AF530530

Winter Vegetables :
Advances and Developments

Winter Vegetables :
Advances and Developments

By :
D.N. SINGH
VISHAL NATH

SATISH SERIAL PUBLISHING HOUSE
403, Express Tower, Commercial Complex, Azadpur, Delhi-110033 (India)
Phone : 011-27672852, Fax : 91-11-27672046
E-mail :info@satishserial.com, hkjain1975@yahoo.com
Website : www.satishserial.com

Published by :

SATISH SERIAL PUBLISHING HOUSE
403, Express Tower, Commercial Complex, Azadpur, Delhi-110033 (INDIA)
Phone : 011-27672852 Fax : 91-11-27672046
E-mail : info@satishserial.com, hkjain1975@yahoo.com

ISBN 978-93-81226-14-8

Typeset at: Harminder Singh Kharb *for* Laxmi Art Creation
Printed at: Salasar Imaging Systems, Delhi-110035

Preface

Vegetable crop assume great significance in providing food and nutritional security. Vegetables are the integral part of our daily diet. The demand of fresh vegetables is in increasing trend due to growth of health consciousness population, enhanced income and living standards. So interest in vegetable farming is being marked significantly among the farmers. As the monsoon favours growing of cereals, farmers mostly give emphasis for raising winter vegetables. But a suitable literature on winter vegetables was in need. To meet the increasing demand of this long felt need of the knowledge on scientific cultivation of these vegetables a book on "Winter Vegetables : Advances and Developments" has been written for use of students, research workers and farmers of the country.

The objective of publishing this book is to provide upto date know how on improved package of practices and post harvest management of commercially as well as lesser known winter vegetables. The origin, distribution, varieties, cultivation methods, INM, IPM, irrigation schedules, harvesting and post harvest practices of the winter vegetables have been given in a lucid manner for easy understanding of the facts. Various books, journals, bulletins, scientific papers etc. related with vegetable growing has been referred while writing this book. The chapters are based on scientific principles and practical experience of the authors.

We may hope, this book will definitely fulfil the long cherished demand of different stake holder like seed dealers, extension workers, farmers, students, teachers etc. those are involved in vegetable production. This book is handy and illustrative. The publication of this book is timely having academic and commercial importance.

We gratefully acknowledge the contribution of fellow research workers, sources of publication from which valuable materials for this book have been utilised. We take this privilege to express out heart felt thanks to our friends and relatives who have supported in preparation of this book.

We also sincerely express our thanks to publisher of this book for his patronage. We still seek the valuable suggestion of the readers of this book to improve our forthcoming editions.

D.N. Singh
Vishal Nath

Contents

Introduction

Diversity in agro-climatic conditions of India favours growing a vast array of vegetable crops. We are the second largest producer of vegetables next to China. Presently, India's share is 13% of total world production which has been achieved from the one third of total area of vegetable production in China. Vegetables are grown in 7.98 million hectares in our country with production of 129.07 million tons. The state wise area and production of vegetables are as follows :

State/UTs	Area in 000 ha.	Production in 000 MT
Andaman Nicobar	3.951	30.823
Andhra Pradesh	324.552	5267.456
Arunachal Pradesh	23.800	110.000
Assam	240.052	2916.691
Bihar	826.872	13385.748
Chandigarh	0.100	1.700
Chhattishgarh	302.563	3041.029
D and N Haveli	0.970	4.480
Daman and Diu	0.162	0.200
Delhi	36.078	617.400

contd...

Goa	5.700	57.600
Gujarat	394.820	6807.121
Haryana	298.430	3893.431
Himachal Pradesh	74.727	1263.915
Jammu and Kashmir	60.661	1023.649
Jharkhand	242.085	3637.029
Karnataka	447.950	7724.883
Kerala	163.570	3509.438
Lakshdweep	0.445	14.120
Madhya Pradesh	291.673	4105.750
Maharashtra	448.300	6368.015
Manipur	16.618	174.263
Meghalaya	44.274	415.794
Mizoram	14.435	114.374
Nagaland	10.380	78.330
Orrisa	672.478	8467.400
Pondicherry	4.451	81.010
Punjab	178.412	3410.315
Rajasthan	125.573	736.701
Sikkim	21.486	98.039
Tamil Nadu	286.277	8693.475
Tripura	25.588	294.712
Uttar Pradesh	987.849	18950.082
Uttarakhand	81.818	1077.550
West Bengal	1323.609	22704.256
Total	**7980.7**	**129076.8**

Source : National Horticulture Board (Year, 2008-2009).

The Present production of vegetable is far below the requirement of rapidly growing population. It is envisaged that our demand by 2020 will be around 250 million tons even @ 250g/caput/day against the daily requirement of 285 g. To achieve such, our area and production must increase as follows :

Year	Expected population (million)	Area in million ha.	Production in million tonnes	Availability/ caput/day in grams
2019-20	1566	9.00	200.00	250
2024-25	1745	10.00	250.00	300

Source : Perspective Plan, PDVR, Varanasi.

Although it seems to be a very hard task, yet we can achieve it. Nature has bestowed us with ample resources to grow vegetables from temperate to humid tropics and from sea level to snow lines. The challenges before us is to produce more from shrinking land and declining water in the scenario of climate change. The solution will come from effective utilisation of diversified gene pool.

Human beings need a wide range of nutrients which give energy, promotes growth and sustain the metabolic functions essential for life. For this we were mostly dependant on cereals. Due to green revolution we could be self sufficient in food grains. Higher yields of modern grain varieties have been achieved largely by increasing the harvest index i.e. the edible portion of the biomass, but this has probably reached its limit (plateau) and under such circumstances, next logical step is to put more emphasis on crops with a higher harvest index, such as those whose vegetative parts are consumed. Thus vegetable crop deserves more attention when one considers its high productivity, nutritive value of the leaves and edibility of roots over a long period. Vegetable crops are equally or more efficient in producing edible dry matters and at the same time providing more minerals and vitamins can definitely be utilised as an alternative towards achieving the nutritional security. Vegetables are immensely important as source of protective nutrients, dieting fibres and folic acids. The demand of vegetables has been increasing fast with a gradual rise in standard of living coupled with development of communication and transport facilities. Therefore, it calls for a major research and development effort to meet the growing demand.

Concentrated efforts made in vegetable research and development in the country have resulted in availability of many improved and hybrid varieties and modern package of practices to augment production and productivity. Further the advances in post harvest management, marketing and value addition has created a new avenue for growing vegetables in a profitable manner.

In our country more than 70 types of vegetables are grown but maximum emphasis has been given on few vegetables leaving behind a number of minor vegetables that are grown in different parts of the country. Although maximum vegetables are being produced in the winter season of the year, recent information on these vegetables are yet to made available in a scientific manner at the hands of different stake holders of vegetable sector.

Hence, an attempt has been made in the succeeding chapters to provide detail information on winter vegetables.

❑❑❑

Chapter-1

Solanaceous Vegetable Crops

1. TOMATO

Botanical Name	:	*Lycopersicon lycopersicum* (L.) Karsten ex. Farw. Syn. *Solanum lycopersicon, Lycopersicon esculentum* Mill.
Family	:	Solanaceae
Chromosome member	:	2n = 2x = 24

Tomato is one of the most popular vegetables, which is grown all over the World. In India tomato has wider coverage and great commercial value in comparison to other vegetables. The fruit is available in cities almost all the year round. The World 'tomath' the Nahua tongue of Mexico. The crop is becoming very popular vegetable throughout the country; earlier some of the old generation people did not like it because of its deep red colour. Tomato is grown throughout the year in almost all home gardens and on large scale in market and truck gardening systems. It is also preferred to raise under protective cultivation system like under green house, polyhouse conditions and soilless culture or hydroponic systems.

ORIGIN

Numerous wild and cultivated relatives of the tomato can still be found in a narrow, elongated mountainous region on the Andes in Peru, Equador and Bolivia as well as in the Galapagos Island.

It is known that cultivated tomato has originated in a wild form in the Peru-Ecuador-Bolivia area of the Andes. Cultivated tomato originated in the new World, while the Andean Zone is likely to centre of origin of wild tomato. The distribution of the allele studies at the Ge locus (Genetic basis) also gives evidence of the origin of cultivated tomato in the Peru-Ecuador region.

It is reported to be a native of tropical America (Thompson and Kelly, 1975) and from where it spread to other parts of the world in the 16th century and became popular in India within the last seven decades. In England, it was popularly known as 'Love Apple' The word tomato not used until 1695 is said to be derived from the Xitomate or Xitotomate.

History and Adaptation

Tomato is one of the most important vegetable in many countries. In England it is known as love apple. Spanish priests introduced it to Europe around 1550. The first written account documenting the arrival of the tomato in the old world appeared in 1550 by the Italian herbalist Pier Andrea Mattioli. In Europe it was known as *Poma amoris* - Amorous apple or love apple. It was also known as *Poma peruviana* - apple of peru. Only as late as 1880, did the British finally concede that tomato is edible. As it belong to night shade (*Solanium nigrum* L.) family Linnacus called it *Solanum lycopersicum* (Lykos-wolf, persicum peach, meaning thereby wolf peach) for the buty of the fruit and its deceitful value. It was robert Gibbon Johnson an ordinary farmer in the USA who first ate tomato on a hot day of August, 1820 to demonstrate its edibility. From then onwards, the tomato spread to different countries.

Even though originated in South America (Peru - Ecuador-Bolivia region of the Andes), Jenkins (1948) regards Mexico particularly veracruz-Puebla region as the major centre of domestication of the cultivated tomato. A large variability existing in Mexico may have given rise to the present day cultivated varieties. From the Andes in South America, Indians (of America) might have carried the material to Central America and Mexico, where the diversification took place (Rick, 1976).

For a long time, tomato was not consumed by human beings due to persistent suspicion of its possible poisonous nature. The Europeans consumed it first and then the Americans. The Italians grew it in about 1550 and might have been the first Europeans to eat it. Its entry into Asia might have been later. Philippines might have received it directly from Mexico or from Spain through trade channels in 16th or 17th century. But in India, British are believed to have introduced it probably in early nineteenth (1928) century and as per the records of Agri-Horticultural Society of Calcutta.

It is an example of fast adaptation to tropical conditions of the country. Probably at the time of introduction, British promoted its cultivation in mid and lower hills in north India. Later in the present century only, the rapid spread took place covering large areas in every state. It has since attained a commercial status with the demand from the processing industry and has undergone a transformation in its production technology with the introduction of F_1 hybrids. The diversity of conditions under which it is grown, is very wide, like growing round the year in Bangalore and other parts of Karnataka, eastern Maharashtra, Ranchi in Chotanagpur plateau, Madanapalli of Andhra Pradesh and tropical Batlagundu (Madurai District) area of Tamil Nadu. It is grown under entirely rainfed conditions (Nov.-Feb.) in parts of Tamil Nadu, Andhra Pradesh, Maharashtra etc. The native selections like 'Annanji' of Tamil Nadu and 'Meeruti' of U.P., from introduced varieties and its derivative 'Pusa Ruby' are examples of adaptation to adverse or hardy conditions. In fact 'Pusa Ruby' a time tested variety for over four decades, is still a popular variety grown in rainfed conditions, covering probably 20 percent of the total area under tomato.

In a typical autonomous crop plant like tomato, diversification and adaptation have been rapid and there has been a steady inflow of new varieties from U.S.A., Europe, Japan etc. during the past four or five decades. Consumption pattern of tomatoes in Indian cuisine partly explains the preference of the farmers to the fresh market varieties of medium fruit size and with slightly higher acidity like 'Pusa Ruby' and 'Annanji'. Processing varieties and thick skinned long distance haulage varieties are of recent origin, especially with the introduction of new F_1 hybrids from abroad and indigenously produced hybrids between imported parents. Specific interest if any, on tomato during Moghul period in Inda, has not been documented and it has not been possible to delineate its course of diversification in India. The natural variability in the introduced germplasm has been rather limited since most of them were bred varieties. There have been some stray pockets in N.E. India and Kashmir where *Lycopersicon pimpinellifolium* has been found in semi-wild state, indicating its preference to cooler conditions.

Tomato production is spread over a wide range of climatic conditions, (even adverse in some areas), which is the main reason for low productivity. A very short ideal period is available for fruit-set, as is prevalent in north western India and other parts. Round the year production poses severe problems of succession of virus, fungal and nematode diseases in some regions.

Tomato cultivation has adapted itself to Indian conditions so much that a virus disease like "Leaf curl virus" not prevalent in other tomato growing countries, has become a serious limiting factor affecting productivity in this country (Seshadri and Chatterjee, 1996).

Importance

Tomato universally treated as protective food is being extensively grown as annual plant all over the world. It is a very good source of income to small and marginal farmers and contributes to the nutrition of the consumers. Tomato is a rich source of minerals, vitamins and organic acid (health acids). These are various types of flavouring compounds found in the fruits, which enrich the taste. A survey made by M.A. Stens indicated that away the main fruits and vegetables tomato ranks 16th as a source of both these vitamins. The total sugar content is 2.5% in ripe fruits and amount of ascorbic acid varies from 16-65 mg/100 g of fruit weight. The total amino acid is 100-350 mg/100 g of fruit weight.

Tomatoes are used directly as raw vegetables in sandwitches, salad, etc. as cooked with mix vegetables form good curry. Several processed item like paste, puree, syrup, juice, ketchup, drinks, whole peeled tomato, etc. are prepared on a large scale. Tomato is very good appetizer and its soup is said to be a good remedy for patients suffering from constipation.

Export Potential

Tomato has been identified as a potential vegetable for export by the Agricultural and Processed Food Products Export Development Authority (APEDA). The specific requirement of tomato for export are round, medium sized and red coloured for Middle East, while cherry tomato is preferred for export in European Countries. Processed product of tomato especially puree and paste have great demand in export. For encouraging export potential of tomato and its products, there is an urgent need of development of suitable varieties, proper packing procedure and long term and consisted export policy. Pre-cooling unit and cold storage for fresh tomato should be established for prolonging the shelf life and minimizing the losses in post harvest handling.

DISTRIBUTION

Tomatoes were distributed to the different parts of tropical America from its centre of origin i.e. Peru-Ecuador-Bolivian area. The first known record of tomato is in the year 1554 in South America and after 200 years it had gradually spread to other parts of the world. The Italians might have been the first (about 1550) European to cultivate tomato and might have been the first European to consume it. After 25 years, it was introduced to the English, Spanish and mid-European garden as a curiosity. In the USA, Thomas Jefferson in virginia first grew the plant of tomato in 1781. Later, the French refugees spread it to Phiadelphia in 1789 and to Massachusettes in 1802 by an Halian Painter, since then tomatoes are being used as food all over the world.

Tomato is one of the most widely grown vegetable crops in India. It is grown across the country as a common vegetable in farm gardens, home gardens, market gardens for fresh consumption and for processing purposes. In India it is grown on about 4.86 lakh hectares land with a production of about 74.42 lakh tons. Maharashtra leads in the production of tomato followed by Karnataka and Bihar. Maharashtra recorded a production of 11.83 lakh tons where as Karnataka recorded 11.43 lakh tons during 2001-2002. (Singh, 2004).

BOTANY

Tomato (*Lycopersicon esculentum Mill*) a herbacious, annual to perennial, prostrate and sexually (occasionally asexually) propagated plant. Tomato is perennial but grown as, an annual for both seed as well as vegetable purposes. It grows up to 2 m in height. Plants have vigorous tap root with extensive fibrous root system. Stems are soft, brittle, round and hairy later on they become angular and woody. The leaves are compound, alternate, spirally arranged pinnatified, oval, irregularly lobed, often with small pinnae between larger leaflets. Flowers are in cymes of 4-12, borne on stem laterally or opposite or between leaves. Flowers are having short pedicel which are constricted at the middle point. Flowers are pendent, perfect and hypogynous. Calyx rotate with a short tube having 6 persistent lobes, pale yellow to yellow in colour. Stamens 5 in number are attached to the base of corolla tube, filament yellow in colour, anther bright yellow, ovary 6 called with a central fleshy placenta. Fruit is fleshy berry with 2-9 loculi, orange, yellow or red when ripe, usually round, pear shaped, smooth or furrowed. The seeds are oval, kidney or pea shaped, hairy, light brown (3-5 mm x 2-4 mm). The test weight (1000 seed weight) is about 2.5 g. The number of seeds - per fruit varies from 150-300 due to variety and climate.

The tomato, *Lycopersicon esculeutum* (Lycos, wolf; persica, peach, esculemtum, edible) is a member of the family *Solanaceae (Syn.Solanam lyeopersicon Mill.)* and it is a herbaceous annual plant with bisexual flowers and the fruit is a true berry. It is a self pollinated crop but cross pollination up to 30 per cent has been reported. According to the growth habit, the tomato plants have been categorized into two: indeterminate and determinate types. The plant of former type terminates in a vegetative bud, whereas that of the determinate type terminates in a flower bud and is appropriately called 'self topping' or 'self pruning' type. Many varieties of determinate type tomato plants do not have adequate foliage to protect their fruit. Some of them fruit very early. The determinate varieties can be harvested in 2-3 harvests while the fruiting period of indeterminate type is prolonged.

Tomato produces a strong taproot but very often it is damaged at the time of transplanting. Consequently, the lateral adventitious roots develop profusely. The adventitious roots develop at the rate of about 33 mm per day penetrating 1m depth and attaining a length of 1.5 m (Ward, 1964). However the root development largely depends upon the cultivar and the growing media and the moisture in it. The stem is soft, brittle and hairy, when young and hard, woody, and capiously branched when mature. Its organisation is in the sequence of two or three leaved sympodia. It is erect to semi-erect. Epidermal hairs with oil glands are present at the early plant growth stage. The leaf is alternate, petiolate and 2.15 inch wide. The leaflets are unequal and odd-pinnate. The apex is narrowed or acuminate and acute irregularly serrate. The number of leaflets varies from four to several and the length of leaf from 6 to 12 inches on the leaves, hairs or trichomes are formed by the outer growth of the epidermal cells. Inflorescence is the extra-axillary cymes with dicotomous or polycotomous branching. It is lateral and arises from the stem opposite to the leaf. It may be terminal or may continue with the vegetative shoot (Muller, 1940). The number of flowers per cluster varies from three to several. The early flowers that open in the cluster are large-sized as in a typical cyme. The abscission of the flowers may take place due to high or low temperature or some other internal and external factors. The flower drop is due to the development of the abscission layer at the mid point of the pedicel. The plant produces bright yellow flowers. The flowers are pentamerous, bisexual, regular, complete ebracteate, and hypogynous. The lateral section of the flower shows the details of sepals, petals, stamens, style ovary and ovules. The pistil has two to several carpels. The anther connate appearing in the throat of the corolla and they dehisce through the pericidal dehiscence. Usually, the style is shorter has anther cone which causes high degree of self-pollination. The fruit is soft berry. Glandular hairs and glands are usually present on the small fruit which degenerate in the advanced stage. The fruits of some varieties or species do not have glandular hairs and gland. The seed is oval, flattened buff to pale brown, 3-5 mm in length. It is derived from the anatropous ovule.

Plant breeding

Nath *et al.* (2008) stated that initially most of the important varieties were introduced from the western countries and some of them became popular among the growers and consumers in different agro climatic regions of this country depending upon their adaptability. The Indian Agricultural Research Institute, New Delhi had introduced important tomato varieties from all over the world which were evaluated during early sixties of last century for various purposes (Mital *et al.*, 1965). Some of the introduced American varieties like 'Sioux', 'Marglobe' etc. were recommended for commercial production.

A significant achievement was made at the IARI, New Delhi, when the new tomato variety Pusa Ruby' was developed by crossing 'Sioux' (a good quality American variety) with 'Improved Meeruti an indigenous hardy variety. The new tomato varieties 'Arka Saurabh' and 'Arka Vikas' from IIHR, Bangalore (Tikoo and Anand, 1983), 'Co-3' from Coimbatore, 'Sel-120' and 'Sel-152' from IARI, New Delhi, 'S-12', 'Punjab Chhuhara' and 'Punjab Kesari from PAU, Ludhiana (Nandpuri *et al.*, 1977), 'HS-10T 'HS-102' from Kalianpur, 'AC-238' and 'Plant T-3' from Pantnagar, and 'Sweet -72' from Gwalior were promising beside some old varieties like 'Pusa Ruby', 'Pusa Eraly Dwarf 'Sioux', 'Roma' and "La Bonita'. The varieties 'Roma' (Singh *et al.*, 1977) 'La Bonita' (Thomas *et al.*, 1976) and 'Dwarf Money Maker' (Thomas *et al.*, 1979) held promise for commercial cultivation in different regions of our country in view of their higher yield and better fruit quality, till late eighties and early nineties of last century. There are some determinate and indeterminate types in tomato. Indeterminate types : Pusa Ruby, Sel-120, T-l, Sioux, Pant Bahar (AC-234), Arka Vikas, Arka Saurabh, PantT-3 (AC-238), Bt-10, Bt-12. Hybrid and determinate varieties: AARTH-3, Pusa Hybrid-2, NA-501 and DTH-4 and ARTH-4, MTH-6, FMH-2 (Arka Vardhan), KT-4, NATH-601, FMH-1, BSS-20 (Indeterminate type). Heterosis breeding is common in tomato to develop Fl hybrids. It is important to note here that till late eighties and early nineties the breeding objective was for higher yield. With the changing agro climate and crop growing stuations, changing pest scenario and value added consumer preference during the middle of the nineties of the last century onwards, additional objectives like breeding for insect pest resistance, breeding for abioty stresses (heat and moisture stress) resistance, breeding for processing qualities etc were added along with breeding for higher yields.

Randhwa *et al.* (1988) evaluated the tomato varieties for processing and found that Punjab Chhuhara tops the yield as well as in product quality for processing. Rai *et al.* (1987) evaluated 20 varieties of tomato for Tripura conditions and variety S-12 gave the best performance under their agro climatic conditions. Supe *et al.* (1989) reported the performance of tomato hybrids in Haryana, and concluded that the Suttons Bragger gave maximum fruit weight and plant height while hybrid MTHH was the best performing hybrid since it had recorded highest yield. Kumar and Irulappan (1990) also studied the seasonal influence of spotted wilt virus resistance in tomato varieties and reported that crops during November had lesser incidence of the disease and yielded 46.7 tons/hectare. Choudhury and Malhotra (2001) have also studied the performance of tomato hybrids developed as a result of line X tester mating design and have reported that the hybrids VTL 33 X Roma to be the best, while Mishra *etal* (2001) reported the performance of four tomato varieties (BT-1, BT-3, BT-10 and Pusa Ruby) to different planting dates in the easteren ghats of Orissa. Verma *et al.* (2001) reported that a tomato hybrid, *viz.*, Pusa Divya had profuse branches and provided 35-45 t fruit/ha. Tiwari *et al.* (2002) studied the suitability of different tomato varieties for processing and found that PR-3 and PR-H and Arka Ahsish were superior. Sharma *et al.* (2002) also reported the suitability of tomato hybrids for the preparation of paste and found JKTH-895 to be superior to others. Tewari *et al.* (2002) described a new variety of tomato named 'Pusa Sadabahar' which set fruits at 30°C of night temperature and suitable for north India. Singh *et al.* (2002) studied the heat stress tolerance in tomato and found Lt-2, and Lt-26 had grown best under stress and non stress conditions where as Lt-10, and Lt-16 were suitable for stress conditions and Lt-34 for non stress conditions. Jasmine *et al.* (1993) evaluated certain F_1 hybrids and open pollinated cultivators of tomato for yield and yield components for Coimbatore tract of Tamil Nadu, and reported that ARTH-4 and ARTH-3 were found suitable. Natarajan *et al.* (2002) reported utility of the COTH-1-A hybrid tomato for high density planting in Tamil Nadu. Rahman *et al.* (2002) evaluated tomato varieties for leaf curl disease of tomato and found that the resistance in varieties Bilah-2, and H-86 should be exploited in tomato improvement programme. Sadashiva (2002) reported the breeding work of tomato for combined resistance to bacterial wilt and tomato leaf curl virus. He reported most hybrids with combined resistance to TLCV and BW were TLBR-3 X IIHR - 2200 (3.88 kg/plant), TLBR-4 X IRHR-2200 (3.8 kg/ plant), TLBR-3 X IIHR-2199 (3.72 kg/ plant), TLBR-6 XIIHR-2202 (3.58 kg/ plant) whereas TLCV resistant hybrid Avinah-2 reported 3.3 kg/ plant in bacterial wilt free soil but succumbed to complete wilt with 1.11 kg/ plant in yield during first and third seasons respectively. This work has resulted in identification of two F_1 hybrids Arka Shreshtha and Arka Abhijit during 1994-95 and Arka Ananya another F_1 hybrid with combained resistance to LCV and BW during 2005. Prasanna *et al.* (2002) reported developing of bacterial wilt resistant tomato hybrids with extended shelf life. High yielding BW resistant F_1 hybrids were IIHR 2188 X IIHR 2052 (2.45 kg/ plant), BWR-IF X nor-1 (2.57 kg/ plant) and IIHR 2199 X IIHR 1136 (2.17 kg/plant). The maximum shelf life observed were in 15 SBSB X IIHR 2050 (32 days), 15 SBSB X Nor -1 (27 days) and BWR IF X Nor-1 (12 days). Kulkarni and Madalangeri (2002), evaluated and reported on correlation in tomato for leaf curl virus. They found that the parents (H-36, H-86 and NDT-VR-60) and three hybrids (H-36 X NDT-VR-73, H-36 XH- 86 and H-36 X L-15) were found resistant as compared with susceptible check Punjab

Chhuhara. Tomato varieties susceptible to these serious diseases and pests have become limiting factors in tomato cultivation in various regions. The Indian Institute of Horticultural Research, Hessargahtta, Bangalore has been engaged in breeding varieties resistant to bacterial wilt, leaf curl, tobacco mosaic virus, root knot nematode, phytophthora fruit rot, septoria leaf spot and fruit borer, while the Division of Vegetable Crops at IARI is engaged in breeding varieties resistant to tobacco mosaic virus. The best source has been recorded from the wild tomato *(L. peruuianwn)*, but it does not cross with cultivated varieties normally. However, this was made possible with special techniques byMajid *et al.* (1968), where few interspecific hybrids were obtained with high resistance to various diseases. As a result of screening 181 varieties for resistance to root knot nematodes, the varieties 'Pelican', 'NMR-1', 'VFN Bush', 'New Yorker', 'Healani, 'Kalohi', 'Small Fry', 'West Virgina', 'LA-121', 'EC-118270', 'EC-118276', 'EC-118277', and 'Hybrid Karnataka' were found to be resistant (Mahajan *et al.*, 1979). There are varieties which are resistant to diseases like Pant Bahar for Verticillium wilt and fusarium wilt; BT-1, BWR-5, LE-79-5 and Arka Abhijit, Arka Shreshta, Arka Alok, BT -10 for bacterial wilt; H-24 and H-36 for TLCV; Pusa 120, Pusa hybrid-2 and Pusa hybrid-4 for root knot nematode and Pusa Uphar for fruit borer. Sabour Suphala is salt tolerant at seed germination stage.

Work is in progress under National Agricultural Research System of the country with regard to fruit yield, appearance, plant parameters, resistance to diseases and pests, salt and drought tolerance etc. Several F_1 hybrids from private companies and research centers have recently come up and they are now gaining popularity due to high yield. They are reported to be resistant to diseases and pests apart from good quality for fresh consumption and processing.

Taxonomy

Tomato belongs to the family solanaceae and gems *Lycopersicon*. The genus includes 12 species, all native to south America. The genus differs from *Solamum* by the absence of spines in the pinnatifid leaves and anther tips are narrow and exhibit longitudinal dehiscence. According to Muller (1940) and Luckwill (1943) the genus is further subdivided into two sub-genera.

Eulycopersicon

Characterised by red-fruited edible species with carotenoid pigmentation and annual. Cultivated tomato *esculentum* and *pimpinellifolium* species are included in this group.

Eriopersicon

Green fruited species with anthocyanin pigmentation and the following species are in this group (Kalloo, 1985, Bose *et al.*, 1993).

a. *pissisi* (Lukwill, 1943)
b. *peruvianum* (Muller, 1940)
c. *hirsutum* (Muller, 1940)
d. *glandulosum* (Muller, 1940)
e. *cheesmanii* (Muller, 1940)

Bailey (1949) classified tomatoes into two species, *L. esculentum* and *L. pimpinellifolium* with five botanical varieties like *L. esculentum* var. *commune* (common tomato), var. *grandifolium* (large-leafed tomato), var. *validum* (upright tomato), var. *cerasiforme* (cherry tomato) and var. *pyriformae* (pear tomato). Rick (1976) divided the genus in to the those species which can be easily crossed with commercial tomato (*esculentum* - complex) and those which can not (*peruvianum* - complex). Species forming the *esculentum* complex are : (a) *L. esculentum* Mill. Mexico is the centre of origin, the cherry tomato (*L.esculentum* var. *cerasiforme* is the direct ancestors; (b) *L. pimpinellifolium* : originated in low elevations in Peru, source of resistance to fusarium wilts, bacterial speck (*Pseudomonas*); (c) *L. cheesmani*, originated in Galapagos island, self compatible; (d) *L. parviflorum* (syn.*L. minutum*); (e) *L. chmielesswskii* and (g) *L. hirsutum* f. *glabratum*, less hairy, resistance to bacterial speck and root-knot nematode.

Tamato can also be classified on the basis of growth habit : (a) Indeterminate-terminal bud ends in leafy bud and there is continuous vegetative growth, (Pusa Ruby, Sweet 72); (b) Determinate-terminal bud ends in a floral bud and further vegetative growth is checked (AC 142, HS 101): and (c) Semi-determinate-in between determinate and indeterminate, number of nodes between two consecutive inflorescence approach one (S 12, Pusa Early Dwarf). On the basis of temperature requirements for fruit set, it can be classified as; (a) Normal temperature requirements-15-20°C night temperature, e.g. Pusa Ruby, Sweet 72 etc; (b) Hot Set-20°C temperature e.g. HS 101, Saladette, Punjab Tropic etc; (c) Cold Set-15°C night temperature e.g. Pusa Sheetal, HS 101, HS 102, HS 110, Cold Set, Precoce, Pusa Early Dwarf, Balkan etc.

Floral biology

The flowers in tomato are borne in cluster on stems. The anthesis starts at 6 a.m. with maximum flower opening between 7 to 8 a.m. Anthesis is followed by dehiscence which is at its peak from 9 to 11 a.m. Pollen grains remain viable for long period. Stigma becomes receptive 16 hours before anthesis and remain the same for 2-3 days after anthesis. Anthesis is much influenced by temperature and soil moisture (Hawthorn and Pollard, 1954). The optimum temperature for pollination is about 21°C. The pollen remains viable for 7-10 days at 20°C and 70% humidity and the stigma remains receptive for 6-8 days. The best fruit set is obtained when the stigmas receive pollen from many different plants. The pollen is most fertile immediately after the opening of the stamens and the stigma is most receptive on the day of anthesis. Pollen grain germination takes place approximately two hrs after pollen has been applied on the stigma. Ovules ripen over two days. Pollen can be stored upto three days under ordinary conditions. Pollen from one days-stored flower provides best seed set. At 0°C, it can store for 2-3 months. While, it remains viable for six months at 5°C in a dessicator. The optimum temperature for pollen germination is 18-25°C. Tomato pollen germinates best in a liquid solution of 20 per cent sucrose and 10 ppm boron. The number of seed produced increases with the quantity of pollen placed on the stigma. The fertilization and fruit formation are maximum when pollination take place at anthesis time. Visible sign of fruit set could be observed after about 96 hours of pollination (Brar, 1975; Tiwari and Choudhary, 1993).

COMPOSITION

Tomato is a rich source of minerals, and organic acid (healthy acids). Tomato is good source of vitamins viz., ascorbic acid or vitamin C, vitamin A, thiamine or vitamin B1 and riboflavin or vitamin B2 in that order. A 100 g of edible portion of tomato contains 94.1 g water, 1 g protein, 0.3 g fat, 4 g carbohydrates, 1100 IU vitamin A, 0.2 mg vitamin B, 0.6 mg Nicotinic acid, 0.31 mg pantothenic acid, 23 mg vitamin C, 0.27 mg vitamin E, 390 mg citric acid, 268 mg potassium, 27 mg phosphorus and 51 mg chlorine (Chatfield, 1959).

Tomato contain important minerals viz. potassium, calcium, sodium, magnesium, phosphorus, boron, manganese, zinc, copper, iron etc. Tomatine is a steroidal glucoalkaloid found in leaves, stems and fruits. The molecular formula of tomatine is $C_{50}H_{83}NO_{21}$ and that of B-tomatine $C_{45}H_{83}NO_{17}H_2O$ (Roddick, 1974). The main site of tomatine biosynthesis is growing regions of the tomato roots (Roddick and Butcher, 1972).

Tomato seeds contain oil (about 24%) which is used as salad oil and in the preparation of margarine (Roddick and Butcher, 1972). Tomato has medicinal importance. Tomato juice promotes gastric secretion, acts as blood purifier, and works as intestinal antiseptic (Hedrick and Booth, 1908).

Uses

Tomatoes are cooked as vegetables preferably mixed with potatoes. Their use in the form of sald is very common. Various products are made from tomatoes are soup, ketchup, sauce, chutney, puree, powder, paste, juice etc. Green tomatoes are also used for pickles and preserves. Tomato juice is used as an appetiser and beverage. It is also used for baking.

Tomato has many medicinal uses. The pulp and juice of the fruit is digestible and a mild aperient, a promoter of gastric secretion and blood purifier. It is also considered to be an intestinal antiseptic as it has a cleaning effect in the enteric portion of the elimentary canal. It is said to be useful in cancer of the mouth, sore mouth etc. Dried tomato juice retains Vitamin C. It stimulate torpid liver and is good in chronic dyspepsia. It is one of the richest vegetable which keeps stomach and intestine in good condition. Tomato protect against cancer of the digestive tract and prostate. Tomato is very useful for the patients of bronchitis and asthma (nadkarni, 1927). People who are 7 or more of raw tomatoes every week had 60 percent less chance of developing colon, rectal or stomach cancer than those who ate less. It is an extraordinary connective for kidneys and is wild and natural stimulant that helps to wash away the poisons which cause disease and contaminate our system.

VARIETIES AND HYBRIDS

Broadly, tomato plants are of two types (1) determinate, and (2) indeterminate.

Table 1 : Differences between types of tomato plants.

Determinate type	Indeterminate type
1. Main stem of tomato plant terminate in flower bud	1. Main stems of tomato plant does not terminate in flower bud instead it possesses vegetative bud
2. Main stem does not grow tall after flowering	2. Main stem remains continue in growth throughout growing season
3. Formation of more laterals	3. Formation of less number of laterals
4. There is self-topping of plant	4. There is no self-topping of plant
5. Flower cluster or truss forms in many cases at every internode	5. Flower cluster or truss forms in many cases at every 3rd internode, separated by three leaves.
6. Comparatively having early maturity	6. Comparatively requires long period for maturity
7. Incidence of sunscalding is less because fruits remain hiding under leaves	7. Incidence of sunscalding is more because fruits remain exposed to sun
8. Plants do not require earthing	8. Plants requires earthing and they are benefitted with it
9. Plants do not need staking	9. Plants need staking
10. Plants habit is bush type	10. Plants growth straight i.e. 'non-bush' type
11. Not much suitable for high planting density	11. Suitable for high planting density

A brief description of commercial tomato varieties and hybrids are given by Singh and Nath, 2011; mentioned below. These are, developed by Agricultural Institutes, Agricultural Universities, State Departments of Agriculture and Private Seed Organizations.

Pusa Red Plum

It is a derivative a cross between the dwarf variety Red Cloud and Merruti. Fruits ripen a week earlier than Pusa Ruby. This variety does well in summer and in winter growing seasons.

Pusa Ruby

It is a derivative of Improved Meeruti and Sioux. It is early maturing variety and matures in 60-65 days after transplanting. It is widely adoptable variety having indeterminate growth habit. Fruits are flattened, grooved, firm and medium size. It is suitable for rainy season for all over the country (Singh and Bhagchandani, 1956).

Pusa Sadabahar

It has been released by IARI, New Delhi. It is highly suitable for growing under wide range of 8-30°C night temperature. It plants are determinated dwarf. This variety can be accommodate more in number per unit area. Very prolific bearer with smooth oval to round attractive fruits. This variety can be grown through out the year in north Indian plains except rainy season (July to early September) (Tewari *et al.,* 2001).

Pusa Sheetal

It has been developed by combining the desired attributes of two cold set lines Balkan from Bulgaria and Temnorrosnij from USSR by hybridization and selection of plants to set fruit at very low temperature in segregating generation was done. This variety was released by IARI Variety Release Committee. Plants of this cultivar are compact with moderate foliage. Fruits are flatish round, smooth, attractive and developed uniformly red colour on maturity. This cultivar is very suitable for raising as an additional early crop in northern India (Tiwari *et al.,* 1990).

Pusa Uphar

It is indeterminate and having higher fruit setting with desirable quality attributes in slightly adverse temperature. It is developed form IARI, New Delhi and released by the Central Variety Release Committee in 1999. Plants are prolific bearers, upright, compact with dark green foliage. Fruits are borne in bunches of 4-5 each. They are attractive round, medium sized (60 g), thick-skinned and uniform in ripening. It is slightly tolerant to fruit borer under filed conditions. It takes 100 days from transplanting to fruit ripening. It is suitable for cultivation in Ganga alluvial and western plains (Tiwari *et al.*, 1999).

Pusa-120

It is well-adopted variety and released from IARI, New Delhi. Plants are semi-determinate, spreading type having dark green foliage with good coverage. Fruits are attractive, round to flattish round, medium to large in size, smooth, uniform red in colour, less acidic and less seeded. It is resistant to nematodes and does well in both summer and winter seasons. Crop duration is 100-125 days.

Pusa Divya

This hybrid has been developed by utilizing closed anther mutant with potato leaf marker gene (functional male sterility) as female parent. This hybrid has been released by the Central Variety Release Committee, especially for cultivation in Uttar Pradesh, Punjab and Bihar (Verma *et al.*, 2001). Plants are indeterminate with profuse branching. Fruits are firm, thick, skinned, round to oval shaped, green with dark green stem end and turn dark red on ripening. Due to thick skin, fruits have longer shelf life with suitability for long distance transportation. Maturity period is 70-90 days from transplanting and 135-150 days from seed sowing in hills, however, this may be reduced in plain. An average yield is 350-450 q/ha. Less infestation to late blight and buckey rot.

Pusa Early Dwarf

It is a derivative of the cross Meeruti x Red Cloud. It is widely adopted variety. It is suitable for rainy season in plains and for spring summer sowing in hills. Plants are typical dwarf (50-55 cm in height) with determinate growth habit. Fruits are roundish, slightly flattish, medium large, uniform red, ribbed, obscure furrows, 5-6 locules, set closely with 30-40 fruits/plant. Fruits mature in 55 to 60 days after transplanting. Yield potential is 250-300 q/ha.

Pusa Gaurav

It has been developed from the exotic segregating generation of a cross between 'Galmour' and 'Watch' and released by the IARI, New Delhi. The plant is dwarf, bushy with moderate foliage cover. The fruits are smooth, elliptical (egg shaped) and borne in clusters. The unripe fruits are firm with thick flesh (0.6 cm) and two well- filled locules that facilitate easy transportation over long distances. Whole fruit is suitable for processing and canning because it is without neck constriction and has higher TSS (6 per cent) and better keeping quality at room temperature (Choudhury and Tewari, 1993).

Pusa Hybrid-1

The plants are compact, determinate with good foliage cover having prolific bearing. Fruits are round, smooth and attractive. It is ready for market from late May to mid July. Pusa Hybrid-1 has an added advantage of providing tomatoes from June to mid July, the lean period for tomato in north India. It can set fruit upto a temperature of 28°C during night. The average fruit weight is 63 g and yielded 320 q/ha (Tewari, 1996).

Pusa Hybrid-2

It has developed by combining desirable commercial characters. The IIVR, Varanasi has recommended it for cultivation in humid western Himalayan regions, sub-humid Sutluj-Ganga alluvial plains and Western plains, semi-arid plateau during 1993. Plants are compact, semi-determinate with good foliage cover having prolific bearing. Fruits are round td flattish round, firm, smooth, attractive and develop a uniform red colour on maturity. It is highly tolerant to root knot nematode. The average yield in co-ordinated trials was 591.9 q/ha (Tiwari *et al.*, 1996).

Pusa Hybrid-4

Plants are determinate, compact with dark green foliage having prolific bearing. Fruits are attractive, round, smooth, medium-sized (70-80 g), 3-loculed, thick pericarp and uniform in ripening. It is highly tolerant to root knot nematodes under field conditions. It takes 130 days from transplanting to fruit ripening in spring season and 120 days in autumn (Tiwari, 1998).

Arka Alok

It is also developed from IIHR, Bangalore. Fruits are determinate. Fruits are large in size (100 g), round in shape having green shoulder. Average yield of this variety is 350 q/ha. It is resistant to bacterial wilt.

Arka Abha

It is developed from IIHR Bangalore. Plants are indeterminate in nature. Fruits are flat, round, green shoulder and average fruit weight is 70-80 g. An average yield is 350 q/ha. This variety is resistant to bacterial wilt.

Arka Abhijit

This was developed at the IIHR, Bangalore. Plants are semi-determinate having 75 cm plant height and takes 55-60 days to first fruit maturity. Fruits are round having 70 g average fruit weight with green shoulder. TSS is 4.3 per cent and 17 days keeping quality at room temperature (from breaker to softening). This yielded 650 q/ ha in 140 days of crop duration. It is resistant to bacterial wilt and suitable for fresh marketing.

Arka Ahuti

It is pure line selection of variety Ottawa-60 and released from IIHR Bangalore. Plants are indeterminate in growth habit. Fruits are oval in shape, pulpy, tough skin and uniform ripening. Total soluble solid is 5.2 per cent. The yield potential is 450 q/ha in 135 days of crop duration.

Arka Meghali

It is a cross of Arka Vikas x IIHR-554, developed through pedigree method of selection and released from IIHR, Bangalore. Plants are semi-Is determinate. Fruits are superior to Pusa Ruby variety in size (63 g), flesh thickness and fruit firmness. This variety is suitable for rainfed condition. An average yield of this variety is 172-182 q/ha (Sadashiva *et al.,* 1996).

Arka Saurabh

It is an improvement over breeding line V-685 from Canada and released from IIHR, Bangalore. Plants are semi-determinate. Fruits are firm, fleshy, deep red and weighing about 70 g. This is a dual purpose variety, suitable for both fresh market and processing purposes. This variety yielded 350 q/ha in 105 -110 days of crop duration. It is resistant to fruit cracking.

Arka Shreshtha

It is developed and released from IIHR, Bangalore. The plant is semi-determinate, 79 cm tall and takes 55-60 days for first fruit maturity. Fruits are round having light green shoulder and 75 g average fruit weight. The TSS is 4.7 per cent. Fruits can be stored for 17 days (from breaker to softening stage) at room temperature. It gives 830 q/ha in 140 days of crop duration. It is resistance to bacterial wilt and suitable for both fresh marketing and processing purposes.

Arka Vardan

It has been developed at IIHR, Bangalore. Plants are indeterminate. Fruits are flat round, firm deep red with green shoulder having average weight of 140 g. It is resistant to nematodes and recommended for cultivation in Jammu and Kashmir, Himachal Pradesh, Uttranchal, Punjab, Bihar, Uttar Pradesh plains, Karnataka, Tamil Nadu and Kerala.

Arka Vikas

It is a selection from American variety TipTop and released from IIHR, Bangalore. Plants are semi-determinate in growth habit. Fruit is oblate, deep red in colour, and each weighing 80-85 g with uniform ripening. It is suitable for fresh market purpose. Yield potential is 350-400 q/ha in 105-110 days of crop duration.

Arka Vishal

It has been developed at IIHR, Bangalore. Plants are indeterminate. Fruits are round, firm, deep red with green shoulder and average fruit weight of 140 g. It is tolerant to fruit cracking and matures in 160 days with an average yield of 750 q/ha. It is recommended for Punjab, Bihar and Uttar Pradesh plains.

Best of All

It is indeterminate and mid season variety. Plants are erect and height ranged from 100 to 200 cm. Fruits are medium in size, scarlet red in colour, round, juicy, small scar, thin skin, three loculed with green shoulder and borne in clusters of four to five fruits.

CO-1

It is developed and released from TNAU, Coimbatore. Plant is determinate and suitable for growing in southern India. Fruits are round with yellow stem end, uniform ripening.

CO-2

It is a pure line isolated from USSR type and adopted in Tamil Nadu State. Plant is 70 cm tall, semi-dwarf, erect in growth habit with moderate branching, well spread and needs no staking. Stem is light green and foliage is dark green in colour. Flowering occurs 65-70 days and first picking starts 100 days after transplanting. Fruits are ovate, medium in size, smooth, green when unripe and bright red on ripening. Fruits are devoid of cracking. Plant contains 4-5 fruits/truss and 20-25 fruits/plant. It gives 410 q/ha fruit in 6-8 harvestings.

CO-3

It is a mutant of Co.1 variety and released in 1981 from TNAU, Coimbatore. Plants are dwarf, erect and determinate in growth habit. Stem is thick with strong primary branches. Leaves are dark green. Fruits are medium in size, borne in clusters (4-5-fruits/ truss), roundish, globular in shape, parrot green when unripe and attractive red on ripening. It is susceptible to mosaic, leaf curl virus and leaf spot diseases. It gives 500 q/ha fruits and 3 q/ha seeds.

COTH-1

It is a derivative of cross between IIHR-709 and LE-812, developed at TNAU, Coimbatore. Plants are determinate and 75 cm tall. Fruits are round to slightly oblong, medium sized (average weight 60 g) and bright red at maturity. First picking starts after 88-94 days. It gives an average yield of 850-950 q/ha in 140 days of crop duration.

PKM-1

It is an induced mutant from Annanji tomato and released in 1878 from TNAU, Coimbatore. Plants are dwarf (45 cm) and less spreading with dark green foliage. Flowering starts 55 days and first harvesting starts 95 days after transplanting. Fruits are flattish with 4-5 light grooves, 4-5 locules, 4-6 fruits/cluster, 30 fruits per plant, dark green when unripe and capsicum red at ripening with green at pedicel end. The mean fruit weight is 55.2 g, fruit length 3 cm, and fruit width 4.6 cm. It is moderately resistant to pest and diseases. Fruit yield is 320 q/ha in 135 days of crop duration. The seed yield is 2.35-2.50 q/ha.

H-24 (Hisar Anmol)

It is a determinate variety resistant to TLCV, derived through backcross-pedigree method from an inter-specific cross L. *esculentum* cv. Sel-7 x L. *hirsutum* f. *glabratum*. Fruits are flattish-round, fleshy, medium and red. It gives an average yield of 400 q/ha. This variety has been most widely utilized as a source of TLCV resistant for developing resistant varieties worldwide. Recommended for release and cultivation in 1997 for Chhattishgarh, Orissa and Andhra Pradesh.

H.S.101

It is a derivative of cross between Sel- 2-3 and exotic cultivar, developed and release from CCSHAU, Hisar. Central Sub-Committee on Release of Varieties in 1977 for Haryana, Uttar Pradesh, Gujarat and Tamil Nadu states releases it. It is suitable for winter season. Plants are 50 cm tall, multibranched, sturdy and determinate in growth habit. Fruits borne in clusters (2-3 fruits/ truss), round small to medium in size, 3-4 loculed and uniform red on ripening. It bears 40-50 fruits/plant and yields 250 q/ha (Kalloo and Singh, 1976).

H.S.102

It is a derivative of cross between S12 and Pusa Early Dwarf and developed and released from CCSHAU, Hisar. The variety is characterized by extremely early, concentrated flowering and fruiting. Fruits are medium to small, round, juicy, and thin pericarp. It is recommended for cultivation in Haryana State (Kalloo and Singh, 1976).

H.S.110

It is a selection from an exotic and recommended for cultivation in Haryana Kalloo and Singh, 1982). Plants are determinate having potato leaf type. Fruits are large, smooth, fleshy and jointless. It is suitable for table purposes.

Hisar Arun

It is derivative of Pusa Early Dwarf x K-l . It is extremely early and high yielding. Plants are dwarf having concentrated flowering and fruiting. Fruits are medium to large, round deep red and recommended for cultivation in Delhi, Uttar Pradesh, Karnataka, Orissa and Maharashtra. It has ability to fruit set at low temperature (Kalloo and Dudi, 1990).

Hisar Gaurav

It is developed and released from CCS-HAU, Hisar. Plants are straight, less branched and required staking. Fruits are larger, flat orange and fleshy. It is recommended for cultivation in leaf curl virus infested area.

Hisar Lalima (Sel-18)

It is developed through hybridization of Pusa Early Dwarf x HS-101 at CCS HAU, Hisar. It is an early maturing and large fruited variety. Fruits are round, flat attractive red, fleshy, with 4-6 locules, a thick pericarp, less prone to cracking and suitable for both the tomato seasons (autumn and spring) in the north India. Harvest index is very high. Because foliage is less and fruits are more in number. An average fruit weight is 75 g.

Hisar Lalit

It is a derivative of HS-101 x Resistant Bangalore. Plant is semi-determinate and early fruiting. Fruits are round and medium to large. It is resistant to root knot nematode and recommended for cultivation in root knot nematode infested areas (Jain *et al.,* 1990).

Sel-32

It is a determinate variety, developed at CCSHAU, Hisar. Fruits are flattish-round. Smooth, large, attractive red and less seeded. Suitable for spring and summer season cultivation. This variety gives an average yield of 320 q/ha. It is recommended for release and cultivation in 1996 for West Bengal, Assam, Rajasthan, Gujarat, Haryana, Delhi, Madhya Pradesh and Maharashtra.

Punjab Chhuhara

It is also derivative of the cross Punjab Tropic x EC55055, released in 1975 from PAU, Ludhiana. It is suitable for summer season. Plants are dwarf (60 cm), bushy and determinate growth habit with dense and luxuriant foliage. Fruits are pear shaped, thick pericarp, fleshy, 2-3 loculed, small to medium in size and well protected under foliage. First harvesting starts after 90 days when planted in February. It has good keeping quality and sets fruits under high temperature conditions. It is good for processing purposes. It is tolerant to root knot nematode and susceptible to early and late blight.

Punjab Kesari

It is a derivative of the cross Punjab Tropic x EC55055, released from PAU, Ludhiana. Plant is about 60 cm tall. Stem is round with green foliage. Fruits are oval round, medium in size (55 g), three loculed and red on maturity.

Punjab Tropic

It is a selection from the introduction of USA and released from 'PAU Ludhiana. Plant is 100 cm tall, indeterminate growth habit with luxuriant growth. Stem and foliage are dark green. Flowering starts 95 days after sowing. Fruits are large (100 g), round, red on maturity, borne in clusters (4-6 fruits) and remain totally covered with foliage. It has late fruit setting with slow development of fruits. Fruit yield is 225 q/ha in 141 days crop duration.

Sel-1-6-1-4

It is a root knot resistant variety, developed at PAU, Ludhiana, derived through selection from a cross between root knot resistant line E24-C8 and susceptible variety Punjab Chhuhara. Plants are determinate with dark green foliage, which protect fruit from sunburn. Fruits are pear shaped and medium (40-50 g). It gives an average yield of 560 q/ha. It is recommended for release and cultivation in 1997 for Jammu & Kashmir, Himachal Pradesh, Uttaranchal, West Bengal, Assam, Punjab, Uttar Pradesh, Bihar, Chhattisgarh, Orissa and Andhra Pradesh.

TH-802

It is derived from a cross between Healani and Castle Rock and developed at PAU, Ludhiana. Plants are determinate, spreading type with broad leaflet and dark green foliage. Fruits are round, medium size (68 g of average weight) with thick pericarp (6.4 mm) and 4.75% TSS. It gives an average yield of 631.8 q/ha in 140 days of crop duration.

TH-2312

It is derived from a cross of VFN-8 and Punjab Chhuhara and developed at PAU, Ludhiana. Plants are dwarf and spreading with dense foliage. Fruits are oval-round, medium size (67.5 g average weight), medium firm with thick pericarp (5.4 mm) and 4.75% TSS. It gives an average yield of 631.8 q/ha (6.46% increase over Rupali) in 150 days of maturity duration.

Utkal Kumari (BT-10)

This is a bacterial wilt resistant and indeterminate variety, developed at OUA& T, Bhubaneshwar. It is derived from the cross Selection 22 x BT-2. Fruits are round, red, medium-large with 5-6 days shelf-life. First picking starts 90-95 days after sowing. It gives an average yield of 404 q/ha in 90 to 100 days (after transplanting) of crop duration. Recommended for release and cultivation in 1995 for Chhattishgarh, Orissa, Andhra Pradesh, Rajasthan, Gujarat, Haryana and Delhi.

Utkal Urbasi (BT-12)

This variety has been developed at OUA&T, Bhubaneshwar. It derived from the cross Punjab Chhuhara x LE-79. Plants are indeterminate (70-85 cm) with good foliage cover, bear 5-8 fruits in cluster. Fruits are pear shaped, red with green shoulder, 2-3 locules, low seed content and thick skin, 10-12 days self-life. First picking starts 100-115 days after sowing. This variety is resistant to bacterial wilt and tolerant to early blight. It gives yield of 300-400 q/ha in 105-110 days of crop duration. Recommended for release and cultivation in 1996 for Jammu & Kashmir, Himachal Pradesh, Uttaranchal, Punjab, Uttar Pradesh and Bihar.

Swarna Baibhav

It is determinate type F_1 hybrid (H- 3 x CHRT-4), developed at Horticulture and Agro-forestry Research Programme, Ranchi, Jharkhand. Fruits are round, deep red colour, firm medium size (150g) and contains high pulp. This hybrid is suitable for long distance transportation. An average yield of this hybrid is 700-800 q/ha in 1 10-120 days of crop duration.

Swarna Lalima

The variety has been developed through pure line selection at Horticulture and Agro-forestry Research Programme, Ranchi, Jharkhand. Fruits are deep red round medium size (125 g) with TSS 4° Brix. This variety is resistant to bacterial wilt. It is recommended for cultivation in Jharkhand, Bihar and adjoining areas where bacterial wilt is a serious problem. The yield potential of this variety is 500-550 q/ha in 1 10-120 days of crop duration.

Swarna Naveen

It is developed through pure line selection at Horticulture and Agro-forestry Research Programme, Ranchi Jharkhand. Fruits are deep red, round, medium size (60 g) with TSS 5° Brix. It is resistant to bacterial wilt. It is recommended for cultivation in Jharkhand, Bihar and adjoining area. An average yield of this variety is 600 q/ha.

Kalyanpur Angoorlata

It is a selection from local germplasm at Kanpur, released in 1982. Plants are creeping nature, 2-3 m tall and indeterminate growth habit. Leaves are medium, light green with spaced leaflets. Fruits are small, oblong, 8-10 fruits/ trusses on average, skin firm and thick, and less juicy. Immature fruits are yellowish green and red on ripening. It is moderately resistant to leaf curl mosaic and resistant to root knot nematode. Fruit yield is 450-500 q/ha. Seed yield is 60-70 kg/ha.

Kalyanpur T-1

It is a selection from local germplasm made at Kanpur and released in 1982. Plants are 100-120 cm tall, erect and indeterminate in growth habit. Foliage is green in colour. Fruit is round, slightly pointed at the stigmatic end, 4-5 loculed, 4-5 fruits/truss, non-cracking, light green when unripe and red on ripening. It is susceptible to tomato mosaic virus. Seed yield is 80-100 kg/ ha.

Keckruth

It is a derivative of the cross Kecki Methi *x* Ruthger. Plant is 100 cm tall, and stem and foliage are green colour. Flowering takes place 93 days after transplanting. Fruits are round, medium to large (70-80 g), deep red on maturity, 3-4 locules and pericarp 0.3 cm thick.

Kechruth Ageti

It is a derivative of Keckruth x Ruthger. Plant is 100 cm tall. Stem and foliage are deep green. Flowering starts 85 days after transplanting. Fruits are round, medium in size (40-50 g) with green shoulder, which disappears on ripening and 3-4 locules.

KS-1

It is a medium early variety having determinate growth habit. Fruits are medium in size, round, flattened and smooth. It is suitable for table purpose.

KS-2

It is also mid early maturing variety having determinate growth habit. Fruits are medium in size, green shoulder, flattened, smooth and semi -fleshy. It is suitable for table purpose.

KS-3

It is a determinate and mid early maturing variety. Fruits of this variety are having green shoulder.

Jawahar Tamatar-99

This variety has been developed from JNKV, Jabalpur by a germplasm collected from USA. Plants are determinate in growth habit. It is an early variety. First picking starts from 100-110 days after seed sowing. Fruits are round, hard and medium size with attractive yellow colour after ripening. This variety is suited for processing due to high TSS (5.3%). An average yield of this variety is 475-500 q/ha.

Sweet-72

It is derived from a cross of Pusa Red Plum x Sioux and developed at the Regional Agricultural Research Institute, Gwalior (Wakankar, 1977). The plants are semi-erect in height and their height is 57 cm. The first picking starts 75-80 days after transplanting. The fruits of Sweet-72 are round oval and somewhat flattened in shape. They are attractive and scarlet red in colour. The fruit is firm, fleshy, possesses fine flavour and sweet in taste. It possesses fewer seeds. The numbers of locules are 4.6. The average weight of fruit is 75 g. Vitamin-C content of tomato Sweet-72 is very high (20.91 mg per 100 ml). It is very sweet in taste with sugar percentage of 1.62. The acidity content is 0.43 mg per 100 ml. An average yield of this variety is 200 q/ha.

Pant T-1

This variety has been developed from GBPUA&T, Pantnagar. Plants are dwarf, straight and 100-120 cm tall. Fruits are round with nipple having 4-6 lowle. Fruiting occure in cluster of 4-5 fruits in each cluster. This variety is tolerant to tomato leaf curl virus.

Pant T-3

It is a pure line selection released from GBPUA&T, Pantnagar. Plant is semi-determinate thick, round with hairy stem. Fruit is medium round and smooth. It is mostly suitable for winter season. It is recommended for Delhi, Uttar Pradesh, Orissa and Maharashtra.

Pant Bahar

It is a selection from germplasm lines maintained at Pantnagar and released by Central Sub Committee on Crop Standards, Notification and Release of Varieties in 1985 for cultivation in both autumn- winter as well as spring- summer season in the plains. Plant is 93 cm tall, bushy, indeterminate in growth habit and branched. Stem is round, medium thick and pubescent. Leaf petiole is purple pigmented at young stage. Fruiting starts 31 days after transplanting and fruits are ready for harvest 75 days after transplanting. Fruits are flattish round, medium in size (polar diameter 4 cm, equatorial diameter 5.3 cm), average fruit weight 58 g, 5-6 loculed, slightly ridged and red on maturity. It is resistant to verticillum wilt and fusarium wilt under field conditions. Fruits can be stored for 5-6 days in winter season and 2-3 days in summer season. Fruits are good for processing purposes.

NDT-108

This variety has been developed from NDUAT, Faizabad. Plants are dwarf. Fruits are like Punjab Chhuhara having good shelf-life. An average yield of this variety is 260-300 q/ha.

NDT-120

This variety has also been developed from NDUAT, Faizabad. Plants are dwarf, harvesting starts 75-80 days after transplanting. Fruits are attractive, big in size and red in colour. An average yield of this variety is 300-350 q/ha.

Azad T-2

It is a derivative of the cross between HB-5 x Kuber, released in 1983 from CS AUA&T, Kanpur. Plants are 50-60 cm tall, compact and determinate. Leaf is curved inward. Fruits are round (Polar diameter 3.5-4.0 cm and equatorial diameter 4 - 4.5 cm), green when unripe and red on maturity with light green netting on dorsal side. Fruits are in clusters, consisting of 6-8 fruits/ truss and 50-60-fruits/ plant. It is resistant to root knot nematode and moderately resistant to leaf curl virus. It is suitable for processing purposes. Fruit yield is 400-500 q/ha.

Junagarh Ruby

It is a derivative of the cross between Inbred Big Silari x Pusa Ruby, released in 1981 from Gujarat state. Plants are semi-spreading in growth habit with purple attractive growth and more foliage. Leaves are medium in size and dark green. Fruits born in cluster (4-6 fruits/truss), round without grooves, four loculed and orange red on ripening with pinkish red flesh. Days to first picking are 85-90 days. On an average, it gives 20 pickings at short intervals. It does not possess complete resistance against any disease and insect/pests.

Marglobe

It is a mid season variety. Plant is 70-80 cm tall, vigorous, erect and indeterminate growth habit. Leaves are large and broad with dense foliage. It has typically 4 or 5 flowers on truss, which bears 3 or 4 fruits. Fruits are nearly round, deep scarlet colour on ripening, 3-5 locules, juicy, mildly acidic, thick walled, smooth, solid and less seeded. Fruits have dark green shoulder. Diameter of ripe fruit is about 7.5 cm and weighing 100-150 g.

Labonita

It is an introduction from USA. Plants are dwarf and determinate in growth habit. Fruits are plum shaped. Plant is heavy bearer and suitable for spring summer season. It is resistant to fusarium wilt (Thomas *et al.*, 1976).

Sioux

It is an exotic variety introduced from USA. It is indeterminate type and produces yellowish green, medium large fruit which becomes uniform red at ripening. It has ability to set the fruits in spring and summer seasons.

Ox Heart

Plant is 70-80 cm tall. Foliage is dark green in colour. Fruit is heart shaped, pink in colour and each weighing 300-400 g. Interior of the fruit extremely solid with thick firm walls and small seed cavities with few seeds. Crop duration is 115 to 130 days.

Roma

It is an introduction from USA. Plants are semi-dwarf, bushy growth habit with luxuriant foliage cover. Fruits are borne in 4-5 clusters, pear shaped with thick pericarp, 7 cm long and 5 cm in girth, red in colour on ripening, free from green shoulder, bilocular, small seed cavity and less seeded. This variety possesses good keeping quality and good for processing.

LE-79 (Shakati)

It was developed through pure line selection at College of Horticulture, Vellanikkara, Kerala. Plants are determinate with pubescent stem and normal leaf Fruits are round, white and of medium size with average weight of 58 g. First picking starts 75 days after sowing. It is resistant to bacterial wilt and fruit cracking It gives an average yield of 435 q/ha in 95-100 days of crop duration. It is recommended for cultivation in the Jammu & Kashmir, Himachal Pradesh, Uttaranchal, Chhattisgarh, Onssa and Andhra Pradesh, Madhya Pradesh, Maharashtra, Karnataka, Tamil Nadu and Kerala.

NDTR-1

It is a derivative from the multiple crosses between *Lycopersicon esculentum* x *L peruvianum* and released in 1978 from Karnataka. Plant is 60 cm tall, spreading type and indeterminate growth habit. Leaves are thick and dark green. Fruits are medium to large (70-80 g), round, 4-6 loculed, 3-4 mm flesh thickness and bright red on ripening. Seeds are light brown colour. Fruit yield is 175-250 q/ha. Seed yield is 1.25-1.40 q/ha in 100 to 120 days of crop duration.

Solan Sagun

This is an early hybrid, the flowering starts 24 days after transplanting. It yields about 580 q/ha of fruits. This hybrid is quite suitable for low and mid-hills of Himachal Pradesh (Sharma and Rattan, 1995).

Solan Gola

This variety has been developed from YS Parmar University of Forestry and Horticulture, Solan. Plants are indeterminate in growth habit. Fruits are round attractive, medium size, red colour and thick pericarp. This variety is suited for transportation to distant markets. An average yield of this variety is 375 q/ha.

Yashwant-2

This variety has been developed from YS Parmar University of Forestry and Horticulture, Solan. Plants are indeterminate growth habit and start fruiting 75 days after transplanting. Fruits are uniform red in colour with thick pericarp. An average yield of this variety is 500 q/ha. This variety is resistant to buckeye rot.

Sonali

It is a high yielding bacterial wilt resistant variety evolved from an exotic collection and recommended for the bacterial wilt prone southern district of Konkan. It matures in 120-140 days and gives 500 q/ha fruit (Patil *et al.*, 1990).

RHRTH-57

It is derived from a cross of 32-2 and 85-1 and developed at MPKV, Rahuri. Fruits are oval, red, medium size and medium stout with 4.1% TSS and uniform ripening. Gives an average yield of 730 q/ha.

RHRTH-92

It is derived from a cross of Dhanashree and RHRT-1 and developed at MPKV, Rahuri. Fruits are round, orange-red, medium sized, smooth surface, very stout with uniform ripening and 3-9% TSS. Gives an average yield of 669.5 q/ha.

Raja Shree

This hybrid is released in 1994. Fruits are medium (90-100 g) firm with high pulp and ascorbic acid (42-50 mg/100 g) content, attractive orange coloured and oval round in shape. It is suitable for long transportation and comparatively less susceptible to the viral diseases (Wagh, *et al.*, 1995).

Balkan

It is an introduction of Bulgaria. Plants are erect and compact with darl leaves resembling as potato. Stems are thick and hardy, plant remains upright throughout the season. Stem internodes are much shorter, resulting in clustering of fruits. It is a prolific bearer (40-50 fruits/ plant), dwarf (35-40 cm) with good foliage cover. The unripe fruits show green stems ends, whereas, the ripe fruits are slightly oval and possess two locules. Fruits are dark reds with no greening or spot and medium to small in size. Plants can withstand in considerable cold. Due to its dwarf, compact and erect habit, more plants/ unit area can be accommodated. Though, the fruits are medium small but attractive and firm, hence suitable for transportation. It is good for cultivation in early winter *(rabi)*. It can be used in hybridization for developing cold and frost tolerant varieties (Sexana and Thomas 1988).

VC 48-1

This variety was brought from Asian Vegetable Research and Development Centre, Taiwan and tested at Assam Agricultural University, Jorhat. Plants are determinate and 80-90 cm tall with heavy foliage. Fruits are medium round and bright red in colour having 5.5% TSS. This variety has been recommended for bacterial wilt infested area.

Krishna Nagar (S-20)

This variety has been released by Department of Agriculture (W.B.). It is a medium to late variety. Fruits for first picking are available 70-80 days after transplanting. Fruits are medium-sized, round in shape and red in colour. It yields nearly 300 to 325 quintals of fruits per hectare.

Floradade

This variety was brought from Florida. Plants are determinate (45-50 cm) tall. Fruits are big in size weighing 150-175 g/ fruit. This variety is early blight tolerant. An average yield of this variety is 400-500 q/ha. This variety has very good keeping quality.

ARTH-3

This hybrid developed at Ankur Seeds Pvt. Ltd., Nagpur. Plants are determinate with medium dark green leaves. Fruits are squarish round and deep red. Flesh is red with 4.5 to 5.0 per cent TSS. It is moderately tolerant to drought. Harvesting starts 80-85 days after sowing and continues up to 160-180 days. This variety gives an average yield of 880-950 q/ha. It is recommended for Assam, West Bengal, Maharashtra, western part of Madhya Pradesh, Karnataka, Tamil Nadu and Kerala.

ARTH-4

This hybrid has also been developed by Ankur Seeds Pvt. Ltd., Nagpur. Plants are semi-spreading and indeterminate with dark green leaves. Fruits are deep red and flat round. Flesh is red with 4-4.5 per cent TSS. It is moderately tolerant to drought. Harvesting starts after 80-85 days after sowing and continues upto 170-190 days which gives an average yield of 1000-1100 q/ha. It is recommended for Punjab, Bihar, Uttar Pradesh plains, Karnataka, Tamil Nadu and Kerala.

DVRT-2

This variety has been developed at IIVR, Varanasi, derived from the cross between *L. esculentum* cv. Sel-7 and *hirsutitm* f. *glabratum* B 6013, following backcross-pedigree selection. Plants are determinate. Fruits are large, spherical, attractive red with 5-6 locules.-Gives an average yield of 600 q/ha, recommended for cultivation and release in 2001 for Rajasthan, Gujarat, Haryana and Delhi.

VRTH-1 (Kashi Hybrid-1)

This is an early maturing hybrid having determinate growth habit developed at IIVR, Varanasi. The average yield is 1100 q /ha. This hybrid shows 87 % heterosis over standard check (ARTH-3). It posses field tolerance to tomato leaf curl virus and early blight.

VRTH-2 (Kashi Hybrid-2)

It is a short duration hybrid, having maximum fruit production of 900 q/ha. This hybrid has been developed at IIVR, Varanasi. This hybrid expresses 54 % heterosis over standard check (ARTH-3).

Manikhamnu

It is dwarf (34.5 cm), early maturing, determinate and multibranched. Fruits are medium size, round, smooth with 2-3 locules, uniform red at maturity and ready for harvesting 70-75 days when planted in September and 85-90 days in case of spring summer planting. Flesh is thick and juicy with 5.4 per cent TSS. It is suitable for processing as well as fresh consumption purposes. It is resistant to fruit cracking and tolerant to leaf curl virus. The yield potential is 533 q/ha. This variety is released from State Variety Release Committee, Manipur (Yadav *et al.*, 2000).

Manileima

It is an early maturing, determinate and multibranched variety. Fruits are medium size, oval, nipple shaped and bilocular. It is suitable for processing and long distance transportation. It is resistance to fruit cracking and tolerance to leaf curl virus. It is also tolerant to moisture stress and suitable for spring summer season. The yield potential of the variety is 630 q/ha. It has been released from State Variety Release Committee, Manipur (Yadav *et al.*, 2000).

Manithoibi

The plants are dwarf (35 cm), determinate and multibranched. It is an early maturing variety and capacity to withstand in moisture stress condition. Fruits are oval, medium, bilocular, good shelf life and uniform ripening. This variety matures in 70-75 days when planted in September and 80-90 days in case of spring summer planting. The yield potential of this variety is 637q/ha. It is released from State Variety Release Committee, Manipur (Yadav *et al.*, 2000).

Gulmohar (MTH-6)

It has been developed at Maharashtra Hybrid Seeds Co. Ltd; Jalna. Plants are determinate. First picking starts 60-67 days after transplanting. Fruits are red, square shape with average weight of 90 g. TSS is 4.17 per cent. It gives an average yield of 900-1000 q/ha. It is recommended for cultivation in Maharashtra, western part of Madhya Pradesh, Karnataka, Tamil Nadu and Kerala.

Meenakshi (BSS-20)

This hybrid has been developed at Beejo Sheetal Seeds Pvt. Ltd., Jalna. Plants are indeterminate. Fruits are dark red, round global with average weight of 80 g. It is resistant to fusarium wilt. It gives an average yield of 600-1000 q/ha.

Madhuri (BSS-39)

It has been developed at Beejo Sheetal Seed Pvt. Ltd Jalna Plants are determinate and early in maturity. Fruits are dark red, flat round with thick skin, average weight of 100 g, crack resistant and can with stand long transportation. It gives an average yield of 500-700 q/ha.

Megha (BSS-98)

It has been developed at Beejo Sheetal Seed Pvt. Ltd., Jalna. Plants are determinate. Fruits are red with thick skin, good shelf life and transport quality. It gives an average yield of 500-700 q/ha.

Manisha (BSS-99)

It has been developed from Beejo Sheetal Seed Pvt. Ltd. It has good self-life and transport quality. It gives an average yield of 800-1000 q/ha.

Hybrids and Varieties Developed by IAHS Co., Bangalore

Mangala

The plant of this hybrid is determinate type. Fruits are medium-sized, good in quality. First picking of fruits is obtained 70 days after transplanting. Average yield is about 750 quintals per hectare.

Sheetal

Plant habit of this hybrid is determinate. Fruits medium-sized (120 g). Fruits are free from cracking. Plant starts yielding about 70 days after transplanting. The harvesting duration, is about 30 days. Average yield is about 750 quintals per hectare.

Vaishali

Plants are determinate type. Fruits are well suitable for processing. Fruits are medium-sized (about 100 g) of high quality. Plants start yielding in about 70 days after transplanting. The harvesting duration is 30 days. Average yield ranges 800-880 quintals per hectare (Yawalkar, 1985).

Rupali

This hybrid is grown on varying climatic condition. It has ability to set fruit under high temperature range. The fruits are round in shape.

Rashmi

The fruits of this hybrid are round in shape and firm. Fruits are suitable for processing. Plants tolerant to fusarium, verticillium wilt, alternaria and stem canker.

IAHS-88-2

This variety produces large fruits. Plants are tolerant to fusarium and verticillium wilt, root-knot nematodes. This is suitable to supply tomatoes fresh in local markets.

Naveen

Fruits are round in shape and medium in size. This is suitable for the supply of fruits fresh in local markets and for processing. This variety is having high yield potential.

CLIMATIC INFLUENCE

Tomato is a warm season plant. It cannot withstand severe frosts. During hot and drought conditions immature blossoms usually drop because of increased transpiration. For plant growth, atmospheric temperature should be in the range of 20 to 28°C. Night and day temperature below 15°C and above 39°C, respectively may cause poor germination and develop off-shaped fruits. A warm sunny weather is most suited for its proper ripening, colour, quality and high yields. Lycopine a red pigment in tomato fruits is destroyed at 40°C and fruits become yellow.

It requires long season to produce a profitable crop. The optimum range of temperature is 21-24°C. The temperatures below 16°C and above 27°C are not desirable. The minimum and optimum soil temperatures for seed germination are 10°C, 30°C and 35°C, respectively. Seed germination maximum occurs at 24°C. Maximum fruit setting occurs at night temperature of 15-20°C. Flowers fails to set fruit below at 13°C and beyond at 38°C. Temperature exceeds 32°C, the formation of lycopene is inhibited. Lycopene is the highest at 21-24°C, while the production of this pigment drop off rapidly above 27°C. Tomato plants are highly susceptible to water logging, hence an areas of high rainfall, if proper drainage is not available, its cultivation is not possible.

It is not only sensitive to frost but it does not thrive at low, non-freezing temperatures. High temperatures, accompanied by low humidity and dry winds, frequently damage floral parts and there is no fruits set. Smith (1932) found that tomato pollen grains germinate best at 29.4°C, poorly at 10°C and very poorly at 37.3°C. Thus both high and low temperatures interfere with the setting of fruit. Tomato plants withstand drought fairly well but fruits are subject to blossom end rot and to growth cracks if moisture supply follows drought. It cannot be grown successfully in regions of higher rainfall due to possible damage by diseases.

SOIL REQUIREMENT

Tomato is grown in many type of soil from sandy to heavy clay. Light soils are good for an early crop, while clay loam and silt loam soils are suited for heavy yields. A well-drained fairly fertile loam with a fair moisture holding capacity is ideal for growing of good crop of tomato.

Tomato do well in soil reaction from pH 6.0 to 7.0. It is moderately tolerant to acid soils (pH 5.5). Application of lime is advisable to improve soil pH when it is below 5.

The field chosen for seed crop should not be grown with any variety or species of tomato as a previous crop. Land should be well prepared by 3-4 ploughings followed by planking once if needed to make friable seed bed before sowing and transplanting (Singh, 1989).

Salt stress do not generally reduce growth of young, growing leaves by decreasing H+-ATPage activity and support the hypothesis that salt stress affects mature leaves and, as a consequence, their ability to supply essential metabolites for growing regions of both shoot and roots (Wilson *et al.*, 1984).

Salt tolerance varies with cultivars but a number of tomato cultivars can tolerate salts up to 8 ds/m without affecting much yield reduction significantly. Fruits ripening is delayed by salts. Cultivars like Marglobe and Pusa Ruby 'are having better salt tolerance capacity (Jaiswal and Singh, 1989).

SOWING

Sowing Time

Tomatoes can be grown almost all the year round. The planting seasons vary from place to place of the country. In northern India, for spring summer crop, the seeds are sown in November and transplanting is done in second fortnight of January. In area, where frost does occur only one sowing is in July - August with transplanted in August -September. For autumn crop in north India, seeds are sown in July-August and transplanted in August-September. In hilly areas, the seeds are sown in March-April with transplants in April-May.

Seed Rate

About 500-800 gram of seed is enough to raise sufficient seedlings for the planting of one hectare area. The quantity of seed varies with season and method of cultivation. Tomato seeds remain viable for four years if stored under good storage conditions. One gram contains about 300 seeds.

Seed Treatment

It is advantageous to treat tomato seeds with IAA in talc powder and to sow the seeds after one months of treatment for better plant growth and higher fruit yields (Singh and Singh, 1991).

Tomato seeds can also be treated osmotically (seeds are soaked in -10 MPa polyethylene glycol (PEG) and are dried before sowing) that results in higher and uniform germination, seedlings growth decreases seed dormancy and improves seed vigour (Lic, 1994).

Raising of Seedlings

The field for raising nursery should be highly porous and fertile. The soil should, be mixed with well rotten compost, or farmyard manure. Raised beds are prepared and the seeds are sown in lines of 5 cm' apart. The seeds are covered with sand. 'Watering should be done with rose can. Nursery beds may be covered with dry Saccharum grass for 3-5 days to induce early germination of seeds because of darkness. This coverage should be removed immediately as soon as seedlings are seen emerging out of the upper soil layer. Seedlings should be protected from direct sun, heavy rains, high winds and frost. Copious irrigation should be avoided to prevent damping off disease. The seed should be treated with Captan or Thiram at the rate of 4 g/kg of seed at the time of sowing.

The seedlings will be ready for transplanting 25-30 days after sowing. Healthy and vigorous seedlings that have produced five leaves should be used for planting in the main field to have a good establishment and to achieve uniform stand in the field.

PLANTING

Tomato seedlings become ready for transplanting when they attain 10-15 cm height and are of 3-4 weeks old. They can be accepted for transplanting up to the age of 40 days, not beyond, irrigation should be withheld for about 4-days. Seedlings should be uprooted carefully from the nursery bed. Transplanting is done either in flat bed (in light soils) or on ridges (in heavy soils) and for taking rainy season crop at the spacing of 60 to 75 cm between rows and 30 to 45 cm between plants within rows. High plant density (60 x 30 cm) results in high total seed yield. High plant density also requires high quantity of nutrients and irrigation water for successful cultivation of tomato crop for fruit and seed yields both. Planting on ridges provides a good support to plants and less attack of fruit diseases. Further, ridge planting with paddy husk results in better survival and plant growth and produces higher marketable fruits (Singh *et al.*, 1995). Winter crop, gives higher fruit yield hence, transplanting is done in September - October.

Planting may be done preferably In the evening. Light irrigation is done immediately after planting.

The planting distance within a row is kept at 30 cm for determinate varieties and 60cm for indeterminate varieties. For F1 hybrids the distance is increased as per the agro-techniques recommended for such hybrids. Initial trial indicated that planting in double row system (30 cm X 30 cm X 1 m) on raised beds provides higher yield of healthy fruits, besides facilitating intercultural operations.

Singh *et al.* (2000) studied the quality and resistance of tomato varieties to population density and training methods and found that quality characters of cytokine, carotene, vitamin C and total soluble solids were best when tomato vines are trained on high trellis. The hybrid Naveen was noticed to be superior to Cv Punjab Tropics in all attributes and lowest plant density (24691 plants per hectare).

Singh (1994) reported the performance of kharif tomato as influenced by cultural practices and found that the transplanting done in second and fourth week of May recorded significantly higher yield as compared to fourth week of June planting. Delaying of transplanting significantly decreased fruit weight and number; whereas crop stand and fruit firmness showed improvements. In both plain bed and ridge furrow staking, the unmarketable yield was significantly reduced, thus producing higher net return and net profit. Cultivation systems had no influence on fruit quality and crop stand. Second and fourth week of May plantings also proved beneficial by recording significantly higher net returns and net profits.

Hooda *et al.* (1998), reported the effect of seeding/planting time and mulching on tomato production. As compared to others treatments, the treatment with 15 December direct seeding with black polythene mulching gave the maximum plant height, number of branches, number of fruits per plant, fruit weight, total marketable yield and cost benefit ratio.

NUTRITIONAL REQUIREMENT

Tomato is a heavy feeder of macro-nutrients (N.P.K.) and therefore these should be applied in right doses in right time and with suitable method so that better growth and development is attained that will ensure higher fruit yield.

Tomato crop is responsive to manorial and fertilizer application. An yield of tomato of about 37 tons/ha removed 80 kg of nitrogen, 20 kg of phosphorus and 130 kg of potassium from the soil. However, the exact requirements of manures and fertilizers would depend on the fertility status of the soil.

Nitrogen is the most important for plant growth flowering and fruiting. Higher level of phosphorus in combination with nitrogen helps in getting higher seed, yield and seed quality (George *et al.*, 1980). Application of potassium is an important as nitrogen and phosphorus. It causes firmness in fruits, development of fruits and improvement in ripeness (Kalloo, 1985). Organic matter helps in improving physical, chemical and biological properties of soil in turn provides favourable crop growing conditions apart from adding several nutrients. Calcium deficiency (falls below 0.2% in plants) leads to a disorder known as 'blossom-end rot' or 'black spot' in tomato. It should be taken care of by spraying of calcium Chloride 0.5 per cent. Magnesium is a constituent of chlorophyll, and maintains the dark green colour of the leaves. Sulphur involves in chlorophyll formation. Manganese activates oxidation and reduction reactions (Kalloo, 1985). Besides, micro-nutrients need to apply through foliar sprays, as and when symptoms are seen. Among micronutrients, boron and zinc are more important. The deficiency of boron in tomato leads to fruits cracking, deshaped fruits and uneven ripening of fruits (Chauhan, 1981). First spray of boron at 0.3 per cent should be done on seedlings in the nursery bed before transplanting; second spray 3 to 4 weeks after transplanting of seedlings; third spray 3 to 4 weeks after the second spray. Availability of Zinc increases ascorbic acid in tomato fruits. Application of zinc sulphate at about 2.5 kg/ha increases seed content. The requirement of tomato crop for macro-nutrients is given in Table 2 (Tandon, 1991).

Table 2 : Recommended doses of nitrogen, phosphorus and potassium and farmyard manure for tomato crop by various states.

State	Recommended doses (kg/ha)			T/ha
	N	P_20	K_20	FYM
Andhra Pradesh	100	50	50	15-20
Assam	45	50	50	10
Gujarat	75	37.5	37.5	10-15
Haryana	100	50	50	25
Himachal Pradesh	100	60	60	-
Karnataka Irrigated	250	250	250	38
Karnataka Rainfed	60	50	30	25
Kerala	75	40	25	20-25
Madhya Pradesh	100	50	50	20
Maharashtra	75	40	40	40-50 carts/ha
Orissa	125	80	110	-
Punjab	138	62	62	25
Tamil Nadu	150	100	50	25
Uttar Pradesh	60	50	50	15-20

However, exact requirement of nutrients depends on the season in which tomato crop is grown, the type of soil, the form of fertilizers and the cultivars grown. The different recommendations made in various regions are 66 to 110 kg of nitrogen, usually in the form of ammonium sulphate, 35 to 60 kg of phosphorus in the form of super phosphate and 35 to 60 kg of potash (muriate of potash) per hectare. In addition, about 20 cartloads of fully decomposed farmyard manure are also applied (Nath *et al.*, 2008).

During rainy season, more quantity of nitrogen is required compared to winter and summer. Tomato crop needs more nitrogen when raised on sandy soils because such soil are subjected to leaching. The splitting doses are superior to basal doses in sandy soils. Further, determinate cultivars require more quantity of nitrogen than interdeterminate and semi-determinate (Kalloo, 1985).

Method and Time of Application

Organic manures are incorporated in the soil before or at final ploughing. Whole quantity of phosphorus and potassium is applied on both the sides of row and mixed with soil. Nitrogen is applied through top dressing along the rows of plants in soil and through foliar spray.

Application of nitrogen is done in two split doses during winter and in three splits during rainy season. The first application is done 3 weeks after transplanting of seedlings and the second application just at beginning of flowering and the third application is done during fruiting stage.

According to Nath *et al.* (2008) the complete doses of farmyard manure, phosphate and potassic fertilizers are applied at the time of field preparation for transplanting the seedlings. The fertilizers are placed in bands 7.5 to 10 cm deep on both sides of the row before making furrows. The nitrogenous fertilizer is applied in two equal split doses, the first to be given before transplanting along with phosphorous and potassic fertilizers and the second dose to be applied around each plant about 45 days after transplanting.

Kamalanathan and Thamburaj (1970) recommended 100 kg of nitrogen, 80 kg phosphorous and 50 kg potash and over 25 tons of FYM/ha for the tomato plants spaced at 75 cm X 45 cm apart. Kamruddin *et al.* (1978) reported that high doses of nitrogen produced significantly more number of flowers and fruits than the lower doses. Six sprays of one per cent urea at weekly interval, starting at the seedling stage, produced significant increase in the yield of tomato (Tiwari and Chhonkar, 1967). In rare cases, deficiency of boron was reported which was corrected by the soil application of 8 to 12 kg/ha of borax (Datta, 1961). Single morphactin treatment significantly checked the abscission and shedding of flowers, thereby increasing the number of fruits per plant. The quantity of fruit was maximum when treated with *GA3* treatments (Mukherjee, 1980).

Singh *et al.* (1995) studied the effect of different levels of nitrogen and phosphorus on yield of hybrid tomatoes. They reported the optimum doses for tomato hybrid TH-2312 up to 100 kg nitrogen and 90 kg phosphorous.

Tomato variety B7-10 was choosen to study the fertigation effect on tomato by Sungadhupe *et al.* (2005) at Bhubhaneshwar. Application of 31 and 69 per cent Nitrogen from 58 kg nitrogen/ha during initial crop growth stages and from flowering to ripening stages respectively resulted in a yield of 41 t/ha and saved 27.5 per cent nitrogen as compared to recommended dose of 80 kg/ha applied in split doses.

For southern and western region the requirement is 100 kg nitrogen, 80 kg phosphorus and 50 kg potassium or 100-120 kg nitrogen, 60-80 kg phosphorus and 100-120 kg potassium. For the states like Punjab, Haryana, U.P, M.P, Bihar and West Bengal, it is 150 kg nitrogen, 60 kg phosphorus and 60 kg potassium.

The hybrid varieties have higher requirements of the nutrients viz 200 kg nitrogen, 100 kg phosphorus and 100 kg potassium. The crop responds well to manurial and fertilizer application. The amount of fertilizers to be added depends on the variety and soil. A crop grown in spring-summer season, require less nitrogen as compared to winter season crop. Late maturing varieties require more nitrogen. In general, the crop requires 80-120 kg nitrogen, 50-80 kg phosphorus and 50-60 kg potassium per hectare.

Inadequate availability of major nutrients cause poor growth and development of tomato plants resulting in decrease in yield potential (Maynard, 1979; Shanmugavelu, 1989).

Nitrogen

The deficiency symptoms of nitrogen in tomato plants are seen by:

1. Light green leaves, lower leaves become yellow and later on dry up. Veins of leaves become deep purple in colour.
2. Stems become hard and purple in colour.
3. Plant growth is retarded.
4. Flower buds turn yellow and drop.

   Excess nitrogen results in laxuriant vegetative growth, and unfruitfulness.

Phosphorus

1. Phosphorus deficient tomato leaves become olive green in colour, underside, surface becomes purplish.
2. Stems become sender.
3. Sparse foliage
4. Growth of plants (both root and shoot) slows down drastically.
5. Maturity period of the tomato fruits is extended.

   Excess of phosphorus affects Lycopene content of tomato fruits adversely.

Potassium

1. The margins of the tomato leaves become yellowish or grayish and later on they become necrotic.
2. Stems become slender and may show necrotic areas.
3. There is uneven ripening of fruits and also they become puffy.

   Excess of potassium reduces the yield in tomato due to salinity.

Calcium

1. Terminal leaves become yellow, brown or purple and become necrotic.
2. Terminal flowers die.
3. Turgidity in plants reduces.
4. Roots become short and much branched.
5. Incidence of blossom-end rot is more.

Magnesium

1. Older leaves of tomato plants develop interveinal chlorosis and necrosis.
2. Petioles tend to downwards.
3. Leaf margins turn up.
4. Stalks become slender.
5. Roots grow longer with less number of secondary and tertiary roots.

Manganese

1. Plants become stunted.
2. Terminal leaves reduce in size and rolled forward and chlorotic.
3. Small dark brown spots are seen on younger leaves irregularly.

   Excess of Manganese may cause iron deficiency since it changes iron into inactive ferric form.

Causes of N Deficiency

1. Initially N content of soil is lacking.
2. Leaching of N which is common in sandy soils.
3. Low content of organic matter in soil.
4. When soil pH is low (below 5.0)

Causes of P Deficiency

1. Acid and alkaline soils release less quantity of P.
2. Prevailing of low soil temperature and / or low soil moisture.
3. Low P content and low availability in peat and muck soil.
4. High rate of fixation of P in the soil.

Causes of K Deficiency

1. When soils are acidic.
2. Sandy soils which are subject to leaching.
3. K is fixed in non-exchangeable form in soil colloids.
4. Irregular and inadequate addition of K fertilizer to crop (Bose *et al.*, 1993).

Use of Chemicals and Growth Substances

Use of plant growth substances in tomato has been found beneficial for yield, quality, earliness, cold and high temperature fruit setting and to develop resistance to TLCV. Seed treatment with B-naphthoxy-acetic acid (NBOA) at 25-50 ppm, Gibberellic acid (GA3) at 5-20 ppm and Chlorophenoxy acetic acid at 10-20 ppm were reported to improve in growth and yield of tomato. Seedlings soaked for 24 hours in NAA at 0.1 ppm showed higher fruit set and increased early and total yield. Foliar application of GA3 at 10 ppm, NAA 1,000 ppm, PCPA at 50 ppm 2,4-D at 0.5 ppm or cytozyme at 1.25% were reported to increase in fruit yield. Spraying of PCPA at 50 ppm, IAA 50 ppm or borax 1% gave better fruit set in summer. The foliar application of PCPA 50-100 ppm at the flowering stage increases the fruit set at low and high temperatures. Spraying is effective in detrerminate varieties having high degree of synchronization in flowering. The spraying at flower cluster is more useful, since the growth of the plant is not affected advessely. (Choudhury and Singh, 1960), reported that application of NAA 25-50 mg/l as as seed treatment increased fruits in tomato. Seed treatment with GA 5-20 mg/l; CIPA 10-20 mg/l and 2-4D, 0.50 mg/l for 8 hrs increased yield, T.S.S. and sugar content in tomato fruits (Mehta *et al.*, 1989). Foliar spray of Ethephon 250 mg/l increased fruit yield (Singh and Singh, 1984). Application of Paclobutrazol 40-150 mg/l as foliar spray increased yield and improved fruit quality (Baruah and Pandita 1994 and Rai *et al.*, 2001). Mixtallol 2 mg/l as foliar spray increased fruit yield (Shukla and Prabhakar, 1989).

Abdul Majeed *et al.* (1991) studied the response of tomato to stage and concentration of DCPTA application at Dharwad. The results showed that DCPTA can be applied either at nursery stage or 15 or 30 days after transplanting at a single spray at 80 ppm concentration to increase the yield of tomato.

Singh and Singh (1993} studied varietal response to 2, 4-dichlorophenoxy acetic acid and methods of its application on summer tomato. Pusa Ruby and Pant T3 cultivars showed positive response to all the formulations of 2,4-D when the seed was completely soaked for 24 hrs in 5 ppm solution, followed by foliar spray at 20,30 and 40 days after transplanting. The 2,4-D also increased the early yield (29.6-33.5%), fruit diameter, volume, earliness, fruits/ plant and reduced the leaf area (38.4-39.7%) and virus incidence (43.3-50.5%) compared with the control.

Barwah *et al.* (1996) studied the effect of Paclobutrazol (PP33) on yield and yield attributing characters of tomato under different nitrogen levels. The treatment combination which helped in improving the yield and other yield attributing characters was 150 kg N/ha +PP33 at 150 ppm which produced maximum number of secondary branches per plant, fruits per truss, total number of fruits per plant and also total yield.

Nainar *et al.* (1997) reported response of tomato accession to fruit cracking. Accession LE812 was found to be promising as it recorded higher yield per plant and resistant to cracking.

Growth regulators have shown useful results in the experimental plots but commercial utilization needs to be explored. Spray of gibberllic acid (50 ppm) at the fruit setting stage increased the ascorbic acid content (Oza and Ranganekar, 1969).

Singh (1999) reported the response of hybrid tomatoes to growth regulators. Plants treated with 500 ppm indolebutyric acid or 500 ppm indolebutric acid +500 ppm naphthalene acetic acid gave maximum yield and proved more economical in both the seasons.

Growth regulators like GA_3 can be sprayed @5 ppm at fruit set. Like wise 2,4D (© 5 ppm) also can be sprayed at 30, 45 and 60 after planting. GA3, has been found effective in hills whereas 2, 4-D in plains (Singh, 2004).

Gupta *et al* (2004) studied the effect of PGR (NAA and IAA) and micronutrient mixture on the ascorbic acid content in tomato products. The plant growth regulators IAA and NAA are applied @ 25 ppm and 75 ppm respectively. Multiplex and Humaur were applied ® 2500 ppm and 2000 ppm along with plant growth regulators. NAA (75 ppm) along with Humaur (2000 ppm) not only enhanced the vitamin 'C' content in fruits but also the ascorbic acid content.

Singh and Lal (2005) conducted an experiment at Pantnagar to study the parameters from the maturity to ripening of fruits. Ripe stage of harvesting was best for seed characters. The highest seed content was in Pusa Ruby (11.5 g/ kg of fruits weight), highest seed, fruit weight was in Pant-T-3 (0.32) while seed: pulp ratio was highest in Pant Bahar (0.010). On the basis of overall performance, Pusa Ruby followed by Pant Bahar have given the best results.

Application of 4000 ppm NaCl to the irrigation water or with spray of 2000 ppm cycocel + 0.25 percent Zn SO4 + 25μM proline at transplanting (El-Oksh *et al.*, 1980) proved beneficial for hardening of seedlings.

Application of Micronutrients

Micronutrients play an important role in tomato production. The application of micronutrients like Zn, Cu and B significantly increased the number of fruits per plant, yield and quality. Application of 20-30 kg/ha of borax and 0.50% Zn is beneficial for yield and good quality of fruit (Dutta, 1961), Zinc deficiency is found to occur in high pH soils. Mallick and Muthukrishnan (1980) found that micronutrients, especially zinc at high doses (500 ppm) increased ascorbic acid content by 2-7 mg/100 g. Maximum increase of 22% was recorded in sugar/acid ratio by using manganese at 10 ppm. The yield was highest when Zinc was applied at the rate of 10 ppm. Das and Das (1981) studied the effect of various micronutrients with and without urea on tomato. They found that plants receiving MO at 0.75% alone gave the highest yield of 296 q/ha as against 221 q/ha with the control treatment. In a study at Almora, U.P., Rawat and Mathpal (1984) found spraying of B, Cu, Fe, Mo or Zn at 30 and 45 days after planting increased fruit yield with Zn 0.5% having the most beneficial effect. Suryanarayana Reddy *et al.* (1985) reported that addition of Zinc, copper and boron significantly increased the number of fruits per plant compared to application of nitrogen, phosphorus and potassium alone.

Sharma (1995) had carried out studies on response of boron and calcium nutrients on plant growth, fruit and seed yield on tomato. Optimum doses of boron as 20 kg and calcium carbonate as 10 kg/ha was found to be most effective when applied alone or in combination and produced the highest yield of fruits and seed per hectare. Verma *et al.* (1995) reported the response of zinc and boron fertilization on yield and quality of tomato fruits at different stages of picking and the higher marketable fruit yield of 285.88 and 275.57 q/ha were recorded respectively. The effect of nutrition source on the yield of tomato has been reported by Sai Reddy *et al.* (2002) based on the results of the experiment. Number of flowers per inflorescence, fruits per truss, fruits per plant, early fruit ripening, fruit size were found to be maximum with foliar spray of borax. Marketable and total yield was maximum with foliar spray. Total yield of 365.4 q/ha was recorded when tomato plants were sprayed with borax @ 5 g/l 40, 50 and 60 days after transplanting. Dube *et al.* (2003) studied the effect of soil application of Zinc sulphate (1 mg, 2.5 mg and 5 mg and 10 mg per kg of soil) on tomato variety Pusa Ruby. It was concluded that though all parameters were high compared to control, yield attributes were highest @ 5 mg/kg of soil treatment. It not only increased the biomass but also increased the fruit yield.

Malkan Singh *et al.* (2003) studied the effect of foliar application of micronutrients on tomato variety Hisar Arun. Application of all the micronutrients @ 5% (0.83 g of each micronutrient Fe, mn, Cu, Zn) at 40, 50 and 60 DAP significantly increased the plant parameters viz plant height, branches/plant, flowers/fruits per plant.

On similar lines, Bhatt *et al.* (2006) studied the effect of foliar spray of Zn, Fe, Cu, Boron @ 100 ppm each and ammomium molybdate @ 50 ppm at Pantnagar. The nutritional status of the tomato fruit improved significantly as it exhibited higher phosphorus (21 mg) and sulphur 910 mg) contents over control.

Boron deficiency in the soil may cause cracking of fruits; hence Borax @ 0.3% can be sprayed at 15 days interval. Spraying of calcium chloride @ 2 g/l of water also reduces the cracking of fruits. A combination of micronutrients can be sprayed @ 0.5% in order to improve the nutritional status of the fruits.

Use of Organic Materials

Sai Reddy *et al.* (2002) have reported the effect of organic farming on the fruit yield of tomato. On the basis of the experiments carried out for two years during Rabi 1999 to 2000 to assess the effect of organic farming on the productivity of tomato at Rajendra Nagar, Hyderabad, the organic manures *viz.,* farm yard manure, vermi compost and their combinations with inorganic fertilizers were evaluated. The study revealed that application of organic manures alone reduced the yields by 29.19 per cent with farm yard manure @ 20 t/ha; 32.76 per cent with the use of vermi compost (5.0 t/ha) and 28.54 per cent with the use of their combination (FYM 10 t/ha +vermi compost 2.5 t/ha). Application of 50 per cent recommended N through FYM + 50 per cent with the application of recommended dose N as chemical fertilizer (urea) gave significantly higher yield of 30.5 t/ha which has 120:60:60 kg NPK/ha of fertilizers with 329.9 q/ha yield.

Lindsay (2005) studied the impact of vermicomposting on tomato @ 2 t/ha along with recommended doses of chemical fertilizers @ 100 kg nitrogen, 40 kg phosphorus and 60 kg potassium. It has shown a beneficial effect on fruit yield of 289 q/ha. It is evident that combination of FYM, vermi compost or green manuring with inorganic fertilizers i.e; integrated nutrient management improved the soil fertility status and reduces quantities of inorganic fertilizers to be applied over a period of time.

IRRIGATION

First irrigation is given just after transplanting of seedlings. The second irrigation should be given on the third day of planting and the subsequent irrigations at 7 days interval depending on the soil conditions and prevailing local climate. Rest of the irrigation should be given regularly so that the soil remains moderately moist. Excessive irrigation induces the plant to vine and dropping off the blossoms. During hot season, the tomato is irrigated once in a week. During winter season, whenever there is a danger of frost, the crop must be irrigated as it helps to maintain temperature above freezing. Imbalance soil moisture may result in fruit cracking (Vadivelu and Srimathi, 1988).

One can save 30 to 40 per cent water through drip irrigation. In this method, the water is applied at a slow rate with desired and required quantity adjacent to the root zone. Fertilizers can also applied with drip irrigation.

In fact, exact water requirement of tomato varies with varieties, but in genera!, at irrigation regime 0.5 IW/CPE ratio at pre-flowering, flowering and fruiting results in higher root length and number of feeder roots. However, irrigation at 0.9\W\CPE ratio results in the higher dry matter production, yield and higher cost benefit ratio (Saraswati and Pappaiah, 1993).

Irrigation should be done regularly so that the soil remains moderately moist. Excessive irrigation induces the plant to vine and dropping off of the blossoms. During summer season, irrigation at every three to four days interval may be necessary, whereas in winter, 1 0 to 15 days interval is sufficient.

Raina *et al.* (1999) carried out experiment on the use of drip irrigation and polyethylene mulch on yield, quality and water use efficiency of tomato variety 'Naveen'. Drip irrigation at 0.8 volume of water (80 per cent evapo transpiration crop) applied, gave significantly higher fruit yield (16.63 t/ha) compared with the surface irrigation (911.7 t/ha). Moreover polyethylene mulch plus drip irrigation further raised the yield to 23.25 t/ha. Fruit weight increased significantly while other analyzed quality characteristics did not differ significantly among treatments. Water use efficiency under the drip irrigation alone, drip irrigation plus polyethylene mulch and surface irrigation was 0.34, 0.48 and 0.16 t/ha, respectively. Drip irrigation, besides giving a saving of 54 per cent irrigation water, resulted in 40 per cent higher fruit yield compared with the surface irrigation.

Hebbar *et al.* (2002) carried out studies on NPK drip fertilization on yield and economics of field grown tomatoes. The fruit yield of tomato was 19.9 per cent higher in drip irrigation (71.8 t/ha) over furrow irrigation (59.6 t/ha). Fertilization with 100 per cent water soluble fertilizers (WSF) increased the fruit yield significantly (79.2 t/ha) over furrow irrigated control and drip irrigation. Sub surface drip fertilization (76.5 t/ha), nitrogen potassium fertilization (76.57 t/ha) and ½ soil and ½ fertilization had given fruit yield similar to WSF fertilization.

The cost of WSF was four times higher than that of normal fertilizers under the recommended rate of 180:150:120 kg, N, P_2O_5 and K_2O/ha. Since the net returns and yields under NK fertilization and ½ soils and ½ fertilization are comparable to WSF fertilization, it is concluded that fertilization with NK fertilizer is most promising and feasible technology for hybrid tomato production.

A record yield of 114.03 t/ha was obtained with Kashi Amrit under drip irrigation systems at spacing of 60 cm x 40 cm at IIVR, Varanasi (Lal, 2006).

Singh *et al.* (1993) reported the performance of tomato cultivars under rain fed conditions of Kandi area (Punjab). Punjab Kesari had maximum number of fruits per plant followed by THL-2312.

INTERCULTURE

Interculture is an essential for keeping down weeds and maintenance of soil mulch. As far as possible, all the intercultural operation should be shallow so that the roots which spread upto a depth of 5 cm below the surface soil may not be injured.

Weeding and hoeing should be done at regular intervals so as to keep the field free of weeds and to help soil aeration and proper root development.

As the plants grow, the intercultural operation should be shallow so that the damage to the roots, which are present five cm below the soil, be avoided.

Spring-summer tomato crop may be provided with shade by raising 3 rows of tomatoes, 1 row of maize that results in increase in percentage of fruit set, number of days from sowing to fruit harvest, number of fruits per/ plant and seed yield per kg fruit (Sharma and Tiwari, 1996).

Staking

Staking of tomato plants has proved to be beneficial. In indeterminate varieties the yield and quality have improved and more so in rainy and winter seasons where the incidence of some of the diseases of fruit and foliage is also reduced.

Individual plant is given support with the help of wooden stick or bamboo stakes. Support prevents plants from bending, fruits do not come in direct contact of soil, insects find comparatively less hiding place, spraying of insecticides become easy.

Rajeshwar and Patil (1979) observed that pruned and staked plants gave significantly more yield/plant than the unstaked, unpruned plants. Training of the tomato plants which includes pruning are practiced in American and European countries, but it has not become popular in India except in some F_1 hybrids, because of inconsistent results at various places and poor economics.

It is evident that staking in tomato reduces the infection by fungal diseases and also increases the yield in both determinate and indeterminate varieties. In case of indeterminate varieties, fruit yield and quality is much more improved that in determinate varieties.

Mulching

The main aim of mulching in tomato crop is to create a barrier to the moisture loss and to alter temperature by a few degrees, and to affect certain attributes pertaining to micro-climatic conditions of the plant particularly near to root zone. Mulching in nursery bed enhances tomato seed germination and plant growth. Mulching results in early flowering and early harvest also enables to earn more profits. It enhances early ripening of tomato fruits, higher yields and quality tomatoes because fruits do not come in direct contact with soil.

The water retention capacity of leaves has been found the maximum with the plant sprayed with 10^5 Molar concentration of Phenyl Mercuric Acetate at 30 and 60 days after transplanting and resulted in higher yields (Das and Panda, 1995).

Mulching materials should be locally available such as dry grasses, dry leaves, saw dust, paddy straw, paddy husk, polythene sheet, etc.

Chakravarti and Sadhu (1994) reported the efect of mulch type and colour on growth and yield of tomato which was carried out during winter seasons of 1991-92 and 1992-93. It was observed that polyethylene mulches irrespective of colour were superior to rice straw or water hyacinth. Higher soil temperature (2-3°C above the control), greater soil moisture conservation (31.5-67.7 per cent) and lower salinity levels (36.7-59.8 per cent) were also observed with polythene mulches but mulch colour had no appreciable effect. Black polyethylene was found more economical. Even paper/cured oil mulches appeared to be effective in suppressing the weeds in tomato under Texas conditions. Biodegradable paper and polymerized vegetable oil mulches also used. Under those conditions only one weed per 1 sqm plot was seen penetrating the paper mulch after 3 months compared to 10 weeds in the bare soil (Rezende *et al.*, 2006). Mulching with black ethylene sheet proved to be superior but pea straw, paddy straw etc. can be used economically. Kumar *et al.* (2005) studied the effect of different mulching materials on tomato. It was found that among mulches, total yield was highest under black polyethylene mulch (412.8 q/ha) followed by clear polythene (379.8 q/ha), sugarcane trash (357.5 q/ha) and rice straw (341.7 q/ha).

Training and Pruning

Indeterminate type of tomatoes respond well to training and pruning. Pruning besides removing the unwanted growth of plants, improves the size, shape and other qualities of the fruits. Pruning also breaks the apical dominance, eliminates the crown set and enhances the plant vigour. Apart from pruning in indeterminate tomato, early and total yield and quality are increased to a greater extent by training.

Clipping and pinching may require sometimes which is attended to carefully otherwise it will reduce total yield.

WEED CONTROL

Weeds affect growth and yield of tomato. They compete for nutrients and water, therefore, it is essential to check them before they become a problem. The early stage of crop is the critical period for weed competition and during this period free environment is needed to ensure good growth and yield. The first hand hoeing and weeding should be attended to on 25th day of planting and the earthing up on 45th day. To control weed 2 lit. Basalin mixed with 600-700 lit of water should 'be sprayed and mixed with soil before transplanting. The best weed control and the highest yield of tomato cv 'Rupali' during rainy season have been achieved with the use of 0.75 kg/ha. Metribuzin after transplanting in Bangalore conditions (Muniyappa *et al.*, 1995).

Gosh *et al.* (2002) reported bioefficacy of metribuzin 70 wp on weed control in tomato. They found that the predominant weeds were *Digitaria sanguinalis, Echinochloa colonnm, Eleusine indica, Cyperiis rotendus, Amaranthus viriis, Solanum nigrum, Digera arverisis, Portulaca oleracea and Phyralis minima.* The maximum reduction in the density and biomass of weeds was achieved under hand weeded (HW) plots during both the years. All the treatments used for weed control, provided significantly higher tomato fruit yield than unweeded control. HW twice recorded 43.98 per cent and 44.00 per cent increase in fruit yield of tomato over unweeded control. The corresponding figures for metribuzin 70 wp @ 600 g/ha was 40.7 per cent and 40.7 per cent. Thus preplanting application of these chemicals can safely replace the costly time consuming and traditional method of handweeding. Treflan at the rate 0.25 kg/ha was effective in controlling the weeds of tomato around Bangalore for about two and half months (Dhurai and Leela, 1970).

Chanda *et al.* (1994) reported the herbicidal effect in preventing weeds in spring summer tomato. In their studies with the application of fluchloralin (basalin) @ 1.0 and 2. Okg a.i./ha one day after transplanting (DAT); sethoxydin (nabu-s) @ 0.06, 0.12, 0.19, 0.39 and 0.75 kg a.i./ha 21 DAT; hand weeding twice (21 and 35 DAT) and unweeded control, sethoxydim

@ 0.75 kg/ha was found to be most effective in controlling grassy weeds followed byO.37 kg/ha. Post planting application of fluchloralin exhibited moderate effect in controlling weeds. Growth and yield of tomato were also influenced by the herbicide. Fruit number, weight and yield were higher in hand weeded plots followed by sethoxydim at higher doses. Better weight of fruit and yield were also obtained with fluchloralin at both doses applied on DAT. Cost benefit ratio indicated parity of the treatments.

Weeds can be controlled by Treflan (0.5-1.5 kg a.i/ha) and Basalin-45 EC as pre planting application, Sencor-70WP (0.35-0.5 kg a.i/ha) as pre and post emergence application. Pre plant incorporation of Basalin-45 EC and EPTAM 2-3 a.i kg/ha is also effective.

Improvement of Fruit and Seed Setting

The spraying of either NAA 50 ppm or Paras 200 ppm spray at flowering and 50 per cent flowering prevents flower drop and improves fruit and seed setting in tomato varieties (Vadivelu and Srimathi, 1988). For seed production of tomato, seed soaking in 2,4-D at 5 ppm for 24 hours has been found to be very effective (Singh and Singh, 1996). Application of Tria cantonal as Miracular in the form of a spray at 7.5 ppm on tomato crop, 4 weeks after transplanting is the most effective in enhancing fruit and seed yields/ha and improving seed quality of tomato cv 'Solan Gola' (Sharma and Rattan, 1995).

PLANT PROTECTION MEASURES

The most common insect pests and diseases attacking to tomatoes and their management are given below :

Insect-Pests

Fruit borer (*Heliothis armigera* Hubner)

These are polyphagus in habits and having a wide range of host plants. The larvae which are greenish in colour feed on the vegetative parts and cut holes on fruits and feed inside. Full grown caterpillars drop down from the plants and burrow in the soil where they pupate.

Control

Collection of affected fruits and picking up of caterpillars by-hands and destroying them proved to be very effective control of fruit borers. Adoption of proper crop rotation and growing of African marigold as a trap crop reduce the population. The most effective results have obtained when a row of African marigold *(Tagetes erecta)* raised on either side and paralled to 14 to 16 rows of tomato crop and 2 sprays of 0.07 per cent Endosulphan applied to the intercropped tomato (Rabinowitch *et al.*, 1974). Other insecticides that should be sprayed are carbaryl at 0.2% or malathion at 0.1%.

Jassids *(Empoasca devastans)*

Jassids are polyphagous in habits. They suck the sap of the leaves. The affected leaves turn yellowish and curl and wrinkle.

Control

Spraying of Phosphamidon at 0.03% or Parathion at 0.02% will control Jassids effectively.

Tobacco caterpillar *(Spodoptera litura Fabr.)*

The major hosts of this caterpillar are tobacco and tomato. The iarvae feed leaves, shoot and fruit at night. During severe infestation, entire crop may be defoliated over night.

Control

- Thorough ploughing of field will expose and kill the pupae in the soil.
- Collect and destroy the leaves having egg-masses and young larvae.
- Spray with 0.05% Dichlorvos or 0.05% Nuvan -100 @ 800 litres per hectare or 0.2% Carbaryl at 10 days in interval.

White fly *(Bemisia tabaci Gen.)*

Whitefties are tiny white insects. They usually crowded in between the veins on ventral surface of leaves. They suck the sap from the plants. The affected parts become yellowish, the leaf curls downward. They exude honeydew which favours the development of black sooty mould which interferes with the photosynthetic activity of the plant resulting in growth retardation. Whiteflies are well known a vector for transmitting viruses and mycoplasma. Rainy season is more favorable for its multiplication.

Control

- Spray the crop with Dimethoate at 0.05% or Metasystox at 0.1%
- On heavy infestation, spray with Oxydenceton methyl or Thiometon at 0.03 per cent.
- Remove alternate host plants like weeds and ornamentals plants.

Root Knot Nematodes *(Meloldogyne arenarla, M. incognita, M. javanica)*

These pests are very harmful to the tomato crop. Affected plants grow slowly, they become stunted and weak, foliage becomes pale yellow, tend to drooping followed by wilting. Invaded roots form galls which vary from spheroid to elongated spindles. The nematodes survive in the root debris and live from one crop season to next in the soil.

Control

- Follow crop rotation i.e. solanaceous vegetables crop should be grown only in rotation of after at least every third year.
- Plough the soil during summer that will expose larvae which are killed at high temperature of 40 to 50°C.
- Grow trap crop like *Sesamum orientale, Nusturtium, Tagetes* sp, Chrysanthemum (Singh, 1977).
- Apply organic amendments like sawdust, neem cake, groundnut cake, mustard cake, mahua cake at 25 quintals per hectare along with fertilizers (100-120 kg N, 60-80 kg P2O5 and 100-120 kg/ha) for economically reduction in nematodes and getting higher yields (Singh, 1985).
- Fumigate the soil with nematicide such as D.D. (dichloro propene dichloro propane) or Nemagon 20 to 25 kg/ha 3 weeks before planting at a depth of 20 cm at 30 cm apart with the help of soil injection.
- Apply Furadon (Carbofuran) or nemacur (Phenamiphos) or Phorate 2.5 kg a.i./ha in the soil immediately or seven day before planting (Singh, 1985) or Basamid GR (Dazornet) for effective nemacidal activities (Verma *et al.*, 1994).
- Drench seedlings in nursery bed with copper oxychloride (0.3%) alone and in combination with Phenamiphos 10 G (Nemacur) or carbofuran 3 G (Furadon 3 G) or Phorate 10 G at 1 kg a.i. per hectare (Gunasekaran *et al.*, 1994).
- Application of leaf extracts of neem (*Azadirachta indica*) and Nilgiri (*Eucalyptus tereticornis*) reduce effectively number of galls and egg masses of *Meloidogyne incognita* (Vats *et al.*, 1995). Seed coating of tomato with Achook, a neem based product improves plant growth, reduces severity of root knot infection, improves plant growth (Akhtar and Mahmood, 1985). Entegration of endomycorrhiza (*Glomus mosseae*) and neem cake control population of nematode. Mycorrhizal seedlings of tomato when transplanted in neem cake amended soil, reported to be the least infected by *Meloidogyne incognita* and there by plants could get desirable rhizopheric changes due to the addition of neem cake to the soil can keep with the effective utilization of *Glomus mvsseae* for the control of *M. incognita* on tomato (Rao *et al.*, 1995).

Diseases

Fungal Diseases

Damping off (*Pythium* Spp., *Rhizoctonia solani, Phytophthora parasitica, Selerotinia* Sp., *Botrytis* Sp. etc.).

Pre-emergence damping off causes killing or rotting of seedlings before coming out of the soil surface. Post emergence damping off causes toppling of seedlings on ground due to invasion of pathogen on cortical tissue of hypocotyle which results in constriction, shrinking of the base and seedlings in the nursery and newly planted one in field collapse. Fungi are soil borne.

Control

- Drench with 0.1 per cent Thiram to reduce pre-emergence damping off incidence.
- Treatment of nursery soil with formalin dust (15 parts of formalin and 85 parts of Charcoal ash) at 100 g/sqm. at 6-8 cm deep.
- Treat the seeds with Agrosan or Captan or Thiram (2.5/kg seed) before sowing.
- Drench the nursery bed with solution of captan or brassicol (2 g/litre of water).

Buckeye rot (*Phytophthora parasitica Dastur*)

It affects lower fruits rather than touching to the soil surface. Brownish circular spots are seen on immature fruits at the blossom-end. The fungus is soil inhabitant and having a number of hosts. Pathogen proliferates in the presence of high soil moisture and soil temperature (18 to 30°C). The disease is both soil and seed borne. Ill drained soil favours to disease. High nitrogen application also favours more incidence of the rot (Sharma and Sohi, 1983).

Control

- Treat the seed with Thiram or Captan (2.5 g/kg seed).
- Spray the crop with Difolaton 0.3 per cent 3-4 times at 10 days interval.
- Avoid raising tomato crop in low laying areas.
- Provide controlled irrigation.
- have provision of staking so that fruits do not touch the soit.
- Apply polythene mulch to obstruct dispersal of soil borne inoculum of *Phytophthora nicotianae* var parasitica and remove affected fruits (Dodan *et al.*, 1994).

- Spray the crop with Captafol or Mancozeb 0.2 per cent (Dodan *et al.*, 1994).
- Locate for buck eye resistant/tolerant varieties like Money Maker, Flat Large Red, Red Cherry (Sharma *et al.*, 1974).

Fusarium wilt of tomato (*Fusarium oxysporum, F. lycopersici,* Snyder and Hansen)

The lower leaves become yellow, wilt, and die. Young leaves show clearing of veinlets and drooping of the petioles. Growth of plants is stunted subsequently the affected plants wilt and die. There are water-soaked lesions on fruits. Warm humid climate favours the disease.

Control

- Use seeds from healthy plants.
- Treat the seed with Agresan GN or Thiram (2.5 g/kg of seed).
- Spray the seedlings with Dithane-Z-78 or Dithane M-45 (0.25%) at 7-8 days intervals.
- Do controlled irrigation.

Early blight (*Alternaria solani* Jones & Grout)

This disease causes damage in both the seasons. All plant parts suffer with the incidence of this disease. The very first symptom starts on leaves with circular to angular spots with dark brown colour. The spots are well recognized by concentric ring. Such spots are also appeared on stems, petioles, and fruits. The mature fruits are comparatively less affected than immature fruits. Usually, fungus fails to germinate on mature fruits. Weak plants are more affected by early blight. Low moisture favours this disease (Rotem, 1969).

Although incidence of disease starts early in the growing crop but its intensity increases with age of the crop. The causal fungus is soil borne. It also survives on the old stock. The disease commonly spreads through air-borne spores.

Control

- Seed treatment in hot water at 50°C for 20-25 minutes.
- Seed treatment with fungicide like Zineb 2g/kg of seed.
- Spray the crop with Dithane-Z-78 or Dithane M-45 at 0.2 per cent.
- Irrigate tomato crop very lightly but at frequent intervals.
- Apply recommended doses of nutrients (NPK).

Late blight (*Phytophthora infestans* (Mont) de Bary)

This is one of the most serious diseases of potato, and also affects tomato considerably. Affected leaves show water-soaked irregular spots on leaves. The disease spreads at extremely fast rate. Within 24 hours it can completely destroy the crop. The stem-end portion of the fruits is affected; spots of varying size are seen. Cloudy weather, intermittent rain and high humidity are favourable conditions for the spread of the disease.

Control

- Avoid planting of tomato crop near to potato crop.
- Controlled irrigations.
- Spray the crop with Dithane M-45 at 0.2 per cent.
- Remove affected plants and their parts regularly and destroy them out of field.
- Follow crop rotation.

Bacterial Diseases

Bacterial wilt (*Pseudomonas solanacearum* Smith) : This is a very destructive disease. Affected plants become stunted, foliage become yellowish in colour and then wilted. Severely wilted plants never recover. Yellow mass exudes when affected stem is cut. This is a seed borne disease.

Control

- Use healthy seeds and seedlings.
- Collect and destroy affected plants.
- Treat the seed with hot water at 50°C for 25 minutes before sowing in nursery bed.
- Apply oil cakes sawdust and adequate nitrogen.

Bacterial canker (*Corynebacterium michiganense* Jenson)

Affected young plants show wilting of leaflets at the extremities of plants. The petioles droop down the typical symptom is wilting of leaflets on one side of the nachis while those on the opposite side appear healthy. Water soaked spots on green fruits appear which soon become surrounded with halo.

Control

- Sterilize nursery soil before raising seedlings.
- Extract the tomato seeds through fermentation of the pulp at room temperature for 72 hours.
- Follow long crop rotation.

Viral Diseases

Leaf curl virus

The white flies (*Bemesia tabaci* Gen) are vector of the causal virus (Tobacco leaf curl virus (TLCV) and *Nicotiana virus* (Dodan *et al.*, 1994). The disease causes dwarfing, puckering and curling of the leaves towards dorsal side. Excessive branching, stunting of plants and partial or complete sterility are major symptoms.

Control

- Apply Phorate or carbofuran (1.5 kg a.i./ha) in the soil at the time of planting.
- Spray Dimethoate 0.05% on seedlings in nursery bed
- Remove all affected plants and destroy them carefully.
- Apply polythene or straw mulch.

Tomato Mosaic Disease TMV or *Nicotiana virus*

The general symptoms are the appearance of chlorotic areas on the leaves, such areas turn yellow and are interspersed with green areas. The edges of the leaves turn downward.

Control

- Treat the seed with hot water at 50°C for 25 minutes.
- Sterilize the nursery soil with formalin dust at 100/sq m at 6-8 cm deep.
- Control aphid vector with spraying of Nuvacron 0.05 per cent.
- Up root host plants like Dhatura, *Solanum nigrum* etc.
- Obtain seeds from acid (HC) seed extraction method (Kalloo, 1985).

Spotted Wilt Virus

This disease has been discovered" in Australia in 1915. Symptoms start with browning of leaves with thickening of veins and stunting of plant growth. Concentric circular spots are also seen on fruits. This disease is mechanically transmitted, but the natural transmission is through main thrips like *Thrips tabaci, Frankliniella schuitzei, F. coccidentalis.* The other thrips are: *Thrips flavus, T. sitosus, F. fusca, Scirtothrips dorsalis.*

Control

- Remove affected plants.
- Control weeds like *Portulaca oleracea* which is most prominent host.

- Spray systemic insecticides like Dimethoate or Endosulfan (0.03%).
- Apply Phorate (1.2 kg a.i./ha) 2.3 weeks after transplanting.
- Use aluminum-surfaced plastic mulch.
- Locate resistant varieties like Pearl harbour (from Hawaii), Manzana (from Argentina) etc.

Physiological Disorders

Physiological disorders are also known as non-parasite diseases. The following are some such disorders causing considerable yield loss in tomatoes.

Blossom-end Rot

Lesions appear at blossorn-end portion of the tomato fruit while it is green. Brown, water-soaked spots appear at the point of attachment of the senescent petals. Slowly and slowly this spot become enlarged. Affected portion of the fruit become sunken, leathery and dark in colour. Such fruits become unfit for human consumption. High moisture, and sudden change in the rate of transpiration especially in reduced moisture favours this disorder. Calcium deficiency (below 0.2 % in leaf) favours incidence of blossom-end rot. Heavy load of trusses on plants may also favour this incidence. Severe moisture deficiency is also associated with this malady. Imbalance nutrients make fruits susceptible.

Control

- Maintain soil moisture balance through light and frequent irrigations.
- Stake the plants if needs.
- Spray the crop with 0.5 per cent calcium chloride at the time of fruit development.
- Remove few trusses when profuse formation of them is occurred.
- Decrase level of nitrogen and increase phosphate.

Sunscalding

Developing fruits are exposed to intense heat (high temperature and high light intensity are fallen on a particular part of tomato fruit), the tissues of that area become dead. The water-soaked spot leads to sunken area that become white or grey colour in immature green fruit and yellowish in mature or red fruits. Sunscald occurs at the temperature of 40°C and above (Rabinowitch *et al.*, 1974).

Control

- Don't prune or stake the plant.
- Grow determinate tye of varieties having dense foliage.
- Grow wind break that can reduce intense sun rays upto some extent.

Fruit Cracking

Tomato fruits do crack when they are approaching to their maturity at the stem end. The cracking has been found four types viz., radial, concentric, burst and cuticular. Former two types are more common than latter types. When fruits crack around the shoulder this type of cracking is known as concentric cracking. While radial cracking is occurred in full ripe fruit causing more on mature green fruit. Exposed fruits to sun develop more cracking in the presence of high moisture on fruit surfaces. Reduced transpiration has more chances of fruit cracking because it increases fruit cell turgidity (Singh and Young, 1970).

Control

- Grow self-topping i.e. profuse foliage bearing varieties that have less cracking.
- Maintain proper soil moisture through light irrigation at regular inteval.
- Apply borax in the soil at 10 to 15 kg per hectare and/or spray borax at 0.25 per cent at fruiting stage.
- Locate for cracking tolerant varieties suitable for local conditions, like Sioux, Marglobe, Roma, Arka Saurabh, etc.

Cat Face

There is a distortion or opening of the blossom-end portion of the fruit. Such fruits have ridges, furrows, indentations and blotches. Abnormal growing conditions cause death of cells in the blossom-end of the ovary and turn dark to form a leathery blotch at the end of the fruit. Poor pollination and fertilization during winters may cause cat face disorder (Kalloo, 1985).

Control

- Balanced applications of nutrients will make free from cat face disorder to tomato fruits.

Puffiness

It is also known as hollowness or pocket. When outer wall of the developing fruit remains continue to develop but internal tissue growth is retarded. The surface of the fruit becomes flattened. Formation of locular jelly or juice is also reduced that results in light weight of fruit. Price et al. (1985) were first to describe Puffiness in "Terracota" variety of tomato and to assign the reason as varietal character. Improper pollination and fertilization due to extremes of temperatures (low as well as high) are some of the important reasons of Puffiness. Low soil moisture is also an important reason. Puffiness is governed by more than one gene. (Palevitch and Kedar, 1963).

Control

- Avoid planting of tomato crop in such period that fruit developing stage does not pass through the period of either very low temperature or very high.
- Maintain soil temperature through regular irrigation at proper intervals.
- Spray 0.3-0.4% borax solution at early plant growth, fruiting and fruit development stages.
- Tomato varieties like Sioux and Marglobe are less affected by Puffiness disorder.

Low Temperature Injury

Tomato is very sensitive to frost. As a maturing crop is exposed to near freezing temperature, the accumulation of sugars may lower the freezing point slightly. When vines are frozen they darken, wither and desiccate. The fruits become soft, water-soaked and full coloured.

Blotchy Ripening

There are diffuse area of yellow or gray in the other wise normal red colour of the ripe fruits. These area has hard white tissues. Viral diseases also may results irregular ripening. Botchy ripening is also caused by lower dose of potassium. Use of balanced fertilizer in the crop prevents the occurrence of blotchy ripening. Adjust planting to achieve favourable light intensity for good fruit development.

Vascular Browning

Vascular browning is caused due to the deficiency of magnesium. When the stem cut from the centre the vascular portion of the affected plants looks like brown colour and kills the tissue at later stage. It can be corrected by using 1.5% Mg as a foliar spray at the time when the symptom appears.

Integrated Pest Management (IPM)

In tomato IPM includes cultural, mechanical and biological control measures.

Cultural Measures

- Crop rotation with French beans reduces bacterial wilt in tomato. With cereals, sesame, mustard, marigold it reduces nematode infestation.
- Adopt raised nursery beds (10 cm) for good drainage to avoid damping off in solanaceous nurseries by preventing soil borne *Pythium, Phytophthora, Rhizoctonia.*
- Soil solarization using 160-100 gauge thick polythene transparent sheets on nursery beds for 15-21 days helps in killing weed seeds, nematodes and resting stages.
- Barrier crops like pearl millet, sorghum avoids TLCV.

Mechanical Control

- Collection and destruction of egg masses and crop debris of previous harvested crop.
- Use of 100 mesh nylon net in nursery beds to prevent the entry of white fly and transmission of leaves curl virus diseases.
- Installation of pheromone traps @ 5/ha *(Helicoverpa armigera* and S. litura).
- Trap crop; Intercropping with a tall variety like African marigold is beneficial. One row of marigold for every 16 rows of tomatoes (25-40 days old) will reduce the incidence the pest (Singh *et al.,* 2004).

Biological Control

Fruit borer (Helicoverpa armigera) : is highly migratory with large number of hosts. The parasitoids such as *Trichogramma* spp; *Eucetatoria bryani: Cotesia marginiventris; Hyposotes didymator; Campoletis flavisineta* were imported in the last 2 decades but certain degree of establishment was with *E. bryani* and *Trichogramma spp.* It was found that *Trichogramma pretiosam* is better than *T. brasiliensis* for the control of *H. armigera* in tomato. The parasitoid was released at weekly intervals from flower initiation on or when the incidence of the pest was observed at the rate of 40,000-60,000 adults/ha.

In Solan region, *T.chilonis* gave even 90-100% parasitism. Project Directorate on Biological Control, Bangalore, has developed endosulfan resistant strain of *T.chilonis* as endosulfan is highly toxic. When the incidence of whitefly is observed at the time of release of parasitoid, spray of insecticide would be necessary. This strain of parasitoid can be recommended where both parasitoid and insecticide (endosulfan) are recommended for use.

For effective control of fruit borer, egg parasitoid, T. *pretioswnand* Nuclear Polyhedral Virus (NPV) of *H. armigera* could be integrated to tackle different stages of pests T. *pretiosum* was released @ 2.5 lakh/ha and HaNPV was sprayed at 250 LE/ha two times. Two sprays of HaNPV would be sufficient with full dosage of parasitoid.

Another parasitoid *Campoletis chlorideae uchida* was also found to be capable of producing 9-40 per cent parasitism in tomato field. It attacks the neonates to first two larval stages of the pest. This can be mass produced like *Eriborus argentiopilosus* for better control of the pest (Krishnamurthy and Mani 1989).

C. chloridae can be at least released in small numbers along with egg parasitoid to take care of both the egg and early larval stages. Acephate and neem are found to be safe for *C. chloridae.* Endosulfan, carbaryl, dichlorovos, phosalone, methyl demeton were less persistent than other insecticides (Reddy, 2001).

HARVESTING

Harvesting is important for transportation, marketing, storage and processing. Harvesting should be done at the right stage which is depends upon the purpose to which the fruits are to be used.

Various Stages of Harvesting

Immature green

Fruits are usually tough leathery and develop poor colour when ipened artificially.

Mature green

Fruits some time used for chutney and curry preparations. Fruits will develop good colour when riped under favourable conditions. They transported to a longer distance or stored for the off season generally picked at this stage.

Turning pink or half ripe

This stage is the best for immediate marketing.

Red ripe

For processing or for extracting the seeds, the fruits are picked at the red ripe stage.

Over ripe

Fully coloured and soft fruits harvested for immediate use on home consumption. Fruits are normally picked at 4 days interval during warm season and at weekly interval when weather is cool. There is 7-10 harvests in a crop life span.

After picking, the fruits are graded and sorted out into cracked, bruised, injured fruits and well matured ripe fruits. For marketing purposes ISI standard is advocated.

YIELD

Average yield of 250 q/ha is expected from the improved tomato varieties but best management practices can produce as high as 400 q/ha. Average yield of hybrids is 600-750 q/ha.

POST HARVEST HANDLING

Grading

The grading of tomato is being followed for specialized city market in India. The Indian Standard Institution has specified four grades viz., super A., super, fancy and commercial for tomato crop. However the grading on size of the fruits are formed (a) small (below 100g) (b) medium (100-255g) (c) large (over 255g) accordingly and quality of fruits for better market prices.

Packing

Fruits are packed in perforated fibre Cartoon or card board boxes for distant marketing. Sometimes fruit are packed in wooden boxes with paper cutting from all sides and fruits are kept layerwise to avoid damage. For local markets fruit are packed in bamboo basket.

Storage

Fruits can be kept in store either, in mature green or turning stage of maturity. Fruits remain firm upto 21 days when kept at 20°C. Fruit develop moderate chilling injury and moderate to severe alternaria rot when stored at 7.2 to 10°C. The green tomatoes can be stored for 4-5 weeks at 12-15°C and 85-90% relative humidity. At the same freezing point, riped tomatoes can be stored for 7-10 weeks at 7-10°C with a maintained relative humidity of 85-90 per cent (Pantastico *et al.*, 1975).

SEED PRODUCTION

Seed production of tomato can be taken in any part of the country. The method of tomato cultivation for seed production is more or less the same as that of the cultivation for fruit production. Tomato is a self pollinated crop and natural cross pollination varies with the cultivars, weather conditions, insect pollinators and isolation.

Following point should be taken into consideration for production of genetically pure seeds

i. Isolation distance
ii. Field insepection and roguing
iii. Harvesting
iv Method of seed extraction

Isolation Distance

Tomato is normally self-pollinated, but there are fair chances of cross-pollination of about 5 per cent due to attractive yellow flowers which attract insects like bees. Seed fields should be isolated from other-variety fields of tomato and fields of the same variety not conforming to varietal purity requirements of certification by at least 50 metres for foundation seed production, and 25 metres for certified seed production (Agarwal, 1980).

Studies conducted at TNAU, Coimbatore reveals that 40-50 m is a safe distance for tomato to prevent cross pollination. It was observed that at 33 m the out crossmg percentage was 5.56 and thereafter no out crossing was recorded up to 50 m (Veeraraghavatham *et al.* 2005). The best time for planting the seed crop is January-February in plains since it is not affected by the viruses at low temperature.

Roguing

It is an important operation in seed crop, where the undesirable plants, off types and diseased plants are removed. At least three to four times roguing should be carried out at different stages of crop growth.

Pre-flowering Stage

The first roguing is done for the plants which bear flowers, plants different in their branch and leaves orientation, leaf colour, plants different in general plant type including determinate and indeterminate habit, should be removed if :hey are different than the grown variety.

Besides, the plants affected severely by known or unknown disease or disorder should be removed.

Flowering Stage

The second roguing should be done at flowering and post flowering stage. In this roguing, the plants which do not bear flowers and plants show different flower colour than the grown variety should be removed from the field. Roguing should also be done for diseased plants affected by early blight and leaf spot from the field.

Immature Fruiting Stage

The third roguing should be done at immature fruiting stage. In this roguing, the plants which show flowering, different fruit size and shape should be removed from the field if they are different than the grown variety. The plants which show ripe fruits should be removed from the field, affected severely by early blight, leaf spot and mosaic (TMV) should be removed from field.

Mature Fruit Stage

The fourth roguing should be done at the time of mature fruit stage. In, this roguing, the plants which show different fruit size and shape should be removed. The plant which show unripe fruit colour or ripe fruit colour other than variety grown in the field should also be removed from the field. Roguing should be done for diseased plants affected by early blight, leaf spot and mosaic (TMV) from the field from time to time as required.

Harvesting and Seed Yield

Tomato fruits are harvested for seed purpose at pink to red-ripe stage. All unwanted fruits (green, rotten, injured, deformed, overripe etc) are sorted. The healthy fruits are subjected to seed extraction by fermentation for 24 to 48 hours that results in the highest quality seeds (pandita *et al.*, 1996). Hybrid seed yield is about 40-50 kg /ha, and 80-100 kg/ha in general.

Seed Extraction

The following any one method may be followed as per convenience (Agrawal, 1980; Singh, 1977; Singh, 1989).

Fermentation method

The ripe fruits are crushed well in a non-metallic container by hand or by any mechanical method to make a slurry. Fruit juice should not be allowed to drain out. The entire material is kept as such till it ferment. When fermentation is completed it shows profuse foam formation on the upper surface of material and tomato flesh separated from seeds completely. The liquid is decanted off and the seeds are washed at least 8 to 10 times with clean- water. Seeds are spread thinly in the sun for drying.

Alkali treatment method

The extracted material with pulp is treated with an alkali mixture (900 g of ordinary washing soda is added to 4 litres of boiling water) in equal volume.

When the alkali mixture is cooled, allow it all to stand overnight in an earthen pot. Next day, all the seeds will settle down at the bottom of the container. The liquid is decanted off. Seeds are washed thoroughly with clean water and allowed to dry in the sun.

Acid treatment method

In the extracted tomato pulp commercial hydrochloric acid. @ 5-6 ml/kg pulp is thoroughly mixed and stirred and washed thoroughly after 30 min with clean water, and seeds are allowed to dry in the sun.

Washing

After extraction the seeds are washed with water to remove the pulp, etc. The water is added to containers with the pulp and seed. It is stirred thoroughly and is drained out along with pulp and other mucilaginous substance. This process is repeated until the seeds are clean.

Drying

After washing the seeds should be dried as rapidly as possible. Seeds may be spread on screen bottom trays, or cioth and placed in the open where a maximum exposure to sun and dry air is available. The seeds should be dried to eight per cent moisture before storage (Agrawal, 1980).

Grading

Whole dry lot of seeds is graded. The seeds are passed through different size of screen preferably above 0.6 mm to 0.8 mm size. However, size of screen or sieve also varies with varieties. For example 'Pusa Gaurav' seeds of the size 1.0 to 1.2 mm thick resulted in maximum field emergence of 75 per cent (Pandita *et al.*, 1996). Large seeds are superior to smaller seeds, hence, grading of seeds is absolutely essential to get maximum seed quality (Palaniswamy *et al.*, 1993).

Packing and Storage

The seeds should be treated with Captan or thiram 75 WDP at 2 g/kg seed. After treating the seeds should be packed in polythene bags then the seeds should be stored with 8-10 per cent moisture in container of moisture-vapour proof. Such seeds will remain viable up to 70-80 per cent at the end of about 25 to 30 months of storage (Vadivelu and Srimathi, 1988).

Heterosis and Hybrid Seed Production

Heterosis has been commercially exploited as the tomato fruit contains large number of seeds. Hybrid vigour in tomato was first reported by Hedrick and Booth in 1908. Since then, a number of investigations from India and abroad have reported heterosis in tomato for yield, its components and other economic traits. It is necessary to utilize the best performing parental lines for one or more characters associated with yield in order to achieve higher gain in the F1 hybrid through exploitation of heterosis (Dev *et al.*, 1994).

"Yield is a multiplicative end product and dependable on the contribution of yield components". There is ample scope for exploitation of hybrid vigour for commercial production as well as isolation of pure lines among the progenies of heterotic crosses for achieving higher yield in tomato. The hybrids with high heterosis have been found to have high inbreeding depression (Singh *et al.*, 1995).

Selection of more fruit weight along with fruit number and large sized fruits as they are also associated with high heritability and genetic advance result in improvement of tomato yield (Ghosh and Syamal, 1994).

Technique of Producing F_1 Hybrid Tomato Seed (Hybridization)

Almost all cultivation practices are similar to the production of commercial tomato for fruits. In addition to this, the following procedure is followed for the production of Hybrids in tomato.

At present, F_1 hybrid tomato seed is in tremendous demand because it has high yield potential with quality fruits.

In Indian Condition, Hybridization Work

In Indian condition, hybridization work is being done under open field, though under protective conditions, it is also not uncommon.

1. Hybridization requires the growing of two separate male and female parents. The male parent provides pollen and the female parent receives pollen resulting in the fruit bearing seed.
2. Usually, male parent is planted 2 to 3 weeks earlier to the planting of female parent so that adequate quantity of pollen is available for pollination and fertilization. Ratio of male to female is 1:5.
3. Operation of hybridization is not started until and unless both the parents are having cent per cent plant true-to-type. That is achieved only by strict roguing of all off types, undesirable and suspect plants.
4. Provide support to both the male and female plants.
5. All the flowers of female parents are emasculated during their late bud stage. It is achieved by removing their anthers (i.e. male organs) with the help of suitable forcep carefully, prior to applying pollen to its stigma.
6. The work of emasculation should be started early in the morning well in advance to anthesis. Male-sterile line does not require emasculation.

7. Pollen is collected from the flowers by detaching from the male parent (line) or by mechanical means like a small 'vibrator', and is applied with a fine brush to each emasculated (or male-sterile) female flower.
8. Pollination should be followed by bagging of flowers avoiding any chances of cross-pollination through bees.
9. Pollinated flowers are marked with tag for ease in identification o! fruits containing hybrid seed at harvest time.

Precautions

- Assure about the perfect removal of off types from both the parental lines.
- Do emasculation at right stage of bud.
- Take out anthers very carefully which are fused in a cone by a pull, using forceps.
- Take care that in no case there is a slight damage in stigma.
- Find out well in advance the right time of emasculation. In fact, the commencement of emasculation depends on local climatic conditions and plant type.
- Removal of unpollinated flowers from the truss and plant should be attended to invariably or tagging follow bagging of pollinated flowers.

REFERENCES

Abdul Majeed, B., Madalageri, B., Dundi, K.B. and Dharmati, P.R. (1991). Response of tomato to stage and concentration of DCPTA application. *South Indian Hort.* 39(5): 292-295.

Agrawal, R.L. (1980). Seed Technology. Oxford and IBH Publishing Co., Pvt.Ltd, New Delhi, pp.193-278.

Akhtar, M. and Mahmood, I. (1985). Control of root knot nematode *(Meloidogyne incognita)* in tomato plants by seed-coating with Achook and neem oil. *International Pest Control,* 37(3): 86-87.

Bailey, L.H. (1949). Manual of cultivated plants, 2nd Ed. Macmillan Company, New York.

Baruah, G.K.S., Arora, A.K. and Pandita, M.L. (1994). *Vegetable Science,* Vol. 21, pp. 93-96.

Borah, P. (1995). Incidence of fruit borer, *Heliothes armigera* Hub. in relation to date of planting on tomato in the hill zone of Assam. *Hort. J.,* 8 (1): 93-94.

Bose, T.K., Som, M.G. and Kabir, J.. (1993). Vegetable Crops. Naya Prakash, Calcutta.

Brar, J.S. (1975). Summer Institute on Vegetable Production Techniques and Method of Vegetable Breeding, at PAU, Ludhiana, Punjab.

Chakraborti, R.C. and Sandhu, M.K. (1994). Effect of mulch type and colour on growth and yield of tomato (*Lycopersicon esculentum* Mill.) *Indian J. Agric. Sci.*, 64(9): 608-612.

Chanda, S., Chattopadhyay, P.K. and Bhattacharya, S.P. (1994). Herbicidal effect in preventing weeds in spring-summer tomato. *Haryana J. Hort. Sci.*, 32(2): 144-148.

Chatfield, C. (1954). Food composition tables minerals and vitamins for international use. FAO Nutritional studies No. 11. FAO, Rome.

Chauhan, D.V.S. (1981). Vegetable Production in India. Ram Prasad and Sons, Agra, pp. 57-69.

Choudhary, D.R. and Malhotra, S.K. (2001). Studies on hybrid vigour in tomato (*Lycopersicon erculentum* Mill.). *Indian J. Agric. Res.*, 35(3): 1760180.

Choudhury, B. and Singh, S.N. (1960). *Indian Horticulture*, Vol. 17, pp. 48-55.

Choudhury, B. and Tiwari, R.N. (1993). *Indian Horticulture*, Vol. 37, p. 14.

Das, R.C. and Das, G. (1981). *Journal of orissa Horticulture*, Vol. 9 : pp. 22-27.

Das, T.K. and Panda, P.K. (1995). Effect of anti-tranparant on water retention capacity and yield of tomato. *Nat. Sym. on Recent Development in Veg. Impr.*, held at Indira Gandhi Krishi Vishwa Vidyalaya, Raipur (M.P.) Feb. 2-5: 1995.

Datta, A.K. (1961). Hunger brings down yield. *Extension*, 3(8): 11-12.

Dev, H., Rattan, R.S. and Thakur, M.C. (1994). Heterosis in tomato. *Hort. J.*, 7 (2) : 125-132.

Dhuria, H.S. and Leela, D. (1970). Control of Weeds in horticultural crops with the aid of modern herbicides. Half-yearly report 1969-1970. Indian Institute of Horticultural Research Bangalore, pp. 40-42.

Dodan, D.S., Shyam, K.R. and Gupta, S.K. (1994). Integrated management of buckeye rot of tomato. *Plant Diseases Res.*, 9 :141-144.

Dube, B.K., Sinha, P., Chatterjee (2003). Effect of zinc on yield and quality of tomato. *Indian J. Hort.*, 60(1): 59-63.

Ei-Oksh II, Ei-Leboudi, A.E. and Taha, E.M. (1980). Res. Bull. Fac. Agric. Ain Shams Univ. No. 1318, pp. 15.

George, RAT., Stephens, R.J. and Varis, S. (1980). The effect of mineral nutrients on the yield and quality of seeds in tomato. In: Seed Production. Hebblethwaite, P.O. ED. Butterworths, London, 561.

Ghosh, P.K. and Syamal, M.M. (1994). Path analysis and correlation studies in tomato (*Lycopersicon escu.'entum* Mill). *Orissa J. Hort.*, 22 (1-2) : 41-45.

Ghosh, R.K., Ghosh, Pritam, Patra, Bikash and Saha, Sankar (2002). Bio-efficicacy of metribuzin 70 WP on weed control in tomato. *Intl. Conf. Veg.*, Bangalore, 11-14 November, 2002. Abstract, pp. 257.

Gunasekaran, C.R., Vinaymurthy, A. and Sundaram, R. (1994). Management of root-knot nematode and damping-off disease in tomato nursery. *South Indian Hort.*, 42(5): 346-347.

Gupta, P.K. and Gupta, A.K. (2004). Studies of PGR and micronutrients mixtures on Vitamin C content in tomato products. *Indian J. Hort.*, 61(1): 102-103.

Hawthorn, L.R. and Pollard, L.H. (1954). Vegetable and Flower Seed Production. The Blakiston Company, New York, 444.

Hebbar, S.S., Ramachandrappa, B.K., Prabhakar, M. and Nanjappa, H.V. (2002). Effect of NPK drip fertilization on yield and economics of field grown tomatoes. *Intl. Conf. Veg.*, Bangalore, 11-14 Nov., 2002. Abst., pp. 217.

Hedrick, U.P. and Booth, N.O. (1908). Mendelian Characters in tomato. *Proc.Am.Soc.Hort. Sci.*, 5 : 19.

Hooda, R.S., Singh, Jitendra; Malite, V.S. and Nirmal, A.K. (1998). Effect of seeding/ planting time and mulching on tomato (*Lycopersicon erculentum* Mill.) production. *Haryana J. Hort. Sci.*, 27(3): 195-200.

Jain, R., Bhutani, R.D., Kalloo, G., Bhatti, D.S. and Singh, K. (1990). *Indian Horticulture*, Vol. 36, p. 24.

Jaiswal, R.C. and Singh, Kirti (1989). Salt tolerance studies in tomato *(Lycopersicon escuientum* Mill). *Hort.J.*, 2 (1) : 33-35.

Jasmine, A.P., Jaya and Ramadas, Seemanthini (1993). Evaluation of certain F1 hybrids and cultivars of tomato for yield and yield components. *South Indian Hort.*, 41(5): 248-250.

Jenkins, J.A. (1948). The origin of cultivated tomato. *Econ. Bot.*, 2:379-392.

Kalloo (1985). Tomato. Allied Publishers, Pvt. (Ltd.), New Delhi.

Kalloo, G. and Dudi, B.S. (1990). Development of Variety for fruit set at low temperature in tomato. *Ann. Rep. Dep. Veg. Crops.* Haryana Agric. Univ., Hisar, p. 16.

Kalloo, G. and Singh, K. (1976). *Haryana Journal of Horticultural Science*, Vol. 5 : pp. 92-95.

Kalloo, G. Singh, K. (1982). *Journal of Horticulture Science*, Vol. 11, pp. : 261-263.

Kamalanathan, S. and Thamburaj, S. (1970). Response of tomato to N.P.K. and plant spacing. *Madras Agric. J.*, 57(11): 526-533.

Kamruddin, M.D., Jain, B.P. and Verma, S.P. (1978). A note on effect of nitrogen and plant growth regulators on tomato (*Lycoperricon exculentum* Mill.). *Indian J. Hort.*, 35(2): 130-132.

Kulkarni, Gururaj P. and Madalageri, M.B. (2002). Evaluation and Correlation studies in tomato for tomato leaf curl virus. *Intl. Conf. Veg.*, Bangalore. 11-14 November, 2002. Abstract: pp. 80.

Kumar, N. and Irulappan, I. (1990). Seasonal influence on the incidence of spotted wilt virus on tomato varieties. *South Indian Hort.*, 38(5): 250-252.

Lal, K. (2006). Drip irrigation in tomato. Kisan World, 33(11): 49.

Liu, Yongqing (1994). Effect of osmotic treatments on tomato seed vigour, *J.Human Agric. College*, 20 (1) : 42-46.

Luckwill, (1943). *Royal Horticulture Society Journal*, Vol. 68. pp. 19-25.

Mahajan, R., Ahuja, S., Kanwar, J.S. and Nandpuri, K.S. (1979). Resistance in tomato against *Meloidogyne incognita* (K&W) Chitwood *Veg. Sci.*, 6(1): 57-60.

Majid, R., Swaminathan, M.S. and Iyer, R.D. (1968). Production and cytogenetic analysis of interspecific hybrids in *Lycoperricon. Indian J. Genet.*, 28(3): 275-286.

Malkan, V.K., Batra, Bhatia, A.K. Virendra; Arora, S.K. (2003). Response of foliar application of micronutrients on tomato variety-Hisar Arun. *Veg. Sci.*, 30(2): 182-184.

Mallick, M.F.R. and Muthukrishnan, C.R. (1980). *South Indian Horticulture*, Vol. 28, pp. 14-20.

Maynard, D.N. (1979). Nutritional disorders of Vegetable crops. *J.Plant Nutrition*, 1 : 1-23.

Mehta, A.K., Singh, R.P. and Lal, G. (1989). *Vegetable Science*, Vol. 16, pp. 1-8.

Mishra, H.N., Mishra, M. and Sahoo, S. (2001). Performance of tomato (*Lycopersicon esculentum* Mill.) genotypes during kharif season in eastern ghat highland zone of Orissa. *Orissa J. Hort.* 29(2): 110-112.

Misra, C.M. and Khanna, K.R. (1977). Heterosis and combining ability studies for some vegetative characters in tomato. (*Lycopersicon esculentum* Mill.). *Indian J. Hort.*, 34: 396-403.

Mital, S.P., Thomas, T.A. and Singh, H.B. (1962). Use of male sterility in the exploitation of hybrid vigour in tomato. *Indian J. Hort.*, 19(1&2): 50-60.

Mital, S.P., Thomas, T.A. and Srivastava, G. (1965). Assessment of Genetic stocks of tomato (*Lycoperricon esculentum* Mill.) and its wild relatives. *Tech. Bull. (Agric.)* 1. Indian Council of Agric. Res. N. Delhi.

Mukherjee, D. (1980). Effect of Morphactin and G.A. on flower formation, fruit growth and biochemical changes in *Lycopersicon esculentum. Indian J. Hort.*, 37(1): 72-76.

Muller, C.H.A. (1940). A revision of the genus *Lycopersicon*, USDA Misc. Publ. 382, 29.

Muniyappa, T.V. Dev, N.P. and Andani, Gowda (1995). Economics of integrated weed control in hybrid tomato (*Lycopersicon esculentum*). *Indian J. Agron.* 40(2): 339-344.

Nadkarni, K.M. (1927). Indian Meteria Medica, Published by Nadkarni & Co., Bombay.

Nainar, P. and Pappiah, C.M. (1997). Studies on the nutrient requirement of direct sown tomato (*Lycopersicon esculentum* Mill.) *South Indian Hort.*, 45(3&4): 120-124.

Nainar, P., Balakrishnan, K., Pappiah, C.M. and Sambandamurthi, S. (1997). Studies on fruit cracking in tomato (*Lycopersicon esculentum* Mill.) *South Indian Hort.*, 45(1&2): 63-65.

Nandpuri, K.S., Singh, Surjan and Kanwar, J.S. (1977). Tomato Punjab Chhuhara. *Indian Hort.*, 22(1); 7&27.

Natarajan, S., Veeraragavathatham, D., Thamburaj, S. and Mageswari, K. (2002). CO-1. A determinate F1 hybrid planting. *Intl. Conf. Veg.*, Bangalore, 11-14 November, 2002. Abstract: pp. 73.

Nath, Prem, Srivastava, V.K., Dutta, O.P. and Swamy, K.R.M. (2008). Vegetable Crops - Improvement and Production. Dr. Prem Nath Agricultural Science Foundation, Bangalore, 34-40.

Oza, A.M. and Rangnekar, Y.B. (1969). Effect of application of gibberellic acid on the ascorbic acid content of tomato. *Indian J. Agric. Sci.* 39(10): 980-83.

Palaniswamy, V., Balkrishnan, K., Karivaratharaju, T.V. and Arumugam, R. (1993). Influence of picking and seed size on seed quality in tomato genotypes. *Golden Jubilee Sym. on Hort Res. Changing Scenario,* Held on May 24-28 at IIHR, Bangalore, India.

Palevitch, D. and Kedar, N. (1963). Puffiness of tomato fruits. *Res.Rep.Sci & Agric,* Hebrew Univ. Jerusalem, pp 436-437.

Pandita, V.K. and Randhawa. K.S. (1993). Seed quality in relation to seed size, grading in some tomato varieties. *Golden Jubilee Sym. on Hort.Res.Changing Scenario,* held on May 24-28. 1993 at IIHR, Bangalore, India.

Pandita, V.K., Randhawa. K.S. and Modi, B.S. (1996). Seed quality in relation to fruit maturity stage and duration of pulp fermentation in tomato. *Gartenbauwissenschaft,* 61 (1) : 33-36.

Pantastico, Er. B., Chattopadhyay and Subramanyam, H. (1975). Post Harvest Physiology, Handling and Utilization of Tropical and Subtropical Fruits and Vegetables (Ed : Er. B. Pantastico), AVI Publishing Co., West Port, Connecticut, U.S.A.

Patil, B.P., Rajput, J.C., Palve, S.B., Hambhole, A.S. and Diwakar, M.P. (1990). *Indian Horticulture,* Vol. 35 : pp. 6-7.

Prasanna, K.M., Sadashiva, A.T., Madhan Reddy, K., Krishna Reddy, M., Rawal, R.D., Balaram, M.V. and Narasimha Prasad, B.C. (2002). Development of bacterial wilt (BW) resistant tomato (*Lycopersicon esculentum* Mill.) F1 hybrids with extended shelf life. *Intl. Conf. Veg.,* Bangalore, 11-14 November, 2002, Abstract : pp. 59.

Rabinowitch, H.D., Kedar, N. and Budowski, P. (1974). Induction of sunscaled damage in tomatoes under natural and controlled conditions. *Scientia Hort.* 2 : 265-272.

Rahman, M.A., Vijaya, M., Neeraja, G., Chiranjeevi, and Reddy, I.P. (2002). Evaluation of tomato cultivars for resistance to tomato leaf curl disease *Intl. Conf. Veg.,* Bangalore, 11-14 November, 2002. Abstract : pp. 271.

Rai, Mathura; Dhandhar D.G. and Varma, S.P. (1987). A comparative performance of tomato (*Lycopersicon esculentum* Mill.). Varieties under Tripura Condition. *Haryana J. Hort. Sci.,* 16(3-4): 253-256.

Rai, N., Yadav, D.S. and Patel, R.K. (2001). Annual Report, ICAR Research Complex for NEH Region.

Raina, J.N., Thakur, B.C. and Verma, M.L. (1999). Effect of drip irrigation and poly-ethylene mulch on yield, quality and water use efficiency of tomato (*Lycopersicon esculentum). Indian J. Agric. Sci.,* 69(6): 430-433.

Rajeswar, S.R. and Patil, V.K. (1979). Flowering and Fruiting of some important varieties of tomato as affected by spacing, staking and pruning. *Indian J. Agric. Sci.,* 49(5): 358-360.

Randhawa, K.S. Saimbhi, M.S. and Gill, B.S. (1988). Commercial evaluation of tomato varieties for processing. *Veg. Sci.*, 15(2): 181-184.

Rao, M.S., Reddy, P.P. and Mohandas, S. (1995). Effect of integration of endomycorrhiza *(Glomus mosseae)* and neem cake on the control of root knot nematode on tomato. *Zeitschriftfur Pflanzen-krankheiten and Pflanzenschutz*, 102 (5) : 526-529.

Reddy, P., Verghese, A. and Kumar, K. (2001). Integrated Pest Management in Horticultural Ecosystems. Capital Publishing Company, New Delhi.

Reddy, P.P. (2002). Impact of Vegetable research, constraints and strategies - An IIHR perspective. Paper presented at the workshop on Impact of Vegetable Research in India. held at IIVR, Varanasi, 1-2 March, 2002.

Rezende, F. and Charles de Arajoo (2006). Biodegradable paper/poly merized vegetable oil mulches for tomato and pepper production. *J. Appl. Hort.*, 61(2): 12-15.

Rick, C.M. (1976). Tomato, In : "Evolution of Crop Plants". Ed. N.W. Simmonds, Longmans London.

Roddick, J.G. and Butcher, D.N. (1972). Tomatine production in cultured excised tomato roots. *Phytochem.* 11 : 2991-2997.

Roddick. J.G. (1974). The steroidal glycoalkaloid tomatine Phytochem. 13 : 9-25.

Rotem, J. (1969). The effect of soil moisture level on the incidence of early blight on potato and tomato plants. *Indian J.Agric.Res.* 19 : 139-141.

Sadashiva, A.T., Madhayi Reddy, K., Krishna Reddy, M., Narasimha Prasad, B.C., Satya Narayana, H.V. (2002). Evaluation of advance tomato (*Lycopersicon esculentum* Mill.) breeding lines for combined resistant to bacterial wilt (BW) and tomato leaf curl virus (TLCV). *Intl. Conf. Veg.*, Bangalore, 11-14 November, 2002. *Abstract*: pp. 71-71.

Sadashiva, A.T., Madhayi Reddy, K., Krishna Reddy, M., Singh, T.H., Balaram, M.K., Narasimha Prasad, B.C., Prasanna, K.M., Satya Narayana, H.V. and Naveen, L.R. (2002). Breeding tomato (*Lycopersicon esculentum* Mill.) for combined resistance to bacterial wilt and tomato leaf curl virus. *Intl. Conf. Veg.*, Bangalore, 11-14 November, 2002. *Abstract* : pp. 47.

Sadasihva, A.T., Deshpande, A.A. and Madhavi, Reddy, K. (1996). *Indian Horticulture*, 41: 6-7.

Sai Reddy, C., Narayanamma, M., Chiranjeevi, Ch. and Prabhakar Reddy, I. (2002). Effect of nutrient sources on the fruit yield of tomato (*Lycopersicon esculentum* Mill.) *Veg. Sci.*, 29(2): 193-194.

Sai Reddy, C., Narayanamma, M., Chiranjeevi, Ch. and Prabhakar Reddy, I. (2002). Effect of organic farming on the fruit ytield of tomato in Andhra Pradesh. *Intl. Conf. Veg.*, Bangalore, 11-14 November, 2002. *Abstract* : pp. 215-216.

Sai, Reddy, C., Narayanamma, M., Chiranjeevi, Ch. and Prabhakar Reddy, I. (2002). Influence of additional application of water soluble fertilizer (Foliar spray) on the yield of tomato in Andhra Pradesh *Intl. Conf. Veg.*, Bangalore, 11-14 November, 2002. *Abstract*. pp 216.

Sarswati, S. and Pappaiah, C.M. (1993). Effect of irrigation regimes on growth, flowering and yield of tomato *(Lycopersico escuientum* Mill) cultivars.Nat.Sym. on Recent Development in vegetable Improvement, held at Indira Gandhi Krishi Vishwa, Vidyalaya, Raipur (M.P.) on Feb. 2-5, 1995.

Saxena, R.K. and Thomas, T.A. (1988). *Indian Horticulture*. 33: 9-10.

Seshadri, V.S. and Chatterjee, S.S. (1996). The history and adaptation of some introduced vegetable crops in India. *Vegetable Science*, 23(2): 127-128.

Shanmugavelu, K.G. (1989). Production Technology of Vegetable Crops. Oxford and IBH Publishing Co., Pvt.Ltd., New Delhi, pp. 184-193.

Sharma, H.R. and Rattan, R.S. (1995). Evaluation of some F_1 hybrids of tomato under mid hill conditions of U.P. *Nat.Sym. on Recent Development in Vegetable Improvement*, held at Indira Gandhi Krishi Vishwa Vidyalaya, Raipur (M.P.) on Feb 2-5, 1995.

Sharma, Meenakshi, Arora, S.K.K., Singh, Jitendar and Dhawan, S.S. (2002). Assessment of tomato hybrids for the preparation of paste. *Intl. Conf. Veg.*, Bangalore, 11-14, November, 2002. *Abstract* : pp. 301.

Sharma, N.K. and Tiwari, R.S. (1996). Effect of shade on growth contributing characters and factors in relation to yield to tomato *(Lycopersicon escuientum)* cv. Pusa Ruby. *Progr.Hort.*, 25 (3-4):184-187.

Sharma, S.K. (1995). Response of boron and calcium nutrition on plant growth, fruit and seed yield of tomato *Veg. Sci.*, 22(1): 27-29.

Sharma, S.L., Sohi, H.S., Chowla, S.C. and Peskin, S.N. (1974). Note on the reaction of tomato varieties in field to the causal organism of the buckeye fruit rot disease. *Indian J.Agric.Sci.*, 4: 323-324.

Sharma, S.R. and Sohi, H.S. (1983). Buckeye rot of tomato as influenced by different levels of N.P. and K. fertilizers *J.Turk.Phytopathol*, 21 : 19-24.

Singh, B., Rawat, P.S. and Pant, P.C. (1979). Heterosis for ascorbic acid in tomato. *Indian J. Agric. Sci.*, 49(3): 177-178.

Singh, Bhangu, Taskiran and Winder Singh, Sukh (1993). Comparative performance of tomato cultivars and rainfed condition of Kandi area (Punjab). *Punjab Hort. J.*, 33(1-2): 123-126.

Singh, D.K. (1999). Response of hybrid tomato (*Lycopersicon esculentum*) to growth regulators. *Indian J. Agric. Sci.*, 69(9): 523-525.

Singh, D.K. and Lal, G. (2005). Effect of cultivars and harvesting stages on yield and quality of tomato seed. *Prog. Hort.*, 37(1): 199-203.

Singh, D.K., Lal, G., Tiwari, R.P. and Jain, S.K. (1995). Effect of method of planting on tomato cultivars during spring-summer. *Annals of Agric. Res.*, 16(1): 125-127.

Singh, D.N. and Nath, Vishal (2011). Varieties and Hybrids of Vegetables. Satish Serial Publishing House, Azadpur, Delhi (India), 181-218.

Singh, Durvesh Kumar and Singh, Raj Pal (1993). Varietal response to 2,4-dichlorophenoxy acetic acid and method of its application on summer tomato (*Lycopersicon esculentum*). *Indian J. Agric. Sci.*, 63(5): 276-282.

Singh, J. and Singh, M.P. (1991). A note to study the effect of IAA and NAA treated seed on growth and yield of tomato (*Lycopersicon esculentum* Mill). *Scientific Hort.*, 1:129-133.

Singh, K. and Young, J.O. (1970). Effect of an anti-transpirant and bagging the fruits or covering the plants on tomato fruit cracking. *J.N. Krishi Vishwa Vidyalaya, J. Res.* 4: 43-45.

Singh, K.P. and Kalloo, G. (2000). AICVIP Recommendations (1991-98), IIVR, Varanasi (ICAR Publication. Largely the results of private research efforts.

Singh, Makhan, Batra, V.K., Singh, Jitender, Bhatia, A.K., Singh, Virendra and Arora, S.K. (2002). Response of foliar application. *Intl. Conf. Veg.*, Bangalore, 11-14 November, 2002, *Abstract* : pp. 176.

Singh, N.P., Bharadwaj, A.K., Kumar, A. and Singh, K.M. (2004). Modern Technology on Vegetable Production. International Book Distributing Company, Lucknow.

Singh, R.D., Seth, J.N. and Kukshal, R.P. (1977). Tomatoes in U.P. hills. *Indian Hort.*, 22(4): 7-8.

Singh, R.S. (1985). Diseases of Vegetables. Oxford & IBH Publishing Co., New Delhi.

Singh, R.V. (1994). Performance of Kharif tomato as influenced by cultural practices. *Indian J. Hort.*, 61(2): 192-196.

Singh, Rajinder, Sandhu, K.S., Singh, Daljit and Sandhu, M.S. (2002). Quality response of tomato to population density and training methods. *Haryana J. Hort. Sci.*, 29(3&4): 235-236.

Singh, S.P. (1989). Production Technology of Vegetable Crops. Agricultural Research Communication Centre, Kernal, Haryana, India.

Singh, S.R (1977). All about tomato. *Farmer and Parliament*, 13: (9):8, 24-25.

Supe, V.S., Kale, P.N., Raijadhav, S.B., Yadav, E.D. and Patil, R.S. (1989). Performance of tomato hybrids. *J. Maharashtra Agric. Univ.*, 14(3): 371-372.

Suryanarayana, Reddy P., Reddy, M.G., Verra, Raghavaiah, L., Subramanyam, K. and Reddy, D.S., 91985). *South Indian Horticulture*, 33: 23-25.

Swarup Vishnu (2006). Vegetable Science and Technology in India. Kalyani Publishers, New Delhi. pp : 656.

Tandon, H.L.S. (1991). Fertilizer Recommendations for Horticultural Crops. A guide-book. 2nd Edition. Fertilizer Development and Consultation Organisation, New Delhi, India, pp. 28-92.

Thomas, T.A., Saxena, R.K. and Hardas, M.W. (1976). La-Bonita - a promising introduction in tomato. *Indian Hort.*, 21(2): 19-22.

Thomas, T.A., Saxena, R.K. and Hardas, M.W. (1979). A new pearshaped tomato. *Indian Horticulture*, 23(4): 9-10.

Thompson, C.H. and Kelly, C.W. (1957). Vegetable Crops. McGraw Hill Book Co., Inc., USA.

Tikoo, S.K. and Anand, N. (1983). Annual Report. Indian Institute of Horticulture Research, Bangalore.

Tiwari, R. and Chhonkar, V.S. (1967). Studies on the effects of foliar application of urea on growth and yield of tomato (*Lycopersicon esculentum* Mill.). *Indian J. Hort.*, 24(3&4):173-179.

Tiwari, R.N. and Choudhary, B. (1993). Solanaceous Crops. In: Vegetables Crops. (Ed. T.K. Bose, M.G. Som and J. Kabir) Naya Prakash, Calcutta.

Tiwari, R.N., Choudhury, B. and Pachauri, D.C. (1990). *Indian Horticulture*, 35: 4-5.

Tiwari, R.N., Pachauri, D.C. and Sharma, R.K. (1996). *Indian Horticulture*, 40: 26-27.

Tiwari, R.N., Pachauri, D.C. and Sharma, R.K. (1999). *Indian Horticulture*, 44: 9-10.

Tiwari, R.N., Pachauri, D.C. and Sharma, R.K. (2001). Pusa Sadabahar : A new tomato variety. *Indian Hort.*, 46(3): 11&24.

Tiwari, R.S., Sadashiva, A.T. and Doreyappa Gowda, J.N. (2002). Stability of different tomato varieties/hybrids for processing into puree. *Intl. Conf. Veg.*, Bangalore, 11-14 November, 2002. *Abstract* : pp. 305.

Vadivelu, K.K. and Srimathi, P. (1988). Tomato seed technology. *Seed and Farms*, 14(11): 22-27.

Vats, R., Nandal, S.N. and Dalal, M.R. (1995). Efficacy of different plant extracts root knot nematode (*Meloidogyne incognita*) on tomato. *Haryana Agric. Univ. J. Res.*, 25(3): 113-116.

Veeraragavatham, D. (2005). Studies on natural cross pollination in tomato. *Crop Res.*, 29(1): 169-170.

Verma, A.C., Anwar, A. and Kumar, A. (1994). Efficacy of Basamid Gr. against *Meloidogyne incognita* in tomato nursery. *Afro-Asian J. Nematology*, 4(1): 73-75.

Verma, S.K., Singh, S.S. and Awasthi, C.P. (1995). Response of zinc and boron fertilization on yield and quality of tomato fruits at different stages of picking. *Veg. Sci.*, 22(1): 5-8.

Verma, T.S., Joshi, S., Sharma, R.K. and Pachauri (2001). *Indian Horticulture*, 6: 1-3.

Verma, T.S., Joshi, S., Sharma, R.K. and Pachauri, D.C. (2001). Pusa Diviya : a new tomato hybrid. *Indian Hort.*, 46(3): 10&24.

Wagh, R.S. Patil, R.S., Kale, P.N., Pujari, C.V. and Kesar, B.G. (1995). Raja Shree - A new high yielding tomato F1 hybrid. *Nat. Sym.* on *Recent Development in Vegetable Improvement*, held at Indira Gandhi Krishi Vishwa Vidyalaya, Raipur (M.P.) on Feb. 2-5: 1995.

Wakankar, S.M. (1977). Tomato Sweet - 72. *Indian Hort.*, 22(3): 11.

Ward, G.M. (1964). *Cannadian Journal of Plant Sciences*, 44: 492-494.

Wilson, C., Clark, R.A. and Shearer, G.C. (1994). Effect of Salinity on the plasma membrane from tomato (*Lycopersicon ersculentum* Mill) leaves. Plant Science, 103(1): 1-9.

Yadav, D.S., Rai, N. and Yadav, R.K. (2000). *Indian Horticulture* (Communicated).

Yawalkar, K.S. (1985). Vegetable Crops of India. Agri. - Horticultural Publishing House, Nagpur.

□□□

2. BRINJAL

Botanical name	:	*Solanum melongena* L.
Family	:	Solanaceae
Chromosome number	:	2n = 24

Brinjal is known as egg plant, garden egg, guinea squash, aubergine or melongene, an important vegetable grown in many of the seasons and in more area compared to tomato and chilli in India. Brinjal can be grown in almost all parts of India except higher altitudes, all the year round. It has a long bearing period when grown under mild climate of southern states but its bearing is shortened under hot summer and cold winter seasons of Sutlej-Ganga alluvial plains of northern India. In hilly regions, it is grown only in summer (Singh and Kalda, 1993). Som and Maity (1988) reported that the actual area under brinjal crop is varying due to its seasonal nature of cultivation. Brinjal is highly productive and usually finds its place as the poor man's crop (Chadha, 1993).

ORIGIN

De Candolle (1904) regarding, brinjal as a native of India because wild form of *Solanum incanum* Roxb. which he regards as belonging to the same species as *Solanum melongena* L. have been found growing in Madras State and Burma.

India is the primary centre of origin of the crop. Bailey (1949) mentioned that the first records of the use of this plant came from India. Vavilov (1951) suggested China as a secondary centre of origin of brinjal. The egg plant was first taken into cultivation in India, where wild plants occur (Purseglove, 1977). More than 16 species, many of which closely related to brinjal, are found to good used in various parts of the country.

DISTRIBUTION

The crop is restricted mainly to south and south east Asia, Southern Europe, China and Japan. Arabs took it to Spain and Persians to Africa. It has now spread throughout the tropics (Purseglove, 1977). The important brinjal growing countries are, China, Japan, Philippines, France, Italy, India, the United States, Indonesia, Bulgaria, Africa, Pakistan and Bangladesh. In India brinjal is grown throughout the year in almost all states except at higher altitude. It is well distributed in Orissa, Bihar, Karnataka, West Bengal, Andhra Pradesh, Maharashtra and Uttar Pradesh.

BOTANY

Brinjal is a herbaceous annual with erect or semi spreading habit. It also behaves like a perennial herb. The plant produces numerous branches. Roots are vigorous, deep growing tap root. Stems and branches are covered with tomentum, stem woody at base, sometimes with violet tinge. Leaves are alternate, simple, singly or in pairs, petioles 2-10 cm long, lamina ovate, sometimes lobed, hairy, pointed at apex, margins undulate. The flowers are large solitary, or in 2-5 flowered cymes; pedicel 1-2 cm long, thickening and lengthening as fruit develops, calyx persistent, gamosepalous approximately 2 cm long, 5-6 lobed, often with purple tinge, densely hairs, enlarging with fruit development. Corolla rotate, persistent with 5-6 with triangular lobes, 3.5 cm in diameter violet or purple within tubes stamens 5-6, 0.5-1 cm, long, erect yellow anther, styles variable in length, short styled do not normally have fertile ovaries, ovary superior, 2-celled, avoid, 2-locular, variable in size (oval or egg-shaped to oblong) and size from 5-12 cm long, 3-6 cm in diameter, smooth and shiny, purple-black green and white colour (Purseglove, 1977). It is a self-pollinated crop but the cross-pollination occur to the extent of 30-40 per cent. There are four type of flowers-long styled, medium styled, pseudo-short styled and true short styled flowers. First two only produce fruit and fruit set ranges 70-86 %. The fruit is berry, born singly or in clusters. Fruit become yellow on ripening. Seeds are small, light brown, numerous, approximately 250 seeds/g. Brinjal seed resemble with those of chillies in colour, shape and appearance but they are comparatively smaller than chillies seeds.

Plant breeding

Nath *et al.* (2008) stated that India is the primary centre of origin of egg plant and has rich biodiversity in Eastern Ghats, North eastern regions and central India including Orissa. The important land races of north central India include Bundelkhand Desi (drought tolerant). Budhiya Baigan cluster bearing, white, Ramnagar Baigan (large fruited green), Jethuwa Baigan (summer season), Kuchbaigan (small, cluster bearing for transportation), Jafrabadi Baigan (deep purple long with less seeds) and Balfahwa Jathua Baigan (tall, cluster bearing good shelf life) suitable for summer season and long distance transportation (Swarup, 2006). The breeding methods adopted in brinjal are pure line selection, pedigree method, backcrossing, heterosis breeding. Brinjal varieties and hybrids are developed exclusively In India. In brinjal, a local cuitivar selection, Nurki is the source for the variety Pusa Purple Cluster which is resistant to bacterial wilt and little leaf diseases. Many local varieties and land races have contributed genes for insect resistance to jassids like Dorli, Benares, Giant, Khandesh Local. Genetic variation with regard to colour, shape, vegetative growth and presence or absence of spines exists among the indigenous materials. Studies on floral biology in long and round cultivars of brinjal were reported (Vijay *et al.*, 1977). Earlier, more emphasis was laid on the selection among the existing strains, which resulted in the release of varieties like 'Pusa Purple Long', 'Pusa Purple Round' and 'Pusa Purple Cluster' at the Indian Agricultural Research Institute, New Delhi. The oblong variety 'Pusa Kranti' was developed at the same institute for North Indian conditions (Choudhury *et al.*, 1971). The high yielding and good quality new varieties, 'Arka Sheel' (dark purple long), 'Arka Shirish' (long green), 'Arka Kusumakar' (small green clusters) and hybrid 'Arka Navaneet' (purple round) have been developed for the southern plains by the Indian Institute of Horticultural Research, Bangalore (Vijay *et al.*, 1971, 1978). Pal and Singh (1949) reported increase in yield up to 490 q/ha to 40-50 per cent In hybrids over the better parent. The hybrids invariably gave large fruits of superior quality. Vijay and Nath (1978) reported hybrid vigour upto 156.9 per cent over the better parent in the cross 'Arka Kusumkar' X 'Supreme'. The hybrid combination IIHR-22-1 X Supreme named 'Arka Navneet' proved to be the best combination for fruit quality and yield (Vijay *et al.*, 1978). Trials involving a large number of *Solanum* species and varieties of *Solatium mclongena* indicated that *S. articulatum* was immune to little leaf virus (Anjaneyulu and Ramakrishna, 1968) which may possibly be used for breeding resistant varieties. Correlation studies in egg plant indicated significant positive association between plant spread and number of branches and between fruit number and yield (Singh and Khanna, 1978). The nature and magnitude of genetic variance involved in the inheritance of some quantitative characters were investigated in a 6 X 6 diallete cross in brinjal (Singh and Khanna, 1979). Mode of inheritance of certain qualitative traits like spineness and purple leaf colour has been reported (Randhawa and Sukhija, 1975). Chadha and Sidhu (1983) studied variability and correlation of newly developed genotypes and observed total yield to be highly correlated with plant height, number of branches and fruit weight. Vadivel and Kannan Bapu (1989) have reported the genotype environment interaction for fruit yield in egg plant. The genotypes Ep 65 and Annamalai were more stable with high fruit yield over the environments. Co-2 fared well under less favorable environment and the genotype Co-1 and EP44 fared well under less favorable environments. Bora *et al.* (1994) reported two-promising strains of brinjal, *viz*, JC-1 and JC-2 which were developed at Assam Agricultural University, Jorhat, through pure line selection from local round varieties. They observed that JC-1 and JC-2 produced average yields of 391.03 q/ha and 408.63 q/ha respectively. Both were found to be resistant to bacterial wilt and tolerant to shoot and fruit borer. The fruits of JC-1 were long with pointed apex, purple in colour where as JC-2 were oblong and some what round in shape with blunt apex. The purple colour of fruits was brighter In JC-2 than in JC-1. The fruit size in both the strains was larger with average fruit weight of 282 g in JC-1 and 330 g in JC-2. These strains have already been recommended for cultivation in Assam. Cupta *et al.* (2000) reported the new brinjal 'Pusa Ankur' a semi erect non-spiny variety. It bears oval round, small sized (60-80 g), and dark purple, attractive fruits. An early bearing variety 'Pusa Ankur' takes 45 days for first picking after transplanting. Prasanth *et al.* (2002), reported performance of hybrids in brinjal. They pointed out that the developing of commercially acceptable brinjai varieties and hybrids with high yield and relatively resistant to shoot and fruit borer has been the objective of many breeding programers. Promising hybrids with high yield and less susceptible to shoot and fruit borer were 190 X Annamalai. 221 X Annamalai (3.363 and 30.6), 193 X MDU-1 (3.281 and 28.5) and 195 X Annamalai (3.122 and 17.1) The major diseases/insect pests affecting brinjal crops are phomospis blight, bacterial wilt (*Pseudomonas solanacearum* E.F. Smith), and fruit and shoot borer. The cultivars 'Annamalai' 'Pusa Purple Cluster', JC-1 and JC-2 were observed to be resistant to bacterial wilt at Assam Agricultural University, Jorhat. Other sources of resistance to bacterial wilt are cv 'Long Green' and SM-81. Sources of resistance to phomopsis blight utilized at IARI, Delhi is the line 'lla'. Hussain *et al.* (2002) screened 15 genotypes of brinjal varieties for black soil of Orissa in kharif season over two years. The variety Black Beauty produced the highest fruit yield (178 q/ha) with moderate fruit weight (70.83 g). The variety Bhawanipatna local recorded the maximum fruit weight with least number of fruits/plant (142.33 g and 4.5 respectively). A moderate yield of 161.83 and 152.67 q/ha was obtained from the genotype BB-6-1 and BB 26, respectively, which were suggested for cultivation in wilt prone areas. The variety Black Beauty was advocated for general cultivation.

Sebastian *et al.* (2002) have studied the characterization of land races of brinjal in Kerala. A collection of land races of brinjal from different parts of Kerala was characterized for morphology, yield attributes and reaction to various biotic stresses *viz.*, fruit and shoot borer (*Leucinodes orbonalis Guen*), bacterial wilt (*Ralstonia solanacearum* Yabuuchi), phomopsis blight (Phomopsis *vexana* (Sacc and Syd. Hartar) and little leaf (mycoplasma). Landraces showed significant differences for all the characters except stomatal distribution and nodes of flowering. S-33 collected from Adoor (Pathanamthitta) was the top yielder. *S. macrantham, S. melongena* var. *insanum* and *S. xanthocarpum* and landraces, *viz.*, SI (Neyya Hinkara, Thiruvantlhapuram), S-13 (Athichanallore, Kollam), S-28 (Thamarakulam, Alappuzha). S-35 (Poonjar, Kottayam), S-36 (Eara, Kottayam) and S-37 (Manarkadu, Kottayam) had resistance against fruit and shoot borer. S-15 (Kundara, Kottam), S-22 (Thirvakkal, Alapuzha), S-23 (Cherthal, Aloppuzha) S-32 and S-33 {Adoor), S-47 Jrinjalakuda, Thrisur) and S-55 (Ozhur, Malappuram) were resistant to bacterial wilt coupled with high yield attributes. Resistance to phomopsis blight was noted in wild semi wild species *viz. S. macranthum. S. melongena* var *insanum* and S. *xanthocarpum* and in cultivated types *viz.*, S-1 (Neyyathihara), S-13 (Kundara), S-159 (Adichanallord), S-17 (Oanavattom, Kollom), S-23 (Cherthala). S-32 and S-33 (Adoor), S-34 and S-35 (Poonjar), S-36 (Eara). S-45 (Moovattupuzha, Emakulam), S-46 (Irinjalakkuda), S-J3 (Wahayar, Palakkad), S-55 (Ozhur) and S-60 (Ambalavayal, Wayanadu). Incidence of little leaf disease was seen only in S-10 (a land race from Nedumanagad, Thiruvanthapuram). Multiple resistances to all the major biotic stresses were noticed in *S. melongena* var. insanum and *S. xanthocorpum* and in land races viz, S-1, S-13 and S-36. Gopalkrishanan *et al.* (2002) have studied the basics of resistance in bacterial wilt resistant hybrids and reported development of three wilt resistant open pollinated brinjal varieties *viz.*, Surya, Swetha and Haritha. Heterosis breeding programme resulted in the development of a purple fruited, high yielding and bacterial wilt resistant F, hybrid, Arka Keshav X SM 16. Considering the bacterial wilt resistance, high yield (51 t/ha) and specific consumer preference due to the attractive fruit colour and large fruit size, the F, hybrid, Surya X SM 116 was recommended by the name 'Neelima'. Somatic hybrids between *S.melongena* and S. *torvum* were obtained by protoplast fusion and embryo culture. The somatic hybrids produced by protoplast fusion between *S. melongena* and *S. sisymbrio* folium were resistant to root knot nematode and *Tetrachynus cinnabainus.* Kanamycin resistant plants were produced by co-cultivation of leaf explants with *Agrobacterium lumifaciens.* In India synthetic Bt toxingenes, Cryl Ab and Cry 1 b were introduced into brinjal for producing transgenes resistant to shoot and fruit borer.

Floral biology

There are four types of flowers in brinjal based on the length of style viz., (i) long styled with big size ovary (25%), (ii) medium styled with medium size ovary (10%), (iii) pseudo short styled with rudimentary ovary (15%), (iv) true short styled with rudimentary ovary (50%). Baksh and Iqbal (1979) reported that the ratio of the three types of flowers varied in different species and varieties, long-styled flowers were always appreciably more in number (52.90%), than the medium-styled (8.5-28%) and short style ones (1.5-25%).

Long and medium styled flowers with swollen ovary at the base produce fruits whereas pseudo short and true short styled flowers with rudimentary ovaries do not set any fruits. However, chances of cross-pollination are more in long styled flowers. Highest percentage of fruit set was found where the stigma is above the stamens. At a temperature of 20-22°C and 50-55 per cent relative humidity, pollen viability is retained for 8-10 days. Repeated pollination with pollen from different plants increases both fruit and seed set. The distribution of different types of flowers on plant varies with variety and season.

Both the anthesis and dehiscence are mainly influenced by the day light, temperature and humidity and cultivar (Sidhu *et al.*, 1980). Anthesis in brinjal flower starts from 7.30 a.m. and continues upto 11.00 a.m. Peak time for anthesis is 8.30 to 10.30 a.m. The pollen dehiscence starts from 9.30 to 10.00 a.m. Stigma is most receptive at the time of opening of flower. The stamens dehiscent at the same time the stigma is receptive so that self pollination is the rule, although there is some cross pollination by itself.

Under natural conditions on an average flowers with long styled set fruits to 50 per cent and flowers of medium styled about 40 per cent while both short styled and pseudo short styled do not set any fruit. With the application of growth regulators (2,4-D, NOA, GA etc) very high percentage of fruit setting (80-95%) is obtained. Even pseudo short styled flowers show fruit setting upto 60 to 70 per cent with the aid of 2, 4-D and NOA, respectively.

The fruit setting in brinjal is increased with the treatment of brinjal seeds by soaking method with 2, 4-D at 5 ppm for 24 hours before sowing or spraying of 2, 4-D at 2 ppm on standing crop of brinjal just at commencement of flowering stage increases yield significantly. Considerable yield improvement is obtained with the application of GA_3 + ascorbic acid each at 250 ppm as root dipping of seedlings before transplanting (Patil *et al.*, 1981).

Classification

According to Peter (1998) botanical classification based on fruit shape and plant spread and they are:

S. melongena var. *esculentum* L.

Fruit is long, round and oval.

S. melongena var. *serpentinum* L.

Snake brinjal, fruit is extra ordinarily long, leaf highly prickly.

S. melongena var. *depressum* L.

Plant is extremely short and dwarf.

S. melongena var. *bulsarensis* L.

Fruit shape

Long Brinjal (Pusa Purple Long, Pusa Purple Cluster), round Brinjal (Pusa Purple Round, T3); and oval Brinjal (Pusa Kranti).

Fruit colour

Purple brinjal (anthocyanin present) (Pusa Purple long); Green brinjal (Chlorophyll present) (Arka Kusumakar).

COMPOSITION

Brinjal has been a staple vegetable in our diet, since ancient times. Brinjal fruits are fairly good source of calcium, phosphorus, iron and vitamins, particularly "B" group. Oblong type rich in total water soluble sugars. For processing high dry matter content and low level of phenolics require. A high anthocyanin content and low glycoalkaloid is essential. Green cultivars have better processing properties than more popular purple cultivars.

The nutritive value of fruit is quite high (Anonymous, 1980). It contains (per 100 g of edible portion) following :

Moisture	92.7 gm	Magnesium	16.0 mg
Protein	1.4 mg	Phosphorus	47.0 mg
Fat	0.3 gm	Iron	0.9 mg
Minerals	0.3 gm	Sulphur	44.0 mg
Fibre	1.3 gm	Chlorin	52.0 mg
Carbohydrates	4.0 gm	Vitamin A	12.4 IU
Calories	24	Thiamine	0.04 mg
Calcium	18.0 mg	Riboflavin	0.11 mg
Oxalic acid	18.0 mg	Nicotinic acid	0.9 mg
Sodium	3.0 mg	Vitamin C	12.0 mg
Copper	0.17 mg		

Uses

Brinjals (unripe fruits) are used primarily as cooked vegetable for preparation of various dishes in different region of the world. Much potential as raw material in pickle making and dehydration industries (Singh *et al.*, 1963). According to Choudhury (1876) white brinjal is said to be good for diabetic patients. Nadakarni (1927) has suggested the different medicinal uses of brinjal i.e., pierced all over with a needle and fried in til oil, the fruit is employed as cure for toothache, an excellent remedy for those suffering from liver complaints. The green leaves of Brinjal plant are the main source of the supply of antiscorbic vitamin C. Its use in Ayurveda as appetizer, aphrodisiac, cardiotonic and beneficial in 'Vata' and 'Kaph' etc. In Unani system of medicine, roots of brinjal are used to alleviate pain, fruit is used as cardiotonic laxative, muturant and reliever of inflammation.

VARIETIES AND HYBRIDS

According to Bailey (1949) brinjal is having three botanical varieties, i.e. var. *esculentum* - fruits are round or egg-shaped; var. *Serpentinum* - fruits are long slender; var. *depressum* - early and dwarf types. Brinjal varieties are grouped on the basis of their colour, fruit shape and size. The selection of varieties depends on the climatic condition and consumers preference. In West Bengal, North UP, Orissa regions green fruited brinjal is more preferred. The preference is vary according to the season also as the white fruited brinjal is mostly used in summer months as there is very less bitterness and fibre at that time. Besides improved varieties, there are promising native eggplants, which are grown in different regions of the country.

Table 1 : The promising native eggplants grown in different regions of the country.

Region		Varieties
Uttar Pradesh and north central region	:	Muktakeshi, Ramanagar Giant, Batia, Bholanath, Sheonath, Bundelkhand Desi, Dindhya Baigan, Ramnagar Baigan, Kucjilachi Baigan, Jafrabadi, Balfahwas Jetliwa, Bhanta.
Bihar	:	Hyderpur, Improved Muktakeshi.
West Bengal	:	Mukta, Lokeshi, Krishnanagar Purple Round, Krishnanagar Green Long, Muktakeshi, Baromashi, Kalyani Green, Nurki.
Maharashtra	:	Manjari Gota, Bourad Local, Borgona, Local Khandesh, Khed Shivapur, Muleti, Malkapuri, Shirur Khandala, Khanpur, Rajangona, Songona Kagal, Tambalwadi, Local, Chikhalgaon Local, Dorli.
Gujarat	:	Junagadh Bhattu, Surti Gota, Gulabi Dorla.
Karnataka	:	Coorg Local, Irangeri, West Coast Green Round.
Madhya Pradesh	:	Nurki Round, Nurki Long, Banjari
Tamil Nadu	:	Wynad Giant, Pattabiram, Annamalai
Punjab	:	Ludhiana Local, Batia
Other cultivars (Mainly from Bihar and Orissa)	:	Changulia, Okhla, Very long, Lamba, Round Bangalore, Satputia Hara, Satputia Kathvatia, Grasevindy, Local Kalimpong, Gola Hara, Long Saharapur, Local oval, Kusumsita, Behda Tupri, Bada Tupri, Raitan Mani, Burupada, Borigumma, Kutiguda, Dhenkanal, Batparsi, Balugaon, Market, Banamalipur, Narayanpur, Nisan Local, Jaganathpur (Adivasi), Malipondi, Anchla-B, Bheda Ramachandrapur, Chekkaguda, Narikiri Tholi, Randha Pathuria, Chitrakonda, Rayagada Baigan, Kharagpuri, Bogma, Bhola, Bara Mahanian, Surajmukhi, Bhanta, Bor Baingna, Pol Baigan.

Source : Swarup, (1998).

Varieties / Hybrids grouped on the basis of shape and colour of fruits

Long purple

Pusa Purple Long, Pusa Purple Cluster, Pusa Kranti, KT-4, DBSR-31, Arka Sheel, Azad Kranti, ARU-1, S-16, Punjab Barsati, Punjab Sadabahar Punjab 129-5, Pant Samrat, H-4, ARU-2C, H-7, NDB-25, BB-26, Sel-4, BB-13, KS-331, JB-15, NDB-28-2, Pusa Hybrid-5, NDBH-6, ARBH-201, PBH-6, Pant Hybrid-1, MHB-1, Long 13, Junagadh Long, Pusa Bhairav.

Round purple

PB-91-2, T-3, K-202-9, H-8, KS-224, DBR-8, AB-1, CHBR-1, D-2-88-6, Pusa Hybrid-6, Vijay Hybrid, NDBH-1, Pusa Hybrid-9, JVH-1, MHB-2, MHB-10, Round 14, Azad Hybrid, Pusa Purple Round, Black Beauty, Jamuni Gola.

Round green

Krishnanagar Purple Round, Pant Rituraj, Shruti Gota, T_3, Banaras Giant, Ramnagar Giant.

Small round

Aruna, DBSR-44, PLR-1, DBSR-91, JB-64-1-2, ABH-1, ABH-2, Phule Hybrid-2.

Long green

Arka Kusumakar, Arka sirish, Krishnanagar Green Long.

Oval or oblong purple

Arka Navneet, Junagadh long, H-4, Pusa Anmol, Suphal.

Round with strips (Purple white)

Manjari, Manjari Gota.

Round white

Swarna Shree.

Brinjal varieties grouped on the basis of their some morphological distinctions

i) *Fruits borne in cluster*: Arka Kusumakar, Arka Neelkanth, Pusa Purple cluster, Pant Samrat, ARU-2C (Almora)

ii) *Spine Present* : H-4, Manjari, Manjari Gota

iii) *Plant grown erect*: Pant samrat, H-4

iv) *Plant growth spreading* : T_3, Pant Rituraj, Surti Gota.

Table 2 : Morphological characters of Brinjal Cultivars / Hybrids described by Singh and Nath, 2011) given below :

Cultivars/ hybrids	Developing Institutes/ centres	Yield (q/ha)	Characteristics
Pusa Anmol	IARI, New Delhi	385.00	It is a hybrid in the attractive dark purple oblong fruits, early and high yielder.
Pusa Kranti	IARI, New Delhi	400.00	Oblong, deep purple glossy fruits, high yielding variety with shining, green calyx and less seeded.
Pusa Purple Long	IARI, New Delhi	300.00	Long purple glossy and tender suitable for summer and autumn crops
Pusa Purple Cluster	IARI, New Delhi	250.00	Fruits, long deep purple born in cluster, moderately resistant to bacterial wilt, suitable for southern and northern hills.
Pusa Bahairav	IARI, New Delhi	275.00	Fruits, deep purple, glossy long early variety resistant to phomopsis fruit rot (Kalda and Choudhury, 1991).
Pusa Purple Round	IARI, New Delhi	325.00	Fruits, attractive round purple, wt. 450/fruit tolerant to little leaf.
Pusa Anupam	IARI, New Delhi	350.00	Fruits, cylindrical, medium long, purple borne in cluster, resistant to bacterial wilt, duration 160 days.
Pusa Upkar	IARI, New Delhi	400.00	Fruits pendent, solitary round medium size glossy, dark purple, average weight 200 g .
Pusa Uttam	IARI, New Delhi	410.00	Fruits pendent, oval medium large sized, glossy dark purple, average weight 250 g.

contd...

Pusa Bindu	IARI, New Delhi	300.00	Fruits pendent, small oval round glossy, violet purple, solitary to partial cluster. About 14 fruits/kg wt.
Pusa Ankur	IARI, New Delhi	337.00	Fruits, small, oval round, dark purple, glossy, 60-80 g av. wt. Early bearer, first picking 45 days after transplanting.
Pusa Hybrid-5	IARI, New Delhi	500.00	Fruits, long, glossy, dark purple, wt. 100g , 80-85 days from sowing to fruit picking.
Pusa Hybrid-6	IARI, New Delhi	650.00	Fruits, round, glossy, violet purple, weight 200 g/fruit. 85-90 days sowing to first picking, early bearing.
Pusa Hybrid-9	IARI, New Delhi	670.00	Fruits, oval-round, glossy, dark purple, weight 300 g/fruit, 90 days from sowing to first picking.
DBR-8	IARI, New Delhi	410.00	Fruit round, glossy, dark purple, medium size, 5 fruits/ kg wt., medium duration (95 days for first picking).
DBR-31	IARI, New Delhi	430.00	Fruits oval-round glossy purple, medium size,4-5 fruits/ kg wt., Medium duration (85 days for first harvest.
DBR-44	IARI, New Delhi	300.00	Fruits small oval round, glossy dark violet purple 14 fruits/kg wt. Medium duration (85 days for first harvest.
Swarnashree	CHES, Ranchi	650.00	Fruits, white round, less seeded, flesh spongy, early variety, moderately tolerant to bacterial wilt and shoot and fruit borer.
Azad B3	CSAUAT, Kanpur	350.00	Fruits medium thick, long, shining purple, glossy attractive high yielding, good for fresh market.
Azad hybrid	SCAUAT, Kanpur	450.00	Fruits, bold, bright purple, early bearer. Cross from Azad B_1 (round) x Kalyanpur T_3 (Round).

contd...

Azad Kranti	CSAUAT, Kanpur	425.00	Fruits, uniform, Thick, oblong (15.20 cm long), dark purple, high yielder, suitable for Zone 1 and IV.
K-202-9	CSAUAT, Kanpur	380.00	Fruits, bright purple, mature in 70-75 DTP. Tolerant to fruit rot, wilt and blight.
T_3	CSAUAT, Kanpur	340.00	Fruits, round, light purple with whitish green colour at stigmatic end. Resistant to little leaf and bacterial blight.
KT-4	IARI, Katrain	370.00	Fruits, cylindrical, tender firm, purple colour, good keeping quality. Variety resistant to wilt.
Arka Sirish	IIHR, Bangalore	384.00	Fruits, thick (2-3 cm), long dark green borne in cluster (6-8 fruits).
Arka Kusumakar	IIHR, Bangalore	490.00	Fruits, small, light green, borne in cluster (5-7), good quality, mature in 110 days after transplanting.
Arka sheel	IIHR, Bangalore	480.00	Fruits, medium-long, tender, deep purple, Maturity 120 days after transplanting.
Arka Nidhi	IIHR, Bangalore	485.00	Fruits, blue-black, glossy, green purple, medium-long in cluster. Excellent keeping and transport quality (25 days at 25-30°C). Resistant to bacterial wilt. Maturity in 120 days after transplanting.
Arka Neelkanth	IIHR, Bangalore	530.00	Fruits short in cluster, glossy, violet-blue, good keeping quality, free from bitter principle. Resistant to bacterial wilt, duration 150 days.
Arka Keshav	IIHR, Bangalore	540.00	Fruits, long in cluster, glossy red purple, free from bitter principle, good cooking and keeping quality. Resistant to bacterial wilt. Duration 150 days.
Arka Navneet	IIHR, Bangalore	700.00	Fruits, deep purple black glossy, wt. 350g/fruit, white flesh less seeded good cooking

contd...

			and keeping quality, mature in 120 days.
Punjab Chamkila	PAU, Ludhiana	500.00	High yielder, developed from local material.
Punjab Barsati	PAU, Ludhiana	350.00	Fruits, dark-purple, 18-20 cm long with 4-6 cm dia, tender, high yielder (Chadha and Sidhu, 1987).
Punjab Sadabahar	PAU, Ludhiana	600.00	Fruits, 20-25 cm long, 3-4 cm dia, dark purplish to blackish in colour. Heavy yielder.
Punjab-8	Punjab (Agril. Dept.)	300.00	Fruits round, medium sized light purple, early bearer, first fruiting 65 days after transplanting.
Jamuni Gola	PAU, Ludhiana	275.00	Fruits round, plumpy, purple and early bearer (65 days after transplanting).
PH-4	PAU, Ludhiana	450.00	Fruits, medium-long and thin, dark purple, flesh light-green, cross of Hyderpur x PPL.
Hisar Jamuni	HAU, Hisar	350.00	Fruits, oblong, bulging at the blossom end, 10-15 cm long, deep bright, purple- blackish, dual season variety, most suitable as ratoon crop in early spring. Product of Aushey round brinjal.
Hisar shyamal	HAU, Hisar	325.00	Fruits, round, fleshy, bright purple, tolerant to little leaf, lesser prone to shoot and fruit borer.
Ks 224	-	400.00	Fruits, round, attractive, purple, wt. 300 g/fruit, tall variety.
Dholi 5	RAU, Samastipur	300.00	Fruits, oval, stocky, dark purple, wt. 150 g/fruit, dual season, early variety, duration 120 days.
Rajendra Baigan	RAU, Samastipur	325.00	Fruits long, light green, slight curve, heavy bearer dual season, tolerant to wilt and phomopsis blight.

contd...

Aruna	-	275.00	Fruits, round, purple, attractive.
Pant Samrat	GBPUAT, Pantnagar	450.00	Fruits, medium long dark purple, borne in cluster. First picking in 70 days after transplanting. Tolerant to phomopsis blight, bacterial wilt, shoot and fruit borers and jassids (PBR-129-5).
Pant Rituraj	GBPUAT, Pantnagar	400.00	Fruits, round, dark purple, shiny and medium sized, field resistant to bacterial wilt. Early variety, first picking 60 days after transplanting (T-3 x PPC).
Ramnagar Giant	VNS, Ramanagar	300.00	Fruits, round, shoft, light green, very big size, less seeded, suitable for Bharta preparation
Junagadh long	GAU, Gujarat	270.00	Fruits, medium, size, pinkish coloured long shaped fruits with tapering ends. Early variety (first picking 50 days after transplanting) cross of Dholi 5 x Panchmahal.
Surti Gota	MS (Agril. Dept.)	200.00	Fruits, medium to big , round with purple colour, soft and watery in taste.
Manjari Gota	MS (Agril.Dept.)	250.00	Fruits, medium, round, rosy colour white strips and very tasty. Resistant to bacterial wilt.
MHB-39 (F_1)	MAHYCO	350.00	Fruits, oval shining, reddish purple, clustered (4-6 fruits), early bearing hybrid, (60-65 days for first picking), duration 130 days, cross of MBA 149 x MBR 1285.
MHB-56 (F_1)	MAHYCO	300.00	Fruits, oval, greenish to yellow, matures in 130 days.
MHB-63 (F_1)	MAHYCO	550.00	Fruits, round, violet with white streaks high yielder
MHB-70 (F_1)	MAHYCO	575.00	Fruits, very long, medium duration, high yielder
MHB-80 (F_1)	MAHYCO	585.00	Fruits, round, reddish-black colour, rough skin, good keeping quality.

contd...

MHB-82 (F_1)	MAHYCO	580.00	Fruits, big and heavy, 150-200 g/fruit, mature in 60-65 days, early hybrid.
MHB-111(3) (F_1)	MAHYCO	565.00	Fruits, long, blackish violet, less seeded.
Narendra Hybrid Brinjal-1 (NDBH-1) (F_1)	NDUAT	570.00	Fruits, round, purple, shining, soft fleshy, less seeded , wt. 150-200 g/fruit, harvested 70-75 DATP resistant to little leaf, cross of PBR91-1 x K 202-14-90-110.
Brinjal-6 (NDBH-6)	NDUAT, Faizabad	600.00	Fruits long-oval, purple soft, fleshy, wt. 150-200g, bear in bunch, moderately resistant to shoot and fruit borer, phomopsis blight and little leaf. Cross of T_3 x Pant Samrat.
Kashi Komal (F_1)	IIVR, Varanasi	720.00	An-excellent long-fruited and prolific bearing hybrid. Fruits are light purple, long, soft textured, picking starts 65-70 days after transplanting. Suitable for growing in all brinjal growing parts of the country (Kalloo *et al.*, 2006).
Kashi Sandesh (F_1)	IIVR, Varanasi	780.00	Fruits, purple, medium-sized, round (12.4 cm long x 10.2 cm diameter) with average weight of 225 g, picking starts 76 days after transplanting (Kalloo *et al.*, 2006).
Kashi Prakash	IIVR, Varanasi	700.00	An open pollinated variety, plants semi-upright, with green stems and leaves , fruits oblong, attractive with light green spots, calyx spiny and average wt. 190 g. The picking starts 80-82 days after transplanting. (Kaloo *et al.*, 2006).
Kashi Taru (F_1)	IIVR, Varanasi	750.00	An open pollinated variety, tall and erect, Fruits long, dark purple, glossy and tender, weigh 75-85 g. Picking starts 75-80 days after transplanting. Suitable for summer and

contd...

			autumn seasons (Kalloo *et al.*, 2006).
VRBHR 1 (F_1)	PDVR, Varanasi	850.00	Fruits, ideal for exports purpose.
PLR-1	Assam	300.00	Fruits, small round, purple and clustered (1-2 fruits).
PPL-47 (F_1)	Golden Seeds	360.00	Fruits, long, cylindrical, deep purple, wt. 90-100 g / fruit.
F_1 321	Golden Seeds	410.00	Fruits, medium long, light purple, wt. 80-100 g/fruit, first harvest 55 to 60 DAP.
Nishant (F_1)	Golden Seeds	450.00	Fruits, elongated oval, dark violet, glossy, clustered (3-4/ cluster), wt. 100-120 g/fruit.
Green Long (F_1)	Golden Seeds	420.00	Fruits, green, long, slender, clustered (3-4/cl), wt. 100-120 g/fruit, resistant to bacterial wilt.
Preeti (F_1)	Golden Seeds	400.00	Fruits, small, round to oval, lustrous purple, clustered (3-5/ cl), 40-60 g each first harvest after 40 DAP.
Krishna (F_1)	Golden Seeds	450.00	Fruits, oval-round, dark purple (blackish), clustered (2-3/cl) wt 80-100 g/fruit, first picking 40-50 DAP potential to set fruit in adverse climatic condition.
New Kiran (F_1)	Golden Seeds	400.00	Fruits, round, bright purple, solitary, weigh 180-200 g each, first picking 62 DAP.
F_1 310 (F_1)	Golden Seeds	420.00	Fruits, large sized, attractive, round, deep black solitary, wt. 200-250 g each, first harvest 60-65 DAP.
Green Moti (F_1)	Golden Seeds	450.00	Fruits, small, oval-round, green with white streaks, weigh 30-40 g each, borne in cluster (3-4 cl) first harvest 40-42 DAP.
Rangeela (F_1)	Golden Seeds	450.00	Fruits, oval-round, light purple with narrow white streaks, borne in cluster (1-3/cl), weigh 100-125 g each, early (55 days first plucking), bearer.

contd...

Rangoli (F_1)	Golden Seeds	350.00	Fruits, oval round, lustrous purple with white streaks, weigh 100-125 g each, borne in cluster (1-3/cluster).
Vijay (F_1)	Golden Seeds	400.00	Fruits attractive, oval-round, light purple with white stripes, weigh 80-100 g each, first harvest 45-50 DAP.
BE 701 (F_1)	Nunhem Seeds	320.00	Fruits, round, purple black, attractive shining, large (300 g), excellent flavoured pulp quality, partial resistant to mites.
Sandhya (257) (F_1)	Nunhem Seeds	330.00	Fruits, round, attractive, purplish black, weigh 250-300 g each, less seeded, good shelf life, early maturing (45-50 DAP).
Black Boy (175) (F_1)	Nunhem Seeds	310.00	Fruits, oval-round, purple black, shining weigh 200-250 g each, good shelf life, partial resistant to sucking pests.
Jaya (1001) (F_1)	Nunhem Seeds	315.00	Fruits, round, deep purple, weight 40-50 g each, borne in clusters, suitable for stuffing, frying and pickles.
Kareena(1002) (F_1)	Nunhem Seeds	320.00	Fruits, glossy, oval- round, blackish, purple, weigh 80-100g each, early maturing (50-55 days), partial resistant to sucking pest and high temperature.
Shilpa (743) (F_1)	Nunhem Seeds	345.00	Fruits, glossy blackish, long, tender flesh, weigh 100-125 g each, partial resistant to sucking pests and bacterial wilt.
Yamini (177) (F_1)	Nunhemn Seeds	335.00	Fruits, oval, blackish, weigh 100-150 g, early maturing (45-50 days), partial resistant to sucking pests.
BE 708 (F_1)	Nunhem Seeds	325.00	Fruits, elongated-oval, purplish black, weight 200-250 g each, high yielder.

contd...

Abhishek(256)(F_1)	Nunhem Seeds	345.00	Fruits, deep purple, round, weigh 300g each.
BE 925 (F_1)	Nunhem Seeds	330.00	Fruits, glossy green, long cylindrical straight, weigh 80-100 g each, partial resistant to sucking pests.
Chaya (3715) (F_1)	Nunhem Seeds	325.00	Fruits, half long, blackish purple, glossy, borne in clusters, resistant to bacterial wilt and partly resistant to sucking pests.
BE 706 (F_1)	Nunhem seeds	335.00	Fruits, long (15-20 cm), glossy purple, early maturing (50-55 days), wide adaptability.
Kanha (091) (F_1)	Nunhem Seeds	315.00	Fruits, small (50 g), oval round, deep purple, borne in cluster, partly resistant to sucking pests.
Mohini (F_1)	Nunhem Seeds	328.00	Fruits, round to oval, violet with white stripes, borne in clusters and less seeded.
Super Mohini (F_1)	Nunhem Seeds	370.00	Fruits, round, pink with white stripes, glossy, firm and attractive, weighing 75-80 g each, resistant to sucking pests.
Kamini (F_1)	Nunhem Seeds	365.00	Fruits, round, pink with white stripes, borne in dusters and spines on calyx, partial resistant to sucking pests.
Harshita(1346)(F_1)	Nunhem Seeds	355.00	Fruits, oval-round with white stripes, glossy, green in colour, borne in cluster (2-3 fruits), medium maturity (50-55 days), high resistant to bacterial wilt.
Pinki (1364) (F_1)	Nunhem Seeds	360.00	Fruits, medium large (15-16 cm), attractive, pink, borne in cluster. Field resistant to bacterial wilt, medium maturing hybrid.
Mahakranti (PK 123) (F_1)	Sungro Seeds	345.00	Fruit, long club shaped, bright purple, wt. 75 g each, first picking 60 days after transplanting, tolerant to fruit borer.

contd...

Pragati (Chu-Chu) (F_1)	Sungro Seeds	675.00	Fruit, oval-round, purple, weight 30 g each, borne in cluster, first picking 45 DAP , high yielder hybrid, tolerant to fruit borers.
Unnati (Chu-Chu) (F_1)	Sungro Seeds	610.00	Fruits, oval-round, black, weigh 40 g each, high yielding hybrid, tolerant to fruit borer, first picking 45 DAP, suitable for stuffing.
Green long (F_1)	Sungo Seeds	520.00	Fruits, long green, weigh 100 g each, borne in clusters, high yielder hybrid, first picking 60 DAP.
Kanhaiya (F_1)	Sungro Seeds	560.00	Fruits, oval-round, uniform, dark -black, weight, 120 g/fruit, tolerant to jassids, first picking 60 DAP high yielder.
Navkiran (F_1)	Sungro Seeds	465.00	Fruits, round, bright purple fruits, weigh 160 g each, first picking 65 DAP, tolerant to fruit borer and jassids, recommended for rainy season.
Mukta (F_1)	Sungro Seeds	475.00	Fruits, oval, dark purple, weigh 100 g each, first picking 65 DAP, do well in rainy season also, continuous fruiting.
PPL-74 (F_1)	Sungro Seeds	485.00	Fruits, cylindrical, dark purple, weight 90 g each, less seeded, first picking 55 DAP.
Ankur 888 (F_1)	Ankur Seeds	440.00	Fruits, round and blackish-purple, solitary, wt. 150-200 g each, first picking 105-115 DAS. Good for bharta purpose.
Ankur-786 (F_1)	Ankur Seeds	430.00	Fruit, long slender with purple colour, cluster fruiting, wt. 60-70 g each, first picking in 100-110 DAS.
Hirak (F_1)	Ankur Seeds	450.00	Fruits, oval and green with light green strips, clusterous fruiting, wt. 50-60 g each, first picking starts 105 –115 DAS.

contd...

Abhimannyu (F_1)	Ankur Seeds	475.00	Fruits, oval-round and purple with white strips, clusterous fruiting, wt. 50-70 g each first picking 110-115 DAS.
Ajay (F_1)	Ankur Seeds	415.00	Fruits, oval, medium size, clusters fruiting, purple with green streaks good shelf life.
Kuldeep (F_1)	Ankur Seeds	435.00	Fruits, oval-round, green colour, medium size, good fruit quality.
Kaustubh (F_1)	Ankur Seeds	385.00	Fruits oval-round, purple with white strips, single to pair fruits.
Utkarsha (F_1)	Ankur Seeds	465.00	Fruits, oval, purple, medium sized, bearing in clusters.
Kushal (F_1)	Ankur Seeds	375.00	Fruits, long, slender, medium, light green colour, fruiting clusterous.
Kirti (F_1)	Ankur Seeds	455.00	Fruits, oval, medium to small size, green with light green stripes, good quality.
Shymal (F_1)	Ankur Seeds	565.00	Fruits, oblong,purple, clusterous fruiting, good cooking quality, excellent yield potential.
Abhishek (F_1)	Ankur Seeds	438.00	Fruits, oval round, purple with white stripes, clusterous fruiting, wt. 55-65 g each.
Ajinky (F_1)	Ankur Seeds	415.00	Fruits, round, green with purple stripes, solitary, wt. 50-70 g each, first picking 100-110 DAS.
Nakul (F_1)	Ankur Seeds	385.00	Fruits, oval, green with light green stripes, wt. 45-60 g. each first picking 90-105 DAS.
NS 509 (F_1)	Namdhari Seeds	425.00	Fruits, dark purple with white stripes, solitary bearing, glossy, attractive, medium sized (70 g), good performer.
Apsara (F_1)	Namdhari Seeds	450.00	Fruits, round to oval, bright purple with white stripes, glossy, medium sized (80 g), clusterous, suitable for growing throughout the year.

contd...

NS 1720 (F_1)	Namdhari Seeds	465.00	Fruits, oval, uniform purple, size 60-70 g each, cluster bearing, early fruiting, profuse yielder.
Nisha (F_1)	Namdhari Seeds	520.00	Fruits, bell shaped, deep purple glossy, uniform, wt. 350 g each, bear solitary.
Sunder (F_1)	Namdhari Seeds	540.00	Fruits, bell-shaped, deep purple colour, shiny, wt.350-400g each, solitary, good adaptability.
BSS 471 (F_1)	Bejo Sheetal Seeds	410.00	Fruits, oval to slightly high oval, purple with green as well as white stripes.
BSS 472 (F_1)	Bejo Sheetal Seed	370.00	Fruits, round to oval, purple with white as well as green stripes, suitable for all three seasons.
Kalyan (F_1)	Bejo Sheetal Seed	425.00	Fruits, reddish purple with white stripes, round to oblong. first harvest 65-70 DAP. Suitable for all seasons.
Arpan (F_1)	Bejo Sheetal Seed	415.00	Fruits, round to oblong, dark purple with white stripes. first harvest 55-60 DAP. Suitable for all seasons.
BSS 331 (F_1)	Bejo Sheetal Seed	421.00	Fruits, oval to oblong, purple, early maturing, very productive, maturity in 50 DAP.
BSS 332 (F_1)	Bejo Sheetal Seed	410.00	Fruits, round to oval, dark, purple, thornless, suitable for growing in all season, mature in 55-65 DAP.
BSS 351 (F_1)	Bejo Sheetal Seed	395.00	Fruits, oblong, black, thorn less, first harvest 55-60 DAP. Suitable for growing in all seasons.
BSS 426 (F_1)	Bejo Sheetal Seed	400.00	Fruits, oval, dark purple, small, cluster- bearing, suitable for all seasons, fruits chu-chu type.
BSS 427 (F_1)	Bejo Sheetal Seed	380.00	Fruits, oval-round, dark purple, first harvest 55-60 DAP, suitable

contd...

			for round the year cultivation. Fruits chu-chu type.
BSS 388 (F_1)	Bejo Sheetal Seed	390.00	Fruits, medium long and dark purple, thick, clusterous, profuse fruiting. Duel season hybrid.
BSS 431 (F_1)	Bejo Sheetal Seed	415.00	Fruits, long to medium long, purple thornless, first harvest 55-60 DAP, suitable for all seasons.
Kokila (BSS 387) (F_1)	Bejo Sheetal Seed	425.00	Fruits, oblong, black, first harvest 55-60 DAP, Thornless chu-chu type, suitable for all seasons.
Sumo (F_1) (BSS 425)	Bejo Sheetal Seed	410.00	Fruits, oblong, black, jumbo sized fruits as big as 1.0 kg, less seeded. First harvest 70-75 DAP. Suitable for all seasons.
Kajol (F_1)	Bejo Sheetal Seed	418.00	Fruit, round, black, medium sized, profuse bearing, first harvest 55-60 DAP.
Kajri(BSS382) (F_1)	Bejo Sheetal Seed	390.00	Fruits, oval, black, medium-biggest size, slightly oblong, less seeded. First harvest 70-75 DAP. Suitable for all round the years.
Nisha (BSS 383) (F_1)	Bejo Sheetal Seed	410.00	Fruit, medium long, dark purple, prolong fruiting, first harvest 55-60 DAP, all season growing hybrid.
BSS 330 (F_1)	Bejo Sheetal Seed	465.00	Fruits, medium long, purple, slightly thick, first harvest 70 DAP, Duel season hybrid.
BSS 434 (F_1)	Bejo Sheetal Seed	460.00	Fruits, medium long to oblong, reddish black to black, first harvest 70-75 DAP. Suitable for round the year cultivation, mid early hybrid.
BSS 465 (F_1)	Bejo Sheetal Seed	470.00	Fruits long, shining, dark black, first harvest 60 DAP, profuse fruiting, duel season hybrid.
Preeti (F_1)	Bejo Sheetal Seed	410.00	Fruits, round to oval, green with white stripe slightly purple

contd...

			shade. First picking 60-65 DAP, duel season hybrid.
BSS 428 (F_1)	Bejo Sheetal Seed	405.00	Fruits, oval, green with white stripes, small sized, first harvest 60-65 DAP, suitable for dual season hybrid.
Raveena (F_1)	Bejo Sheetal Seed	435.00	Fruits long green thorn less, first harvest 65-70 DAP, duel season hybrid.
Greenlong (F_1)	Century Seed	445.00	Fruits, long green, firm glossy, large sized, borne in bunches (2-3), early and prolong bearing, excellent flavour of pulp.
Kokila (F_1)	Century Seed	425.00	Fruits, long, deep purple – black, contrasted, less seeded.
Vardhan (F_1)	Century Seed	385.00	Fruits, attractive, purplish, clusterous, wt. 40-50 g, high yielding.
Mahadeva (F_1)	Century Seed	415.00	Fruits, globular, purplish-black, less seeded wt. 300 g each a good cropper.
Shiva (F_1)	Century Seed	395.00	Fruits, round, medium size, purplish black weight 220-260 g, excellent flavour less seeded.
Hybrid-1 (F_1)	Century Seed	425.0	Fruit, cylindrical uniform, attractive shiny purple black, good and continuous fruiting for longer period.
Nisha (F_1)	Century Seed	405.00	Fruits, attractive, bluish purple, medium sized, clusterous, very early maturing hybrid.
Choo-choo (F_1)	Century Seed	370.00	Fruits, elongated small, glossy, purple black, clusterous (3-5), profuse bearing, continuous over a long period.
Nisha improved (F_1)	Century Seed	395.00	Fruit, medium size, slow maturing adapted to monsoon cultivation.
Kamdhenu (F_1)	Century Seed	415.00	Fruits, small, purple with white stripes, A prolific cropper.

contd...

Sanyog (F_1)	Century seed	428.00	Fruits, dark blackish crimson, wt. 45-55 g each, good flavoured, less seeded, clusterous.
Angora (F_1)	Sandoz Seed	895.00	Fruits, oval-round, firm , fleshy, deep, purple black, wt 200-400 g, good shelf life, ideal for long distance transport and export.
Harit (F_1)	Sandoz Seed	550.00	Fruits, oblong-ovate, brilliant green with creamy streaks, wt. 80-100 g, early bearer (45-50 DAP).
Green Bunch (F_1)	Sandoz Seed	800.00	Fruits, oval, round, green with creamy streaks, wt. 75-90 g, clusterous, early maturing.
AUB -132 (F_1)	Sandoz Seed	1000.00	Fruits, oval-round, brilliant black, less seeded, thin skinned and solid flesh.
Kanha (F_1)	Sandoz Seed	1100.00	Fruits, long glossy, violet skin, wt. 90-110 g each, vigorous plant.
Manju (F_1)	Sandoz Seed	1050.00	Fruits, small, oval round, light purple with white stripes, wt. 80-100 g each.
Mew - Mew(F_1)	Novartis Seed	800.00	Fruits, small, round, violet, wt. 60-80 g each, lustrous violet, clusterous, medium maturing.
Sanju (F_1)	Novartis Seed	750.00	Fruits, oval round (8-10 cm long), clusterous wt 90-100 g each, medium maturing.
Green Star	Sreemaa Seed	350.00	Fruits, long cylindrical, dark purple, wt. 150-200 g each, tolerant to long distance transport and bacterial and root knot wilt, clusterous.
Green Star Long	Sreemaa Seed	410.00	Fruits ,oval-long green with white striped wt. 150-200 g each, resistant to bacterial wilt.
Cross-99	Sreemaa Seed	380.00	Fruits, oval, whitish green, clusterous (4-5), wt. 200-250 g each, suitable for long distance transportation, resistant to bacterial and nematode wilt.

contd...

Uphar	Sreemaa Seed	375.00	Fruits, long (15-20 cm), cylindrical, dark purple, wt 15-20 g each resistant to long distant transportation, bacterial and nematode root knot wilt.
Sudha	Sreemaa Seed	700.00	Fruits oblong, dark purple, smaller (100 g) size, clusterous and profuse bearer.
Saptarathi	Sreemaa Seed	500.00	Fruits, long, small, light purple, clusterous (8-10 fruits), resistant to bacterial and root-knot nematode wilting. Suitable for summer cultivation
Vijaya	Sreemaa Seed	385.00	Fruits oblong, green-white, striped, wt. 100-150 g each, clusterous, tolerant to shoot and fruit borer, resistant to wilt.
Blue Star	Sreemaa Seed	395.00	Fruits, oval, glossy, purple colour, wt. 200-250g each , wide adaptability, resistant to wilt, cultivated throughout the year, heavy cropper.
Chandrika	Sreemaa Seed	390.00	Fruits, round, dark green, white striped, wt. 200 g each, resistant to bacterial and nematode wilt.
Panchasakha	Sreemaa Seed	375.00	Fruits oval, blue, clusterous (8-10), suitable for summer cultivation.
Mukta	Sreemaa Seed	345.00	Fruits, oval, purple, glossy, wt. 100 g each, resistant to bacterial and nematode wilt.
Purnima	Sreemaa Seed	350.00	Fruits, cylindrical, whitish green, Clusterous (6-7) wt. 100-200 g, suitable to long distance transportation resistant to bacterial and nematode wilt .
Komal (F_1)	Unicorn seeds	318.00	Fruits, long, dark purple, wt. 80-150 g each medium size, solitary.

contd...

Kajal (F_1)	Unicorn Seeds	327.00	Fruits, oval-round dark purple, wt. 300-400 g, first picking 60-65 days from transplanting.
Vijay (F_1)	Unicorn seeds	328.00	Fruits, oval, firm glossy, black purple attractive, wt. 300-450 g, uniform size.
Ghanshyam (F_1)	Unicorn seed	325.00	Fruits, oval, medium to big, dark purple skin, wt. 300 g each, tolerant to fruit borer and early bearer.
Shyam (F_1)	Unicorn seed	320.00	Fruits, round-oval, purple, small 30-45 g, uniform and early bearing, clusterous suited for stuffing, Ist picking 45 days after planting.
Prabha (F_1)	Unicorn seed	342.00	Fruits, round-oval, violet with white streaks. Wt. 80-100 g, early prolific bearer.
Gowri (F_1)	Unicorn seed	335.00	Fruits, long, shiny green, medium size wt. 40-50 each clusterous (6 fruit).
Pratibha (F_1)	Unicorn seed	340.00	Fruits, round, violet with white streaks, size 100-125 g each.
Divya	Unicorn seed	302.00	Fruits, round, shiny purple, wt. 250-300 g, first picking in 65 days after transplanting.
Kiran	Unicorn seed	325.00	Fruits, round, shiny green with white spots, borne in clusters, wt. 40-50 g each.
BJT 103	Unicorn seeds	360.00	Fruits, long, black purple, wt. 100-130 g each.
BJT 146	Unicorn seeds	345.00	Fruits, round-oval, purple, wt. 500-600 g each.
Long-13 (F_1)	IAHS	450.00	Fruits, long, dark purple, early maturing hybrid.
Round-14 (F_1)	IAHS	500.00	Fruits, round, light purple, medium in size.

Table 3 : Varieties resistant to different pests and diseases.

Varieties/Hybrids	Resistant to Pests and Diseases	Reported by
West Coast Green, Round 112-18	Bacterial wilt	Gopinath and Madal ageri (1986)
S. indicum and *S. sisymbrifolium*	Little leaf diseases	Anjaneyulu and Ramakrishnan (1968)
Pusa purple cluster	Little leaf and bacterial wilt	Chakrabarti and Choudhury (1974)
Nurki, Bourad Local No.4 and Chikhalgaon Local No.1	Moderately resistant to little leaf	Mayee and Munshi (1973)
Hisar Shyamal and H-10	Little leaf and bacterial wilt (tolerant under field condition)	Kalloo (1990)
S. xanthocarpum, S. indicum, S. gilo, S. khasianum, S. nigram and S. sisymberifolium, S. melongena line, cultivated line 11a and 264	Phomopsis blight	Khalda *et al.* (1977)
Vijaya and Black Beauty	Tolerant to root knot nematode	Yadav *et al.* (1975)
Pusa Purple Long	Tolerant to shoot and fruit borers and Jassid	Gill and Chadha (1979)
Punjab Chamkila and Pusa Purple Cluster	Tolerant to shoot and fruit borer and jassid	Gill and Chadha (1979)
Tambalwadi Local, Gopuri, Pusa Purple Round, Jalgaon Local, Hingana Dorla, Baramasi, Punjab Bahar	Highly resistant to shoot infestation	Kale *et al.* (1986)
Junagadh Bhattu, Nurki, Pusa Purple Cluster, Gulabi Dorla,	Highly resistant to fruit infestation	Kale *et al.* (1986)
SM 17-4	Field resistant to fruit and fruit borer	Bajaj *et al.* (1989)
Dorli, Jumbli, Malayalum, Black Beauty, Manjri Gota, Ramnagar Giant, Vijay Hybrid	Resistant to Jassid	Mote (1978 and 1982)
Arka Keshav, Arka Neelkanth, Arka Nidhi, Utkal Tarini, SM-6-6 (Soorya), JC-1 & JC-2	*Resistant to Bacterial wilt (Pseudomonas solanacearum)*	Anonymous (2000)
Pusa Bhairav	Phomopsis rot *(Phomopsis vexans)*	Anonymous (2000)

Note : (i) Higher glyco alkaloid content, peroxidase and polyphenol oxidoase conferring to eggplant fruits towards pests.

(ii) Hairyness of leaves was the main cause of resistance to Jassids which hinders the oviposition.

CLIMATIC INFLUENCE

Brinjal show a wide climatic tolerance and can be grown in the open wherever there is more than 3 months of frost-free weather. Brinjal is well adapted to both wet and dry season cultivation but excessive rainfall will check both vegetative growth and flower formation. It requires a long and warm growing season. It is susceptible to forst. The plant should not be transplanted in the field until the daily mean temperature reaches 18.3°C to 21.1°C. The optimum growing temperature is 22 to 30°C and growth stop at temperature below 17°C. Pollen deformity increases at temperature of 15-16°C. High soil temperature is injurious to the root system (Shukla and Naik, 1993).

Temperature range of 17-25°C is suitable for growth, fruit set and fruit yield. It is very susceptible to frost. Late round fruited varieties are more tolerant to frost than the early long fruited varieties. The cool season affects adversely plant growth., fruit quality in respect to size and colour development. The planting of brinjal should be done when the daily mean temperature reaches 13°C to 22°C. The seeds germinate well at 22°C to 25°C. Brinjal is well adapted to both wet and dry season cultivation. Elevation upto 1200 m is satisfactory for cultivation.

SOIL REQUIREMENT

Brinjal is a hardy plant and can be grown on wide range of soils but does best in silt loam or clay loams. A pH range from 5.5-6.8 is considered suitable. The root system is sensitive to excess soil moisture and higher soil temperature. Since brinjal is a long duration crop, deep cultivation prior to planting is preferable.

For successful cultivation of brinjal well drained, fertile soil is necessary. Sandy or sandy loam soils are good for early crop. Higher yields are obtained when crop is raised in clay loam and silt-loam soils. The brinjal plant is moderately tolerant to acidic soil which is also favourable for its better growth and development.

SOWING

Brinjal is mainly propagated through seed, but axillary shoots can be rooted if required. Where bacterial wilt is prevalent in Puerto Rico, eggplant is some times grafted on *Solanum torvum* Swartz, which is resistant to the disease (Purseglove, 1977). The seeds are sown in the nursery and seedlings are transplanted in the field.

Seed Rate

In order to raise brinjal crop in one hectare area, about 400-500 g seed is sufficient to provide seedlings. Brinjal seeds have 70-80 per cent germination.

According to Purewal (1957), brinjal seeds are light in weight having an average germinability of 75-80 per cent. About 500 to 750 g of seeds are sufficient for raising seedlings for one hectare of land.

Seed Treatment

Thiram @ 4 g/kg of seeds or ceresan or agrosan GN @ 3 g/kg of seeds is sufficient for seed treatment. Seeds are also treated with 30 ppm IAA before sowing to get vigorous seedlings.

Nursery Management

Nursery bed should be prepared according to the season. In the rainy season raised beds are prepared but in the winter season flat beds are suitable. Size of beds should be 3.0 m x 1.0 m and 15 cm high. 20-25 beds of such size are sufficient to raise seedlings for one hectare area.

Generally raised beds are prepared for raising seedlings of brinjal. The top soil of the nursery bed should be mixed properly with well rotten farm yard manure or compost or leaf mould.

Nursery beds are treated by drenching the soil with 0.2% brassical or captan. Nursery beds can also be sterilize with 40% formaline and half litre of formaline is sufficient for 1 sqm. bed. Generally 10% formaldehyde is used for fumigating the dug up soil and soon after fumigation beds are covered with polythene for 24 hrs. The seeds are sown only after 5-6 day of fumigation.

Seed beds are finally prepared by mixing of 20 kg compost/FYM, 150 g single super phosphate, 50 g muriate of potash, in each bed. Seeds are sown 5 to 10 mm deep in rows of 5 cm apart. Sowing of seed is followed by covering it with a fine layer of soil and compost. It is advisable to cover the nursery bed with any locally available mulching material like paddy straw or dry grasses like *Sachharum spontanium* till the seeds germinate. Brinjal seeds germinate in 8 to 10 days. The germination is epigeous.

Under good nursery management, brinjal seedlings become ready for transplanting in 20-25 days or so after sowing of seeds. Brinjal is less sensitive to damping-off than tomato. Seedlings of 8 to 10 cm height with 2-3 true leaves are ready to transplant. Verma *et al.* (1980) obtained the highest yield of brinjal from seedlings transplanted at 5 weeks of age. The seedlings should be hardened by withholding watering for 4 to 6 days before transplanting. Chaugule and Pandey (1958) reported that the plants hardened by irrigating at 7th or 11th day gave an average of 56 per cent higher yield than those irrigated every third day (no hardening). Desai (1962) also found that even the maximum hardening of seedlings (irrigation on every 12th day) increased fruit yield. The varieties also differ in their response to hardening. Desai (1962) reported that Surti Gota and Manjari Gota responded very well to hardening of seedlings.

However, nursery bed should be softened with water, if soil become too hard, so that roots are not injured when the seedlings are uprooted. Crumbling of roots should be avoided while transplanting.

The following are main season during which brinjals are cultivated.

In Plains

Season	Sowing Time	Transplanting time
1. Autumn-Winter	June-July	July-August
2. Spring-summer	Nov.-Dec.	Jan.-Feb.
3. Rainy season	March-April	April-May

In Hills

Brinjal seeds are sown in April and transplanting is done in May in hills. In many places, brinjal is also sown in September and transplanted in october or in early November.

Land Preparation

The soil is well prepared. Salinity affects plant growth and yield of brinjal. It is more pronounced beyond 4 ds/m Ece, however, yield reduction depends on variety under cultivation (Sharma and Malik, 1983).

As the crop remains in the field for a number of months therefore, the soil should be well prepared by ploughing and harrowing 4 to 5 times before transplanting the seedlings. Bulky organic manures should be incorporated into the soil during land preparation. When land is sufficiently prepared and levelled, the whole plot is divided into the beds of suitable dimensions before transplanting (Chauhan, 1981).

Planting Distance

The distance of transplanting for non-spreading brinjal varieties would be 50 to 60 cm both ways, and for spreading varieties, the row to row distance should be 75 to 90 cm and for plant 60 to 70 cm (Singh, 1989).

Rainy and summer crop transplanted at spacing of 75 cm x 60 cm (round varieties). Spring and summer crop at spacing of 60 cm x 45 cm (With long fruited varieties). Spreading types are generally planted at spacing of 75cm x 75 cm from row to row and plant to plant. High yielding and hybrids should be placed at 90 cm x 90 cm.

PLANTING

In plain brinjal crop can be grown in all three season *viz.*, Feb-March, June-July and Oct.-Nov. The planting is done on flat beds of convenient size for irrigation. In high rainfall region crop is grown on raised beds for proper drainage. The spring summer crop may be raised on ridges for efficient use of water. The seedlings are ready for transplanting when they attain a height of 15 cm with 4-5 leaves, stocky, healthy seedlings which are free from disease and shoot and fruit borer infestation are selected for planting. Watering of the seedling is withstand for four days before planting. This result, in better establishment of the seedlings.

Seedling should be transplanted in afternoon hours to avoid mortality. Light irrigation or pot watering should be given just after transplanting.

NURITIONAL REQUIREMENT

Brinjal is a long duration crop and occupy the land for nearly 6-8 months, besides it is a heavy yielding and removing large quantities of plant food in one cycle of plant growth. A crop yielding about 60 q/ha removes about 190-25-159 kg N, P_2O_5, K_2O/ha (Gnanakumari and Satyanarayan, 1971).

For better fruiting and fruit yield, application of balanced nutrients are essential. They are supplied both through organic manures and fertilizers (Table 4).

Table 4 : Fertilizer recommendations for brinjal crop in various states (Tandon, 1991) NPK (kg/ha), FYM (t/ha).

State	N	P_2O_5	K_2O	FYM t/ha
Andhra Pradesh	100	60	60	20
Assam	50	50	50	10
Bihar	160	88	90	20
Gujarat	100	50	50	10-15
Haryana	100	50	25	25
Himachal Pradesh	45	60	40	-
Karnataka	125	100	50	25
Kerala	75	40	25	20-25
Madhya Pradesh	100	60	25	20
Maharashtra	90	40	0	50
Orissa	125	80	110	-
Punjab	125	62	30	25
Rajasthan	80	80	60	15
Tamil Nadu	100	50	30	25
Uttar Pradesh	100	50	50	20
West Bengal	120	50	50	15

Fertilizer dose for hybrids is generally higher than the improved varieties. Pal *et al.* (2002) reported that application of 187.5 kg nitrogen and 90 kg phosphorus produced maximum fruits and fruit yield of hybrid brinjal in loamy sand soil of Punjab.

Method and Time of Application

Both method and time of application of nutrients in brinjal crop are important. Basal application of one fourth of the total nitrogen alongwith whole quantity of phosphorus and potassium should be applied within 24 hours before transplanting of seedlings. The half dose from the remaining quantity of nitrogen should be applied 25-30 days after transplanting and the second application should be during flowering (45-50 days after transplanting). Yawalkar *et al.* (1962) suggested three split doses of nitrogen. The first dose is to be given 1½ month after transplanting, the second dose one month after the first dose and the final dose 3½ months after transplanting.

Top dressing of nitrogen fertilizer is done in two ways (i) band application i.e., fertilizer is placed either in shallow furrow opened along the row of brinjal plants or (ii) fertilizer is placed on ground along the row of the plants and shallow digging suffice mixing of it. Top dressing can also be done by foliar spraying. The concentration of solution should not exceed 2% otherwise nitrogen may cause leaf scorching or leaf burning. The absorption efficiency of nutrition is also reduced when spray solution is not having optimum concentration. Foliar feeding of nitrogen is also advantageous in the way that one can also apply pesticides or fungicides simultaneously there by reducing cost of spray (Singh, 2001).

Use of Macro and Micro Nutrients

The micronutrient deficiencies in soils and crop has increased in the recent years due to the factors like intensive cropping, loss of top soil by erosion, leaching, decreased use of farm yard manure, increased purity of chemical fertilizers and liming of acid soils. Increased in quality and yield of vegetables from the application of micronutrients have been reported in many parts of the world. The micronutrients malnutrition is seriously damaging the health of over 40 % of the worlds population (De and Rai, 2006). Micronutrients serve mainly as constituents of prosthetic groups in metalloproteins and as activators of enzyme reactions. Adequate ranges of micronutrients in dry matter of crop will be an important aid to diagnose the deficiency and toxicity. The Brinjal crop shows adequate concentration ranges (ppm) of Fe (40-200), Mn (30-125), Zn (20-60), Cu (5-15), B (25-75) and Mn (0.2-0.8). Very low analytical values below the "adequate range" means acute deficiency with visual symptoms , very high analytical values indicate uneconomic nutrition (luxury consumption) or toxicity symptoms.

The micronutrients deficiencies can be corrected by adding the micronutrient that is indeficit or by correcting the factor for its unavailability. Micronutrients can be applied in three different ways: (i) soil application, (ii) applied as a single foliar spray at just before flowering stage, suspected mild deficiency and (iii) applied as a foliar spray at critical crop growth stages (i.e., active growth, flowering and fruit setting suspected severe deficiency. Nair and Mehta (1959) reported that deficiency of Zn caused chlorosis, area near the leaf margins and between the viens become necrotic, producing a scorching effect and the lower leaves began to wither away. Flowering was delayed and flowers failed to set fruits and withered. Foliar application of 3000 to 5000 ppm of Zn correct the deficiency.

Spraying of Cu was found to increase the number of flowers and fruits while Zn and Mn increased fruit weight (Nagarajan and Shanmugasundaram, 1961). Aliev (1968) from his experiment with B, Mn, Cu and Zn concluded that all these micro elements, particularly B and Mn, stimulated photosynthesis and increased yield by 11 to 23%. The highest yield was, however, noted with Cu and all the micro elements resulted in an increased in sugar drymatter and vitamin C contents.

Mehrotra *et al.* (1973) identified deficiencies symptoms of different nutrient in brinjal plants are given below : (i) N deficiency: general paling and stunted growth , (ii) P deficiency: dirty grayish green leaves and premature shedding, (iii) K and Fe deficiency: interveinal chlorosis of young leaves followed by yellowing and premature shedding, (iv) Ca deficiency: light green colour of young leaves along with necrotic spotting, (v) Mg deficiency : inverted 'V' shaped intervinal chlorosis, (vi) B deficiency : death of apical bud coupled with browning of roots. Das *et al.* (1974) reported that spraying of boron 105 ppm, molybdenum 325 ppm, calcium 21 ppm and planofix 200 ppm on brinjal plant 35 days after transplanting significantly improved fruit quality of brinjal. Vaganov *et al.* (1975) observed that B, Mn, Zn or Mo at 0.01% increased the leaf water retaining capacity during almost all growth phases except flowering when sprayed 3 times. Shatrova (1977) indicated that foliar application of Mn or B, each at 0.05 to 0.15% during bud formation and flowering decreased the incidence of wilt by 2-3 times. Mn was lightly more effective than B. Reddy and Reddy (1986) reported that fruit yield increased by spraying with KNO_3 at 2000 ppm or by root dipping. Treatment with Zn and B each at 500 ppm also showed increased yield. Pratima *et al.* (2001) determined the boron requirement of brinjal plant as 0.0033 mg/litre acute, 0.033 mg/litre moderate and 0.33 mg/litre adequate. Pratima *et al.* (2002) observed that calcium deficiency effect on enzyme activity of brinjal at 0.05 meq/litre, however, 8.0 meq/litre of calcium was noted as adequate level for brinjal.

Use of Biofertilizer

Biofertilizers are culture of microbial cells which benefit the plant by providing nitrogen or phosphorus to bring about rapid mineralization of organic materials. They help enhance productivity by biological nitrogen fixation or solubilization of insoluble phosphate or producing hormones, vitamins and other growth factors required for plant growth (Bhattacharyya *et al.*, 2000; Kanaujia and Narayan, 2003). Among these, nitrogenous and phosphatic biofertilizers are being widely used in vegetable crops. Nitrogenous biofertilizers are mostly bacterial biofertilizers viz., Rhizobium, Azotobacter and Azospirillum. They are economical in saving 25-50 per cent of the recommended dose of nitrogenous fertilizers and increase the yield 15-50 per cent in vegetables. Phosphatic biofertilizers are bacterial and mycorrhizal and inoculated in vegetable crops. *Pseudomonas* and *bacillus* possess the ability to bring insoluble phosphate in soil into soluble by secreting organic acids. These organisms can grow on insoluble phosphatic sources such as tricalcium phosphate, magnesium phosphate, rock phosphate, bonemeal and convert in soluble form.

Biofertilizers are insolubled by seed inoculation, seedling inoculation, foliar spray and soil inoculation. About 200 g of biofertilizer is required to treat 10-14 kg of seeds. Suspend one packet of 200 g in about 400 ml water and mix thoroughly. Pour this mixture on seeds and mix with hands to obtain uniform coating on each and every seed. Spread the seeds in shade for drying for 10-15 minutes then sow them immediately. For seedling treatments 1.0 kg culture dissolved in 10-15 litre of water and it is sufficient to treat the seedlings for one acre. Make small bundle of seedling and keep the root portion of seedling in solution of culture for 10-15 minutes. Transplant it immediately. The ratio of inoculant and water should be 1:10 i.e., 1 packet (200 g) in 10 litres of water. For soil application prepare the mixture of 2-3 kg of biofertilizer in 40-60 kg of soil/compost. Broadcast the mixture in one acre of land either at sowing time or 24 hrs before sowing. Jose *et al.* (1988) reported that organic form of nitrogen (biofertilizers) were effective to increase the fruit number and yield of brinjal. Nanthakumar and Veeraragavathatham (2000) reported beneficial effect of biofertilizers in increasing the yield and yield attributes of brinjal. Devi *et al.* (2002) revealed that combined application of organic manure, biofertilizers and 50% reduction in chemical fertilizer increased the growth, yield attributes and fruit yield of brinjal as compared to sole application of chemical fertilizer or chemical fertilizer at low dose in association with biofertilizer. The treatment 50% N (60 kg/ha) + 25% poultry manure (1.675 tonne/ha) + 25% biofertilizer (1.875 kg/ha) recorded a maximum fruit yield of 27.5 tonnes/ha. Kamili *et al.* (2002) recorded maximum fruit yield of 184.40 q/ha with combined application of Azotobacter + 100 kg N/ha.

Mycorrhizal fungi are also useful in increasing the yield and yield attributes of Brinjal crop. These fungi are most common fungal association among angiosperms. The vesicular-arbuscular mycorrhizal (VAM) are formed by the non-septate phycomycetes fungi belonging to *Glomus, Gigaspora, Acaulospora* and *Sclerocystis* in the family Endogonaceae of the order Mucorales. They produce vesicles and arbuscules inside the root system. Arbuscules are highly branched fungal hyphae while vesicles are the bulbous swellings of these hyphae. The VAM fungi are beneficial to the plant which they colonize. They make more nutrients available to the plants, improve soil texture, water-holding capacity, improve disease resistance and help in better plant growth. Besides, mycorrhizae are also helpful in the biological control of root pathogens (Asokan *et al.*, 2000). Brinjal crop is inoculated with VAM fungi in the nursery-beds. The inoculum consisting of spores of fungus mixed with soil and infected root bits of the host plants on which they are raised is either spread uniformly in the nursery beds or placed in rows over which seeds of brinjal are sown. The seedlings are colonized by the fungus 45 days after transplanting. Inoculated plants show better growth and improved nutrient content. Addition of inorganic phosphates should be reduced by about 25-50% depending upon the soil P content. Mycorrhizal fungi help in control of nematodes. Mycorrhizal fungi inoculated brinjal plants show more resistance to nematode attack compared to non-mycorrhizal plants.

Use of Growth Regulators

Application of triacontanol 2 ppm + boron 4 ppm at 15 days after transplanting and again at the flowering give maximum fruit yield. Para-chloropheonxy-acetic acid (4-CPA), 2, 4-dichlorophenoxy acetic acid and napthalene acetic acid (NAA) improve fruit set and yield in brinjal. Spraying of 2, 4-D @ 2 ppm at flowering stage at interval of one week over a period of 60-70 days from the commencement of flowering, induce parthenocarpy, increase fruit set, advance fruit maturation and significant increase in total yield. (Muthukrishnan and Srinivasan, 1963).

Pre-sowing treatment of seed with IAA at 50 mg/litre, GA at 40 mg/litre (Sadawarte and Gupta, 1968; Das and Prusty, 1972) or soaking of seedlings root in NAA at 0.2 mg/litre and ascorbic acid at 250 mg/litre has been reported to produce highest fruit yield (Shirasat *et al.*, 1981). Dipping of seedling roots in solution of NAA at 0.1 or 0.2 mg/litre for 24 hours has been most effective treatment in brinjal (Sambasiva Rao *et al.*, 1980). Application of 2, 4-D at 0.00025% in a linolin paste to cut ends of styles or as foliar sprays to freshly opened flower clusters has been reported to induce parthenocarpy in brinjal (Muthukrishnan and Srinivasan, 1966). In pot experiment on brinjal, Das and Prusty (1972) reported enhanced seed germination with GA at 10 mg/litre. Bisaria and Bhatnagar (1978) obtained higher fruit yield with foliar sprays of IAA at 20 mg/litre. Hooda *et al.* (1985) reported higher fruit yield with cytozyme seed treatment at 10% and foliars sprays at 1.25%. Foliar sprays of mixtallol at 4 mg/litre four weeks after transplanting have been found to give highest fruit yield (Shukla and Prabhakar, 1988). 25 kg N/ha as 1% urea as foliar spray at transplanting + a basal dressing. 25 kg N/ha as 1% in form of foliar spray at transplanting + a further 25 kg N/ha as 1% urea at flowering is highly beneficial (Rajeeva and Rao, 1980).

IRRIGATION

Irrigation is essential for brinjal cultivation, being a shallow rooted crop need irrigation at frequent intervals. Irrigation should be given according to local need, insufficient water results in leaf fall and flower bud abscission. Hence, timely irrigation is necessary for high yield. Irrigation is to be given once in 4-5 days during summer and once in 10-14 days during winter.

Immediately after transplanting of brinjal seedlings, light irrigation is utmost essential for survival and establishment of seedlings. Subsequent irrigations are needed according to season. During the summer, brinjal crop requires irrigation at every fourth or fifth day whereas 10 to 12 days during the winter (Yawalkar, 1969). Higher yields are obtained when crop is irrigated at right time of its requirements. Drip irrigation saves 30 to 40 per cent water. About 100-110 cm of irrigation water are required for successful brinjal crop (Choudhury, 1979).

The Brinjal crop is managed to prevent from moisture stress by seed soaking in 1% KH_2 PO_4 solution for 72 hours. The crop is given stress for about 15 days during flowering by withholding irrigation, and such stressed crop sprayed with 1% KCP, significantly recorded higher yield because sprayed plants maintained higher photosynthesis (19.6 mg CU_2/dm_2/Hrx, proline content (160 U_2 mg/g) coupled with low transpiration rate (1.97 mg H_2/cm^2/Sc) which might have led to higher yield (Balkrishnan *et al.*, 1993).

INTERCULTURE

Brinjal is initially slow growing crop, incapable of offering any competition to the weed. Therefore weed should be effectively controlled. During early stage of crop growth shallow cultivations will be sufficient to provide good environment for better root development and well also help in keeping down the weeds.

Two to three hoeing and earthing (on 20, 30 and 60 DAP) are required to raise a good crop. After the hoeing, earthing up is essential by mounting the soil around the main stem with the help of spade to provide the plants a strong base to bear the fruit load.

Mulching

Brinjal responds well to the mulching operation with the use of black plastic or black polythene. Mulching materials help in stimulating growth and improving earliness ultimately plastic mulch also reduces the injection rates of verticillium wilt and increases the brinjal yield. The other mulch materials which are locally available may also be used (*Guinea grass, Panicum maximum*) at 4 tonnes per hectare (Singh, 1989). Marked increased in yield by mulching the brinjal crop with Sarkanda grass (*Erianthus arundinaceus*) (Singh *et al.*, 1985). Awasthi *et al.*, (2006) reported that fruit quality parameters were significantly influenced by mulch treatments. Application of 75 m thick black polyethylene film and 75 m thick white polyethylene film and lasoda (*Cordia myxa*) leaf (10 tonnes/ha) mulch recorded maximum fruit size, fruit yield and better quality fruits.

WEED CONTROL

The weed should be controlled as soon as they are seen either by traditional method of hand weeding and hoeing or by application of herbicides. The most serious weed is the *Orobanchae* sp and few other weed plants like *Amaranthus viridis, Acanthospermum hespidum, Cyperus rotundus, Digitaria marginata, Orobancha spp., Portulaca oleracea, Triantheme portulacastrum etc.* (Ali, 1985; Singh, 1997) are commonly seen growing with Brinjal crop. They can be controlled by deep ploughing, crop rotation, summer ploughing etc. Following weedicides are also effective to control weeds in Brinjal crop. They are: Stomp, a weedicide @ 1.5 kg a.i./ha may be sprayed 3-4 days after transplanting followed by one hand weeding. Trifluralin @ 1 kg/ha give the best yield with a good weed controlled for 8 weeks. Pre-emergence/planting application of stomp 30 EC @ 1 kg a.i./ha followed by hoeing/weeding 35-40 days after planting and second 60 days after planting is advisable in brinjal to escape the critical period of crop-weed competition (Anonymous, 1995). Raghav (1995) reported that Fluchloralin @ 1.5 kg a.i./ha (pre-plant) + one hand weeding at 6 WAT gave maximum fruit yield (266.52 kg/ha). Singh *et al.* (1996) observed that pendimethalin at 0.56 kg a.i./ha + hand weeding and oxyfluorfen at 0.1 kg a.i./ha and one hand weeding gave the best weed control and freuit yield. But for summer crop oxyfluorfen at 0.1 kg /ha and 0.12 kg/ha + one hand weeding gave the best weed control and fruit yield. Pre-plant application of Fluchloraline @ 1.5 kg a.i./ha + one hand weeding at six weeks after transplanting, effectively controlled weeds and gave the maximum fruit yield and net profit with a cost benefit ratio (Raghav, 1995). Pre-emergence application of stomp 30 EC @ 2.5 l/ha and one hand weeding was effective to control the weeds (Pal *et al.*, 2002).

PLANT PROTECTION MEASURES

The important insect pests along with their control and management (Butani and Jotwani, 1984) are given below :

Insect Pests

Shoot and Fruit borer (*Leucinodes orbonalis* Guen)

It is the most serious insect pest of Brinjal. The larvae are light pink in colour and about 15-18 mm long. In young plants, the larvae bore into the petioles, midribs of large leaves and young shoots resulting in withering of the terminal portion. In later stages, they bore into the flower buds and fruits. Consequently such flower buds dry up and the fruits are rendered unmarketable. The infested fruits emit foul smell. They can be controlled in following ways. (i) Avoidance of continuous cropping of Brinjal in the same land. (ii) Removal and destruction of the infested shoots and fruits regularly. (iii) Spraying the crop with 0.1 % carbaryl (Sevin 2 g/l), fenvalerate (Sumuicidin 0.5 ml/l), cypermethrin (Ambush 0.2 ml/l), decamethrin (Decis 0.5 ml/l), quinalphos (Ekalux 2 ml/l), monocrotophos (Nuvacron 1.25 ml/l) or endosulfan (Thiodan 1.5 ml/l) at fortnightly interval (Bagle, 1993). (iv) Use of pheromone trap for adult females. (v) Growing less susceptible varieties like, Pusa Purple Long, H-4, Pusa Kranti, SM-41, Pant Samrat, Punjab Barsati, Pusa Purple Round, Punjab Neelam, HE-12, etc.

Stem borer (*Euzophera perticella* Ragonot)

This pest is widely distributed in the Indian subcontinent . The larva is light brown in colour and 16 to 18 mm long. The larvae bore through the main stem and hamper the growth of the plant which ultimately affects flowering and fruiting. The attacked plant may wilt and wither. Control measures are same as for fruit and shoot borer.

Epilachna or Hadda beetle (*Epilachna vigintioctopunctata* Fabricius)

It is a polyphagus coccinellid beetle and causes tremendous damage to the plants. The beetle is bronze to red, small spherical and mottled with black spots. The grubs are yellowish and stout bodied with stiff hairs on the bodies. The adults and grubs feed voraciously on the leaves by scraping in a characteristic manner leaving the veins intact. As a result, skeletonised patches develop on the leaves which later dry away. Epilachna beetle can be control by : (i) Hand picking of grubs, adults and eggs from the infested leaves in the early stages of attack. (ii) Dusting the crop with 5% BHC @ 20 kg/ha. (iii) Spraying the crop with 0.05% endosulfan (Thiodan 1.5 ml/l), 0.05% monocrotophos (Nuvacron 1.25 ml/l) or 0.2% carbaryl (Sevin 4 g/l). (iv) Growing less susceptible varieties like Arka Shirish.

Leaf roller (*Eublemma olivacea* Walk)

It is a common foliage pest of brinjal. The caterpillars roll the leaves and feed on the green tissue while remaining inside the fold and thus lead a concealed life. The rolled leaves ultimately wither and dry up. The caterpillars also bore into the shoot and feed inside resulting in withering of the plant. To control leaf roller following measures should be taken: (i) Collection and destruction of the folded leaves along with the caterpillars. (ii) Spraying the crop with 0.05% endosulfan (Thiodan 1.5 ml/l), 0.08% malathion (Cythion 2 ml/l) or 0.01% carbaryl (Sevin 2 g/l).

Jassids or Leaf hoppers (*Amrasca biguttula* biguttula Ishida and *Hishimonas phycitis* Dist.)

Nymphs and adults of these Hemipteran species suck the sap from the undersurface of the leaves. The infested leaves curl upwards along the margin like a bowl which become sick and may show burnt up patches. Apart from direct damage they ooze out a sugary liquid called 'honeydew'. Sooty mould which develops on these honeydews impair photosynthesis. These jassids also transmit mycoplasma and virus which cause little leaf and mosaic disease, respectively. They should be checked by: (i) Spraying the crop with 0.05% dimethoate (Rogor 2 ml/l), 0.08% malathion (Cythion 2 ml/l), 0.02% carbaryl (Sevin 4 g/l) or cypermethrin (Ripcord 0.4 ml/l) at 10 days interval. (ii) Application of granular insecticides like Furadon 3G (30 kg/ha) or Thimet 10 G (25 kg/ha) at the time of transplanting followed by 3 fortnightly sprays with cypermethrin (Ripcord 0.4 ml/l) or 0.2% carbaryl (Sevin 4 g/l). (iii) Growing field tolerant varieties like H-4, T-3, Aushey, Pant Samrat.

Aphids (*Myzus persicae* (Sulzer), *Aphis gossypii* Glover)

These polyphagus pests are small soft, yellowish green or greenish-brown and is found in colonies of hundreds on the tender shoot and the undersurface of the leaves. The nymphs and adults suck the sap from the leaves which become curled and finally dry up. Black shooty moulds which develop on the honeydews excreted by the aphids binder photosynthesis and devitalise the plants Control measures are same as for jassids.

Red spider mite (*Tetranychus neocaledonicus* Andre, *T. cinnaarinus* Boisduval, *T. telarius, Parateiraayclus indicus*)

These are very small, polyphagus pest found in large colonies on the lower surface of the leaves covered with fine, silky webs. The protonymphs and adults suck the sap from the leaves causing characteristic blotching of leaves. Red spider mite can be controlled by: (i) Dusting the crop with fine sulphur @ 20-25 kg/ha (ii) Spraying the crop with Dicofol (3ml/l) or phosalone (Zolone 2 ml/l) at 10 days interval. (iii) Spray sevisulf (a mixture of carbaryl and sulphur) at the rate of 0.1% (Kumar *et al.*, 1989).

DISEASES AND THEIR CONTROL

Fungal Diseases

Phomopsis blight

Caused by *Phomopsis vexans* (Sacc. and Syd) Harter. This is a serious seed as well as soil-borne disease of brinjal (Walker, 1952). The fungus attacks all the aboveground parts of the plant. Dark brown spots of irrigular size develop on the leaves which later spread over the entire area. On fruits the disease symptom appears as minute, sunken, dull and dusky spots, which later merge to form rotten areas. In severely infected fruits, the entire flesh may rot. In seedling infection, it may cause wilting of the plants. It can be controlled in following ways:

(i) Care should be taken at the time of seed extraction. Calyx are to be removed from the healthy fruit and the fruits be dipped in mercuric chloride solution (25 g in 35 litres of water) for 10 minutes. After extraction, the seeds should be dried properly before storage. (ii) In the seed production field, repeated painting of the peduncle, calyx and adjacent fruit region with blitox 15-20 days after fruit set. (iii) Use of only bold, shiny and disease free seed. (iv) Hot water treatment of seeds at 50°C for 30 minutes (1 litre of boiling water may be mixed with 1 litre of normal water) to get the required temperature of water. (v) Seed treatment with bavistin @ 1 g/kg of seed or with thiram or captan @ 2.5 g/kg of seed. (vi) Spraying the seedlings in the nursery bed and in the main field with 0.2% captan or 0.1% bavistin or mancozeb @ 0.3% (Das, 1995).(vii) Seed treatment with bavistin @ 1 kg/kg of seed followed by seedling dip in 0.05% bavistin solution for 15 minutes and 2 sprays of 0.2% bavistin at 30 and 50 days after transplanting. (viii) Growing field resistant varieties like Pusa Bhairav, Florida Market, Florida Beauty, SM 6, Pant Samrat, K-202-9, etc. (ix) Adoption of long crop rotation excluding brinjal.

Damping Off

This serious nursery disease is caused by *Pythium debaryanum, Rhizoctonia solani, Verticillium albo-atrum* and *Phytophthora* spp. Both pre- and post emergence damping off is common in brinjal. In the initial stage, the germinating seeds are attacked by the fungus. The infection later spreads to the hypocotyls, basal stem and developing root. The affected seedlings look pale green and brownish lesions are found at the collar region. With the advancement of the disease, collar rotting occurs and the seedlings topple over the ground. To check damping-off following control measure should be taken : (i) Seed treatment with ceresan, thiram and captan @ 3 g/kg of seed or with bavistin 2 g/kg of seed. (ii) Soil sterilization of the nursery bed by drenching with 1% formaldehyde. (iii) Soil drenching before sowing with 0.1% Aretan or brassicol. (iv) Spraying the seedlings with 0.2% captan, difolatan-80, 0.1% bavistin or 0.3% blitox. (v) Seed treatment and soil drenching twice with 0.1% bavistin, first after germination and second 20 days after germination is the best treatment.

Wilt

A number of fungi are associated with this disease. Among them, *Fusarium solani* (Mart.) Snyder and Hansen and *Verticillium dahliae* Kleb, are most common causal organisms. In *Fusarium* wilt, the underground stems become dry and brown owing to cortical decay while roots may have a soft and water-soaked appearance. Stunting of plants withering of immature fruits, yellowing of lower leaves, drooping of the apical portion, browning of vascular bundles and ultimate drying of the whole plant are the characteristic symptoms.

In Verticillium wilt, the first symptom is noticed on the lowest leaves. The affected leaves turn yellow and then brown between the veins followed by wilting. The infected plants become stuntied and generally do not flower. In bearing plants, the flowers and fruits become deformed, flaccid and finally withered. Root systems are damaged and vascular tissue of the stem show dark discolouration. Following control measure should be taken: (i) Seed treatment with thiram @ 2.5 g/kg of seeds or with 0.2% solution of ziram for 30 minutes. (ii) Application of potassic fertilizer to the soil. (iii) Avoidance of excess application of nitrogenous fertilizers. (iv) Soil drenching with 0.3% lonacol or 0.2% brassicol. (v) Spraying the crop with 0.2 % captan or bavistin or 0.15% benlate at weekly interval to check further spread of the disease. (vi) Adoption of long crop rotation excluding solanaceous crops.(vii) Growing field resistant/tolerant varieties like KT-4, K 202-9, etc.

Powdery mildew

Caused by *Laveillula taurica* (Lev.) Arn, and *Phyllactinia moricola* (P. Henn.). This disease appears in dry weather. In the initial stages the infected leaves show chlorotic spots which enlarge and coalesce. Later, white powdery of dusty coatings appear on the leaves, stem and other succulent plant parts due to development of numerous conidia. Plants remain stunted, leave curl, become yellow, wither and drop off. This disease can be controlled by (i) Spraying the crop with 0.05% topaz or 0.1% moroside, morestan or karathane at fortnightly intervals. (ii) Application of sulphur dust @ 15-20 kg/ha.

Cercospora leaf spot

Caused by *Cercospora solanimelongenae* Chupp. and *C. solani* de Thuemen. The leaf spots are characterized by chlorotic lesions, angular to irregular in shape, later turning grayish brown with profuse sproulation at the centre of the spot. Severely infected leaves drop off prematurely. Leaf spot can be controlled in following ways: (i) Collect leaves and destroy. (ii) Regular spray of dithane Z-78 or dithane M-45 or fytolan or blitox @ 0.2% (Yadav and Singh, 1995). (iii) Spraying the crop with 0.1% bavistin or benlate 3-4 times at fortnightly interval commencing 45 days after transplanting. (iv) Grow brinjal varieties like, Junagadh Long, P-8, Manjari Gota, Black Round , Pusa Purple Cluster, H-8, Pant Brinjal-7 and Pant Rituraj (Pandey *et al.*, 1995).

Bacterial Diseases

Bacterial wilt

Caused by *Pseudomonas solanacearum* (E.E. smith). It is a serious disease of brinjal and appears in almost all states. The pathogen is soil borne. Vascular bundles are affected and turn brown in colour. Drooping of leaves is followed by wilting and finally drying of the plants are characteristic to this disease. The freshly cut stem pieces when placed in water, a white milky stream of bacteria oozes out (Yawalkar, 1985) which is the diagnostic symptom for bacterial wilt. It causes rapid and complete wilting of the plants.

It can be controlled by (i) Application of neem cake, groundnut cake, cashew nut shell powder, sawdust to the field which reduce the disease incidence. (ii) Seed treatment with hot water at 52°C for 30 minutes. (iii) Crop rotation with maize, cowpea, cabbage, other cereals and grasses. (iv) Growing field resistant/tolerant varieties like Banaras Giant, Marvel of Banaras, Improved Muktakeshi, Pusa Kranti, Pusa Purple Cluster, Arka Sheel, Arka Shirish, Arka Kusumakar, Arka Nidhi, Arka Keshav, Arka Neelkant, T-3, SM-6, BB-7, ARU-2C, Pant Raturaj, etc.

Viral and Mycoplasmal Diseases

Mosaic

Alfalfa mosaic virus causes mosaic disease in brinjal. This virus is sap transmissible but aphids also transmit the virus in nonpersistent manner. This disease is characterized by chlorotic vein banding, oak leaf pattern, yellow blotches, slight puckering of leaves and stunting of the entire plant. Mosaic can be checked by : (i) Roguing of the affected plant as soon as possible. (ii) Avoidance of mechanical contact between diseased and healthy plants as far as possible. (iii) Control of aphids by systemic insecticides like 0.05% dimethoate (Rogor 1.5 ml/l) or 0.05% methyl demeton (Metasystox 1.5 ml/l) at 10 days interval. (iv) Growing field tolerant variety like, Pusa Purple Cluster.

Little leaf

Caused by a mycoplasma and transmitted by the jassids, *Hishimonus phycitis* Dist. and *Amrasca biguttula* biguttula Ishida. The early symptoms are seen in the young leaves near the tip of the plant which turn light yellow in colour and are thinner than healthy leaves. The flower buds take an upright position instead of being pendulous. As the disease progresses all the freshly produced flowers become phylloid, leaves become progressively smaller. The stimulated axillarybuds grow out into axillary shoots which in turn produce secondary and tertiary axillary branches. In severe cases, excessive crowding of short branches and production of smaller leaves give the plant a rosette appearance. The affected plants hardly bear any fruit. Control measures are: (i) Roguing of the affected plants as soon as possible. (ii) Control of insect vectors in the nursery bed and in the field by spraying 0.05% methyl parathion (Metacide 2 ml/l), 0.05% phosphamidon (Dimecron 1.5 m/l) or 0.05% endosulfan (Thiodan 1.5 ml/l) at 10 days interval. (iii) Dipping of seedling root at the time of transplanting in 0.1% tetracycline solution followed by 3 sprays of 0.05% tetracycline hydrochloride at weekly interval 4-5 weeks after transplanting check the disease. (iv) Dipping of seedling root in 0.2% carbofuran (Furadon) solution for 24 hours helps to keep the seedling away from the initial attack of jassids. (v) Growing field tolerant varieties like, Pusa Purple Cluster, Pusa Purple Long, Pusa Purple Round, T-3, Hisar Shyamal, etc.

Nematodes

Root-knot nematode (*Meloidogyne incognita* (Kofoid and White) Chitwood and *M. Javanica*) and Rentiform nematode (*Rotylenchus reniformis*): Infestation of these plant parasitic nematodes is common in brinjal. These celworms infest the roots and cause root galls. The infested plants become stunted and the leaves show chlorotic symptoms. Fruit production is greatly hampered. Root knot nematodes can be controlled by (i) Ploughing the field in summer months to expose the soil to high temperature as the nematodes cannot survive on surface soils under hot summer condition. (ii) Incorporation of neem cake or castor cake in the field and dipping of seedling root in 5% aqueous extract of neem cake. (iii) Growing antagonistic and non-host crops like marigold, wheat in rotation. Brinjal followed by garlic also reduces nematode population in the soil. (iv) Soil fumigation by DD mixture @ 70-75 l/ha. (v) Application of granular insecticides like Thimet 10 G 925 kg/ha) or Furadon 3 G (35 kg/ha) in the field 10-15 days after transplanting. (vi) Planting 6 week old marigold seedlings in alternate rows of brinjal, (vii) Growing comparatively less susceptible varieties like, Black Beauty, Banaras Giant, T-2, etc.

Physiological Disorders

Calyx withering

This physiological disorder generally occurs unsightly between mid-February and mid-April. The symptoms of this disorder become apparent 7 days before harvest. The affected fruits become reddish brown in colour which lack in normal luster and thus marketability of the fruits is reduced. The affected fruits have much higher calcium and nitrate contents than healthy ones.

Poor fruit set

In brinjal, four types of flower according to length of style, are borne irrespective of the variety namely, long-styled, medium styled, pseudoshort-styled and short styled. Pseudoshort and short styled flowers do not normally set fruit but their number in a plant are normally higher than long and medium styled flower which produce fruits. In brinjal, flowers are normally borne solitarily as well as in clusters. The solitary flowers are mostly long or medium styled whereas mixtures of short-styled, pseudoshort styled and medium/ long styled flowers are found in the cluster. Obviously ratio of short-styled flowers to long/medium styled flowers in the cluster is much high (3-4:1). So, fruit set depends on the number of productive flowers (long/medium-styled) in the plant.

Natural drop is also a problem in brinjal which maybe as high 50% in medium styled flower and 30% in long-styled flower particularly in high temperature condition. Preponderance of unproductive flower types and flower drop sometimes result in poor fruit set in brinjal. Following control measures should be taken. True short-styled and pseudo short styled flowers are actually sterile flowers which cannot be induced to set fruits. However, application of growth substances on medium-styled and long-styled flowers induce higher percentage of fruit set. (i) Spraying the plant with 2 ppm 2-4-D at flowering stage when few flower clusters appear. (ii) Spraying the plant with 60 ppm NAA (naphthalene acetic acid) or 50 ppm PCPA (parachlorophenoxy scetic acid) at full bloom stage. Application of growth substances also produce earlier and more uniform fruits.

HARVESTING

The harvesting of the fruits should be done as soon as it attain a good size and colour. A heavy crop will produced if the fruits are removed before they reach full size. The fruit should not loose its bright and glossy appearance. When a fruit look dull, it is an indication of over maturity and loss of quality. The maturity of the fruits can be tested by pressing the thumb against the side of the fruits. If the pressed portion springs back to its original shape, the fruit is too immature (Chadha, 1993).

YIELD

The average yield varies from 200-500 quintals per hectare. However, hybrid may yield upto 950 q fruits/ha.

POST HARVEST HANDLING

Packing

Usually fruits are packed in bamboo and wicker baskets of different shapes and sizes for market. Some times for local market farmers pack the fruits in gunny bags. For shipping fruits are wrapped in paper before packing. They are also packed loose in bulk or packed in containers for trading and transport.

For transport fruits pack in loosely woven gunny bags or net bags or wooden crates. Mangal *et al.* (2001) reported that packing of fruits individually in polybag (150 gauze) and placing them in cardboard boxes reduced the post harvest losses during storage.

Grading

Before packing fruits are usually graded somewhat to separate the size and to cull out inferior ones. The Indian standard institution has recommended three grades of brinjal viz. Super, fancy and commercial grade. A good size of well develop colour with bright and glossy appearance fruits are kept in "super" grade. Medium size fruits with bright colour and free from any spots and blemishes are graded into fancy grade. Fruits with general appearance, dull colour over or undersized, free from any deformity and spots are kept in commercial grade.

Storage

Storage extends the consumption period of vegetables. Brinjal fruits can be stored for 1-2 days in summer and 3-4 days in winter provided they are kept in shade under ordinary conditions (Chauhan, 1981). It was found that fruits can be stored for 7-10 days in fairly good condition at 7.2 to 10°C with 85-95% relatively humidity (Purewal, 1988).

SEED PRODUCTION

Brinjal is partially self and cross pollinated crop. The extent of natural crossing depends upon insect activity and has been recorded from 0 to 48 per cent. In order to obtain high quality seed for next season crop, it is very essential to produce seeds of brinjal. Besides all cultural operation which are needful for commercial production, the following points should be taken into care. (i) Maintain proper isolation distance between two varieties. (ii) Grow seed crop during autumn- winter (spring-summer give less seed). (iii) Do field inspection for roguing three times, (a) before flowering, (b) at flowering and fruiting, (c) fruit maturity stage. (iv) Take proper plant protection measure (specially phomopsis fruit rot and little leaf). (v) Harvest fruit when they are usually ripe yellow. Extract the seeds by removing outer skin and by cutting the flesh into slices and soak over night in water. Separate seed by washing in running water and sieving. Dry seed in partial shade.

Isolation Distance

Isolation is used to prevent contamination by cross-pollination with a different but selected variety and to prevent mechanical mixing of the seed during harvesting. It is achieved through isolation distance. But it can also be obtained by enclosing plants or group of plants in cages, enclosing individual platns, or removing male flower parts and then artificial pollination.

For pure seed production, seed field must be isolated from other variety of brinjal fields, and fields of same variety not conforming to varietal purity requirements of seed certification at least 200 metres for foundation seed production, and 100 metres for certified seed production (Salunkhe *et al.*, 1987; Dhamdhere and Mathur, 1994).

Roguing

In order to maintain genetic purity of a variety, it becomes essential to attend to roguing operations carefully. The minimum of three roguings suffice the purpose. They are done at different stages of crop growth.

Before Flowering

The first roguing is done before flowering. In this roguing, the plants which are in flowering process or have already bore flowers should be removed. The plants showing different type of branch orientation, leaf colour, presence or absence of pubescent should be removed if they are different than the grown variety. Plant may be erect type and/or spreading type but when such plants show differences from the variety under cultivation, are removed. Besides, the plants affected severely by known or un-known disease or disorder should be removed.

At Flowering and Fruiting

The second roguing is done at the time of flowering and fruiting stage. In this roguing, the plants which are not flowered and the plants which are in flowering process should be removed. The plants showing different flower colour, flower orientation and leaf colour should be removed if they are different than the grown variety. At the time of fruiting stage, plants showing different fruit colour, fruit shape and those plants which are flowered should be removed if they are different than the grown variety. The plants affected by diseases and insect pests should be removed from the field.

At Maturity Stage

The third and final roguing is done at the time of maturity. In this roguing the plants which are having immatured or over matured fruits should be removed from the field. Check fruit for size, shape, colour and external features. The plants affected with phomopsis blight, little leaf, bacterial wilt etc., diseases should be removed from the field and destroyed.

Fruit Set

Four types of flowers have been described in brinjal. This grouping is made on the basis of the style. They are long styled , medium styled, pseudo-short styled and true short styled. Only long and medium styled flower marked by a swollen ovary at the base, bear fruits.The other two types have rudimentary overies which do not develop into fruits. The fruit set can be increased by use of plant regulators. The seed can be soaked for 24 hours in a solution of 2,4-D at 4 to 5 ppm before sowing or spraying the whole plant with 2, 4-D solution at 2 ppm when the first few flower clusters appear. These treatments give about 50% earlier and higher fruit set (Pal, 2001).

Harvesting

Brinjal fruits are allowed to become mature and fully ripe appearing shining yellow in colour. In ripe fruits an abscission layer develops between the proximal end of the fruit and the calyx resulting in fruit, fall on the ground; such fruits are collected for seed extraction. It is not necessary to wait that fruits fall on the ground and even before they can also be harvested. Fruits ripe in 75-80 days in early varieties and 110-125 days or so in late varieties after transplanting. Jayabarathi *et al.* (1990) reported that the seed obtained from the fruits harvested at completely yellow colour stage recorded the highest fruit and seed yield/ha, seed recovery, percentage, germination and vigour index. Delay harvest reduced the yield and quality of seeds.

Seed Yield

The seed yield varies with the varieties, climatic conditions, plant density etc. The highest yield is obtained at the close spacing of 60 x 60 cm (Singh and Syamal, 1995). Satisfactory seed yield of brinjal is about 100-120 kg/ha. The 1000 seed weight is approximately 4 to 5 g (George, 1985). Hybrid seed yield is 50-60 kg/ha.

It has been found that seed content and seed quality vary with different four portions of brinjal fruit (cv. Arka Neelkanth). The second half of the fruit as against 27% in the first half of the fruit (on the stalk side), contained over 73% of the total seeds of the fruit. The seed content is the least in the first quarter (50%), the second and the third quarters contain the maximum quantity of seeds (26 and 52%, respectively). The last quarter contains 17% of the total seeds. The test weight of the seeds has been maximum in the seeds from the first quarter (5.48 g/1000 seeds) and it decreased successively till the third quarter and at par with fourth quarter. There has not much variation in the germination and vigour of the seeds among the first three quarters. While the seeds from the last quarter having less germination as well as vigour (Naik *et al.*, 1993).

Seed Extraction

The outer skin of the healthy brinjal fruits is peeled off. The flesh in which large number of seeds are imbibed, is cut into pieces or crushed. The fruit flesh is also pulped in electric pulp and is allowed to soak in water overnight. The seeds are separated out by washing in running water and sieving.

In other method, seed extraction is achieved when commercial HCI is added in pulp at 1:4 ratio (25 ml Hcl/kg pulp). The pulp is stirred for 25-30 minutes and then seeds are separated out by washing and sieving.

During washing, some seeds float, would be taken out and rejected because such seeds either dead or having very por germinability. Recovery of seed is 5 per cent or so.

Drying

The seeds just after extraction are dried in partial shade to moisture content of 8 to 10% (Agrawal, 1980).

Packing and Storage

Dried seeds are packed in glass jars, polythene boxes, polythene bags. The seeds are packed in cloth bags or in tinwares. Packed seeds are stored in dry clean and air tight containers in cold storage at 10°C.

Production of Hybrid Seed

The improvement in brinjal crop largely depends on the magnitude of genetic variability and the extent of which desirable characters are heritable. A good amount of variability (phenotypic and genotypic) has been observed for the characters like weight of fruit, size of fruit, number of fruits/plant and early yield. Heritability in broad sense has been maximum for fruit weight followed by early yield, size of fruit, number of fruits/plant. High genetic advance (more than 75%) is exhibit by number of fruits/plant, size and weight of fruit and early yield. Days to flowering having lowest value of genetic advance (14.45%) as recorded by Choudhary and Mishra, (1993).

F_1 hybrids are high yielders, better quality in respect of uniformity in fruit size, shape, taste and maturity, and high degree of tolerance against pests and diseases. The cost of F_1 seed is very high but production technique is easy.

The crossing work is not difficult between two parental lines through hand emasculation and pollination. A fruit contains large number of seeds. Moreover, the seed requirement per hectare is only around 500 g. Thus the cost of hybrid seed productionis not high.

Both the lines (male and female) are seen carefully and off types are strictly removed. Emasculation is done one day advance to anthesis. It is occurred between 7 a.m. to 11 a.m. Maximum pollen grains are available at 9.30 a.m. and stigma is most receptive at the time of opening of flower. The pollen is collected and placed properly on receptive stigma with the help of brush. Pollinated flowers are bagged and unpollinated flowers are removed. The best period is October and November. The success of fruit setting is more in fore-noon pollination as compared to afternoon pollination. There is positive correlation between fruit setting and pollination in fore-noon and negative correlation between fruit setting and wind velocity (Mandal *et al.*, 1995).

Fully ripe fruits are harvested and all operations done accordingly as mentioned in the text.

Precautions

i) Emasculated flowers should be pollinated soon

ii) Bagging and tagging are helpful in confirming the fruits obtained from hybridization

iii) Always avoid cloudy and windy weather during pollination.

REFERENCES

Agrawal, R.L. (1980). Seed technology. Oxford and IBH Publishing Co., New Delhi.

Ali, AM (1985). *Amaranthus polygamous, Flaveria australasica* and *Parthenium* control in brinjal . In Abstr. Papers, ISWS. Coimbatore, pp.23.

Aliev, D.A. (1968). Trudy Azerb. Nanc-Issled. *Inst. Ovasc.* 2: 157-68.

Anjaneyulu, A. and Ramakrishnan, (1968). Reaction of *Solanum species* to little leaf of brinjal. *Madras Agril. J.* 55: 142-143.

Anonymous, (1980). Nutritive value of Indian Food. National Institute of Nutrition. Indian Council of Medical Research, Hyderabad.

Anonymous, (1996). Effect of herbicides treatments on brinjal cv. BB1, Vegetable improvement project, Annual Progress Report, 1996 (ICAR).

Anonymous, (2000). AICVIP Recommendatins 1971-98, IIVRE, Publications, varanasi.

Asokan, R., Mohandas, S. and Anand L. (2000). Biofertilizers and biopesticides for horticultural crops. *Indian Horticulture*, 45(1): 45-47.

Awasthi, O.P., Singh, I.S. and Sharma, B.D. (2006). Effect of mulch on soil hydrothermal regimes, growth and fruit yield of brinjal under arid conditions. *Indian J. Hort.* 63(2): 192-194.

Bagle, B.G. (1993). Efficacy of fenvalerate on avoidance losses in brinjal as affected by shoot and fruit borer (*Leucinodes orbonalis* Guen). *Scientific Hort.*, 3:125-128.

Baily, L.H. (1949). Manual of cultivated plants. Mac-Millan Co., New York.

Bajaj, K.L., Singh, D. and Kaur, G. (1989). Bio-chemical basis of relative field resistance of egg plant (*Solanum melongena*) to the shoot and fruit borer (*Leucinodes orbonalis* Guen). *Veg. Sci.*, 16: 145-149.

Baksh, S. and Iqbal, M. (1979). Crossability relationship in some non-tuberous species of *Solanum*. *J. Hort. Sci.*, 54: 163.

Balkrishnan, B., Thandapani, V., Prakash, M., Palaniswamy, V. and Ramugam (1993). Physiological management of moisture stress in brinjal. *Golden Jubilee. Symposium on Hort. Res. changing Scenario*, held on May 24-28 at IIHR,, Bangalore, India.

Bhattacharyya, P., Jain, R.K. and Paliwal, M.K. (2000). Biofertilizers for vegetables. *Indian Horticulture*, 45(2): 12-13.

Bisaria, A.K. and Bhatnagar, V.K. (1978). Effect of growth regulators on growth, fruit and yield in brinjal (*Solanum melongena* L.). *Indian J. Hortic*, 35: 381-383.

Bora, G.C., Shadeque, A., Bora, L.C. and Deka, B.C. (1994). 'JC-1' and 'JC-2' : Two promising strains of brinjal. *Haryana J. Hort. Sci.*, 23(2): 148.152.

Butani, D.K.and Jotwani, M.G. (1984). Insects in vegetables. Periodical Expert Book Agency, D-42, Vivek Vihar, Delhi-32.

Chadha, M.L. (1993). Improvement of Brinjal. In Advances in Horticulture vol.5-Vegetable Crops: Part.1 (Eds.: K.L. Chadha and G. Kalloo). Malhotra Publishing House, New Delhi.

Chadha, M.L. and Sidhu, A.S. (1983). Variability and correlation studies in brinjal. *Indian J. Hort.*, 40(1&2): 60-64.

Chadha, M.L. and Sidhu, A.S. (1987). Punjab Barsati, variety of brinjal (*Solanum melongena* L.). *J. Res. Punjab Agric. Univ.* 24(2): 337.

Chakrabarti, A.K. and Choudhury, B. (1974). Effect of little leaf disease on metabolic changes in susceptible cultivar and resistant allied species of brinjal. *Vegetable Science*, 1:12-17.

Chaugule, B.A. and Pandey, S.G. (1958). Hardening of seed for moisture in relation to growth, yield and irrigation requirement of Brinjal fruit. *Indian J. Agron.*, 3: 1-14.

Chauhan, D.V.S. (1981). Vegetable production in India. Ram Prasad and Sons, Agra.

Choudhary, D.N. and Mishra, G.M. (1993). Studies on genetic variability in eggplant (*Solanum melongena* L.) *Adv. Hort. and For.*, 3:109-116.

Choudhary, P.K. and Saraf, R.K. (1996). Influence of growth regulators and insecticides on growth and yield of brinjal (*Solanum melongena* L.) *Orissa J. Hort.*, 20(2): 42-47.

Choudhury, B. (1976) Eggplant. In Evolution of Crop Plants. (Ed: N.W. Simmonds). Longman, London.

Choudhury, B. (1976). Vegetables. National Book Trust, New Delhi.

Choudhury, B.; Kalda, T.S. and Bhagchandani, P.M. (1971). 'Pusa Kranti' a new name in brinjal. *Indian Hort.,* 16(1): 11-12.

Das, R.C. and Prusty, S.S. (1972). Growth regulator effect on seed treated brinjal plant. (*Solanum melongena* L.) with relation to the vegetative development. *Indian J. Hortic.,* 29:334-335.

Das, R.C., Mahapatra, P. and Panda, N. (1974). Post harvest studies on the effect of mineral nutrients and growth substance on Brinjal (*Solanum melongena* L. cv. Muktakeshi). *The Orissa J. Hort.* II(1&2): 15-18.

Das, SR (1995). Bioefficiency and economics of fungicidal control of phomopsis leaf blight/ fruit rot of brinjal in Orissa. *Orissa J. Hort.,* 23(1 & 2): 9-12.

De Candolle, A. (1904). Origin of cultivated plants. London.

De, Nirmal and Rai, Mathura (2005). Micronutrient disorders and management in vegetable crops: A review. *Vegetable Science.* 32(1): 1-10.

Desai, S.N. (1962). Effect of varying levels of moisture stress on the growth and yield of three varieties of brinjal, M.Sc.(Ag). thesis, University of Poona.

Devi, J.H., Maity, T.K., Paria, N.C. and Thapa, U. (2002). Response of Brinjal (*Solanum melongena* L.) to different sources of nitrogen. *Vegetable Science.* 29(1):45-47.

Dhamdhere, S. and Mathur, S. (1994). Efficacy of some foliar insecticides against *Aphis gossypii* Glover infesting brinjal. *J. Ento. Res.* 18(3): 283-285.

George, R.A.T. (1985). Vegetable Seed Production. Longman, London, New York.

Gill, C.K. and Chadha, M.L. (1979). Resistance in brinjal to shoot and fruit borer. *Indian J. Hort.,* 36:67-71.

Gnanakumari and Satyanarayan, G.(1971). Effect of N, P and K fertilizer at different rate on flowering, Yield and composition of brinjal (*Solanum melongena* L.). *Indian J. Agric. Sci.,* 41: 554-558.

Gopalkrishnan, T.R.; Singh, P.K., Narayanankutty, C. and Peter, K.V. (2002). Bacterial wilt resistant hybrids in brinjal and basis of resistance. *Intl. Conf. Veg.,* Bangalore. 11-14, November, 2002. *Abstract* : pp. 123-124.

Gopinath, G. and Madalageri, B.B. (1986). Bacterial wilt (*Pseudomonas solanacerum* E.F. Smith) resistance in eggplant. *Veg. Sci.,* 13: 189-195.

Gupta, S.S., Kalda, T.S., Singh, Narendra and Kumar, Ravinder (2002). New brinjal : Pusa Ankur. *Indian Hort.,* 45(2): 17-18.

Hooda, R.S., Sidhu, A.S. and Pandita, M.L. (1985). Effect of seedling treatment and foliar application of cytozyme on growth and yield of tomato, brinjal and chillies. *Haryana Agric. Univ. J. Res.,* 15: 329-331.

Hussain, M.M.; Mohanty, B.K. and Prusti, A.M. (2000). Screening of brinjal varieties in black soils of Orissa. *Orissa J. Hort.,* 28(2):81-83.

Jayabarathi, M., Palaniswamy, V., Kalavathi, D. and Balamurugan, P. (1990). Influence of harvesting conditions on the yield and quality of brinjal seeds. *Vegetable Science*, 17(2): 113-118.

Jose, D., Shanmugavelu, K.G. and Thamburaj, S. (1988). Studies on the efficacy of organic vs. inorganic form of nitrogen in brinjal. *Indian J. Hort.*, 45(1-2): 100-103.

Kalda, T.K. and Choudhary, B. (1991). Note on the evolution of 'Pusa Bhairav 'a resistant variety to Phomopsis fruit rot disease in egg plant. *Scientific Hort.*, 1: 145-150.

Kale, P.B.; Mohod, U.V., Dod, V.N. and Thakare, HS. (1986). Screening of brinjal germplasm (*Solanum* spp.) for resistance to shoot and fruit borer (*Leucinodes orbonalis* Guen) under field conditions. Veg. Sci., 13:376-382.

Kalloo, G. (1990). Reaction of egg plant varieties to little leaf disease at Hisar (unpublished data).

Kalloo, G., Rai, M., Kumar, R., Prasanna, H.C., Singh, M., Kumar, S.; Singh, B., Ram, D., Pandey, S., Lal, H.,Pandey, P.K., Rai, S., Pandey, K.K. and Satpathy, S. (2006). New vegetable varieties from IIVR, Varanasi, *Indian Horticulture*. 51(3): 16-22.

Kananjia, S.P. and Narayan, Raj (2003). Biofertilizers for sustainable vegetable production. *Intensive agriculture*. 41(5-6): 15-17.

Krishnamurthi, S and Subramanian, D (1962). Effect of 2,4 Dichloro phenoxy acetic acid on *Solanum melongena L. Curr. Sci.*, 23: 125.

Kumar, S., Jain, P.C. and Kumawat, S.R. (1989). Field efficacy of various insecticides against brinjal mite (*Tetranychus macfarlanei* Baker and Pritchard). *Indian J. Entom.*, 51(3): 253-255.

Mandal, S.C., Singh, Y.V. and Harihar Ram (1995). Hybrid seed production in brinjal (*Solanum melongena* L.) under Tarai conditions of U.P. *Nat. symp. On Recent Dev. In Vegetable Impr* held at Indira Gandhi Krishi Vishwa vidyalaya, Raipur (MP) on February 2-5.

Mangal, J.L., Kumar, J., Batra, V.K. and Singh, J. (2001). Effect of cultivars, packing types and waxing on shelf life of Brinjal (*Solanum malongena* L.) *Vegetable Science*. 28(1): 43-44.

Mayee, C.D. and Munshi, C.D. (1973). Mycoplasma a new threat to vegetables. *Punjab Veg. Grower*. VIII: (2): 12-14.

Mehrotra, O.N., Saxena, H.K. and Misra, P.H. (1973). *Progressive Hort.*, 4: 93-100.

Mehta, Y.R. (1959). Vegetable Growing in Uttar Pradesh, Bureau of Agric. Information, Lucknow.

Mishra, D.N. (1995). Bioefficiency of some insecticides and acaroinsecticides for the control of *Aphis gossypii* Glover infesting *Solanum melongena* L. at West Bengal conditions. *Environment Ecology*. 14(2); 279-281.

Mote, U.N. (1978). Studies on the varietal resistance of brinjal (*Solanum melongena* L.) to jassid (*Amrasca biguttula* Ishida) under field conditions. *Veg. Sci.*, 5: 107-110.

Mote, U.N. (1982). Varietal susceptibility of brinjal (*Solanum melongena* L.) to jassid (*Amrasca bitltula* Ishida). *J. Maharashtra Agric. Univ.*, 7: 59-60.

Muthukrishnan, C.R. and Srinivasan, K.M. (1963). A note on increasing yields of brinjal (*Solanum melongena*) by application of whole plant repeat sprays of growth regulators. *Indian J. Hort.*, 20(2): 150-51.

Muthukrishnan, C.R.and Srinivasan, K.M. (1966). Some observations on parthenocarpic fruits of egg-plant (*Solanum melongena* L.) due to treatment with plant growth regulators. *South Indian Hortic.*, 14: 68-70.

Nadkarni, K.M. (1927). Indian Meteria medica . Published by Nadkarni and Co. Bombay.

Nagarajan, S.S. and Shanmugasundaram, S. (1961). Studies on micronutrients on crop plants. I. Foliar nutrient studies on Brinjal. *Madras Agric. J.*, 418(2): 22-23.

Naik, L.B. , Prabhakar, M. and doijode, S.D. (1993b). A study on quality and content of seeds in different portions of fruit in brinjal. *Golden Jubilee Sym. on Hort. Res. Changing Scenario,* held on May 24-28 at IIHR, Bangalore (India).

Naik, L.B., Prabhakar M. and Doijode, S.D. (1993a). Influence of fruit size on seed quality and content in brinjal. *Golden Jubilee Sym. on Hort. Res. Changing Scenario,* held on May 24-28 at IIHR, Bangalore (India).

Nanthakumar, S. and Veeraragavathatham, D. (2000). Effect of integrated nutrient management on yield and yield attributes of brinjal (*Solanum melongena*) cv. PLR-1. *South Indian Hort.*, 48:42-48.

Nath, Prem; Srivastava, V.K.; Dutta, O.P. and Swamy, K.R.M. (2008). Vegetable Crops - Improvement and Production. PNASF, Bangalore, 49-52.

Pal, B.P. and Singh, H.B. (1949). Hybrid brinjal gave increased yields. *Indian Fmg.*, 10: 375.

Pal, Netra (2001). Seed production technology of brinjal and Okra. In *advances in seed production and technology of vegetable crops.* Centre of Advanced studies, Horticulture (Vegetables), held at 1st-21st December, 2001 at Y.S.P.U.H & F. Nauni-Solan (H.P.).

Pal, S., Saimbhi, M.S. and Bal, S.S. (2002). Effect of nitrogen and phosphorus levels on growth and yield of brinjal hybrids (*Solanum melongena* L.) *Vegetable Science.* 29(1): 90-91.

Pandey, K.K., Pandey, P.K. and Vishwakarma, S.N. (1995). Varietal reactions on brinjal against *Alternaria alternata. Nat. Sym. on Recent Dev. in Vegetable Impr.* held at Indira Gandhi Krishi Vishwa vidyalaya, Raipur (MP) on February 2-5, 1995.

Patil, B.N., Sontakey, P.V. and Shiraat, S.P. (1981). Effect of root dipping treatment with growth regulators on growth and yield of *Solanum melongena* L. *Indian J. Plant Physiol.,* 24: 145.

Peter, K.V. (1998). Genetics and Breeding of Vegetables. Indian Council of Agricultural Research, New Delhi.

Prasanth, D., Natarajan, S. and Thamburaj, S. (2002). Performance of F1 hybrids in brinjal (*Solanum melongena* L.) *Intl. Conf. Veg.*, Bangalore, 11-14 November, 2002, *Abstract*: pp. 105.

Purewal, S.S. (1988). Vegetable cultivation in North India, *Farm. Bull.*, ICAR, New Delhi No.36.

Purseglove, J.W. (1977). Tropical crops: Dicotyledons. English Language Book Society, Longman, London.

Raghav, M. (1995). Weed management studies in kharif Brinjal. *Vegetable Science.* 22(2): 90-91.

Rajeevan, P.K. and Rao, S.N. (1980). Effect of soil and foliar application of nitrogen on growth and yield of brinjal (*Solanum melongena* L.) under rainfed condition. *Agri. Res. J. Kerala.* 18:45-50.

Ramu, N. and Muthuswamy, S. (1964). Preliminary studies on foliar application of nitrogen on brinjal. *Madras Agric. J.*, 51:80-81.

Randhawa, K.S. and Sukhija, B.S. (1975). Studies on the mode of inheritance of some quantitative characters in brinjal. *Haryana J. Hort. Sci.*, 4(1&2): 64-68.

Reddy, K.B. and Reddy, E.N. (1986). Effect of plant nutrient on growth and yield of brinjal (*Solanum melongena* L.). *South Indian Hort.*, 34: 100-104.

Sadawarte, K.T. and Gupta, P.K. (1968). Effect of seed treatment with plant growth regulators on germination, growth and yield of brinjal. *Punjab Hort. J.*, 8: 95-99.

Salunkhe, D.K., Desai, B.B. and Bhatt, N.R. (1987). Vegetable and Flower Seed Production. Agricole Publishing Academy. 208 defence colony, Flyover, New Delhi.

Sambasiva, Rao, G., Rao, D.V.S. and Suryanarayana, V. (1980). Effect of starters and naphthalene acetic acid in brinjal. *Veg. Sci.*, 7:129-132.

Sebastian, S., Rajamoni, L. and Vohab, M.A. (2002). Characterization of land race of brinjal (*Solanum melongena* L.) in Kerala. *Intl. Conf., Veg.*, Bangalore, 11-14 November, 2002. Abstract : pp. 16-17.

Sharma, N.K. and Malik, Y.S. (1983). The influence of salinity on the yield of brinjal genotypes. *Haryana J. Hort. Sci.* 12(3-4): 231-236.

Shatrova, G.L. (1977). *Referativny i-hurnal*, 10: 55-611.

Shirasat, S.P., Soutarey, P. Y. and Patil, B.N. (1981). Effect of root dipping treatment with a growth regulators on growth and yield of *Solanum melongena* L. *Indian`J. Plant Physiol.* 24: 145-149.

Shukla, V. and Prabhakar, B.S.(1988). Effect of mixtallol on production potential of Brinjal. *Indian J. Hort.*, 43: 403-407.

Shukla, Vishnu and Naik,L.B. (1993). Agro-Techniques for solanaceous Vegetables. In Advances in Horticulture, vol.5-Vegetable Crops. Part 1. (Eds.: K.L. Chadha and G. Kalloo). Malhotra Publishing House, New Delhi.

Sidhu, A.S., Kalloo and Pandita, M.L. (1980). Studies on some important aspects of floral biology in vegetable crops. A review. *Haryana J. Hort. Sci.*, 9:207-217.

Singh D.N. and Nath, Vishal (2011). Varieties and Hybrids of Vegetables. Satish Serial Publishing House, Azadpur, Delhi (India). 126-147.

Singh, B. and Khanna, K.R. (1978). Correlation studies in egg plant. *Indian J. Hort.*, 35(1): 39-42.

Singh, B. and Khanna, K.R. (1979). Inheritance of some quantitative characters in brinjal. *Indian J. Hort.*, 36(1): 80-83.

Singh, K. and Sandhu, D.S.(1970). Effect of soil and foliar application of nitrogen on growth and yield of brinjal (*Solanum melongena* L.) *Pubjab Hort.*, 10:103-110.

Singh, N., Pandita, M.L. and Vashista, R.N. (1985). Effect of mulching on growth and yield of brinjal under rainfed conditions. *Haryana J.Hort. Sci.*, 14: 285-286.

Singh, Narendra and Kalda, T.S. (2001). Brinjal. In Vegetables, Tubercrops and Spices. (Eds.: S. Thamuraj and Narendra Singh). Indian Council of Agricultural Research, New Delhi.

Singh, S., Krishnamurthi, S. and Katyal, S.L. (1963). Fruit culture in India. Indian Council of Agricultural Research, New Delhi.

Singh, S.P. (1989). Production Technology of Vegetable Crops. Agriculture Research Communication Centre, Karnal, Haryana, India.

Singh, S.P. (1989). Studies on mulching in vegetable crops A review. *Adv. Hort. and For.* 2:115-143.

Singh, S.P. (1997). Principles of Vegetable Production. Agro Tech Publishing Academy, Udipur, Rajasthan (India).

Singh, S.P. (2001). Seed Production of Commercial Vegetables. Agrotech Publishing Academy, Udaipur, 313002.

Singh, V.N. and Syamal, M.M. (1995). Effect of nitrogen and spacing on yield and quality attributes of brinjal (*Solanum melongena* L.). *J. Res. Birsa Agric. Univ.*, 7(2): 137-139.

Sinha, P., Dube, B.K. and Chatterjee, C. (2002). Changes in brinjal (*Sonalum melongena* L.) Physiology of calcium deficiency. *Indian J. Hort.*, 59(4): 411-415.

Sinha, P., Dube, B.K., Singh, N. and Chatterjee, C. (2001). Boron deficiency and physiological disorders in brinjal. *Indian J. Hort.* 58(4): 354-359.

Som, M.G. and Maity, T.K. (1988). Brinjal. Vegetable Crops in India (Eds. T.K. Bose and M.G. som), Naya Prokash, Calcutta.

Swarup, V. (1998). Genetic Resources of Vegetable Crops in India. National Symposium on Emerging Scenario in Vegetable Research and Development. Organized by Indian Society of Vegetable Science held at Project Directorate of Vegetable Research, Varanasi from 12-14 December.

Swarup, Vishnu (2006). Vegetable Science and technology in India. Kalyani Publishers. New Delhi. pp. 656.

Tandon, H.L.S. (1991). Fertilizer Recommendations for Horticultural Crops. A Guidebook (2nd Edition). Fertilizer Development and Consultation Organization, New Delhi, India.

Vadivel, E. and Kannan Bapu, J.R. (1989). Genotype x environment interaction for fruit yield in egg plant (*Solanum melongena* L.). *South Indian Hort.*, 37(3): 141-143.

Vaganov, A.P., Kulik, N.J. and Vinichenko, N.V. (1975). *Referativny i. Zhurnal*, 2. 55. 762.

Vavilov, N.I. (1951). The origin , variation, immunity and breeding of cultivated plants . *Chron. Bot.*, 13(1/6): 1-366.

Verma, J.P., Rathore, S.V.S. and Sharma, B.R. (1980). Performance of Brinjal (*Solanum melongena* L.) as influenced by age of seedlings and date of transplanting. *Haryana J. Hort. Sci.*, 9: 218-219.

Vijay, O.P. and Nath, Prem (1978). Studies on heterosis and development of hybrid variety in brinjal (*Solanum melongena* L.). *Indian J. Hort.*, 35(3): 229-234.

Vijay, O.P.; Dutta, O.P. and Nath, Prem (1971). A note on the improvement of brinjal. *Indian J. Hort.*, 29(2): 67-69.

Vijay, O.P.; Dutta, O.P. and Nath, Prem (1972). Three new varieties in brinjal. *Indian Hort.*, 16(3): 21-22.

Vijay, O.P.; Nath, Prem and Jalikop, S.H. (1977). Studies on floral biology in long and round type brinjal (*Solanum melongena* L.). *Indian J. Agric. Sci.*, 47:288-291.

Vijay, O.P.; Nath, Prem and Jalikop, S.H. (1978). Combining ability in diallel cross of brinjal. *Indian J. Hort.*, 35(1): 35-38.

Walkar, J.C. (1952). Diseases of Vegetable Crops. Mc Graw Hill Book Co, INC, New Delhi, India.

Yadav, B.S., Nandwane, R.P., Lal, A. and Verma, M.K. (1975). Evaluation of certain brinjal varieties to root-knot nematodes *Meloidogyne incognita*. *Indian J. Mycol. Pl. Path.*, 5:17.

Yadav, F.D. and Singh, HB. (1995). Management of Cercospora leaf spot disease of brinjal by fungicides in Faizabad. *J. Liviing World*, 2 (1) : 23-26.

Yawalkar, K.S. (1985). Vegetable Crops of India. Agri-Horticultural Publishing House, Nagpur, Maharashtra, India.

Yawalkar, K.S., Jakate, P.N. and Srivastava, M.M.P. (1962). Commercial Fertilizers in India. Agric. Hort. Publ. House, Nagpur.

□□□

3. CHILLI

Botanical name	:	*Capsicum annuum* L.
Family	:	Solanaceae
Chromosome number	:	2n = 24

Chilli is one of the most important solanaceous vegetable and cash crop in India where it occupies a prominent place in the dietary habit of the people. Chillies are very useful in our daily diet. They are rich source of vitamins A and C, and rutin which is invaluable substance for pharmaceutical purpose (Purseglove, 1977). The pungency in chillies is due to an alkaloid 'Capsaicin', pungent varieties belong to *Capsicum frutescens* i.e., hot chillies.

ORIGIN

The native home from of the chilli is considered to be tropical America, especially Brazil where it is still found growing in a wild state. (Thompson and Kelly, 1957). They are known from pre historic remains in Peru, were widely cultivated in Central and South America in early times and were unknown in Europe prior to discovery of America. Columbus carried seed to spain in 1493 (Boswell, 1951) and from there the cultivation of peppers spread rapidly in Europe. The centre of diversity of the common cultivated pepper (*Capsicum annuum* L.) is probably Mexico with secondary centre at Guatemala (De Candolle, 1986). Already by 1542 three races were recognized in India. The Spainards and the Portuguese in their search for pepper were quick to seize this new and much more pungent pepper from the new World and to distributed it widely.

It is commonly believed that *Capsicum annuum* is having its origin place Mexico and Guatemala and *Capsicum frutescens* in South America (Brazil). In India chilli was brought, by Portuguese from Brazil in 17th Century (Chauhan, 1981; Singh, 1989).

DISTRIBUTION

Chilli (*Capsicum annuum* L.) is an important crop being cultivated extensively throughout the world. India, Mexico, Japan, Ethiopia, Uganda, Nigeria, Thailand, Turkey, Indonesia, China and Pakistan are the major chilli growing countries. The important states growing chillies are Andhra Pradesh, Maharashtra, Tamil Nadu, Karnataka, Orissa, West Bengal and Rajasthan.

Both the types are grown all over India. However, Andhra Pradesh, Maharashtra, Karnataka, Tamil Nadu and Orissa are most important states, and having three forth of the total area followed by other three states like

Madhya Pradesh, Punjab and Bihar. Pickle chillies are grown on large scale in Uttar Pradesh (Chauhan, 1981). Besides India the other chilli growing countries are china, Ethiopia, Hungary, Indonesia, Japan, Kenya, Mexico, Nigeria, Pakistan, Spain, Sudan, Thailand, Turkey, Uganda.

BOTANY

Capsicum annuum L and *Capsicum frutescens* are mainly two species of Chillies cultivated commercially.

Capsicum annuum: (Chillies, Red or, Sweet peppers) belongs to solanaceae family, is an annual or short lived perennial herb upto 1.5 m in height plant is having a well developed tap root, often with many lateral roots. Stems branched, erect or semi prostate, fleshy, often woody at the base, round or slightly angular, growth normally indeterminate. Leaves - alternate, simple ovate to lanceolate margin entire, tip pointed variable in size upto 12 cm long 7.5 cm wide. Flowers are borne singly and are terminal. Pedicels upto 1.5 cm. calyx campanulate, small 5 toothed, yellow green, petal 5 to 6 'white' bell shaped or rotate upto 15 mm in diameter, stamens-5, ovary bilocular, becoming unilocular above, style-slender, stigma-capitate. Fruits indehiscent, many seeded berrry hollow, (Purseglove, 1977), rectangular to long with 2 or more locules, 1-5 cm long, not pungent, 1-4 cm diameter. Seeds are flattened kidney shaped and pale yellow in colour. 250 seeds/g. the test (1000 seed weight) wieght is 5.5 g.

Capsicum frutescens : (Bird Chillies) : A branching perennial herb, upto 1.5 m in height, having vigorous tap root. Stems woody at base, leaves unequal in shape, oval or oblong acute appex 1.5 to 10 cm in length 0.5 to 5 cm wide. Calyx small 5 toothed yellow green; petal yellow or green white, rotate, 0.5 to 1 cm in diameter, stamens -5; ovary-2 celled; style-slender, stigma-capitate. Fruits small narrow, berry variable in size upto 7.5 cm long, 1 cm in diameter, orange yellow or red when ripe highly pungent. Seeds are 250 per gram.

Plant breeding

All the species of Capsicum both wild and cultivated are diploids, 2n=2x=24. Chillies and sweet peppers can be crossed easily and viable seeds can be obtained in F_1 and F_2 generations. The germplasm is maintained by NBGPR and ICAR centers at All India Coordinated Vegetable Improvement Project, Lam (Guntur, Andhra Pradesh), Kovilpatti (Tamil Nadu) and Kalianpur (Uttar Pradesh). Several local varieties of chillies are grown by the farmers (Nath *et al.*, 2008).

A great genetic variation exits in chillies and sweet pepper with regard to shape, size, pungency and pigmentation of the fruits. Most important strains have been reported from various parts of the country where it is intensively grown for dry as well as green chillies. The important centres of growing chillies are Andhra Pradesh, Maharashtra, Karnataka and Tamil Nadu. Madhya Pradesh, Rajasthan, Bihar and Punjab. Efforts were made by the breeders to evolve suitable strains which are being widely cultivated. The local varieties of 'Byadgi' of Karnataka and 'Sankeshwar' of Maharashtra are grown during rainy season. In Andhra Pradesh, there are several local cultivars like Nallapadu, Warangal, Lanka chilli, Seema Mirapa, Golcanda Mirchi etc. are commonly grown. Like wise, different states have different cultivars that are adapted to local conditions. (Swarup 2006) In some of these chilli growing areas, there has been a set back because of the serious incidence of the virus disease. However, the insect pest could be controlled by chemical measures. Attempts have been made to breed high yielding varieties coupled with resistance to diseases, and high vitamin C content (Pal *et al.*, 1941; Cheema *et al.*, 1944; Deshpande, 1944; Parthasarthy and Kedarnath, 1950; Jeswani *et al.*, 1956; Jagadish *et al.*, 1964; Singh and Singh, 1970; Singh *et al.*, 1977). The probable source of resistance to virus disease was reported by Jayarajan and Ramakrishnan (1969). Source of resistance to anthracnose, cercospora leaf spot, bacterial leaf spot and powdery mildew have been identified by Cheema (1982), Sontakke (1981). Raju (1979) and Cheema (1982) also suggested different breeding methods for the improvement of chillies.

Deshpande (1933) demonstrated hybrid vigour in chillies with regard to total number of fruits and total weight of dry, produce, Sontakke (1981) observed heterosis and heterobeltiosis for yield, branches, earlincss, plant height, carotenoids, ascorbic acid, fruit size, etc. Similarly, Pal (1945) observed increased yield and early maturity in hybrid chillies. Singh and Singh (1978) said it may be worthwhile to subject the-parents to diallele selective mating for fostering greater recombination if simultaneous improvement of developmental characters related to productivity is sought in chilli. Raju (1979) and Usha Rani (1981) studied variability, heritability and gene action in chillies for various traits including ascorbic acid, pigment and pungency. Several variants and mutants have been observed which are mainly useful for genetic studies. However, use of closed flower mutant (cf) and male sterile (ms) mutants can be used for hybrid seed production (Deshpande *et al.*, 1983a, B. Pathak *et al.*, 1983a,b). Use of synthetic or composite varieties and need for reciprocal recurrent selection for simultaneous improvement for various characters have been advocated by Usha Rani (1981) and Cheema (1982). Natarajan *et al.* (1988) reported the evaluation of chilli genotypes under semi dry conditions at Paramakudi. Among the types, CA42 was adjudged to be the best recording the highest yield of 82 g of dry fruits per plant with moderate fruit size. CA 419 followed by CA 349 were found to be superior in fruit characters. Dambe and Kawarkhe (1998) reported a new variety of chilli, *viz.*, Jayanti (AKC 86-39). It was developed by pure line selection and resulted in higher yield potential, bigger size and red colour fruits. In Vidarbaha region of Maharashtra state this strain recorded an yield of 86.03 q/ha of wet red ripe chillies under irrigated conditions with an increase of 48.81 per cent over CA 760 and CA 960 chilli varieties. It had a total duration of 215 days from transplanting to last picking of fruits.

Verma and Pant (1999) evaluated the chilli germplasm for utilization in Uttar Pradesh hills. Evaluation revealed that P-32, P-1399, P-2072, EC-362903-A, EC-362916-B, EC-362930 and KN-613 have elite types amounting to 119 accessions. Most notable genotypes include P-2072 for maximum fruit length and uniform fruiting and appealing colour and having disease tolerance.

Chandra and Pandey (2001) reported a new variety of chilli, viz., 'Jawahar Mirch 283' which is suitable for areas around Jabalpur of Madhya Pradesh. The plants of 'Jawahar Mirch 283' are dwarf (40-50 cm) with bushy habit. It produces 7-9 primary branches with dense foliage cover. The fruits are straight, firm, medium in length, thin (2.5-3.5 cm girth in the middle) and dark green in colour. Since its fruits are very attractive with high consumer preference, they fetch a premium price in the market.

Kameswari *et al.* (2005) evaluated genotypes of chiili for salt tolerance at Bapatla, A.P., Among chilli varieties, LCA 235 was found to be salt tolerant followed by LCA 64, LCA 235 withstood the salinity up to 8 EC. Variety LCA 334 was found to be sensitive and its plant height, leaf area and dry matter production was reduced by 46.3%, 33-7% and 58.7% respectively with increase in salinity from 0-8 dsm-1. Chilli local strains were identified and evaluated under Kashmir valley conditions. The highest yield of dry chilli was recorded in CC-84 (262 g/plant) followed by CC-167 (227 g/plant) and CC-39 (202 g/plant). The longest fruit was recorded in CC-151 (15,3 cm) followed by CC-132 and CC-,126(15.1 cm) (Anon., 2006).

Overall, the breeding methods of improvement include, pure line selection, pedigree method of selection followed by hybridization and backcrossing. Many improved varieties are developed by pure line selection in land races or heterogenous local cultivars such as Aparna, CO_1, CO_2, Musalwadi, Arka lohit and Kalyanpur Red. Several varieities like Andhra Jyoti, Pusa Jwala, Pusa Sadabahar, Pant-C-1, Pant-C-2, Punjab Lal etc. were developed by hybridization. Backcrossing is useful in transferring a single gene controlled character into a desired variety particularly disease resistance (Swarup, 2006). Transgression of useful genes for biotic and abiotic stresses, for disease and pest resistance is possible through utilizing resistant genes from wild and weedy species of capsicum species.

Taxonomy

According to early taxonomists the genus capsicum resulted in more than 100 species and botanical varieties (Irish, 1989). The modern taxonomists recognizing the extent of variability have consolidated the cultivated *Capsicum* into following species (Pickersgill, 1980).

Capsicum annuum L.; syn. *C. purpureum, C. grossum, C. cerasiformoe, C. frutescens* L.; Syn. *C. minimum.*

C. chinense Jacq.. syn. *C. luteum, C. umbilicatum, C. sinense, C. baccatum* L.; *syn. C. pendulum, C. microcarpum, C. angulosum, C. pubescens,* R. and P.

Based on starch gel electrophoresis Mc leod (1978) classified 14 taxa into 5 biological species, with *C. cardenasii, C. eximum* and *C. pubescens* representing a single species, *C. baccatum var. baccatum* and *C. baccatum* var. pendulum and *C. practermissum* representing another and the wild and cultivated *C. annuum, C. frutescens,* and *C. chinense* yet another *C. chacoense* and an unnamed species are considered valid wild species. Slightly deviating from the above classification, based on the electrophoresis of peroxidase isozyme extracted from functional leaf, Wang and MA (1987) assigned 8 *Capsicum* species to 4 groups.

Group I. *C. annuum, C. frutescens* and *C. chinense*
Group II. *C. chacoense* and *C. pubescens*
Group III. *C. practermissum*
Group IV. *C. baccatum* and *C. eximium*

Horticultural Classification

Smith *et al.* (1987) classified chilli cultivars on the basis of fruit shape, colour and usage:

I. Fruit Large Smooth, Thick Fleshed

A. Bell group

Fruit large, 7.5-12.5 cm long, blocky, blunt, 3-4 lobed, square to rectangular, or tapering in longitudinal section. Colour usually green when immature, red at maturity. Mostly non pungent, although a few pungent forms are known.

1. Non pungent

(a) Green, turning red (or rarely orange-yellow) when ripe, (California Wonder, Yolo Wonder' Key Stone giant', 'Early Calowonder, "Tambel 2, 'Golden California Wonder).

(b) Yellow, turning red when ripe (golden Bell, Roumanian')

B. Pimento group

Fruit heart shaped but pointed, 3.75-12.5 cm long, smooth, thick walled, non-pungent, (Pimento', Pimento perfection, Pimento L.).

II. Fruit broad, smooth, thin walled

A. Ancho group

Fruit 10-15 cm long, heart shaped but pointed, somewhat flattened, sweet to mildly pungent.

1. Dark green turning red at maturity. (Mexican Chilli, Ancho, Poblano).
2. Turning brown at maturity (Mulato).

III. Pods long, slender

A. Anacheim chilli group: (long green/long red chilli)

Fruit medium to dark green, smooth, 12.5-20cm x 3.2-5 cm tapering to point flesh medium thick, moderately pungent to sweet, e.g.

1. Moderately pungent (Sandia, New Mexico 9);
2. Mildly pungent, (Mild California);
3. Slightly pungent (Mild California, New Mexico 6, Tam Mild chile 2);
4. Non pungent (Paprika).

B. Cayenne group

Fruits slender 12.5-25 x 1.9–2.5 cm, medium green, characteristically wrinkled and irregular in shape, thin walled and highly pungent.

1. Mature fruit red, (Cuban, Cubanelle, Aconcagua, Italian E1, 'Pepperoncini).

IV. Fruit elongated to 7.5 cm long, green when immature

A. *Jalapeno group*: Fruit 3.75 – 5 cm wide, 5-7.5 cm long, rounded cylindrical shape, thick walled, dark green, smooth (Jalapeno, Mild Jalapeno).

B. *Serrano group*: Fruit slender, cylindrical , often slightly constricted near middle, tapering to abrupt point, highly pungent, 1.25 x 5-6.25 cm (Serrano).

C. *Small hot group:* Fruit slender, medium to thin walled < 7.5 cm long, highly pungent, (Red Chilli, Chile de arbol, Japanese Chilli, Santaka, Hontaka).

V. Fruit small to 5 cm globular to oblate, thick flesh

A. Cherry group

1 Non-pungent (Sweet Cherry):

2 pungent, (Large Red Cerry, Small Red Cherry).

VI. Fruit yellow when immature

A. Small wax group

Fruit 7.5 cm or less in length.

1. Pungent, (Floral Gem, Cascabella, Calora);
2. Non-pungent, (Petite Yellow Sweet Tam, Rio Grande Gold).

B.Long wax group

Fruit 8.8 cm or more in length pointed or blunt.

1. Pungent (Hungarian Yellow Wax).
2. Non pungent, (Sweet Banana, Hungarian Sweet Wax, Long Yellow Sweet).

VII. Fruit slender, yellow turning red at maturity, 2.5-3.75 cm long, highly pungent, of the species *C. frutescens*

A. Tabasco group (Tabasco, Green Leaf Tabasco)

Floral biology

Chilli is a self pollinated but cross pollination also takes place to some extent. Flowers are bisexual. (protogynous). Anther dehisces only half to 5.5 hr after stigma becomes receptive. Anthesis occurs between 6.00 and 9.00 hrs. Receptivity of stigma was the highest at the day of flower anthesis. Flower opening takes place between 5 and 6 a.m. Anther dehiscence take place from 9 to 11 am. Maximum fruit setting takes place when pollination is done at the time of opening of flower (Padda and Singh, 1971). Flowers remain open for 2-3 days. The percentage of fruit setting is 40-50. Since anthesis is completed by 8.00 a.m. and anther dehiscence starts from 9 am, emasculation and pollination can be done simultaneously. Saxena *et al.* (2003) observed that stigma remain receptive for 24 hours after anthesis, it is necessary to pollinate the flower when they are completely open or latest 12 hours after opening to get maximum fruit set and hybrid seed yield. Bees, ants and thrips are the possible agents of pollination (Peter, 1998).

COMPOSITION

Chilli besides imparting pungency and red colour, is a rich source of Vitamin A, C and E. Recently Russian Scientists have identified Vitamin P in green Chillies. Which is considered to be important as it protects from secondary irradiations injury. The pungent principles of chilli is due to an alkaloid capsaicin, $C_{18}H_{27}NO_3$, the decylenic acid derivative of vanillylamine, which is present in the placentas; commercial samples contain about 0.1 per cent. Green chillies contain about 83 per cent moisture, 0.6 per cent fat, 3 per cent protein, 6 per cent carbohydrate and 7 per cent fibre. Capsanthin, $C_{40}H_{48}O_3$ is the most important pigment of paprika, Dried Chillies contain about 10 per cent moisture, *C. annuum* contains 50-230 mg/100 g ascorbic acid and *C. frutescens* 2-50 mg/100 g of ascorbic acid. The former contains 100-1200 IU Vitamin A and the latter 200-20,000 I.U. of vitamin A.

The detail composition of green chillies is given below: (per 100 g of edible portion (Aykroyd, 1963).

Moisture	85.7 gm	Phosphorus	80.0 mg
Protein	2.9 gm	Iron	1.2 mg
Fat	0.6 gm	Sodium	6.5 mg
Minerals	1.0 gm	Potassium	217.0 mg
Fibre	6.8 gm	Copper	1.55 mg
Other carbohydrates	3.0 gm	Sulphur	34.0 mg
Calories	29	Chlorine	15.0 mg
Calcium	30.0 mg	Vitamin A	292 I.U.
Magnesium	24.0 mg	Thiamine	0.19 mg
Riboflavin	0.39 mg	Nicotinic acid	0.9 mg
Oxalic acid	67.0 mg	Vitamin C	111.0 mg

Uses

Chilli is the universal spice and used as an essential ingredient of Indian curry, which is characterised by tempting colour and titillating pungency. In curry chilli is used as paste, powder, broken splits or whole form. There are Indian pickles, especially with tender mango in which chilli is added lavishly to form a thick paste with bitting sensation at the end of curing. In all Indian homes, chilli is used by the poor and rich alike. Oleoresin, sauce and essence are prepared from chillies.

Capsaicin has high medicinal value and used as chemical ingredient in many, pain balm, linaments, carminative tonics etc. It also prevent the heart diseases by dilating blood vessels. According to Nadkarni (1927) chilli paste is externally used as a rubefacient & as a local stimulant for the tonsils in tonsilliis. In diptheria its application is said to hasten the separation of false membranes. Internally, it is irritant and large doses produce gastro-enteritis. When made into lozenge with sugar and tragacanth, it is a remedy for hoarseness. With asafoetida and camphor it is used in the form of pills in cases of cholera. The decoction of the fruit, with the addition of opium and fried asafoetida is also given with equal success in cholera. Chilli has a powerful action, on the mucus membrane and in sore throat, and a putrid throat a gargle made of chillies (4g inone bottle of boiling water) is found beneficial. All its essential qualities can be preserved in vinegar. The whole plant steeped in milk is successfully applied to reduce swellings and hardened tumours. An infusion of the chilli with cinnamon and sugar is a valuable drink for patients suffering from delirium for it satisfies the craving in dipsomaniacs. In the west Indies it is used to relieve the sinking at the epigastrium felt by drunkards (Chauhan, 1981). According to Siddappa and Bhatia (1954), chillies can be canned in 2% brine processed for 75 minutes at 4 kg pressure. Canned product is generally of good quality, having the characteristics pungent taste and aroma.

According to Singh (1989), chillies are also used in following ways:

1. Chillies increase the palatability and taste of cooked vegetables and several other food items.

2. Chillies are prepared into pickles
3. Dry chillies are made into powder for the use as condiments. They are also used to flavour soups, sauces, ketchup, essences, oleoresin etc.
4. Green chillies are used in the preparation of salads. It is reported that green chillies contain an anti-cancer compound.
5. Oleoresin has several medicinal uses in pain balms, vapourbs, lineaments etc., because capsaicin serves as an effective counter-irritant.
6. Chilli seed oil can be used as edible purpose, chilli seed cake as fertilizer or as an animal feed.
7. Pungency free chilli colour can be used for colouring foods.
8. Chilli seeds are used for raising crop for pod as well as seed purposes.

VARIETIES AND HYBRIDS

Many Varieties of Chilli

There are many varieties of chilli according to shape size, colour and pungency, grown in the country. Important varieties are given below:

Pusa Jwala, Pusa Sadabahar, Jwahar mirch-283, Resham Patti, Bharni, Japani Laungi, Surya Rekha, J-218, Musalwadi, X-235, K-1, K-2, Pant C-1, Pant C-2, WP 46 -A, BR-red, Sindhur (CA-960), CA-1068 (Aparna), X 200 (Kiran), CO-1, CO-2, MDU-1, Andhra Jyoti (G-5), Bhagyalakshmi (G-4), G-3, LCA-306, HC-44, HC-28, Utkal Rashmi, Utkal Ava, Utkal Ragini, Arka Lohit Arka Abir, Punjab Lal, Punjab Surkh, Punjab, Guchhedar, Ujwala, AAUM-1 and AAUM-2, Kt-pl 19.

Hot types

Sanauri, Patna Red Sirhindi , NP-34, NP-41, NP-46 and NP-61, G-1, G-2, CA-451-1.

Foreign varieties grown in India. *Hot types* : Tobasco, Maxican chillies, Anaheim chillies, Hungarian wax etc.

Besides, these improved and introduced genotypes , some local varieties are also grown in different regions of the country (Swarup, 1998). They are :

Andhra Pradesh : Oosimilagi, Lavang Mirche, Golgonda Micle, S4, Mundi, Sadasivpet, Shivali, Nallapadu, India Dundicut, Warangal, Lanka Chilli, Tomato Chilli, Seema Mirapa, Gundu Type, India Long, Bagyalakshmi, Samba.

Tamil Nadu : Salem, Ramanand, Samba Kodakal, Sattur, Samba, Kovilpatti.

Karnataka : Javari, Javagoal, Coilbil, Danur, Gowridumr, Bijadagi, Dubba Bijadagi, Kaddi, Sankeshwar Type, Chincholi, Kollegal Type, Arsikera, Madhugiri.

Kerala : Kandhari, Achar, Vella Khandhari, Aana Mulagu, (Elephant chilli).

Maharashtra : Musalvadi, Shankeshwar, Sindhur, Seema Mirapa, Dandicha, Achalpuri, Malkapuri, DPLC, Bombay-742.

Gujarat : Gholar, Patta.

Rajasthan : Mathania, Mandoria

Madhya Pradesh : Elichip, Kalipeeth.

Punjab : Sirhindi, Sanauri Red, Sanauri Yellow, Patna Red, Punjab Lal, Rajpura Long Red, Jullundur, Shokti, Longi.

Uttar Pradesh : Kalianpur Red, Kandhari, Suryamukhi

Bihar : Rahri, Sity, Patna Red, Sabour, Angon, Sabour Anal.

Orissa : Puri Red, Puri Orange

Assam :Latabh, Tiger, Baghi

Sikkim : Bird's eye.

Table 1 : Key characteristics of different varieties and hybrids described by Singh and Nath (2011) are given below :

Varieties/ hybrids	Souce	Yield Potential (q/ha)	Characters
Pusa Jwala (NP46-A x Puri Red)	IARI	250.00	Fruits long thing green, shrunken skin, usually curved, medium maturity, high capsaicin and red at maturity. Tolerant to virus and root knot nematodes, suitable for green chilli and powder also , highly pungent.
Pusa Sadabahar	IARI	280.00	Fruits medium , long, green, clusterous (6-14), capsaicin 7 %, oleoresin 12 %, good for green and powder. Tolerant to mosaic and leaf curl disease (Pruthi, 1993; Tewari, 1991).
Pant C-1	GBPUA&T	210.00	Fruits medium long (6-7 cm), erect, green, smooth, blunt apex, highly pungent, Ist picking 60 DAP, tolerant to leaf curl and mosaic disease.

contd...

Pant C-2 (NP-46 x Local)	GBPUAT	205.00	Fruits medium long, pendant, green tolerant to leaf curl virus good for fresh market.
Sindhur (CA-960)	Maharashtra	200.00	Fruits long (8.2 cm), thick, green, deep red at maturity, capsaicin 0.81 mg /g of fruits, duration 180 days (Murthy and Bavaji, 1980). Suitable for green purpose.
Kiran	Lam	335.00	Fruits long, light green, thin , dry red on ripe, tolerant to thrips, mite and aphids
Aparna	Lam	330.00	Fruits medium long blunt tip , green, turn to yellow on ripening, good for capsaicin extraction, tolerant to salinity (Singh and Singh,1996)
Arka Lohit	IIHR	280.00	Fruits long, straight, dark green, smooth, highly pungent, deep red on maturity, duration 180-210. days. Tolerant to moisture stress.
JCA-154	JNKVV	195.00	Fruits long, dark green, turn to bright red when ripe, duration 115-120 days. Most suitable for pickle making.
Bharni	JNKVV	270.00	Fruits long (10x7 cm), deep green, straight, bright red on ripe, suitable for all purposes.
Japani Laungi	JNKVV	210.00	Fruits long, green, clusterous (10-20), upright "Jhumka", flowering 60-65 DAP. Tolerant to pest and diseases.
Surya Rekha	JNKVV	150.00	Fruits long, light green , single, erect, upward, flowering 45-50 DAP, maroon on ripe, red on drying, strong pungent, suitable for duel season.
Jawahar Mirch -218	JNKVV	220.00	Fruits long, thin, light green, erect, purplish ting, bright, red

contd...

			and glossy on ripe, highly pungent, early maturing (Chandra *et al.*, 1987).
CO-1	TNAU	225.00	Fruits medium long (7.3 x 1.5 cm), green, samba type, high capsaicin, bright red at ripening, duration 210 days.
CO-2	TNAU	280.00	Fruits oblong, green, bright red on ripening, capsaicin 0.56 mg/g duel purpose i.e. green and dry use . Suitable for all season cultivation.
K1	TNAU	290.00	Fruits medium long (6.6 x 3.8 cm), green, good capsaicin, shiny red at ripening, duration 215 days, suitable for rainfed area (Pruthi, 1993).
K2	Tamil Nadu	235.00	Fruits , medium long (7.3 x 5.5 cm), green, good in capsaicin (0.49 mg/g), bright red at ripening, duration 210 days. Tolerant to thrips and fruit rot.
MDU1	TNAU	210.00	Fruits long (8 cm) , green clusterous (4-9), good capsaicin (0.70 mg/g), shiny red at ripening, developed through induced mutation (K1 by gamma irradiation).
Agni (F_1)	Sandoz	305.00	Fruits long (10 x 1.5 cm), green attractive glossy, pungent, good for long transportation, tolerant to TMV, PVY.
Angare (F_1)	Sandoz	310.00	Fruits long (12 x 1.5 cm), cylindrical, pale green, moderate pungency.
Bijlee (F_1)	Sandoz	312.00	Fruits long (13 x 2.0 cm), Conical, thick walled, green medium pungent, red at maturity early bearing, longer shelf life and good for long distance transportation tolerant to TMV.

contd...

Roshni (F_1)	Novartis	314.00	Fruits long (14 x 2.0 cm), cylindrical, dark green glossy, mature to dark red, medium pungent, thin skin, suitable for both fresh and dehydration.
Sholay (F_1)	Novartis	317.00	Fruits long (13 x 1.5 cm), dark green smooth, thin walled and uniform cylindrical, moderate pungency.
Agni Rekha (F_1)	Novartis	315.00	Fruits medium long (11 x 2.0 cm), faint green, lustrous, smooth, thick walled, uniform, solitary, medium pungency, deep red at maturity, good for long distance transport, and fresh market, bearing at 55 DAP.
Flame Hot (F_1)	Novartis	318.00	Fruits small (8 x 1.5 cm) long , smooth, lustrous dark green, strong pungent, deep red on maturity, start bearing at 55 DAP good for long transportation and dry use.
Hot Line (F_1)	Novartis	315.00	Fruits long (14 x 1.5 cm), medium thick walled, smooth, lustrous faint green, deep red on maturity. Strong pungent good for long transportation and dry use and powder.
Picador (F_1)	Novartis	312.00	Fruits long (20 x 8 cm), glossy, thick walled, green taste and flavour of capsicum, bearing start 45-50 DAP, good keeping quality, tolerant to TMV, suitable for pickling, salad and fry.
Parbhani			
Tejas	MAU, Parbhani	260.00	Fruits long, green, smooth, red at maturity, good quality and use both for green and dry purpose high yielder (Sontakke *et al.*, 1995).

contd...

Bhagya Lakshmi	RARS, Guntur	240.00	Fruits medium long (8.8 cm), green, firm attachment, bright red on ripening, good capsaicin, suitable for both green and dry use.
Andhra Jyoti			
(G-5) x (G2 x 1331)	Andhra Pradesh	275.00	Fruits short (5.1 x 6.3 cm), "gundu type" green smooth, bright, red on ripening, good capsaicin (0.65 mg/g). suitable for both green and dry purpose.
NP 46-A		120.00	Fruits long (10.5 cm), thin , pendant, wrinkled, green samba type, red on ripening, good capsaicin (0.53 mg/g), suitable for both green and dry use.
Arka Abir	IIHR	220.00	Long fruits, thin and green, red at maturity, paprika variety for oleoresin industry, red chilli yield 20 q/ha.
LCA 304	-	180.00	Fruits (7.8 cm) long, medium, stout, green mature fruit red, tolerant to virus, suitable for rainfed crop.
LCA 324	-	175.00	Fruits long thin, tolerant to virus diseases, suitable for irrigated condition.
LCA 334	-	185.00	Fruits medium long and stout, early maturing, ideal for green and red chillies.
HC 28	Assam	209.00	Fruits smooth, clustered, upright, straight, medium long (8-9 cm), thin , dark green, highly pungent, bright red on ripening, slight blackish tinge on fruit surface under low temperature, a marker character. First picking in 63 DAP for green and 82 DAP for ripe fruits, suitable for dry powder. Highly resistant to

contd...

			TMV,CMV, PVY and leaf curl, free from fruit rot, powdery mildew and diback disease and suitable for dry powder, good for rainy, spring season.
HC 44	Assam	327.00	Fruits, erect, clustered, medium long (8-9 cm), light green, red glossy at maturity, high pungency, rich in capsicin, oleoresin, dry matter, pigment, suitable for dry powder and duel season variety.
J 218	JNKVV	215.00	Fruits long (10 x 3 cm), thin, light green with purplish tinge, ripe fruits light bright red, glossy, suitable for green and dry use.
Musalwadi	MPKVP, Rahuri	178.00	Fruits, medium long (6.86 x 0.95 cm), dark green, black patches, on ripening blood red, resistant to leaf curl.
X-235	Lam	185.00	Fruits, medium long green, highly pungent, suitable for pickling and export, early maturity. Tolerant to thrips.
Jayanti	PDMKV Akola	186.00	Fruits, long, green, bigger size, good ascorbic and capsaicin content, duration 215 DAP.
Punjab Surkh	PAU	242.00	Fruits, medium long, dark green, red at maturity rich in capsaicin, oleoresin and dry matter, red at ripening. Resistant to mosaic wilt and die back.
Akash,	IAHS	369.00	Long, bright green, glossy, good size, tolerant to foliar diseases long, bright textured and shape, tolerant to foliar diseases
Utpan	IAHS	327.00	
Ruchi	IAHS	325.00	
No.6	IAHS	318.00	Long medium, green glossy, good shape and size, tolerant to leaf curl, die back diseases.

contd...

No.8	IAHS	320.00	
No.9	IAHS	335.00	
Ujwala	KAU	220.00	Plant height medium (56 cm), vigorous branching, (11), bears maximum fruits (86/plant). Good plant canopy (Mini and Joseph, 1998)
Phule suryamukhi	AICVIP, MPKV	280.00	Fruits medium, green, clusterous, erect, red at maturity tolerant to leaf curl and powdery mildew, duel type suitable for kharif and summer planting.
Phule Jyoti	MPKV,	285.00	Fruits medium long green, clusterous, pendant, red at ripening, duel purpose, suitable for kharif and summer planting, tolerant to leaf curl and powdery mildew., resistant to lodging.
Utkal Rashmi	OUAT	295.00	Fruits long, medium girth, straight, green, good pungency deep red on ripening, pendant, suitable for dry chilli, mature in 130-135 days, Resistant to bacterial wilt, fruit borer.
Utkal Ava	OUAT	220.00	Fruits medium size, green solitary, highly pungent, mature in 125-130 days, deep red on ripening good for both green and dry use . Resistance to backterial wilt.
Utkal Ragini	OUAT	230.00	Fruits medium, small, culusterous, green, highly pungent upright fruiting, mature in 140-145 days, sutiable for green chilli.
BC-24	OUAT	250.00	Fruits long, bold, green, attractive, pendant, deep red on ripening,. Mature in 125-135 days , good transport quality, suitable for dry chilli purpose.

contd...

Punjab Lal (Perennial x long red)	PAU	245.00	Fruits medium long (4.25 x 0.79 cm), erect, dark green, turning dark red at maturity, rich in capsaicin (0.7 %), oleoresin, dry matter and red pigment, resistant to TMV and CMV.
Resham Patty	MP	120.00	Fruits long, thick and green and glossy, deep red on ripening. Mild pungent, less seeded, suitable for dry purpose.
Kashi Anmol	IIVR	250.00	Plants determinate, bushy (60-75 cm), umbrella type. Fruits pendant, attractive, compact, dark green at maturity, pungent, first picking at 50-55 DAT, early bearer, duration 150-160 days and suitable for green chilli and long distance transportation (Kalloo *et al.*, 2006) .
Kashi Vishwanath (F_1)	IIVR	200.00	Plants tall (120 cm), dull green stems and bear pendant, long (10-12 cm x 1.0 -1.2 cm) fruits, highly attractive, yellowish green and turn bright red at maturity, pungent with wrinkled surface. First picking of ripe red fruits at 90 DAT, duration 220 days . Suitable for dry fruit production (Kalloo *et al.*, 2006).
Kashi Early (F_1)	IIVR	300.00	Plants tall (110 cm), dull green stem, fruits pendant, long (8-9 cm x 1-1.2 cm), highly attractive, dark green and turn bright red at maturity, pungent with smooth surface. First picking 45 DAT, duration 220 days. Suitable for green as well as dry fruit production (Kalloo *et al.*, 2006).
Azad Mirch 1	CSAUAT	70.00 (Ripe)	Plants dwarf bushy (35-45 cm), 8-10 branches, bear erect and

contd...

			clusterd green fruits and red at maturity. 12-14 clusters per plant having green 8-12 fruits in a cluster. First picking of green fruits 50 DAT. Takes 135-140 days from transplanting to full maturity (Singh and Srivastava, 2006).
Divya Jyoti (F_1)	Unicorn Seeds	375.00	Fruits medium, green, long , slender, highly pungent solitary downward.
Shama (F_1)	Unicorn Sseed	380.00	Fruits long (20 x 2.0 cm), waxy light green, thick skin, suitable for fresh market, pickle , and Bhaji .
Asha Jyoti (F_1)	Unicorn Seeds	387.00	Fruits long (14 x 1.3 cm), medium thick, slender.
Usha Jyoti (F_1)	Unicorn Seeds	370.00	Fruits long (16 x 1.3 cm), slender medium green, pungent, shiny red at maturity, very smooth, good for transport, suitable for both fresh market and delay duration.
Nav Jyoti (F_1)	Unicorn Seeds	367.00	Fruits long (15 x 1.2 cm), slim, dark green, very pungent, turn bright red at maturity, good for fresh market and drying.
Red flame (F_1)	Unicorn Seeds	389.00	Fruits long (16 x 1.8 cm), medium thick, light green, medium pungent, excellent fruit setting ability.
Maha Jwala (F_1)	Unicorn Seeds	372.00	Fruits long (15 x 0.8cm), slender, light, green and highly pungent.
Wonder Jyoti	Unicorn Seeds	375.0	Fruits long (16 x 1.5 cm), slender, medium thick skin, shining, excellent for both fresh and dry market.
CHT-367 (F_1)	Unicorn Seeds	372.00	Fruits medium long, green, upright, clustered highly pungent, excellent, upright hybrid.

contd...

CHT 298 (F_1)	Unicorn Seeds	376.00	Fruits long (13 x 1.3 cm) light green to green, high pungency suitable for fresh market.
CHT 347 (F_1)	Unicorn Seeds	379.00	Fruits long (14 x 1.3 cm), light green to green, high pungency suitable for processing.
CHT 380 (F_1)	Unicorn Seeds	377.00	Fruits long (13.5 x 0.75 cm), green, high pungency good for fresh market.
CHT 365 (F_1)	Unicorn Seeds	365.00	Fruits (14.5 x 1.5 cm), green, high pungency, good for fresh market.
CHT 399(F_1)	Unicorn Seeds	372.00	Frits long (15 x 2.0 cm), light green, medium pungency suitable for fresh market.
CHT 248 (F_1)	Unicorn seeds	370.00	Fruits (10 x 1.0 cm), green, high pungency, good for fresh market and processing.
MHP 58 (F_1)	MAHYCO	310.00	Fruits long, deep green, pungent, first picking 75 DAP, suitable for both season.
MHP 59 (F_1)	MAHYCO	305.00	Fruits long, deep green highly pungent, first picking, 75 DAP duel type.
Agnishikha (F_1)	MAHYCO	315.00	Fruits medium long (10 x 2 cm), thick skin, green, solitary, first picking 55 DAP, suitable for green and dry purpose.
MHP-62 (F_1)	MAHYCO	332.00	Fruits medium long (12.5cm), light green, wrinkles, pendent, moderate pungency, first picking 72 DAP, red at maturity, ideal for green chilli, summer cultivation.
MHP-63 (F_1)	MAHYCO	330.00	Fruits long (13 cm), light green shiny, wrinkles, pendent, moderate pungency, red at maturity, first picking 70 DAP, suitable for green chilli and for summer cultivation.

contd...

MHCP 155 (F_1)	MAHYCO	315.00	Fruits long (13 cm), shiny dark green, smooth, pendent, highly pungent, first picking 72 DAP, deep red at maturity, duel use i.e., green and red purpose.
MHCP-157 (F_1)	MAHYCO	320.00	Fruits medium long (10.5 cm), shiny dark green, smooth, pendent orientation, low pungency, first picking 70 DAP, deep red at maturity
MHCP-158 (F_1)	MAHYCO	330.00	Fruits medium long (10-11 cm), light green, medium pungency, first picking 65 DAP, light red at maturity.
Tejaswini (F_1)	MAHYCO	375.00	Fruits long, green to deep green, first picking 75 DAP, highly pungent.
Sujata (F_1)	MAHYCO	364.00	Fruits long, light green, medium pungent, first picking, 70 DAP, duel type.
Surya (F_1)	MAHYCO	385.00	Fruits long, green, upright, highly pungent, first picking 75 DAP.
MHCP-159 (F_1)	MAHYCO	337.00	Fruits medium long (10-11 cm), green, pendent, highly pungent, first picking, 78 DAP, red on maturity.
CCH-1 (F_1)	Century	340.00	Fruits, medium size (6.6 cm), green, pungent, deep red on ripening, good VHC, capsaicin, moderate resistant to mosaic, leaf curl, wet rot, resistant to die-back.
CCH-2 (F_1)	Century	335.00	Fruits long (12 x 1 cm), deep green, upright clusters , extremely pungent, mature 70 DAP, scarlet red at maturity, early-medium type.
Agnivaan (F_1)	Century	345.00	Fruits long (12 x 1 cm), medium-green colour, early maturing (65-70 DAP), excellent quality crimson on ripening.

contd...

Agnidev (F_1)	Century	342.00	Fruits medium long (7 x 1 cm), excellent deep green colour and very high pungency, solitary pendent fruits, good for dehydration, maturity (68-72 DAP).
Agnipath (F_1)	Century	338.00	Fruits long (10 x 1.6 cm), green, smooth, tight skin, medium-thick wall, high pungency, medium maturing (68-70 DAP), red on maturity, suitable for fresh and dehydration purpose.
CCH-791 (F_1)	Century	358.00	Fruits long (8.0 cm), thin, light green, medium maturing (70-75 days). Red on ripening, good for fresh use and dehydration.
CCH-792 (F_1)	Century	348.00	Fruits long (8-9 cm), medium thickness, light green colour, maturity at 75-80 DAP, deep red at maturity.
Anaheim TMR 23 (F_1)	IAES	325.00	Fruits long (19 x 5 cm) bright green, medium pungent, turning red at maturity, smooth pods, resistant to T.M.V., Mature in 74-76 days.
Cayenne large red thick (F_1)	IAES	335.00	Fruits medium, dark green, thin wall wrinkled (15 x 2 cm) two celled, dark red at maturity, medium thin wall, pendent, mature in 73-76 DAP, excellent for processing and fresh market.
Hungarian yellow wax hot (F_1)	IAES	321.00	Fruits long (14 x 4 cm), light yellow to orange red , smooth conical, upright, pendant, medium thick wall, early type, good for processing and home garden.
Cayenne long Slim (F_1)	IAES	330.00	Fruits long (13 x 1 cm), dark green to red, thin walled, 2-

contd...

			celled, slim, wrinkled, pendant, early-medium maturing (65-70 days) good for home garden.
BSS 269 (F_1)	Bejo sheetal Seeds	359.00	Fruits long (16 x 2.0 cm), less wrinkled, light green, pungent, first picking 75 DAP, dark red at ripening, Hungarian wax type, medium thick, skin, good for fresh market.
BSS 356 (F_1)	Bejo Sheetal Seeds	348.00	Fruits long (15 x 30 cm), rough, green, mild pungent red at ripening, first picking 80 DAP, Hungarian wax type, thick skin, good for all seasons crop.
BSS 377 (F_1)	Bejo Sheetal Seeds	385.00	Fruits long (14 x 1.4 cm), green, smooth, pungent, first picking 85-90 DAP, dark red at ripening, tolerant to virus complex , good for fresh and dry use.
BSS 345 (F_1)	Bejo Sheetal Seeds	415.00	Fruits long (13 x 1.1 cm), smooth, green, medium pungent, first picking 65 DAP good for dry purpose.
BSS 219 (F_1)	Bejo sheetal Seeds	379.00	Fruits long (14 x 1.5 cm), green to light green, thick and smooth surface, medium pungent, first picking 70-75 DAP, dark red at ripening, suitable for both green and dry use.
BSS 308 (F_1)	Bejo Sheetal Seeds	435.00	Fruits long (14 x 1.8 cm), smooth, green to light green, pungent, thick skin, first picking 75 -80 DAP, used for green.
Laxmi (F_1)	Bejo Sheetal Seeds	395.00	Fruits long (14 x 1.5 cm), smooth blackish green, mild pungent, medium, thick skin, dark red at ripening first picking 75 DAP, good for fresh and processing use, tolerant to virus.
BSS 411 (F_1)	Bejo Sheetal Seeds	379.00	Fruits long (15 x 1.5 cm), smooth, green, medium

contd...

			pungent, dark red at ripening first picking 75 DAP, good for both fresh and dry purpose.
BSS 413 (F_1)	Bejo Sheetal Seeds	359.00	Fruits long (18 x 1.4 cm), smooth dark green, medium pungent, dark red at ripening, first picking 70 DAP, excellent for dry chilli purpose.
BSS 346 (F_1)	Bejo sheetal Seeds	430.00	Fruits long (14 x 1.2 cm), smooth, dark green, mild pungent, first picking 75 DAP, dark red at ripening, good for dry purpose.
Arjun (F_1)	Bejo Sheetal Seeds	438.00	Fruits long (16 x 1.6 cm), less wrinkled, green medium pungency, thick skin, red at ripening, first Picking, 60 DAP, good for fresh use
BSS 213 (F_1)	Bejo Sheetal Seeds	387.00	Fruits long fresh use (16 x 3 cm), rough, green, mild pungent, first picking 75 DAP thick skin, suitable for all season crop. dark red at ripening, Hungari an wax type.
BSS 414 (F_1)	Bejo Sheetal Seeds	365.00	Fruits long (10 x 1.2 cm), light green to green, medium smooth, highly pungent, first picking 75-80 DAP, suitable for all seasons.
BSS 140 (F_1)	Bejo Sheetal Seeds	368.00	Fruits long (10 x 1.2 cm), green to light green, very pungent, medium smooth surface, first picking 60-70 DAP, red at maturity early maturing, suitable for rainy season for fresh market.
BSS 202 (F_1)	Bejo Sheetal Seeds	355.00	Fruits long (10 x 1.5 cm), green to dark green, medium group, highly pungent, first picking 70-75 DAP, dark red at maturity, good for rainy and summer season.

contd...

BSS 344 (F_1)	Bejo Sheetal Seeds	360.00	Fruits long (10 x 1.0 cm), yellowish green wrinkled, highly pungent, first picking 70 DAP, thin skin, red at maturity, good for fresh as well as dry purpose.
BSS 375 (F_1)	Bejo sheetal Seeds	367.00	Fruits long (13 x 1.2 cm), yellowish green smooth, highly pungent, first picking 75-80 DAP, dark red at ripening tolerant to virus and other diseases good for all season crop.
Garima (F_1)	Bejo Sheetal Seeds	378.00	Fruits long (15 x 1.2 cm), yellow green, smooth, highly pungent first picking 70-75 DAP, tolerance to virus complex good for all season cultivation.
BSS 141 (F_1)	Bejo Sheetal Seeds	367.00	Fruits long (132 x 1.2 cm), dark green, less wrinkled, pungent, thin skin, first picking 70-75 DAP, dark red at ripening good for green as well as dry chilli.
BSS 376 (F_1)	Bejo Sheetal Seeds	359.00	Fruits long (12-14 cm), green, smooth, medium pungent, dark red at ripening, suitable for dry chilli.
Anmol	Bejo Sheetal Seeds	362.00	Fruits long (16 x 1.3 cm), dark green, smooth, pungent. first picking 75 DAP, dark red at ripening, thin skin, excellent for dry chilli.
BSS 305 (F_1)	Bejo Sheetal Seeds	364.00	Fruits long (14 x1.4 cm), medium wrinkled, dark green, medium pungent, first picking 80-85 DAP, thin, skin, good for dry purpose, tolerant to virus.
BSS 306 (F_1)	Bejo Sheetal Seeds	360.00	Fruits long (18 x 1.2 cm), attractive dark green, tough surface, medium pungent, good for dry purpose and good tolerance to virus complex.

contd...

BSS 355 (F_1)	Bejo Sheetal Seeds	425.00	Fruits long (16 x 1.2 cm), rough, dark green, pungent. first picking 70 DAP, dark red at ripening, good for dry purpose.
Suraj (F_1)	Bejo Sheetal Seeds	395.00	Fruits long (14 x 1.2 cm), wrinkled , light green to green pungent, first picking 75 DAP, bright red at ripening thin skin, can also used for dry purpose.
Savitri (F_1)	Bejo Sheetal Seeds	380.00	Fruits long (10 x 1 cm), wrinkled, yellowish green, high pungent, thin skin, red at ripening, first picking 70 DAP, high pungent, good for green and dry use.
BSS 267 (F_1)	Bejo Sheetal Seeds	375.00	Fruits long (10 x 1 cm), smooth to slightly wrinkled, light green to green, highly pungent, first picking 75 to 80 DAP, good for dry chilli.
BSS 268 (F_1)	Bejo Sheetal Seds	364.00	Fruits long (10 x 1 cm), smooth, green, highly pungent, first picking 75 DAP, good for green as well as dry purposes.
Gagan (F_1)	Golden Seeds	280	Fruits small, green upright, very high pungency, first picking 65-70 DAP, suitable for both fresh and dry purposes. Tolerant to mosaic and leaf curl virus diseases.
Alankar (F_1)	Golden Seeds	300	Fruits light green, pendant, high pungency, Ist picking 70-75 DAP, prolific bearer, suitable for both fresh and dry use.
Byadagi (F_1)	Golden Seeds	290	Fruits long light green, pendant, prominently wrinkled, medium pungent, medium early maturing. Suitable for both green and dry fruit. Tolerant to virus diseases.
A.K. 47 (F_1)	Golden Seeds	320	Fruits long, green, and pendant. Pungency medium to high, suitable for fresh and market.

contd...

			first picking 70 DAP, adaptation to dry and wet conditions, do well in hot regions and in long rainy period. Tolerant to mosaic and leaf curl diseases.
Farmers No.1 (F_1)	Golden Seeds	320	Fruits long, dark green, pendant, 12-15 cm x 1.0 - 1.2 cm size, high pungency, first picking 68 DAP, suitable for fresh and dry purpose. Tolerant to viruses and anthracnose diseases.
Farmers Choice (F_1)	Golden Seeds	300	Fruits long green shiny, 12-14 cm x 1.3 x 1.3-1.5 cm size, pendant, medium pungency, mid early maturity, suitable for both fresh and dry use. Tolerant to virus diseases.
GC-39 (F_1)	Golden Seeds	290	Fruits large, 12-14 cm x 1.8 - 2.0 cm size, light green, pendant, light to medium pungent, mid early prolific bearer, suitable for dressings, stuffing, and pickles. Tolerant to phytophthora root rot and bacterial diseases.
S-16 (F_1)	Sungro Seeds	270.00	Fruits pendant, dark green, bright red at maturity, high pungency, skin thick, length 8-9 cm. Tolerant to thrips and mites.
S-86235 (F_1)	Sungro Seeds	275.00	Fruits long (10-12 cm), dark green, pendant, pungency medium, thick skin. Compact umbrella type.
New Vardan (693)	Nunhems Seeds	245. 00	Fruits long (12.5 x 1.1 cm), pendant, light green, thin skin , very pungent, early maturing, strong, suitable for fresh and dry use. Cayenne type, partial resistant to PVY, mites and stress.

contd...

Sona (F_1)	Nunhems Seeds	325.00	Fruits long (8 x 0.8 cm) , light green, pendant, thin skin, very pungent, suitable for fresh and dry use, Cayenne type resistant to PVY
Mashal (F_1)	Nunhems Seeds	345.00	Fruits long (15 x 1.2 cm), light green, mild pungent, early maturing, pendant, good for fresh use, Cayenne type, Strong and stall.
Red star (F_1)	Nunhems Seeds	435.00	Fruit long (14 x 1.0 cm), dark green, mild pungent, good for fresh and dry use, early maturing, pendantly cayenne type, resistant to CVMV.
Sankranti (695)	Nunhems Seeds	430.00	Fruits long (13 x 1.2 cm), pendant, dark green, mild pungent, good for fresh and dry use, early maturing, strong framework, cayenne type, resistant to CVMV.
Dushhera (F_1)	Nunhems Seeds	415.00	Fruit long (11.5 x 1.2cm) green, pungent, medium, early maturing, pendant, good for fresh and dry use, cayenne type, good for processing and storage.
Devanur Deluxe (F_1)	Nunhems Seeds	427.00	Fruits long (12 x 1.3 cm), green pendant, medium, skin, thickness, very pungent, good for fresh and dry use, cayenne type, resistant to CVMV.
Soldier (F_1)	Nunhems Seeds	435.00	Fruits long (10 x 1 cm), green, thin skin, pendent, very pungent, good fresh & dry use, cayenne, resistant to CVMV.
Deepali (F_1)	Nunhems Seeds	417.00	Fruits long (8 x 0.9 cm), green, thin skinned, very pungent, medium early pendant, good for fresh and dry use cayenne type.
Green Wonder (F_1)	Nunhems Seeds	422.00	Fruits long (15 x 1.7 cm), green, thick skinned, mild pungent,

contd...

			good for fresh and dry use, medium - early maturity, pendant, anaheim type, partial resistant to high temperature.
Rudra (F_1)	Nunhems Seeds	427.00	Fruits long (12 x 1.4 cm), light green, thick, skinned, mild pungent, suitable for fresh and dry use, early maturity, pendant, anaheim type.
Kranti (F_1)	Nunhems Seeds	445.00	Fruits long (13.5 x 1.4 cm) light green, thick skinned, mild pungent, early maturing, pendant, good for fresh and dry use, anaheim type.
Veeru (F_1)	Nunhems Seeds	425.00	Fruits long (12 x 2 cm), green thick skinned, pendant, mild pungent, good for fresh and dry use, early maturing, anaheim type.
Shikha (F_1)	Nunhems Seeds	435.00	Fruits long (8 x 0.8 cm), green, thin skinned, very pungent, early maturing, upright, pendant, cluster bearing, good for fresh and dry use cayenne type.
Anjula (F_1)	Nunhems Seeds	427.00	Fruits long (6 x 1.2 cm) , dark green, medium skin, very pungent, early maturing, pendant, good for processing, Jalopeno type.
Sindoor (F_1)	Nunhems Seeds	437.00	Fruits long (13 x 1.2 cm), green medium skin, pendent, early maturing, mild pungent, good for fresh and dry use, cayenne type, partial resistant to mites.
ARCH -236 (F_1)	Ankur Seeds	430.00	Fruit, elongated , undulated, pointed and green, turn red at maturity, good shelf life and good for fresh market, high capsaicin content, duration 110-140 days.
ARCH-226 (F_1)	Ankur Seeds	415.00	Fruits long, medium, green shining, turn deep red at

contd...

			maturity. Stable red colour, suitable for summer cultivation.
ARCH-228 (F_1)	Ankur Seeds	415.00	Fruits medium long, attractive dark green colour, later deep red, high capsaicin content in the good colour value.
ARCH-001 (F_1)	Ankur Seeds	422.00	Fruits medium long, dark green smooth, deep red at maturity, suitable for green and red dry fruits, good colour, high capsaicin content.
ARCH-112 (F_1)	Ankur Seeds	416.00	Fruits attractive, lustrous light green, medium long, bright red colour at maturity and good pungency.
ARCH-162 (F_1)	Ankur Seeds	410.00	Fruits, medium, long attractive, light green wrinkled, dark red at maturity, suitable for fresh market green chillies.
ARCH-9 (F_1)	Ankur Seeds	405.00	Fruit medium size, green, compact, deep red at maturity, highly pungent, excellent flowering and fruiting, good transportation.
ARCH 55 (F_1)	Ankur Seeds	435.00	Fruits, lustrous green, medium long, bright red at maturity, medium capsaicin content and high oleoresin, suitable for green and red chilli.
ARCH -313 (F_1)	Ankur Seeds	425.00	Fruits bold triangular shaped, blunt end attractive green, dark red at maturity, medium pungent, good for green chillies.
ARCH-83 (F_1)	Ankur Seeds	405.00	Fruits long, dark green, lustrous, deep red at maturity, medium pungent, good for red chilli.
CHILLI-82 (F_1)	Ankur Seeds	422.00	Fruits medium long, light green, lustrous, smooth, medium pungency, good for fresh green chilli.

contd...

NS 101 (F_1)	Namdhari seeds	401.00	Fruits long (12 x 15 cm), dark green, thick skinned, medium pungent, maturity 70 days (DAP) suitable for green and dry chilli purpose.
NS 1101 (F_1)	Namdhari Seeds	415.00	Fruits medium long (8 x 1 cm), light green, thin skinned, very high pungent, dark red at maturity, mature in 70 DAP, tolerant to virus and thrips.
NS 170 (F_1)	Namdhari Seeds	407.00	Fruits medium long (8 x 0.8 cm) light green, thin skinned very high pungency, dark red at maturity, fruits mature in 75 DAP. Tolerant to viruses and thrips, suitable for all seasons.
NS 555 (F_1)	Namdhari Seeds	375.00	Fruits long (13 x 2.25 cm), light green, skin medium thick, medium pungent, dark red at maturity, early bearer.
NS 999 (F_1)	Namdhari Seeds	390.00	Fruits long (14 X 2 cm), light green, medium thick skinned, medium pungency, dark red at maturity, early bearing.
Balwant (F_1)	Namdhari Seeds	385.00	Fruits long (13 x 1.1 cm), light green, thick skin, dark red at maturity, highly pungent, early bearer, tolerant to virus.
Angara (F_1)	Namdhari Seeds	365.00	Fruits long (14 x 1.4 cm), green, highly pungent, thin skinned, dark red at maturity, early fruiting.
NS 3074 (F_1)	Namdhari Seeds	370.00	Fruits long (13 x 1.2 cm), dark green, thin skin, highly pungent, wrinkled, mature in 70 DAP and deep red at maturity.
Goli (F_1)	Namdhari Seeds	368.00	Fruits long (11 x 1.2 cm), straight, dark green, glossy, pungent, thin skinned mature in 70 DAP, dark red at maturity.

contd...

Akash (F_1)	Namdhari Seeds	395.00	Fruits medium long (6 x 0.8 cm) light green thin skinned, very high pungent, solitary bearer, red at maturity and mature in 95 DAP.

CLIMATIC INFLUENCE

Chillies require a warm and humid climate for its best growth and dry weather during the maturation of fruits. Chilli crop come up well in tropical and subtropical regions, due to their wide range of adaptability and can withstand heat and moderate cold to some extent. The crop can be grown through a wide range of altitudes from sea level upto nearly 2100 meters. It is generally a cold weather crop but does not withstand a low temperature in winter. In general a frost free period of four months with maximum temperature ranging from 20-30°C. and minimum temperature not below 10°C is optimum for raising the crop. Fruits colouring is delayed considerably temperature below 15°C At high temperature fruit drop and poor fruit set is a serious problem in chilli.

Chilli thrives in areas having a moderate rainfall within the range of 60-120 cm. Excessive rain however brings about defoliation and rotting of the plant. Similarly, supply of low moisture to the soil during blossom development and fruit formation results in abscission of buds, flowers and small fruits. Winds at the time of flowering and fruiting cause the flowers and fruit drop.

Plants die under extremes of high and low temperatures. When low temperature (8-10°C) exists during night before anthesis is useful for the production of market fruits. The most suitable temperature range for cultivation of chillies is 20-25°C. High temperature at about 38°C in the presence of low humidity cause formation of small-sized fruits, poorly shaped, off-type fruits and reducing the seed production significantly (Hawthorn and Moguruder, 1938). Heat stress causes ethylene production in capsicum flowers that results in flower abscission and abscission is enhanced further when flowers are sensitive to ethylene (Aloni *et al.*, 1994). A seed crop essentially requires a dry, warm climate particularly during ripening of fruits (Thakur and Gill, 1985). Dry winds are harmful particularly during flowering and fruiting periods as they cause flower and fruit drop.

SOIL REQUIREMENT

Chillies are grown in almost all type of soils, well drained loam soils, rich in organic matter are best suited for it. Rainy season crops require black soils which retain moisture for long period. Well drained soils which retain moistrue but are not susceptible to water logging condition are also suitable for rain fed crop. The ideal soil type is rich heavy loam soil containing adequate lime. Sandy soils can also be used in chilli cultivation with sufficient addition of organic manure and frequent irrigations. The pH of soil may vary from 6.0 to 7.5 where chilli crop can be grown successfully. Quality of fruits are better in light soils than heavy soils. The plant produce best at soil pH 7.6.

Ill drained soils are lacking in sufficient supply of oxygen which is very unfavourable to chilli plants because they are susceptible to oxygen deficiency. Inadequate drainage may cause leaf shedding and impair physiological activities in plant.

Good preparation of land is needed that is fulfilled by 4 or 5 ploughings and each is followed by planking for achieving a fine tilth.

SOWING

Time of Sowing

The seeds are sown in December-January and May-June (northern plains) in May-June and September-November (southern plains) and in March-June in hills.

Generally, the crop is raised by transplanting the seedlings in the field. However in some area direct seedling in the field is also practised in case of commercial crop grown for dry ripe chillies. For raising seedling nursery management is most important.

Seed Rate

For planting one hectare area of chilli crop about 1.5 to 2.0 kg seed is required (Thakur and Gill, 1985) for 250 square metres nursery area.

Seed Treatment

Before sowing seed should be treated with Thiram 4 g or bavistin 2 g per kg of seeds. Seed can also be treated with captan 75% dust (2.5 g/kg of seed). Some times seeds are rubbed between the palms with a little kerosene oil in order to keep off ants which often carry them from nursery beds. According to Thomas (1961) seeds should be rubbed between the palms and washed in a water few times so that their pungency is removed. It is believed that the reduction of pungency facilitate germination and also reduces the attack of ants.

Nursery Management

Nursery beds are made near partially shaded areas. First of all field is deeply dug and all plant debries, stones etc are collected. Levelling is done if needed. Whole plot is divided into convenient length of beds with about 1.5

m width. Surface of the bed should be full pulverized. Well decomposed farm yard manures and fertilizer is applied at the rate of 10 kg FYM, 150 g SSP, and 50 g MOP per 3 m^2 area. Raised beds of one metre width and 30 metre length in one or more bits according to the availability of land surrounded by drainage channels of 30 cm width are to be formed. Generally, 250 square metre area of nursery bed is sufficient to raise seedlings for one hectare land. 100 g of phorate granules. are applied in nursery area to avoid damage by ants. Raised beds are prepared during rainy season whereas flat beds are prepared during winter and/or spring season.

It will be advantageous if nursery beds are also treated with formalin and covered with polythene sheet for 96 hours and kept open for at least 72 hours before placement of seeds.

The seeds are sown thinly in lines spaced 5 cm apart and 2 cm deep and well decomposed compost are sprinkled over the seeds. Mini and Joseph (1998) reported that seed sown in nursery at 4 cm spacing within row produced vigorous seedling than conventional method of sowing.

As soon as sowing of seeds is over, the seeds should be covered with a thin layer of well screened FYM followed by sprinkling of water to provide soil moisture for seed germination. Watering should be done once in a day till germination.

A fine mulching with paddy straw can given soon after sowing. The mulch is removed as soon as the seeds starts germinating. Drenching of nursery beds with blue copper or fytolan @ 0.25 g/litre of water at fortnightly interval is done against damping off diseases. The seed germinate in 6-10 days.

PLANTING

When seedlings become 4-5 weeks old (10-15 cm height) are ready for transplanting. However, by all means transplanting should be completed as soon as seedlings become 6-7 weeks old. The distance of planting 30-60 cm within and 45-60 cm between row either in flat beds or on ridges, is followed. Watering of seedlings is withheld for four days before transplanting to make them harden. The seedlings are transplanted in well prepared field at given spacing. Pot watering should be given at least for 3-4 days for proper establishment of the seedlings in the field.

Planting on ridges is advisable during rainy season and when crop is grown on sandy or poor soils and in other seasons planting on flat bed is followed. Transplanting is done in the evening. Before transplanting seedlings are treated with the solution containing 4g carbofuran + 1 g bavistin + 1 g plantomycin per litre of water to avoid disease incidence upto one month after planting.

Raising of chilli crop by direct sowing of seeds in the field is also not uncommon. This method is also known as 'Chuki method" (5-6 kg/ha). In this method 4-5 seeds are hand dibbled at desired distances. The sowing of seeds is followed by light irrigation. Germination of seeds takes place in 15-20 days. When seedlings attain 10-15 cm height, they are thinned to 1 plant per hill (Pruthi, 1993).

NUTRITIONAL REQUIREMENT

Chilli respond well to application of fertilizer both under rainfed and irrigated condition. Heavy application of organic manures is followed in irrigated area. A basal dose of 10 tonnes per ha of farm yard manure is applied in June. Sheep penning @ 2500 to 3000 sheep per hectare is done, if available and neem cake @ 3-4 q/ha preferably mixing with fertilizers at the time of final ploughing. Green manuring is practised by sowing billipesara, cowpea or sunhemp with early rains and incorporating it after 40 days growth.

Depending on fertility status of the soil, fertilizer applications are varied, nitrogen 60-150 kg, phosphorus 50-110 kg and potassium 50-98 kg per hectare are sufficient for higher yield. Gunewardhane and Pereira (1975) recorded the highest yield of capsicum with 90 x 30 cm spacing and 80-100-60 kg N, P_2O_5 and K_2O/ha. Under semi arid condition the close spacing of 45 cm x 30 cm with the balanced fertilization of 150:120:60 kg of N:P:K per hectare is recommended for higher yields with better quality fruits for irrigated green chilli crop of the variety Jwala (Chaugule and Mahajan, 1979). Application of 92:52:30 kg N, P_2O_5, K_2O/ha gave highest yield of chilli (Joseph and Pillai, 1985). Sharma *et al.* (1995) indicated that 150 kg N/ha (125 kg N on soil + 25 kg N on foliage) gave the highest fruit yield of chilli. Singh *et al.* (1999) found that combination of 104.93 kg nitrogen and 50.97 kg potassium/ha is optimum for economic yield of chilli in clay loam soils of Orissa. Palaniappan *et al.* (1999) suggested fertilizer schedule of 150:70:70 kg N, P_2O_5 and K_2O/ha for a maximum yield of chilli. Anu *et al.* (2005) obtained maximum growth and yield of chilli by application of 93.8:50:31.50 kg N:P:K/ha with variety Jwalasakhi. Singh and Srivastava (2006) advocated the fertilizer dose of 120 kg N, 60 kg P_2O_5 and 60 kg K_2O/ha to achieve maximum yield with Azad Mirch 1.

The approximate quantity based on an average requirement which is being applied in various states is given in Table 1. Chilli growers are advised to decide economic doses of fertilisers to be added based on soil test i.e., initial fertility status of the soil because lot of variations are existed even plot to plot.

Table 2 : Doses of nitrogen phosphorus and potassium (kg/ha) and farm yard manure (tonnes/ha) recommended for chilli crop by various states (Tandon, 1991).

STATES	N	P_2O_5	K_2O	FYM
Andhra pradesh				
Rainfed	60	30	50	25
Irrigated	160	90	75	25
Assam	70	40	60	10
Bihar	140	80	90	9
Gujarat	100	50	50	10,15
Haryana	60	30	30	25
Himachal pradesh	75	75	50	-
Karanataka Irrigated	250	250	250	25
Rainfed	100	50	50	25
Kerala	75	40	25	20, 25
Madhya Pradesh	150	60	40	20
Maharashtra				
Rainfed	60	30	50	25-30 carts
Irrigated	150	60	50	20-30 Carts/ha
Orissa	110	70	75	-
Punjab	62	30	30	-
Tamil Nadu	75	35	35	25
Uttar pradesh	80	40	40	25

Apply ½ N, full P_2O_5 and ½ K_2O in furrow at the time of transplanting of seedling. Rest ½ N should be given in two split doses i.e., at 30 and 60 days after transplanting. Another ½ of K_2O should be mixed with 2nd split of nitrogen. At each applications of fertilizer some quantity of farm yard manures is to be incorporated for proper utilization of nutrients. If growth is not rigorous two more split dose of N @ 20 kg/ha is given.

Use of Biofertilizers

Application of bio-fertilizer like azotobacter, phosphate solubilizing micro-organism (*Bacillus polymyxa, Aspergillus awamori, Pencillium digitatium* etc.) and mycorrhizal fungi (Vesicular Arbuscular Mycorrhizae (VAM), play significant role to increase the pod yield of chilli and also save 25-30 per cent nitrogenous fertilizers (Bhattacharyya *et al.*, 2000; Meghwal, 2002). They can use in following manner.

Seed Treatment

About 200 g of biofertilizer is required to treat 10-14 kg of seed. Suspend one packet of 200 g in approximately 400 ml water and mix it thoroughly.

Pour this mixture on seeds and mix with hands to obtain uniform coating on each and every seed. Spread the seeds in shade for drying for 10-15 minutes then sow them immediately.

Seedling Treatment

Mix 1 kg of culture (Azotobacter + PSM) in 10-15 litres of water. Get seedlings required for 1 acre and make small bundles of seedlings in the suspension for 1-20 minutes. Transplant these immediately. Generally the ratio of inoculant and water should be 1:10 approximately, i.e., 1 kg packet in 10 litres of water.

Soil Application

Prepare the mixture of 2-3 kg of biofertilizer in 40-60 kg of soil/compost. Broadcast the mixture in one acre of lands either at sowing time or 24 hours before sowing. The application of phosphate solubilizers is very common.

VAM Fungi

These VAM fungi are beneficial to the plant, which they colonize. They make more nutrients available to the plant, improve soil texture, water holding capacity, improve disease resistance and help in better plant growth. Besides, mycorrhizae are also helpful in the biological control of root pathogens. Chilli crop is inoculated with VAM fungi in the nursery beds. The inoculum consisting of spores of fungus mixed with soil and infected root bits of the host plants on which they are raised is either spread uniformly in the nursery bed or placed in rows over which seeds of chilli are sown. The seedlings are colonized by the fungus 45 days after transplanting. Inoculated plants show better growth and improve nutrient content (Asokan *et al.*, 2000). Application of biofertilizers i.e., fluorescent pseudomonas and Azospirillum significantly improved the growth and yield of chilli var. Jwalasakhi (Anu *et al.*, 2005).

Use of Growth Substances

Application of CCC (cycocel) 10 days after transplanting, NAA (Planofix) at flowering had the most beneficial effect on plant growth; 10 ppm of planofix at flowering and three weeks later increased the growth. Triacontanol soil drenching @ 1 ppm at 18 and 43 days after planting under greenhouse cultivation improved the growth of chilli plants. Chilli plant height was decreased but branching increased with application of ethrel (300-500 ppm) and CCC (500-2000 pppm). Similarly spraying with chloroflurenol (50-200 ppm) at 4 weeks after planting increased the number of secondary branches. Fruit set in chilli can be improved by application of growth regulatiors like GA(10-100ppm), NAA (20-200 ppm) and CCC (1000 ppm). The fruit and seed yield can be improved by an application of triacontanol (miraculan) at 5 ppm spray at 4 weeks after transplanting (Sharma, 1993).

i) Two spray of planofix @ 50 ppm (NAA/or 0.5 ml planofix per 44.5 litre of water) for better fruits set, one at the time of flowering and second spray 10-20 days after first spray. It will check the flower drop and improve the fruit quality and fruit yield. Application of Abscissic acid (ABA)/cycocel/2, 4-D @ 5 ppm reduces plants height and helps in accommodating an increased number of plants per unit area, giving higher yield per hectare, besides reduction in transpiration rate it enables the plants to utilize available water efficiently. It is of immense utility in drier areas especially during the summer, when the yield loss and water shortage is more.

ii) Seed soaking treatment with 70 ppm NAA caused significant increased vegetative growth, flowering and fruit set in chilli (Napoleon and George, 2003).

iii) Application of napthalene acetic acid (NAA) and triacontanol induces earliness in chilli. Spraying of NAA @ 15 ppm and triacontanol at 2.5 ppm , 15, 30, 45 and 60 days after transplanting results into an advance in flowering by about 10 days with NAA and week with triacontanol. They induced earliness, ensures higher early yield, avoiding a total yield loss due to drought in the later stages.

iv) Application of NAA @ 15 ppm is commercially used to control flower fall. During the summer the incidence of leaf-curl enhances the flower fall and results into poor yield. Application of triacontanol @ 2.5 ppm also control, the flower drop during summer. Besides it checks the incidence of leaf-curl and increase fruit set up to 77 per cent (Usha and Peter, 1988).

v) The plants treated with triacontanol (2.5 ppm) give a higher yield during the summer. In the rainy season, NAA (15 ppm) treated plant also give higher yield (Usha and Peter, 1995).

vi) Application of CCC @ 100 ppm during kharif is also increase fruit yield.

Use of Micronutrients and Chemicals

i) Micronutirents serve mainly as constituents of prosthetic groups in metalloproteins and activators of enzyme reactions.

ii) Micronutrients can be applied as soil application, foliar spray at different stages of crop growth.

iii) Mineral nutrient contents range in dry matter of chilli is : Fe 30-300, Mn 30-1000, Zn 20-60, Cu 8-15, B 40-80 and Mo 0.2-0.6 ppm (De and Rai, 2005).

iv) Application of micronutrients in any form improve the growth, check the flower and fruit drop and increase the fruit yield of chilli.

v) Application of urea + single super phosphate + muriate of potash (2:1:1) in liquid form as starter solution at 300 ml/hill soon after transplanting gave the highest yield.

vi) Foliar application of 1 % urea for three times (45, 78 and 108 days) after transplanting increase fruit yield.

vii) Soil application of zinc sulphate @ 25 kg/ha and also foliar spray at 2 g/litre of water using 1250 g of zinc sulphate/ha correct the zinc deficiency.

viii) Tracels-2 @ 4 kg/ha as basal application before transplanting increase fruit yield.

ix) Agromin @ 5 g/1.5 litre of water as foliar spray, 3-4 times, commencing 30 days after transplanting at interval of 15 days increase yield and quality of chilli pods.

x) Foliar spray of multiplex @ 1.5 g/litre of water, thrice at 10 days interval at flowering and fruitset increase yield and fruit quality of chilli.

xi) Foliar spray of Dhanzyme 1 ml/litre of water at flower initiation and 20-40 days later in crease fruit yield and improve pod quality.

xii) Application of Dhanzyme granules @ 20 kg/ha, first dose on 10-15 days after planting and second dose 20-30 days later. It should be uniformly broadcast by hand in field having sufficient moisture. It increase fruit yield and fruit quality.

xiii) Application of "Action-100" @ 1 ml/litre of water keep the plant vigorous and free from diseases.

xiv) Foliar application of plantaid @ 2 ml/litre of water reduce flower drop and increase fruit yield.

IRRIGATION

Chillies are cultivated both as rainfed and irrigated crop. Plants grown with sufficient moisture produce more flowers and set more fruits than the plants growing under drier conditions. Drought may have little effect on fruit set although the number of flowers produced is considerably reduced. Chilli is more drought resistant than other solanaceous crop. But best yield is dependent upon ample moisture supply and well distributed rains, during crop period. The maintenance of uniform soil moisture is essential to blossom and prevent fruit drop. Choudhury (1983) reported that water requirement of chilli is 450 mm on sandy loam soil at Delhi. It was found that chilli plants irrigated with water having EC 0.58 to 1.59, increased flowering, fruit set and fruit yield (Venkatachalam, 1982). Suryanarayana *et al.* (1983), found that irrigation after every 12 days alongwith anti transpirant spray of alchlor was most economical for sandy loam soil. The seasonal consumptive use was 347.5 mm. Wankhade and Morey (1985) at Akola obtained maximum yield of chilli with irrigations at 0.6 CPE which recorded total consumptive use of 926 mm. Generally 8 to 9 irrigations are given, depending upon rainfall, soil type, humidity and prevailing temperature. However, irrigation must be given at flowering and fruiting (critical)stages. If irrigations are not given at these critical stages, the seed yield will be reduced drastically (Agrawal, 1985). First irrigation, is given soon after transplanting and subsequent irrigations are given at 5-7 days interval during summer and 10-15 day in winter depending upon the retentiveness of the soil and weather conditions (Hegde, 1993). Sivanappan *et al.* (1978) observed that drip system proved to be much beneficial for yield and water saving in chilli. It is reported that drip irrigation produced a high yield about twice as high as sprinkler irrigation. Sprinkler irrigation is most suitable for light sandy clay soil and irrigation by furrow and pipes for heavy meadow soil.

INTERCULTURE

In order to make soil loose, around root system, enhance aeration and to keep down the weeds, shallow intercultural operations should be done during early stages of plant growth.

Under rainfed condition frequent intercultivation are done to keep the weed under check and conserve the soil moisture. Usually two to three hoeings are carried out before application of fertilizer. Earthing up operation proves beneficial to chilli crop so it is given after each top dressing and it also keep plant stable and strong. In commercially cultivated irrigated crop, inter cultivation is given either by junior hoe or light plough after each irrigation and followed hand weeding to check the weed growth.

In the direct sown crop, blade harrow is worked starting from 30th day of sowing. Four interculture are needed at 10 days intervals alternated with blade harrow and tyned harrow or junior-hoe. Final intercultivation is given by the country plough.

WEED CONTROL

Weeds affect plant growth and development, and ultimately results in poor yields. Two to three hoeings and manual weedings are necessary to keep the chilli crop free of weeds. There are some weeds which usually found in chilli crop are : *Ageratum conyzoides, Blumea spp., Cannabis sativa, Chenopodium album, Cirsium arvense, Convolvulus arvensis, Cynodon dactylon, Cyperus rotundus, Euphorbia hirta, E. microphyylla, Leucas aspera,, Orobanche cernua, Portulaka oleracea, Trianthema monogyna, Vicia hirsuta, V. sativa.* Herbicides control weeds effectively upto 4-5 weeks period by the time majority of vegetables develop sufficient canopy and provide smoothening effect on weed species. Pre plant incorporation (10 days before transplanting) of EPTC @ 3.75 kg per hectare followed by application of nitrofen @ 1 -2 kg a.i./ha or alachlor at 2.5 kg a.i. per hectare provided the best control of weed without any adverse effect on chilli (Saimbhi and Randhawa, 1976; Rajagopal *et al.*, 1976). Spraying of Tok E-25 @ 2 litres a.i./ha with one hand weeding are effective to control the weeds. Lasso @ 2.0 litre per hectare as pre-plant spray with one hand weeding are effective in controlling the weeds. Pre-planting application of trifluralin (Treflan) @ 1.0-1.5 kg a.i. per hectare effective to control the weeds. Singh *et al.* (1992), obtained highest fruit yield when pendimethalin plus oxy-flufen at 1.00 + 0.15 kg/ha was applied as pre-emergence (one day before transplanting) and it increased 77.6 and 41.4 per cent yield over weedy check and hand weeding 20 and 40 DAT respectively.

PLANT PROTECTION MEASURES

Insect Pests and their Control

Cutworm (*Agrotis ipsylon*)

Cut worms generally cut the young plants at ground level. It can be controlled by regular interculturing and stirring up of soil around plants. Soil application of aldrin 30 kg/hectare before sowing.

Thrips (*Scirtothrips dorsalis* Hood)

It is a polyphagus pest having a wide host range and affect both the places at nursery and fields. The nymphs and adults are tiny, slender, fragile and yellowish-straw in colour.The adults have grey coloured heavily fringed wings. Reproduction is sexual and parthenogenetical as well.Female lays eggs into the vein (40-48) and completes life cycle in 10-20 days (Butani, 1976). Both nymphs and adults lacerate the tender leaf tissues and feed on the oozing sap which results in curling and crumbling of the leaves. The buds and flowers are also attacked which become brittle and even drop down. Crop growth is stunted. They also transmit leaf curl virus. Infestation of thrips become severe in rainless, dry weather condition.

Thrips can be controlled by application of granular insecticide like, aldicarb 10 G @ 15 kg/ha at the time of transplanting followed by 4-5 sprays of 0.05% monocrotophos (Nuvacron 1.25 ml/l) at fortnightly intervals the best method of controlling this pest. Spraying the crop with 0.05% monocrotophos (Nuvacron 1.25 ml/l) 0.03% dimethoate (Rogor 1.5 ml/l) or 0.2% carbaryl (Sevin 4 g/l) or 0.3% phosalone (Zolone 2 ml/l) at fortnightly interval. Grow thrips tolerant varieties like K-2.

Fruit borer (*Longitarus nigripannis, Spodoptera litura*, Fabricius, *Heliothis armigera*)

The grubs bore the fruits and feed inside while the adult scrape the leaves and fruits (Thakur and Gill, 1985). Spray quinalphos 25 EC (4 ml in 1 litre of water) or carbaryl 50 WP (3 g in 1 litre of water) to control the pest.

Aphids (*Aphis gossypii* Glover, cotton aphid and *Myzus persicae* (Sulzer), peach green aphid)

Both nymphs and adults are harmful. Small, ovate, soft-bodied nymphs and adults are found in large number on undersurface of tender leaves and shoots sucking the cell sap from the tissue which causes curling and ultimately drying up of the leaves. Black sooty moulds which is developed on the honeydew secretion of the aphids hinder photosynthetic activity and thereby retard growth and production. They also transmit different viruses like PVY,CMV,TEV and PVMV. Cloudy weather is conducive to their rapid multiplication. Aphids can be controlled by (i) Spraying the crop with 0.03% dimethoate (Rogor 1.5 ml/l) or 0.05% methyl demeton (Metasystox 1.5 ml/l) or 0.05% monocrotophos (Nuvacron 1.25 ml/l) or 0.02% fenvalerate (Sumicidin 0.3 ml/l)at fortnightly interval. (ii) Application of granular insecticides like thimet 10G, furadon 3 G aldicarb 10G is also effective in controlling aphid infestation, (iii) To check aphids grow border crop with *Agathi, Sesbania grandiflora* (Nelson and Natarajan, 1994).

Mites (*Polyphagotarsonemus latus* (Banks), *Tarsonemus translucens* (Green) and *Tetranychus cinnabarinus* (Boisduval)

These tiny accarids are found in large numbers on the ventral surface of the leaves under a protective cover of fine webs (Butani, 1976). Both nymphs and adults suck the cell sap and devitalise the plants. Severely affected leaves show glossy brown or reddish brown colouration on the leaves. Spraying the crop with dicofol (3 ml/l) or 0.3% phosalone (Zolone 2 ml/l). Dusting the crop with fine sulphur @ 20 kg/ha is effective to control mites.

Flea beetle (*Longitarsus nigripennis* Motschulsky)

It is a serious pest of sweet pepper and also found damaging to chillies. The eggs are laid on fruits. The grubs after hatching bore inside the fruit and feed within the fruits while the adult beetles scrap the green tissues of the leaves and fruits. Removal and destruction of the affected fruits with the grubs and spraying the crop with 0.2% carbaryl (Sevin 4 g/l) or 0.05% endosulfan (Thiodan 1.5 ml/l) or 0.05% malathion (Cythion 1.5 ml/l) from bloom stage at weekly interval are suitable control measure.

Pepper weevil (*Anthonomus eugenii*)

This weevil is black with grey or yellow markings and particularly severe in sweet pepper. Eggs are laid beneath the fruit surface and the grubs after hatching, feed within the fruits. The affected fruits become misshaped and discoloured. Control measures are same and given for flea beetle.

Pepper maggot (*Spilographa electa*)

It is a serious pest of sweet pepper. The maggots feed on the small fruits from inside, so that their presence may not be detected from outside. The damaged fruits may be destroyed by the secondary infection of soft rot.Control measures are same as given for flea beetle.

Diseases and their Control

Fungal Diseases

Damping off

A serious nursery disease caused by *Pythium aphanidermatum* (Eds. Fitz.) and *Rhizoctonia solani* Kuhn. These fungi are soil borne. The seeds may rot or the seedlings may be killed before they emerge through the soil. Collar rotting and subsequent toppling over of the young seedlings are the most common symptom (Singh, 1985).

Control measures

i) Seed treatment with thiram or captan or captaf @ 2.5 g/kg of seed or with bavistin @ 2 g/kg of seed.

ii) Preparation of raised nursery bed and partial sterilization of the soil by burning straw and other trash materials on the surface of the nurserybed.

iii) Soil of the nursery bed may be sterilized by drenching with 1% formaldehyde solution up to depth of 10-15 cm. The treated soil should be covered by polyethylene sheet for 4 days and after removing the soil cover, the soil should be air dried for 7 days.

iv) Soil drenching with the solution of captan or aretan @ 4 g or with bavistin or brassicol @ 1 g per litre of water or with mixture of maxcozin (0.3%) and bavistin (0.2%).

v) Spraying the young seedlings with 0.2% captan or difoltan or 0.1% bavistin.

vi) Seed treatment with captaf @ 2.5 g/kg seed + soil drenching with 0.5% captaf twice, first after germination and second 20 days after germination is very effective.

vii) Application of bio-agents like *Trichoderma* @ 5 kg/ha decrease the pathogen load in soil.

Ripe fruit rot and die back

This disease, also called anthracnose, is caused by *Colletotrichum capsici* Syd.) Bulter and Bisby. It is a very serious disease of chilli and causes severe loss of mature fruits in the field and in transit and storage. The fungus is externally seed borne. Elongated, ash-coloured, angular spots develop on the veins and ventral surface of the leaf, lesions on the stem are dark brown and scattered. Leaves fall down and the branches die back from the tip. In the ripe fruits, the lesions are usually circular and sunken with black margin.The sunken spots are covered with a pinkish mass of fungal spots. As the disease advances, the lesions spread forming concentric marking with dark fructifications. Fruits may drop-off or may rot in transit or storage (Walker, 1952).

Control measures

i) Seed treatment with thiram or captan or ceresan @ 2.5 g/kg of seed.

ii) Spraying the crop with fungicidal solution containing 1 g benlate + 2 g captan or difolatan per litre of water.

iii) Spraying the crop with 0.1% benla te or 0.2% difolatan or 0.4% blitox or 0.1% topsin or 0.2% bavistin or 0.25% dithane M-45 four times starting from the appearance of the disease at an interval of 15 days.

iv) Spraying the crop with 0.1% bavistin followed by 0.3% blitox after removal of infected twigs at 10 days interval.

Powdery Mildew

Caused by *laveillula taurica* (Lev.) Salmon. This disease comes in advanced stage of the crop in the winter months. White floury coatings appear on the leaves, stems and other succulent parts of the leaves which cause yellowing of leaves and subsequently rapid defoliation of the plants. Powdery mildew can be controlled by spraying the crop with 0.1% karathane or 0.2% moroside or 0.1% bavistin or 0.1% calixin or 0.1% topaz 3-4 times at an interval of 10 days, Dusting the crop with sulphur @ 15-20 kg/ha also helpful to check powdery mildew.

Leaf spot or frogeye leaf spot

This fungal disease is seed-born and caused by *Cercospora capsici* Heald, and Wolf. The leaf spot is characterized by chlorotic lesions, angular to irregular in shape which later turns grayish brown with profuse sporulation at the centre of the spot. Severely infected leaves drop-off prematurely and thus the yield is reduced to a great extent. Four sprays of 0.1% bavistin or benlate at 10 days interval effectively control the disease. Spraying the crop with 0.2% dithane M-45 or difolatan also control this disease.

Fusarium wilt

It is a serious disease caused by *Fusarium solani* (Mart.) Snyder and Hansen and *F. annuum* Leonian. These fungi are both seed and soil borne. Upward and inward rolling of the leaves which turn yellow with the drooping of apical portion. Ultimately, the whole plant wilts due to cortical decay while the roots may have a soft water soaked appearance.

Disease can be controlled by (i) Adoption of long crop rotations excluding solanaceous crop. (ii) Seed treatment with 2.5 g thiram or 1 g bavistin per kg of seed. (iii) Soil drenching with 0.2% fytolan or brassicol may give protection. (iv) Spraying the crop with 0.2% captan or 0.1% bavistin or benlate at weekly interval check further spread of the disease.

Fruit rot

Caused by *Phytophthora capsici* Leonian, the disease is particularly severe in sweet pepper and mostly appears during rainy season. The symptom appears in the form of dark green water-soaked spots which enlarge rapidly to cover the entire fruit causing rotting. In the early stages of crop growth, leaf blight may occur where leaf becomes water-soaked with a border of white layer of the fungus. Control measure are same as given for ripe fruit rot.

Bacterial Diseases

Bacterial leaf spot

This seed borne disease caused by *Xanthomonas campestris* pv. *Versicatoria* (Doidge) Dye. Induces serious injury to the leaves and fruits of both chilli and sweet pepper. Moderate temperature (21-31°C and high relative humidity (85-90%) coupled with continuous drizzling and cloudiness is highly favourable for development of the disease. On the young leaves the spots are yellowish green while on older leaves they are dark coloured and water soaked. The spots later appear with straw-coloured centres and dark margins. In severe leaf spotting, premature defoliation occurs.Fruits are also infested with the appearance of small blister like watery spots. The bacterial leaf spot can be controlled by (i) Growing field tolerant varieties like, CA 960, CO1, G5, X197, G4, NP23, Pusa Jwala, G2,IC 9110 etc. (ii) Use of disease-free seed, (iii) Hot water treatment of seeds at 50°C for 20 minutes. (iv) Preventive spraying with a combination of 100 ppm agrimycin-100 + 0.3% blitox during rainy season. (v) Spray paushamycin 200 ppm + copper oxychloride 0.15% two spray at 15-20 days interval (Pruthi, 1993).

Bacterial wilt

Caused by *Pseudomonas solanacearum* E.E. Smith. This disease has now assumed serious proportion in chilli growing areas. This typical soil-borne bacteria invade the plant through root injuries or at the point of emergence of root hair and lateral roots. Vascular discolouration and wilting of the plants are the characteristic symptoms. The wilting is rapid and complete damage of plants. Bacterial wilt can be controlled in following ways. (i) Application of neem cake, cashew nut shell powder to reduce the disease incidence. (ii) Seed treatment with hot water at 50°C for 30 minutes. (iii) Growing tolerant varieties like Pant C-1,Long Red Kandhari, Pungent Pride, Cherry Red, Vattal, etc. (iv) Long crop rotations excluding solanaceous crops. (v) Application of bleaching powder @ 15 kg/ha before planting.

Soft rot

It is a common post-harvest disease of sweet pepper caused by *Erwinia carotovora* (Jones) Bergey *et al.* Fruit decay usually starts at wounds in the carpels or peduncles, which reduces palatability as well as marketability. Soft rot can be checked by (i) Avoidance of injury to the fruits during harvesting. (ii) Storage of fruits in low temperature condition.

Viral Diseases

Leaf curl

This serious disease is caused by a strain of tobacco leaf curl virus. This virus is transmitted by white fly (*Bemisia tabacci* Gennadius), thrips (*Scirtothrips dorsalis* Hood and mites (*Polyphagotarsonemus latus* (Banks)), *Tarsonemus translucens* (Green) and *Tetranychus cinnabariuus* (Boisduval). Inward and outward curling of leaf margin and the leaves become rough. Plant growth is stunted which gives a bushy appearance. In severe cases, defoliation occurs. New leaves become small. Flowers and fruits are also few and small with curling ends. Leaf curl can be controlled by following way:

i) Growing field tolerant varieties like Pant C-1, JCA-198, JCA-248, JCA 218, Pusa Jwala, etc.
ii) The plants showing symptoms should be burnt or buried immediately to avoid further infection.
iii) Application of granular insecticides like furadon 3 G, temik (Aldicarb), disyston in the seedbed at the time of sowing.
iv) Application of granular insecticides like furadon 3 G (30 kg/ha), aldicarb 10 G (15 kg/ha), thimet 10 G (20 kg/ha) at transplanting followed by 2-3 foliar sprays with 0.05% phosphamidon (Dimecron 2 ml/l) or 0.03% methyl demeton (Metasystox 1.5 ml/l) or 0.05% monocrotophos (Nuvacron 1.25 ml/l) at 10 days interval.
v) Spraying of 2% neem oil to the plants checks the spread of the virus, through control of vector.

vi) Instalation of yellow/sticky trap @ 10 acre reduces the vector population.

vii) Combined use of border or barrier cropping of maize, jowar and bajra and 2-3 foliar sprays of insecticides. Five to six rows of border crops around the chilli plot should be sown at least 50-60 days before transplanting of chilli.

viii) Mulching with yellow, white black or blue polyethylene sheet because reflecting sunlight from the polyethylene mulch restrict the movement of the white fly.

Mosaic

Several viruses cause mosaic symptoms in chilli viz., tobacco mosaic virus (TMV), potato virus X (PVX), potato virus Y (PVY), cucumber mosaic virus (CMV), tobacco-etch virus (TEV), pepper veinal-mottle virus (PVMV) and tomato spotted wilt virus (TSWV). TMV and PVX are sap transmitted; TSWV is transmitted by thrips while other viruses are transmitted by aphid. Yellow green mottling on the leaves, reduction in leaf size and stunted growth. The plant bears less number of flowers and fruits.

Following control measures should be taken to check mosaic :

i) Use of healthy seeds. Seeds should be collected from virus free plants.

ii) Careful removal of the infected plants from the seedbed and main field.

iii) Washing of all tools and implements with 3% solution of trisodium phosphate (1.5 kg/50 litre water) before using it again because TMV and PVX are transmitted by contact.

iv) Tobacco users should wash their hands thoroughly with soap and water before handling the plants because TMV may spread through smoking and chewing tobacco.

v) Spraying the crop with 0.05% monocrotophos (Nuvacron 1.25 ml/l) or 0.05% methyl parathion (Metacide 1.5 ml/l) to control the vectors population.

vi) Application of granular insecticides in the seed bed followed by 1-2 foliar sprays of 0.05% monocrotophos (Nuvacron 1.25 ml/l) check the aphid population and thus reduce the disease incidence.

vii) Viral infection can be reduced by growing border crops like sunflower, sesame, sorghum around the chilli plot.

viii) Soil application of carbofuran 3 g @ 1.00 kg a.i./ha + covering the nursery bed with nylon net of 400 mesh followed by three sprays of neem oil at 15 and 35 days after transplanting is also effective.

Pepper ring spot

Caused by a strain of tobacco rattle virus and transmitted by aphids. Brown to slate-grey, elongated and sunken necrotic tissues develop on the stems and leaf veins.These strips may run parallel to one another on the stems. The leaves present a curly or crumpled appearance owing to collapse of veins. Control measure is same as for leaf curl and mosaic disease.

Disease-pest complex

Chillies are often attacked simultaneously by the diseases like fruit rot, die back and leaf curl and insect pests like mites and thrips. Thus a severe disease-pest complex is formed. It can be checked by spraying 3-4 times with a combination of 2.5 g dithane M-45 + 3 ml dicofol + 1.5 ml metasystox in one litre of water at fortnightly interval starting from the appearance of the malady.

Nematodes (*Meloidogyne* spp., *Rotylenchus* spp.)

These root-knot nematodes are polyphagus pests with a wide host range. The nematodes attack the roots and produce tiny galls there. The attacked plants show symptoms of withering and wilting of leaves. Plant growth is stunted and fruiting is adversely affected.

Control measures

i) Ploughing the field to expose the soil in summer months reduces the load of nematodes in the soil

ii) Incorporation of neem cake/castor cake/neem leaf/castor leaf/ Pongamia leaf @ 500 g/m^2 helps in reducing the infestatio

iii) Growing antagonistic and non-host crops like cereals and marigold in rotation or as an intercrop

iv) Application of granular insecticides, like aldicarb 10 G (20 kg/ha), furadon 3G (35 kg/ha). Thimet 10 G (20 kg/ha) in the field at transplanting considerably brings down nematode population

v) Growing less susceptible variety like, G-2

vi) Follow crop rotation excluding other solanaceous crops (cowpea-cauliflower-chilli, okra-pea-chilli; cabbage-chilli-cowpea; carrot-chilli-cucumber).

Physiological Disorders

Flower and fruit drop

It is one of the major disorder in chilli cultivation. Flower and the fruit drop may occur due to (i) low humidity and high temperature condition which result in excessive transpiration and water deficit in the plant causing abscission of buds, flowers and small fruits. (ii) decreasing light intensity, (iii) short day and high temperature and (iv) high temperature during early flowering stages. Flower and fruit drop can be checked by : (i) Irrigation at flowering and fruit set stage helps in reducing blossom and fruit drop. (ii) Foliar application of 50 ppm NAA at full bloom stage effectively controls the drop. (iii) Foliar application of 20 ppm NAA at first flower opening followed by two sprays of NAA at an interval of 30 days increase fruit set. (iv) Application of triacontanol also markedly reduces flower and fruit drop.

HARVESTING

In chilli, flowering begins 1-2 months after transplanting and it takes another month for green fruits. Maturity of the fruits varies according to the age of the plant and fruits mature within 2 to 2½ months while in south the crop is ready for harvesting in about 3 to months after transplanting (Chauhan, 1981).

The stage of maturity for picking depends on the purpose for which they are grown. Chillies used for vegetable purposes are generally picked at full grown green stage. For pickles picking is done at green or ripe stage. For drying fruits are picked at red ripe stage. Fruits are picked by snapping off the brittle stems by hand. They are sold green, red or mixed according to demand and domestic need. There after ripe fruits are picked at interval of 1-2 weeks and harvesting continues over a period of about 3 months. The number of picking varies from 6-10 depending upon the season, cultivar and cultural practices (Muthukrishnan *et al.*, 1986). Since chilli is a cash crop and some portion of the total production of dried chilli is exported so its timely harvesting, curing and proper grading is very important (Shukla and Naik, 1993).

YIELD

The average yield of green chillies range between 100-150 q/ha, where as dry fruits are produced at the rate of 20 to 25 q/ha.

POST HARVEST HANDLING

Grading

After harvesting, green chillies are kept in a cool place. Whole fruits with glossy appearance, good texture and free from blemishes are kept separately for marketing. Damaged, broken and deformed fruits collected in different bags. Dry chillies are graded as per "Agmark" grade specifications for chillies. The quality factors considered under these specifications are colour, length of the pods, broken, damaged and discoloured pods, loose seeds, foreign matter, moisture content, pods with or without stalks etc. The chillies graded under these standards, shall be dried ripe fruits and of current year crops. Separate specification schedules exist for important commercial varieties like "Sannam", Mundu, , Rari, Gospurea and a general one for all other varieties produced in India. These grades are (a) Sannam special (SS)- light red shining and 5 cm and above, pods without stalk of more than 80 by weight (b) Sannam general (SG) light red shining, above 3 cm below 5 cm, pods without stalk not more than 10 % by weight. (c) Sannam fair (SF) - blackish dull red, above 3 cm and below 5 cm. (d) Non specified (NS) –it is not a regular grade. It provided to meet such specific requirements of the buyer which are not covered under any of the regular grades.It shall be allowed only for export grading against a specific order from the buyer indicating the quality and quantity required.

Packing

Usually the green chillies are packed in gunny bags, bamboo basket and plastic tray and send to market for sale, whereas dry chillies are packed in bags, basket, and platic bags to transport distant market for sale. Packing of dry chilli is done after the removal of difective and decolourised pods in gunny bags, jute boras or palmyrah baskets. It was found that if fresh red chillies are packed in perforated plastic bags of one kg capacity and further packed in to bamboo baskets, cartoons, and plastic bags and transported to distant market, even after 17 hours transport,weight loss and damage are least and quality remained best. In American countries chillies for market are shipped largely in 1-bushel tub baskets, but in other countries such as ½ and 1-bushel hampers are used, especially for local market.

Doijode (1996) reported that, red ripe fruits of chilli can be packed successfully in paper bags and cloth bags for longer period.

Storage

Green chillies are sold in fresh condition soon after picking. They are sold as early as possible. During glut period or even during high production period green chillies are be stored for about 40 days at 0°C and 98% relative humidity. Dried chillies can be kept for months together in dry places well protected from insect pests.

Drying

Sun drying is the common practice in India. The preparation of drying floor differ from tract to tract in places where sand belts are available, they are used for drying chillies. Sand is also carted and spread as a thin layer on drying floor to get uniform and quick drying. In places where sand is not available, the drying floor is prepared by working a blade harrow or a spade.

The grass patches are removed and the land is leveled and compacted, if necessary after sprinkling water, the pods are spread on such drying yards in layers of 8-10 cm. From fifth day onwards, the produce is initiated on alternate days so that the pods in the lower layers are brought up to ensure quick and uniform drying.

Since the produce is exposed to sun for 10-15 days, spreading on the open yards, it is not only subjected to contamination with dust and foreign matter, but the development of colour also is not upto the mark, due to the bleaching effect of sun rays. The work on mechanical drying has shown that the produce can be dried within a period of 18 hours with the aid of air blown drier keeping the temperature at 44-46°C . This method not only saves and avoids watch for 10-15 days, but also imparts deep red colour and glossy texture to the fruits which are liked in foreign trade and fetch higher premium than that of sun drying (the moisture content of dry pods is to be kept at 8-10%). The cost of mechanical drying worked out to 45 paise per kg of dry fruits.

Processing

Dry chillies, powdered or ground into a paste and even green chillies are used for curries, sambar, rasam and savoury dishes to impart pungency and colour to food items. The great processed form is oleoresin, which permits better distribution of colour and flavour in food. The food industry prefers to use highly coloured and less pungent chillies for preparation of oleoresins.

The principal colouring pigment of dried chillies is carotenoid capsanthin. This is accompanied by capsorubin, zeaxanthin, cryptoxanthin, lutein, and few unidentified xanthophylls. The total pigment content in capsicum is between 4 to 5 grams per kilogram of dried fruits and sum total of these carotenoids provide shades from light yellow, orange, red, deep red, brick red, blood red and dark red incidentally providing a good content of carotenoids as precursor of vitamins A. Oleoresin give cleanliness and uniform flavour. Oleoresin is used in preparation of processed product and also incorporated into a number of pharmaceutical formulations.

Marketing and Export

Chillies are mostly grown in kharif and harvesting of red ripe chilli started from October-November and continue upto February-March. They send to marketing centre by head load, cart and trucks. There are scores of important assembling and distributing and terminal markets of chilli in India. Some of the important assembling centres are Byadagi, and Hubli in Karnataka, Guntur and Warangal in Andhra Pradesh, Madurai, Virudhanagar in Tamil Nadu and Amaravati and Dondaiacha in Maharashtra. The distributing and terminal markets are Mumbai, Calcutta, Chennai, Tuticorin, Virudhnagar, Guntur, Calicut, patna, Kanpur, Nagpur, Delhi, Panipat, Pune, Solapur, Hyderabad, Vijayawada, Salem, Warangal and Solan.

The present system of marketing chilli is through regulated market committees. The agricultural produce market committees are operating market yards in all the major markets in Andhra Pradesh and Karnataka. Commission agents organize closed auctions, in Tamil Nadu and elsewhere in the country.

The total export of chillies from India is on an average only 4% of the total production. This is mainly because of high domestic consumption. China has emerged as a principal exporter and is the main competitor in international market. The other main producers and exporters of chilli are Pakistan, Morocco, Mexico.,Turkey and Bangladesh. Chillies are exported in the form of fresh chillies, stalkless dry chilli, green chillies, pickles, chilli powder and also oleoresin both colour as well as pungent. The importing country such as USA insists on STA (American Spice Trade Association) cleanliness specification. It is enforced by FDA (Food and Drug Administration). The importing countries insist on freedom of products from mould and insects, cleanliness, safe level of mycotoxic and pesticide residues (Deshpande, 2003).

SEED PRODUCTION

Chilli is a self pollinated crop but cross pollination also takes places to some extent. Flowers are bisexual. Bees, ants and thrips are the possible agents of pollination. Anthesis takes place some times after the flower have opened. Flowers remain open for 2-3 days. The percentage of fruit setting is 40-50. The flowers open in the morning between 2 to 10 am. The anther normally dehisce after the flower opening. During cold as well as cloudy days, the opening is delayed.

Source of Seed

Good healthy and well-developed fruits should be selected for producing seeds or obtain nucleus/breeder's/foundation seed from source approved by a seed certification agency.

Time of Sowing

(i) Kharif crop - June to July (ii) Spring crop - February-March.

The seed rate is 1 to 1.5 kilograms per hectare.

Isolation Distance

Chiliies are often-cross pollinated crops. Cross-pollination is mainly by insects. The extent of cross pollination from 7 to 36 per cent has been recorded in the USA, and upto 68 per cent in India. The seed crop must be isolated from other variety and also the same variety growing for commercial purposes following 400 metres distance for foundation seed and 200 metres for certified seed production (Agrawal, 1985).

Other cultural operations and agronomic practices are same as given in crop production section.

Roguing

Agrawal (1985), George (1985) and Singh (2001) observed that it is utmost important to attend roguing operation very carefully thrice to obtain pure and true to the type seed. It is mainly done on the basis of varietal characters of the plants and fruits at the main roguing stages.

Pre Flowering Stage

First roguing should be done at pre flowering stage. In this roguing, plants which bear flowers should be uprooted. Plants show, different flowering habit, different type of branch orientation, should be removed. Those plants which show different leaf colour and leaf arrangement other than the grown variety should be uprooted.

At the Time of Flowering and Fruiting

This is the second roguing which is done at the time of flowering and fruit setting. In this roguing, the plants which do not bear flowers should be removed from the field. Plants bearing flowers of different colour than the grown variety would be removed from the field.

At the immature fruit stage plants bearing no fruits should be removed. Plants show different fruit shape and size should be removed if they are different than the grown variety. It is essential to remove all affected plants by leaf blight, anthracnose and virus disease, from the seed crop.

At the Time of Maturity Stage

The final roguing is done at the time of maturity stage. The roguing should be done on plant basis but not of individual fruit basis. Plants show different type of shape and size than the grown variety should be rogued. Those plants which show different fruit colour than the grown variety should be uprooted. Plants affected by diseases particularly by seed born diseases such as leaf blight, fruit rot, anthracnose, wilt, bacterial wilt, brown rot, viral and mosaic diseases, should be removed from the seed field.

Harvesting

Chillies become red or orange or crimson or scarlet in colour when ripe depending on varieties. Fruits are picked up when fully ripe. Fruits can be heaped under shade for a couple of days. The fruits which are injured during harvesting or during transportation are removed avoiding the spoilage the whole seed lot by rotting. Selection of fruits can also be done at this stage to obtain high quality seeds.

In India, chillies are harvested manually. Fruits are easily separated from the plants by hands.

Seed Extraction

Seeds from chilli fruits are extracted manually without any difficulty. There are two main methods of seeds extraction i.e., a wet and a dry. Drying of long pungent chillies can be subjected to drying in open sun or in electrically operated drying cabinet. Solar cabinet dress are also available, in which within 3 to 5 days fruits are completely dried upto the moisture content 8-10 per cent or below.

Seed Yield

The average seed yield varies from 45 to 120 kg per hectare (Thakur and Gill, 1985) depending on variety grown, climate condition and management practices.

Drying and Packing

Seed extraction is followed by drying of seeds till moisture percentage comes down to below 8. It will be far better if seed moisture is kept at minimum level of 6 to 8 per cent. The moisture percentage is tested with moisture tester. Properly dried seeds are graded and then tested for germination.

Dried seeds are packed in polythene bags or glass containers or any other suitable containers those can be made air tight.

Seed Storage

After packing the seeds in suitable containers as mentioned above, are stored in cool and dry place.

Chilli seeds dried to a moisture content of 8 per cent with an initial germination of 85 per cent recorded 73 per cent germination after 30 months storage under ambient conditions. Slurry treatment with captain 75% WP at 2 g/kg of seed maintains germination percentage at higher value in comparison to untreated seeds. Seed germination decreases as storage period increases (Thiagarajan, 1994).

Germination, field emergence, seedling growth and vigour index decrease with increase in drying air temperature. Chilli seeds maintain very high germination (92%) when they are dried under sun and seeds are extracted from fully ripe red coloured fruits followed by fully ripe red coloured fruits dried at an air temperature of 40°C (Dhanelappagol *et al.*, 1994).

REFERENCES

Agrawal, R.L. (1985). Seed Technology. Oxford and IBH Publishing Co., New Delhi.

Akroyd, W.R. (1963). The nutritive value of Indian foods and the planning of satisfactory diets. ICMR special Rept. 42: New Delhi.

Aloni, B., Karni, L., Zaidonan, Z., Riov, Y., Huberman, M.and Goren, R. (1994). The susceptibility of pepper (*Capsicum annuum*) to heat induced flower abscission: Possible involvement of ethylene. *J. Hort. Sci.* 69:923-928.

Anonymous. (2006). Annual Report 2005-06, Indian Institute of Horticulture Research, Bangalore, pp. 7-8.

Anu,V., Syriac, K.E., Yadav, P.I.P. and Geethakumari, V.L. (2005). Growth characters of chilli (*Capsicum annuum* L.) as influenced by varying levels of fertilizers in combination with fluorescent pseudomonos and Azospirillum. *Vegetable Science*, 32(2): 150-153.

Asokan, R.; Mohandas, S. and Anand, L. (2000). Biofertilizers and biopesticides for horticultural crops. *Indian Horticulture*, 45(1): 44-47.

Bhattacharyya, P., Jain, R.K. and Paliwal, M.K. (2000). Biofertilizers for vegetables. *Indian Horticulture*, 45(2): 12-13.

Boswell, V.R. (1951). Garden pepper, both a vegetable and condiment. *Natl. Geogr. Mag.* 96: 166-167.

Butani,D.K. (1976). Pests and diseases of chillies and their control. *Pesticides*, 10(80):38-44

Chandra, A., Singh, K., Nair, P.K.R. and Bhalle, Usha (1987). Jawahar Mirch 218: A new bright red pungent chilli. *Indian Horticulture* 31(4):7-8.

Chandra, Amar and Pandey, R.P. (2001). Jawahar Mirch 283: a summer-set chilli. *Indian Hort.*, 46(1): 20-21.

Chaugule, A.B. and Mahajan, P.R. (1979). Effect of varying levels of plant population, nitrogen, phosphorus and potash on growth and yield of chilli (*Capsicum annuum* L.) *Vegetable Science*, 6: 73-80.

Chauhan, D.V.S. (1981). Vegetable Production in India. Ram Prasad and Sons, Agra.

Cheema, G.S. (1982). Studies on improvement of chillies for resistance to anthrocnose and cereorpora leaf spot diseases. *Ph.D. Thesis*, submitted to UAS, Bangalore.

Cheema, G.S., Nazareth, B. and Dhaneshwar, S.R. (1994). Improvement of Chilli (Capsicum annum L.) by selection in the Bombay province. *Indian J. Hort.*, 11(2): 49-61.

Choudhury, B. (1983). Vegetables. National Book Trust, New Delhi.

De, Nirmal and Rai, Malhura, (2005). Micronutrient disorders and management in vegetable crops; A review, *Vegetable Science*, 32(1): 1-10.

De. Candolle, A. (1986). Origin of cultivated Plants. Hafner Publishing Co. New York.

Deshpande, A.A. (2001).Chilli and capsicums (*Capsicum annuum* L.). In Vegetables, Tuber Crops and Spices (Eds: S. Thamburaj and Narendra Singh), ICAR, New Delhi.

Deshpande, A.A., Pathak, C.S. and Singh, D.P. (1983). Variants in Chilli pepper (*C. annum* L.). *Capsicum Newsl.*, 2: 81.

Deshpande, R.B. (1933). Studies in Chillies III. The inheritance of some Characters in C. *annum* L. *Indian J. Agric. Sci.*, 3: 219-300.

Deshpande, R.B. (1944). Inheritance of bunchy habit in Chilli. *Indian J. Genet.*, 4(1): 54.

Dhanelappagol, M.S., Shashidhara, S.D. and Kulkarni, G.N. (1994). Effect of stages of harvesting and methods of drying of chilli fruits on seed quality. *Karnataka J.Agric. Sci.*, 7(1): 36-39.

Doijode, S.D. (1996). Effect of packaging and fruit storage on seed viability, vigour and longevity in chilli (*Capsicum annuum* L.). *Vegetable Science*, 23(1): 36-41.

George, Raymond, A.T. (1985). Vegetable Seed Production Longman, London and New York.

Gunewardhane, S.D.I.E. and Pereira, B.D. (1975). Agronomic investigations with chilli in the wet zone. Studies on interaction between fertilizer levels and planting density. *Trop. Agril.*, 131-41.

Hawthorn, L.R. and Moguruder, R. (1938). Weather influence on pepper fruits (size, shape and seed production respond to weather conditions). National seedsman, April, 1938.

Hegde, D.M. (1993). Water Management in Vegetable Crops, In Advances in Horticulture Vol.5. Vegetable Crops Part-1 (Eds: K.L.Chadha and G. Kalloo). Malhotra Publishing House, New Delhi.

Irish, H.C. (1989). A revision of the genus Capsicum with special reference to garden varieties. *Report, of Missouri Botanical Garden*, 9: 53-110.

Jagadish, C.A., Chandrasekharan, P.; Ponnaiya, B.W.X. (1964). Studies on the genus Capsicum. *Madras Agric. J.*, 51(8): 36.

Jayarajan, R. and Rama Krishnan, K. (1969). Potato Virus on Chilli (Capsicum annum L.) in Tamil Nadu. *Madras Agric. J.*, 56(12): 761-766.

Jeswani, L.M., Deshpande, R.B. and Joshi, A.B. (1956). Inheritance of some fruit characters in Chilli. *Indian J. Genet.*, 16(2): 138-143.

Joseph, P.A. and Pillai, P.B. (1985). Effect of N, P and K on the growth and yield of chilli variety Pant C-1. *Agr. Res. J. Kerala*, 23(1): 75-82.

Kalloo, G., Rai, M., Kumar, R., Prasanna, H.C., Singh, M., Kumar, S., Singh, B., Ram, D., Lal H., Pandey, P.K., Rai, S., Pandey, K.K. and Satpathy, S. (2006). New Vegetable varieties from IIVR, Varanasi, *Indian Horticulture*, 51(3): 16-22.

Meghwal, R.R. (2002). Increasing horticultural produce through VAM inoculation. *Indian Horticulture*, 47(3): 29-30.

Mini, C. and Joseph, A. (1998). Effect of seed spacing in nursery on seedling vigour and field performance in chilli (*Capsicum annuum* L.). *Vegetable Science*, 25(2): 124-126.

Murthy, N.S.C.and Bavaji, J.N. (1980). Sindoor - a new dual purpose chilli strain, Indian Cocoa, *Arecanut Spices, J.*, 4(2): 36-38.

Muthukrishnan, C.R., Thangaraj, T. and Chatterjee, R. (1986). Chilli and Capsicum. In Vegetable Crops in India (Eds: T.K. Bose and M.G.Som). Naya Prokash, Calcutta-6.

Nadkarni, K.M., (1927). Indian Meteria Medica, Published by Nadakarni and Co., Bombay.

Napoleon, Ann and George, Annamma (2003). Morphological characters of chilli (*Capsicum annuum* L.) as affected by vegetable picking and seed soaking with NAA. *Vegetable Science*, 30(2): 185-186.

Nelson, S.J. and Natarajan, S. (1994). Influence of barrier crops on pests of chilli under semi dry condition. *South Indian Hort.*, 42(6): 390-392.

Padda, D.S. and Singh, J. (1971). Studies on some important aspects of floral biology in chillies. *Indian Journal of Agricultural Research*, 5(3): 217-8.

Pal, B.P. (1945). Studies in hybrid vigour II. Notes on the manifestation of hybrid vigour in gram, sesamum and chillies, *Indian J. Genet.*, 5: 106-121.

Pal, B.P.; Ramanujam, S. and Raju, A.B. (1941). Colchicine induced polyploidy in crop plants-chilli. *Indian J. Genet.*, 1: 128-139.

Palaniappan, S.P.,Jeyabal, A. and Chelliah, S. (1999). Response of tomato and chilli to foliar application of speciality fertilizers. *Vegetable Science*, 26(2): 198-200.

Parthasarathy, N. and Kedarnath, S. (1950). Varietal differences in the breeding behaviour of colchicine induced auto-tetraploids of chilli. *Indian J. Genet.*, 10(1): 14-20.

Pathak, C.S.; Singh, D.P. and Deshpande, A.A. (1983a). Male and Female sterility in Chilli pepper (*C. annum*). *Capsicum Newsl.*, 2: 102.

Pathak, C.S.; Singh, D.P. and Deshpande, A.A. (1983b). Closed flower mutant in *Capsicum annum*. *Capsicum Newsl.* 2: 106.

Peter, K.V. (1998) Genetics and Breeding of Vegetables. *Indian Council of Agricultural Research*, New Delhi.

Pickersgill, B. (1980). Some aspects of interspecific hybridization in Capsicum. Fourth Meeting of the EUCARPTIA Capsicum Working Group, held during 14-16 October, 1980, at Wageningen, the Netherlands.

Pruthi,J.S. (1993) Major spices of India. Crop Management and Post harvest Technology. *Indian Council of Agril. Research*, New Delhi.

Purseglove, J.W. (1977). Tropical Crops : Dicotyledones. Longman, London.

Rajagopal, A., Mohideen, M.K., and Mehta, V.A. (1976). Weed control studies in chilli. *Madras Agric. J.*, 63: 470-473.

Raju Narsimaha, D.N. (1979). Correlation and path coefficient analysis in capsicum (*C. annum* var. *grossum*). M.Sc. Thesis, submitted to UAS, Bangalore.

Saimbhi, M.S. and Randhawa, K.S. (1976). Herbicidal control of weeds in transplanted chilli. *Punjab Hort. J.* 16:146-149.

Saxena, Ashish, Hundal, J.S. and Dhall, R.K. (2003). Floral biology studies in chilli (*Capsicum annuum* L.) *Vegetable Science*, 30(2): 173-175.

Sharma, B.R., Chadha, A.P.S., Naidu, A.K. and Verma, A.C. (1995). Studies on the determination of doses and methods of nitrogen application on chillies (*Capsicum annuum* L.). *Vegetable Science*, 22(1): 16-18.

Sharma, S.K. (1993). Effect of triacontanol on capsicum seed production. Golen Jubilee symp. on *Hort. Res. Changing Scenario*, held on May 28 at IIHR, Bangalore.

Shukla, V. and Naik, L.B. (1993). Agro-Techniques for Solanaceous Vegetables. In Advances in Horticulture Vol.5-Vegetable Crops: Part 1, (Eds: K.L. Chadha and G. Kalloo). Malhotra Publishing House, New Delhi.

Siddappa, G.S. and Bhatia, B.S. (1954) Canning of Chillies; *Bull of C.F. Tech. Reser. Insti.* Mysore, 4(1).

Singh, A. and Singh, H.N. (1978). Line x tester analysis of yield in chilli. *Indian J. Genet.*, 38(1): 52-56.

Singh, A.; Singh, H.N. and Bajpayee, N.K. (1977). A new strain of chilli with loose calyx. *Indian Hort.*, 22(4): 13.

Singh, D.N.,Sahu,A. and Parida,A.K. (1999). Response of chilli (*Capsicum annuum*) to applied nitrogen and potassium in clay loam soils of Orissa. *Indian Journal of Agricultural Sciences*, 69(4): 787-8.

Singh, N.B. and Singh, B. (1970). Interrelationship, heritability estimate and genetic advance in yield and other characters in chillies. *Madras Agric. J.*, 57(7): 369-373.

Singh, N.P., Johri, A.K. and Singh,A.K. (1992). Weed Management in bell-pepper (*Capsicum annuum* L.). *Indian J. Weed Science*, 24(3 & 4): 56-59.

Singh, S.K. and Srivastava, J.P. (2006). A new clustured chilli: Azad Mirch 1, *Indian Horticulture*, 51(5): 8.

Singh, S.P. (1989). Production Technology of Vegetable Crops. Agricultural Research Communication Centre, Karnal Haryana (India).

Singh, S.P. (2001). Seed Production of Commercial Vegetables. Agrotech Publishing Academy, Udaipur.

Singh, V.B. and Singh, Kirti (1996). Spices. New Age International Publishers, New Delhi.

Singh, R.S. (1985). Diseases of Vegetables. Oxford and IBH Publishing Co., New Delhi.

Sivanappan, R.K., Aruna, R. and Palaniswami, D. (1978). Response of chillies to drip irrigation. *Madras Agric. J.*, 65: 576-579.

Smith, P.G., Villalon, B. and Villa, P.L. (1987). Horticultural Classification of peppers grown in the United States. *Hort.Science*, 22: 11-13.

Sontakke, M.B. (1981). Genetical studies in chillies. *Ph.D. Thesis*, submitted to UAS., Bangalore.

Sontakke, M.B., Gunjkar, S.N., Purandare, N.D., Anserwadakar, K.W. and Shinde, N.N.(1995). A new rainfed chilli variety 'Parbhani Tejas' for Marathwada region. *Nat. Symp. on Recent Development in Vegetable Improvement* held at Indira Gandhi Krishi Vishwa Vidyalaya, Raipur (MP) on February 2-5, 1995.

Suryanarayana, V., Raja, K.T.R.and Rao, D.V.S. (1983). Effect of irrigation frequencies and antitranspirants on chilli. *Veg. Sci.*, 10:77-83.

Swarup Vishnu (2006). Vegetable Science and technology in India. Kalyani Publishers, New Delhi. pp. 656.

Swarup, Vishnu (1998). Genetic resources of Vegetable Crops in India. Presented on Dr. Harbhajan Singh Memorial Lecture during *National Symposium of Indian Society of Vegetable Science*, 12-14 December, 1998 at Varanasi-221 005.

Tandon, H.L.S.(1991). Fertilizer Recommendations for Horticultural Crops. A Guidebook. Fertilizer Development and Consultation organization. New Delhi, India.

Tewari, V.P. (1991). A multipurpose perennial chilli, Pusa Sadabahar. *Indian Hort.*, 35(4): 29-31.

Thakur, P.C. and Gill, H.S. (1985). Produce quality seed of Sweet pepper. *Seeds and Farms*, 11(12): 25-27.

Thiagarajan, C.P. (1994). Storability of chilli (*Capsicum annuum*) Seed. *Madras Agric. J.* 81(80): 442-445.

Thomas, K.M. (1961). No need to be Charry of Chillies. *Indian Farming*, pp.27-28 July.

Thompson, H.C. and Kelly, W.C. (1957) Vegetable Crops. Tata McGraw-Hill Publishing Co., Ltd., New Delhi.

Usha Rani, P. (1981). Varietal difference in chillies for pungency, pigmentation and ascorbic acid contents. *M.Sc. Thesis*, submitted to University of Agricultural Sciences. Bangalore.

Usha, P. and Peter, K.V. (1988). Control of flower fall through stimulants and regulators in chilli, *Vegetable Science*, 15: 185-89.

Usha, P. and Peter, K.V. (1995). Exogenous chemicals for increasing chilli production. *Indian Horticulture*, 40(1): 44-45.

Venkatachalam, R. (1982). M.Sc. (Hort.) Thesis, Tamil Nadu Agricultural University, Coimbatore.

Verma, S.K. and Pant, K.C. (1999). Evaluation and Utilization of Chillies (*Capsicum* sp.) germplasm in Uttar Pradesh hills. *Haryana J. Hort. Sci.,* 28(1&2): 80-81.

Walker, J.C. (1952). Diseases of Vegetable Crops. Mc. Graw Hill Book Co., Inc. New York, London.

Wang, M.and M. A., D.H. (1987). Study of analytical electrophoresis on capsicum germplasm. *Capsicum News letter,* 6:25.

Wankhade, B.N. and Mony, D.K. (1985). Water use studies in chillis (*Capsicum annuum* L.). *PKV Res., J.,* 9:20-25.

□□□

4. CAPSICUM

Botanical name	:	*Capsicum annuum* L. var. grossum (L) Sendt.
Family	:	Solanaceae
Chromosome number	:	2n = 24.

Capsicum , bell pepper, sweet pepper (*Capsicum annuum* L.) is an important vegetable of recent introduction into India from USA, Italy, Brazil, Hungary, Bulgaria, Czechoslovakia and USSR. It is widely used as food and flavour additive in various delicious preparation all over the world.

ORIGIN

Capsicum is originated in the new world tropics and subtropics (greenleaf, 1985). The original home of capsicum is thought to be tropical south America especially, Brazil and introduced in India by the British in Himachal Pradesh and Tamil Nadu (Neelgiris).

DISTRIBUTION

Capsicum, sweet pepper or paprika as it is called today, appeared in Europe in the early 16th century and the spread elsewhere took place from Portugal (Seshadri and Chatterjee, 1996). Capsicum is widely distributed in central and south America, Peru, Bolivia, Costa Rica, Mexico in almost all the European countries, Hong Kong and India. The spread to India has been ascribed to the Portuguese in the early 16th or 17th century (Somos, 1984). In India it is cultivated commercially in Tamil Nadu, Karnataka, Himachal Pradesh and in some part of Uttar Pradesh. In northern India it is commonly known as 'Simla mirch' and also in Kalimpong sub-division in Darjeeling district of West Bengal. However, due to change in food habits and advent of hybrids large quantities of bell pepper are now also grown in other states such as Jammu and Kashmir, Andhra Pradesh, Uttar Pradesh, West Bengal, Maharashtra, Karnataka, Tamil Nadu, Bihar and Madhya Pradesh.

BOTANY

The capsicum genus is a member of solanaceae family. It is generally regarded as C. *annuum* var. grossum. Green leaf (1985) recognised 13 horticultural group in C. *annuum*. Capsicum is an annual or short lived perennial herb upto 1.5 m in height. Plants are sub-shrub woody at base, erect, branched root-taproot; shoot radial but lateral branches, leaves variable, simple. Flowers borne singly and are terminal. Fruits large 3.5" x 2.4", smooth, thick, fleshy, blocky of various shapes, more like a bell and hence named bell pepper non-pungent, wild, thick-fleshed type used as vegetables so named as sweet pepper. Fruit colour green red to yellow. Fruits are "berry", botanically, the fruits have many seeded berry with 200, more locules (Purseglove, 1977).

Taxonomy

It is considered that chilli and sweet pepper are intercrossable belonging to the same species annuum. The large fruited bell peppers of today originated from tinny, pungent and pointed fruited wild species *C. annuum* var. *aviculare*. Besides, *C. annuum* which encompasses both bell and chilli peppers, the other four cultivated species are *C. frutescens* L., *C. chinense* Jacquin; *C. pendulum* var. baccatum and *C. pubescens* Ruiz and Pavon. The native American Indian cultivators-breeders, probably women developed the range of fruit shapes and sizes from tiny pungent fruits to large sweet bell types. The modern pepper breeders owe a lot to the natives who created wide variability within this horticultural group. This development is thought to have occurred during the last 4000 years (Pickersgill, 1969). During this process, sweet or partially sweet or less pungent types possibly got elliminated, since they require cooler temperature for fruit setting.

Sweet types were later introduced into India during British days and their cultivation developed with preference towards green blocky types than to elephant trunk types; cooler regions like Bangalore, Hyderabad, Poona, Ranchi, Solan, Kshmir valley etc., are now growing sweet pepper. Now a days, sweet pepper production has spread to some parts of north India and eastern plains, where cooler winters are suitable for their growing. The sweet pepper consumption in India is increasing and is mostly confined to upper strata of society in urban localities or cities. Wider adaptation to warmer conditions has not been possible in sweet pepper types, because introduced varieties were from typically cooler countries of Europe (Seshadri and Chatterjee, 1996).

Floral biology

In sweet peppers, anthesis starts at 7.15 a.m. and continues upto 11.15 a.m., with the peak at 7.15 a.m. (Vijay *et al.*, 1979). Anthers dehisce after 30 minutes of anthesis. The stigma becomes receptive from the day of anthesis and remains receiptive upto 2 days after anthesis. Fruit set was maximum on the day of anthesis with around 80 % set and pollen fertility around 90 %. However, it was feel that considerable variability exists for time of anthesis and dehiscence. Fruit set in flower buds pollinated a day prior to anthesis has been observed to be good. In fact, using buds, selected a day prior to anthesis may be desirable since dehiscence of anthers takes place before anthesis in some of the genotypes.

Table 1 : Morphological characteristics of domesticated *Capsicum* species and their wild relatives

Species	Wild/ cultivated	Clayx teeth	Corolla	Flowers/ node	Seeds
C. annuum annuum	Cultivated	Present	White large	1	Yellow smooth
C. annuum var. aviculare	Wild Progenitor	Absent	White small	1(3-A)	Yellow smooth
C. baccatum var. pendulum	Cultivated	Present	White with yellow/green spot	1	Yellow smooth
C. baccatum var. baccatum	Wild Progenitor	Present	White with yellow green spot	2-3	Yellow smooth
C. chinense	Cultivated	Absent	Dull-white	2-3	Yellow smooth
C. frutescens	Cultivated	Present	Greenish-white	2-3	Yellow smooth
C. pubescens	Cultivated	Absent	Purple	1(2-3)	Rough
C. cardenasii	Wild progenitor	Absent	Purple campanulate	2-3	Yellow smooth
C. eximium	Wild progenitor	Absent	Magenta to white with yellow/ green spots	2-3	Yellow smooth

Deshpande (2001)

COMPOSITION

Capsicum is very rich in ascorbic acid and about 400 g of raw pepper contains enough ascorbic acid to supply the daily requirements of nine persons. The red capsicum/bell peppers have different types of pigments, which details are given below:

Water	92.4%	Nicotinic acid	6-10 mg
Capsanthin	36%	Cryptocapsin	4%
β-carotene	10%	Vit. C	321 mg
Violaxanthin	10%	Vit B_1 and B_2	0.02-0.1 mg
Cryptoxanthin	6%	Folic acid	1.3-2.9 mg
Food value per 100 g edible portion			
Energy	29 calories	Vit. A	870 IU
Protein	1.2 g	Ascorbic acid	175 mg
Calcium	11 mg	Thiamine	0.06 mg
Riboflavin	0.03 mg	Niacin	0.55 mg

Uses

Capsicum, green or red, may be eaten cooked or raw, sliced in salads. In stews a little capsicum imparts a novel flavour. Capsicum are also used for pickling in brine, baking and stuffing. Diced green or red capsicums are some times mixed with sweet corn or other vegetables.

VARIETIES AND HYBRIDS

Most of the cultivars adopted by the growers are introduction from foreign countries. The important varieties and hybrids are California Wonder, Yolo Wonder, King of North, Early Giant, World Beater, Neapolitan, Early Calwonder, Windsor A, Harris Early Giant, Vinedale, Bullnose, Russian 449, Chinese Giant, Ruby King, Tara, California Arka Basant, Arka Mohini, Arka Gaurav, Bharat, Oshkosh,Hungarian Wax, Elephant's Trunk, R 449, Golden Bell, Domino etc.

Detail characteristics of some improved varieties and hybrids described by Singh and Nath (2011) are given below:

Varieties/ hybrids	Developing Institute/ Centre Agencies	Yield (q/ha)	Remarks (Special Features)
Arka Mohini	IIHR	200.00	Fruits blocky, 3-4 lobed, dark green, thick flesh, non pungent, good aroma, fruit 80-100 g each, suitable for both season of cultivation.

contd...

Arka Gaurav	IIHR	230.00	Fruits dark green, blocky, erect, 3-4 lobed, medium sized (70-80 g), good for both season crop. Tolerant to bacterial wilt diseases.
Arka Basant	IIHR	210.00	Fruits creamish white, conical, erect, thick, flesh and prolific continuous bearing, excellent keeping and cooking qualities. Good aroma and crisp texture, good for stuffing. Suitable for both season good export value (Nath *et al.*, 1987).
California Wonder	IARI, RS, Katrain	285.00	Fruits smooth blocky, 4 lobed , dark green, wt. 90-150 g, maturity 70-80 DAP, 3-4 picking, thick fleshy layer, most popular variety in India. It takes 170-195 days for maturity from seed to seed stage (Thakur & Gill, 1985).
California Supreme	Introduction	450.00	Fruits dark green, blocky, thick and juicy flesh, 3-4 lobes, dark, red at ripening, long, shelf life, tolerate long distance shipping, tolerant to tobacco mosaic virus.
Kt-1 (F_1)	IARI	320.00	Very early and heavy yielder with yellowish green fruits pointed at the tip end. Suitable for growing in humid western Himalayan region.
Chinese Giant	Foreign Introduction	145.00	Plants vigorous and prolific bearer, fruits dark green, 3-4 lobed, medium flesh thickness and vigorous (Thomas *et al.*, 1983).
World Beater	Foreign Introduction	125.00	Plant upright, fruit 4 lobed with medium thick flesh, mild and sweet.
Yolow Wonder	IARI, RS, Katrain	120.00	Fruits dark green , drooping, flesh medium thick 3-4 lobed fruit, fleshy sweet and fine flavour, turn bright crimson on maturity (Thakur and Gill, 1985).

contd...

Bharat (F_1)	IAHS	350.00	Fruits are 4-lobed with thick flesh, good taste.
IAHS-P-2	IAHS	300.00	It is relatively heat tolerant, early, compact hybrid. Fruits thick walled, tri lobed, dark green, slightly elongated; light green in colour. Fruit matured in 65 days after transplanting. Average fruit weight, 140 g.
Pusa Deepti (F_1)	IARI, RS, Katrain	380.00	Fruits smooth, erect, conical, light green, thick flesh, long (10 x 4 cm), early to medium maturing (60-70 DAP), suitable for commercial purpose, tolerant to bacterial leaf spot and anthracnose under field condition.
NS 436(F_1)	Namdhari Seeds	300.00	Fruits dark green, good size, very firm, attractive, 4-lobed and blocky, early bearer.
Sindhoori (F_1)	Namdhari Seeds	320.00	Fruits lustrous green , turn yellow to maturity, blocky, 8 x 8 cm size, firm and very smooth.
Ratan (F_1)	Namdhari Seeds	310.00	Fruits attractive, shiny dark green, blocky, 8 x 8 cm size, firm, smooth red at maturity, very good transport quality.
Arun (F_1)	Namdhari Seeds	315.00	Fruits shiny, dark green, blocky, 8 x 8 cm size and firm, red at maturity. Prolific bearer.
Gagan (F_1)	Namdhari Seeds	370.00	Fruits beautiful light green, blocky, 8 x 6 cm size., medium firm and thin walled, turn red at maturity.
Vikas (F_1)	Golden Seeds	327.00	Fruits large smooth, uniform, blocky, medium green and four lobed, wall thick and firm, av. fruit wt. 150-200 g each. Suitable for outdoor cultivation.
Anupam (F_1)	Nunhem Seeds	255.00	Fruits, bright green, smooth, 3-4 lobed, thick, walled, wt, 75-80 g, early (55 days) maturing.

contd...

Alankar (F_1)	Nunhems Seeds	270.00	Fruits bell shaped, blocky, excellent firmness, partial resistance to PVYs TMV.
Harit (F_1)	Nunhems Seeds	265.00	Fruits firm, green blocky, smooth and thick wall, suitable for processing.
Nutan(F_1)	Nunhems Seeds	278.00	Fruits, lobed , firm, green and smooth surface, turn yellow at maturity. Recommended for green house cultivation.
Tanvi (F_1)	Unicorn Seeds	210.00	Fruits, 4 lobed, blocky, thick walled, firm, wt. 150-200 g each, good keeping quality, tolerant to tobacco mosaic virus.
Tanvi Plus(F_1)	Unicorn Seeds	215.00	Fruits, dark green, elongated blocky, thick walled, wt. 200-250 g each, long pendency.
BSS 89 (F_1)	Bejo Sheetal Seeds	225.00	Fruits green, blocky, 4 lobed, wt. 100 g each, first picking 65 days after transplanting, suitable for all seasons growing.
BSS 399(F_1)	Bejo Sheetal Seeds	245.00	Fruits green, blocky, 4 lobed, wt. 100 g, each, thick, skin first picking 55 DAP, good for long transportation, very early type and can grown in all seasons.
BSS 401(F_1)	Bejo Sheetal Seeds	240.00	Fruits green, 4 lobed , wt 120 g each, mature in 85-90 DAP, vigorous.
Green gold (F_1)	MAHYCO	250.00	Fruits dark green, blocky, on maturity dark red, medium thick skin, 3-4 lobed, pendent, first picking 60-65 DAP.
Sungold (F_1)	MAHYCO	240.00	Fruits long (10-12 cm), shiny dark green, blocky, medium thick skinned with 3-4 lobes first picking 65-70 DAP.
Bombay (F_1)	Novartis, India Seeds	700.00	Fruits dark green, thick walled glossy wt. 140 g, 3-4 lobes, longer shelf life, ideal for long

contd...

			distance transportations, tolerant to TMV, PVY and early bearer.
Indra (F_1)	Novartis, India Seeds	550.00	Fruits dark green, thick walled and glossy, wt. 170 g, 3-4 lobes, fruiting starts 55 DAP, longer shelf life. Ideal for long distance transport. High export potential, tolerant to TMVS and PVY.
Lario (F_1)	Novartis, India Seeds	800.00	Fruit medium sized, dark green, thick walled, fleshy and blocky type, wt. 150 g, 3-4 lobes, first picking starts 55 DAP, long shelf life. Ideal for long distance-transport, tolerant to TMV.
Portos (F_1)	Novartis, India Seeds	600.00	Fruits blocky, 4 lobed, excellent long distance transportability, tolerant to TMV and PMVY.
Tara	Novartis, India Seeds	385.00	Fruits medium large, 4-lobed, thick wall, wt. 250 g, dark green fruits turn to golden yellow on maturity.
Zenith	Tropica (Green Co.)	375.00	Fruits dark green to dark red at full maturity, size (10 x 8 cm), 3-4 lobes, thick flesh, ideal for export.

CLIMATIC INFLUENCE

Capsicum is a cool season crop with narrow genetic base (Ado and Samarawira, 1988) and cultivated during winter or rabi season. But it is highly sensitive to light frost and grow poorly in cool temperature (5-15°C). It can not tolerate frost. Temperature at 23.5°C or above affects, developments of fruits adversely. The optimum temperature for seed germination is 18-24°C while for vegetative growth 24-27°C. It can be grown throughout the country, but regions having mild climate are more suitable. In case of availability of disease free seed in temperate region, the chances of viral diseases are lesser. High temperature and low relative humidity at the time of flowering increase the transpiration pull resulting in abscission of buds, flowers and small fruits. Higher night temperature is responsible for the higher capsaicin content

(Rylski, 1973). Fruit development is adversely affected at temperature 37.8°C or more. The average cropping period in summer season (Hill region) is 103 ± 2 days and temperature range 18-36°C. During winter season average the cropping period is 132 ± 3 days and average temperature range (8-28°C). These give maximum fruit yield of capsicum. High relative humidity is good for the crop growth, fruit set and increase crop yield. Providing shed to sweet or bell peppers through poly or net houses during summer and only shed nets in open during other seasons is beneficial.

SOIL REQUIREMENT

Although capsicum or bell pepper can be grown in almost all types of soil, well-drained clay loam is considered as ideal. Sandy loam soil holding moisture with liberal supply of organic matter is also suitable for its cultivation. Bell pepper performs best in red loams and alluvial soil with slightly acidic pH and having good drainage and aerations. It is sensitive to acidic soils and can grow well under a pH of 5.5-7.0. Plants produce best when soil pH is 6.0-6.5 (Joshi and Singh, 1975).

SOWING

Sowing Time

The sowing time for northern plain and low hills are July, November-February and March. In the mid hills seeds can be sown from March-May. At higher altitudes in Himachal Pradesh and Nilgiris seed sowing is done in March-April, while in tropical plains seeds are sown in September-November. Two successive crops can be taken in area having very mild climate.

Seed Rate

Seed rate is 750 g/ha for high yielding (improved) varieties and 350 g seed is enough for one hectare to transplant hybrid varieties. Seed should be treated with thiram or captan @ 3g kg of seeds.

Nursery Raising

The site for nursery bed should be sunny and well drained; about 10-12 beds are prepared to raise seedling for one hectare land. The size of each bed should be 8.0 m long, 1.20m wide and 15 cm high. Apply 15-20 kg of well-rotten FYM and 500 g of 15:15:15 NPK fertilizer during preparation to each bed. Treat the nursery beds with formalin solution (1 part formaldehyde and 48 parts water) for raising healthy seedlings. Cartaf (0.3%, @ 5 litre solution per sq.m.) can be used alternately for drenching. The soil surface should be covered with an alkathene sheet for 48 hours and then open for 48-72 hrs. before sowing. Seed should be sown 2.5 mm deep in rows 5-7 cm apart. Seed beds are then covered with dry grass and watering is done with fine shower or irrigated with sprinkler to maintain optimum moisture till the seeds germinate (Arya, 1983). On germination mulch/grass is removed.

Regular light watering in the morning and evenings with a rose can or micro sprinklers is advisable. To ensure healthy seedlings it is better to cover the nursery area with 40 mesh nylon nets which protect the seedlings from virus carrying vectors. Seedlings become ready 30-35 days after sowing. Seedlings in the nursery may be allowed for about 40-50 days. Clipping of capsicum seedlings about 10 days prior to transplanting helps in establishment of transplanted seedlings and also accelerates the growth of auxillary buds resulting in better branching. Hardening of seedlings should be started a week before transplanting by regulating water to the nursery (Deshpande, 2001).

PLANTING

Land Preparatin

The land should be thoroughly prepared by ploughing 3-5 times. Planting can be done in either flat bed or raised beds. The beds of suitable size (3 m x 3 m or 3m x 4m) are prepared before transplanting, apply 3 g furadan in each furrow by mixing well with the soil to avoid damage of cutworm.

Planting Distance

In capsicum planting distance and plant population appears to be the most important for getting higher yields. Maximum fruit yield was obtained at a spacing of 45 x 45 cm (Rastogi *et al.*, 1980). A spacing of 60 x 30 cm with 55,000 plant/ha gave highest yield of 12.3 t/ha in cv. California Wonder. Sharma (2001), recorded maximum and economic seed yield in paprika with 60 x30 cm spacing and 100 g N/ha.

Usually seedlings are transplanted in rows about 60 cm apart either in plain beds or on the ridges depending on the type of soil and irrigation facilities available. The plants in the row are kept about 30 cm apart. The seedlings having 4-5 leaves should be transplanted. The nursery bed should be irrigated before lifting of seedlings. Transplanting should be done in evening or during cloudy day followed by irrigation.

NUTRITIONAL REQUIREMENT

Capsicum responds well to application of fertilizers both under irrigations and rainfed condition. The nutrient removal of capsicum cultivar Yolo Wonder showed that for production of 419 q/ha, the N, P, K, Ca and Mg uptake were 132.9, 43.9, 147.8, 124.6 and 19.9 kg/ha, respectively (Grain fenberg *et al.*, 1983). Which indicated high nutrient uptake by the plants. Balance application of both organic and inorganic nutrients were found to increase the growth and yield.

Apply 20-25 tonnes well rotten farm yard manure while preparing the land. Depending on the fertility status of the soil fertilizer requirement vary. However, fertilizer dose of 120:60:120 kg/ha N, P_2O_5 and K_2O should be given for high yield. Sharma (1995) observed the highest yield of capsicum hybrid cv. Pusa Deepti at 240 kg N and 60 kg P_2O_5/ha alongwith common dose of 80 kg K_2O/ha. Sharma (1999), suggested the application of N, P, K @ 150, 75 and 60 kg/ha to get higher yield from improved varieties of capsicum under agroclimatic condition of Himachal Pradesh. Kumar *et al.* (2000) found that application of fertilizer (100: 60:60 kg/ha NPK) + 4 ml/l biozyme sprayed at 50 day after transplanting gave highest yield of capsicum cv. California Wonder. Chauhan *et al.* (2005) got the highest yield of hybrid PRC-2 with 120 kg N and 60 kg P_2O_5/ha. Full dose of phosphorus, half potash and 1/3 nitrogen are to be applied at planting as basal dose. Another 2/3 N and ½ K_2O should be given in two splits i.e., at 25 and 45 days after planting as top dressings. The response of capsicum to different spacings and fertilizers doses are summarized in Table 2.

Use of Micronutrients

The beneficial effects of micronutrients like boron, zinc, magnesium, iron and manganese on vegetative growth, flowering, fruiting and fruit quality of capsicum have been observed by several authors (Kim, 1975; Alla, 1984; Sakr, 1992; Manuel *et al.*, 1999; Tadesse *et al.*, 2002).

(i) Spraying with EDTA, chelates of Cu, Zn, Mn and Fe five times at 15 days interval increase the yield of pepper (ii) The colour of fruits is improve by spraying with B, Cu and zn or Cu and zn. (iii) Treatment of seeds with Zn chloride or Zn sulphate (25, 58, 75, and 100 mg per litre of water) for 20-24 hours before sowing and plant spraying with similar solution at two leaf stage, give one to three time more vegetative growth. (iv) Application of dhanzyme granules @ 8-12 kg/ac in two splits i.e., first at 10-15 days after planting and second 20-30 days later increase fruit yield, fruit quality and resistance to pests, diseases and climatic stress. (v) Application of dhanzyme liquid @ 1 ml/litre of water (180 ml/ac) as a foliar spray, with high volume sprayer at flower initiation and 20 and 40 days later increase fruit yield, quality and resistance to pests diseases and climatic stress. (vi) Foliar spray of plantaid @ 3 ml/litre of water twice at interval of 3-4 weeks to the crop showing specific symptom of micronutrient deficiency, correct the deficiency, ensure healthy growth, better flowering as well as better fruit and seed formation. (vii) Spraying of Mn and Fe each at 0.05 per cent 30, 50 and 70 days after transplanting was found effective in bringing about earliness and higher yield in cvs. California Wonder and Ruby King (Elabdeen and Metwally, 1982). (viii) Rawat and Mathpal (1984) reported that spraying of B, Cu, Fe, Mn or Zn 30 and 45 days after transplanting significantly increased the total sugar content of fruits and all the nutrients were also found effective in increasing the yield. (ix) Application of borax 20 kg and calcium carbonate 20 kg/ha proved best in recording the highest fruit and seed yield of capsicum (Sharma, 1999). (x) It was revealed that foliar application of 1000 ppm boron or zinc increased physical attributes whereas 2000 ppm boron enhanced the quality traits when sprayed thrice at 10 days intervals commencing from 50 days after transplanting (Table 3). However, Mg SO_4 (1000 ppm) increased ascorbic acid content upto 130.87 kg/100 g fresh weight (Pal *et al.*, 2004).

Use of Growth Substances

Foliar spray of planofix (NAA 20 mg/l) followed by celmone (NAA) gave highest fruit yield in capsicum (Sinha and Pal, 1980). Foliar spray of biozyme @ 4 ml/l at 50 days after transplanting enhanced the yield by making the plants photosynthetically more efficient (Kumar *et al.*, 2000).

Use of Biofertilizers

Bhattacharyya *et al.* (2000) reported the beneficial effect of biofertilizers in vegetable production. They suggested the use of Azotobacter and phosphate-solubilizing biofertilizers in different ways.

Azotobacter is free living nitrogen fixing bacteria, fixing nitrogen equivalent to 25-30 kg N/ha. It also produces hormones like, Indole Acetic Acid (IAA) and Gibberellin, vitamins like biotin, folic acid and different B-group. The application of azotobacter supported by judicious use of organic matter, ensures good seed germination, increasing productivity.

Phosphate-solubilizing biofertilizers play a significant role in solubilizing insoluble phosphate. Around 95-99 per cent of the total soil phosphorus are insoluble which are not directly available to plants. The P-solubilizers containing bacteria or fungi may convert insoluble form of phosphate to soluble form by producing organic acids in general. About 15-25 per cent of insoluble phosphate can be solubilized, saving chemical fertilizer significantly. *Bacillus polymyxa, aspergillus awamori, penicillium digitatum* etc., are important such microorganisms, they are used in following ways :

i) *Seed treatment* : About 200 g of biofertilizer is required to treat 10-14 kg of seed. Suspend one packet of 200 g in approximately 400 ml water and mix it thoroughly. Pour this mixture on seeds and mix with hands to obtain uniform coating on each and every seeds. Spread the seeds in shade to dry for 10-15 minutes then sow them immediately.

ii) *Seedling treatment (Root dipping)* : Prepare the suspension by mixing 1 kg of culture in 10-15 litre of water. Get seedling required for 1 acre and make small bundles of seedling in suspension for 15-20 minutes. Transplant these immediately. Generally the ratio of inoculant and water should be 1:10 approximately, i.e., 1 kg packet in 10 litres of water.

Table 2 : Response of capsicum varieties/hybrids to different spacing and fertilizers

Varieties/ Hybrids	Spacing (cm)	Fertilizer dose (kg/ha) N	P_2O_5	K_2O	Yield (q/ha)	Reported by
California Wonder	45 x 45	-	-	-	135	Rastogi *et al* (1980)
Arka Basant	50 x 50	-	-	-	97	Srinivas (1982)
California Wonder	50 x 30	-	-	-	134	Srinivas (1982)
California Wonder	-	150	-	-	111	Srinivas and Prabhakar (1982)
California Wonder	50 x 50	-	-	-	Increased dry matter and leaf area/ plant	Srinivas and Hegde (1984)
California Wonder	-	180	-	-	Increased fruit weight and dry matter and thickness	Hegde (1986)
HS-2	-	160	160	160	182	Markovic (1984)
California Wonder	40 x 20	160	-	-	115	Manchanda and Singh (1988)
California Wonder	-	180	50	-	-	Shukla *et al* . (1987)
Pusa Deepti	60 x 45	240	60	80	752	Sharma (1995)
California Wonder	60 x 45	150	75	60	244	Sharma (1999)
California Wonder	50 x 45	100	60	60	223	Kumar *et al* . (2000)
KT-PL-19	60 x 30	100	-	-	3.4 (seed yield)	Sharma (2001)
PRC-2	50 x 30	120	60	80	262	Chauhan *et al* . (2005)

The vesicular-arbuscular mycorrhizae (VAM) are formed by the non-septate phycomycetes fungi belonging to *Glomus, Gigaspora, Acaulospora* and *Sclerocystis* in the family Endogonaceae of the order Mucorales. They produce vesicles and arburcules inside the root system. These VAM fungi are beneficial to the plant which they colonize. They make more nutrients available to the plant, improve soil texture, water-holding capacity, improve disease resistance and help in better plant growth (Asokan *et al.*, 2000). VAM fungi are inoculated in the nursery -beds before sowing the seed in the beds. The inoculum consisting of spores of fungus mixed with soil and infected root bits of the host plants on which they are raised is either spread uniformly in the nursery bed or placed in rows over which seeds of capsicum is sown. The seedlings are colonized by the fungus 45 days after transplanting. Inoculated plants show better growth and improve nutrient content. Application of 2 kg VAM for a bed size of 90 x 30 x 10 cm give good effects. Addition of inorganic phosphates should be reduced by about 25-50 % depending upon the soil P content (Asokan *et al.*, 2000).

IRRIGATION

The first irrigation should be given just after transplanting. Later, the field should be irrigated as and when required. Generally irrigation is given 7-15 days intervals depending upon weather conditions. More frequent irrigation, however decreased the yield. Therefore, no need of irrigation at the on set of the rainy season. The drainage of excessive water should properly be done to prevent the crop losses from disease complex. Sprinkler irrigation in the field and mist irrigation in plastic greenhouse are effective to reduce the incidence of *Leveillula taurica*.

Economic feasibility of adopting drip irrigation in sweet pepper cv. 'California Wonder' given in Table 4, indicates higher fruit yield and gross returns and water use efficiency. The mulch crop (rice straw mulch at 5 tonnes/ha, application is done 20 days after transplanting) increases gross return to 16-44 per cent in comparison with the crop without mulch. System of irrigations have no much differences. However, the levels of irrigation influence gross return considerably. Irrigation at 0.8 CPE (cumulative pan evaporation) results in maximum gross returns due to higher yield (Narayan, *et al.*, 1994).

Capsicum cv. Arka Gaurav grown in micro-irrigation systems, which produced around 20% more yield (18 to 20 t/ha) than yields under furrow method of irrigation. On sandy clay loam soils of Bangalore, bell pepper gave maximum yield by providing irrigation at 60% soil moisture depletion level in the top 30 cm soil profile. This required 15 irrigations with a consumptive use of 446 mm. Hegde (1989) found that alternate furrow irrigation or irrigation in widely spaced furrows with double planting could save irrigation water to the extent of about 40 per cent without any adverse effect on

Table 3 : Effect of micronutrients on physical properties of capsicum cultivars (Pal *et al.*, 2004).

	Control	$MgSO_4$ (ppm)		Boron (ppm)		$ZnSO_4$ (ppm)		Mean*
		500	1000	1000	2000	1000	2000	
Length (cm)								
California Wonder	5.90^{abc}	5.90^{abc}	5.0^{abc}	5.90^{abc}	5.90^{abc}	5.90^{abc}	5.90^{abc}	5.90
Golden Bell	6.50^{bc}	6.67^{c}	5.57^{abc}	6.20^{abc}	8.10^{e}	5.60^{abc}	5.57^{abc}	6.32
Yolo Wonder	6.80^{cd}	8.00^{dc}	6.43^{abc}	6.83^{abc}	5.30^{ab}	6.20^{abc}	5.20^{a}	6.39
Mean*	6.40	6.86	5.97	6.31	6.43	5.90	5.56	
Base Diameter (cm)								
California Wonder	6.50	6.73	5.17	6.63	5.13	6.57	5.20	5.99^{d}
Golden Bell	7.40	8.83	7.70	7.93	5.97	8.23	6.50	7.51^{f}
Yolo Wonder	6.80	6.50	6.07	7.10	5.93	7.90	6.20	6.64^{e}
Mean*	6.90^{bc}	7.35^{c}	6.31^{ab}	7.22^{c}	5.68^{a}	7.57^{c}	5.97^{a}	
Pericarp thickness (cm)								
California Wonder	0.50^{bi}	0.55^{ij}	0.50^{hi}	0.55^{ij}	0.50^{hi}	0.50^{hi}	0.35^{g}	0.49^{e}
Golden Bell	0.52^{hi}	0.50^{bi}	0.50^{bi}	0.68^{k}	0.50^{hi}	0.50^{hi}	0.48^{h}	0.53^{f}
Yolo Wonder	0.55^{ij}	0.60^{i}	0.55^{ij}	0.55^{ij}	0.50^{hi}	0.55^{ij}	0.50^{hi}	0.54^{f}
Mean*	0.52^{bc}	0.55^{c}	0.52^{bc}	0.59^{d}	0.50^{b}	0.52^{bc}	0.44^{a}	
Average fruit weight (g)								
California Wonder	65.38^{gh}	66.06^{gh}	46.30^{f}	85.57^{jklm}	68.49^{ghi}	77.25^{hijk}	76.48^{ghij}	69.52^{c}
Golden Bell	68.82^{ghi}	101.04^{n}	88.37^{jklmn}	95.37^{mn}	72.34^{ghi}	91.33^{klmn}	58.05^{hg}	82.19^{e}
Yolo Wonder	60.37^{fg}	89.35^{jkimn}	64.27^{gh}	81.08^{ijkl}	75.71^{ghij}	94.531^{mn}	62.67^{g}	75.43^{d}
Mean*	64.86^{a}	85.48^{b}	66.31^{a}	87.71^{b}	72.18^{a}	87.70^{b}	65.73^{a}	

productivity of bell pepper as compared to every furrow irrigation in Bangalore condition. This can be used for medium to heavy textured soils with good movement of water and it will give more profit.

Table 4 : Effect of mulch systems and levels of irrigation on fruit yield, gross returns of sweet pepper and water use efficiency (Narayan *et al.*, 1994).

Treatment	Total green-fruit yield (tonnes/ha)	Gross return (Rs/ha)	Water-use efficiency (kg/ha-cm)
Mulch			
M_0, No mulch	10.82	32,460	277
M_1, Mulch	12.95	38,850	269
CD (P=0.05)	0.26	-	5
System of irrigation			
S_1, Emitter	11.93	35,790	249
S_2, T-tape	12.16	36,480	254
S_3 Furrow	11.55	34,650	242
CD (P = 0.05)	0.23		
Level of irrigation			
I_1, 0.4 CPE	9.61	28,830	263
I_2, 0.6 CPE	12.53	27,690	258
I_3, 0.8 CPE	13.51	40,630	223
CD (P = 0.05)	0.23	-	12

CPE, cumulative pan evaporation

INTERCULTURE

Intercultural operations are done frequently to keep the weed under check and conserve the soil moisture. Usually two to three hoeing are carried out i.e. just after establishment of seedlings, at 25 days stage and at 45 days stage in main field. After each hoeing earthing is given to keep plant stable and strong. In commercially cultivated irrigated crop inter cultivation is given either by junior-hoe or light plough after each irrigation soil compactness.

WEED CONTROL

Two weedings i.e., 30 and 60 days after planting led to the highest yield of sweet pepper. Hoeing and weeding should be done regularly to keep the crop weed free as well as to facilitate better aeration in the root zone. The following weedicides are effective to control weeds in capsicum. Diphenamid @ 4-8 kg/ha, dinitramine at 1 kg/ha and Pendimethalin at 1.25 kg/ha alone or in mixture with EPTC at 3 kg/ha incorporated prior to transplanting gave satisfactory weed control of sweet pepper.

Singh (1989) reported that pre-emergence application of pendimethalin at 1.25 kg/ha and manual weedings at 20, 40 days after transplanting significantly produced more number of fruits, prolonged duration of marketable fruits and more fruit yield. Singh *et al.*, (1992) recorded highest fruit yield when pendimethalin plus oxyfluorfen at (1.00 + 0.15 kg/ha) was applied as pre-emergence (one day before transplanting).

Maximum return (Rs. 18,379) and cost benefit ratio (1.47) was also obtained when pendimethalin + oxyfluorfen (1.00 + 0.15 kg/ha) as per emergence was applied (Singh *et al.*, 1991).

PLANT PROTECTION MEASURES

Insect Pests

Some important insect pests alongwith their management are given below

Thrips, mites, aphids, fruit borer and hairy catterpillers are most common pests of capsicum. Spraying of endosulfan @ 0.12 per cent are effective to control these pests. Spray should be given thrice at 10-15 days intervals.

Pepper weevil (*Anthonomers enginii*)

This is a serious pest in capsicum. The adult of pepper weevil is a black beetle with grey or yellow margins. The larva hatching from eggs laid beneath the surface of the pod feeds within the pods, infested large fruits are likely to be mis-shapen or discoloured. Dusting the crop with folidol dust at every 10 days interval beginning at the time of fruit setting control the pest.

Pepper maggot (*Spilographa electa*)

The larva is about 1.5 cm long when full grown an infest the pods of peppers and some times causes serious losses in some area . Spraying of endosulfan @ 0.12 per cent are effective to control the pest. In fruiting stage malathion 50 EC @ 2 ml/litre of water should spray to check the pest attack.

Cutworm (*Agrostis* sp.)

It is a problem to capsicum in different parts of the world now. It can be control by application of monocrotophos @ 2 ml/litre of water at vegetative stage and nuvan 1.5 ml/litre or dunnet 2.5 ml/litre of water at fruiting stage.

Diseases

Fungal Diseases

Phytophthora leaf blight and fruit rot

Phytopththora capsici Leon has been reported to cause this disease in most of the localities, but in some areas *P. nicotianae* var. nicotianae Waterhouse has also been found responsible for causing the disease. The pathogen survives in infected crop debris or in the soil or is also carried in and on the seed. Warm temperature in combination with high humidity is quite favourable for disease development. The zoospores are splashed by rain and carried by running water to different plant parts and fields.

All the plant parts are attacked by this disease. The first infection begins on green fruits as water soaked, dull green spots that expand rapidly to encompass the entire fruit which later become mummified. Whitish grey mould appear on the rotten fruits under humid conditions. Leaf lesions are dark green, water soaked, round or irregular which expand rapidly. Later on, these lesions become dry and turn light tan. Eventually, infected leaves absciss from the plant. Collar rot and wilt phase is characterized by a dark brown stem discoloration expanding upward from the soil line, accompanied by sudden wilt of the entire plant without foliage yellowing.

With the onset of monsoon, fortnightly sprays with copper fungicides have been found quite effective in minimizing the fruit rot phase of disease (Bhardwaj and Sharma, 1985). Metalaxyl, oxadixyl, propamocarb and dithianon are highly effective in limiting the spread of pathogen. Metalaxyl alone applied as drench has been shown to control root rot more effectively and persistently. Five foliar sprays of mancozeb were found tobe the most effective in controlling fruit rot (Sharma, 1995).

Adoption of cultural measures such as crop rotation, timely roguing of diseased plant parts avoiding contact of fruits and leaves on ground soil, use of disease free seed, mulching with straw, earthing up of plants after transplanting, intercropping with sesame or peanut or onion, ginger and peas (Lee *et al.*, 1991) has been found effective in suppressing the disease.

Trichoderma harzianum, Pseudomonas cepacia, Bacillus polymyxa and *Streptomyces violaceoniger* are effective biocontrol agents against *P. capsici*. Capsicum germplasm such as PI 201232, PI201234, BGH 176, BGH 2678, BGH 3032, BGH 3036 and BGH 3056, CNPH 286, root stock LS 279, sweet pepper cultivar Smith-5, cultivar Phytostop (151-I x P51) etc have been reported to possess resistance to different phases of pathogen infection. Cuiyu (F_1 hybrid) is reported tolerant to blight in China. Lines AF 19141, 1916L, 1947L and SCM-334 showed resistance reaction in Brazil.

Anthracnose or Dieback or Ripe Rot

The disease is caused by *Colletotrichum capsici* (Syd.) and Bisby (Tel: *Glomerella cingulata* (Stonem) Spauld and Schrenk) though number of other *Colleotrichum* spp. such as *C. gloeosporioides* (Peng.) Sacc., *C. acutatun* Simmonds and *C. coccodes* (Waller). Hughes (1988) have been found associated with the disease in different geographical areas. The pathogen over winters in crop debris and seeds of infected fruits. Conidia present in a gelatinous matrix in acervuli on plant debris are disseminated by rain splashes. The optimum temperature for germination of conidia varies from 28 to 30°C with 100 per cent relative humidity.

Ripe rot symptoms first appear on mature fruits as small, water soaked, sunken lesion that rapidly expand, range from dark red to light tan with varying amounts of visible dark acervuli of the fungus. Pale buff to salmon spore masses develop in concentric rings or in a scattered fashion on these lesions. The pathogen also infects healthy growing plants and cause necrosis of twigs from tip background. The lesions are watersoaked grayish white covered with minute dot like fruiting bodies of the fungus.

Chemicals such as captan, carbendazim, benomyl, nabam, zineb, captafal and mancozeb at 10to15 days interval after transplanting have been recommended by various workers to control the disease and increase the yield. Seed dipping in thiram has also been reported to control fruit rot effectively.

Crop rotation with non-host, eradication of weed hosts and cultivation in well drained soils are helpful in preventing the appearance of disease.Varieties/lines such as C1, B7-9, BG1, Loai and Perennial have been found resistant to C. *capsici.Azospirillum* sp. and Azotobacter sp. have been found effective biocontrol agents in the control of dieback of chillies. Garlic and neem extract minimize the ripe fruit rot.

Cercospora Leaf Spot

The fungus Cercospora capsici Heald and Wolf is responsible for causing leaf spot. The fungus subsists on seed and in association with crop residues. The conidia are disseminated by wind, rain or transported on tools and clothes to other plants or field. The fungus sporulate abundantly at 20 to 30°C. The optimum temperature for spore germination is 20 to 30°C.

Small, necrotic lesions develop on the surface of leaves, which later coalesce giving an irregular and blightened appearance. These lesions are typically brown, circular with small light grey centers and brown margins. They may enlarge and center may fall out giving shot holes. When numerous spots appear the entire leaf turn yellow and drop off.

Spraying either carbendazim (0.1%) or captafal (0.3%) 5 to 6 times at fortnightly intervals or triadimefon (0.1%) applied at 50, 65, 80 and 95 days after transplanting has been recommended for the control of this disease (Sharma and Sohi, 1984).

Destruction of crop debris, use of healthy seeds, rotation with non-host crops and good drainage are the general recommendations for checking spread of this disease. Small fruited varieties do possess resistance against the disease.

Powdery Mildew

Powdery mildew is caused by *Leveillula taurica* (Lev.) (Ana: Oidiopsissicula scalia). Cultivated and wild hosts ensure the survival of the fungus between seasons. Low relative humidity i.e. less than 50 per cent cause high disease spread. The optimum and maximum temperatures for *L. taurica* development are 20° and less than 30°C, respectively.

First disease symptoms are noticed on the older leaves which progress to younger leaves. Chlorotic spots becoming necrotic are noticed on upper surface of leaves. The undersurface corresponding to these lesions is covered with white to grey powdery growth of the fungus. With the passage of time, the lower leaf surface also turn necrotic. Heavy infection leads to leaf shedding resulting in heavy losses in yield. These necrotic lesions gradually turn brownish black with the appearance of fungal fructifications.

Fungicides such as wettable sulphur, carbendazim, potassium bicarbonate, triadimefon, tridemorph and triforine have been used to control powdery mildew of capsicum.The recommendations for the control of disease were spraying with dinocap (0.1%) and thiovit (0.5%) or wettable sulphur (0.25%) at 10 to 15 days interval, starting with mildew appearance.

Varieties such as B 15, R 7, Padasali, Tinwari, Vrias de California, Du Danube, Minis 27, Vidra 9 and Golden Superior have been reported as moderately resistance to Litaurica. The C. frutescens line IHR 703 and capsicum cultivar *viz.*, S-32, Pondaicha, K. Red and Local were highly resistant.

Stem or White Rot

The disease is caused by *Sclerotium rolfsii* Sacc. The Sclerotia have a long but variable survival time and germinate eruptivelly to infect the plants directly. High soil moisture and temperature favour disease development, although symptom expression may be more severe during dry conditions following a wet period. The fungus is locally disseminated by irrigation water and intercultural operation tools.

The disease is characterised by sudden wilt of individual plants scattered around the field. Initially, there is foliar discolouration but later the leaves may turn yellow. The cortical tissues at the base of the stem turn brown and decayed above and below the soil line. White mycelial growth usually is visible at the base of the stem and on the soil around the base of the plant. Sclerotia about the size of mustard seed that are tan to brown when mature are produced in the mycelial mat.

The combination of carboxin plus thiram and quintozene have been found most effective in preventing sclerotial formation and mycelial growth. Burning of crop debris, deep summer ploughing, use of mulches and solarization of fallow soil in raised beds has been found to significantly reduce disease incidence in pepper crop. Dual inoculation of *Glomus fasciculatum* and *G. macrocarpum* suppressed the pathogen more effectively. The combined inoculation of *G. macrocarpum* and *Trichoderma harzianum* was most effective in suppressing *S. rolfsii*.

Capsicum chinense viz., line PI 224428, *C. frutescens* cultivars Green leaf Tabasco and Mcllhenny Tabasco, and *C. annuum* cultivars Golden California Wonder and Santaka have shown resistance to southern blight.

Fusarium Wilt

Fusarium oxysporum Schlecht. Ex. Fr. F.sp. capsici has been reported to cause this disease. Initial symptoms are characterised by a slight yellowing of the foliage and wilting of upper leaves followed by permanent wilt of infected plants. Vascular system of plant is discoloured particularly in the lower stem and roots. The disease appears in patches in the field which can be recognized from distance.

High temperature and soil conditions favour disease development and more were in poorly drained soils.The organism grows and sporulate best at 20° to 30°C and 5 to 8 pH. Carbendazim, maneb or carboxin seed treatment has been reported to eradicate seed borne infection of pathogen. The most resistant accessions with less than 20 per cent infection tested under artificial inoculation identified as 412, 403 and 313. *Capsicum baccatum* accession also possess resistance against *Fusarium oxysporum* sp. *capsici*.

Bacterial Diseases

Bacterial leaf spot

Bacterial leaf spot of capsicum is caused by *Xanthomonas vesicatoria* (ed. Doidge). The bacterium persists in crop debris and infested seeds. The bacterium enters through the stomata and injuries and lesions develop more frequently on wounded fruits as compared to unwounded fruits. Wind and wind generated injuries through sand particles have been reported to play a vital role in infection of pepper leaves. The pathogen requires high moisture and temperature for infection. Heavy, continuous and prolonged periods of rainfall favour quick disease development. The optimum temperature lies between 22° to 34°C.

The disease affects leaves, fruits and stems. Initially on the lower surface, lesions are watersoaked, circular or irregular that become necrotic with brown centres and thin chlorotic borders. Generally, these lesions are slightly depressed on the upper surface and raised on the under surface. Severe infection leads to large chlorotic and necrotic areas ultimately resulting into blight symptoms. Such leaves turn yellow and drop down causing defoliation.

If the spots appear on the edge of leaf, they may break away, leaving serrated appearance or twisted leaves. Fruit lesions are 2 to 5 mm in diameter, raised with a roughened and cracked surface. Lesions are first pale green but soon turn brown. On stem, narrow, elongated lesions or streak develop. The centre of stem finally becomes roughened and light brown in colour.

Spraying the plants with a mixture of streptomycin and basic copper sulphate, Agrimycin 100 alone or with copper oxychloride, copper sulphate and mancozeb have been found promising for the control of bacterial leaf spot. Spraying with oxytetracycline or with copper oxychloride, mancozeb, zineb or maneb and Zn ion or maneb plus copper have been found effective in controlling the disease.A three to four year rotation with non-solanaceous crop and use of disease free seed have been found effective in reducing disease. Line 271-4 is resistant to race 1 and race 2. Sweet pepper varieties Medalla-de-Ovo, True Heart 27A, True Heart 2B and Florida XVR 3-25 have shown tolerance or resistance to leaf spot disease. Red pepper cultivars Nilum and Jalapeno are also resistant to this disease.

Bacterial Wilt

The disease is caused by *Ralstonia solanacearum* (Smith). It is soil inhabitant surviving for long periods. Diseased plants debries also help in the overwintering of bacterium. Initial infection comes through the wounds created during uprooting of seedlings from seed beds. Mechanical/insect injury also provides avenues for the entry of wilt organism. Secondary spread of bacterium occurs by irrigation water, infested soil and diseased seedlings. The optimum temperature for the growth of bacterium is 35° to 37°C. High soil moisture favours the disease development. The pathogen grows over a wide range of pH ranging from 6.2 to 7.4.

The disease appears in field as well as in nurseries in scattered pattern or in patches. Initially slight wilting of lower leaves in older plants is noticed. In young seedlings the upper leaves show the wilting symptoms first. With the progress of the disease the initial wilt is succeeded by a sudden and permanent wilt of entire plant, with a slight or no leaf yellowing. Infected young plants die rapidly whereas older plants may first show leaf dropping and discolouration or one side wilting and stunting, before the plant wilt permanently and dies. The vascular tissues of stem and roots turn brown.

Crop rotation with non solanaceous crops, good drainage, liberal use of nitrogen fertilizers, selection of healthy seedlings, avoiding injuries to roots during uprooting are of immense value. Nursery should be raised in disease free areas. The infested soil should be either steamed or treated with formaldehyde before seed sowing to ensure protection from disease at least in nursery beds.

CA 364 was found resistance in Thialand to wilt pathogen.The local chilli cv. Kandhari has been identified resistant under field conditions and cluster fruited plants (CA 219 and CA 33) have showed more resistance. Cultivar Manjeri has been recommended for cultivation in wilt prone areas. Sweet pepper cv. Cholo has been identified as tolerant. Takeda *et al.* (1996) have identified Kaala - a bell type pepper and Waialua - a Jalapeno type pepper resistant to wilt for cooler season and all types of elevations, respectively.

Viral Diseases

Mosaic

Mosaic is induced by more than seven viruses viz., TMV, CMV, PVY and its strains pepper vein banding virus, pepper mottle virus, pepper severe mosaic virus.These viruses mainly belong to three taxonomic groups namely tobamovirus, eneumovirus and potyviruses. All mosaic causing viruses are easily sap transmitted. The members and strains of cucumoviruses and potyviruses are mostly transmitted through aphid vectors viz., *Myzus persicae, Aphis craccivora, A. gossypii, Rhopalosiphum maydis* etc. in a non-persistent manner. Temperature affect both virus multiplication as well as build up of the vector population.

Potato virus Y and its strains generally induce vein-banding followed by necrosis of the veins, petioles and stems of chillies. Pepper vein mottle virus causes mosaic mottling, distortion and filiform of the leaves with bushy appearance of the infected plants. Symptoms developed by pepper mottle virus include systemic mottling and deformities of the infected leaves. The virus also causes systemic necrosis of the leaves resulting in their death. Pepper severe mosaic virus produces mosaic mottle, vein clearing and deformation of infected dark green gein banding, cupping, laminal necrosis of leaves, brown elongated lesions on branches and ring line pattern on fruits. Cucumber mosaic virus causes mosaic mottling, reducing in leaf lamina and disortion of leaves, subsequently the leaves become filiform and plants remain stunted.

The aphid borne viruses are non-persistent in nature, thus can be managed by using practices which affect the vector activities and transmission efficiency. Berries (maize, pearl millet, sunflower, snap bean, marigold, sesame and sorghum) have proved quite useful for the control of PVY and CMV in chillies and bell peppers. Punjab Lal, Gauhati Black and perennial possess resistance against different viruses infecting capsicum. Jawala,Agronomico 9 and Agronomoco 10 have been reported resistant to different strains of several isolates of CMV. Application of insecticides viz., diazinon, metasystox or malathion or endosulphan at 10 to 15 days intervals, starting from early stages of plant growth have been found effective in reducing the disease incidence.

Physiological Disorders

Blossom end rot

It is a disorder of sweet pepper due to Ca deficiency (Walker, 1952), where water-soaked spots first appear on the blossom end of the fruit. The spots soon become light brown and pappery as the lesions dry out. Heavy irrigation after a period of low soil moisture condition and heavy application of nitrogenous fertilizer may cause this disorder. High day time relative humidity lowers this disorder. It can be checked by : (i) Judicious and timely irrigation. (ii) The crop should be grown in the soil having good water holding capacity. (iii) Crop load should be well balanced. (iv) Drip irrigation at night lowers this disorder.

Skin cracking

A disorder of sweet pepper where cracking occurs around the shoulder of the fruits. It is often associated with fluctuations in temperature and humidity. High day temperature and average relative humidity increase the incidence of cracking.

Sunscald

A disorder of sweet pepper, occurs when the fruits are exposed to scorching sunlight (Choudhury, 1983). This disorder arises as a soft, light coloured area in fruit that becomes slightly wrinkled. Later on, the discoloured area becomes slightly sunken and papery. Frits of healthy and vigorous plant having enough foliage cover are less prone to this disorder.

HARVESTING

The large capsicum usually are picked while they are still green in colour but full grown when sold in marlet. Harvesting starts 90-100 days after sowing Fruits are picked with an upward twist which leave a piece of stem attached. Young immature capsicums are rather soft and yield readily to mild pressure of the fingers. Green fruits ready for harvest are relatively firm and crisp. Plants give 4-5 picking and 10-15 fruits/plants. Fruits attain marketable maturity 60-85 days depending upon cultivars.

YIELD

A good crop of capsicum may yield 200-300 quintals per hectare.

POST HARVEST HANDLING

Grading

Grading should be done on basis of uniformity in size, shapes, colour quality and freedom from blemishes. Generally three gradings i.e., Super, fancy and commercial are followed as per Indian standard. In super grade, uniform, well developed size, colour and quality fruit, are arranged. Medium size fruits with uniform shape, developed colour and free from any spots are kept in fancy grade to fetch higher price in the market. In commercial grade, small fruits with dull colour and free from blemishes are arranged separately to sale in the market as a whole. Fruits are graded to get better price and dispose quickly in the market.

Packing

Generally, capsicums are packed in wooden boxes and bamboo baskets in India. Ventilated corrugated fibre board boxes (CFB) are also used for packing for export purposes. In large scale, peppers for market are shipped largely in 1-bushel tub baskets, but other container, such as ½ and 1-bushel hampers, are used to pack sweet papers for local market in U.S. countries.

Storage

Capsicum can be stored in good condition for at least 40 days at 0°C and RH of 95-98 per cent. Shrinkage of the fruits stored under these conditions was only 4% in 40 days. Respiration was found to be lower and carotenoid synthesis higher in darkness than in light.

SEED PRODUCTION

Capsicum annuum are generally self pollinated but considerable outcrossing does occur between and within cultivars of chillies and bell pepper respectively. Natural crossing in bell pepper in different parts of the world has been reported to vary from 7 to 68 per cent (Murthy and Murthy, 1962). The high degree of occasional cross pollination is due not only to the large population of pollinating insects but also to the dehiscence of the anthers upto two or three days after individual flowers open. In a study on chilli, Ahmed (1994) indicated that direction of planting has negligible influence on outcrossing rate but a minimum isolation distance of 600 m and 400 m respectively must be maintained for the production of foundation and certified seed, respectively. Natural cross pollination is mostly carried out by bees (58%), followed by flies (13%), ants (8%), wasps (5%) and others (16%). Bees visit more frequently with maximum activity between 8 am to 12.30 pm.

Climate

Bell peppers are frost sensitive and require warm weather and a long period for good seed production. Mean day temperature of 20-25°C are ideal, plant growth is improved when night temperatures do not exceed 20°C. At low temperatures plants and fruits are susceptible to chilling injury. Flowers are not fertilized at temperatures below 16°C or above 32°C because of poor pollen production. Pollination and fertilization are optimum between 20°C and 25°C. Capsicum produce parthenocarpic fruits or fruits containing very few seeds when the night temperature is relatively low. The effect of low temperature is greater before anthesis than afterwards (Rylski,1973). Seed yield can be considerably less for pollinated flowers under low temperature conditions.

Land Requirement

The land in which crops are grown for seed production must be free of objectionable weeds and free of seeds or plant remains of previous seed crop. Ensure that the land is free of volunteer plants and the soil of selected fields should be well drained and aerated.

Isolation Distance

As already mentioned, bell pepper is an often cross pollinated crop, therefore, the seed fields must be isolated from other cultivars of bell pepper and chilli pepper by 400 metres for foundation seed production and 200 metres for certified seed production.

Brief Cultural Practices

Sowing time

Bell pepper seeds can be produced either in mid hills or in glasshouses elsewhere.The sowing time for northern plain and low hills are July, November-February and March. In the mid hills sowing time is from March-May. However, for raising seed crop March-April is the optimum sowing time because bell peppers require long growth period from sowing to seed harvest. Moreover, lower temperatures may lead to development of parthenocarpic fruits.

Seed

The seed should be obtained from an approved seed certification agency and should be of the required purity level, that is breeder or foundation seed as the case may be. About 500 g seed is sufficient for raising one hectare seed crop.

Spacing

The crop should be spaced 60 cm apart between rows and 45 cm within rows to facilitate easier cultivation operations.

Manures and fertilizers

The fertilizer requirements may vary with soil type, heritability, cultivar, spacing and other factors. However, under Himachal Pradesh conditions, the recommended dosages level are 10 tonnes FYM, 400 kg CAN, 475 kg SSP and 90 kg muriate of potash. Of these, full quantities of FYM, SSP and muriate of potash and half of CAN is added at the time of land preparation. The remaining half of CAN is added in two splits at monthly intervals after transplanting (Anonymous, 1996).

Irrigation

Over irrigation should be avoided especially after fruit maturity to reduce the risk of fruit decay. Crop water requirements in bell pepper usually range from 400 to 1000 millimeters which should be uniformly supplied during growth. Periods of even intermittent moisture stress can significantly reduce plant growth and ultimately seed yield.

Interculture

Crop should be regularly weeded to keep it completely weed free otherwise the yield may be affected significantly.

Roguing

Plants should be rogued at least three times. Before flowering, roguing should be carried out for general plant form and vigour, leaf size, shape and colour. At the flowering and fruiting stage check flower colour, fruit size, shape and colour. When fruits are mature, check fruit size, shape and colour. At all stages check carefully for anthracnose and leaf spot.

Field standards

In the field, a standing seed crop must meet certain standards regarding trueness-to-type and freedom from objectionable weeds, diseased crop plants and other crop plants. The field standards recommended for bell pepper are given in Table 5. The values are related to the seed crop after final roguing.

Table 5 : Recommended field standards for bell pepper.

S. No.	Particulars	Maximum percentage permitted	
		Foundation seed	Certified seed
1.	Off types		
	a) Definite	Nil	0.5
	b) Doubtful	0.3	3.0
2.	Other crop plants	Nil	Nil
3.	Objectionable weeds	Nil	Nil
4.	Crop plants affected by seed borne diseases	Nil	0.05

Table 6 : Seed and fruit yield of capsicum cultivar during summer and winter

Hybrids/ parents	Summer seed yield (g/plant)	1000-seed wt (g)	Winter seed yield (g/plant)	1000-seed weight (g)	Summer fruit yield (kg/plant)	Winter fruit yield (kg/plant)
DARL-2002	6.15	7.38	7.50	9.20	2.200	1.800
DARL-204	6.30	7.56	8.66	9.35	1.950	1.500
CW-51	8.15	4.07	10.10	5.50	1.350	1.110
Yolo Wonder	9.80	4.90	12.25	6.10	1.275	0.975
Arka Basant	11.00	4.75	13.00	6.25	1.000	0.890
CD at 5 %	2.70	0.96	1.18	0.70	0.478	0.637

Source : Ahmed et al., 2005.

Seed standards : The seed standards for bell pepper are given below.

S. No.	Particulars	Percentage	
		Foundation seed	Certified seed
1.	Purs seed (min)	98	98
2.	Other crops (Max.)	Nil	0.1
3.	Other cultivars (max.)	Nil	Nil
4.	Inert matter (max)	2.0	2.0
5.	Weed seeds (max).	Nil	Nil.
6.	Germination (%)	60.0	60.0
7.	Moisture content	7.5	7.5

Harvesting and Yield

Fruits should be harvested only when red and fully ripe (Shrinking stage) to obtain high quality seeds. Fruits are crushed, cut or macerated. Seeds are washed free of pulp and skins. After washing, seeds are dried in the sun to below eight per cent moisture content before storage.

Bell pepper seed yields are lower as compared to chilli and approximately 50-80 kg seed can be obtained per hectare.

The average seed yield was 105 kg/ha with a maximum of 225 kg, in Kulu valley with cultivars California wonder (Singh *et al.*, 1964). It was reported that winter crop gave higher seed yield than the summer crop (Table 6). During winter season the seed yield both qualitatively and quantitatively was excellent (Ahmed *et al.* 2005).

Hybrid Seed Production

All the hybrid seed of bell pepper is produced by emasculation and hand pollination in glass houses in India. Information from United States Seed Growers indicate that the manual procedure is the usual production technique and the primary production areas for F_1 hybrid seed are Taiwan, South Korea and Guatemala.

Emasculation could be avoided if male sterile forms were used. At least three non-allelic male sterile genes are known (Shifriss, 1973). Cytoplasmic male steriles have been found in cayenne and California Wonder (Novak, 1971). Shifriss and Frankel (1971) found a CMS system in two pungent types which carried only restorer genes and for which the sweet types, Yolo Y, Naharia and Vinedale were mixtures for maintainers or restorers. This CMS in certain genetic backgrounds was highly unstable and 28-62% selfing occurred in CMS form. Attempts to use these male sterile forms in hybrid seed production have not been successful enough to establish the method on a commercial scale.

REFERENCES

Ado, S. G. and Samarawira, I.(1988).Estimates of genetic parameters of yield components in pepper (*Capsicum annuum* L.). *East African Agril. and Forestry J.*, 52: 130-140.

Ahmed, Z., Tiwari, H. C., Pandey, V. and Kumar, N. (2005). Fruit and seed yield of capsicum lines and hybrids as influenced by environment. *Vegetable Science*. 32(2): 192-193.

Alla, M.A., Abed, T.A. and Shatshak, N.S. (1984). The response of summer sweet pepper plants to micronutrients foliar spray. *Ann. Agric. Sci. (Moshtohor)* 21(3): 897-912.

Anonymous. (1996). Himachal Pradesh Mein Sabji Utpadan. Directorate of Extn. Edu., UHF, Nauni, 21pp.

Arya, P.S. (1983). *Indian Farming Digest*, 16: 13-16.

Asokan, R., Mohandas, Sukhada and Anand, Lalitha (2000). Biofertilizers and biopesticides for horticultural crops. *Indian Horticulture*, 45(1): 44-47.

Bal, S.S., Singh, J. and Dhanju, K.C. (1995). Genetics of resistance to mosaic and leaf curl viruses in chilli (*Capsicum annuum* L.). *Indian J.Virol.*, 11: 77-79.

Bhardwaj, S.S. and Sharma, S.L. (1985). Chemical control of fruit rot and leaf blight of bell pepper in India. *Agric.Ecosystems Env.* 14: 151-154.

Bhattacharyya, P., Jain, R.K.and Paliwal,M.K. (2000). Biofertilizers for vegetables. *Indian Horticulture*, 45(2): 12-13.

Chauhan, Vijay, Laxmi; Singh, R.V. and Raghav, Manoj (2005). Optimum nitrogen and phosphorus fertilization in hybrid capsicum. *Vegetable Science*, 32(2): 200-202.

Chowla, S.C. and Sharma, P.N. (1990). Management of bell pepper mosaic disease complex in Himachal Pradesh .*Indian Phytopath.*, 43: 349-351.

Deshpande, A.A. (2001). Chilli and Capsicums. In Vegetables, Tuber crops and Spices. (Eds.: S. Thamburaj and Narendra Singh). DIPA-ICAR, New Delhi.

Elabdeen, A.Z. and Metwally, A.M. (1982). *Agril.Res. Rev.*, 60: 143-64.

Graifenberg, A., Petsas, S. and Giustiniani, L. (1983). Informator di Ortiflorofrutticoltura, 24: 5-10.

Greenleaf, W.H. (1985). Pepper Breeding; In Breeding vegetable crops.(Ed: Mark J. Bassett).The AVI Publishing Company, INC, West Port, Connecticut, USA.

Gupta, Meenu. (1997). Studies on ripe fruit rot of chilli caused due to *Colletotrichum capsici* (Syd.) Butler and Bisley. M.Sc. Thesis UHF, Nauni,Solan, pp.58.

Hegde, D.M. (1986). Fruit development in sweet pepper (*Capsicum annuum* L.) in relation to soil moisture and nitrogen fertilization. *Singapur J. Primary Industry*, 14(1): 64-75.

Hegde, D.M. (1989). Effect of method and volume of irrigation on yield and water use of sweet pepper (*Capsicum annuum* L.) *Indian J. Hort.*, 46: 225-229.

Joshi, M.C. and Singh, D.P. (1975). *Indian Hort.*, 20: 19-21.

Kim, C.S. (1975). The effect of fertilizers, growth regulators and air temperature on fruit set in red peppers. *J. Kor. Soc. Horti. Sci.*, 16(2): 167-184.

Kumar, Pramod; Dube, S.D. and Chauhan, V.S. (2000). Photosynthetic response of bell pepper to biozyme in relation to fruit yield. *Vegetable Science*, 27(1): 54-56.

Lee, H.V., Kim, C.H. and Nam, K.W. (1991). Suppression of Phytophthora blight incidence of red pepper by cropping system. *Korean J. Plant Pathol.*, 7: 140-146.

Manchanda, A.K. and Singh, Bhopal (1988). Effect of plant density and nitrogen on growth and fruit yield of bell pepper (*Capsicum annuum* L.). *Indian J. Agron.*, 33(4): 445-447.

Manuel, J.G., Uwe, N.and Steven, J.S. (1999). Paprika (*Capsicum annuum*) Oleoresin extraction with supercritical carbon dioxide. *J. Agric. Food Chem.*, 47: 3558-3564.

Markovic, V. (1984). *Sevremena Poljoprivreda*, 32: 37-45.

Murthy, N.S.R. and Murthy,B.S. (1962). Natural cross pollination in chilli. *Andhra Agric. J.*, 9(3): 161-165.

Naik, L.B., Hebbar, S.S. and Doijode, S.D. (1995). Influence of fruit size on seed content and quality in Capsicum (*Capsicum annuum* L.). *National symposium on recent development in Vegetable Improvements* held at Indira Gandhi krishi Viswa Vidyalaya, Raipur (M.P.) on February 2-5, 1995.

Narayan, T.A., Thimmegowda, S., Bandi, A.G. and Devakumar, N. (1994). Economic feasibility of adopting drip irrigation in sweet pepper (*Capsicum frudescens* var. grossum) .*Indian J. Agron.*, 39(1): 150-1511.

Nath, P., Velayudhan, S. and Singh, D.P. (1987). Vegetables for the tropical region. ICAR, New Delhi.

Novak, F. (1971).Cytoplasmic male sterility in sweet pepper II. Tapetal development in male sterile anthers. *Z. Pflanzenzuecht*, 65: 231-232.

Pal, R.K., Behera, T.K., Sen, Nita and Singh, Manoj (2004). Response of bell pepper (*Capsicum annuum* L.) cultivators to foliar application of magnesium, boron and zinc. *Vegetable Science*, 31(1): 40-44.

Pickersgill, B. (1969). The archeological record of chilli peppers (*Capsicum* spp.) and the sequence of plant domestication in Peru. *Amer. Antiq.*, 34: 53-61.

Purseglove, J.W. (1977). Tropical Crops- Dicotyledons, Longman, Londan.

Rastogi, K.B., Korla, B. N. and Saini, S.S. (1980). Effect of different levels of nitrogen and spacing on fruit yield of bell pepper. *Punjab Hort. J.*, 20: 88-90.

Rawat,P.S.and Mathpal, K.N. (1984). *Sci. and Cult.*, 50: 243-4.

Rylski, I. (1973). Effect of night temperature on shape and size of sweet pepper (*Capsicum annuum L.). J. Amer. Soc. Hort. Sci.*, 98(2): 149-152.

Sakr, M.T. (1992). Physiological aspects of sweet pepper plant affected by certain micronutrients and growth substances. *J. Agric. Sci. (Mansoura Unia).* 17(8): 2958-2963.

Seshadri, V.S. and Chatterjee, S.S. (1996). The history and adaptation of some introduced vegetable crops in India. *Vegetable Science*, 23(2): 114-129.

Sharma, A.K. (1995). Control of fruit rot in off-season crop of capsicum(sweet paper) in Kumaon Hills through safer and less hazardous chemicals. *Vegetable Science*, 22(2): 126-127.

Sharma, S.K. (1995). Response of nitrogen and phosphorus on plant growth and fruit yield in hybrid sweet pepper cultivar Pusa Deepti, *Vegetable Science*, 22(1): 19-21.

Sharma, S.K. (1999). Effect of boron and calcium on seed production of bell-pepper (*Capsicuum annuum* L.). *Vegetable Science* 26(1): 87-88.

Sharma, S.K. (2001). Effect of nitrogen and spacing on seed production of paprika. *Vegetable Science*, 28(2): 175-176.

Sharma, S.R. and Sohi, H.S. (1984). Control of Cercospora leaf spot of chillies with Bavistin. *Pesticides,* 18:21-23.

Sharma, Satish, K. (1999). Bacterial and fungal diseases of Chillies and Bell Pepper.In: Diseases of Horticultural Crops –Vegetables, Ornamentals and Mushrooms. (Eds: L.R. Verma and R.C. Sharma). Indus Publ. Co. New Delhi, p.370-409.

Shifriss, C. (1973). Additional spontaneous male sterile mutants in *Capsicum annuum, Euphytica,* 22: 527-529.

Shifriss, C. and Frankel,R. (1971): New sources of cytoplasmic male sterility in cultivated peppers. *J. Hered.,* 62: 254-256.

Shukla, Vishnu, Srinivas, K. and Prabhakar, B.S. (1987). Response of bell pepper to nitrogen, phosphorus and potash fertilization. *Indian J. Hort.,* 44: 81-84.

Singh, D.N. and Nath, Vishal (2011). Varieties and Hybrids of Vegetables. Satish Serial Publishing House, Azadpur, Delhi (India). 148-153.

Singh, D.P., Anand, N. and Deshpande, A.A. (1993). Improvement of Bell pepper. In advance in Horticulture vol.5-Vegetable crops: part-I. (Eds: K.L. Chadha and G. Kalloo). Malhotra Publishing House, New Delhi.

Singh, H.B., Bhagchandani, P.M. and Thakur, M.R. (1964). Vegetable seed production in Kulu Valley. 5. Summer vegetables. *Indian J. Hort.,* 21(3 & 4) : 221-31.

Singh, N.P. (1989). Effect of herbicides, Spacing and plant nutrients on fruit characters in bell-pepper . Ph.D. thesis 1-177 pages. Dr. Y.S. Parmar UHF., Nauni (HP).

Singh, N.P. , Johri, A.K. and singh, A.K. (1992). Weed management in bell-pepper (*Capsicum annuum* L.). *Indian J. Weed Sci.,* 24(3 & 4) : 56-59.

Singh, N.P.; Johri, A.K. and Singh, K.K. (1991). Economics of herbicides in bell-pepper (*Capsicum annuum* L.) *Veg. Sci.,* 18(2): 130-133.

Singh, R.S. (1985). Diseases of vegetables. Oxford and IBH publishing Co., New Delhi.

Singh, S.P. (2001). Seed Production of Commercial Vegetables. Agrotech publishing Academy , Udaipur - 313002.

Sinha, M.M. and Pal, R.S. (1980). Studies on the effect of plant growth regulators on regulation of growth and yield of *Capsicum annuum* (L.).var. *Bullnose, Prog. Hot.,* 12(2): 35-38.

Somos, Andras. (1984). The paprika, Akademiai kiado – Budapest.

Srinivas, K and Prabhakar, B.S. (1982). Response of capsicum to nitrogen fertilization. *Veg. Sci.,* 9: 71-74.

Srinivas, K. (1982). Response of capsicum cultivars to plant spacing. *Maysore J. Agric. Sci.,* 16: 396-399.

Srinivas, K. and Hegde, D.M. (1984). Growth analysis of capsicum (*Capsicum annuum* L.) cultivars in relation to plant spacing. *Singapur J. Primary Industry,* 12: 152-158.

Tadesse, T., Hewett, E.W., Nicholas, M.A. and Fisher, K.J. (2002). Changes in physicochemical attributes of sweet pepper cv. Domino during fruit growth and development. *Scientia Horti*, 93: 91-103.

Takeda,K.Y., Tanaka, J.S., Sekioka,T.T. and Hamilton, R.D. (1996). Kaala and Waialua peppers. *Hort.Sci.*, 31: 1054.

Thakur, P.C. and Gill, H.S. (1985). Produce quality seed of sweet pepper. *Seeds and Farms*, 11(12): 25-27.

Thomas, T.A., Singh, R. and Prasad, R. (1983). Genetic resources for improvement of vegetable crops. *Indian Hort.*, 31: 59-73.

Thompson, H.C. and Kelly, W.C. (1983). Vegetable crops. Tata Mc. Graw Hill Publishing Co. Ltd., New Delhi.

Valdez, J.A.and Wolfenbarger,D.A. (1995). Yellow traps and insecticides for control of a strain of sweet potato whitefly and associated virus incidence on pepper. *J. Ento. Sci.*, 30: 342-348.

Vijay, O.P., Singh, D.P. and Yadav, I.S. (1979). Studies on flower development, fruit maturity and floral biology in sweet pepper (*Capsicum annuum* L.) *Veg Sci.*, 6: 5-10.

Walker, J.C. (1952). Diseases of Vegetable crops. McGraw Hill Book Co., Inc. New York.

❑❑❑

5. POTATO

Botanical name	:	*Solanum tuberosum* L.
Chromosome number	:	2n = 4x = 48
Family	:	Solanaceae

Potato some times called the European or Irish potato to differentiate it from the sweet potato; is one of the world's great food plants especially in Europe and America. It is cultivated both in larger tracts and in home gardens, providing a cheap and nutritious food. The crop is unique and different from other crops in that the food materials are stored in underground stem parts called tubers.

Most of the cvs of *solanum tubersum* grown in the Andes are tetraploids (2n = 48), but diploid and triploid forms are also cultivated. It is considered that *S. tuberosum* arose either directly from an ancestral form of *S. stenotomum* Juz & Buk. by a process of simple chromosome doubling, or as spontaneous amphidiploid hybrid between the more ancient diploid cultigen *S. stenotomum* and the diploid wild species *S. sparsipilum* Juz. & Buk. Brucher favours *S. vernei* Juz. & Buk. (2n = 24) as the ancestral species.

The wild species of potatos are distributed more or less continously from the South of the United States to southern Chile. The indigenous cultivated potato has a narrower range. The main centre of diversity is in South America in the Andes between 10°N and 20°S at altitudes above 6500 ft with maximum variability on the altiplano around Lake Titicaca in Bolivia.

Many wild species still occur in South America, specifically in the Andes of Peru and Boliva. Two contrasting forms were observed one at chile and the other at Andean region of Peru and Bolivia. The former is botanically called *Solanum tuberosum* L. (*S. tuberosum* var. *tuberosum*) and the later *S.andigensum* (*S. tuberosum* var. *andigenum*).

ORIGIN

Potato is a native of the Andes mountains of Peru and Ecuador in South America and had been cultivated in northern chile and Peru by the Inca Indians before the coming of the white men. They are also native to Colorado (MacGillivray, 1953). Evidences indicate that potatoes were cultivated for centuries by South American Indians and the tubers were used as a common article of food. The Spaniards, during invasion found this economic plant worthy of introduction into Europe. The Spaniards brought the potato from Peru to Spain in 1565. They were probably brought to England about 1586 by

Sir Francis Drake. Potato was introduced in to India in early 17th century only about 40 years after its first introduction in Europe (Pushkarnath, 1964). The first mention of the potato in India occurs in Terry's account of a banquet at Ajmer given by Asaph Jhan to Sir Thomas Roe in 1615. By 1675, it seems to have established itself in the gardens of Surat and Karnatak. In India it was first cultivated on large scale in the Niligiris in 1822. Since then its cultivation has extended to other parts of India (Venkateswarn, 1961). India is fourth largest potato growing country.

DISTRIBUTION

The wild species of potatoes are distributed more or less continuously from the south of the United States to Southern Chile. The indigenous cultivated potato has a narrower range. The main centre of diversity is in South America in the Andes between 10°N to 20°S at altitudes above 65000 ft with maximum variability on the altiplano around lake Tritica basin in Peru Bolivian region in high mountains.

At present potato is cultivated in Russia, Germany, France, the USA, Britain, Spain, Belgium, Netherland, Sweden and Italy. The highest yield (75 t/ha) is obtained in Belgium, the Netherland and Sweden due to favourable climatic condition.

1570 (Chile) (*S. tuberosum*) → Europe → India

↑

South America

↓

Andean region (*S. andigenum*) → Europe → India

Details of the spread of crop are indicated as in 1570 by Spaniards

1570 (Spaniards)

Southern spain → Italy → Australia, Belgium

↓ ↘ Asia

South America 17th Century → China, Japan, India

↓ ↗

England → Ireland, Scotland, Wales, North Europe

The major potato growing countries of the world are China, Russian Federation, Poland, Germany, France, USA, England, Brazil, India, Romania and Netherlands. India ranks fifth in area in the world. Potato is grown in 130 countries of the world. Though it is cultivated throughout India yet Uttar Pradesh, Bihar, West Bengal, Assam and Gujrat are the leading potato producing states. The average yield of potato in India is the highest (282 q/ ha) in Gujarat. Other states like Himachal Pradesh, Karnataka, Maharashtra, Meghalaya, Orissa and Punjab also have increase in area and production as well as yield under potato cultivation.

Economic Importance

Potato (Aalu or Batata) is one of the four major food crops of the world, after rice, wheat and maize. Among tuber vegetable it is the most important commercial cash crop being grown almost all the states except in few states and territories in which it is stayed in kitchen gardens. Potato is not only grown as vegetable but also has become staple food crop in many countries. It can supplement the food needs of the country in a substantial way. It has proved worthy feeding the nations in emergency. It is an important crop for the high population area of Asia because it produces more dry matter food, well balanced protein and more calories from unit area of land and time than other major food crops. It plays an important role in Indian diet. Potato is a nutritious food. It contains practically all the essential dietary constituents like cereals, carbohydrate which are the major constituents of Potato.

The potato supplies about two and a half times more calories as wheat and rice per hectare. It produces the highest yield of energy and protein per hecatre per day. The potato by virtue of its inherent potentialities to solve the tremendous food problems, demands its due place in the agricultural set up. About 50% potato produced in the world is utilized as human foods. It is used as staple diet in many of the countries specially in the West.

BOTANY

Potato belongs to the genus *Solanum, S. tuberosum* L. The genus *Solanum* comprises approximately 2000 species. Of which 150-200 tuber-bearing and belong to the section Tuberarium. The tuber bearing *Solanum* falls into a polyploid series having a basic chromosome number of 12 and containing forms with 2n = 24, 36, 48, 60 and 72. The genomes is the *Solanum* spp. appears to be relatively similar and genome evolution has not progressed far. Pentaploids (2n = 60) are grown in South America. Since they all have similar and common genetic system, they are all included under *S. tuberosum* (Peter, 1998).

Potato is a perennial but as a crop it is treated as an annual. The plant is herbaceous, branched annual 0.3 to 1.0 m high succulent bushy with a mass of fine, fibrous, adventitous root and swollen stem tubers. It may be erect or spreading type. The stem is of a branching type, the (branches arising laterlly along the whole length of the stem). The stem is more or less triangular with prominent wing like growth along its side. A special feature of potato is the presence of underground stems known as stolons. At the ends of stolons are borne the tubers. The tuber is an enlargement of the terminal end of the stolon. The bud end of the tuber, represents the growing point of the original stolon. The tuber is an adaptation to overcome the advise seasonal conditions. According to Pushkarnath (1964), in nature the prolonged dormancy of two to three months helps to preserve the seed tuber during periods of unfavourable for growth. The buds on the tubers, placed as they are within the eyes (often deep-set in the flesh in several varieties), are also protected. All these characteristics make a potato tuber an efficient organ for perpetuation of the species. Potato plants grown from sprouted tuber develop adventitous roots, where as the seedlings grown from true seeds have a tap root in the beginnings which owining to extensive development of lateral roots, turns into a fibrous toor sytem. The roots are generally confined to 20-25 cms top soil. Potato leaf is compound, irregular pinnate. The petiole is small, between two leaflets a leaf like structure is present which is knwon as foliole. The size of leaflet depend on cultivar. Leaves arise along the stem in a spiral arrangement. The potato tuber is a modified stem. The attachment end of the tuber to the stolon is termed as "heel end" and the other "rose end". Spaced on the surface of tubers are eyes, each of the them containing several buds. The flowers are borne on a stalk known as the "floral axis". The potato flower is complete bisexual, hypogynous. Fruit is berry.

Plant breeding

The cultivated Potato *Solanum tuberosum* L. is tetraploid (2n = 4x = 48). Potato is a long day plant for flowering and photoperiod of 12-14 hours and temperature of 15-20°C favours flowering and berry setting (Nath *et al.*, 2008). There are just a few genotypes which may flower under short day conditions of plains. Though both pollen and ovule sterility can occur, pollen sterility ranging from partial to complete absence of pollen grains is common in potato. About one third of the potato cultivars derived from S. *tuberosum* ssp. *tuberosum* do not form berries. Since potato is prorogated vegetatively, it is highly heterozygous for all the characters. Parents for hybridization are selected on the basis of characters to be incorporated and then after hybridization seed is collected. Desired clones are selected from the progeny. Genetically distant crosses with wild species are avoided, as these result in progenies with wild characters which are difficult to eliminate. Assessment is based on unreplicated plots of just a single plant per genotype in the seedling (F_1) generation and a few hills per genotype in first (F_1C_1) and second clonal (F_1C_2) generations. Gopal and Kumar (2003) have suggested following procedure for an early and safe elimination of unproductive genotypes:

- Seedlings of poor vigour may be rejected prior to transplanting in the field.

- Clones with undesirable tuber colour, tuber shape, eye depth and tuber cracking may be rejected from the seedling stage onward.
- No rejection should be done on the basis of tuber yield, average tuber weight or number of tubers in the seedling generation.
- Negative selection (rejection of the poor phenotypes) for tuber yield and tuber weight can be initiated from the first clonal generation onwards. Whereas number of tubers can be considered for the rejection of undesirable types from second clonal generation onward.
- In the subsequent generations, for simultaneous improvement in tuber yield and its components, average tuber weight should be used as selection criteria after fixing a standard for the minimum number of tubers required in the selected types.

All later generations have to be tested in replicated trials at multi locations.

Potato is a vegetatively propagated crop. Clonal selection is effective in maintaining any desirable genotypes from selections made in the segregating progenies followed by hybridization which is between two selected individuals from genetically divergent sources. The breeding strategy followed in potato is through use of dihaploids obtained by pseudogamy in 4x cultivated potatoes. The dihaploids are crossed with diploid species that may be cultivated or wild for transgression of their useful genes like resistance to pests and diseases or for abiotic stresses. The dihaploids are maintained by vegetative propagation.

Dihaploids are generally produced by their parthenogenetic development in 4x X 2x crosses between S. tuberosum X S. phureja (pollinator). Anther culture has been used to produce dihaploid in potatoes. But, due to lack of response of many genotypes to anther culture, its usage is limited. Dihaploid of *tuberosum* cross readily with many diploid wild species and thus provide the opportunity to introgress useful traits from alien sources to cultivated background. The final goal of breeding at the diploid level is to develop 2x clones with specific attributes, good breeding value for agronomic traits and 2n gametes. Production of 2n (unreduced) gametes by diploid-dihaploid hybrids is important to revert back to tetraploid level through 4x X 2x or 2x X 4x (unilateral) or 2x X 2x (bilateral) polyploidization. The natural production of 2n gametes in many tuber bearing *Solanum* sp is well established. This would help in ihe development of good quality tubers. The other techniques worth exploitation in potato breeding programme are: somaclonal variations for creating useful variability, somatic hybridization for over coming barriers to normal crossing, molecular markers assisted selection and genetic transformation for transfer of single, specific genes from any source to the existing varieties (Joseph *et al.*, 2003).

Biotechnological approach in breeding

Development of varieties with specific characteristics such as high protein content, resistant to pests and diseases through genetic engineering is also in progress in different parts of the world. During last decade 277 patent applications on genetically modified potatoes have been filed in the world (Chakrabarti *et al.* 2003). Field tests of transgcnic potatoes have been reported in the United States of America, Canada, Europe, Argentina, Brazil, Egypt, India. Mexico, Peru, Russia, South America, Ukraine. Transgenic potato lines resistant to tuber moth, developed at CPRI, Shimla, are being evaluated under limited field trials. Efforts are also on at CPRI to develop late blight and virus resistant transgenic potato lines. Nutritional quality of potato is also under improvement by transferring a protein gene from Amaranthus.

In potato, the technique for transfer of genes *Agrobacteriun* strain C58-PB1121 has been standardized and production of transgenic plants has been confirmed by the GVS assay. A genetic engineering target to introduce CP gene (for virus resistant) and BT gene (for insect resistant potato) has been set up by CTRI. For developing transgenics, there is a need to change in specific carbohydrates proteins and oils in order to have improved seed quality. Through transgenes, in potato, there were experiments which proved useful in developing resistant transgenes for Colorado potato beetle, two viruses, *Verticillium* (fungi), bruising resistance and allied carbohydrate metabolism (Gopal *et al.,* 2003).

Floral biology

The flowers in potato are borne in cluster on stems. Flowering take place in long day condition. For seed setting it requires 16 hours day length, low temperature and high humidity. In Indian at higher altitude Potato blooms profusely and set seeds.

Flowers of potato are garmopetalous, hermaphrodite and complete. They are highly self pollinated but due to self-incompatibility some cross pollination takes place. Due to absence of pollens most commercial varieties fails to set berries, but on pollination with their own pollen (if the variety is not self sterile) or on cross pollination berries can be readily secured (Pushkarnath, 1964). Fruit is small, globular and purplish yellowish in colour. Fruit contains about 300 seeds. To induce flowering in plains under short day or day neutral conditions GA_3 is used to extend day length requirements. Hybrid TPS has shown superiority over open-pollinated TPS. After location specific cobiners and after inducing flowering, 2-3 pollination are done at 8 hr interval for good fruit set. The TPS can be sotred safety at ordinary room temperature and under low humidity for 2-3 years. Freshly extracted TPS has dormaney for 5-6 months. Dormany can be broken by dropping the TPS in 1500-2000 ppm GA_3 for 48 hrs (Peter, 1998).

COMOSITION

Potatoes contain much greater amounts of water and lower quantities of carbohydrate than the cereals and pulses we generally consume. The major constituents of potato, besides water, are the carbohydrates (starch and sugar) and nitrogen containing compounds (crude proteins).

According to Aykroyd (1941) potato contains 74.7% water, 22.9% sugar, 1.6% protein, 0.1% fat, 0.6% mineral maters (0.01%, calcium, 0.03% phosphorus and 0.007% iron, while according to Mae Gillivray (1953), it contains 77.8% water and food value per 100 g of edible portion is energy 85 calories, protien 20 g, calcium 13 mg, Vit A 40 I.U, ascorbic acid 12 mg, thaimine 0.11 mg, riboflavin 0.06 mg, and niacin 1.18mg, and according to Chatfield (1949 and 1954), raw potatoes also contain in 100g edible portion : Na3 mgm, K 410 mgm, Ca 14 mgm, Mg 27 mgm, Fe 0.8 mgm, Cu 0.16 mgm, P56 mgm., S 29 mgm, and Cl 35 mgm. The potato tuber may contain an alkloid (Solanine) if it is exposed to light. It is poisonous compound but 70% of it is removed when potatoes are peeled. This alklaoid may cause sickness or death both to live stock and the human beings.

Table 1 : Major constituents of Potato tubers.

Water	75.80%
Carbohydrate	16.20%
Crude proteins	2.5-3.0%
Fibre	0.6%
Fat	0.1%
Minerals	1.0%

Table 2 : Essential Amino Acid Content (g/100g protein).

Arginine	4.6
Cystine	0.8
Histidine	1.5
Leucines	10.4
Lysine	6.2
Methionine	2.8
Phenylalanine	4.4
Threonine	4.8
Tryptophan	1.6

Potato is low calory food with good quality protein. The glycoalkaloid, solanine are present in potato. The exposure of tubers to light (the tubers may become the green) and low temerature, the quantitiy of glycoalkaloids increases. If it becomes more than 20 mg /100g, the tubers could taste bitter. About one percen tof the weight of potato tubers consists of mineral matter : calcium, iron, phosphorus and potassium are the important minerals, nutritionally, occur in potatoes. Potato tubers contain vit 'C' and vitamins of 'B'-group. They are considered to be a good source of vitam 'C'.

Table 3 : Minerals in Potato.

Minerals	Content (mg/100g of fresh weight)
Calcium	7.7
Copper	0.15
Iron	0.75
Magnesium	24.2
Phosphorus	40.3
Potassium	568.0
Sodium	6.5

Table 4 : Vitamins in Potato.

Vitamins	Content (per 100 g fresh weight)
Vitamin c (mg)	14.0-25.0
Thiamine (mg)	0.04-0.06
Riboflavin (mg)	0.02-0.04
Nictotinic (mg) acid	1.0-2.3

Potatoes also contains some β-carotene (pro-Vitamin A); Yellow-fleshed varieties may contain upto 100 mg, β-carotene/100g of fresh weight. Most of the one percent of the fibre in potatoes is accounted for by the skin and is removed during peeling. As such potatoes do not contribute any significant amount to the dietary fibre. Potatoes contain very small quantity of fat out it has been found that a major proportion of it consists of unsaturated fatty acids which are nutritionally superior.

Uses

Potato is the most important vegetable in the World today, the tubers being cooked in various ways. In India potato is used only as a vegetable. Sometimes it is also mxed with other vegetables. It is also used for making a chips, halwa, gulabjamun, rasgulla, murabba, kheer, guzia and barfi etc.

Potato in terms of total production of energy for human consumption, is one of the five major food crops in the world, the other are wheat, rice and barley. Potato is utilized in the form of stock-feed, processed potato, seed, waste and as fresh food in major producing countries like, Netherland, Japan, China, Poland, Germany. Potato is a relatively staple part of the diet of European and North American people. It is also used for stock feed. Starch, spirits and industrial alcohol are prepared from it.

According to Nadkarni (1927) potato also has different medicinal uses. It is anti-scorbutic. Persons with neurotic and liver dyspepsia digest it well. It is also employed as an aperient, diuretic and galactagogue, nervous sedative and stimulant in gout. The leaves in the form of extract are employed as an antispamodic in chronic cough producing effects as those of opium etc. Potato ground into a paste is applied as plaster to burns caused by fire with good results. The juice of the potato is a dark coloured liquid due to the presence of citric, and succinic acids. Potatoes may be mealy or non-mealy. Mealy Potatoes are those in which the cells, on cooking, separate from one another so that they may be piled up in a loose flour with air between them instead of remaining in a solid mass with water in the small intercellular spaces. But this is absent in immature potatoes due to low quantity starch.

Mealiness and high starch are associate with high specific gravity. According to Covdon and Boussard (1897), disruption of potatoes in cooking is partly due to turgidity of tissues - the more turgid they are the more easily do they fracture and permit the ready penetration of boiling water. According to mealiness, potatoes are classified into

1. Non -mealy potato if $\frac{\text{Albuminoid}}{\text{starch}} \times 100$ is between 8.5-14.0.
2. Slightly mealy if it is between 6.6-8.4.
3. Mealy if it is between 4.3-6.5.

Nitrogen matter content in potato x $\frac{100}{1}$ is also an index of good quality in potato. (1) If it is between starch 20-25, it is very good. (2) If it is between 17-19, it is good. (3) If it is above 16, it is moderate. (4) If it is below 16, it is poor. Ashby (1947) reported that potatoes of good quality have a higher water content and lower nitrogen content, particularly the protein nitrogen.

VARIETIES AND HYBRIDS

There are many varieties of potatoes either introduced into or bred in India. Varieties of the potato described on the basis of their maturity period viz. early, medium and late maturing. Knowledge of maturity period helps in getting higher prices in the markets, it becomes easy to follow or accomodate variety in intensive cropping systems. Further, varieties of different maturity periods can escape easily from incidence of inset-pests, diseases and disorders upto some extent.

Early maturing varieties
Gola
Up-to-date
Kufri Alankar
Kufri Bahar
Kufri kuber
Great scot
Kufri Chandramukhi
Kufri Jyoti
Kufri Lauvkar
Kufri Navatal
Kufri Ashoka
Kufri Chamatkar
Kufri Khasigaro
Kufri Jawahar

Medium maturing varieties
President
Criag's Defiance
Kufri Neelmani
Kufri Moti
Kufri Anand
Kufri Kanchan
Kufri Himalini
Darjeeling Red Round
Kufri Badshah
Kufri Lalima
Kufri Muthu
Kufri Sheetman
Kufri Sherpa
Kufri Swarna
Kufri Megha
Kufri Pukhraj
Kufri Sutlez
Kufri Giriraj
Kufri Chipsona-1
Kufri Chipsona-2

Late maturing varieties
Kufri Thenmalai
Kufri Red
Kufri Kumar
Kufri Kundan
Kufri Kisan
Kufri Safed
Kufri Neela
Kufri Naveen
Kufri Sindhuri
Kufri Dewa
Phulwa
Kufri Jeevan

The detail characteristics of varieties and hybrids described by Singh and Nath (2011) are given below:

Gola

The commercial 'Gola' is a mixture of three varieties of which 'Gola B' is dominant and has large, pear-shaped, white tubers, with deep-set eyes. The flesh is cream-coloured this is an eary variety.

Up-to-date (Disease-free clone)

It is an early maturing variety (90 days) and is adapted to a wide range of environmental conditions and is particulary suited for Uttar Pradesh, Himachal Pradesh, the Punjab, Madhya Pradesh, Bihar and Maharashtra. Its tubers are of a large size, have a smooth surface, an oval and flattened shape, fllet eyes and white flesh. It is one of the most popular varieties in cultivation in India.

Phulwa

It is probably an old European potato variety which has lost its identity. It has been under cultivation in this country for more than a century. It produces a good yield under a wide range throughout North India. The tubers are small to medium, skin smooth, colour white, shape round, eyes moderate in number, deep flesh, colour yellow. It keeps well. It is a late maturing variety.

Darjeeling Red Round

It is slightly earlier then phulwa. It has also a wide adaptability. The tuber are medium-sized, round with deep red-coloured eyes. It does not keep as well as phulwa.

Kufri Red

This clonal selection from Darjeeling Red Round was first released in 1957. This variety gives excellent yields in the plains and is particularly suited for U.P., Bihar and West Bangal. Its tubers are of medium size, have a smooth and red skin, round shape, yellow flesh adn medium deep eyes. It has a good keeping quality under ordinary conditions.

Kufri Kuber

This variety is suited for both the plains and the hills of Maharashtra, the Punjab, and Bihar. It is also suitable as an early and main crop variety. It yields fairly under long days, but tuberises better under short-day conditions. The tubers are of medium size, have smooth surface, oval, flattened shape, fleet eyes and white flesh and skin.

Kufri Kisan

This is a late maturing variety and gives heavy yields in the plains. It is particularly suitable for Uttar Pradesh, Madhya Pradesh, Delhi and the Punjab. The tubers are of medium size, with a smooth surface, deep eyes, white skin and round shape. It has a very good keeping quality.

Kufri Kundan

It is an alternate variety to up-to-date. This is particularly suitable for hills areas, heavy and wet soils. It performs wet under long-day conditions in the hills. It has tubers of medium size, having a smooth surface, round shape, white skin, fleet eyes and yellow flesh. It has excellent keeping qualitities.

Criag's Defiance

This variety was introduced from the United Kingdom in the year 1953. It is suited for culture in the hill areas of Himachal Pradesh and the Punjab. The tubers of this variety are of a large size, have a smooth surface, long shape, white skin colour, fleet eyes, and white flesh.

Kufri Safed

A late (4.5 to 5.5 months) variety. It produces small, round white tubers with deep eyes and a light, creamy-yellow flesh. A clonal selection of phulwa, it is an excellent substitute for the latter. It is a hardy variety, suitable for a poor soil, with meagre facilities of irrigation. It has an excellent keeping quality. It is recommended for the plains of Uttar Pradesh.

Kufri Kumar

A late maturing (4.4 to 5.5 monts) variety, with large, white oblong tubers. It is good yielder both in the hills and on the plains and possesses a high degree of field resistance to late blight. It has, therefore, been recommended for the hills, particularly where late blight is a problem.

Great Scot

Introduced from England by the Madras Department of Agriculture; It is an early (3 to 3.5 months) variety with oval, white tubers. It is very popular in the Nilgiri Hills because of its ability to give high yields and rapid bulking under conditions prevaling there. It is however, susceptible to late blight, mosaic,a dn leaf-roll viruses and other diseases.

President

Introduced from England, it is a medium maturing (2.3.5 to 4.5 months) variety with round, white tubers. This variety is also grown in the Nilgiri Hills. It is field resistance to blight. A clone of this variety has been selected at the central Potato Research Institute. This clone has given a better performance than the parent variety and has been named as the president-Kufri clone.

Kufri Neela

A late maturing (4.5 to 5.5 months) variety with round, white tubers. As a substitute for Great Scot, it has several virtues. It answers the trade needs and possess Great Scot type of tubers. Besides, it yields about 70 per cent higher than Great Scot. The greatest value of this variety is that it is resistant to late blight.

Kufri Sindhuri (Kufri Red x Kufri Kundan)

Late maturing (3.5 to 4.5 months) with round, light red medium-sized tubers. It is suitable for cultivation as a main crop variety in the plains, as a replacement to Kufri Red and other late varieties. The average yield about 350 to 400 q/ha is obtained in 120 days period. Bigger tubers have tendency to develop hallow heart disorders. This variety is suitable for cultivation in north Indian plains. It is released in 1967.

Kufri Chandramukhi (Sd 4485 x Kufri Kuber)

Early maturing (matures 10 days earlier than Up-to-date), with attractive oval, white tubers. It degenerates slowly and keeps well in storage. It gives high yields (higher than up-to-date) and cooks easily and does not degenerate rapidly in the plains. It is suitable for cultivation in North Indian plains and plateau regions of peninsular India. It is released in 1968.

Kufri Khasi-Garo (Tobrky x Sd. 698-D)

The plants are medium tall and spreading type. Tubers are reound-oval in shape, skin white, eyes deep and flesh pale yellow in colour. It is an early to medium maturing variety, suitable to grow in the hilly areas of Assam, Meghalaya etc. It is moderately resistant to early and late blight but susceptible to charcoal rot, wart and virus disease. It is released in 1968.

Kufri Naveen

A fertilizer-responsible variety for the Khasi Hills region of Assam; possesses a high degree of field resistance to late blight, resistance to wart and moderate resistance to cercospora leaf spot and early blight.

Kufri Chamatkar (Ekishirazu x Phulwa)

This is a medium maturing variety (110-120 days). Tubers are medium, round white eyes medium deep. Flesh colour is pale yellow with uniform - sized, shining and smooth tubers; suitable for the punjab, Haryana, Uttar Pradesh and Madhya Pradesh. It is released in 1968.

Kufri Neelmani

Suited for the Nilgiri Hills in the south. It has given 100 per cent more yields than the variety Great, Scot, which has long been popular in the Niligiri area. This variety can be grown in all the three potato crop seasons in that area.

Kufri Sheetman

A fost-resistar variety, suitable especially for the Punjab, Rajasthan, Haryana and Uttar Pradesh as also for Madhya Pradesh Pradesh and West

Bengal. It can be planted late in the northern plains owing to its high degree of resistance to frost.

Kufri Jyoti {3069 (4) x 2814 a (1)}

The plants are tall, erect and compact. The tubers are large white and oval in shape. The eyes are fleet and flesh is white. A widely adaptable, fertilizer-responsive variety, possessing a high degree of field resistance to late blight disesase in the foliage and also a good degree of tuber resistance to the same disease. This variety is also resistant to the wart and moderately resistant to Cercospora leaf spot. It is recommended for Himachal Pradesh and Kumaon Hills of Uttar Pradesh.

Kufri Alankar (Kennebec x 0. N. 2090)

A very early-tubering (75 days), photo-insensitive variety capable of giving high yields, suitable for cultivation in the plains of northern India. This variety, by virtue of its earliness, can be included in the multiple cropping patterns which are coming into vogue with the introduction of the short-duration, dwarf, high-yielding varieties of rice and wheat. It is relased in 1968. Tuber is oval to oblong and tapering towards heel. Flesh is white. It is suitable in north Indian plains.

Kufri Jeevan (M 109-3 x 698-D)

The plants are tall, erect and medium compact. The tuber are oval-shaped, eyes fleet skin white and flesh colour is pale yellow. A late-maturing, high-yielding variety, possessing a high degree of field resistance to late blight, resistance to the wart and Cercospora leaf spot disease. It is recommended for cultivation in Himachal Pradesh.

Kufri Moti

It is suitable for growing in the Niligiri Hills. The Tubers are oval white and round and takes 110-120 days to mature. It is field immune to race 1 of late blight.

Kufri Lauvkar (serkov x Adina)

The plants are tall erect and medium compact. The leaves are dark green having prominent folioles. It is an early maturing variety (Plateau-70; Plains 100 Days) with round white tubers and light cream and firm flesh. It is suitable for growing the in Deccan Plateau in Maharashtra. It can be grown in borh rabi and kharif seasons. It is released in 1972.

Kufri Dewa (Criags Defiance x Phulwa)

Its plants are tall, erect and open, it is recommended for growing in the Tarai area of Western Uttar Pradesh. The tubers are oval round and white with yellow flesh and takes about 130 days for harvest in the plains and 150-160 days in the hills. It is widely grown in Tarai area of western U.P and Bihar and released in 1972.

Kufri Ashoka ((EM/C-1021 x CP 1468)

Its plants are medium-tall and spreading with glossy-green foliage. White skinned variety, an early maturing (75-80 days after planting). Tubers are oval to oblong and eyes are shallow with attractive appearance and large sized. Its yield is 25-30% higher than Kufri chandramukhi.

Kufri Badshah

Suitable for growing in the plains of Uttar Pradesh. Plants tall, erect, medium compact. Tubers medium to large, oval, skin white, eye shallow, flesh dull white, floury in texture. The variety matures in 90-100 days in the plains and yield 250-500 q/ha. It is tolerant to frost and late blight.

Kufri Bahar (Kufri Red x Gineke)

Suitable for cultivation in plains of U.P., Haryana and Punjab. Plants tall, medium size. Tubers white, medium in size, round to oval with medium deep eye. Flesh white, texture floury. The variety matures in 90-100 days and yield 200-250 q/ha. It is released in 1980.

Kufri Lalima {Kufri Red x AG 14 (Wis x 37)}

Plants tall, erect, compact, vigorous, foliage dark gray-green. Tuber size medium, shape round, skin red, smooth, eyes medium deep, flesh white. It matures in about 90-100 days. The variety is moderately resistant to early blight. The average yield is about 200-300 q/ha. It is released in 1982.

Kufri Sutlej (Kufri Bahar x Kufri Alankar)

This variety, notified in 1996, has been recommended for cultivation in the western central Indo-Gangetic plains. The variety gives high yield (30-35 tonnes/ha) in about 90-100 days and is moderately resistant to late blight. It produces very attractive large, white, oval and shallow-eyed tubers. The tubers are white fleshed and have good dry matter content of about 19%, it has been found quite suitable for production of french fries. Its early tuberizations and bulking character makes it more popular than other varieties and helps in the production of good yield in 60-70 days. In this regard, it is a stiff competitor for Kufri Chandramukhi, Kufri Bahar and Kufri Ashoka, the three early maturing /bulking varieties.

Kufri Pukhraj

This early - bulking, medium maturing variety was relased in 1998, has andigena blood in it. It is very high yielder, producing 35-40 tonnes/ha in about 90-100 days. The variety is field resistant to late blight and produces attractive yellow, oval, shallow-eyed tuber, with light yellow flesh. The tubers have a good keeping quality, with dry matter content about 18%.

Kufri Anand (PJ-376 x 698-D)

The plants grow up to 60-70 cm height, smaller foliage. This is the latest of the releases (1998) and has been recommended for cultivation in Uttar Pradesh by the Uttar Pradesh State Variety Release Committee. It is a medium maturing variety and has higher degree of resistance against late blight and some frost hardiness. It takes about 100-110 days for maturity. The variety is a higher yielder (35-40 tonnes/ha) and produced very attractive, white long oval, flattened, shallow-eyed tubers with white flesh. The tubers have about 19% dry matter and posses good keeping quality. Like Kufri Sutlej, this variety too has been found very suitable for French fries. Though the variety is recommended for Uttar Pradesh, it is finding wide acceptance in surrounding states, viz. Punjab, Haryana, Rajasthan and Gujarat.

Kufri Swarna

The nematode-resistant culture 'SON 73/110' was released for general cultivation in the hills of Tamil Nadu as "Kufri Swarna". It is medium maturing variety. The plants are tall and erect. Foliage dark green in colour. Tubers are medium sized, oval to round in shape. The colour of both skin and flesh is white. The eyes are fleet. Its tuber yield is 30.9 tonnes/ha compared with 23.8 tonnes/ha of "Kufri Jyoti". It is resistant to early and late blight also. It is released in 1985.

Kufri Jawahar (Kufri Neelmani x Kufri Jyoti)

The plants are tall, erect growing, flowers white. Kufri Jawahar is an early maturing variety (80 days). It is recommended for cultivation as a main season variety in Punjab, Haryana, Uttar Pradesh, Gujarat and Karnataka. Its plants are short, erect and compact with light green foliage. Its tubers are oval, medium sized with fleet eyes. This is moderately resistant to late blight. Its tubers have good keeping quality. Due to compact foliage and erect habit this is suitable for intercropping and mechanized cultivation. It is suitable for growing in plateau region as it requires less number of irrigation due to its early maturity. It yields 25-31 tonnes/ha.

Kufri Giriraj (MS/85-45)

Kufri Giriraj released for hills of Himachal Pradesh, however, it performs well even in south and north-eastern hills. It is a medium-maturing (130-135 days) variety producing medium-tall, semi-erect, medium compact and vigorous plants with light purple flowers. Tubers are medium to large, white, oval, with fleet eyes and unlike Kufri Jyoti do not show cracking. It is field resistant to late blight and immune to wart, average yields being 25-30 tonnes/ha.

Kufri Megha

Kufri Megha released for north-eastern hills, however, it could not become so popular among farmers because of its longer maturity period (140-145 days). It produces medium-tall, semi-erect, medium compact plants with white flowers. Tubers are medium, round-oval, white with fleet eyes. It is highly resistant to late blight and yields on an average 22-28 tonnes/ha.

Kufri Himalini

It has maturity period of 110-120 days and also adapted to subtropical plains. It provides a yield advantage of about 10 per cent over Kufri Jyoti and Kufri Giriraj in the plains and its keeping quality is better than all the other cultivars developed so far hills. It has been recommended for cultivation in north-western and north eastern hills of the country during summer season.

Kufri Kanchan

Kufri Kanchan has been released for north Bengal hills and Sikkim, it is a medium-maturing (110-135 days in spring and 90-100 days in autumn) variety, producing medium-tall, semi-erect, medium-compact, vigorous plants with red purple flowers. Tubers are medium-large, oval-oblong and red with fleet eyes. The variety is field resistant to late blight, resistant to wart and slow rate of degeneration with an average yield of about 25-28 tonnes/ha.

Kufri Chipsona-1 (MP/90-83)

This variety is released in 1997 specially for processing (chip making) as having high dry-matter content (23%). This is medium in maturity. Tubers are white skinned, medium, smooth and round-oval with shallow eyes. Keeping quality is good. The variety is resistant to late blight; average yield is 250 q/ha.

Kufri Chipsona-2 (MP/91G)

This variety is also excellent for processing and keeping quality is good. The content of tuber dry matter is high (23%). This variety is resistant to late blight and released in 1997. It is suitable for growing in U.P. and Bihar.

Kufri Chipsona-3 (MP 97/583)

It is a superior chipping hybrid, has high dry matter content (22%) and robust during storage and produces superior quality chips i.e. crispy low oily chips (Anon., 2005).

Kufri Thenmalai (D/79-56)

This hybrid has been released by Tamil Nadu variety release Committee. Suitability has been recorded for cultivation in the part of Tamil Nadu hills. The colour of sprout is red. It is late maturing and high yielding variety. Moderately resistant to early and late blight.

For processing purpose, varieties like white potato, Burbanle, Russet and Katahdin were introduce from U.S.A.

The duel (Table and processing) purpose varieties are Russet, Round White, Long White, Round Red, Yellow Flesh, Blue and purple are the exotic varieties.

BSS296

Tuber shape-round to oval; Skin colour-yellow; Flesh colour-yellowish; Depth of eye-fleet; Maturity-90 days; Disease resistance-field tolerance to late blight; Dry matter-21.9% (approx). It is good for processing.

CLIMATIC INFLUENCE

The potato has a wide range of seasonal adaptability. It is a cool season crop and is moderately tolerant to frost. Temperature is an important factor in the production of potato. The young plants grow best at a temp. of 24°C later growth is favoured at temp of 18°C. Tuber production is maximum at 20°C and decreases with the rise of temperature. At about 30°C the tuber production stops totally. Temperature influences starch-sugar contents of potatoes. Under low temperature (5-6°C), potato starch converting into sugar due to activation of starch enzyme and at normal temperature starch is formed at the expense of sugars. Short days are beneficial for tuber production. The early maturing varieties are favoured by relatively large days, where as late varities are better adapted to short days during the tuber development. In the Indo-Gangetic plains, it is grown under short day conditions during October to March. In the Himalayan hills, it is grown during the long day summer and autumn months of April to September. In Peninsular India, parts of north-eastern India, it is grown, during June-July to March-April. In the temperate hills, only about 10° north of the equatorial line and well fixed in the tropic cancer, potato can be cultivated almost throughout the year (Mandal and Chaterjee, 1993). Thus it is possible to see a crop of potato in every month of the year in one part or other part of India. In Europe, potato is being grown under long day conditions during April to September.

SOIL REQUIREMENT

Though potato can be grown on various types of soils, yet the best yields are obtained from sandy loam or silt loam rich in organic matter. The soil should be well drained, well aerated, deep and having a pH range of 5.2 to 6.4. Alkaline soil create favourable conditions for scab disease. Excellent potato crops can be grown under the river bed system of cultivation.

The following points should be kept in mind before soil selection. (1) duration of the crop, (2) friability of the soil, (3) soil pH and (4) its water holding capacity. The aeration of soil has great effects on tuber set and tuber development. The vegetation on land, which has been last for 3-4 years, is cut, dried and burnt before crop cultivation. In Shilong, Meghalaya, burning of 20 kg dried twigs/4m^2 has given an average tuber yield of 16.66 t/ha compared with 7.25 t/ha without soil burning. Application of N+P at 60 + 52 kg/ha has resulted the yields of 25.83 and 33.27 t/ha on the burnt soil and 14.25 and 12.77 t/ha on unburnt soil respectively. Soil burning has a highly significant residual effect on potatoes in the subsequent season, while fertilizers have no residual effects (Sahota and Virk, 1987).

Increasing soil salinity from 5.0 to 12.5 ds/m causes considerably delay in sprouting, decline in number of stems per hill, height of plants, number of tuber and yield of tuber. There has been no significant reduction in yields of potato genotypes like kufri Chandramukhi, Kufri Sindhuri, P-3349, B S/C-287 and JB-808, upto the salinity level of 5.0 ds/m (Malik *et al.*, 1985). Initial potato tuber yield starts declining at soil salinity of 1.7 ds/m Ece and there is about 12 per cent yield reduction with per unit increase in salinity (Mangal *et al.*, 1990).

Soil Treatment

Soil should be treated with Thimet 10 G granules at the rate of 10 kg per hectare. It should be applied in furrows at the time of planting to avoid early attack of sucking insects, and potato cut worms.

PLANTING

Seed Tuber Preparation

Tubers can be planted whole or cut. It is better to plant whole tubers of size 2.5 cm in diameter with weight 25-50 g. If bigger tubers are used, the same should be cut into pieces having atleast 2 eyes. They should be cut at right angles to the main axis to remove apical dominance. Whole or cut tubers before planting should be allowed to sprout. Cut whole tuber longitudinally from rose end to heel end. Cover cut tuber with wet gunny bag for night to be dry & harden of the cut end, known as suberization.

Seed Treatment

Use Emisan 3g/Bavistin @ 1g, Dithahe M-45 1g and Thiram 2g per litre of water for 30 minutes and thoroughly dried in the shade before planting.

Seed treatment with 3% Borc acid or 0.25%. Emisan-6 for 25 minutes, followed by shade drying before storage.

Propagation

Potato is propagated through tubers that are commonly spoken of as seeds, which have a dormancy of 8-10 weeks. Eyes on surface of tubers have axillary buds. Sprouted tubers when used for planting put up fast and vigorous growth. To increase sprouting, tubers are taken out from cold storage 10-15 days before planting and kept in pre-cooling chamber for 24 hours. It is then spread in cool and shady places to sprout.

Thirty main stems/m^2 is optimum for maximum yield in potato. This is achieved by adjusting size of tubers and spacing. Planting of tubers of 40g size at 60 x 20 cm spacing results in optimum stem number. If tuber size is increased or decreased, intra row spacing is adjusted accordingly.

Seed accounts for nearly 40% cost of potato cultivation. Use of large tubers is more expensive and hence, cut tubers are also used for planting purpose.

Breaking of Dormancy in Seed Potato

Monsoon seed potato procured from hilly area need to be treated with 1% Thiourea + 1 ppm GA (For 20 lit. sol. 200g thiourea + 20mg GA are required) for half hour and allowed to sprout before planting.

Seed Requirement

Seed requirement is depend upon the size of seed tubers to be planted. Details are given below :

Tuber Size	Seed requirement per ha
Large size whole tubers	15-20 quintals
Medium size tubers	10-15 quintals
Small size tubers	8-10 quintals
Cut potato seeds	6-8 quintals

Selection of Tubers

Tuber should be true to the type. Select only medium size from 3.5 to 4.0 cm diameter unless known to be disease free. Small size tubers produce virus infected plants, while large size (more than 4 cm diameter) may not always produce them sufficiently large to offset the extra weight of the seed required. There is always danger of exposure of extra surface. Tough surface appearance may be helpful yet injured or damaged tubers with disease lesions should be rejected. Internal discolouration may be due to environmental or pathological causes. If there is any such suspicion, discard all such tubers for seeds. Weak sprouters are not good for seed. Use the tubers after carefully grading them into large size tubers, normal, medium, and the smallest size tubers (Chauhan, 1972).

Whole vs Cut Seed

When the supply of small potato tubers is inadequate, cut pieces may be used. According to Mac Gillivray (1953), cut pieces of about 30 to 65g give greater yields than 30 to 50g pieces or the smaller ones. Each seed piece should have at least 2 or 3 eyes or buds. As the apical end of the potato usually has the largest number of buds, so it should be cut to make more than one piece. Size of potato tubers also affects the position of the out. The cut surface should be sealed over by suberisation which proceeds most rapidly at 68°F and 90% relative humidity. Cut potatoes are planted only in December-January, otherwise whole potatoes are used for planting. Sprouting of tubers may be checked or delayed by the use of Tuberito, Resposive, Belvitank, Agermine and Sodium molybdate etc. (Singh *et al.*, 1957).

Sprouting of Tubers

After selection of tubers, it should be spread under diffused light in cool place preferably underneath thatched shed for about 6-8 days. The seed tubers should have multiple sprout at the time of planting. Such sprouts are healthy and strong, capable of early emergence and vigorous growth.

When hill potatoes are used as seed, treat them with Gibberellic acid (GA3) at the rate of one ppm for 5 to 10 minute then dry in shade and fill in small gunny bags (of 30-40 kg capacity) and keep the bags in vertical position for 10-12 days for sprout initiation in dark ventilated place. The number of sprouts per tuber varies due to variety, physiological age of tuber, temperature etc. The temperature 10° to 12°C is favourable for sprouting.

Monsoon seed potato procured from hilly area required treatment for breaking of dormancy. Seed tubers should be treated with 1% thiourea + 1 ppm GA (for 20 lit. solution 200 g thiourea + 20 mg GA are required) for half hour and allowed to sprout before planting.

Planting Time

Planting time of potatoes varies from region to region in the following manner.

(1) In Plains

Monsoon crop : July to August (only in some parts of maharastra, Mysore and Orissa).

Early winter crop : Mid September to Mid October (U.P. and Bihar)

Main crop : Mid October to Mid November (North and South)

Seed crop : December-January (U.P. and Punjab)

(2) In Hill area

Early crop : Jan-Feb. Main crop - March-April. Late crop- July-August.

Planting is done in mid March in Mukhteswar. In Nilgiri at Ooty (southern hills), the planting for main crop extends from March to April.

Plateau areas : In the southern plateau at Rajgurunagar near Pune, where winter is mild and very short, the optimum time for autumn planting is the second fortnight of November. For rainy season crop early July. At Ranchi, optimum time of planting was found between 20th October to 5th November. In Karnataka crop cultivation is possible almost throughout the year.

When optimum time of planting is not followed then low yields are resulted in both early and late plantings, prebably associated with fewer tubers per hill and lower LAI, and most the potato varieties doing well when they are planted in the month of October to November (Krishnappa, 1991; Ezekiel and Bhargava, 1992 and Noorudin and Mehta, 1995).

Land Preparation

Prepare a land to fine with tilth which is very helpful for subsequent development of tubers. At the time of land preparation incorporate 25-30 tonnes fym/ha and mix well in the soil three to four week before planting. Hot weather cultivation by ploughing field deep and giving one or two turning of soils in the summer months of April-June in the plains and keeping it fallow helps in checking the problem of soil-borne pathogens and also the perennial weeds. After burying of green manure, the fields are properly levelled and proper provision for good drainage is provided in the fields. Before planting, fields are ploughed with mould board plough or disc harrow followed by one or two tillings with tillers.

Planting Method

Potato is planted by different methods in various parts of the country. However, ridge and furrow method is the most popular method which is widely accepted. The crop is planted manually as well as mechanically.

Manual Planting

Planting of tubers in shallow furrows followed by ridge making

In eastern plains of India, planting is done in shallow furrows drawn with curved narrow-blade spade followed by fertilizer application and covering, spreading and dibbling of seed tuber on the north side of the ridge (Patna method).

Planting of tubers in flat bed followed by ridge making

After lay-out of the field, lines are marked at inter row spacing. Fertilizers are placed on both sides of lines as continous bands. Tubers are placed on these lines and tubers are covered with soil by making ridge.

Double row method of planting

In double row method of planting lines are marked at a distance of 70-80 cm apart. Fertilizers are placed on these lines as a continous bands in rows. Tubers are placed on soil surface on both sides of the band about 10-15 cm apart, followed by ridging.

Planting of tubers behind deshi plough

Field is prepared well after pre-planting irrigation. Cut tubers or small size whole tubers are placed in the furrows by pora manually, leaving the alternate furrow vacant. This is followed by planking and the field is irrigated after full emergence.

Dibbling of tubers in ridges

Dibbling of tubers in ridges are made at definite plant to distance. Tubers are dibbled in the centre of the ridges at a depth of 5-7 cm from the top with khurfi.

Mechanical planting

A tractor drawn 2 or 4 row potato planter is used. All the operations i.e. furrow opening, fertilizer application, tuber placement and ridging are combined together simultaneously. It saves time and reduces soil compaction by tractor wheels compared with the method where these operations are done separately in two trip. Granular fertilizer should be used to avoid blocking of the fertilizer tube. In second device tractor drawn 2 or 4 row marker-cum-fertilizer drill is used for making furrow and placing of fertilizers in the furrows followed by planting with 2 or 4 row potato planter-cum-ridger. In third device furrow opening, fertilizer application, insecticide application are done by tractor drawn implement and tubers are placed in furrows manually followed by ridging with tractor drawn implement and tubers are placed in furrows manually followed by ridging with tractor drawn ridge. In fourth device, ridges are made with the tractor drawn ridges after application of fertilizer. Tubers are dibbled 5-7 cm deep on top of ridges manually.

Care must be taken that tubers, particularly cut seed tubers do not come in direct contact of fertilizers and should remain 4-5 cm above the zone of fertilizer application.

Potato is planted either on ridges or in flat bed. But the planting on ridges given higher yield, though it is a more labourious method than the planting in flat beds. Generally monsoon crops are sown on the ridges to avoid rotting of tubers and staganation of water in the field to prevent from wilting. Before planting the tubers, the land should be divided into strips of 2 to 2.5 metre with channels for irrigation. Inside the strips, furrow should be drawn and tubers should be planted in furrow according to given spacing. The tubers should be covered by soil about 2.5 cm thick after planting. Furrow opening and planting of potato can be done efficiently by tractor operator semi autometic potato planter.

Spacing

The closer spacing causes the individual plant to suffer from competition for water, nutrients and space for growing results in poor growth and finally less yield. On the other hand, the wider spacing causes increase in the size of tubers and yield per plant, but the total yield per hectare may be decreased because of less plant population. Thus both wider and closer plant spacings are not desirable.

Plant spacing appears to have a far greater influence on the tuber size. When plants spaced at 15 cm within a row and 60 cm between rows, higher tuber yield is obtained as compared to 10 cm and 40 cm spacings. However spacing between each row is kept 15-22 cms, while the spacing for ridges may vary from 45-60 to 75 cm row to row, and 15-22 cms from plant to plant, according to the variety and time of sowing. Generally spacing for flat bed method is to be given 45 cm from row to row and 20 cm from seed to seed and for ridge method the spacing should be row to row 60 cm and plant to plant 20 cm. For the production of seed potatoes spacing should be adjusted according to the seed grades (Table 4).

Table 4 : Suitable planting distance in relation to potato seed grades.

Diameter of tuber from longer axis	Planting distance (Row x seed)
2.5-3.5 cm	50x20 cm
2.5-3.5 cm	60x15 cm
3.5-5.0 cm	60x25 cm
5.0-6.0 cm	60x40 cm

With these spacings (Table 4) mechanized cultivation becomes easy and also larger seed tuber (5-6 cm) could also be used with advantage by adopting wider spacing (60x40 cm). Most of the favourable conditions available during bulking period are resulted in big size tubers and higher yield. If available soil moisture level is inadequate due to drought or irregular irrigations; the physiological disorders like hollow heart, tuber cracking, etc., may develop in large tubers. Thus to restrict exceptional increases in the size of tubers, close spacing and optimum moisture supply are good measures. After completion of rapid buking rate, removal of foliage will control formation of large tubers. High proportion of seed tubers can only be expected from high density planting by maintaining row distance of 50 cm (Singh *et al.*, 1990 and Singh 1994).

NUTRITIONAL REQUIREMENT

Potato is a heavy surface feeder being a shallow rooted crop, the potato gets most of its nutrients requirement from surface soil. As a heavy crop is produced within a limited time, this crop has a high nutrients requirement. A yield of 25 tonnes per hectare, of tubers removes 110 kg of notrogen, 50 kg of Phosphate and 225 kg of Potassium. In light soil and places where organic manure is not easily available, green manuring proves beneficial.

It is better to apply fertilizer as per soil test recommendation if the soil test is not done, the fertilizer may be applied @ 120 kg N, 60 kg P2O5 and 120 kg K2O per ha. However, the fertilizer requirements of potato vary according to region as per soil type, soil fertility, crop rotation and season of planting. The requirements of major nutrients depending on soil are given below :

Soil type	N (kg/ha)	P (kg/ha)	K (kg/ha)
Alluvial	180	80	110
Black	115	45	50
Red	120	115	135
Hill soils	115	135	95

Source : Shekhawat, 2005.

Additional application of nitrogen is helpful for the long duration varieties. Application of extra phosphorus helps bulking in short duration varieties. Apply one third of nitrogen, full dose of phophorus and half of potash as basel dressing in furows and mix well with compost and then tubers should be planted. Rest of nitrogen may be applied into two splits during first and second earthing, when plant will be about 30 days and 45 days of sowing respectively. The remaining quantity of potash should be given alongwith second split of nitrogen at the time of second earthing up. In case of early variety, the split should be reduced to two i.e. basal and first earthingup. Calcium ammonium mitrate and ammonium sulphate are better sources of N than urea. Phosphorus increases number of medium sized tubers while potassium enhances size of individual tubers. Potassium sulphate is better than muriate of Potash. Similarly single super phosphate is good sourse of phosphorus. Application of farm yard manure @ 30 t/ha meets the entire requirement of phorphorus and potassium.

Use of Micronutrients

Micronutrients have a very high efficiency value, that is, very small quantities are sufficient to produced optimum yields. On the other hand, even slight deficiency or slight excess can also cause severe damage. Therefore, it is essential that micronutrients should be applied only if their deficiency is known in the soils.

Potato Response to Micronutrients

Potato response to micronutrients application was recorded at a number of places when NPK was applied as a basal dose. However, the magnitide of response varied to a great extent depending upon the nature of the soil, micronutrient availability, climatic conditions and the yield potential. The increase in tuber yield due to zinc, copper, iron, manganese, boron and molybdenum application in different soil type is established. The results indicate that zinc was the most deficient micronutrient in potato growing soils, followed by iron, copper, manganese, boron and molybdenum. In general, the response to these micronutrients except molybdenum was more in plain soils than the hill soils but reverse was true in case of molybdenum. It was obeserved that the micronutrients can be a limiting factor for potato production in many soils and they should be applied along with major nutrients.

Variety is an important factor for potato response to micronutrients. Marked differences in potato varieties in regard to their sensitivity to micronutrients exist. It was found that in a soil deficient in Zn, one variety may suffer from Zn deficiency and develop deficiency symptoms but the other variety may not show any such synptoms and its yield may also not be affected to any marked extent. the implications of differential behaviour of varieties are that certain varieties can be grown successfully on zinc deficient soil without any addition of zinc.

Application of micronutrients also improves the quality of potato tubers. The application of zinc, copper and manganese has positive effect on the starch, ascorbic acid and protein contents of tubers. Molybdenum application increases the protein and ascorbic acid content of the tubers. Application of boron also has positive effect on starch, ascorbic acid and dry matter content of the tubers.

Diagnosis of Micronutrient Deficiency

The indiscriminate application of micronutrients can adversely affect the plants growth and can lead to micronutrient toxicity in the soil. Diagnostic techniques can help in determining the application of micronutrients. Micronutrient deficiencies in potato may be diagnosed by visual symptoms and by plant and soil analysis.

Visual Symptoms

Diagnosis of micronutrients deficiencies in potato plant could be done by visual symptoms of malnutrition which are characterstic of each nutrient and appear on specific leaf or leaves of the plant.

Zinc

Its deficiency symptoms usually appear on the new growth. Deficient plants show severe stunting and bronzing or yellowing of the foilage, usually around the leaf margins, starting from the tips. Youngest leaves are cupped upwards and rolled to such an extent that the terminal growth resembles that of ferns. Leaves of affected plants are smaller and their upper internodes are shorter.

Iron

The iron deficiency symptoms are slight chlorosis in the young leaves. The discoloured tissues become a clear pale yellow and in extreme cases turn almost pure white. The principal veins and the leaves that develop before the symptom appears retain their green colour.

Manganese

Defecient plants are characterised by small upper leaves, with a tendency to roll upward and light green chlorotic areas in the interveinal tissues of the upper leaves. Later, numerous like brown patches develop along with veins and these dead areas increase in both size and number. During mild deficiency, upper parts of the plants become somewhat chlorotic but do not develop dead spots.

Boron

The first boron deficiency symptoms on potato is that growing points die and tips of terminal shoots are stimulated into a characteristic growth of lateral buds. Internodes remain short and give the plant bushy appearance. Leaves thicken and margins, roll upward. The leaf points and margins especially those of the older leaves die prematurely. In case of mild boron deficiency, slight upward curling of the margins of the older leaves is visible.

Copper

In the event of its deficiency, the young leaves are wilted, terminal bud bends to droop when flower buds are developing and drying of leaflet tips occurs in advance stages.

Molybdenum

The symptoms of molybdenum deficiency are marked chlorosis, associated with reduction in growth and yield.

Plant and Soil Analysis

Apparently easy, the diagnosis of micronutrient deficiencies by visual symptoms requires a special skill and constant touch with the deficient plants. Moreover, the deficiency symptoms appear only in extreme cases of deficiency and in general there are hidden hunger signs which can be detected only by chemical analysis of soils and plant tissue. Experiments conducted at Central Potato Research Institute, Simla revealed that zinc supplementation may increase the yield even when no specific visual symptoms are observed. Therefore, soil and plant tissue samples should be got analysed to find the micronutrient deficiency in the crop.

Correction of Micronutrient Deficiency

The micronutrient deficiencies can be corrected by their application through soil, spray or seed treatment. The experiments conducted at various places in India revealed that the application of micronutrients to potato through soil, spray or seed treatments was quite effective. Soil appliction of micronutrient fertilizer is the widely practised corrective method. In case the deficiency is discovered in a standing crop, the most frequent remedy is to spray the foliage with a solution containing the deficient nutrient. The difficulty with soil applied micronutrients is that they are strikingly inefficient because of their fixation by the soils. The low efficiency of micronutrients in soils of high fixation capacity can be improved by applying them through seed treatment which is particularly suitable for potato because of its high seed rate (30 q/ha) and large seed surface. For soil application, the recommended dose of micronutrient is applied in furrows below the seed at the time of planting. In spray application recommended dose is sprayed twice 40 and 60 days after planting in the plains and 60 and 80 days in the hills. In case of seed treatment, seed tubers are soaked in the recommended micronutrient solution for four hours and then dried for 24 hours in the shade before planting them in the field.

Use of Biofertilizers

Biofertilizers are natural fertilizers containing carried-based micro-organisms which help to enhance productivity by biological nitrogen fixation or solubilization of ensoluble phosphate or producing hormones, vitamins and other growth factors required for plant growth. For Potato Azotobacter and Phosphate solubillizers are important biofertilizers.

Azotobacter

Agotobactor is free-living nitrogen-fixing bacteria, fixing notrogen equivalent to 25-30 kg N/ha. It also produces hormones like Indole Acetic

Acid (IAA) and gibberralin. Vitamin like biotin, folic acid and different B-group are also produced. The application of Azotabacter, supported by judicious use of organic matter, ensures good seed germination, increasing productivity.

Phosphate Solubilizing Micro-organism

These biofertilizers play a significant role in solubilizing insoluble phosphate. Around 95-99 per cent of the total soil phosphorus are insoluble which are not directly available to plants. The P-solubilizer containing bacteria or fungi may convert insoluble form of phosphate, to soluble form by producing organic acids in general. About 15-25% of insoluble phosphate can be solublized, saving chemical fertilizers significantly. *Bacillus polymyxa, aspergillus awamori, penicillium digitatum* etc. are important such micro organisms.

Methods of Application

There are four methods i.e. seed treatment, cut piece set treatment, seedling treatment and soil application.

Seed Treatment

About 200 g of biofertilizer is required to treat 10-14 kg of seed. Suspend one pkt of 200 gram in approximately 400 ml water and mix it thoroughly. Pour this mixture on seeds and mixed with hands to obtain uniform coating on each, and every seed. Spread the seeds in shed for drying for 10-15 minutes then so them immediately.

Cutting/set Treatment

Prepare a culture suspension by mixing one kg of culture in 50-60 litre of water. The cut pieces of planting material required for 1 acre kept immersed in the suspension for 10-15 minutes. Than bring out these cut pieces and allow to dry for sometimes before planting. Cut pieces method is suitable for crops like potato.

Seedling Treatment

Prepare the suspension by mixing 1 kg of culture in 10-15 litres of water. Get seedlings required for 1 acre and make small bundles of seedlings in the suspension for 15-20 minutes. Transplant these immediately. Generally the ratio of inoculant and water should be 1:10 approximately, i.e. 1 kg pkt in 10 litre in water.

Soil Application

Prepare the mixture of 2-3 kg of biofertilizers in 40-60 kg of soil/compost. Broadcast the mixture in 1 acre of land either at sowing time or 24 hour before sowing. The application of phosphate-solublizers is very common. Generally 4.0-5.0 kg of N biofertilizer is require for one hectare of potato planting and it can be used on soil/seed treatment.

Use of VAM Fungi

The vesicular - arbuscular mycorrhizae, batter known as VAM fungi, are beneficial soil microorgansm, which form as symbiotic association with fine roots of plants. They colonize root cortical tissues and extend their hyphae into sorrounding soil, forming an important linkage between plants and its below the ground environment. The fungus obtain photosynthates and other growth factors from its host. In turn, it extends and increases the absorbing area of root system enabling plants to more fully tap soils for nutrients and water. Mycorrhizal plants frequently take greater uptake of phosphorus and free metals particularly when those nutrients are less soluble.

The plant materials like seeds, seedlings cutting or micro propagated plants can be inoculated, about 1-2 kg of bulk soil inoculums (5,000-10,000 progagates) in one m^2 of seedbed could be appropriate. For increasing VAM efficiency different horticultural practices should be well coordinated. Soil disinfection by chemical treatment or other means is a pre-requisite for application of mycorrhizal biotechnology, appropriate fertilizer regime must be carefully established with regards to quantities, nutrients composition and application schedule.

IRRIGATION

One irrigation before planting is advantageous for ensuring uniform germination. If pre-irrigation has not been given, then the first irrigation may be given the day after planting. This post planting irrigation should be light. Potato being a shallow rooted crop need irrigation at frequent intervals. Irrigation should be given at 6-10 days interval in sandy loam soil and 10-12 days in heavy soils. The intervals of irrigation is decreased when tubers are formed. The irrigation frequency is depend on the nature of the soil and weather condition. Larger the plant size more will be its water requirement. During winter season (between December-January), when temperature comes down, water requirement of the plant reduces. Scheduling irrigation at 20 mm CPE or 0.25 bar soil moisture tension or at the depletion of soil moisture by 15 to 30 per cent to that of field capacity has been found to be optimum in potato for obtaining high yields. In this way about 350-500 mm of water is needed through furrow method of irrigation. Water stress at tuber initiation has been found to be most detrimental (Nath *et al.*, 2008). In no case the plant should be allowed to wilt. Over irrigation should be avoided. The safe limit is to allow irrigation water in the furrow to the extent of 2/3 of the height of the ridges. Irrigation is withheld before a fortnight of harvest.

INTERCULTURE

It should be done after first irrigation, when 75 per cent plant emergence is seen in the potato field. The gap filling should be done immediately with sprouted tubers to maintain proper plant population. The first hoeing should be done after 20-25 days of planting, when the potato plants are 8-10 cm high then earthing up should be done by taking soil from the spaces in between the line. After 40-45 days, the second earthing is done, taking care that the tubers already formed are not disturbed. Before earthing the field should be lightly hoed for weeding and at the time of earthing the remaining weeds are removed.

Mulching

Mulch helps in conserving soil-moisture, reducing soil-temperature and inducing quick germination. It also suppresses weed growth. Plant material such as paddy straw, wheat husk, maize or jowar stalks or farm refuse acts as good mulch and is applied on ridges. In hilly regions, local available material such as pine needles or leaf litters are quite effective in controlling run off loss and conserving moisture.

WEED CONTROL

The use of weedicides in potato crop, in general is not so essential because earthing operations destroy almost all weeds, if somehow, weeds plants are growing on ridges, they may be pulled out by hands. However, due to non availability and costly labour, weed control becomes difficult manually. Moreover, on a large scale of potato cultivation, earthing operation may be done by tractor driven implements and weeds may be controlled by pre-emergence and post-emergence herbicides. The pre-emergence herbicide is sprayed in potato crop just after planting on ridges as well as in furrows. Any one of the pre-emergence and post-emergence herbicide given in table may be used in solution form at 800-1000 lit/ha. Care should be taken while spraying post-emergence herbicides that they should not come in contact to potato plants. For this, on the nozzle of sprayer, the protecter is fitted at the time of spray of paraquat/propanil and inter-row spraying is done (Singh, 1977 and Singh and Hooda, 1982).

Singh *et al.*, 1982; Singh, 1992; Jaiswal and Lal, 1993 and Singh *et al.*, 1999 reported that following herbicides are effective for control of weeds in potato crops.

Herbicide	Trade name	% active ingredient	Rate by a.i./ha
(i) Pre-emergence herbicides			
Nitrofen	TOK-E-25	25	1.0
Metribuzin	Sencor	70	1.0
Azuron	Tribunil	70	2.0
Linuron	Lorox or Afalan	50	0.5
Alachlor	Lasso	50	2.0
Simazine	Tafazine	50	0.5
(ii) Post-emergence herbicides			
Paraquat	Gramezone	20	0.5
Propanil	Stam-M-34	35	1.0

Crop Rotation / Cropping-system

Being a short duration crop, potato easily accomm dates in intensive cropping system. Farmers can easily adopt several system but potato-moong-okra cropping system give higher net return (Singh and Singh, 1955). Potato can be harvested at any time after 60-70 days after planting till 100-110 days and sold in the market. Its adaptability to wide range of climatic conditions, soil types and wide flexibility in planting times make it deal for including in intensive cropping systems. The crop fits in well with several crops in different cropping sequences (Saini, 1996), which are given below:

1. Potato - Wheat - Maize
2. Potato - Wheat - Paddy
3. Potato - Tomato - Bhindi
4. Potato - Bhindi - Soyabean
5. Potato - Moong - Maize
6. Potato - Bajra - Groundnut
7. Potato - Greengram - Groundnut
8. Potto - Mung - Paddy
9. Potato - Summer paddy - Paddy
10. Potato - Cowpea - Soybean
11. Potato - Pumpkin - Maize
12. Potato - Onion - Paddy
13. Potato - Okra - Cowpea
14. Potato - Onion - Maize

15. Potato - Onion - Maize
15. Potato - Onion - Okra
16. Potato - Cowpea - Paddy
17. Potato - Bittergourd - Amaranth
18. Potato - Amaranth - Cowpea - Radish
19. Potato - Dolichos - Cauliflower

Potato can also be profitably intercropped with several other crops. The net returns from various companion crops depend on the crop involved. For example :

i. Two rows of wheat can be sown between the two ridges of potato.
ii. Potato can also be grown with linseed, mustard i.e. three row of linseed or one row of mustard can be planted in place of every fourth row of potato.
iii. Sugarcane and potato have been found to be highly paying crop combination.
iv. Potato and Fennel and Potato and Onion are profitable companion crops.

PLANT PROTECTION MEASURES

Being rich source of carbohydrates it considered as paradise for pests and diseases, yield losses to the tune of 75-80% have been observed by late blight, bacterial wilt and viral diseases. Therefore, management of all the diseases and pests is very important to harvest a bumper crop, which details are given below :

Insect Pests

Aphids

The important species of aphids associated with potato crops are *Myzus persicae, aphis gossypii* and *A. fabae*. However, *Myzus persicae* is the principal vector of aphid - transmitted potato viruses. Both adults and nymphs suck the sap from the foliage and cause the leaves to curl downwards. If the population is very high, infected leaves turn yellow and die. However, the main damage caused by these insects is by the dissemination of viral disease in the crop. Spray metasystox 25 EC (Methyl demeton) or roger (dimethoate) 30 EC @ 1.2 litre & 800 ml in 125 litre of water for one ha. to control the pest. soon after germination and continue at 15 days interval.

Leaf hoppers

The leaf hoppers are tiny insects having slender bodies. When disturbed they jump often several feet. Several species of leaf hoppers i.e. *Allebroides nigroscutulatus,* Dist. and *Seriana equata* cause damage to potato crop by sucking sap from the leaves. The infected leaves turn yellow in severe cases dies. They are responsible for transmitting the mycoplasmal diseases, viz. Witch broom, purple top roll and marginal flavescence. Soil application of phorate 10 G @ 1.0 kg ai/ha followed by 1-2 spray of Fenitrothion 50 EC @ 500 g ai./ha.

Cut worms

Agrotis ipsilon and *Agrotis segetum* are mainly responsible for causing damage to potato crop. The former is more common in plains and later in hills. Larvae of moths cut the stems or the leaves of potato plant just above the ground and thus affect their growth, vigour and yield. They also feed on tubers by boring into them and affect their market value. They are active only at night. Soil application of 50% steptachlor or 5% aldrin dust @ 45 kg/ha at the time of planting gives effective control. Chloropyriphos 20 EC @ 500 g ai/ ha. is also beneficial to control pest.

Epilachna beetle (*Epilachna ocellata, E. vigintiocto-punctata*)

These beetles are important leaf eating pests of potatoes. Epilachna beetles are small insects of yellowish brown colour. The young grubs of the beetles cause damage by feeding on the foliage.

The grubs and adults of epilachina beetles can be effectively controlled by spraying with sevin 50 W.P. 2 kg/ha or Thiodon 35 EC @ 1.5 litre/ha dilute in about 700 litres of water per/ha.

Potato tuber moth (*Gnorimoschema perculella*)

Infestation of this pests generally starts in the fields, but it causes heavy damage to potato tubers in the country stores in plains. The larva of the moth bores into the tender leaves, petioles, young shoots and exposed tubers. The tuber moth in the potato fields can effectively be controlled by spraying with sevin 50 WP 4 kg per hectare respectively. In the stores, pest can be controlled by keeping the potato covered with thin (2 cm thick) layer of dry leaves of *Lantana* or *Eucalyptus.*

Mites (Hemitarsonemus latus)

The nymph and adult mites feed by sucking the sap from leaves which roll inwards, showing oily spots on upper surfaces and shooty appearance on lower surfaces. The leaves become bronze in colour and weather and the plant become stunted and finally die. Spraying the plant with kelthane 10% (2.0 litres/ha) at an interval of 15 days control the pest.

Wire worms

Wire worms are widely distributed. Some of its species are of great economic importance in certain areas. In some years and some places they can cause a lot of damage. Severely affected tubers are not marketable. Major wire worm species in India are *Drasterius* sp., *Agonichis* sp. and *Lacon* sp. Wire worms are often a problem, when potatoes follow cereal crops or planted in fields taken out of sod, pasture or a grass. Wire worms may bore into potato seed pieces and developing shoots. These feeding sites often become infected with fungi and bacteria, which may result in weak shoots and thin stands. Sometimes damage is so great as to necessitare replanting. Later in the season, wire worms bore into tubers as they are bulking. Feeding holes in tubers may be shallow or deep, and damaged tubers are often misshapen, which lowers their quality and value. Sometimes tubers still contain wire worms when they are lifted; stored potatoes rarely contain them.

Legume crops are good rotational choices in fields prone to wire worm infestation, so long as they are kept weed-free. If potentially damaging populations of wire worms are prevent in a field in which potatoes are to be planted, application of soil insecticides is the only viable management option. Insecticides for wireworm control can be broadcast before planting or they can be sideressed after potato shoots being to emerge.

Diseases

Late blight

The disease was introduced in India from Europe through the infest of potato seed tubers. In india it was first recoreded in Nilgiri hills sometimes between 1870-1880. The disease is caused by *phytophthora infestans*. The disease affects the foliage, stem and tubers of the potato plant. The disease appears on the leaves as brown water-soaked patch which ultimately turn brownish black. Later on black areas develop from the margin of leaves and ultimately whole plants become blighted. The stems get soft and weak. The affected tubers show light brown patches on the surface and rusty brown necrosis of the underneath flesh. To control the disease takeup following measures:

- Plant resistant/tolerant varieties
- Apply mancozeb 75% NP 2.5 kg/ha or apply metoxyl 8% + mancozeb 64% WP combination formulation @ 2.5 kg/ha.

Early blight

It is caused by *Alternaria solani*, in dry weather with intermittent rains, the disease may appear in an epidemic form. The disease first appears on the older leaves of potato plants as necrotic spots of angular, oval or circular shape. It can control in following ways:

- Plant resistant varities like Darjeeling Red Round, Kufri Sindhuri, Kufri Naveen and Kufri Jeevan.
- Spray mancozeb 75% NP @ 2.5 kg/ha or copper oxychloride 50% WP @ 2.5 kg/ha or chlorothalonil 7.5% WP @ 1.25% kg/ha.
- Spraying of bordeaux mixture, at the occurrence of intermittent rains in dry weather, 3-4 times at the interval of 10 days check the disease.

Black scurf

It is caused by *Rhizoctonia solani* Kuhn. The pathogen may attack sprout, stem, tubers, stolans and roots. It causes black scurf on tubers, stem canker, aerial tuber and top rosette. Chocolate brown specks or scurf are produced on the surface of tubers. These are sclerotia of the fungus which are attacked to the skin of tubers by the hyphae of the pathogen and are easily removable. Older plants may get the disease near the ground level, producing brown canker at the collar.

The seed tubers after harvest may be treated with 3% boric acid for 30 minutes and shade dried before storage or treat the tubers in 1% acetic acid + 0.5% zinc sulphate for 15 minutes.

Common scab

As the name indicates common in potato growing areas. It causes economic loss to the growers as infected tubers fetch low price. The disease is caused by *Streptomyces scabies* (Jhaxter) Waksman and Henrie. The affected tubers show more rusecting or deep pitted lesions. Mainly three types of lesions have been commonly observed on potato tubers. These are (a) slight obrsion or rusetting of the skin (b) raised (1-3 mm) and rough corky scab upto 1 mm in diameter and (c) pitted or deep scab upto 12 mm in diameter and 1-3 mm deep. All the three types of lesions may be present on the same tubers. It can be controlled by dipping of infected tubers in 3% boric acid for 30 minutes or application of PNCB @ 30 kg/ha in soil is enough to manage soil-borne infection.

Powdery scab

It is caused by *Spongospora subteranea*. It is common in higher hills and attacks all the underground parts, but the tubers show the worst effect. Formation of small brown spots on young developing tubers, which enlarge and become raised as pimples. Black brown powdery mass is released by breaking skin. The disease can be controlled by use of disease free healthy seeds. Maintain good drainage and field sanitation. Crop rotation with non-solanaceous crop. Seed treatment with agallol-3-or dithane M-45 (0.5% solution) for 10 minutes before planting or keeping in cold storage.

Brown rot (Bacterial wilt)

The disease is caused by *Pseudomonas solanacearum*. Seed tubers and infected soil are sources of inoculum. The bacterium enters the plant through injured roots and rapidly multiplies in the vascular tissues secreting a slime. The bacterial cells and slime fill the conducting tissues and blocking the supply of water to the aerial parts. Hence the plant starts wilting. If the infected tubers or stem is cut transversely and pressed, white cream coloured droplets (bacterial cells + slime) are seem oozing from the vascular region. The brown rot can be controlled by following ways : Use disease free seed and apply stable bleaching powder 12 kg/ha in soil will be adequate for disease control. Application of bleaching powder can be substituted with 2 year rotations with crops like wheat, barley, finger millet, cabbage, cauliflower, knol-khol, carrot, onion, garlic etc.

Wart disease / Black wart

The disease is caused by *Synchytrium endobioticum*. Prominent Warty protuberances resembling cauliflower on a bunches of "cocks combs" like proliferated out growth on tubers are observed. Sometimes small greenish yellow warty growth on the stalks may be observed near the ground level. Buds on the stem (near soil level), stolens and tubers are the centers of infection, roots are not affected. Under high humidity, intermittent drizzling and low temperature 12°C, the disease symptoms are observed on aerial parts and malformation of floral buds, axillary buds, leaves and leafy out growth on leaves. Do not use seeds from affected area. Plants immune varieties Kurfi Sherpa, Kufri Kanchan and Kufri Jyoti. Soil sterilization by steam, mereuric chloride, copper sulphate or 5 per cent formalin is useful to check wart.

Fusarium wilt

This disease is caused by *Fusarium oxysporum* f. sp. tuberosi and *Fusarium solani* f. sp. *cumertii*. In irrigated crop it attacks on seed spieces and the infection then progresses to the underground part of the stem and roots. The yellowing of leaves and wilting is usually unilateral on one side of the plant in few stems. In non-irrigated crop, the pathogen first attacks roots and progresses slowly. *Fusarium solani* attacks the plant mostly through roots. Interveinal chlorsis of young leaves is accompanies by development of small brown spots in these areas on the dorsal surface. It progresses rapidly in dry weather. It can be controlled by the use of disease free seeds, resistant varieties and long rotations.

Soft rot

The disease is caused by *Erwinia carotovora*. It affects the tubers in the fields and also during storage. The general symptom of the disease is softening of tissues. In advanced stage, infection often be accompanied by a watery ooze and an unpleasent smell from the rotting tubers. The infection takes place through wounds, bruises or other diseases spots. The organism cannot penetrate in normal healthy tissues. The disease spreads from decaying tubers to healthy ones. The disease can be controlled by taking sanitary and cultural practices. Harvest tubers at full maturity. Use healthy seed, material. Disinfect all the articles related with storage, with formaline (10%). Treat seed tubers with Agallol-3 @ 0.5% solution for 10 minutes and dry before storage.

Charcoal rot

The disease is caused by fungus *Macrophomina phaseolina* which lives in soil. High soil temerature above 28°C at crop maturity is conducive for the development of the fungus. It causes 20-80 per cent rot of the tubers during storage. The fungus infects the tubers in the soils through proliferated lenticels and injuries. The symptoms appear at the eyes and lenticels which ultimately enlarge into black patches extending deep into the tuber flesh. The infection can be checked by early harvesting of the crop before the soil temperature reaches 28°C. Use disease free, healthy and certified seeds. Frequent irrigation to lower the soil temperature.

Viral Diseases

Potato viruses causing degeneration of seed. Virus infected seed tubers look almost similar to healthy ones but the disease ones give rise to weak and sick plants with symptoms like mosaic, rugusity, crinkling, necrosis, vein banding, rolling etc. in their foilage. The virus infested plants are also prone to many soil borne infections like wilts, blights and suffer greater damage due to insect pests.

There are two types of viruses affecting the potato crop viz., mechanically transmitted and the insect transmitted. The heavy losses in yield of the potato crop are caused due to the attack of the insect vector transmitted viruses. The vector transmitted viruses are potato virus Y (PVY). Potato virus A (PVA), and potato leaf roll virus (PLRV). Mechanically transmitted viruses are potato virus x (PVX) and potato virus S (PVS or PVM). Most potato viruses (except PLRV) cause mesaics. Few important virus diseases are described below:

Leaf roll

It is caused by potato leaf roll virus. It is also called as potato phloem necrosis virus, leaf curl virus. Rolling of younger or top leaves which turn yellowish and some develop red margins. Secondary or proceeding crop infection causes rolling of lower or older leaves. The rolled leaves are thick, leathery and brittle. Plants produce rattling sound when distrubed. The virus is transmitted through infected tubers and aphid vectors mainly *Myzus per sicae* and *aphis gossypii.*

Streak and severe (Rugose) mosaic

It is caused by potato virus Y. It is the major virus of potatoes because it spreads easily. PVY is spread mechanically and also aphid transmitted ones (*Muzus persicae*). Virus Y causes mild to severe (prominent) green - chlerotia mosaic with or without veinal necrosis, blistering rugosity and stunting of infected plants. Primary infection of PVY causes leaf drop streak and secondary infections result in mild to severe mosaic and even chlorsis/death of susceptible varities; 50-70% yield losses can occur due to PVY.

Simple mild mosaic

Virus A and M/S causes mild green to yellowish mosaic. When present singly, they do not induce any marked stunting or distortion of plants.

It can be control by cultivation of virus resistant varities like Kufri Lalima, Kufri Sindhuri, Kufri Jyoti, Kufri Ashoka because they have inbuilt resistance (or tolerance) to PVX and PVY and also to PLRC. Kufri Badshah, Kufri Sutlej, Kufri Jawahar and Kufri Pukhraj also have some degree of virus resistance to mild and severe moisac and suffer less yield losses.

To check the spread of contagious viruses i.e. potato virus X and S/M, frequent movement in the field should be avoided. Vector transmitted viruses can be controlled by spraying insecticides like metasyatox or rogor @ 1.5 ml/ litre of water or tinox (0.1%) at frotnightly intervals. Apply thimet granules to the seed plots at 15-16 kg/ha in the soil. Rogue out the affected plants and cut the haulmn before the aphids multiply and attack.

Purpe top roll

The pathogen is transmitted by the leaf hoppers (*Orosius albicnetus* and *Alebroides nigroscutellatus*). Rolling of basal parts of the top younger leaves usually accompanied by light to intense pink or purple pigmentation takes place. Tubers produced by such plants are smaller and about 6 to 20 per cent of such tubers develope hairy sprouts. Control insect vector by spray of insecticides i.e. rogor or metasystox @ 1.5 ml/liter of water.

Marginal flavescence

The Mycoplasma like organism (MLO) is transmitted through tubers and by the leaf hoppers (*Orosinus albicinctus* and *seriana equata*). The affected plants develop intense chlorosis on the top leaves. Leaf blades become slightly thick, rough, flucherea and plant grown stunted. The disease badly affects the tuber production and the tubers often develop hairy sprouts. Yield losses between 35 to 80 per cent have been recorded in northen hills. The disease can be controlled by use of disease free tubers and control insect vector through spray of Rogor or Metasystox @ 1.5 ml/litre of water.

Witch's broom

The pathogen MLO is transmitted through both tubers and the leaf hoppers, identified as *Alebroides dravidancus*. Plants develop many axillary and basal shoots simulating a bushy broom like appearance bearing small round, simple, chlorotic leaves. Affected plants either do not produce tubers or only very small sized tubers are produced most of which develop hairy sprouts. It can be controlled by use of disease free tubers and insect vector through spray of rogor or metasystox @ 1.5 ml/litre of water.

Potato phyllody

The pathogen is transmitted through tubers. The affected plants develop flattening of stem, hairyness with wavy wings and margins of the entire foliage, rough lreaf lamina and intense chlorotic leaves with enlarged petioles or without petioles. The flower parts turn into leafy structure. It can be controlled by use of disease free tubers.

Potato spindle tuber virus

This is a viroid disease caused by Solanum virus 12 smith, synonymous with potato spindle - tuber virus (PSTV) Schultz & Folsum. The eyes are more numbers and shallower than those of nealthy potatoes. The tuber becomes lighter in colour.

Nematodes

Though as many as 27 different nematodes have been observed around the root zone of potatos, only two, namely cyst forming (*Globodera* species) and root knot (*Meloidogyne* species) are the most serious known nematodes.

Root knot nematodes (Meloidegyne species)

This nematode is prevalent in all parts of the country where potato or any of its host plant grown. The plants infested with root knot nematodes show stunted growth with small and poor foliage and their premature yellowing, shedding and have tendency to wilt during hot period of the days, poor and shallow root system. The roots become swollen and gall formation takes place and give a beaded appearance the tubers possess pimple - like protuberances on its surface.

Control

Follow crop rotation of atleast 3 years with cereal crops like wheat, barley and oats, millet crops like sorghum and bajra. Wherever feasible cropping or inter-cropping with marigold (*Sagetes patula*) improves the crop yield and reduces nematode population. Use nematode free seed. Apply carbofuran 3 G @ 2.6 kg ai/ha as soil application. Apply 1-2 sprays of dicofol 18.5 EC @ 550 g ai/ha against mites.

Cyst forming nematodes

Globodera rostochiensis and *G. Pollido* also known as the golden nematode, has been found to infest potato crop in and around Ootacamund in Nilgiri hills Tamil Nadu. The persistant stage of the nematode found the soil is called cyst. Plants become stunted with dull and unhealthy. Application of aldicarb 10G or carbofuran 3G @ 2-3 kg ai/ha (50% at planting + 50% at the time of earthing up) prove effective against potato cyst nematode and root knot nematode.

HARVESTING

The time of harvest is very important in potato. In case of early crop, harvesting should be done when the prices are remunerative. Generally potatoes are harvested at physiological maturity when leaves start yellowing. The late crop should be harvested before the temperature rises about 20-28°C to avoid tuber rottage from soft rot/charcoal rot. Harvesting is to be done carefully with minimum damage of tubers. It is better to harvest the crop with bullock drawn potato digger or a tractor operated potato digger-cum-elevator. After harvest, tuber should be allowed to dry for 24 hour to get surface dried and kept in heap for 10-15 days in shade for curing of skin. All damaged and rotten tubers should be removed. In order to get better returns the produce should be graded and packed in gunny bags before marketing. The bags should be kept in cool place. Harvesting of immature tubers results in lower yield and more damage to the skin which reduce the keeping quality. The tubers should not be exposed direct sunlight, as otherwise they will become green. Such potatoes are bitter in taste, difficult to cook and are poisonous in nature.

YIELD

The yield varies from variety to variety, however, the average yield of the early maturing varieties is about 200 q/ha. Late varieties yields about 300 q/ha. Yields as high as 500-600 qt/ha but Krishi Pandit has produced more than 1000 q/ha.

POST HARVEST HANDLING

After harvesting potatoes are kept under shade for curing of skin. All damaged and rotten tubers shortedout. For better marketing and to get higher price from produce, it is necessary to put them in proper grading and packing before sending to market for sale.

Grading

Grading makes orderly marketing of crops. This results in higher price to the producer as most of the poorer grades are not marketed. Properly graded potatoes make purchasing easy without any inspections. Potatoes generally graded into four sizes, "A"Grade >100g (above 100g); "B" grade = 50-100g; "C" grade = 25-50g and "D" grade = <25g (below 25g). Grading is also done to obtain seed size tubers. A single grade from 2.5 to 5.5 cm is being made to present by hand grading. While grading, the shape, colour, depth of eyes, etc. of tuber should be crtically examined and off types are discarded.

Packing

Potatoes are pack in gunny bags generally for storing in cold storage and for transportation to different places. For home consumption it kept on floor or in bamboo baskets. Before bagging the tubers with cuts, bruises, cracks or otherwise mechanically damaged, should invariably be removed. After sorting and grading the potatoes should be put in gunny bags.

Storage

Potato being a semi perishable commodity it can not be stored under ordinary conditions particularly in plains and tropics where high temp and dry weather prevail after harvest. Therefore to sustain increased production, proper storage facilities are a must. Sprouting is a serious problem in storage. A pre-harvest spray of maleic hydrazide (500-2500 ppm) checks sprouting of tubers in the storage. Earlier some non-refrigerated storage structure were developed which included, sand pits diffused light storage rooms, thatched mud wall rooms, passive evaporative cool stores etc. These structures are suitable to hold potatoes upto the month of June-July. Obviously they can not be used for storage of seed particularly that of new high yielding varieties. The cold storage of potato is most appropriate for holding them for long period. Some local and indigenous method are quite common for storing of potato. These are (i) Room storage (ii) Pit Storage.

Room Storage

The room should have a high plinth and good ventilation. The doors and ventilators should be fitted with insect proof wire-nets. Potatoes can be stored in single layers on sand. Frequent inspection is necessary to discard tubers showing rotting symptoms.

Pit Storage

In this system the tubers are stored in pits 60-75 cm deep and 2.5 m long and 1.0 metre wide. Pits are made in some cool shady place. Water is sprinkled inside the pits to cool it. After two days neem leaves, dry grass or sugarcane trash is lined all round the pit inside. Bamboo chimneys of 1.5 metre length are placed inside the pit 1 metre apart for facilitating evaporation of the moistrue deposited due to transpiration of stored tubers. Pits are then filled with tubers leaving 15 cm on top followed by a one foot layer of dry grass. A thatch is provided over the pit as protection from rain and sun.

Cold Storage

Temperature and humidity are regulated in such storage. Best temperature and humidity condition for potatoes are 2.2°C to 3.3°C and 75 to 80% relative humidity. Potatoes stored less than 0°C suffer from internal breakdown known as "Black Heart". Potatoes stored over the shelf in small gunny bags. Tubers stored in cold temperature should be kept for 12 to 24 hours at 15°C (in precooling) before been exposed to atmospheric conditions. This ensure proper germination.

Physiological Disorders

Internal Brown Spot

This disorder of minor importance that appears particularly in light sandy soils when not irrigated regularly. It is characterized by irregular, dry-brown spots scattered through the flash of tubers. The market value of tubers is reduced. Craige Defiance is suscaptible against this disorder.

Black Heart

This disorder is caused due to sub-oxidation i.e. lack of oxygen. When potato tubers are stored at high storage temperature (35°C-40°C), the inner tissues break down and become black due to accumulation of CO_2 at the centre of the tuber, which give the appearance of black heart.

Hallow Heart

Formation of an irregular cavity in the centre of tubers without decay and discolouration is the character of this disorder. Closer spacing of plants, removal of foliage when hallow heart starts and avoiding excessive use of fertilizers is the control measures. The quick bulking varieties are more prone to this disorders.

Greening

Exposure of tubers to direct sunlight favours accumulation of solanin, a slightly poisonous green pigment, is termed as greening. Proper earthingup to prevent the exposure of the tubers is common practice to check the greening.

Chilling Injury

Prolong storage of tubers at 0°C temperature cause chilling injury, which results in discoloured blotches in the flesh of tubers, diffused brownish black patches on skin and reduced or completely inhibited sprouting on planting.

Freezing Injury

Ice formation in potato tuber when exposed to temperature below 1°C cause freezing injury, which results in discolouration of the fibrous vascular tissue. Briefly frozen tubers show a blue-black discontinuous ring in the vascular region, whereas severely injured tubers show blue-black network in the pith alongwith necrosis in the vascular region when cut in half. Entirely frozen tubers shows diffused areas of black discolouration, but on thawing the tissue breaks down into slimy, watery masses.

Marketing

Potatoes are semi-perishable and bulky and, therefore, their transport to long distances is a problem. Often potatoes rot during transit because of high temperature at the time of transport. The problem is compounded further due to shortage of transporting wagons. The total cold storage capacity in the contry is adequate only to store nearly 35-40% of total potato production. Their functioning is many times not upto the mark and is marked by various malpractices. The markets in potato producing belts in northern India are also not properly integrated with major consuming markets of metropolitan cities in other regions. Because of these factors marketing of potato is complicated process and potato prices are prone to high fluctuations and often glut situations.

Physiological Changes During Storage

Rapid Periderm Formation

If the tubers are kept at 15-20°C after harvest, periderm formation is rapid. It is affected by the variety, temperature, humidity and maturity of tubers. Usually effective periderm formation is complete within the first two weeks. Immediate storing after harvest at 5°C delays the periderm formation. So far long storage, the tubers should be kept at 15-20°C for the first two weeks.

Starch-Sugar Balance

It is affected by maturity, season, variety and storage conditions. High starch and lower sugar content are desirable. When a tuber is kept at 10°C, it will have a low suger content till the storage of tubers is terminated by sprouting. If the temperature is lowered, more sugar is formed.

Sprouting

Sprouting is often a serious problem in storage of potatoes. It is affected by temperature, humidity, variety and maturity. A preharvest spray of maleic hydrazide controls the sprouting.

SEED PRODUCTION

There are two ways of seed production.

Hill grown : In hills 2500 metre above mean sea level.

Plain grown : In low aphid infestation areas of plain.

Technique of Seed Production

Generally seed plot technique is adopted for potato seed production which details are given below:

Seed Plot Techniques

Potato being a vegetatively propagated crop, the perpetuation of viral, mycroplasmal and other soil borne pathogen brings down the quality of seed stocks and results in poor yield in subsequent multiplication of such seeds. The major vector responsible for spread of virus disease is aphid, which is almost absent during crop season in hills, therefore initially hills were the major source of healthy seeds till 1959.

The potato in northern hills is harvested in September or October, which is also the time of planting of the main crop in north western Indian plains, but being dormant or partially dormant, these cannot be used immediately for seed purpose and hence they are stored in cold for the next crop season as early or main crop. In addition to this the seed tuber produced in the hilly areas are not sufficient for the production of potato in all over country.

Keeping in view problems of hill grown seed and the limited supply, efforts were made to produce healthy seed in plains. Surveys were made by CPRI from 1952 onwards, in important potato growing areas and the report was made that the aphid population remained very low in North-Western and central Indo-Gangetic plains, during October to December. Taking the advantage of very low population coupled with the use of improved cultural practices, it was possible during the year 1968 to produce healthy seed in plain also.

The technique of seed production by growing a crop during low aphid period, using healthy seed, taking precautions of use of insecticides and periodic rouging of diseased plants, and finally dehaulming the crop before the aphid reach the critical level of 20 aphids/100 compounds leaves, was developed by CPRI, Shimla in 1959 which is called as "Seed Plot Technique". Thus quality seed production was possible under this technique in sub-tropical plains by advancing the date of planting from December end to October first week.

The essentials/Characteristic features of seed plot technique :

1. Seed should take from reliable source i.e. disease free seed to stocks.
2. Selection of aphid-free location for tuber production.
3. A minimum isolation distance of 20 m for foundation class and 5 m for certification seed class should be provided.

Roguing and inspections of crop should be done in proper time (thrice i.e. at 25 & 60 days after planting and final roguing just before cutting the foliage).

Good crop management is required for maximum production of seed size tubers are : (a) Timely planting (25th Sept. to 15th October), proper seed rate (30 q/ha), good seed tubers size (4-6 cm) and optimum spacing (60 cm x 20 cm). (b) Raising crop on not to rich soil (Judicious manuring). (c) Use pre-sprouted tubers and spacing relatively close. (d) Ristrict irrigation after tuberization well withhold irrigation (10-15 days before haulam cutting). (e) Haulam cutting should be done in December and harvesting is to be done 15 days after haulm cutting. Other cultural operation are same as mentioned in crop production section.

Plant Protection Measures

The main diseases are black scurf, blights, mosaic and leafroll and the main harmfull insects are aphids, jassids, and cutworms. Soil application of thimet 10G (phorate) 16 kg/ha should be applied before planting to safeguard against cut worm, grube and other soil insect. For control of black scurf treat the seed tubers with 3% boric acid for 30 minutes before sowing. For prevention against blights, mosaic, leafroll disease and aphid, jassids insects apray the crop with dithan M-45 or difoltan (2 kg per ha) + rogor or metasystox 750 ml/ha or dimecorn 190 ml/ha from first week of November at 15 days interval.

Haulms Killing

All haulms must be cut at close to the ground as possible. Four litres of gramexon (parquet) in 1000 litres of water may be sprayed on the crop for haulms killing. Do not allow any regrowth to come up Regrowth may be checked by spraying of 1% copper sulphate solution in the field. This practice allows skin hardening and virus transmission through aphids is avoided.

Harvesting

After haulms killing, tuber should be left underground for 10-15 days for skin hardening. Digging of seed potatoes should preferably be done by spade or by specially devised plough to avoid injury to the tubers at the time of digging. Do not expose tubers to direct sun.

Curing and Grading

After harvesting, the tubers should be kept in sheds for about a week or so at roomtemperature so that their skin may become thick. Grading is done to obtain seed size tubers. A single grade from 2.5 to 5.5 cm is being made at present by hand grading. While grading, the shape, colour, depth of eyes etc. of tubers should be critically examined and off types are discarded.

Packing

The tubers with cuts, bruises, cracks or otherwise mechanically, dameged, or showing visual symptoms of late blight, dry rot, wet rot, scab, blackscurf, etc. should invariably be removed. Late blight affected tubers can be recognized by sunken purple spots on the surface and the flesh has brown necrosis. In dry rot affected tubers, brown patches are observed on the surface of the tubers. In charcoal rot affected tubers, black areas initially appear around eyes and lenticels. On cutting the tubrs, the internal tissues show black patches beneath the affected areas. Black scurf affected tubers have chocolate coloured sclerotial bodies are formed on the tubers. After sorting and grading the seed potatoes should be put in gunny bags (40 kg size) and the bags appropriately labelled.

Yield

About 300 to 350 q/ha of seed tu‹ ers are obtained under good crop management condition. However after s‹ rting and grading the yield of seed size potato varies from 125-175 q/ha.

Storage

Soon after packing the seed potato should be shifted in cold storage. Best temperature and humidity conditions for potato storage are 4-6°C temperature and 85-90 per cent relative humidity.

Field Standards

The crop should meet the following standards

Factor		Maximum permissible limits (%)		
	Inspection	Foundation I	Foundation II	Certified
Off types	I and II	0.05	0.05	0.10
Mild Mosaic	- do -	1.00	2.00	3.00
Severe Mosaic, Leaf Roll and yellows	- do -	0.50	0.75	1.00
Total virus	- do -	1.00	2.00	3.00
Brown rot	- do -	None	3 Plants/ha	10 Plants/ha

Seed Standards

The maximum tolerance limits of tubers showing visible symptoms caused by diseases are as follows :

Late blight, dry rot or charcoal rot : not more than 1.00% by number.

Soft rot : nil

Common scab incidence from single tuber in a lot upto 5% is acceptable.

Black scurf incidence upto 5% (with 10% scurfed area) are in permissible limits and beyond this the tubers should be treated with glacial acetic acid 1.0% + Zinc Sulphate 0.05% for 15 minutes or 3% boric acid.

Root knot nematode infected tubers : nil

Total number of diseased tubers should not be more than 5% in any case.

Seed size grade specifications for hill and plains

Hill Seed	Medium = 40 to 60 mm Large = above 60 mm
Plains Seed	Small = 25 to 35 mm Medium = above 35 to 45 mm Large = above 45 to 65 mm

True Potato Seed (TPS) Technology

The TPS technology is newly developed in this country, which can bridge the gap of inadequate supply of healthy seed tubers at an affordable cost. TPS is the matured ovule developed in berry of the potato plant as a result of sexual reproduction/fertilization. This concept was first realized to raise commercial crop in India by Dr. S. Ramanujam, the first Director of CPRI in early fifties then at Patna. The advantages conceived by him were : (i) The TPS can be produced in all potato growing regions. (ii) Freedom of TPS from viral diseases, which is common in seed tuber crops. (iii) Requirement of TPS in small quantity and diversion of tubers meant for seeds towards human consumption. (iv) Transport and storage cost is lesser than seed tuber. (v) Elimination of storage losses in seed tubers, and (vi) TPS provide a low cost potato production technology.

A regular TPS research programme was started at CPRI, Shimla in 1976 and the researches during 1977-1984 yielded superior TPS populations. Later on research work on TPS has been jointly conducted by the CPRI, the International Potato Centre and the All India Coordinated Potato Improvement Project resulting in development and identifications of high yielding TPS populations, viz., TPS C3 and TPS C17 from Patna and HPS 1 / 13, HPS2/67 and 1 HPS7/67 from Modipuram.

In case of crop raising through TPS, there are two methods in use viz., (i) using seedlings as planting material and (ii) using seedling tubers as planting material.

The TPS technology involves two major steps via, (a) Production of hybrid seeds, and (b) Its use as planting material for raising the commercial crop.

Production of Hybrid TPS

The hybrid TPS can be produced as summer crop in the Northern hills. In the plains where crop is grown in winters extra light of 4-5 hrs at the end of day by 150 W sodium vapour lamp has to be provided to induce flowering in the crop. For good flowering and seed setting at moderate temperature are desirable.

Prepare, the field as for ware potato crop by adding FYM 20t/ha in the soil. Male and female lines are planted in two separate but adjacent plots in the ratio of about 1:4. Male block is planted about a week before planting female block at a spacing of 60 x 20 cm. In the female block, prepare beds of three rows each at a spacing of 40-50 x 15 cm having 80 cm walking space between two adjacent beds. Use about 30 g size seed tubers or the seed pieces for planting female block. After germination, only single stem per plant is maintained in the female block. Follow all other cultural and plant protection practices as for ware potato crop.

When the potential material comes to flowering, trim the flower bunches in the female plants to retain only 6-8 large size buds/bunch. Old flowers and berries if any should be removed from the bunches prepared for pollination next day. Freshly opened flowers/large sized buds which would open next day are collected from male parent in the evening preceding the day of pollination and they are spread on a sheet of paper on a table. Next morning pollen are extracted in a small dish by shaking anthers a motorized mini-mixer, electric buzzer or manually using a pair of foreceps. Female flowers are pollinated applying pollen to the stigma using a brush in the morning followed by repollination next day.

Well-developed berries 45-50 days after pollination are collected and allowed to ripe at room temperature for 2-3 weeks. Ripe berries are crushed (in screw type juice extractor/low speed food blender) and seed and pulp are treated with 10% HCl with continuous stirring for 20 minutes. Seeds are washed thoroughly with water and dried in shade in the plains or in mild sun in the hills. Dried seeds are kept in moisture proof containers and stored at low temperatures.

Crop production using TPS

There are tow methods : (i) using seedlings as planting material. (ii) Using seedling tubers as planting material.

The first method is successful in regions where the winters are mild eg. Assam, Bihar, eastern U.P., Gujarat, Karnataka, Madhya Pradesh and West Bengal, whereas, the second method is successful in all the potato growing regions. The yield is higher when the crop is raised using seedling tubers as planting material.

(i) Crop from transplanted seedlings

For transplanting in one hectare about 100-150 g of TPS to raise seedlings and a nursery bed area of 50-60 m^2 are required. The width of the nursery bed is kept at 1 m. as it has been found to be most convenient for management and cultural operations. The beds may be separated by 80 cm wide pathway The nursery bed is bordered using bricks or any other suitable and easily available local material upto a height of 10 cm. A substrate is filled in the bed to a thickness of 7-8 cm. The substrate consists of sterilized soil and well rotten FYM or compost or biogas slurry in ratio of 1:1 + 4-5 g N, 6-8 g P_2O_5 and 10 g K_2O/m^2. Cover the top of nursery bed with 2-3 cm thick layer of sieved FYM. The TPS soaked in 1500-2,000 ppm $GA_3$for 48 hrs. to break dormancy in freshly extracted seeds. Seeds after drying in shade are sown in the early rabi season in the nursery bed in furrows (0.5 cm deep) 10 cm apart or seeds can be broadcasted @ 2-3 g/m^2 followed by covering with about 0.5 cm thick layer of sieved FYM. Soil is kept moist with light spray irrigation. After germination seedlings are sprayed every 2-3 days with 0.1% urea till they are ready for transplanting i.e. 4-5 leaf stage.

For transplanting the seedlings, the field is prepared in usual way and 20 cm high ridges 40-45 cm apart are made in East-West direction. Before transplanting, the furrows are irrigated up to 8-10 cm depth and seedlings (one seedling/hill) are transplanted after irrigation at 10 cm intra-row distance at the watermark on the northern side of the ridges. Second irrigation and transplanting are successive operations and are to be done row by row whereas, later on irrigations are given as per requirement. After 30-35 days, earthing along with required nitrogen is done in such a way that the seedlings come to lie in the center of the ridge. All subsequent cultural operations and take plant protection measures as per standard cultivation practices for potato crop in the region are adopted. After harvesting, tubers are graded and 10-40 g size tubers are used as seed next year and the large size (>40g) tubers may be disposed off for table purposes.

(ii) Crop from seedling tubers

This method of raising commercial crop from seedling tubers. Seedling tubers are produced either in primary beds (10 x 10 cm) or seedlings are raised in primary beds and subsequently transplanted in secondary/

production nursery beds for tuberization. About 40-50 g TPS and 300-375 m^2 nursery bed areas are required for producing enough seedling tubers for planting one-hectare area the next year. The requirement of seedling tubers to plant one ha area is as given below:

Seed size (g)	Manual cultivation		Mechanized cultivation	
	Spacing (cm)	Approximate seed rate (q/ha)	Spacing (cm)	Approximate seed rate (q/ha)
20-40	45-50x20	30-33	60x20	25
10-20	45-50x15	20-22	60x15	17
5-10	45-50x10	15-17	60x10	13

Follow all the cultural and manurial practices for raising the crop as when the crop is raised using seed tubers of a standard variety (Singh, 2004).

Processing quality

Potato is a very important items of food for human consumption. It is used in culinary preparation in variety of product made from fresh and from its processed forms. In the USA 54% of total consumption is in the form of processed products. In the U.K. about 25% of potatoes are processed. Processing is becoming increasing by popular in the European countries too. In India, apart from about 1000 tonnes of potatoes which are dehydrated, a small quantity which is canned, and a very small quantity which is dehydrated by housewives, the rest is used directly in culinary preparations. Potatoes may be processed into many forms.

In USA Ppotatoes are processed as frozen Fench fries and other frozen products (25.16%) like chips (10.5%) and dehydrated potatoes (8.7%).

In West Germany and Natherland the frozen products are more preferred forms. In U.K. frozen products and chips, canned potato are used.

Potato Chips

The potatoes are thoroughly washed after removing the sprouts part, they are peeled with a knife in a small-scale unit. For a large-scale plant, alrasise peeler may be used. The peeled potatoes are kept in water to prevent brown cabinet size. The spices are dried to a minimum level of 2-3 per cent moisure content. Then chips are packed in polyethylene film of 400 g LDP (Low density polyethylene). In such packing it may be kept satisfactorily up to 6 month at room temperature and up to 9 month at low temperature (7°C).

Potato Flour

To avoid browning generally fresh potatoes are used to prepare plants flour. Pieces can be ground in to flour in a power mill or stone mill (Conamom Chakki). The powder is sieved through a 30-mesh sieve to get a fine flour and packed in same way as potato chips. Several products such as potato tikki, burfi, potato Jamun, potato gujiya, potato halwa can be prepared from potato flour. It can also be used in the preparation of chapati and soup. It is recommended that potato product should be used in the preparation of different edible processed products. It may be taken commercially to solve the problem of surplus produce.

Processed Potato Products

a. *Fried Potato Products :* Alu Sakarpara, Potato Lachha.

b. *Dehydrated Potato Products :* Potato chips, Potato waris, Potato sticks, Potato-sago papads, Potato chiclets, Potato pickle, Potato lollipops.

c. *Non-Fried Potato Products :* Potato Jam, Potato sweet buttons/murabba, sweet and sour potato lachha, potato flakes, potato eclairs, potato sweet pie, potato hajmola candy, potato custard powder, potato soup mix, potato biscuit, potato cake.

Potatoes stored at low temperatures accumulate reducing sugars rendering them emfit for processing. This sugar accumulation is due largely to vacuolar invertase activity. An invertase inhibitor (inhh) gene (accession AY 145781) from tobacco has been cloned and integrated in T-DNA of pBin AR. This construct will be used for transformation of both table and processing potatoes to reduce sugar accumulation during storage.

REFERENCES

Abrahm, E.V., Padmanaban, M.D. and Mohandass, Alfred. (1972). Control of potato cut worms. *Indian Agric. Sci.* 42: 418-420.

Accationo, P. and Malagmba, P. (1982). Potato production from true seeds. *Centra International de la papa,* Lima, Peru.

Agrawal, R.L. (1985). Seed Technology. Oxford & IBH Publishing Co., New Delhi. pp. 247-52.

Anand, J.C., Mini, S.B. and Diwan, Brijesh. (1982). Export potential for popular chips. *Indian Horticulture.* 27(1): 19-20.

Anonymous (2005). Annual Report 2004-05. Indian Council of Agricultural Research, New Delhi. pp. 10-15.

Anonymous. (1971). Potato in India. CPRI. *Tech. Bull No. I.*

Aykroyd, W.R. (1941). Nutritive Value of Indian Foods. Publ. by Nut. Res. Lab., Koonoor.

Bejo Sheetal. (2000). Hybrid Seeds. Bejo Sheetal Seeds Pvt. Ltd., A-3, Old M.I.D.C. Jalna-43, 1203, Maharashtra, (India).

Chakrabarti, S.K.; Kaushik, S.K., Chimote, V.P., Pattanayak, D. and Naik, P.S. (2003). Molecular markers. In: S.M. Paul Khurana, J.S. Minhas and S.K. Pandey's (eds.). The Potato, Production and Utilization in sub-tropics. Mehta Publishers, New Delhi, p. 239.

Chandla, V.K. and Chandel, R.S. (2003). Managing tuber damaging pests of potato. 48(2): 15-17.

Chauhan, D.V.S. (1972). Vegetable Production in India. Ram Prasad and Sons, Agra.

Chick, H. and Slack, E .G. (1949). Distribution and the nutritive value of nitrogenous substance in potato. *Bio Chem.*, 7(45): 211-221.

Choudhuri, R.P. (1953). Control of cut worms in potato fields, *Indian J. Ent.* 15: 203-206.

Choudhury, B. (1967). Vegetables. National Book Trust India. New Delhi, pp.25-39.

Choudhury, N.C. (1914). Potato cultivation in Bengal. Bihar and Orissa. *Agric. J Bihar, Orissa* 2: 89-101.

Choudhury, N.C. (1965). Control of epilachna beetles attacking potato in Shimla Hills *Indian potato J.*, 7(2): 108.

Ezekiel, R. and Bhargava, S.C. (1992). Nitrogen distribution within the potato plant in relation to planting date under short day conditions. *Indian J. Plant Physiology*. 35(2): 130-139.

Ezekiel, R., Singh, Brajesh, Kumar, N.R. and Khurana, Paul, S.M. (2003). Storing potato scientifically. *Indian Horticulture*. 48(2):28-30.

Fricker, A. (1970). Fatty acid composition of lipids in potatoes. *Z. Lebensm. Umiters Forsch.*, 142:14-30.

Gopal, J. and Raj Kumar (2003). Strategies for successful potato breeding. In: The Potato, Production Utilization in sub-tropics, p. 239. S.M. Paul Khurana, J.S. Minhas and S.K. Pandey (eds.) Mehta Publishers, New Delhi. pp. 61-68.

Gould, W.A. (1985). Quality assurance manual for the manufacture of potato chips and snack foods. *Potato Chips / Snack Food Association,* Alexandria VA.

Grewal, J.S. and Trehan, S.P. (1981). Potato needs micronutrients. *Indian farming* Nov. 1981, 3-5.

Harris, P.M. (1987). Water in the potato crop. Harris PM. ed.. Chapman and Hall. London, pp. 245.

Hawkes, J.G. (1990). Potato: Evolution, Biodiversity and Genetic Resources, Smithsonian Institution Press, Washington DC.

Hukkeri, S.N., Chauhan, D.E. and Dastane, N.G. (1970). Effect of soil moisture stress at different stages of growth on the yield of potato. *Indian J. Agric. Sci.* 40:318-325.

Jaiswal, V.P. and Lal, S.S. (1993). Effect of herbicides on weeds and tuber yield of potato (*Solanum tuberosum* L.) In: Abstr, Papers, Proc. ISWS, 3: 202-203 (Hisar), India.

Joseph, A.A., Roy Choudhury, R.N., Indiramma, K., Rao, M. Narayan, Swaminathan, M. Sreenivasan, A. and Subrahmanyan, V. (1963). Amino Acid composition and nutritive value of the proteins of different potato varieties. *Food Sci.,* 12: 255-257.

Kallo, G. (2003). Research and Developmental priorities in potato. *Indian Horticulture* 48(2): 3.

Kaul, H.N. and Mehta, Ashiv. (1991). A risk less sprout suppressant for ware potatoes: Maleic Hydrazide. *Indian Horticulture.* 36(1): 10, 11 & 13.

Kaushik, S.K., Singh, H.P., Joseph, T.A., Vinay and Singh B.P. (2003). Late blight resistant potato varieties for hills. *Indian Horticulture.* 48(2): Cover III.

Kouwenhoven, J.K. (1970). Yield, grading and distribution of potato in ridges in relation to planting depth and ridge size. *Potato Res.,* 13:59-77.

Krishnappa, K.S. (1991). Effect of planting time on the yield and yield attributes of potato in eastern karnataka. *Curr. Res.,* 20(11): 239-240.

Kumar, Raj, Kang, G.S. and Pandey, S.K. (2003). Early potato for Indo Gangetic plains. *Indian Horticulture.* 48(2):12-13.

Kumar, Shiv, Singh, H.P., Garg, I.D. and Khurana, Paul, S.M. (2003). Integrated Management of Potato Diseases. *Indian Horticulture.* 48 (2): 25-27.

Lal, S.S. (1998). Managing weeds in potato. *Indian Horticulture.* 43(1): 16-19.

Lampitt, L.H. and Goldenborg, N. (1940). The composition of the potato. *Chemistry and Industry.* 18:748-761.

Mac Gillivray, J.H. (1953). Vegetable Production. McGraw-Hill Book Company, Inc. New York.

Mahajan, S.V., Mogal, B.H. and Chudmurwar, R.D. (1976). Chemical control of the potato tuber moth. *Phthorimaea opercullella* (Zeller) (Lepidoptera: Gelchiidal). *Pesticides* 10(7): 50-51.

Malagamba, P. (1984). Agronomic management for transplanting TPS seedlings. Innovative methods for propagating potatoes. Rep. 28th planning. *Conf. International potato center.* Lima, Peru. pp 63.

Malagamba, P. (1988). Potato production from true seed in tropical climates. *Hort. Science,* 23,495.

Malik, Y.S., Singh, S.S. and Pandita, M.L. (1985). Response of Potato genotypes to several levels of salinity. *Haryana J. Hort. Sci.,* 13(1-2): 69-71.

Mandal, S.S. and Chatterjee, B.N. (1993). Potato In: Vegetable Crops in India (Ed. Bose, T.K., Som, M.G. and Kabir, J. Naya Prokash, Calcutta, pp. 382-469.

Mangal, J.L., Srivastava, V.K. and Karwasra, S.P.S. (1990). Salt tolerance in vegetable crops. *Tech. Bull.* 1., Haryana Agric. Univ. *Res. J. Hisar. Bull* 1. Haryana (India).

Marwaha, R.S. and Sandhu, S.K. (2003). Enjoy finger-licking potato products *Indian Horticulture.* 48(2):20-24.

Mishra, P.K., Singh, D.N. and Sontakke, B.K. (1988). Adaptabiiity of some released and pre-released polato varieties in mid-hill of Orissa. *Orissa Hort., Vol.XVI* (1&2):46-49.

Mishra, S.S. and Chandla, V.K. (1995). Manage potato tuber moth. *Indian Horticulture,* 40(1):30-35.

Nadkarni, K.M. (1927). Indian Materia Medica Published by Nadkarni & Co., Bombay.

Nagaich, B.B. and Giri, B.K. (1973). Potato Top Roll disease of potato. *Am. Potato J.,* 7.50:79-85.

Nagaich, B.B. (1971). Virus disease of potato in India and their control. *ICAR tech. Bull (Agric.)* No.28. New Delhi.

Nagaich, B.B. (1974). Viral and mycoplasmal diseases of potato. *Indian farming,* 26(6):33.

Narula, K.K. (1962). Insecticidal control of the spread of leaf roll and virus y in potatoes. *Indian potato Journal* 4(1):1-6.

Nath, Prem; Srivastava, V.K., Dutta, O.P. and Swamy, K.R.M. (2008). Vegetable Crops – Improvement and Production. PNASF, Bangalore, 215-217.

Noorudin, A. and Mehta, A.N. (1995). The effect of shifting planting dates and irrigation levels on tuber size and yield of potato cv. Kufri Badshah under middle agroclimatic zone of Gujarat. *J. Indian Potato Assoc.,* 22(1-2): 17-72.

Pal, B.P. and Pushkarnath. (1951). Indian potato varieties. *Misc. Bull. 62. Indian Conn. Agric. Res.,* New Delhi.

Peter, K.V. (1998). Genetics and Breeding of Vegetables. ICAR., New Delhi.

Purohit, M.L., Khatri, A.K. and Sinde, C.B. (1973). Control of potato cut worms. *Agrotis ypsilon.* (Hfm) *Noctuidae-Lepidoptera. Pesticides* 7:25.

Pushkarnath (1964). Potato in India. ICAR, Publication, New Delhi, pp. 18-20.

Puskarnath. (1967). Seed potato production in the subtropical plains of India. *Am. Potato J.,* 429-41.

Sagar, V.R. (1997). Potato flour is a cheap home made product. *Indian Horticulture.* 41(4):52-53.

Saha, S.N., Murli, G.S.R., Banerjee, V.N., Purohit, A.N. and Singh, Mukhtar. (1974). Effect of night temperature on growth and development of Indian potato varieties under short day conditions. *Indian]. Agric. Sci.,* 44:376-382.

Sahota, T.S. and Virk, M.S. (1987). Effect of soil burning on fertilizer requirements on potato in the soil of the khasi hills. *Indian J. Ecology,* 14(1): 78-82.

Saini, G.S. (1996). A Text Book of Vegetable Production. Aman Publishing House, Meerut (U.P.), India.

Salman, N.R. (1949). The History and Social Influence of the potato. Cambridge Univ. Press, Cambridge.

Saxena, A.P. and Rizvi, S.M.A. (1974). Insect pest problems of potatoes in India. *J I P A* 1 (1-2):45-50.

Sen, K. and Dasgupta, M.K. (1982). Control of root-knot nematodes in India. *Pesticide.* 16(1):1-17.

Singh, D.N. (1998). Effect of cut and whole tubers on vegetative growth and tuber yield of potato. *Environment & Ecology.* 16(3): 745-747.

Singh, D.N. and Nath, Vishal (2011). Varieties and Hybrids of Vegetables. Satish Serial Publishing House, Azadpur, Delhi (India), 95-103.

Singh, D.N. and Parida, A.K. (1999). Response of potato varieties to different dates of planting haulm cutting and spacing under mid hill and coastal plain of Orissa. *Environment & Ecology,* 1 7(4):845-847.

Singh, D.N., Parida, A.K. and Sahu, A. (1994). Evaluation of different seed size and seed rate in potato under coastal plain of Orissa. *J.Indian Potato Assoc.,* 21(3&4): 240-242.

Singh, D.N., Jena, S.N.; Behera, A.K. and Sahu, G. (1999). Weed management in potato under West Central Table Land Zone of Orissa. *ISWS*: VIII *Biennial Conference.* Dept. Agronomy, Inst. Agric. Sci., BHU, Varanasi. 5-7 Feb. p.90.

Singh, D.N., Mishra, P.K., Sahu, S.K. and Sontakke, B.K. (1986). Response of potato to varying levels of nitrogen and potash under Eastern Ghat Land Zone of Orissa. *Journal of Indian Potato Association.* 13(1&2):55-64.

Singh, D.N., Nandi, A. and Parida, A.K. (1994). Effect of Urea and CAN with or without neemcake on growth and yield of potato. *Orissa Journal of Horticulture.* Vol. 21(1&2): 7-10.

Singh, D.N., Nandi, A. and Tripathy, P. (1993). Effect of different seed rate and seed size on growth, yield and seed size tuber production in potato under coastal plain of Orissa. *J. Indian potato Assco.,* 20(1): 182-1 83.

Singh, D.N., Sahu, S.K. and Dhal, J.K. (1989). Evaluation of different nitrogenous fertilizers and their methods of applications on potato under mid-hill and coastal plain of Orissa. J. *Indian potato Assoc.,* 16(1 & 2):34-37.

Singh, D.N., Sahu, S.K., Mishra, P.K. and Dhal, J.K. (1988). Effect of micronutrients application on potato under coastal and hill region of Orissa. *The Andhra Agric. J.,* 35(1&2):131-132.

Singh, K., Kumar, V. and Pandita, M.L. (1973). Influence of sprinkler and surface method of irrigation and different levels of nitrogen on the growth, yield and quality of potato *(solatium tuberorum* L.). *Haryana J. Hart. Sci.,* 2:101-111.

Singh, K.P. (2001). Potato: A stable vegetable. National Training Course on Vegetable Production Technology. Indian Institute of Vegetable Research, Varanasi. pp. 79-92.

Singh, M.; Gandhi, R.T. and Destana, M.G. (1961). Methods of irrigation in India and suggestions for their improvement and evaluation. *Indian J. Agron.,* 4:155-166.

Singh, M.V. and Singh, A.P. (1995). Studies on potato based crop rotations in eastern Uttar Pradesh. *J. Indian Potato Assoc.,* 22(3-4): 156-158.

Singh, N. (1994). Yield responses of potato cv. Kufri Swarna to varying seed sizes and spacing. *J. Indian potato Assoc.,* 21(3-4): 231-233.

Singh, N. and Sood, M.C. (1994). Water and nitrogen needs of potato under modern irrigation methods. *Nat. Symp. March,* 1-3, 1994 held at Shimla.

Singh, P.K. (2004). Text Book on Tuber Crops. Published by N.D. University of Agriculture & Technology Narendra Nagar (Kumarganj), Faizabad-224229 (U.P.), India.

Singh, R.S. (1973). Plant diseases. Oxford and IBH Publishing Co., New Delhi.

Singh, S.K. and Rai, R.P. (2003). Producing quality potato seed. *Indian Horticulture,* 48(2):8-11.

Singh, S.P. (1977). Effect of farmyard manure and nitrogen on potato (*Solanum tuberosum* L.) variety kufri Sindhuri (C-140). *Madras Agric. J.,* 62(8): 504-507.

Singh, S.V., Pandey, S.K. and Gaur, P.C. (1994). Stability performance of new potato hybrids in Western Uttar Pradesh. *J. Indian Potato Assoc.,* 21(3-4): 201-206.

Singh, U.P. and Bhan, V.M. (1999). Herbicidal Weed Control in Potato *(solanum tuberosum* L.*)*. VIII *Biennial Conference,* Dept. of Agronomy, Inst. Agric. Sci., BHU. Varanasi. 5-7 Feb. p.88.

Singh, V.P. and Hooda, R.S. (1982). Weeds and their control in potato field in India. A critical Review. *Agric. Rev.,* 3(2): 83-84.

Singh, Vijay; Singh Ganesh and Raghav, M. (1990). Effect of variety of potato. *Narendra Deva J. Agric. Res.,* 5(1): 29-33.

Smith, Ora. (1977). Potatoes: Production, storing, processing. 2nd Ed., Avi Publishing Co. Inc. Westport. Connecticut, USA, pp.107-115.

Sontakke, B.K., Singh, D.N. and Mishra, B. (1989). Seasonal abundance of potato pests in relation to environmental factors. *Environment & Ecology.* 7(2):391-394.

Srivastava, B.K. and Khan, R.M. (1962). Cutworms of potato and their control in Rajasthan. *Indian Potato J.,* 4(2); 88-92.

Swaminthan, K. and Pushkarnath. (1962). Nutritive value of Indian Potato Varieties. *Indian Potato J.,* 4:76-83.

Thakur, K.C., Pandey, P.C. and Pandey, S.K. (2003). Producing potato through TPS. *Indian Horticulture.* 48(2):32-34.

Trehan, S.P. (2003). Managing micronutrients in Potato. *Indian Horticulture,* 48(2):18-19.

Van dar Zaag, D.E. (1976). Potato production and utilization in the world. *Pot. Res.,* 19:37.

Venkateshwaran, P.A. (1961). Agriculture in South India, Published by Directorate of Extension, Ministry of Food and Agriculture, New Delhi.

Verma, S.C. (1945). Potatoes for processing. Quality parameters. *Indian Horticulture,* 40(1):37-41.

Verma, S.C., Sharma, T.R. and Joshi, K.C. (1973). The sugar content of potato tubers. *Indian J. Agric. Sci.*, 43:643-745.

Wiersema, S.G. (1986). A method of producing seed tubers from true potato seeds. *Potato Res.*, 29: 225.

Wiersema, S.G. and Cabello, R. (1986). Comparative performance of different - sized tubers derived from true seed. *Am. Potato J.*, 66: 109.

□□□

Chapter-2

Cole Crops

1. CAULIFLOWER

Botanical Name	:	*Brassica oleracea* Var. *Botrytis*
Chromosome number	:	2n = 18
Family	:	Brassicaceae (Cruciferae)

Cauliflower is one of the most important cole crop grown as vegetable throughout the world in temperate countries from 11°N to 35°N latitude. India is largest producer of cauliflower in the world. In India the area and production of cauliflower was 3 lakh hectares and 52 lakh tonnes in 1999-2002 (Swarup, 2006). There is hardly any house where it is not regularly used as vegetable. With the development of tropical types in addition to the temperate types, it has now become possible to grow this vegetable alomost throughout the year particularly in the western and central part of India. Cauliflower growers may be benefited by growing this crop to a great extent near large cities or by sending the produce to distant places where they can fetch better price. Cauliflower has high quality proteins. Very little quantity of Vitamin is lost after cooking which is a peculiar character of Cauliflower. It also retain considerable quantity of Vitamin B. It is rich source of minerals, such as potassium, sodium, iron, phorphorus, calcium, magnesium etc.

ORIGIN

Cauliflower (*Brassica oleracea* var. *botrytis* L.) is belivered to have originated in cyprus or somewhere in southern part of Europe around the Mediterranean coast. According to Bosewell (1949), it is originated in the iseland of cyprus from where it moved to other areas like Syria, Turkey, Egypt, Itlay, Spain and north western Europe. Hyam (1971) considered Syria is a place of origin for cauliflower. In the middle of the 16th century, the first-illustration and description of cauliflower was presented by the herbalist Dodoens (1544). It was after this period that the crop become more commonly cultivated, particulary in the beginning of the 18th century. It has descended through mutation and selection from common ancestor, the wild cabbage. (*Brassica oleracea* L. var. *sylvestirstis* L.) Which is still found growing in western and southern Europe and north Africa. The different types of cauliflowers like cornish, Northern, Roscoff and Angers and Erfurts originated from the Italians independently indifferent regions viz. Cornish and Northern in England. Roscoff and Angers in France and Erfurts or Snowball in Germany and the Netherlands. Cornish type, perhaps the first to be introduced in India, has itself gone out of cultivation after coutributing many genes to Indian varieties like resistance to black rot, self-incompatibility, curd flavour, open plant habit, exposed yellow loose curds, etc. (Swarup and Chatterjee, 1972). Indian cauliflower is an early maturing annual type, tolerant to high temperature and humid conditions.

History

The cauliflower (named from the latin caulis (cabbage) floris (flower), var. botrytis (meaning budding). It has been developed from plant still found in wild state in various localities from Greece to Great Britain. Cauliflower was first described as a *Brassica florida botrytis* in 1576 and as *Brassica cauliflora* in 1583. It was probably introduce into Italy by Genoese from Levant or Cyprus probably around 1490. It was grown for seed along the Gulf of Naples. At the beginning of 17th century it reaches Germany, France and England. Cauliflower was introduced in India in 1822 when Dr. Jemson, a botanist from Kew, took charge of the company Bag (United Provinces, Saharanpur in the northern plains) to carry out of some horticultural experiments during the period of the East India Compnay (Swarup and Chatterjee, 1972). The Royal Agri-Horticulatural Society also introduced seeds of cauliflowers. After that it has undergone acclimatization and selection as a result of which the Indian cauliflower today has attained a characteristically different from compared to other temperate types of the world. The main differences are tolerancence to high temperature and rainfall and earliness. Its further development and improvement were achived in Noth and Noth-Western Europe extending its cultivation up to 60°N latitude. The development on Indian cauliflower types made it possible to extend its growing area in the tropics and sub-tropics of the world. The crop is presently cultivated from 11°N to 60°N latitude.

DISTRIBUTION

Cauliflower is grown throughout the world in temperate countries from 11°N to 35°N latitude. After originating in Ciprus, the cauliflower distributed to other areas like Syria, Turkey, Egypt, Italy, Spain and North Werstern Europe. Seeds were also imported from South Africa and England and were distibuted to farmers in North India. During a period of about one hundred years (1822-1929), these introduced cauliflower varieties underwent selection by local growers when seed production was attempted by them in North Indian Plains. Selection were made for early maturity and wider adaptability to hot and humid weather conditions. These types are commonly known as Indian or tropical cauliflower which are good for early sowing and ealry harvest. In later season, temperate types are grown which are known as snowball or late cauliflower or main season cauliflower. Now India is the leading producer of cauliflower in the world with a production of 48,91 lakh metric tonnes from an area of 2.7 lakh hectares. West Bengal, Bihar, Orissa, Assam and Jharkhand are the leading cauliflower producing states of India.

BOTANY

The cauliflower belongs to the species *Brassica oleracea* L., below the species level a generally accepted treatment is to classify cauliflower as *B. oleracea* L. convar *botrytis* (L.) Alef. var. *botrytis* L. The edible portion of cauliflower is the white curd like mass composed of a close aggregation of abortive flowers, developed on thick bunches of the inflorescence. This edible potion is called curd surrounded by leaves which are longer but narrower than those of cabbage. Cauliflower is herbaceous annual for vegetable production and biennial for seed production. Cauliflower has long and linear leaves, curd is the early stage of inflorescence development without any axillary sprouts. The edible protion of plant is the curd which is made up of numerous divided hypertrophic branches which terminate the main stem of the plant and is highly suppressed with no part of flower appernt there (Kato 1964).

Plant breeding

Cauliflower is corss pollinated in which pollination is by insects and honey bees. To combat self incompatibility bud pollination procedure is adopted for self pollination. Cytoplasmic male sterility technology is same as that of cabbage. A few hybrids have been developed by utilizing self incompatibility and cytoplasmic male sterility. The F_1 hybrids cover 13% of the total area of cauliflower (Nath *et al.*, 2009).

'Eealy' and 'Main crop Patna' and 'Early' and 'Main crop Banaras' were the first four listed Indian varieties, perhaps the earliest in the world and have greatly contributed for developing improved varieties adapted to hot-weather conditions in different countries, viz. 'Pua Kea' in Hawaii, 'compinus' in Brazil. 'Improved Japanese' and 96-D in Israel and 'Extra Early' in Taiwan. Singh (1971) and Swarup and Chatterjee (1972) classified cauliflowers into four distinct maturity groups and four plant types, Extensive genetics and plant-breeding work was initiated by Singh *et al.* (1966) at IARI on India cauliflowers (Singh, 1971 and Singh *et al.*, 1975; 1976a,b; 1981b). The same authors developed for the first time the synthetic varieties in Indian cauliflowers of maturity group II and III (Singh, *et al.*, 1981a). An excellent genetic and breeding work at IARI has resulted in the development of some very high-yielding varieties like 'Pusa Deepali'. Early Synthetic'. 'Synthetic II', 'Line 328', 'Line 12C', 'Synthetic III', etc. Work taekn up later at Pantnagar and Ludhiana is also quite appreciabel. The other informations on Indian cauliflowers and snowball type are also available (Chatterjee and Mukherjee. 1964. 1965; Pal and Swarup, 1966: Sandhu *et al.*, 1977; Kale *et al.*, 1979).

Floral biology

The flowers are borne in racemes on the main stem and its branches. There are 4 sepals, 4 petals, 6 stamens and two carpels. The carpels form a superior ovary with false septum and two rows of camylotropous ovules. The androecium is tetradynamous i.e., these are two short and four long stamens. The pollens are 30-40 μ in diameter and have three germination pres. The bright yellow petals become 15-25 mm long and about 10 mm wide. The sepals are erect. The buds open under the pressure of the rapidly growing petals. This process start in the afternoon, and usually the flowers become fully expanded during the following morning. The anthers open a few hours later, the flowers being slightly protogynous. The flowers are pollinated by insects, particularly bees, which collect pollen and nectar. The nectar is secreted by two nectaries situated between the baris of the short stamen and the ovary. Situated outside the basis of pairs of long stamens are a also two nectaries but these are not active. The anthesis of flowers starts at 8.00 a.m. and continue upto 11.30 a.m. The pollen fertility is considerably high at the time of anthesis and one day before anthesis. The fertility is reduced one day after anthesis. The stigma becomes receptive two days before and remains upto two days after anthesis and thus facilitates selfing by bud Pollination even in the self incompatible group (Kanwar, 1996 and Nieuwhof, 1969). There is an absence of a central main stem above the point where branching begins (Salunkhe *et al.*, 1987). The inflorescence may attain a length of 1-2 m but the slender pedicels are only 1.5-2 cm long. The fruits are glabrous siliques, 4-5 mm wide and sometimes over 10 cm long with two rows of seeds lying along the edge of the replum (false septum and out growth of the placentae). A silique contains from ten to thirty seeds. Three to four weeks after the opening of the flower from which it is formed, the silique reaches its maximum length when it is ripe, dehiscence takes place through the two valves breaking away from below upwards, leaving the seeds attached to the placentas (Nieuwhof, 1969).

Classification

(I) According to Swarup and Chatterjee (1972) cauliflower varieties have been classified in different group. The details of classification are give below :

Cauliflower types	Country of origin	Probable period of first cultivation	Characters
Italians or original	Mediterranean	16th Century	Plants short; leaves erect broad with rounded tips, bluish green; curds good not protected.
Cornish	England	Early 19th Century	Plants vigous; long stalked; leaves loosely arranged, broadly wavy; curds flat; irregular, loose, not protected, yellow, highly flavoured.
Northerns	England	19th Century	Leaves petioles, broad, very wavy, serrated; curds good, well protected.
Roscoff	France	19th Century	Plants short; leaves long erect, slightly wavy with

contd...

			pointed tip, midrib prominent, bluish green; curds white or creamy, hemispherical and well protected.
Angers	France	19th Century	Leaves very wavy, serrated, greyish green; curds solid and well protected.
Erfurt and snowball	Germany and Netherlands	18th Century	Pants dwarf; leaves short, erect, glaucous green; curds solid and well protected.
Indian cauliflower	India	Late 19th Century	Plant short, long stalked; leaves loosely arranged, broadly wavy; curds flat, somewhat loose, yellow to creamy, not protected and highly flavoured.

(II) The cauliflower varieties have also been classified in different groups according to their phyllogeny as given by crisp (1982). The details are given below:

Group name	Characteristics	Common Types
Italian	Very diverse, include both annual and biennials and curds with peculiar conformations colours.	Jezi, Naples (autumn Giant), Romanesco and Flora Blanca.
North European annuals	Developed in Northern Europe for at least 400 years. Origin unknown, perhaps Italian Eastern Mediterranean.	Alpha, Mechelse, Erfurt and Danish.
North West European biennials	Eveloped from Italian material within 300 years.	Angers, old English Roscoff St. Malo, Walcheran
Asian	Recombinants of European annuals and biennials, developed within last 250 years, adapted to tropics.	Four maturity groups are recognized by Swarup and Chatterjee (1972).
Austrilian	Recombinants of European annuals and biennials and perhaps Italian stock, developed during the last 200 years.	Not yet been categorized.

(III) The cauliflower varieties can also be classified on the basis of maturity. Most of the early Indian verieties of cauliflower are called by the name of the month in which the curds are mature and become ready for harvest, viz 'Kunwari (September-October maturity), Katki' (October-November maturity), 'Agahani' (November Maturity), 'Posi' (December maturity), 'Maghi' (January maturity). Most of these verities are highly heterozygous for maturity, plant growth and curd traits.

(IV) The cultivars/varieties of cauliflowers are also classified into five group on the basis of curd maturity, which details are given below:

Maturity	**Varieties**	**Temperature requirement for curd initiation and development**
Extra-Early	Early Kunwari	20-27°C
Early-September maturity	Pusa Early sysnthetic	
(Mid September Mid October)	Pant Gobhi-3	
October maturity	Pusa Deppali	20-25°C
Medium-(Mid Oct.-Mid Nov.)	Pant Gobhi-2	
Mid Early	Improved Japanese	16-19°C
November maturity	Pusa Hybrid-2	
(Mid November-Mid December)	Pusa Hybrid-3	
	Pusa Sharad	
	Pant Gobhi-4	
Mid Late	Pusa Synthetic	12-16°C
December maturity	Pusa shubira	
(Mid December- January)	Pusa Himjyoti	
	Punjab Giant-35	
Late Snow ball	Pusa Snow ball -I	10-16°C
	Pusa Snowball K-I	

COMPOSTITION

Cauliflower is a source of Vitamin A and C. (Nath, 1976), it also contain minerals like potassium, sodium, calcium, iron, phosphorus, magnesium. According to Conew (1959), Cauliflower contians 92.7 percent water and food value per 100g of edibale portion is as follows: energy 31 calories, protein 2.4 g, calcium 22mg, vitamin A 40 I.U., ascorbic acid 70 mg, thiamine 0.2 mg, riboflavin 0.1 mg and niacin 0.57 mg. Sulphur containing compounds viz.; hydrogen sulphide-methanethiol, ethanethiol, propanethiol and dimethyl sulfide in addition to acetaldehyde and 2-methyl propanol have been identified in cooked cauliflower.

Uses

Cauliflower is used as a cooked vegetable and a variety of delicious dishes are prepared out of 'Curd'. It is also used in combination with pea, potato etc. and is liked by both poor and rich. Cauliflower seedlings are used for salad and green. The curd is used in curries, soups, pakoras and Pickles. In abundant areas of production, cauliflower curd is cut in to pieces, dried and preserved for off-season use. The therapeutic effect of cauliflower has been well documented. The inflorescence extract has been used in the treatment of scurvy, as a blood purifier and as an antacid. The seeds have contraceptive properties. Cauliflower extract has been reported to be effective in the inhibition of initiation and promotion of carcinogensis in *in-vitro*. Potential chemopreventive agents include ascorbic acid, carotenoids, tocopherols, isothiocyanates, indoles and flavonoids.

VARIETIES AND HYBRIDS

The cauliflower varieties are very responsive to temperature and photoperiod. So it requires specific temperature for their curd formation and development. According to temperature requirement varieties have been classified in 5 groups.

Detail characteristics of important commercial varieties and hybrids of different groups described by Singh and Nath (2011) are given below:

Extra Early Group

Temperature requirement for curd initiation and development 20°-27°C.

Early Kunwari

It is the earliest cultivar released by the Punjab Agricultural University. It has bluish-green leaves with waxy blooms. Early maturing, semi-erect, partial blanching habit curd, colour creamish. Field tolerant to alternaria blight and black rot. Curds are not so compact and sensitive to riceyness.

Pusa Early synthetic

It is developed from IARI, New Delhi, curd is yellowish white, medium sized and moderately compact. Suitable for planting in the end of June or first week of July. Average curd yield 110 q/ha. Suitable for sub-Humid Sutlej-Ganga alluvial plains and humid to semi arid western Ghats, Karnataka plateau and Orissa.

Early selection

It is a early cauliflower. Seeds are sown in nursery during. May-June then after 25 days old seedings are transplanted in main field. It mature within 70-75 days after transplanting.

Pusa Meghna

It has been developed from heterozygous material collected from Hazipur (Bihar), using recurrent selection method. Belonging to early maturity (September) group of cauliflower. It has semi-spreadings plants with 5-7 cm long stalks and 45-48 cm height. It has medium sized, light green, smooth, glabrous, ovate, petiolate and non-blistered leaves with entire margin waving outward. It produces retentive white, compact, medium-sized curd weighing 350-400 g. It takes about 90-100 days for maturity and temperature range for curd development is 20-27°C. It is ready for harvesting, from early September.

Ati shighra (F_1) (NS 1351)

Early summer cauliflower with open plant habit. Curds are medium white in colour and semi dome in shape with good firmness. Each curd weighs 0.5 kg and matures in 50 days. Comes up well in North and North East India (22-30 degree latitude).

Parijat

An extra early variety, maturing in 50-60 days after transplanting snow white curd. Suitable for early spring in western and southern region. Summer sowing in northern region.

Himdhara

It is extra early type and cultivated in rainy season. It matures within 65-75 days after planting. Curd size bigger i.e., 700-800 g, curd uniform and compact, required narrow spacing for higher return.

Himangini

It is extra early type and cultivated in summer and rainy seasons. It matures within 55-60 days after planting. Curds are more compact and can be procured for more days in the field.

Himlata

It is tolerant to high temperature and grown in summer. It matures within 50-55 days after planting. It has highly compact curd and has permanent white colour. It is popular and well known to throughout country.

Himdev

It is extra early type grown in summer and rainy seasons. It mature within 35-40 days after planting. It requires low spacing for high return. Curd size is medium and size varies from 250 g to 300 g.

Summer King

It produce 400-500 g on an average curd weight. Whitish round shaped curd and ready for harvesting with 65 days after transplanting. It can be grown during summer season in north plain.

Basant (F_1) (NS 245)

A good hybrid with open type plants. The hybrid has good white curd colour with semi dome shape and good firmness. Each curd weighs 1.0-1.5 kg with a maturity of 55-60 days. It is suitable for late summers and can perform well in the Southern states (latitude range of 13-26 degree).

NS 60N (F_1)

A typical summer cauliflower with early maturity. The hybrid has open canopy with medium vigour. Each curd weighs 0.75-1 kg with good white colour and firmness. The curds are dome shaped with maturity of 55-60 days. It has wide adaptibility.

Early group

Temperature requirement for curd initiation and development 20-25°C.

Pusa Katiki

One of the early released variety from the IARI, New Delhi, having medium-sized plants and bluish-green leaves, medium sized curds.

Pusa Deepali

Released by IARI, New Delhi, developed through inbreeding from the local collection. Early season, crop in 100-120 days. It is a hot weather cauliflower and temperature requirement is 20-25°C during curd initiation and development. The plants are medium tall, erect, leaves a waxy green with rounded apex and inner leaves covering the curd moderately in early stages of development. The curds are somewhat round, very compact, medium sized, weighing between 500-700 g and white to creamy-white colour. Due to its erect habit and narrower frame (40-50 cm) it is possible to accommodate 35-40% more plant per unit area. It is highly tolerant to riceyness, fuzzyness when grown in proper climate.

Arka Kanti

It is developed from IIHR, Bangalore. It is a early variety with medium plant height. Heads compact white and mature in 60 days. Average curd yield is 25 tonnes/ha.

Pant Gobhi-2

Early maturing, compact curds, composite; creamy white curds, ready for harvest in 125 days after planting.

Pant Gobhi-3

Early maturing; curds medium sized, solid white, nursery sowing upto 15th May. Curds ready for harvest in 110 days after planting.

Summer Special-10

Early maturing hybrid suitable for sowing after mid may to July in northern India and during February to April in southern region. Curds mature in 45-50 days after transplantation. The plants are erect with medium frame. Curds are white, domeshaped, compact and weigh 800 to 900 g and have good self blanching habit.

Rimzim

Early maturing hybrid. Curds can be harvested approximately in about 55-60 days after transplantation. Curds are well protected by inner folding leaves, white in colour, semi-dome shaped, and weigh 700-1000 g; excellent adaptability to different agroclimatic regions. Suitable for rainy season cultivation. It is recommended to plant at wider spacing to obtain uniform curds of larger size.

Sharavani

Early maturing hybrid, mature in about 60 days after transplantation. Curds are pure white and of excellent quality and dome shaped, weigh 800-1000 g. Upright leaves of the plant provide good protection to the curds. Tolerant to soft rot (*Erwinin* sp.) and suitable for rainy season cultivation.

Hybrid-212

Plant green, maturity 45 days after transplanting, curd white, flattish dome, maximum curd weight 900 g, suitable for early sowing (tropical condition).

Pawas

It is most suitable early hybrid, which can be grown from middle May to August. It produces 700-800 g average curd weight and ready within 60 days after transplanting.

Medium group

Temperature requirement for curd initiation and development 16-19ºC.

Improved Japanese

It is an introduction from Israel with long, erect, narrow leaves and medium-sized curds. A mid season variety, crop in about 90-95 days.Stem long. Curds medium to large, compact and white. The average curd yield is 145 q/ha.

Pusa Himjyoti

A product of selection at the IARI, Regional Station, Katrain. It has erect bluish-green leaves with a waxy coating. It is pure line selection from 'MG 2-4' (Maghi Sri Ganga Nagar Selection), give stable yield in both the sowing (April and June). It has erect bluish green leaves with waxy coating. The leaves cover the head tightly in the initial stages. Curds, are pure white and they retain their colour even after exposure. The average curd weight is 500-600 g. It takes 62-75 days from transplanting to curd formation. A good crop may yield up to 150-170 q/ha.

Pusa Sharad

It has semierect and open type plants with small stalk. Leaves are glabrous, bluish-green, long, petiolate oblong with narrow apex, wavy margin, prominent midrib and ear like the lobes at the base of the lamina. Its marketable curd becomes ready 85 days after transplanting. It is about two weeks early in maturity over the Improved Japanese. Curds are retentive white, knobby, very semidome shaped and about 750-1000 g in weight. Average yield is 26 tonnes/ha. The optimum temperature for its curd initiation and development is 16-20°C. Seed should be sown by the end of July to early August depending upon the locality and climatic conditions.

Pusa Hybrid-2

It is a combination of a self-incompatible line (female parent) and sel. 1-3-18-19 (Male parent). Plants are semi erect with bluish-green upright leaves, resistant to downy mildew. The curd is a creamy white, very compact, weighing on an average 907 g. It matures form mid Nov to mid-Dec.

The curd-plant ratio is 10 per cent higher because the curds are very compact and with lesser number of leaves compared with "Improved Japanese". The yield is 23 tonnes/ha. Temperature between 16-20°C is ideal for curd formation. Optimum sowing time is late July.

Pant Shubra

Developed at Pantnagar. Outer leaves semi-erect, inner leaves partially cover the curd, compact, slightly conical, non-ricey, slightly creamish white curds. The average curd yield varies from 185-225 q/ha.

Pant Gobhi-4

It has medium long stem, sparse, semi erect leaves, semispherical creamish-white, medium, compact, non-ricey curds. This is a November maturity group, takes 115 days for marketable curds.

Swarna

It can be grown from September to December in plain and round the year in the hills. Curds are white, compact and ready within 80-85 days after transplanting. It yielded 1.0 to 2.5 kg on an average curd weight.

Shweta

A medium maturing hybrid (70-75 days after transplanting) suitable for tropical and subtropical regions. Plants have thick leaves which provide protection to the curd. Solid and tender curds are white in colour, dome shaped weighing approxmately 500-600 gms.

Chandramukhi (OP)

Plant vigorous and upright, first picking 75-80 days after transplanting. Curd set at temperature 15-17°C, dome shaped, snow white, average weight 1.5 kg, typical self branching type and keeps the curd white for longer period.

Basanti (OP)

Plant erect and vigorous, maturity 85-90 days after transplanting, curd set at temperature 20°C, shape dome, snow white, average weight 2.0 kg, most suitable for medium planting in North India.

Somersat

It has got a better heat tolerance, matures in 65-70 days with curd weight of 2.0 kg each. Curds are very well protected with leaves. Recommended for August to December sowing. High leavel tolerant to fusarium wilt.

Amazing

It is an uniform mid early cauliflower which gives non hairy white curd, mature in 70-80 days after transplanting. Compact with self blanching leaves, which gives snow white curd with weight of 1.0 to 1.5 kg. Recommended for August to November sowing in the plains and tolerant to fusarium wilt.

Hybrid 218

Plant deep green, maturity 60 days after transplanting, curd white, flattish dome, average weight 650 g, suitable for areas having milder climate.

NS 60 (F_1)

It is a very popular hybrid with wide adaptability. It is a mild season tropical cauliflower with semi blanch. The hybrid has very good attractive curds with compact dome shape and good firmness. Curds are milky white in colour and weigh 1.25-1.5 kg maturing in 60-65 days.

NS 50

A mild maturity hybrid. Heads are compact, uniform and deep dome shaped. Holds shape very well even when harvesting is delayed. Uniform heads with tolerant to downy mildew and black rot. Suitable for processing, fresh market and home garden.

Puspanjali

An early-mid variety, leaves are medium sized upright, greenish blue colour, curds are solid dome shaped and fine texture, weight 0.8 kg., mature in 70 days after transplanting.

Mid-Late group

Temperature requirement for curd initiation and development 12-16°C.

Pusa Synthetic

Developed at the IARI, New Delhi, synthesized from the lines with good general combining ability. Plants are erect, very vigorous, 70-90 cm tall, bluish-green to green, leaves, numbering 22-26 at the time of curding; curds some what self balanched, colour varies from white to creamy white, fine-textured and compact. It takes about 130 days from sowing to 50 percent maturity. The most ideal temperature for curding is 12-16°C. Sowing and transplanting time are end of August and end of September.

This is the first-synthetic variety, recommended so for in the vegetable crops. It can set seeds in plains without sacrificing yields. The cauliflower is tolerant to curd-blight and farmer can produce their own seed for 3-4 generations, vigour and yield are hardly affected.

Pusa Shubra

A selection form a triple cross (MGS 2-3', '15-1-1' and D 96') developed at the IARI, New Delhi. Plants erect with somewhat long stalk and light bluish-green leaves. Light bluish-green twisting back wards from the middle. Curds are of medium size flat, compact, white, each weighing 700-800 g. Once ready, they should be harvested without delay. They have a tendency to loose, yet least prone to ricyness. The best temp, for curd formation is 12-16°C. In plains sowing in late august and transplanting by late September are consider ideal. The curd yield is about 25 tonnes/ha.

Marghashira

It has pure white and big size curd. Seeds are sown in nursery during September. 25 days old seedings are transplanted in the main field. It matures within 80-90 days after transplanting.

Hisar 1

It is suitable for mid late season under Haryana conditions. Medium to large sized white heads. It yields 250 q/ha.

Amarabati

It is highly profitable variety and mature within 65 to 70 days. Curds are white and compact. Seed sowing should be done in September. It is medium group cauliflower.

Aparajita

It has big white and compact curd. Seeds are sown in nursery during September and curds are harvested in December. It is medium type cauliflower.

Goodman

Mature in 65 days after transplanting, head weight 2.0 kg and compact, snowwhite curd. Seeds sown in Sept.-Oct. and tolerant to fusarium, self blanching leaves structure. The leaves are spreading and cover the head very nicely.

Fastman

It mature in 60-65 days after transplanting. Very compact and snowwhite curds with weight 1.0 to 1.5 kg. Suitable for cultivation during winter season in plains. Resistant to fusarium. It is an early maturing cauliflower which gives nice white, non-hairy curds.

Lateman

It mature in 80-90 days after transplanting. Very compact self blanching white head, with upright growing leaves. It gives good heads having attractive white colour with average weight of 2.0 kg. Recommended for September to November sowing.

NS 66 (F_1)

Mid late season cauliflower with excellent uniformity and curd quality. Plants are very vigorous with medium blanch. Curds are compact and smooth. They are dome shaped and have excellent firmness and density. Each curd weighs 1.5-1.75 kg with excellent attractive pure white colour and mature in 70-75 days.

NS 75 (F_1)

This hybrid cauliflower is very much suitable for early winter season. It has very vigorous plants with dark coloured foliage and an excellent blanch covering the curds. Each curd weighs 1.5-2.0 kg, which are dome shaped and admirable snow white colour. Curds are compact with good firmness and density. Matures in 80-85 days.

NS 76 (F_1)

This cauliflower has very vigorous plants with good blanch. Curds are compact and dome shaped with excellent milky colour, firmness and density. The curds weigh 1.5-2.0 kg each. It is mild season cauliflower maturing in 75-80 days.

Romanesco

It is a fractal in structure, with conical head formed of small peaks of buds. It has slightly waxy, large, grey green leaves, bigger in curd size (0.6-0.75 kg each) complex, symmetrical head features whorls of pointy, vivid green florets, self blanching type. A good source of vitamin C for antiageing, wound healing, decreasing blood cholesterol. It takes 120-125 days for head formation after transplanting.

Cassius

This is vigorous growing hybrid, which gives a strong and healthy plant. It is having very nice self-blanching leaves which gives snowwhite curds. Mature in 65-75 days, good comapact head (curd) with average weight of 2.0 kg. Recommended for August to December sowing. Tolerant to *Fusarium* and other diseases.

Anarkali

Upright plants. Sown in September-October, average weight 1.5 kg. curds are being harvested within 65-70 days, curds are white, dome shaped.

Puspagiri (F_1)

Suitable for western and southern region of india, maturing in 55-60 days after transplanting, curds are very compact medium sized and of snow white colour.

Himsagar (F_1)

Suitable for western and southern region of India. Maturing in 65-70 days after transplanting, curds are very compact, medium sized and of snow white colour.

Late group

Temperature requirement for curd initiation and curd development 10-16°C.

Pusa Snowball-1

Selection made at Katrain from an inter-varietal cross EC 12012 X EC 12013. Straight, upright leaves, tightly covering the curd. Curds are compact, medium in size and snow-white in colour. Suitable time for sowing in northern India is the first fortnight of October.

Pusa Snowbatt-2

It has upright leaves covering head tightly and slightly puckered at the margins. The curds are solid, white with very good staying power and slightly raised in the center. It takes 110-135 days from sowing to curd harvesting. It yielded 150-120 q/ha in hills and 400-500 q/ha in plains.

Snowball-16

The leaves broad, semi-erect, darkgreen with waxy coating. The curds are snow-white compact, medium to large sized. Its average yield of curds is about 200 quintals per hectare.

Pusa Snowball K-1

Field resistance to black-rot. Selection made at Katrain from an exotic material. It has upright, wavy, light green leaves, self-blanched, snow white, hi-dome curds. Becoming increasingly popular with the growers because of its curd quality and late maturity. Period of fresh availability is extended by at least 15 days.

Pusa Snowball K-25

It has been developed by mass pedigree method of selection from cross of heading broccoli EC 103576 and Pusa Snowball-1. Its curds are very solid, compact, medium, sized, retentive shining white with very good staying power. Inner leaves cover the curd very tightly. Leaves are waxy, slightly bending towards inner sides with puckered margins. It takes 175-180 days from transplanting to first harvesting at katrain, and 105 to 130 days in different parts of the country. It is a very late variety, average yield being 175-300 q/ha.

Dania

This variety developed at Kalimpong, successful in the eastern region, more particularly the eastern hills. It has very sturdy plants with wavy leaves and medium-deep curds. It is sensitive to fluctuating environment.

Dania Kalimpong

More commonly grown in Eastern parts of India. Plants sturdy, wavy leaves; curds medium deep; it is less sensitive to fluctuations of environment.

Punjab Giant-26

Main season variety, foliage green, plant height 57 cm with a spread of 75 cm. Curds solid, white, medium in size, developed by P.A.U., Ludhiana.

Punjab Giant-35

The leaves erect and taller than Punjab Giant-26. The curds are snow white in colour, compact and large sized. The yield potential is higher (about 325 q/ha).

Himyug

It is late type and seed is sown during late Nov.-Dec. in nursery and 25 days old seedling are transplanted in main field. It matures within 70-80 days after planting. Curds are bigger in size i.e. 1100 g. Leaves are dark green and curds are more white and compact.

Garima

Early maturing vigorous hybrid, curds can be harvested approximately 45-50 days after transplantation. Leaves are bluish green coloured. Curds are dome shaped, white and weigh 700-800 g, suitable for planting from middle of February to end of May. Tolerant to black rot and heat.

Sharad safedhi-70

Top quality hybrid in early snowball group. Curds are snow white, dome shaped and relatively large weighing 1200 to 1500 g, upright leaves provide excellent protection to the developing curds. The hybrid has flexible adaptability to different cliamatic conditions.

Mahima-80

Late maturing hybrid of mid snowball group. Curds ready for harvest in about 80-85 days from transplantation. The plants are vigorous, upright and have excellent self blanching habit. Curds are dome-shaped, pure white and weigh 2000 to 2500 g and fairly tolerant to black rot disease. The bluish green coloured waxy leaves provide some protection from insect pests also. This hybrid is recommended for winter season cultivation.

Sharmilee-85

Vigorous growing hybrid of late snowball group. Curds can be harvested in about 85-90 days after transplantation. Curds are snow white, high dome-shaped and weigh 1750 to 2000 g. This hybrid is recommended for winter cultivation in the plants and for summer in the hills.

Shivangi (F_1)

Snow ball type cauliflower with very vigorous plant habit. It is suited very well for late season sowings. This hybrid has very good blanch giving very good protection to the curds. Curds are very compact and dome shaped with excellent firmness. Each curd weighs 1.5-2.0 kgs and is of milky white colour. Matures in 90-95 days.

NS 84 (F_1)

This hybrid is of winter type late cauliflower with vigorous plant habit and an excellent bloom. Plants are uniform with compact curds and superior blanch. Each curd weights 2.0 kgs and is of a superb pure white colour and deep dome shape. Matures in 85-90 days.

NS 90 (F_1)

It is a late season cauliflower with large heads bred for winter season sowing in areas with long winters. It has very vigorous plants with excellent blanch giving good tolerance to cold. Curds are compact, deep dome shaped with excellent pure white colour. Each curd weights 2.0-2.5 kgs and matures in 90 days.

A few F_1 hybrids of 3-5 months maturity have also been developed by Australian Company. Some of these hybrids are Flanca, Sebal, Gija and Nedcha.

CLIMATIC INFLUENCE

Varieties of cauliflower are very much sensitive to temperature and photo periodic requirements. It is therefore essential to choose a proper variety to be shown at the proper time. Cauliflower generally require a cool and moist growing season, it can not withstand so low temp or so much heat. Dry weather and low humidity cause curd to be small and hard. Varieties of early group require higher temperature and longer day length.

In principle, cauliflower now can be grown in wide range of climate but during the transformation from vegetative to curding and curd development stage, the climatic factor play very important role. Some deviation may also be attributed to light condition and also to water regime. At higher altitude the longer day length promotes much faster growth of the plant. (Salter 1960). The cauliflower has a wide range in adpatation to diverse climatic conditions prevailing in temperate, subtropics and tropical parts and therefore, the cultivation of cauliflower is wide spread in many contries of the world. The optimum temperature for growth of young plants is around 23°C, but in later stages 17-20°C are most favourable. The tropical cultivars show growth even 35°C. In temperate regions, the growth of young seedlings may be ceased, when temperatures are slightly above 0°C, while, the early cultivars gown in plains of north India and other tropical parts can grow even at 35°C or still higher temprature (Nieuwhof, 1969). Lower temperatures ranging from 5°C to 28-30°C are needed for trasition from vegetative to curding phase. The termperature higher or lower than the optimum required for curd formation of the cultivars may cause physiological disorder viz. riceyness, leafy curd and blindness. The optimum temperature for most of the temperate cultivars is the 15°C to 20°C for growth. In tropical cultivar 20°C temperature transform vegetative to the curding stage.

SOIL REQUIRMENT

Cauliflower can be grown in all types of soil with good fertility and good regime. In light soil, the plants are more sensitive to drought and therefore, adequate moisture supply is important. For ealry crops, the light soils are preferred, while, loamy and clay loam soils are more suitable for mid season and late maturing types. Cauliflower is relatively more sensitive to deficiency of boron and molybdenum, and it has high rquirement of magnesium. The deficiency of magnesium may quickly appear in acid soils. High pH reduces the availability of boron. It requires on an average pH range of 5.5 to 6.6 and below 5.5 pH it would be necessary to do liming at the rate of 4 to 8 tonnes/ha.

SOWING

Sowing should be done according to variety and their climatic requirements. Generally a cauliflower crop can be grown from April-May to Dec-Jan i.e. near about 8-10 months in a years.

Sowing Time

The optimum time of seed sowing in the nursery varies greatly depending upon climate, varieties and their temperature requirement for curd formation. The nursery for mid season crop can be raised successfully by protection of beds from heavy rains during July-August is essential. The seedlings for mid late and late crop can be conveniently raised, because of favourable temperature conditions. Depending upon curd maturity of the varieties, the following sowing times have been recommended which details are mentioned below:

Group	Time of sowing of seeds	Time of curd maturity
Early maturity group		
I (a)	May-June	Sept.
I (b)	First fortnight of July	Oct. to mid Nov.
Mid season maturity group II	July-August	Late Nov. to mid Dec.
Mid late maturity grup III	September	Late Dec. to mid Jan.
Late maturity group (Snowball type)	October	Mid january onwards.

(Swarup and Chatterjee, 1972)

Seed Rate

For cultivation of cauliflower in one hectare, 200-500 g seed would be needed, though, in the early crops, the rquirement may be more to the tune of 1kg.

Seed Treatment

Seed may be treated with 2.5 captan or 3g thiram + 1.5g streptocyclin per kg of seed. To avoid becterial rot, treat the seed with hot water at 52°C for 10 minutes.

Nursery Bed Preparation

Nursery soil should be deep, well drained comparatively rich and free from stones and pables is considered to be a good garden soil. Beds should be 3 m in length and 1 m wide and 15 cm high from the ground.

20-25 beds of this size are sufficient to produce enough seedling to cover one ha. land.

The nursery beds should be prepared by addition of well rotten farmyard manure or compst @ 10kg/m^2. The soil of the beds is made friable by thorough digging/ploughing and harrowing and breaking the clots. The sterlization of soil by drenching, the nursery beds with formalin @ 1:49, about 15-20 days before seed sowing is beneficial for preventing the attack of the fungal diseases. After drenching, the beds should be kept covered with polythene for a week and then beds are again dug and left open for 5-6 days. So that, there is no injurious effect of formalin ot the seeds. The nursery must be raised about 15 cm high from ground surface for efficient drainage.

Sowing of the Seeds and Care of Seedlings

Sow the seeds in shallow furrows at 10cm apart. The depth furrows may be 1.5-2 cm. After sowing the seed, the furrows are covered with a mixture of sieved fine manure and soil and the beds are gently leveled at the surface and press a little by palm followed by watering. For better and early germination, the beds may be covered with straw mats which will conserve moisture, regulate temperature and protect the seeds and sprouts from heavy rain specially in early season nursery. Within four days the sprouts will come up after which the mats should be removed and palced at about 1 feet above the seed beds to protect the young seedling from heavy rain and strong mild day sun. After germination, exposure to one hour morning sun and one hour evening sun is beneficial. Gradually the light hours should be increased during both the morning and evening hours. The two weeks old seedlings should be fully exposed to sun and open air conditions. Watering must be done carefully, hoeing, weeding should be done in the morning and the soil should be kept loose throughout the day time and water capiously in the evening. During third week the seedling need to be hardened to bear the shock of uprooting. It is done by curtailing the water supply and allowing the plant to just survive with minimum water requirement to develop woodiness. It can only be done by keeping the beds dry by gradual withholdings of water supply and allowing the plants just to survive. This shock treatment is to be given 2-3 times within a week time and one week before transplanting. To protect the seedling from insect pests and disease attack a common insecticide i.e. Rogor @ 1.5 ml/l and Dithane Z-78/Blitox- 50 @ 3g/l. may be sprayed at weekly intervals.

Weeding and interculture of nursery beds should be regularly done for raising healthy seedlings. Generally, the seedlings are ready for transplanting in 3-6 weeks after seeds sowing, depending upon variety, temperature and soil fertility. In case of early crop, 5-6 weeks old seedlings have better establishment and less mortality in the field. while in mid-season and late varieties 3-4 weeks old seedlings may be transplanted.

PLANTING

Treat the seedlings with starter solution before transplanting. Dissolve 4kg. 5 : 10 : 10 : N_1, P_2O_5, K_2 O in 225 litter of water or 24 g (8 g : 16 g : 16 g : N, P_2O_5, K_2O) in 2.25 l. of water for preparing starters solution.

Planting distance

The planting distance may vary according to the variety, climate and soil. The seedlings of early Kharif virieties are transplanted on the ridges of 20 cm high are made in the prepared field at 50 cm distance row in a well drained soil. The seedings are transplanted 30 cm apart in the row with a little compost in normal pits. The main season and late varieities are planted in a flat bed on spacing of row to row 60-75 cm and plant to plant 40-50 cm. Older seedlins should not be transplanted as they may result in buttoning.

Singh *et al.* (1977) observed that transplanting of seedings of 'Snowball-16' variety by 15th October proved beneficial over other dates of trasplanting in North-East Indian conditions. It requires and average of 500 to 650 g of seeds for the early varieties and 400 to 500 g for medium and late varieties to raise the nursery for planting of one hectare land. Srivastava *et al.* (2002) reported the effect of date of planting and growing environment on the plant survival, growth and yeld of early cauliflower in rainy season at Pantnagar during 1998-99 on the variety Pant Gobhi-4. The maximum yield of 194.56 q/ha and 378.72 q/ha were found due to the planting on 10th August and under polyhouse planting respectively. Ghanti and Mallik (1992) reported the influence of planting time on stem growth and curd compactness in early cauliflower. August transplanted crop produced the highest length of stem and june transplanted crop recorded the lowest length. Varieties like Patna No.1 and Kartika were adjudged as best, whereas Early Market as least so far as curd compactness was concerned. Singh *et al.* (1997) evaluated the yield and quality of cauliflower cv. Best Early as affected by different dates of planting. It was observed that transplanting between 25th November to 5th December during 1995 and 1996 took minimum days for 50 per cent flowering, produced curds of maximum weight and diameter and ultimately between 25th November to 5th December was found to be optimum for getting a quality yield. Choudhary *et al.* (1999) studied the curd yield of mid season cauliflower as influenced by dates of planting and plant density. The planting date of 16th July and plant spacing of 45 cm X 45 cm proved to be the most promising for getting highest curd yield in mid cauliflower cv. Pusa Deepali.

Sharma and Choudhary (1996) studied the time of sowing and plant density on growth and curd yield in early cauliflower (*Brassica oleracea* L. var. *botrytis*) cv. Early Kunwari. Tooo early/late plantings resulted in decreased yield and other horticultural traits. The planting date of 12th May and plant spacing of 30cm X 30 cm proved to be the most promising for getting optimum curd yield. Gautam *et al.* (1998) reported the effect of sowing dates and veriety on growth and yield of early cauliflower at Jorhat (Assam) during 1993-94. Sowing on July 15th gave the maximum curd yield (78.48q/ha). which significantly declined with each delay in sowing time. Among the varieties 'Heavy Silver Plate' gave the highest curd yield with and average yield of 81.15 q/ha of marketable curd. Srivastava *et al.* (2002) reported the response of early cauliflower to dates of planting in open and protected environments during rainy season at Pantnagar. Maximum plant survival was achieved in 10th August planting and polyhouse environment as against the minimum of 17.97 per cent observed in open field flat be planting. It is revealed that cauliflower has early, mid-early mid late and late sowing seasons, thus varieties are specifice according to the season. Appropriate sowing season gives maximum curd yield. The planting space varies from 45-60 X 30 cm for early varieties and 60 X 45 cm for mid and late season varieties (Nath *et al.*, 2008).

NUTRITIONAL REQUIREMENT

Cauliflower is a heavy feeder. An average crop of cauliflower giving 16 tonnes/ha of marketable curd removes 46kg nitrogen, 16kg of phosphorus and 55kg of potash.

Manure and fertilizers requirements in cauliflower are depend upon fertility status of the soil, which is determined by soil testing. If soil testing is not possible, then 150-200 q/ha farmyard manure should be applied in the soil before 15-20 days of transplanting. It should be mixed thoroughly in the field. The quantity of nitrogen, phosphorus and potash should be given after determining the fertility status of soil. However, research carried out under All India Coordinated Vegetable Imporvement Project for different state of the India, following recommendation for quantity of nitrigen phosphorous and potash have been given (Anon, 2000).

- A dose of 40 kg N/ha as basal + 40 kg N/ha as top dressing + 30 kg N/ha foliar spray has been recommended for variety Snowbal-16 under Sabour conditions. However, for Pantangar conditions, 60kg N/ha as basal + 30 kg as foliar spray in three splits has been recommended for the same variety. Further, at Kalyanpur, the best response was 60 kg N/ha as basal + 60 kg N/ha as top dressing in two equal splits.
- Application of 150 kg N/ha +120 kg P_2O_5 per hectare has been recommended for obtaining economic yield under Kalyanpur conditions.
- For variety Pant Shubra, recommendation of 100kg N, 60 kg P_2O_5 and 60 kg K_2O per hectare has been made for Faizabad conditions.

In addition to NPK, borax @ 15 kg/ha and ammonium molybdate @ 1.5 kg/ha should also be applied in deficient soil of boron and molybdenum, respectively. Half doses of nitrogen and full doses of each phosphorus and potash are applied at the time of transplanting. Remaining quantity of nitrogen is applied 30-35 days after transplanting.

Depending on the climate and soil conditions following recommendations have also been given: FYM 20-25 tons/ha. nitrogen 100-120 kg/ha. phosphorus 60-80 kg/ha and potassium 80 kg/ha (Swarup, 2006). The complete dose of FYM should be applied, and thoroughly mixed in the soil at the time of field preparation and the complete dose of phosphorus and potash is applied before transplanting. Half the dose of nitrogen may be applied alongwith the potash and phosphorus before transplanting and the other half in bands around the well established plants one month after transplanting followed by irrigation. Chhonkar and Sharma (1966) reported that the application of starter containing ammonium sulphate and single superphosphate (1.2:1.8 kg of mixture in 225 litres of water) in liquid form in combination with 0.1 ppm IBA was most effective in increasing the total yield of cauliflower.

Lawade *et al.* (1991) reported the closer spacing of 60 X 45 cm and application of 225 kg N/ha seemed to be profitable proposition to boost the yield of cauliflower cv. Snowball-16. Balyan *et al.* (1988) reported the maximum yield at 160 kg N, 50 kg P_2O_5 and 20 kg $ZnSO_4$/ha in the cauliflower variety snowball-16. Kotur (1998) reported on the standardization of foliar spray of boron for correction of brown rot and for increasing yield of cauliflower in Bihar plateau. Foliar application of sprays of 0.125 per cent boric acied was optimum for high curd yield (23-46t/ha. 76 per cent higher than the control), low curd rot (5%)

negligible residual hot water soluble B in soil (0.11 mg/g) and high net ruturns per rupee invested (Rs. 4.47). Arora *et al.* (1989) reported the effect of cytozyme and miraculan on yield of cauliflower, cabbage and knol-khol. Trials were conducted druing 1981-82 to study the effect of cytozyme and miraculan on yield of cauliflower, cabbage and knol-khol. Early harvest was obtained in all these crops when sprayed with cytozyme @ 0.2 per cent, weight of curd/ head/knob and yield were improved under the same treatment. Miraculan treatments performed poor as compared to cytozyme but were better than control. Kotur (1998) reported on synergistic interaction of lime, boron and molybdenum on curd rot and curd yield of cauliflower. Lime application to soil combined with three foliar sprays of 0.1 per cent boric acid and 0.1 per cent ammonium molybdate together is optimum for high curd yield, low curd rot and high net returnes for cauliflower grown on Alfisols of Bihar plateau. Singh (2003) reported that application of borax @ 5 kg/ha as soil application + 0.25% foliar spray at 45 and 60 days after planting gave highest curd yield with highest cost: benefit ratio (4.20) of cauliflower var. Pusa Syntheitc grown in lateratic soil of western Orissa.

Use of Bio-fertilizers

Use of biofertilizer is also helpful in increasing the curd quality and curd yield of cauliflower and it also reduces the requirement of nitrogenous fertilizer and efficiently increases the use of phosphorus to the plant. It can be applied in different way. Azotobacter is free living nitrogen fixing bacteria, phosphate solubilizing micro organism (PSM) and vescicular arbuscular mycorrhizae (VAM) play significant role in increasing the curd yield and its quality.

Seed treatment

About 200g of bio fertilizer is required to treat 10-14 kg of seeds. Suspend one packet of 200g in approximately 400 ml water and mix it thoroughly. Pour this mixture on seeds and mix with hands to obtain uniform coating on each and every seed. Spread the seeds in shade for drying for 10-15 minutes then sow them immediately.

Seedling treatment

Prepare the suspension by mixing 1 kg of culture in 10-15 litres of water. Get seedling required for 1 acre and make small bundles of seedlings and keep the root portion in the suspension for 15-20 minutes. Transplant these immediately. Generally the ratio of inoculant and water should be 1:10 approximately, i.e. 1 kg packet in 10 litre of water. Usually 1.5-2.0 kg of N bio fertillizer required for one hectare.

VAM Fungi

Nursery beds are inoculated with VAM fungi inoculum consists of spores of the fungus with the soil and the infected root bits of the host plants in which they are raised in either spread uniformly in the nursery bed or placed in rows on which seeds are raised. The seedlings get colonized by the fungus by 45 days after which they are transplanted to the field, 200-500 g of inoculum could be added depending upon the concentration of VAM spores alongwith FYM and neem cake.

IRRIGATION

First irrigation is given just after transplanting. Further irrigation will depend upon weather, soild types and variety. However, regular maintenance of optimum moisture supply is essential during both growth and curd development stage. In areas having rainfall, the planting is done on ridges, furrow irrigation should be given. Even in some cases pot watering is also suggested for 3-4 days for proper establishment of seedlings. The early and mid season crops require only protective irrigation depending upon the monsoon showers. However moisture during both growth and curding phase should be adquate to maintain an even growth and proper development of curd. It may be given to the crop on every 5-6 days to the early planting stage and 8-10 days for late crop. The yield and quality are highest with single or double planting method in drip system of irrigation. However more commonly furrow irrigation followed.

INTERCULATURE

Shallow and frequent cultivation's should be given in the cauliflower field to kill young weeds and to provide soil mulch. Cauliflower being shallow rooted crop, most of its roots are within 45-60 cm of the top soil. Deep cultivation should, therefore be avoided. Hoeing, weeding and light earthing should be given for 3 times starting from 3rd week of transplanting upto one and half month after transplanting. Earthing provide stability to the plants. Usually, in medium heavy and clay soils, there are crust fromation soon after transplanting. These crust must be broken, otherwise, water and air pernetration in root system are hindered, which affect plant growth. During rainy season, the roots of plants in ridge planting may be exposed, which requires adequate earthing. When earthing or ridging is done, the side dressing of nitrogenous fertilizer should also be done for healthy growth of the crop. During monsoon, the roots may be exposed after every showers, specially on ridge planting and in such cases earthing more than the usual number may also be necessary.

Blanching

In cauliflower, for getting quality curd, blanching is an important operation to protect the curds from yellowing due todirect exposure to sun (Mehta, 1959; Srivastava, 1960). The curds may also loose some of their flavour, because of this exposure. This problem generally occurs in such varieties of early and mid season maturity group, which have spreading and open plant type. In some varities, the curd remains naturally protected and surrounded by inner whorls of leaves. These are called self-blanching types. Drawing and tying the tips of leaves when curds are fully developed may do the blanching. The other way to place a leaf cauliflower over the curd 4-5 days prior to harvest.

Mulching

Polythene mulch had beneficial effects on yield while straw had adverse effect 10 cm thick mulching with paddy husk had been found to be beneficial in increasing the growth and yield of cauliflower (Patel and Jyotishi 1969). Mulching and long intervals of irrigation are found to be as good as short intervals of irrigation and no mulching. Mango leaves mulch is highly economic and beneficial and it give maximum return.

Crop Rotations

Land has to be utilized in such a way that maximum return is obtained because it is a scarce commodity in present days due to ever increasing population. Maximum returns were obtained from cauliflower-tomato-okra (Randhawa and Singh, 1977) and also from late cauliflower when it was intercropped with spinach. Randhawa (1990) and Saini (1996) have suggested the following crop rotations that have been found to be beneficial under Indian conditions.

i) Cauliflower - Tomato - Okra
ii) Eggplant - Cauliflower - Bottle gourd
iii) Cauliflower (Early) - Radish - Chilli
iv) Okra - Cauliflower (main crop) - French bean
v) Maize - cauliflower (mid-season) - Brinjal
vi) Cauliflower (Early) - Onion
vii) Cauliflower (Early) - Carrot (Seed crop)

viii) Cauliflower (Early) - Radish (Seed crop)
ix) Cluster bean - Cauliflower (Main crop) - Bittergourd
x) Greengram - Cauliflower (Late) - Okra
xi) Blackgram - Cauliflower (Late) - Cowpea
xii) Sponge gourd - Cauliflower (Late) - Okra
xiii) Cowpea - Radish - Cauliflower (Late)
xiv) Cauliflower - Chilli - Jackbean
xv) Cauliflower - Potato - Dolichos
xvi) Garlic - Turnip - Cauliflower

WEED CONTROL

Weeding should be started as soon as plants are set in the field. Generally three hand weeding is done starting from third week of transplanting up to one and half month after transplanting to keep plot weed free. Some times one or two shallow cultivation or hand weeding and pre-plant application of Fluchloralin at 1.2 kg per hectare or Nitrofen at 2.0 kg per hectare controls most of the weeds effectively (Randhawa *et al.*, 1987). According to Singh *et al.* (2005), Pendimethalin @ 1.25 kg/ha as pre-plant application recorded the lowest uptake of nitrogen, phosphorus and potassium in the crop. In commercial weed control, basalin @ 2-3 kg a.i./ha + 1-2 hand weeding or tenoran 50 WP @ 1 kg a.i./ha can be used.

Use of Micronutrients

Micronutrients like Zinc, Boron, Moybdenum and Sulphur are essential for good yield of cauliflowers. (i) Folair spray of Borax @ 0.25% (2.5g/lit of water) thrice at 10 days interval starting from 30 days after planting will improve the curd quality and yield. (ii) Application of 10-12 kg/ha Borax at field preparation in boron deficient soils will help to improve the curd yield and quality of curds. (iii) Tracel-2 which contains Molybdenum, Suphur, Zinc and Boron is applied to the soil @ 10-12 kg/ha or its 0-4% solution is spray on the twice at an interval of 15 days improve curd quality and increase yield. (iv) Spraying of 0.4% Borax solution twice on the seedling and increase yield. (iv) Spraying of 0.4% Borax solution twice on the seedling to correct boron deficiency is effective. (v) Application of Dhanzyme granules @ 8-16 kg/ac in two split first at 10-15 days after planting and second 20-30 days later increase curd yield and curd quality. (vi) Application of Dhanzyme liquid@ 1ml/lt of water (180 ml per acre) as a foliar spray, with high volume sprayer at 6-8 leaf stage and 15-20 days later improve curd yield and curd quality. (vii) Application of plantaid as foliar spray 3ml/lit.

of water twice at an interval of 3-4 weeks when plant shows symptom of deficiency. It improve the plant growth and give better flowering as well as better curd seed formation. (viii) Incorporation of plantaid (S) @ 10 kg/ac in soil in a single or split dose increase vegetative growth, flowering, curd yield and quality.

Use of growth requlators

Use of NAA 10 ppm as starter solution for the treatment of couliflower seedlings has been found effective in respect of plant stand in the field and the vegetative growth. Application of GA_4 + GA_7 at the rate of 80 mg/litre of water shortened the period from transpanting to the harvest. The quality of the curd was not affected by this treatment (Booij, 1990). In another study Reddy (1989) obtained maximum plant height, curd diameter and yield of cauliflower by two applications of GA_3. Synergistic effect of mineral nutrition and growth regulators have been noticed for plant growth and curd yield of cauliflower. Besides GA and NAA have also proved to be effective in increasing the yield (Muthoo *et al.*, 1987). Various recommendations on use of plant growth regulators (PGRs) are given below Table 1 :

Table 1 :

Name of PGRs	Concentrations (mg/l)	Methods of application	Attributes affected	References
IBA	10	Seedling treatement	Increases yield	Chhonkar and Sharma, 1966
GA+NAA+Mo	100+120+0.2 per cent	Foliar spray	Increases yield	Muthoo *et al.*, 1987
GA + Urea	50+1 per cent	Foliar spray	Increases yield	Amarender and Reddy, 1989
GA_3	50	Foliar spray twice at 20 & 40 days after transplanting	Increases yield	Vijay Kumar and Rai, 2000

PLANT PROTECTION MEASURES

Some important insect pests, and diseases alongwith their control measures are given below :

Insect Pests

Mustard saw fly. (Athalia sp)

Adult is minute black fly that lays eggs inside the leaf tissue. The caterpillar feeds on the leaves of young seedlings. It can be controlled by spray of malathion @ 2 ml/l of water twice at 7 days intervals.

Mustard aphid (brevicoryne sp)

The greenish or black aphids damage the leaves by sucking the cell sap. The affected leaves get curled and plant wither away and die. Spray malathion @ 0.5% twice at 7 days inteval to control the pest.

Cabbage butterfly (pieris sp)

The young green ceterpillars feed on the surface of the leaves and skeletonise them. In severe case leaves, tender shoots, flowers and fruits are completely destroyed resulting in the rugged apperance of the attacked plants. Spray nicotinic sulphate @ 0.05% to check the infestation.

Semi looper (Plusia sp)

The green caterpillars which are loop while moving attack the crop and cause serious damage. Spray nicotinic sulphate @ 0.05% or thiodan @ 0.15% effectively control the all bove pest of cauliflower.

Diamond back moth (Plutella maculipennis)

This, causes damage to the plant and curd. Spray endosulphan @ 2 ml/l. of water in pre-curding stage to control the infestation. Spray malathion @ 2ml/l of water at harvesting stage and keep 7 days waiting period.

Diseases

Damping off (Pythium Sp.; Rizoctonia Sp.; Fusarium Sp.)

It is common in nursery seedlings. Rotting starts in the collar region of the seedlings. Drench the nursery bed with 0.1% solution of Brassicol/ Captan/ Fytolon. Treat the seeds with agrosan GN or Ceresan @ 2g/kg or Thiram 3g/kg of seed at the time of seed sowing in nursery.

Black rot (xanthomonas or Pseudomonas sp)

The margin of leaves turn yellow. It becomes dark and vascular region of the main stem becomes discoloured. The curd gets discoloured. (i) Follow crop rotation. (ii) The hot water treatment of seeds at 48.9°C for 30 Minutes give good control. (iii) Use resistant varieties.

Leaf spot or Blight (Alternaria sp)

The small dark coloured spot, appear on the leaves which spread rapidly and form circular lesions. In severe cases the cauliflower develop brown colour. The hot water treatement of seeds at 48.9°C for 30 minute give effective control. Treat the seed with Bavistin @ 1g/kg and apply three sprays of Indofil M-45@ 0.2% at an interval of 15 days soon after appearance of disease.

Club root (plasmodiophora brassicae)

In severe cases malformed roots present a clubbed appearance. Foliage wilt in sunny days and recovers towards the evening. It is prevalent in acidic soil. The infected areas should be abandoned for growing cruciferous crops. Treat the seedlings with mercuric chloride solution 250 ml of 0.05 to 0.01 percent solution for control. Also roots are treated with 4 percent calomel paste before transplanting. Application of 250 ml or 0.05 percent PCNB solution per hole. Properly liming of the soil is required.

Black leg (Phoma lingam)

The young plant in nursery bed are affected usually. The fungus attacks at the base of stem and roots and the plant wilts. Seed treatment with Bavistin @ 1.5g/kg seeds. Hot water treatments of seed at 50°C for 30 minutes. Follow proper crop rotation.

Physiological Disorders

Buttoning

Development of small curds with inadequate foliage in cauliflower is known as buttoning or premature heading. On planting an early variety late, the growth remains stunted due to low temperature which leads to buttoning. Slow plant growth in the nursery, over- crowding, insufficient water and poor nitrogen supply, water logging, hot dry weather, lack of weeding, bad condition of the soil, excessive salt concentration, low lying areas or fields with shallow and poor top soil may also cause buttoning. Vigorously grown nursery plants with lenky stem and sessile foliage, may also induce a tendency of button formation in field. Moreover, it is observed that any check in vegetative growth of seedling/plant may induced buttoning.

Riceyness

Premature elongation of peduncles bearing flower buds, rendering curds become granular, loose and poor quality for marketing referred as riceyness in cauliflower.

Normally riceyness in curds develops during the warm weather when the crop is close to maturity and harvesting has been delayed. If late varieties are planted early., riceyness develops due to the prevalence of high temperature. However, it can also appear at low temperature if present as an inherited trait. Besides, rampant growth, a heavy dose of nitrogen differently to riceyness due to genetic differences. Cultivation of genetically pure seeds and appropriate varieties with recommended cultural practices help keep this disorder at its minimum level.

Fuzziness

Fuzziness appears as a flower pedicels of velvety curds elongate. It is also caused due to the poor seeds, uneven temperature, over matured curd etc. Cultivation of cauliflower varieties out of their normal season encourages fuzziness. Sowing quality seed in the right season under proper cultural pracitces minimize fuzziness.

Leafiness

Extremely small green leaves in between the curd segments in other words when the bracts grow out the curds called leafiness due to heritable or non-heritable factors. Prevalence of high temperature during vegetative growth and curding phase aggravates leafiness. Selection of varieties according to their adaptability prevents this disorder. It can be controlled by selection of varieties according to their adaptability.

Blindness

When the growing point collapses at an early stage and the terminal bud fails to develop, a plant become blind. It may be due to low temperature, frost, cutworms and insect pests. Leaves that develop are large, dark green, thick and leathery owing to the accumulation of carbohydrates. But blindness may also be attributed to low temperature experience by small plants. It can be controlled by avoiding young plant from low temperature exposure and care seedling. While planting and handling, seedlings are avoided damage from insect-pests.

Whiptail

In cauliflower the leaf blades fails to develop properly and remain strap like and extremely savoyed. The leaves are ruffled and distored. In severe conditions only the midrib develops, and hence the name whiptail. Growing points are severely deformed and do not produce marketable curds. Sprouting near the base of the plants is also stimulated in extreme cases. Deficiency of molybdenum causes whiptail syndrome especially in highly acidic soils due to high manganese concentrations, in such soils hinders the uptake of molydenum which seldom occurs when the soil pH is 5.5 or higher. Drenching nursery bed with 2g of sodium/ammonium molybdate/sq.m two weeks before transplantings controls whiptail. Spraying of 0.01-0.1 per cent molybdate solution @ 250 g/ha is also effecive.

Browning (brown-rot or red-rot)

Browning of curd is caused by boron deficiency which is influenced by soil pH. The availability of boron decreases at neutral soil reaction.

Young leaves of boron deficient plants become dark green and brittle, the old leaves puckered, chlorotic and often drop off. Sometimes the downward curling of older leaves is followed by development of blisters when boron deficiency is severe, the leaves remain small and the growing point may die. Water soaked, light brown to dark brown spots formed on the stem and branches may ultimately lead to the formation of cavaties and a hollow stem. Curds may also show irregular water-soaked spots, which later change to a rusty brown colour. In seed crop, sometimes flowers and siliquas collapse. A topdressing of 10-15 kg/ha of borax (Sodium tetraborate) can control boron deficiency effectively. Spraying of 0.2-0.4 per cent solution of broax @ 2 kg/ha depending upon growth, soil reaction and extent of deficiency can also control boron deficiency.

Chlorosis

Cauliflower has a high magnesium requirement but its deficiency causes chorosis when grown on highly acidic soils. Chlorosis shows an interveinal, yellow mottling or lower older leaves. The affected leaves turn bronze in colour and become stiff. In severely deficient plants, abscission of the lower leaves occurs and results into smaller curd formation. The disorder could be avoided by applying magnesium oxide @ 300 kg/ha. Liming the soil with dolomite limestone to bring the soil pH to 6.5 is an effective control measure. Use of a fartilizer containing soluble magnesium, keeps it under control.

Hollow Stem

In heavily fertilized soils, particularly with nitrogen, rapidly, growing plants of cauliflower and broccoli develop hollow stems and curds. In broccoli it is maximum at the time of first harvest, which declines with successive harvests. Hollowness caused by boron may be identified by water soaked and discoloured tissue, where as hollowness cuased by nitrogen, the stem is perfectly clear. Stems do not show any apparent signs of disintegration. Hollow stem may be controlled by spacing the plants closer and with optimum use of nitrogenous fertilizers. Hollowness is controlled by spraying of borax at 0.1 to 0.3 per cent or soil application of borax @ 15-20 kg/ha, if hollow stem is caused by boron deficiency.

Premature Bolting

Development of seed stalks before the formation of sufficient edible part is known as premature bolting. It may be due to early sowing, warm open winters, cold winters, extreme fluctuations in temperature, poor growth, poor seeds, poor-soil and other cultural practices, affecting the growth rate. Exposure of plants to continuous low termperature (below 10°C) for 6-8 weeks result into premature bolting.

Frost Injury

In cauliflower leaves of young seedling turn yellowish white on both the surfaces. Petioles become flaccid and white, midrib alongwith adjacent parenchyma and stem may also be injured. Fully grown curds of cauliflower are more sensitive to frost damage than the smaller ones.

Frost injury can be minimised by irrigating the field on anticipating the danger of frost and by raising the field temperature by creating smoke.

Pinking

Sometimes curd show pink tinge, this appears due to the exposure of curds to high light intensities. Under this condition, anthocyanin form and gives rise pink colour curds. This disorder is not so common.

HARVESTING

Cauliflower should be harvested when the curd has attained the proper size, bright colour and compactness and reaches prime condition, if there is any doubt about the maturity, it is better to harvest early than late. If the harvesting is delayed, the curds become over mature, whose quality is deteriorated. Such curds may turn loose, leafy, ricey or fuzzy. The over mature curds should be sorted out while sending the produce to market. For harvesting, the curds are cut off stalk well below the curds with a sharp cutting knife or sickle or khurpi, so that the stub thus left protects the curd during transport. The leaves should be trimmed away until only sufficient jacket leaves remain to protect curd. When transporting loose, more jacket leaves should remain than when packing in crates.

YIELD

Cauliflower produces an average of 20 to 30 tonnes curds containing 18,000-24,000 curds per hectare depending upon the maturity group of the varieties. Early varieties generally produce lower yield than the mid-season or late-season varieties. The average yield of an early variety is 10-12 t/ha and the yield of mid and late season varieties is around 20-30t/ha.

POST HARVEST HANDLING

After harvesting, it is desirable to sortout compact, well developed quality curds and loose, leafy, ricey and fuzzy curds seperately before formation of grades (Singh and Singh, 2007).

Grading

Cauliflower is separated into three grades, depending on size, colour quality and variely. Grade "A" (super) has big size head with, bright colour fully compact and in prime condition. Grade "B" (fancy) medium size curd as for varietal characterisitcs, compact with full shining and without any spot on the curd. Grade "C" (Commercial) medium to small heads with general curd condition and free from any blemish curd should be sorted out from the whole lot.

Packing

Handle the cauliflower curds (heads) carefully because they are tender and damage easily. The usual method of packing for transport is in big nets and then loaded in a trucks. For distant market curds are packed securely in baskets and huge quantity in Jhallies. For local market it is packed in loose gunny bags and upper portion cover with grasses and stitched properly. For export purposes it is packed in crates and wire bound boxes. Cellophane and transparent films are also used to pack cauliflower which are hydrocooled to remove field heat. Tight packing is essential to prevent shifting and bruising.

Storage

The edible curds of cauliflower with leave may be stored in open for 2-3 days during winter season. The curd can also be stored for 3 to 4 days at ordinary temperature. But it is difficult to store as such in summer. However good sound curd with leaves attached can be stored for 30 days at 0°C with 85 to 90% relative humidity. Provision of ventilation should be there to increase storage life.

Processing Quality

(i) Drying of cauliflowers curd for lean period to use product during off season is also in practice in rural and urban areas. (ii) Cauliflower curds are also used for preparation of different type of vegetable pickles. (iii) Cauliflower seeds are also processed for edible oil. Variety 'Pusa synthetic-1' has oil content of 37.1%, which contains Palmitic acid 3-9%, Oleic acid 13.8% Linolenic acid 10-7%, Linolenic + Eicosenoic acid 14.6% and Erucic acid 53.6%; these are fatty acid. Punjab Giant 26' contain 39.3% oil which has palmitic acid 4.1% oleic acid 13.3% linolenic acid 10.5%, linolenic+ eicosenoic acid 16.1% and erucic acid 54.%. All these comes under fatty acid group.

SEED PRODUCTION

Cauliflower is a cross pollinated crop. Pollination is normally done by bees. It is possible to produce the seeds of the varieties of first three maturity groups. i.e. Indian cauliflower varieties in different parts of the plains. The usual method is to leave the palnts with good curd in the field which later produces flower stalks and flowers and the seed ripens from March to May. But this method occupies lot of area becuase plants for seed production are left here and there in the field. It is recommended that the selected plants with best curds may be uprooted carefully and replanted in a compact block for seed production. In this process, the curd of the stecklings before transplanting or at the time of transplanting is given scooping or incision in the middle of the curds to facilitate the growth of the side seed stalks which permit better quality seed production. High yield of seeds, however, has been obtained from the pants which were left in situ and not transplanted (Choudhury and Ramphal, 1961). While transplanting, the stecklings may be spaced at 75 cm X 75 cm. It is necessary that the field is well prepared and manured before transplanting. The bees activities for efficient pollination should be encouraged. Use of insecticide spray at the flowering time should be avoided. The seed gets ready from March to May depending upon the maturity group. It should be cured, threshed, cleaned, graded and stored before the onset of rains. Whereas mid late and late varieties need much cooler period and seed production is done in the hill.

There are two methods of seed production. (i) In situ method. (ii) Transplanting method. In *in situ* method the plant with good true to the type selected and left as such in the field which later produce flower stocks and flowers. For good flower stocks and flower pod, the scooping is done by removal of some curd, with help of knife before flower stock arises. Flower pods after maturity ripen from March to May. In transplanting method the selected plant with best curds uprooted carefully with an earth ball and replanted in compact block for seed production. In this process the curd of the stecklings before transplanting or at the time of transplanting is given scooping or incission in the middle of the curd to facilitate the growth of the side seed stalks which permit better quality seed production. While transplanting the steckling may be spaced at 75cm X 75 cm. It is necessary that field is well prepared and manured before transplanting. The bees activities for efficient pollination should be encouraged. Use of insecticide spray at the flowering time should be avoided. The seed get ready from March to May depending upon the maturity group. It should be cured, threshed, cleaned graded and stored before the onset of rains.

Since it is a cross pollinated corp, it is necessary to maintain the isolation distance of 1000 m between the varieties to avoid contamination (Anon., 1971). In early and mid season varieties (up to III maturity group) excellent seed set is observed in the plains and on an average, seed yield of 500-650 kg/ha is not uncommon. On an average about 220 kg of seeds/ha in 'Snowball' variety were produced in the hills (Singh *et al.*, 1959). Early and mid season varieties like. Pusa Meghna, Pusa Early Synthetic, Early Kunwari, Pusa Deepali, Pusa Sharad, Improved Japanese, Pant Gobhi-2, Pusa Himjyoti, Arka Kanti are recommended for seed production in plains. Late varieties like Pusa Snowball, Indian Snowball. Snowball-16, Dania, Daniakalimpong. Ooty-1, are recommended for seed production in hilly regions.

Seed Production of Main Season Cauliflowers

For successful seed production of main season varieties a temperate climate is needed. The place like Kashmir, Katrain are suitable for seed production. The seed production methods are given below:

Isolation Requirement

1600 meter for foundation seed class and by 1000 metres for certified seed classes.

Sowing

The seed sown in the nursery by the last week of August and transplanting should be completed by the end of September. The curds of these plants will be completed by the 1st fortnight of March when the temperature of 44°F to 52°F. Both earlier and later sowing than the above time is not suited for proper seed production as they result either in rotting of the curd or delayed flowering. In-situ method is most successful for seed production.

Soil, Manuring and Cultural Methods

Medium loam soils are well suited for it. The soil should be made fertile by mixing a sufficient quantity of well rotten FYM to it. The nitrogen requirement of cauliflower for seed production is very high so it must be supplemented by at least two application, one during Oct.-Nov. and other during Feb.-March to produce a vigorous crop.

Promotion of Flowering in Cauliflower

Without exposure to low temperature after curd formation, peduncles do not elongate properly and flower bud development is suppressed at the stage of primary protuberance formation, then the curd withers. The extent of chilling requirement for flower bud initiation in cauliflower is thought to be greater than that for curd formation. When the growing temperature is lower than the optimum temperature for curd development, riciness occurs probably resulting from the precedence of flower bud development over curd development. In contrast, when the growing temperature is higher than the optimum for curd development, fuziness, resulting from the development of bracholes, is observed, possibly due to the partial dieversion of curd development to the vegetative phase. Leafy heads in which green leaves grow through the surface of the curd occurs when cauliflower curds are exposed to temperatures higher than that including fuzzy heads. These leaves develop from axillary bracts of primary peduncles (Fujime and Okuda, 1996).

Generally, cauliflower curds after passing their maturity start loosing the compactness of the cruds that makes favourable condition for easy emergence of floral stalks. Sometimes, a cross cut is made on the curd like cabbage at the maturity that will help in formation of floral stalks and allow to draw away the rain water and to provide the developing branches better aeration. It will be beneficial if Blitox or Copper oxychloride (0.1%) is sprayed after the cutting the curd to prevent decay (Salunkhe *et al.,* 1987).

Roguing

Cauliflower varieties vary in their morphological characters viz, curd size, depth, colour, compactness, uniformity time of maturity etc. These characters are taken into consideration while roguing operations are done. Any character which is not conform with the variety grown does mean that plant is off type$_s$ and should be uprooted. Requing should be done at following stages.

1. Before normal curding period

All those curds which are very small button types, such plants need to be removed. Leaves should also be checked for their size (broad or narrow), spreading type or erect, colour (bluish green, dark green, light green), waxy coating or non-waxy leaves, leaf number, shape and extent of crinkle as per type should be checked.

2. Curd at market maturity

This stage of roguing will more helpful to restore genetic purity of the growing variety. Curds are checked for their colour, shape, size, absence of bracts, absence of riciness, compactness (loose curds are undesirable), coiour (variation than the true to type is undesirable), leaf protection of cured as per growing variety.

3. Maturity stage

At this stage, one can judge on the basis of pod maturity whether a particular plant(s) is early, or late maturity type. Based on type of variety taken for seed production the plants are removed.

The field standards for roguing of off-types are presented in Table 2. These standards also include maximum limit in percentage of seed borne disease plants.

4. Harvesting

Cauliflower pods tend to shatter their seeds. This necessitates, harvesting of mature pods before reaching to the stage of shattering. Mature plants become orange brown in colour. Most of the leaves become yellow and start shedding on the ground. Mature seeds also become brown which can be seen from a few pods after opening them; this indicates the proper stage for harvesting. Whole plant is harvested above ground and kept on cemented flower (threshing floor) or on tarpaulin for drying.

Threshing and Cleaning

Fully dried material is subjected to threshing by beating with wooden stick, trampling by bullock rolling under roller, threshing in mechanically operated threshers.

Table 2 : Field standards for cauliflower seed production (maximum permitted in per cent).

Factors	Foundation	Certified
Off-types	1.0	0.5
Other Corp Plants	None	None
Objectionable Weed Plants	None	None
Diseased	0.1	0.5

Threshed seed material is forked and cleaned through' blowing against wind for chaf, small and light seeds. Seed should meet the standards given in Table 3.

Drying

Cleaned seed is allowed to dry in open sun for about 6-7 days. Drying can also be done in electrically operated drier. The moisture percentage of seed should reach to 7.0 per cent.

Grading

Small, light, dead, split seeds are separated out from the heaithy seed lot by a spiral separator.

Packing

Graded seed lot is packed in suitable containers of standard size based on market demand and stored under dry, cool and sufficiently ventilated places for next season use.

Table 3 : Seed standards for cauliflower seed production (in percentage).

Standard	Foundation	Certified
Pure Seed (minimum)	98.0	98.0
Inert matter (maximum)	2.0	2.0
Other Corp seed (maximum)	0.05	0.10
Total weed seeds (maximum)	0.05	0.20
Objectionable weed seed (maximum)	None	None
Germination (minimum)	65.0	65.0
Moisture (maximum)	7.0	7.0

Seed Yield

The seed yield of cauliflower varies greatly due to variety, climate, soil type plant density, time of planting and management practices. Early maturing varieties have comparatively high yield potential than the late varieties. About 500-800 kg seed per hectare is obtained in early maturing varieties about 300 to 400 kg seed per hectare in late varieties. Higher seed yield is obtained in seed to seed method (*in situ*) compared to head to seed and stump methods.

REFERENCES

Amarender, Reddy, (1989). Effect of foliar application of urea and gibberellic acid on cauliflower (*Brassica oleracea* L. var. *botrytis*). *J. Res. APA V.*, 17: 79-80.

Anonymous (2000). AICVIP *Recommendations* (1971-98), IIVR, Varanasi, India.

Anonymous (1971). Indian minimum seed cerification standards. Central Seed Committee, Ministry of food and Agriculture, Community Developement and Co-operation, New Delhi.

Arora, S.K., Vashistha, R.N. and Pratap, P.S. (1989). Effect of cytozyme and muraculan on yield of cauliflower, cabbage and knol-khol. *Haryana J. Horti. Sci.*, 18 (1-2) : 94-98.

Balyan, D.S., Dhankar, B.S., Ruhal, D.S. and Singh K.P. (1988). Growth and yield of caulflower variety, Snowball-16 as influenced by nitrogen, phosphorus and zinc. *Haryana J. Hort. Sci.*, 17(3-4) : 247-258.

Bose, P.C. (1966). Things to know for growing cauliflower. *Ind. Hort.*, 11: 21.

Bosewell, V.R. (1949). Our vegetable travellers. *Natural Geog.*; 96: 145-217.

Chatterjee, S.S. and Mukherjee, S.K. (1964). Dania Kalimpong matures early and yields high. *Indian Hort.*, 8(4): 16.

Chatterjee, S.S. and Mukherjee, S.K. (1965). Selection and maintenance of cauliflower (*Brassica oleracea* L. Var. *botrytis*). *Indian J. Hort.*, 22(1) : 60-68.

Chhonkar , V.S. and Sharma, S.B. (1966). Studies on the effect of starters and plant-growth regulators on growth, yield and quality of cauliflower. *Indian J. Hort.*, 23 (1&2) : 81-87.

Choudhary, D.R. (1999). Curd yield of mid season cualiflower (*Brassica oleracea* L. var. *botrytis*) as influenced by date of planting and plant density. *Haryana J. Hort. Sci.*, 28 (1&2) : 106-107.

Choudhury, B. and Ramphal. (1961) Seed production with regard to yield and quality in early cauliflower. *Indian J. Hort.*, 18 (2) : 152-155.

Conev, T. (1959). *Horticulture abstract*, Vol. 48: 519.

Crisp, P. (1982). The use of an evolutionary scheme for cauliflowers in the screening of genetic resources. *Euphytica*, 31: 725-734.

Fujime, Y. and Okuda, N. (1996). The physiology of flowering in Brassica especially about cauliflower and broccoli. *Acta Hort.*, 407: 247-254.

Gautam, B.P., Shadeque, A. and Saikia, L. (1998). Effect of sowing dates and varieties on growth and yield of early cauliflower. *Veg. Sci.*, 25(1):1-4.

Ghanti, P. and Mallik S.C. (1992). Influence of planting time on stem growth and curd compactness in early cauliflower. *Orissa. J. Hort.*, 20 (Special) : 52-59.

Halappa, G. and Srinivas, L. (1973) Effect of black polyethylene film mulch on soil conditions and yield of cauliflower (*Brassica oleracea* L. var. *Botrytis*). *Mysore J. Agric. Sci.*, 6:510-519.

Hyam, E. (1971). Plants in service of man, London, pp. 33-81.

Jyotishi, R.P. and Bhalla, B.L. (1967). Study on the effect of some kinds of mulches under different intervals of irrigation on the growth and yield of cauliflower. *JNKVV. Res. J.* 1: 147-152.

Kale, P.N.; Chatterjee, S.S. and Swarup, V. (1979). Heterosis and combining ability in Indian cauliflower. *Veg. Sci.*, 6(1) : 11-20.

Kanwar, J.S. (1996). Effect of Sowing date on seed yield of cauliflower. *Seed Res.*, 22(1): 67-70.

Kato, T. (1964). On flower bud formation and development in cauliflower. I. Ecological studies on flower bud formation and development. *J. Jap. Soc. Hort. Sci.*, 33: 316-326.

Kotur, Shivaputra C. (1998). Standardisation of foliar spray of boron for correction of brown rot and for increasing yield of cauliflower (*Brassica oleracea convar botrytis* var. *botrytis*) in Bihar plateau. *Indian J. Agric. Sci.*, 88(4) : 2: 8-221.

Kotur, Shivaputra C. (1998). Synergistic interaction of lime, boron and molybdenum on curd and curd yield of cauliflower (*Brassica oleracea convar botrytis* **var.** *botrytis*) on an Alfisol. *Indian J. Agric Sci.*, 68(5): 268-270.

Lawade, K.E. Khaire U.A. and Bhore, D.P. (1991). Effect of various levels of spacing and NPK on yield of cauliflower. *J. Maharashtra Agric. Univ.*, 16(1) : 112-113.

Mehta, Y.R. (1959). Vegetable growing in Uttar Pradesh. Bureau of Agricultural Information, Uttar Pradesh.

Muthoo A.K., Kumar, S. and Maurya, A.N. (1987). Studies on the effect of foliar application of GA_3, NAA and molybdenum on growth and yield of cauliflower (*Brassica oleracea* L. var. *botrytis*), *Haryana J. Hortic. Sci.*, 16: 115-120.

Nath, P. (1976). Vegetable for the tropical region. ICAR, Low Priced Book Series No.2. ICAR, New Delhi.

Nath, Prem, Srivastava, V.K., Dutta, O.P. and Swamy, K.R.M. (2008). Vegetable Crops - Improvement and Production. PNASF, Bangalore. 135-138.

Nieuwhof, M. (1969). Cole crops-Botamy, cultivation and utilization. World Crop Books, Leonard Hill, Londan.

Pal, A.B. and Swarup, V. (1966). Gene effects and heterosis in cauliflower II. *Indian J. Genet.*, 26 (3) : 282-94.

Pandey, U.C., Lal, S., Mangal, J.L. and Singh G. (1981). Effect of date of transplanting on yield of cauliflower, variety Hissar-1. *Haryana Agric. Univ. J. Res.*, 11(3): 379-383.

Patel, R.M. and Jyotishi, R.P. (1969). Effect of paddy husk used as mulch on cauliflower under two intervals of irrigation and two levels of nitrogen. *Madras Agric. J.*, 56(4): 181-84.

Randhawa, K.S. and Singh, D. (1977). Studies on production potential of certain vegetable crops in multiple cropping in Punjab. *Indian J. Hort.*, 34: 264-269.

Rehman, F. (1962). Effect of mulches, levels of nitrogen and soil moisture condition on growth and yield of cauliflower and certain soil properties. Ph. D thesis Agronomy Division. IARI, New Delhi.

Saini, G.S. (1996). A Text Book of Vegetable Production. Aman Publishing House, Meerut (U.P.), India.

Salter, P.J. (1960). The growth and development of early summer cauliflower in relation to environmental factors. *J. Hort. Sci.*, 44: 129-140.

Salunkhe, D.K., Desai, B.B. and Bhat, N.R. (1987). Vegetable and Flower Seed Production. *Agricole Publishing Academy*, New Delhi.

Sandhu, J.S., Thakur, J.C. and Nandpuri, K.S. (1977). Investigation on hybrid vigour in cauliflower (*Brassica oleracea* L. var. *botrytis*). *Indian J. Hort.*, 25 : 430-34.

Sharma, D.K. and Choudhury, D.R. (1996). Time of sowing and plant density on growth and curd yield in early cauliflower (*Brassica oleracea* L. var. *botrytis*) cv. Early Kunwari. *Veg. Sci.*, 23(2) : 151-144.

Singh D.P.; Swarup, V. and Chatterjee, S.S. (1981). Development of synthetic varieties in Indian cauliflowers. *Indian J. Hort.*, 38(1&2) : 79-82.

Singh Ram; Gill, H.S. and Chatter jee, S.S. (1997). Breeding Pusa Early Synthetic cauliflower. *Veg. Sci.*, 24 (1) : 23-25.

Singh, D.B., Suryanarayanan, M.A. and Swaroop, Kishan (1997). A note on yield and quality of cauliflower var. Best Early as affected by different dates of planting . *Veg. Sci.*, 24 (1) : 75.

Singh, D.N. (2003). Effect of boron on growth and yield of cauliflower in lateratic soil of Western Orissa. *Indian J. Hort.*, 60 (3) : 283-286.

Singh, D.N. and Nath, Vishal (2011). Varieties and Hybrids of Vegetables. Satish Serial Publishing House, Azadpur, Delhi (India). 32-41.

Singh, D.N. and Singh, S.P. (2007). Production Management of Underutilized Vegetables. Agrihortica Publications. Junagadh (Gujarat) India. 42.

Singh, D.P. (1971). Genetical studies in Indian cauliflowers Ph.D. Thesis submitted to IARI, New Delhi.

Singh, D.P.; Swarup, V. and Chatterjee, S.S. (1975). Genetical studies in Indian cauliflower, heterosis and combining ability in maturity group-I. *Veg. Sci.*, 2 (1-2) : 1-7.

Singh, D.P.; Swarup, V. and Chatterjee, S.S. (1981). Note on self compatibility and incompatibility nature in Indian cauliflowers. *Indian J. Hort.*, 38 (3&4) : 229-231.

Singh, D.P.; Swarup, V. and Chatterjee, S.S. (1976). Genetical studies in Indian cauliflower: heterosis and combining ability in maturity group-II. *Veg. Sci.*, 3(1): 41-46.

Singh, G.; Bhan, V.M. and Tripathi, S.S. (1983). Weed control in cauliflower (*Brassica oleracea* L.). *Indian J. Weed Sci.*, 15: 1-6.

Singh, H.B.; Thakur, M.R. and Bhagchandani, P.M. (1959). Vegetable seed production in kullu valley-II seed production of cabbage, cauliflower, Knol Khol and chinese cabbage. *Indian J. Hortic.*, 16 : 153-61.

Singh, K., Tewari, G.N. and Pandita, M.L. (1978). Effect of methods of planting and plant population in cauliflower variety Hissar-1 *Vatika*, 1 : 87-90.

Singh, R.A., Jain, P.C. and Singh, S.M. (1958). Studies on transplanting of vegetable crops, effect of age and number of seedlings with and without starter solution and hardening on cauliflower. *Indian J. Hort.*, 17 (4) : 217-224.

Singh, Ram, Chatterjee, S.S.; Swarup, V.; Trivedi, B.M. and Sen, Bineeta (1993). Evaluation of Pusa Subhra-a resistant cauliflower to black rot and curd blight diseases. *Indian J. Hort.*, 50 (4) : 370-372.

Singh, Ram, Gill, H.S. and Chatterjee, S.S. (1997). Breeding Pusa Early Synthetic cauliflower, *Veg. Sci.*, 24(1): 23-25.

Singh, S.J.; Sastry, K.S. and Sastry, K.S.M. (1977). *Indian J. Hort.*, 21: 221-231.

Singh, V. (1955). Cauliflower seed production. *Indian J. Hort.*, 12: 19-21.

Singh. D.P.; Swarup, V. and Chatterjee, S.S. (1976). Genetical studies in Indian cauliflower: heterosis and combining ability in maturity group - III. *Veg. Sci.*, 3(1) : 47-53.

Srivastava, R.C. (1960). Steps to a bumper cauliflower crop. *Indian Fmg.*, 10(11):8-9.

Swarup, Vishnu, (2006). Vegetable Science and Techonology in India. Kalyani Publishers, New Delhi. pp. 656.

Swarup, V. and Chatterjee, S.S. (1972). Origin and genetic improvement in Indian cauliflower. *Econ. Bot.*, 26 : 381-393.

Tripathi, S.N. and Singh, R.N. (1969). A note on the method of seed raising in cauliflower. *Indian J. Hort.*, 26: 184-186.

Vijay, Kumar and Ray, N. (2000). Effect of plant growth regulators on cauliflower cv. Pant Subhra. *Orissa J. Hortic.*, 28(1): 65-67.

□□□

2. CABBAGE

Botanical name	:	*Brassica oleracea* var. *Capitata*
Chromosome number	:	2n = 18
Family	:	Brassicaceae (Cruciferae)

Cabbage is one of the most economically impotant member of cole crops. It is the most popular vegetable around the world in area production and availability, almost round the year. In India, it ranks second next to couliflower in area and production. Traditionally, it is grown to a large extent in the southern, eastern and coastal areas of India. But after the development of hybrids and some tropical lines it is produced almost round the year throughout the country barring 1-2 months (during very high and low temperature). It is consumed throughout the country by every class of people as a fresh leafy vegetable or raw as salad. Cabbage is distinguished by its swollen heads which is formed by thickening of edible bud with thick tightly packed overlapping leaves manifesting a large head. The shape of the head may be round, conical, oblong, flat, or savoyed depending on the variety. It covers 4.5% areas under vegetable crops in our country. The increase in area has been possible due to the introduction of heat-tolerant varieties and F_1 hybrids which give more yield per unit area as compared to traditional cultivars in the region of mild winters. The hybrids occupy 22.9% of the area under cabbage (Swarup, 1998).

ORIGIN

Cabbage is a native of western Europe and nothern shores of mediterranean sea (region). Cabbage and kale were first to be domesticated and have probably orginated in Europe. Cabbage belong to the cole group of vegetable (genus - *Brassica*) which originated from a wild ancestor *Brassica oleracea* L. var. sylvestris L. as wild cabbage cliff cabbage or cole wort, through mutation, human seleetion and adaptation. Cabbage evolved in Germany (Helm, 1963) and both red and white cabbage known in Germany by 1150 A D and by 14th century in England. Cabbage is one of the earlier cultivated forms of cole crops (*Brassica oleracea* L.) and was probably in general use, around 2000-2500. B.C. Snogerup (1980) was of the view that cabbage originated in western Europe. The genus *Brassica* includes about 100 species. Majorities of which are native to the mediterranean region. The crop is attributed to mediterranean centre of origin. It is also considered that the real headed cabbage was evolved in Germany. The savoy cabbage originated in Italy and spread to France and Germany in the sixteenth and seventh centuries. It was mationed that (Snogerup, 1980) savoy cabbage may be the result of introgression with other cole crop. Gates (1953) was of the view that white cabbage might have orginated from *oleracea* in western Eruope. Cabbage is one of the earlier cultivated forms of cole crops (*Brassica oleracea* L.) and was probably in general use, around 2000-2500 B.C.

History

The wild plant of cabbage does not form head and is very similar to collared, hence it is consumed by the surrounding populations. In India cabbage was introduced in 15th century from Portugal and for the first time it was cultivated on the banks of malabar and later it spread in whole country.

The character of wider adaptability to fluctuating conditions of climate, has not been exploited in cabbage, to the some extent as in cauliflower, in India. However, some stray reports do suggest that in some varieties, head formation and seed production can take place under higher temperature. The time of introduction itself of cabbage into India, could not be precisely estimated. There is a stray mention of cabbage in the Ain-i-Akbari (Anon) which gives the details of cuisine of Moghul rulers (16th centry A.D.). It is not known whether cabbage seed production had been attempted during the Moghul period. The only recorded evidence of organized seed production of temperate vegetables, is in the early forties of this century, when experiments were conducted in Baluchistan (Quetta) and later in Kashmir Valley which later extended to Himachal Pradesh. This was necessitated by the stoppage of seed imports from Europe during World War-II. Whether Moghul rulers regularly imported cabbage seeds and seeds of other temperate vegetable like carrot, turnip, beet root etc., or these were being presented by the European traders during their visits to the Moghul courts, is neither known nor could be established.

No doubt Japanese Sea Trade were the first to develop tropical cabbage varieties and the classical "KK" and "KY" crosses of 1950s spread the cabbage cultivation to the tropical regions of Philippines, Thailand and other parts of S.E. Asia. A similar effort was started in Darjeeling hills in the fifties but the work was never pursued. The current interest in the imported cabbage hybrids especially in penisular, eastern and western India, is for the tolerance to high temperature and the hard compact quality of heads which have good keeping quality in long distance transportation (Seshadri and Chatterjee, 1996).

It is therefore evidence that potentialities for adaptation to tropical conditions were available in cabbage and Japanese stole a march over the Indians, in contrast to cualiflower. There has been a lack of co-ordination between vagetable growers in the plains and seed growers in the hills to promote directed selections towards better adaptation, in cabbage. It is however significant to record that cabbage may no longer be considered a typical temperate vegetable crop requiring cool temperature of north Indian winters. The present day hot weather hybrids form compact heads under tropical conditions of day temperatures of 30-35°C. Especially the imported hybrids like "Green Boy" "Green Express" etc. have been accepted by the farmers for growing in warmer conditions of east, south and western India.

However, it should not be overlooked that seed production of hot weather cabbage requires cool temperature of north Indian hills for brief periods to ensure good seed yield. In contrast in cauliflower, acclimatisation of seed production at higher temperatures. of Bihar and estern U.P. conditions, could be achieved. In cabbage however, we still have to depend upon imported seeds of cabbage hybrids suited to high temperature growing - that too mainly from Japanese Seed Trade. There have been sporadic reports of the efforts made in Darjeeling hills in eastern India (Chatterjee and Mukherjee 1957). Some low chilling types were selected in exotic introduction like "Eclipse Drumhead", "Danish Ballhead etc., which work was not bee pursued further.

DISTRIBUTION

Basically, cabbage is a crop of temperate zones but it has spread to both subtropical and tropical regions of the world. At present cabbages are grown in caribbean countries, Indonesia Malasyia, Central East and west America, the phillppines and tropics in general.

Cabbage is an inroduced vegetable crop in India, has adapted it self well and is now grown all over the country. It is the most common vegetable crop available during winter throughout India. In India the traditional cabbage growing areas are the southern and south-eastern coastal regions where it is cultivated on a large scale. The other cabbage growing states are U.P., Orissa, Bihar, Karnataka, W.B. and Assam. However with the introduction of new open pollinated/hybrid varieties both in temperate and heat-tolerant types, its cultivation is geared up around the country by adjusting their sowing and planting time.

The area under cabbage was 2.4 lakh hectares. In our country, West Bengal occupies first place with a production of 1843. million tonnes (2001-2002), followed by Orissa 1238.9 million tonnes and Bihar 561.4 million tonnes (Singh *et al.*, 2004).

The head vary from flat-topped to long oval. Preference is generally given to varieties with compact, round heads though pointed head varieties are also grown.

Classification

Bailey (1930) classified the cultivated *Oleracea* in to 7 divisions.

(A) The leaf-races grown for the open foliage

(i) Branching shrubby kales-*B. oleracea* var. *fruticasa.*

(ii) Common biennial mostly unbranching kales - *B. oleracea* var. *acephala.*

(iii) Low cabbage like Portuguese Kales - *B. oleracea* var. *tronchuda.*

(B) The head races grown for the compact leaf-head or buds.

(iv) Edible product a terminal head, the cabbage-*B. oleracea* var. *capitata.*

(v) Edible product lateral small heads or buds, the Brussels sprouts-*B. oleracea* var. *gemmifera.*

(C) The inflorescence races grown for the modified thickened flower shoots of malformed flowers.

(vi) Flowers only partially if at all modified-*B. oleracea* var *italica.*

(vii) Flowers thickened and malformed into a head or several head-*B. oleracea* var *botrytis.*

Cabbage can also be calssified in various ways

1. *Place of origin* : (a) *B. oleracea* var. *capitata* ssp. *mediterranea* e.g. Mediterranean cabbage (b) *B. oleracea* var. *capitata* ssp. *orintalis,* e.g. Oriental cabbage (c) *B. oleracea* var. *capitata* ssp. *europea.* e.g. European cabbage.

2. *Botanical colour and form of Head* : (a) *B. oleracea* var. *capitata* L.*f. alba,* e.g. white cabbage (b) *B. oleracea* var. *capitata* L.*f. rubra,* e.g. red cabbage (c) *B. oleracea* var. *sabauda* e.g. savoy cabbage.

3. *On the Basis of Head Shape (Choudhary, 1976)* : (a) Round shape head or ball head types (Golden Acre, Pride of India. Copenhagen Market and Mammoth Rock Red (b) Flat head or drum head type (Pusa Drum Head) (c) Conical head type (Jersy Wakefield) (d) Savoy type ('Chieftain').

4. *Classification Based on the Size and Shape of the Head (Allen, 1914)* : (a) Wakefield and Winnings Kadt group-head are small, conical and very early (Jersy Wakefield, Charleston Wakefield). (b) Copenhagen Market Group-round head, early large head, (Copenhagen Market, Golden

Drumhead, Globe, Bonanza). (c) Flat head or Drumhead Group-heads flattened from both ends (Pusa Drum head, Early Round Dutch, All Head early, Succession, All Seasons, Sure Crops). (d) Savoy Group-leaves are wrinkled heads are of very high quality, limited commercial use (Drum Head Savoy, Improved American Savoy, and Chieftain). (e) Danish Ball Head Group-leaves are thin, heads compact solid, medium sized, good keeping quality, fine texture (Danish Ball Head, Hollander, Wisconsin Ball Head). (f) Alpha Group-the earliest group head, very small, solid head limited commercial uses, (Miniature Marrow). (g) Volga Group-thick leaves, shining blue coloured leaves, loose bottom (Volga). (h) Red Cabbage Group-similar to Danish Ball Head except red colour of leaves (Red Rock, Red Dutch, Large Red).

BOTANY

The common cabbage grown in Inda is white cabbage. The red cabbage has the same botanical name except form is *rubra* (L). Thell. Plants are polymorphous, annual or biennial, erect, 30-150 cm high and much branched in the upper part. Stem is obtusely quadrangular or subterete, pruinose and glabrous. Leaves are variable at the base of the stem and 2-10 cm long. Petiole is lyrate or obovate, subentire or undulate, more or less deeply and irregularly lobed. Termainal lobe is orbicular or ovate and apex rounded. Leaves are mostly sessile, semi-amplexicaul, ovate-oblong or linear-lanceolate, entire or subentire. All leaves are variable as to shape, colour and size, green, yellowish green or red, fleshy, thick, glabrous, more or less distinctly dotted. Inflorescence in terminal, elongate, lax, racemiform or paniculiform, 10-50 cm long and racemes 20-40 flowered. Flowers are rather large, buds raised for above the expanded flowers. Pedicels are 0.5-0.8 cm long. Sepals are ecrect, obtuse, 1 cm long, the outer ones ovate-oblong and narrow, saccate at the base, the inner ones more ovate or broadly ovate oblong. Petals are oblong-spathulate, rounded at the tips, light yellow and 1.5-2.0 cm long. Stamens are subequal, crect and 1-1.3 cm long. Pods on stalks are 1.5-2.0 cm long, more or less patent, thin, cylindrical or linear elongate, with a long beak with many seeded. 5.0-9.5 cm long and 0.3-0.5 cm thick. Seeds are seriate, globose and smooth (purseglove, 1974).

Plant breeding

Cabbage is basically a temperate crop which does not produce seeds in the plains and hence no work either on breeding or on seed production has been takent up in the plains. It can produce only the edible heads in the plains, the seeds for which are produced in the temperate hilly regions. Their inheritance and hybrid vigour studies have been made on available genetic stocks (Swarup and Singh, 1963; Bhagchandani *et al.*, 1977 and Gill *et al.*, 1979). Watson as Baker (1969) have given an account of the possible gene centres for resistance in the genus *Brassica* to *Plasomodiophora brassica*. The resistance to black rot, *Xanthomonas campestris* was observed to be dominant and governed by polygenes (Tewari *et al.*, 1979).

The cytoplasm of Ogura male sterile radish has been successfully utilized in cabbage the nucleus of cabbage was introduced in to the Ogura cytoplasm of radish by back crossing. Protoplast fusion of Ogura cytoplasm, from broccoli is transferred to cabbage to produce hybrid seeds. The other two sources of CMS were from *Brassica nigra* (sterile) cytoplasm and sterile Anand cytoplasm from *Brassica rapa*. The hybrids developed CMS systems are now available for commercial cultivation in cabbage cauliflowers. Dania variety was developed as self compatible line for late season. Male sterility has been reported which may be used in the production of hyrid seeds. Swarup and Gill (1964) suggested the use of purple pigmentation in hybrid seed production of cabbage. In cabbage, almost all hybrids are exotic. The area under F_1 hybrids is about 85% of the total area of cabbage in the country. Seeds of the exotic varieties Green Express and Green Boy are marketed by National Seeds Corporation (Nath *et al.*, 2008). The round head varieties mature the earliest followed by the conical varieties, whereas drumhead varieties with flat head and Savoy varieties are usually late. Other varieties of white cabbage grown in India are 'Golden Ball' and Early Drumhead.

Taxonomy

Cabbage Belong to the genus *Brassica, B. oleracea* L. convar *capitata* (L.) Alef, var. *capitata* of the family Cruciferae. Inflorescence is a raceme; Flowers are actinomorphic and bisexual calyx gemosepalous with 4 petals arranged opposite to each other in a square, cross 6 stamens of which 2 are short. Pollination is entomophilous Cabbage is highly cross-pollinated due to sporophytic self-incompatibility. Cabbage is diploid with 2n = 18. It crosses easily with other horticultural form of *Brassica oleracea*.

Nieuwhof (1969) recognized the following three cultivated forms :

(i) *Brassica oleracea* var. *capitata* L. f. alba 1 DC white cabbage.

(ii) *B. oleracea* var. *capitata* f. rubra (L) Jhell Red cabbage.

(iii) *B. oleracea* var. *sabauda* L. Savoy cabbage.

Place of origin : There are three sub-species :

(a) *B. oleracea* var. *capitata* ssp. *mediterranea* e.g. mediterranean cabbage.

(b) *B. oleracea* var. *capitata* ssp. orientalis, eg. oriental cabbage.

(c) *B. oleracea* var. *capitata* spp. *europea* eg. European cabbage.

According to Allen (1914) Mayer's grouped cabbage as per size and shape of head.

(a) Wakefield and winnings Kadt. group heads are small conical and very early. (Jersey wake-field, charleston Wakefield).

(b) Copenhagen market group round heads, early, large head, (Copenhagen market, Golden Drum Head, Globe, Bonanza).

(c) Flat head or drum head group = Heads, flattened from both end, (Pusa Drum Head, Early Round Dutch, All Head Early Succession, All Seasons, Sure Crops).

(d) Savoy group = leaves are wrinkled, heads are of very high quality, limited commercial use (Drum Head Savoy, Improved American Savoy, chieftain).

(e) Danish Ball Head group= leaves are very thin, heads compact, solid, medium sized, good keeping quality, fine texture (Danish Ball Head, Hollander, Wisconsin Ball Head).

Choudhary (1976) placed cabbage into different groups according to head shape :

(a) *Round Shape* = Head or Ball Head type, Golden Acre, Pride of India, Copenhagen Market and Mammoth Rock Red.

(b) Flat Head Drum Head type = Pusa Drum head.

(c) Conical head type = Jersey wakefield.

(d) Savoy type = Chieftain.

Floral biology

Cabbage flower has four sepals, four petals, six stamens in tetradynamous condition (two short and four long stamens) and a bicarpellary ovary which is superior and has a false septum. Ovules are attached on both the sides of septum. Two active nectaries are located between the bases of short stamens and ovary. The buds open under pressure of rapidly growing petals and become fully expanded in about 12 hrs. Flowers are slightly protogynous and cabbage is naturally cross pollinated due to sphorophytic self-incompatibility. Pollination is brought about by bees and flies. Bud pollination is effective to achieve selfing. For cross-pollination flower buds expected to open within 1-2 days are emasculated, and are pollinated immediately with desired pollen, using a brush flower stamens.

COMPOSITION

It is a rich source of Vitamin A, B, and C, it also contains phosphorus, Potassium, Calcium. Sodium and Iron. It is said to help in digestion. Cabbage leaves are low in calories (27 per cent), fat (0.1 per cent) and carbohydrates (4.6 per cent). It is good source of protein (1.3 per cent) which contains all essential amino acids, particularly sulphur containing amino acids. Cabbage proteins are comparable to pea proteins, in biological value, digestibility and NPU. Cabbage is an excellent source of minerals such as calcium (39 mg), iron (0.8 mg), magnesium (10 mg), sodium (14.1 mg), potassium (114 mg) and phosphorus (44mg). It has substantial amounts of β-carotene (provitamin-A), ascorbic acid, riboflavin, niacin and thiamine. Ascorbic acid content varies from 30-65 mg per 100g fresh weights. Cooking or shredding of cabbage and leaving it exposed to air can result substantial loss of ascorbic acid. Red cabbage is a rich source of anthocyanin pigments. Both mono and diacylated cyanidin derivatives have been identified in red cabbage. Cabbage contains thermo labile sulphur compounds which evolve sulfurous odours. The major volatile sulfur compounds (dimethy disulfide and hydrogen sulfide) of cooked cabbage are derived from a precursor of L-S-methyl-cysteine sulfoxide, a free amino acid. The presence of 2- propeny1-isothiocyanate in both white and red cabbage and 3-butenyl isothiocyanate in red cabbage has also been reported. The nutritive value of cabbage is given (Aykroyd, 1963) below in the table (per 100g of edible portion).

Moisture	91.9g	Phosphorus	44.0 mg
Protein	1.8g	Iron	0.8mg
Fat	0.1g	Sodium	14.1mg
Minerals	0.6g	Potassium	114.0mg
Fibre	1.0g	Copper	0.08 mg
Other Carbotydrates	4.6g	Sulphur	67.0mg
Calories	2.7g	Vitamin 'A'	2000 I.U.
Calcium	39mg	Thiamine	0.06 mg
Magnesium	10mg	Riboflavin	0.03 mg
Oxalic Acid	3 mg	Nicotinic Acid	0.4mg
		Vitamin 'C'	124 mg

Uses

Cabbage is used a salad, boiled vegetable, cooked in curries, pickling as well as dehydrated vegetable. It is used more raw as processed vagetable. Cabbage soup is good and nourishing. Nice pickle is prepared from it. Cabbage is also used for salad, mixed in tomato, green chllies, beetroot etc. Flavour in cabbage leaves is due to the glucoside sinigrin. Cabbage contains goitrogens which cause enlargement of thyroid glands. The thioglucosides with their associate enzyme (s) impart the desirable culinary flavour to cabbage, broccoli and cauliflower. The thioglucosides (sinigrin) are non-goitrogenic which depend upon enzymatic hydrolysis in presence of the enzyme myrosinase yields glucose, bisulfate and allyl thiocyanate-a goitrogen. The thioglucosides. (Progoitrin and epi-orogoitrin) are the precursors of an anit-thyroid compound called goitrin. The goitrin is formed subsequent to hydroysis and then through cyclization of an unstable isothiocyanate containing hydroxyl group. The total glucosinolate content of white cabbage cultivars ranged from 299-1288 ppm, whereas-the goitrin content varied from 1.2 to 26.0 ppm. Although the enzymatic hydrolytic products from the glucosinolates are goitrogenic, some of these breakdown properties of cabbage could be lost during cooking. Cabbage is well known for its medicinal properties. In ayurvedic medicine, cabbage leaves are prescribed for cough, fever, skin diseases peptic ulcers, urinary discharge and hemorrhoids. Fresh cabbage juice is reported to contain a heat labile anti-peptic ulcer component. A clinical study has indicated that concentrated cabbage juice is significantly effective in healing peptic ulcers. Cabbage is one of the best sources of a sulphur containing amino acids. Fresh cabbage juice is reported to have anti bacterial activity and has been shown to inhibit the growth of various strains of lactic acid bacteria. Cabbage is also reported to have significant anti cancer activity. Model studies and epidemiological data in humans have confirmed the protective role of cabbage on the development of cancer.

VARIETIES AND HYBRIDS

Detail characteristics of varieties and hybrids described by Singh and Nath (2011) are given below :

Early Maturity group (45-75 days)

Golden Acre

A selection from the 'copenhagen market', the heads are small, round, weighing 1.0 to 1.5 kg. It matures in 60-70 days after transplanting. The outer leaves are few and cup-shaped. The harvesting should not be delayed beyond 10 days of head-formation, otherwise the heads have a tendency to burst open.

Pride of India

The plant type of this variety is similar to golden Acre but it is about a week later in maturity. It is available about a week later than the 'Golden

Acre'. Head size 1.5 to 2.0 kg, slightly bigger frame than 'Golden Acre'. An introduction recommended by YSPUHF, Solan.

Sel 8

A selection from an intervarietal cross of EC 10109 X EC 24855 made at Katrain. The variety is characterized by a short-stalk, medium-sized leaves, light green with wavy margins and slightly puckered leaf-blade. Heads slightly flattish round weighing 1.5 to 2.0 kg. also available a week later than the 'Golden Acre'. It is resistant to black-rot.

Copenhagen Market

In most of the areas it has been replaced by 'Golden Acre'. But it is still grown on a limited scale in some pockets. , Head size is 2.0 to 2.5 kg. heavy outer frame. Shape of leaves and outer whorls are same as in 'Golden Acre'.

Express

'Golden Acre' is marketed by some of the seed companies under this name. It resembles 'Golden Acre' in all respects except it is said to be slightly earlier in maturity.

Sri Ganesh Gol

Being marketed by Maharashtra Hybrid Seed Company (Mahyco). Its plant type is like that of 'Golden Acre' but it matures a week early. Its head size and shape also resembles that of 'Golden Acre'. It has got a much better staying ability.

Pusa Mukta

It is a cross between EC 24855 and EC 10109, both exotic parental lines. It has a short stalk, medium sized puckered light green leaves with wavy margin. Heads are flattish-round, weighing 1.5-2.0 kg. It is ready for harvest 65-70 days after transplanting, a week later than 'Golden Acre'. The flat and compact headed 'Pusa Mukta' yields 58.41 tonnes/ha. It is a field resistant to black-rot and is almost free from insect-pest attack.

Pusa Ageti

It has been developed from TKCBH-28 (F1 hybrid) procured from Taiwan. Selection were made F2 generation onwards by using simple recurrent selection. It is the first heat tolerant variety which can produce seed in subtropical climate. It takes about 70-90 days from transplanting to harvesting. Plant has medium sized frame, small stalk, upright growing habit and wax coated grey-green foliage broad lamina with prominent mid rib and wavy margin. It has flattish-round, compact, non-cracking heads with large wrapper leaves. Head weight varies from 600-1200 g yielding 110-380 q/ha depending upon the time of transplanting. Optimum temperature for its growth and head development is 18-20ºC but it can grow and form head when the mean temperature is between 15-30ºC and day temperature should not be more than 35ºC, the crop can be planted in March, August and October.

Pusa Synthetic

Plants with short stalk, cup shaped leaves, early maturity in 60-70 days after transplanting. Heads are very compact and good quality. Average weight about 1.0 kg and it is a high yielding.

Bajrang (BSS-50)

It has been developed by Beejo Sheetal Seeds Pvt. Ltd, Jalna. Heads are dark green, smooth, highly compact, average head weight 1 kg ; outer leaves are upright; mature in 65 days but can stand on the field for 45 days after maturity; tolerate high temperature upto 36°C; resistance to *fusarium wilt.* Female parent (BH-8,20) Early in maturity and has compact round head. Male parent (NORT-14) Early maturity and has highly compact round head.

Pusa Sambandha

An early maturity variety with wide adaptability. Heads round and compact. Suitable for close planting.

Green Express

Marketed by the National Seeds Corporation (NSC). Even before the NSC started marketing the variety, it had become popular in West Bengal, where its seeds had been smuggled from Bangla Desh. Medium sized plants with few outer leaves. The head size depends upon the time of planting. Those planting earlier can attain head size up to 3 kg while in the subsequent plantings the head size decreases.

Kalimpong Eclipse Ball

This variety is also developed at the HRS, Krishnanagar and the VRS, Kalimpong. This is an early variety. Heads are ready for harvesting 60 to 80 days. Heads are oval in shape, big sized, and compact.

Green Boy

Marketed by National Seeds Corporation. An introduction from Japan. Early in habit, produces medium sized round heads.

CH-21

Extra early dark green ball headed hybrid with solid texture. Heads weigh 750 g to 1.0 kg with very short core. Inner leaves are very dense and outer excellent wrappers. Excellent field holding ability enables harvest at will to catch best market. Tolerant to cabbage yellow and Tip burn.

Quisto

It is a high yielder hybrid and has ability to stand over severe hot conditions. Heads can stand in the field conditions upto 70 days after maturity. Heads are compact round and very solid with dark blackish colour. It is good for tropical climate.

Kranti

It is ready for harvesting within 60 days after transplanting. It is good for close space planting. Heads are 0.8-1.2 kg weight, round and compact.

Mitra

An early hybrid with excellent head to plant ratio. Heads are fresh green uniform, compact, ball shaped with weight around 0.8 to 1.2 kg each. Good field retention ability.

Early Drum Head

An early variety, foliage light green, heads flat, leaves medium to large curving inward enclosing the head loosly. Leaves forming the head fold over each other at the centre.

Sudha (BSS-115)

It has been developed by Bejo Sheetal Seeds Pvt. Ltd. Jalna. Heads are highly, compact, good interior structure, matures in 75 days and can be grown round the year; average weight of head is 2-3 kg; have resistance to *Fusarium* wilt. Female parent (FRICO-11): Medium late maturity; leaves are bluish and plant structure is good; has flattish round and medium heads. Male parents (LAWIC): Early in maturity; bluish leaves, compact, round and medium heads.

BSS-44

It has been developed by Bejo Sheetal Seeds Pvt. Ltd. Jalna Heads are roundish, average weight 2.5 Kg; matures in 75 days but can stand for 20 days in field after maturity; grows well both in cold and hot weather; resistance to *Fusarium* wilt. Female parent (BOL-20)- Medium early in maturity, bluish leaves; compact, round and big heads. Male parent (BI X 12)- Early in maturity, bluish leaves; round and compact heads.

Hara gola

This early maturing hybrid produces globe shaped head with extra fine quality crispy sweet flesh. Fowits weigh around 1.0-1.25 kg.

Vaibhav

Early maturing hybrid with excellent field holding ability makes it best choice. Globe shaped compact heads are green and weigh 1.0-1.5 kg, very easy to grow.

Evergreen

Very promising and early maturing hybrid with relatively very good field standing ability. Head is round, firm, compact and weigh about 1.0-1.2 kg.

Super cross

Very promising, early maturing hybrid with good field holding ability. Head is semi-flat, very attractive and weighs about 1.3-1.7 kg, easy to grow in hot and wet conditions.

Harini

Early maturing hybrid with excellent adaptability for upland/high altitude areas. Head is semi flat, compact and weighs around 1.2 kg. This variety has good field holding ability and is suitable for long distance shipping.

Red Green

This newly introduced hybrid cabbage produces red colour round shaped head which are firm and compact. Slightly upstanding in habit. Matures in 70 days from transplanting. Highly tolerant to wet conditions in the field. It is suitable for fresh market and processing.

Primero

It is a red cabbage. Mature in 65-70 days after transplanting. Attractive round head, very compact. Its standing ability is 30-40 days after maturity. Head weight 1.0-1.2 kg and used for salad decoration. It gives attractive dark red colour heads.

Benson

This is an early hybrid and good for intercropping in sugarcane. Crop mature in 45-50 days after transplanting. Head round with less outer leaves, medium compact. It can stand one week in the field after maturity. Average head weight is 800 g and suitable for August to December sowing.

Fresco

It produce uniform round, compact head, mature in 70-75 days after transplanting. It can stand 25 days in the field after maturity. Head weight vary from 2.0-3.0 kg, it can be planted in all three season and gives good heads even in the summer season. It gives good taste and an attractive head.

Gonzales

It produces very attractive, round, compact head weighing 0.8-1.0 kg each, which have less outer leaves. It matures in 50 days after transplanting and has standing ability 15-20 days after maturity. Suitable for August-December sowing in plains.

Morris

It is 65-70 days maturing, round shape and compact hybrid. Head weight is 1.0 to 1.5 kg and is recommended specially for sowing in summer season.

Fieldman

It is very durable hybrid, can give head in all seasons. Head is round with less leaves and very compact with green leaves. Head weight is 1.0 to 1.2 kg and has standing ability 30 days after maturity. Crop mature in 60-65 days after transplanting. Suitable for kharif, rabi and summer crop and good for long transport.

General

This is a 60 days cabbage with 1½ kg head weight. It is medium compact with 15 days standing ability and tolerant to *fusarium wilt.*

Invento

It produces attractive round, compact with short core head, which mature in 60-65 days after transplanting. Average head weight is 1.0 kg and has standing ability of 30 days after maturity. It is suitable for August to December sowing in the plains. It is good for transport and resistant to *fusarium wilt.*

Bronco

This is a very uniform, high yielding hybrid which can be grown round the year and specially recommended for sowing in summer and rainy season. It produces, very compact, round and firm head weighing 1.5-2.5 kg each. It mature in 70-75 days after transplanting and can stand 30 days in the field without bursting. It is resistant to *fusarium* and highly tolerant to *xanthomonas.*

Governor

This is 75 days maturing, round shape hybrids. Average head weight is 1½ kg and it can stand in the field for 30 days after maturity. Having disease tolerance and can be sown round the year.

Rotonda

It is round, very compact hybrid, which matures in 75 days after transplanting. Average head weight 1½ kg with good field standing ability. It has good tolerance to *fusarium* and can be grown round the year.

Pragati

An early maturing hybrid with compact round heads and green outer leaves. Late bursting and excellent performance under spring and fall season and partial resistant to *fusarium* yellow.

Green Ball

An early maturing hybrid, compact, round heads with green leaves, wide adaptability. Tolerant to black rot and bursting. Average head weight is 600-700 gms.

Green Gold

It is green type and head mature within 55 to 65 days after planting. Head weight varies from 1.5 to 2.0 kg.

Green Diamond

It is round type and early variety. It mature in 60 days after planting. Head are compact and uniform in maturity and size. It is highly beneficial for farmers and greenish type head with weight from 1.0 kg to 1.5 kg.

Green Dimond (Imp.)

An extra early variety maturing in 50-55 days after transplanting, white coloured, suitable for early spring, sowing in western and southern region of India. Summer sowing in northern region of India.

Green-Samrat (F_1)

A mid season hybrid weighing 2.5-3.0 kg of medium large, round and compact headed. Ready for harvesting 70-75 days after transplantation. An excellent material for strong and distance transportation. Tolerant to black-rot, *fusarium wilt.*

Gold-Star (F_1)

An excellent hybrid of cabbage. Heads are very solid and don't burst even after several days of maturity. Dark green coloured leaves have a protective layer. Maturity 60-65 days after transplanting. Average head weight is 1.5 kg.

Super Ball-50

Strong and vigorous early maturing hybrid with excellent head and plant ratio. Heads are compact, round with gray green outer waxy leaves and weigh 1.0-1.5 kg. Outer waxy leaves provide good protection against the pest diamond back moth. Excellent field holding ability and wide adaptability makes this hybrid quite popular. Tolerant to black rot, black leg, tip burn and diamond back moth.

Gray Ball-60

Early maturing vigorous hybrid ready for harvest in about 55-58 days after transplantation. Heads are very compact, round, gray green and weigh 1.0-1.2 kg. Good field holding capacity and tolerant to black rot and tip burn diseases.

Goody Ball

An excellent hybrid with bluish green leaves and compact round heads, weighing 0.8-1.2 kg. Field holding capacity is excellent and heads do not burst for a long time after maturity. Suitable for planting from July to January in northern India and throughout the year in southern parts.

Big Ball-75

A relatively late maturing hybrid, ready for harvest in 70-75 days after transplantation. The heads are very compact, dark green, predominently globe-shaped and weigh 1.5-2.0 kg. The hybrid has excellent field holding capacity and very good field tolerance to black rot and fungal diseases.

Quick Ball-45

Extremely early maturing hybrid, ready for harvest after 45 days of establishment of transplanted seedling. The plants are medium sized and heads are round, compact, dark green and weigh 0.8-1.0 kg, suitable for planting from the second week ot July and onwards and up to December in the northern parts of the country and throughout the year in southern parts. Suitable for high density planting. Tolerant to black rot and tip and burn diseases.

Golden Ball

Early maturing hybrid, ready for harvest after 50-55 days of transplantation. The heads are uniformly round, compact, green weigh 1.0-1.2 kg with dark green wrapper leaves. Field holding capacity is very good. The hybrid has very good field tolerance to major fungal diseases and black rot.

Kopen Hegen Market Improrted

It is cultivated in more acerage in Orissa. Its heads are medium in size and varies from 1.0 to 3.0 kg; oval, green and compact head and seed are sown in nursery during August and Sept. It mature within 75 days after planting.

Pashan

It is early type and head mature within 55 to 65 days. It is resistant to root rot and tolerate rain, so it can be cultivated in rainy seasons.

Priya (F_1)

A tropical type hybrid, medium vigorous plant with very uniform growth and maturity, round and very compact. Heads are of 1-1.5 kg in size. Ready to mature in 60-70 days after transplanting. It yields 700-1050 q/ha.

Gold Star (F_1)

It mature within 60 days after planting of seedlings. It is dark green and compact head type. Head are very solid and do not burst after several days of maturity. Average head weight is 1.5 kg. It yields, 200 to 250 q head/Acre.

Red Baby

A medium maturity hybrid, compact heads can be harvested in 65-70 days after transplantation. Heads are round, uniformly dark red with strong waxy outer leaves and weigh 1.0-1.2 kg. Inside colour uniformly dark red and. layers finely arranged with a short core. Possess good storage ability. Tolerant to black rot disease and head bursting.

Mahima (F_1)

Categorised under round early group of cabbage. Good head firmness. Round shaped, light green coloured and compact heads weighing 0.8 to 1.0 kg. approximately. Head maturity in 60-65 days after planting with a field holdng capacity of 8-10 days.

Simran (F_1)

Categorised under round early group of cabbage. Good standing ability. Round shaped, green coloured and compact heads weighing 1 to 2 kg. approximately. Head maturity in 60-65 days after planting with a field holding capacity of 12-15 days.

Sumithra

Early maturing hybrid and heads can be harvested in about 58-60 days after transplantation. Campact heads are round, dark green and weigh 0.9-1.2 kg, tolerant to black rot disease.

Bobby

A medium early maturing vigorous hybrid. Heads can be harvested in about 70-75 days after transplantation. Heads are round, compact with grey green leaves and weigh 2.0-3.0 kg. Its versatile nature makes it ideal for long distance transportation. Tolerant to black rot, tip burn and diamond back moth.

NS 22 (F_1)

Tropical type hybrid, vigorous plants, uniform growth and maturity. Heads are round with very good firmness and density. The heads weigh 1.5-

2.0 kgs each with good field holding capacity and is suitable for spring and summer sowings, matures in 70 days.

NS 25 (F_1)

This early hybrid has upright habit, uniform compact heads and good wrapper leaves. The heads are round and weigh 1.0-1.25 kg each. It matures in 55 days. This hybrid is suitable for close planting because of small compact heads. It has very good field holding capacity.

NS 26 (F_1)

Very good hybrid with vigorous plants and a green foliage with good wrapper leaves. The round heads are very compact with good density and firmness. They weigh 1.5-2.0 kgs each with good field holding capacity. This hybrid is suitable for spring and summer sowings with a maturity period of 65-70 days.

Mansi

An early maturing hybrid has compact, round heads with green leaves. Wide adaptability. Intermediate level of partial resistance to black rot, bursting and high temperature. Aaverage head weight is 900-1250 g.

Bharat (45)

This promising hybrid is round in shape and early maturing (55-60 days). Medium sized heads are bluish green in colour, partial tolerant to diamond back moth, high temperature and weighing about 1.25-1.50 kg.

Ria (105)

Medium maturing hybrid (65-70 days). Round heads are compact, dark green in colour with good taste. Good field holding ability. Average head weights 1.5 to 2.0 kg.

Summer Queen (F_1)

Plants are strong, compact with 10-12 auxiliary leaves. Heads are attractive green, semi-flat, weighing about 2-0 to 2.5 kg. Heads will be ready in 55-60 days after transplanting. It is suitable for growing round the year, especially for summer. This is a good hybrid for hot and humid climate. It is tolerant to soft-rot and cracking.

Vishwas (F_1)

It is very early maturing hybrid. Leaves deep green, fleshy-broad, open. Heads are medium size, moderately compact, round and uniform green coloured, weighing 0.8 to 1.0 kg each. Head is ready in 45-50 days after transplanting. Uniform maturity which make single harvest possible. Matured head does not split up to one week. It is suitable for close planting (70-80 thousand plants/ha), could be plant densely. Yield 50-60 mt/ha under optimal conditions. Tolerant to *fusarium wilt*.

Combinor (F_1)

Compact, vigorous, strong plant with 8-10 auxiliary leaves. Heads round, green, compact, weighing approximately 1.0-1.25 kg each. Ready for harvest 60-65 days after transplanting. Most suitable for cool and dry season. Tolerant to head bursting. Excellent standing ability of 45-60 days after the crop is ready.

Equatoria (F_1)

Strong plant with vigorous growth, 8-10 auxiliary leaves. Head flattened attractive green coloured, globe-shaped head with height 14-16 cm, diameter 20-22 cm and weight 2.5-3.0 kg, head is ready within 70-75 days after transplanting. Tolerant to head bursting and low temperature.

Green Globe (F_1)

Plant vigorous, compact with 8-10 auxiliary leaves. Head is round, attractive green, weight is 1.0-1.25 kg. Extremely early maturing. Head is ready within 50-55 days after transplanting. Suitable for growing round the year. Tolerant to cracking, makes the harvest period more adjustable.

KS-Cross (F_1)

Plants strong, sturdy vigorous with 10-12 auxiliary leaves. Head is green, semi-flat, 13-15 cm in height and 19-21 cm in diameter, weighing about 1.0-1.5 kg with attractive appearance. Very early maturing and ready within 56-60 days after transplanting. Tolerant to bacterial soft rot, black rot and suitable for growing round the year and specially for hot and humid climate.

S-107

It comes to maturity in 70-75 days after transplanting, head flattened globe, green, very good firm, average weight 2.0 kg, excellent holding capacity.

Hybrid 123

It matures in 45 days after transplanting, head ball type, deep green, solid excellent firm, average weight 1.2 kg, good field holding capacity.

Hybrid 124

Plants come to maturity 50-60 days after transplanting, head flattish round, light green, good solid, average weight 1.2 kg, moderate field holding capacity.

Hybrid 125

Plants mature in 65 days after transplanting, heads round, light green, good solid, average weight 1.4 kg, good field holding capacity.

Hybrid 126

Plants come to maturity 65 days after transplanting, heads round, green, good solid, average weight 1.6 kg, very good field holding capacity.

Hybrid 133

Plants mature in 65 days after transplanting, heads flattish round, bluish green, solid, average weight 1.5 kg, excellent field holding capacity.

Super Hero

It mature in 65-75 days after transplanting head semi flat green, firmness excellent, average weight 2.0 kg, tolerant against pests and diseases.

S-101

It come to maturity in 55 days after transplanting, head ball shaped, glossy green, very compact, average weight 1.2 kg, and extra earlv hybrid with excellent taste, tolerant to diseases.

Improved Savitri (OP)

It mature in 65 days after transplanting, heads round, green solid, average weight 1.2 kg, suitable for tropical condition.

Medium Maturing group (76-90 days)

Pusa Drum Head

Selection made at Katrain, and an introduction from Japan (EC 6774), the variety is the earliest in this group. The heads are flat, slightly loose weighing 3-5 kg. The outer leaves are light green with prominent midrib. It has field-resistance to black-leg disease (phoma lingam). Heads are ready in 80-90 days after transplanting.

Bahar

Medium early maturing hybrid. Medium sized heads, round and compact with short core and bluish outer leaves. Excellent performance under low land summer fall cropping. Tolerant to black rot, soft rot, and least damage from diamond back moth.

Unnati

A late maturing hybrid (75-85 days), deep green leaves with large plant frame, flattish heads are compact, good for storage and long distance transportation. Excellent field holding ability. Average head weight 3.0-4.0 kg.

Geetanjali

It mature within 80 to 90 days after planting. Head weight vary from 2.5 to 3.0 kg. It is highly suitable for cultivation in Orissa and Bengal.

Konark

Extremely good material for North East region, West Bengal and Orissa. A mid-season hybrid with flattish-round solid head weighing 2.5-3.0 kg with large frame and strong outer leaves. Have good holding ability. More adapted for sowing during rainy season in these regions. Head is crisp, tasty and excellent for fresh market.

Raj-2

A dark-bluish green ball headed hybrid suitable for fresh market and long trans-shipment. Maturity 80-85 days, medium large head grows upto 4.0 kg depending on field conditions. Leaves are thick and core short. Tolerant to black rot. This hybrid has been bred for a wide adaptability.

Meenakshi

It is a deep green and head size is up to 5.0 kg and can kept in field for more days without any damage. Generally head size varies from 2.5 to 3.0 kg. Mature in 70-80 days.

Hari Rani

It is also marketed by Mahyco and is very popular in the Southern States, especially in the Nilgiri hills. It has a medium-sized frame with light green foliage. The heads are flattish round weighing 2.5 to 3.0 kg.

Varun

A medium maturing hybrid, medium compact globe shaped head is medium green in colour and weigh 1.0-1.5 kg, excellent field holding ability is also adaptable for high density planting and long distance shipping.

Hema II

An outstanding hybrid produces flattened globe shaped heads of weight 1.5-2.5 kg, good tolerance to bursting gives an extended harveting time.

Bhima

A versatile late maturing hybrid ready for harvest in about 75-80 days after transplantation. Heads are very compact and solid, short cored, flattened globe in shape, weigh 2.5-4.0 kg. and covered with dark green wrapper leaves. Tolerant to black rot disease. Excellent field holding capacity allows flexible harvest. Ideally suited for all season cultivation in southern India, especially Karnataka.

Vijay

Medium maturing hybrid (80-85 days). Semi round, heads are compact, dark green in colour with good taste. Excellent field holding ability. Ideal for processing and fresh use. Tolerant to yellow and black rot.

Dublin

This is a new medium sized variety with a fresh medium green colour. The head is filled quickly and it generates a firm fair sized ball, which is well closed at the base.

Hero

It mature in 80-85 days after transplanting, head flat, bluish green, firmness excellent, average weight 2.0 kg, excellent field retention ability and very strong against pests and diseases.

Qutsts (F_1)

Plant strong, self supporting, sturdy, erect compact and well built. Bluish-green, fleshy, broad leaves with serrated collars, 10-15 auxiliary leaves. Head compact, dark green, heavy, round and uniform. Ready for harvest in 75-80 days of planting. Short, tender core with fine internal structure. Head weigh 1.0-3.0 kg. Matured head does not split for 90 days. Very testy and good for table, cooking or for process-industry. Tolerant to fusarium wilt. Yield 60-100 mt/ha under optimal conditions.

Gideon

It gives very compact, bluish green, round head with pointed upward leaves. Maturity 75-80 days after transplanting and standing ability is 35 days after maturity. Average head weight is 2.0 kg. Good for round the year production and suitable for long storage. It is resistant to *fusarium* and *xanthomonas*.

Ramada

It is suitable for long storage and transport. It has good vigour and suitable for round the year planting. It produces flattish round, very compact head with 2.0-3.0 kg weight. It matures in 80-85 days after transplanting and can stand 25-30 days in the field after maturity. It is resistantjto *fusarium wilt*.

Shristi

Late maturing hybrid, extremely suitable for planting in southern plains. Heads are relatively large, compact, semi-globe to flat shape with attractive outer dark green leaves and weigh 2.5 to 3.5 kg. Heads mature in 80 -85 days and have excellent field holding capacity. Can be grown throughout the year in the plains of southern parts of the country. Tolerant to black rot and black leg diseases.

Melisa

It matures in 75-80 days after transplanting. It has wrinkled leaves with dark green outer leaves and a dense interior structure and has good taste. It is healthy with very good disease tolerance. It is suitable for August-November sowing.

Drumhead Savoy

It has a dark green foliage which is much blistered or wrinkled. The quality is superior but it is not commercially acceptable in our country.

Red cabbage

All the red cabbage cultivars are tolerant to diamond back moth insect which is serious pest of cabbage. It has distinct coat of wax and produces a head of 1-2 kg, taking about 90 days from transplanting to head formations.

Field rocket

It gives flattish round, compact head, with well covered leaves, mature in 85 to 90 days, after transplanting. It can stand 25 days in the field after maturity. Head weight is 2.5-3.0 kg. Resistant to *fusarium wilt.* It can be sown round the year except very hot summer. It is good for transport.

Jersey Wakefield

A pointed head cultivar is preferred by some growers but it has very limited market. It takes about 55 to 60 days for head formation. The outer leaves are few and slightly crinkled. Average weight is 1.0-1.5 kg. It has got better staying power. This is also preferred for raw eating due to better taste.

Savoy group

This group is also not popular in India. All the cultivars have blistered leaves. The group has all the shapes like pointed, round and flat. The commonly catalogued cultivar is Chieftain. It has fairly light green foliage and large, rather smooth, flat round head.

Green Samrat (F_1)

A mid seasons hybrid, has the head weight of 2.5 kg to 3.0 kg with bigger size. It has dark green and compact medium large round head. Heads mature within 80-90 days after planting. Tolerant to black rot and *fusarium wilt.* An excellent hybrid for sowing and distance transportation. It yields 700-1000 q/ha.

MCV-1

Heads become ready for harvesting after 80 days from transplanting. It has round shape, bigger head (1.5-2.5 kg) and compact. The foliage is dark green with waxy margins.

NS 21 (F_1)

Very vigorous hybrid with an excellent plant stand. Foliage is dark bluishgreen in colour with strong uniform bloom cover giving tolerance to insect pests. This hybrid has very uniform heads which are flat in shape with a weight of 1.5-2.0 kgs. Produce, very compact dense heads of very good firmness and short core. It matures in 75-80 days.

S-103

It matures in 80-85 days after transplanting, head ball shaped bluish green, excellent firm, average weight 2.5 kg, suitable for long distance transportation, possesses outstanding field holding ability.

601

A medium- early maturing hybrid is suitable for fresh market and long transportation, Ball shaped heads are 1-1.5 kg in weight, very firm, short cored with attractive bluish green colour and excellent field holding ability.

Bejo 1857

It is a 85-90 days maturing hybrid. The head is flattish round and weighs 1½ to 2 kgs, with attractive green colour. It is very compact and can stand for 30 days in the field without bursting. It has a long storage ability i.e. it can be stored at room temperature after harvesting for 25 days. It is recommended for round the year sowing, as it has good tolerance to *fusarium wilt.*

Late Maturing group (90-120 days)

Late Large Drum Head

Marketed by different seed companies. It matures in 110 to 115 days. The heads are large flat, somewhat loose.

Kalimpong Eclipse Drumhead

This variety is developed at the Horticultural Research Station, Krishna Nagar and at the Vegetable Research Station, Kalimpong, Darjeeling. This variety is suitable for West Bengal. It is comparatively late variety, heads become ready for harvesting in 100 to 110 days.The heads are large, flat and compact.

September

An introduction from the German Democratic Republic and a most popular open pollinated variety in the 'Nilgiri hills. It is characterized by a large frame, bluish-green leaves with serrated margins. It has a long stalk and the plants fall on one side after the head-formation has completed. The heads are flattish oblong, very tight and weigh 4 to 6 kg. Highly susceptible to black-rot. It takes 96 to 100 days from planting to head formation.

Hari Rani Gole

It is a good hybrid for medium late maturity. It produces medium sized, ball shaped dark blakish green, solid heads with good wrapper leaves. Each head weigh 1.5-2.0 kg. It can be harvested after 95-100 days of transplanting.

Ratna

An outstanding hybrid with medium late maturity. Big size heads are dark green in colour and very compact. Heads weigh approximately 1.5-2.5 kg.

Suvarna (BSS-32)

It has been developed by Beejo Sheetal Seeds Pvt. Ltd., Jalna. Heads are round, average head weight 3 Kg; strong smooth outer leaves; matures in 110 days. Female Parent (ET-2-2) -Medium late in maturity blue leaves, vigorous growth ; round and medium heads. Male parents (BOL-13): Late in maturity blue leaves, flattish round and big heads.

CLIMATIC INFLUENCE

Cabbage is a cool season crop which thrives best in a relatively cool moist climate. It can with stand extreme cold and frost relatively better than cauliflowers. It losses its flavour in dry warm weather. The optimum seed germination is obtained at 12.8°C to 15.6°C soil temperature. In mild condition do not form tight-heads resulting in high percentage of loose and/or opend head with European cultivars. The optimum temperature for growth and heading is in betwen 15°C to 20°C. Above 25°C the growth is arrested. For tropical cultivar temperature an average is 25°C. The minimum temperature for growth of cabbage is just above O°C. The temperate cultivars grow as biennial having two distinct phases-vegetative and generative. During the first phase plants, develop vegetatively to form heads and these heads after exposure to chilling temperatures enter into reproductive phase when flowering and fruit set occurs. The intensity of flowering depends upon the age of the plants and the period for which they are exposed to low temperatures. It is reported that after cold treatment at 4°C to 10°C for 7 to 9 weeks the plants bolt sooner and flower more abundantly than when exposed to such temperatures for shorter period but before this temperature treatment, plants must have passed juvenile phase (Nieuwhof, 1969). The plants which are not exposed to chilling temparatures for required duration depending upon the cultivrs, remain vegetative and become perennial also. The same plant may form two heads in same year. It may be occur in tropical and sub-topical regions.

SOIL REQUIREMENT

Cabbage can be grown in all types of soils from light to heavy soils. In clay loam or silt soils larger yield may be obtained from the late crop. It does not grow well in a highly acidic soil and the maximum phosphorus availability to the plants is between pH 5.5 and 6.5. Higher acidity can be reduced but it should not be reduced beyond 6.5 pH. Most cabbage cultivars are moderately tolerant to salt. In saline soils, the plant show die back of leaf margins and dark foliage and become susceptible to disease like black leg. (Lal, 1993).

SOWING

The '*in situ*' seed sowing of cabbage is followed in some limited parts of India to get early crop. But there are certain disadvantages of this method, e.g., uneven germination or washing of seed by rains which results in gaps in the field or poor crop stand.

Sowing Time

Seed-sowing time varies in different parts of India. In hills, especially, in high hills, seeds are sown in May-June for summer/autumn crop. In the hilly areas which receive heavy rains, the summer and autumn crop is rather limited and sowing is undertaken in autumn to harvest them in late spring in early summer by over watering them. In the plains of northern India, sowing *in situ* or in seedbed starts from early August and continue till November for the late cultivars. In eastern India, sowing is generally started from mid to late September. Sowing is delayed further for areas in the southern hills except in the western and southern peninsular, where the crop can be growth almost round the year by selecting proper cultivars/hybrids.

Seed Rate

On an average 500 to 750g seeds is required for planting one hectare of land. For late crop 375g seed is suffirent to cover an area of one hectare, whereas for hybrid seed crop requirement is 250g per hectare.

Nursery Raising

In India, the seeds are sown in a seedbed. In the plains, the seed beds for early crop require cover to save the small seedlings from rains, while in the high hills, it may be better to grow them under glass frames or poly tunnels. However, the seedlings of mid season or late cultivar may be raised in the open. The soil of nursery bed should be well prepared and free from disease of organisms. The sieved well rotten farmyard manure or compost @ 2-3 kg/m^2 must be added in the seedbed. Before sowing, the seed should be treated with any one of fungicide like thiram, ceresan, agrosan GN or bavistin @ 3 g/kg seeds. The optimum spacing between rows in the nursery bed is 10 cm and the depth of sowing should be 1.5-2.5 cm. For planting one hectare 200-300 m^2 seedbed at the rate of 1 to 2g per sq.m is required. One-gram seed would give rise to about 1000 seedlings. After sowing, the seed is properly covered with a thin layer of mixture of fine manure and soil. The beds are leveled gently. A regular and good moisture supply is needed for rapid germination of seed and the optimum growth of seedlings. To keep the upper soil of seedbed moist, a thin layer of dry grass is spread on the beds and watering is done in the form of shower either with watering cane or with the help of a sprinkler. But, the seed bed cover should be romoved as soon as the emergence of young seedlings above the ground starts. If there is over crowding of seedlings due to thick sowing, the extra seedlings should be thinned out. Attention should also be given for control of diseased and insect pests in the nursery "Damping off" caused by different fungi viz. *Pythium* spp, *Rhizoctonia* spp. is common and sever disease in nursery. The spray of blitox-50 @ 0.3 per cent is effective against these diseases.

PLANTING

Field Preparation

The field is prepared by 3 or 4 ploughing. 21-25 days old seedlings when they are 12-15 cm tall, are transplanted in well prepared field.

Spacing

The planting distance may be varying according to cultivar, planting season and soil conditions. The following distances are generally recommended on the basis of maturity of cultivars viz: for early 45×45. cm or 60×30 cm, mid 60×45 cm and late 60×60 cm. The planting is done on the flat land, ridges or in furrows depending on climate and soil conditions. For early planting, ridge method will be more suitable, especially, in areas where the rain coccurs at the time of planting. In saline soils, planting should be done in furrows.

Dave and Bambhaniya (1993) reported the performance of cabbage varieties viz. Express, pride of India and Golden Aere under south Gujarat conditions with five plant spacings of 30 cm x 30 cm, 45 cm x 30 cm, 60 cm x 30 cm, 45 cm x 45 cm and 60 cm x 45 cm. Pride of India produced maximum plant spread, significantly higher head diameter, highest head weight per plant and highest yield. Closer spacing of 30·

cm × 30 cm gave the highest yield of 35.11 t/ha while under wider spacing of 60 cm × 45 cm gave the biger head weight with less yield/ unit area. The plant spacing for early varieties is recommended as 45 cm × 45 cm and 60-70 cm × 45-50 cm for main season varieties. Planting at closer spacing gives more yield but yield parameters showed improvement in wider spacing. The plant spacing is adjusted depending on the variety and the prevailing conditions.

Transplanting of Seedlings

Generally, 4-6 weeks old seedlings are ready for transplanting but older seedlings up to 8 weeks age can also be transplanted. However, older seedlings have adverse effect on the establishment and growth of young plants. The optimum time of transplanting varies in different climatic conditions. The planting in plains of India can be done from August to December and in hills from April to August. Early matruring cultivars like "Golden Acre" does well in the plains because of shorter cool period, while, "Drum Head Late", "Pusa Drum Head" and other late maturing varieties perform better in hills, where, cool termperature are available for longer duration. However, cultivars of all matruity groups can be grown in hills of north India by planting on suitable altitudes at optimum time.

NUTRITIONAL REQUIREMENT

Cabbage is a heavy feeder of nutrients especially of nitrogen and potash. The nutrient removal by a crop of 700q heads, 350q leaves and 50q stalks per hectare is recorded as 230 kg nitrogen, 85 kg phosphorus, 320kg potash and 426kg. calcium (Bauer, 1932). The recommendation made by Yawalkar *et al.*, (1962). FYM 20-50 t/ha, 50-70 kg nitrogen, 50-120 kg phosphorus and 90 kg potassium per ha. The complete dose of FYM should be incorporated in the soil at the time of field preparation. The entire dose of phosphorus and potash and half the dose of nitrogen should be mixed in the soil before transplanting. The remaining dose of nitrogenous fertilizer should be top dressed around the plants one month after tranplanting.

Chhnkar and Jha (1993) recommended the application of starter solution to seedlings immediately after transplanting to help increased production. They reported that the starter solution containing urea (1 part), potassium sulphate (1 part) and single superphosphate (2 parts) applied immediately after transplating the seedlings at the rate of 0.25 litre and again after 15 days of first application at the same rate gave the highest yield of 700q/ha of heads. Khurana *et al.* (1987) studied the effect of introgen and spacing on cabbage cv. pride of India. Nitrogen at 60 kg/ha gave highest head yield and was most economical too. Average head weight was also highest in this teatment. Spacing of 70 cm × 30 cm recorded highest head yield while average head weight was not influenced significantly. Wange *et al.* (1995) studied the response of cabbage on microbial inoculents and incremental levels of introgen. The application of ntrogen @ 220 kg/ha recorded significantly more (84%) cabbage yield, highest monetary returns and more benefit cost ratio over control. Inoculation with cultures showed an increase from 15-20 per cent in yield with benefit cost ratio between 4.6 to 6.5 over uninoculated control.

Rajanna *et al.* (1995) also reported the effect of different levels of NPK on yield of cabbage cv. prode of India. They found that application of 150: 100: 125 kg NPK/ha was optimum in red lateritic soils of Bidar. Mohanty and Hussain (1998) reported that 120 kg N and 30 kg P_2O_5 may be applied to cabbage for obtaining the higher yield with optimum return. Medhi and Borthakur (1999) studied the methods and rate of application of nitrogen on growth and yield of cabbage. The application of 40 kg N/ha plus 0.5 per cent urea spray at 15, 30 and 45 days after transplating had the highest benefit cost ratio while the lowest were recorded 40kg N/ha plus 1.0 per cent urea spray at 15 days after transplanting. Recommendations of quantity of nitrogen, phosphorus and potash in cabbage for different states of India (anon, 2000) is given below:

State	Nutrients (Kg/ha)			Variety
	N	P	K	
Madhya predesh (Jabalpur)	180	50	50	Pride of India
Uttar Pradesh (Kanpur)	180	60	60	Pride of India
Bihar (Sabour)	150	-	-	Pride of India

Half quantity of nitrogen and full quantity of each phosphorus and Potoash is applied at the teime of transplanting. Remaining quantity of nitrogen is applied at 30-45 days of transplanting. Sharma and Arya (2001) inferred that the application of 120kg. N and 20 tonnes farm yard manure/ha was optimum for higher yield of cabbage under dry temperature conditions of Himachal Pradesh with cv. Pride of India. Sharma and Kumar (2002) studied the effect of organic source in cabbage production at Nauni. They revealed that soil moisture was maximum with pine needles 10 t/ha, where as soil temperature was maximum with FYM 10 t/ha + 100 percent NPK. Nutrient uptake, yield per plant, per plot and per hectare was recorded maximum in treatment FYM 10 t/ha + 100 per cent NPK. Spray of 1% urea as top dressing is also recommended as cabbage responds well to nitrogen. Also a major nutient requirement of 120-180 kg nitrogen, 75-80 kg shosphorus and 60-75 kg potassium ideally suits the cabbage crop.

Use of Micromtrients

(i) Application of dhanzyme granules @ 8-16 kg/ac in two split; first at 10-15 days after planting and second 20-30 days later increase curd yield and curd quality. (ii) Application of Dhanzyme, liquid @ 1 ml/lt of water (180 ml per acre) as a foliar spray, with high volume sprayer at 6-8 leaf stage and 15-20 days later improve curd yield and curd quality. (iii) Application of plantaid as a foliar spray 3 ml/lit. of water, twice at an interval of 3-4 weeks when plant shows symptom of deficiency. It improve the plant growth and better flowering as well as better fruit formation. (iv) Incorporation of plantaid (s) @ 10 kg/ac in soil in a single or split dose increase vegetative growth, flowering, fruit yield and quality. (v) Foliar application of molybdenum @ 20 mg/litre of water at seedling stage 4-5 weeks after transplanting, increase weight and volume of head and number of outer leaves. (vi) Soil application of sodium molybdate @ 2.5kg/ha to the acid soils increase the yield of cabbage.

Prakash and Bhradwaj (1965) reported that the foliar application of molybdenum and cobalt increased the production of cabbage head. Dixit and Singh (1997) Studied the effect of boron and calcium on yield and mutrition of cabbage at Agra and concluded that the application of calcium increased the cabbage yield significantly. Cabbage yield also increased with 0.5 mg Ca/ kg boron and thereafter it showed a decline at 2 mg Ca/kg boron level. However, interaction between B and Ca affected the yield significatly and maximum mean yield was recorded at 0.5 mg Ca/kg B and 40 mg Ca/kg B treatement. Application of borax (0.5%) recorded the highest-yield (43-47 t/ ha) in cabbage. Spraying of manganese sulphate (0.5%) recorded higher chlorophyll content of 0.61 mg/g (Sarma, 2005).

Use of Bioferilizers

Application of biofertilizers like *azotobacter*, phosphate solubilizing micro-organisms (*Bacillus Polymyxa, Aspergillus awamori, Penicillium digitatum* etc.) and micorrhizal fungi *Vasicular-arbuscular mycorrhizae,* (VAM) play significant role to increase the yield of cabbage and also save 25-30% nitrogenous fertilizers. They can be used in following manner.

Seed treatment

About 200g of bio fertilizer is required to treat 10-14 kg of seeds. Suspend one packet of 200g in approximately 400 ml water and mix it thoroughly. Pour this mixture on seeds and mix with hands to obtain uniform coating on each and every seeds. Spread the seeds in shade for drying for 10-15 minutes then sow them immediately.

Seedling treatment

Mix 1 kg of culture (Azotobacter + PSM) in 10-15 litre of water. Get seedling required for 1 acre and make small bundles of seedlings in the suspension for 15-20 minutes. Transplant these immediately. Generally the ratio of inoculate and water should be 1:10 approximately, i.e. 1 kg packet in 10 lit. of water.

Soil application

Prepare the mixture of 2-3 kg of biofertilizer in 40-60 kg of soil/compost. Broadcast the mixture in one acre of land either at sowing time or 24 hr before sowing. The application of phosphate-solubilizers is very common.

VAM Fungi

These VAM fungi are beneficial to the plant which they colonize. They make more nutrients available to the plant, improve soil texture, water holding capacity, improve disease resistance and help in better plant growth. Besides, mycorrhizae are also helpful in the biological control of root pathogens. Cabbage crop are inoculated with VAM fungi in the nursery beds. The inoculum consisting of spores of fungus mixed with soil and infected root bits of the host plants on which they are raised is either spread uniformly in the nursery bed or place in rows over which seeds is sown. The seedlings are colonized by the fungus 45 days after transplanting. Inoculated plants show better growth and improve nutrient content.

Use of Growth Regulators

Chhonkar and Jha (1963) reported that two sprays of GA at 5 ppm to the seedlings gave the maximum increase in yield and quality of cabbage heads. Dipping of seedling roots in solution of NAA at 0.1 or 0.2 mg/l for 24 hours has been most effective treatments in cabbage (Chhonkar and Jha, 1965). In cabbage seedling treatment with Ga at 5 or 10 mg/l was better than NAA at 0.1 mg/l (Sinha, 1977).

Recommended Plant Growth Regulators (PGRs) in Cabbage

Name of PGRs	Concentration (mg/l)	Mathod of application	Attributes affected	Reference
NAA	0.1	Seed treatment	Improves head	Chhonkar and Jha, 1963
IBA GA_3	0.4 5-10	Foliar spray	Size and yield	Chauhan Bordia, 1971 and Sinha, 1977.
NAA + Chelated Zn	100+0.2 (per cent)	Foliar spray at 45 days after transplanting	Increases yield	Mishra *et al.*, 1984.

Cropping System

Cabbage can be grown in rotation, succession cropping or intercropping to utitize land and labour efficiently especially in a kitchen and market garden. Cabbage fits, into different cropping systems. It is grown in multiple cropping systems in Kornataka as sunflower-cabbage-Tomato. Multiple cropping also include intercropping. Palak and Methi intercropped in cabbage, spinach and beet in cabbage and mustard in cabbage gives high returns. For success in these cropping patterns, heavy fertilization and good cultural practices are required. In Ranchi (Jharkhand), the vegetable based cropping system that found to be profitable was cabbage-coriander and cabbbage and fenugreek. In poplar plantation, during spring and summer season, cabbage can be grown with partial shade to decrease the effect of high temperatures. Patil *et al.* (2002) concluded that the intercropping system of cabbage + radish was found to be the best system for getting maximum returns as well as highest net profit per hectare followed by the inter cropping system of cabbage + methi and cabbage + palak, during kharif season under the Pune conditions of Maharashtra. Gawade *et al.* (2004) experimented at Poona, on the cropping systems with cabbage, radish, palak and methi. Cabbage intercropped with palak recorded highest net returns of Rs. 1,01,760/- and yield of main crop was 243.14 q/ha (Cabbage) and intercrop (Palak) was 178.19 q/ha.

Any plan of cropping system should not be made permanent but it should be flexible depending on individual or market needs, change in weather condition, disease and pests infestation. Any kind of combination can'be worked out to meet·the needs of the individual grower and the market. Cabbage can be intercropped with lettuce from

November- March followed by clusterbean and french been from March - October as succession crops in kitchen gardens. Cropping pattern where in cabbage (late) can be intercropped with lettuce (October-February) followed by Okra (Marh-June) and radish (July-September) as succession crops or cabbage (early) intercropped with knol-khol (September-December) followed by bitter gourd (January-May) and cowpea (June-August) as succession crops. But while doing intercropping care should be taken that cabbage crop get proper light, moisture and be weed free, otherwise, lengthy seedlings growth may not be able to form uniform heads. Prabhakar *et al.* (1985) reported that Sunflower-Cabbage-Tomato was found remunerative under varied weather conditions around Bangalore.

To achieve higher returns and proper utilization of resources (land labour, water and famine) some suitable crop rotations given by Saini (1996) mentioned below:

1. Okra - Cabbage - Brinjal
2. Bitter Gourd - Cabbage - Okra
3. Sponge gourd - Cabbage - Cowpea
4. Cluster bean - Cabbage - Chilli
5. Amaranthus - Cabbage - Bottlegourd
6. Maize - Cabbage - French bean
7. Cowpea - Cabbage - Bittergourd
8. Green gram - Cabbage - Muskmelon
9. Black gram - Cabbage - Okra
10. Spogegourd - Cabbage - Water melon
11. Cabbage - Cowpea - Fenugreek
12. Cabbage - Capsicum - Cowpea
13. Cabbage - Brinjal - Cucumber

IRRIGATION

Cabbage is very sensitive to soil moisture and maximum growth and yield can only be obtained when a plentiful supply of water is available to the plants throughout growth. The first irrigation should be given immdiately after transplanting. As Cabbage is succulent and leafy, its water need is high. Though it requires continuous supply of moisture for proper development, yet timely irrigation is also necessary for it. It may be irrigated after every 8 to 10 days. This interval may be short for late varieties and hot season.

Jadahav and Sreenivas (1968) writing on black clay loam soil during during winter season, observed that irrigation at 8 to 16 days interval did not significantly influence the cabbage yield. Hence, 8 irrigations of 50 mm each applied at 16 days interval were adequate in Pune condition. Som *et al.* (1976), working on sandy loam soils at kalyani (WB), observed that irrigation at a soil moisture tension of 0.25 bar at 15cm depth was optimum for cabbage with a consumptive use value of 354 mm. At Hisar, on sandy loam soils, Mangal *et al.* (1982) obtained highest yield of cabbage when irrigated at ID/CPE. (Irrigation depth/cumulative pan evaporation) ratio of 1.25 when the depth of irrigation was 55 mm. This required about 4 irrigations. Studied carried out at Bangalore indicated that irrigations at 30 mm cumulative pan evaoration was optimum (Gupta, 1987).

Studies conducted on method of irrigation in cabbage by Lal *et al.* (1986) showed that maximum biological and economical yield was obtained where only canal water irrigations were done. There was 20% reduction in yield when irrigations were done alternatively with canal and brackish water. Singh *et al.* (1990) reported that crop yield decreased with decrease in seasonal depth of irrigation under trickle and surface method. Crop yield and water expense efficiency under filter capillary, subsurface irrigation were highest though this method proved problematic. Trickle and sub. surface irrigation produced lower yield and water expense efficiency as compared to other methods.

A full grown crop has the greater need for water and the effect of irrigation is often greatest at this statge. Soil water stress prior to heading reduces plant size and yield of marketable heads. The period from head formation to harvest is most critical for soil moisture. Maintenance of uniform, constant soil moisture is very essential at this stage and moisture deficit during this stage causes biggest reduction in yields. Cabbage heads are subjected to bursting when dry weather is followed by rains or when irrigation applications cause uneven soil moisture. Excessive irrigations in early stages cause superficial rooting and washing down of nutrients. Heavy irrigation should also be avoided when the heads have formed as it will result in splitting of heads.

INTERCULTURE

Post planting care mainly concentrates on checking the weed growth and to keep proper moisture condition of the soil to promote sustained growth. Rugular intercultures are necessary for proper aeration of root system and control of weeds to promote healthy plant growth. Normally, very shallow hoeing is done in order to remove the weeds and loosen the soil for better aeration. In order to produce solid heads the plants are earthed up after 5 to 6 weeks of transplanting. Once the foliage has covered the soil, it is better to stop hoeing since it may damage the roots.

WEED CONTROL

Cabbage needs at least two weeks of weed-free period after emergence to prevent yield reduction as reported by Zimdahl (1980). Two-three manual weedings are required to keep the crop weed free. Weed competition in transplanted spring cabbage was found to reduce yield upto the two thirds on unweeded compared with weed free plots as observed by Lawson and Wiseman (1978). It was found that tenoran was effective herbicide for cabbage (Gupta and Singh, 1965). It was reported by Gautam *et al.* (1976) that at least two hand weedings were necessary for good yield of cabage. They tried herbicides like trifluralin 1.0 kg/ha and fluchloralin at 0.5 kg/ha as soil incorporated using a rotary hoe for controlling weeds in cabbage var. Golden Acre. They also used desmetryne 0.5 kg, nitrofen 2.0 kg and alachlor and butachlor at 2.0 kg/ha as pre-emergence pre-plant spray. Mani *et al.* (1973) found that cabbage (var. Golden Acre) yields were enhanced due to herbicidal treatments, viz. trifluralin 1.0 and 2.0 kg/ha, fluchlorlin 2.0 kg/ha., both applied as pre-plant incorported spray's and desmetryne 0.5 kg and nitrofen 2.0 kg/ha as pre emergence sprays.

Leela (1882, 1985) found that oxyflurofen at 0.25 kg a.i./ha controlled all dicot and monocot weeds including *cyprus* spp. and *cynodon doctylon* upto a period of 60 days as measured by the dry matter production of weeds. Fluchloralin, although was very effective against these two grasses, controlled all broad leaf weeds and the rest of the grasses well for a period of 50-55 days. The yield of cabbage was at par with the hand weeded control in case of oxyfluorfen and fluchloralin treatments. It was found that herbicides like trifluralin @ 0.5 kg/ha (Soil incorporation) and fluchloralin @ 0.5 kg/ha (Soil incorporation) is effective for weed control in cabbage.

PLANT PROTECTION MEASURES

Important insect pests and diseases alongwith their control measures are mentioned here with :

Insect of Pests

Mustard Sawfly (Althalia lugens proxima Klug*)*

The blackish caterpillar that attacks all cole crops and feed on leaves of young seedling in the early stages. The infested leaves become curl and falls on the ground when touched. If low infestation and number of attacked plants are few, hand picking of caterpillar, is the best method of controlling the larvae. Dusting the plant with sevin@ 20-25 kg per hectare or spraying the same @ 5 g per litre of water is also effective to control this insect.

Mustard Aphids (Lipaphis erysimi Kalt)

This is a greenish white small insect that attacks cabbage and other cole crops, when there is a cloudy weather. They suck the sap from the plants. The affected leaves get curled and plant wither and die. Aphid can effectively be controlled by spraying the crop with any one of the insecticides like methyl demiton (metasystox-25 EC, hexasystox-35 EC etc.) or dimethoate (rogor 30 EC, methoate 30 EC, sumidon, demidon etc.) @ 1 ml per litre of water or phosphamidon (demicron-100 EC, sumidon, demidon etc.) @ 0.5 ml per litre of water, starting just after the germination of seed and subsequently at an interval of 15 days. Soil application of phorate 10 G @ 15 kg/ha or thimate 10 G @ 12 kg per hectare in two doses i.e. half at the time of planting and other half at the time of earthing up has also been effective against aphids.

Diamond Back Moth (Plutella exylostella L.)

In India it was first recorded in 1914 (Fletcher, 1914) on cruciferous vegetables. A small slender pale green catrpillar that feeds on leaves and makes holes in them. The infested crop becomes unsuitable for marketing. The pest can be control by (i) Foliar spray twice or thrice cypermethrin (0.2 kg a.i.) or fenavalerate (0.1 kg a.i./ha) or sulprofos (1.0kg a.i./ha) or prothiophos (0.75kg a.i./ha). (ii) Under All India Co-ordinated Vegetable Improvement Projects on control of diamond back moth (DBM) results revealed that against DBM and leaf webber, three sprays of fenavalerate @ 50 g.a.i./ha was found to give maximum cost benefit ratio at IIHR, Bangalore (Anon, 2000). The use of mustard as trap crop proved quite effective in controlling diamond backmoth incidence. Hence, this practice has been recommended to control diamond backmoth throughout India. Mustard crop should be sown 15 days before cabbage planting. Two to three sprays of dichlorvos @ 1 kg. a.i./ha in mustard and two sprays of cartap hydrochloride @ 500 g a.i./ha or 5 per cent neem seed kernal extract in cabbage, starting from primordial stage of the crop at 15 days interval has been recommended.

Cabbage Borer (Hellula undalis F.)

A white brown caterpillar that mines the leaves and feed on shoot later, it burrows into the stem of cabbage. The affected plants wilt, remain dwarf or may even die. Chemicals recommended for diamond back moth and leaf webber control are also effective for cabbage borer.

Leaf Webber (*Crocidolomia binotalis* Zeller)

Adult moths are small and light brown coloured Eggs are laid clusters on the under surface or leaves and held together by gelatinous glue. (i) Most of the insecticides recommended for diamond back moth control are also effective against leaf webber. (ii) In biological control, major natural mortality factors for leaf webber was recorded parasitization by *Bracon hebetor* Say and *Palexorusta solennis* (Wilk) in the late larval instars (Singh. 1978).

Cabbage Aphids (*Brevicoryne brassicae*)

Both nymphs and wingless adults suck cell sap from the leaves, stem and pods. Leaves become curly and pods are not produced. Tolerant varieties are all season, Red Drum Head, Sure Head, Express Mail. In cabbge the larvae of stemborers (*Hellula undalis*) bore into central shoots. The infested plant does not produce the heads. Insecticidal spray of endosulfan/Thiodan (0.2%), fenvalerate (0.05%) or quinalphos (0.2%) at 10-15 days interval is recommended. Cabbage caterpillar (*Pieris brassicae*), cabbage semiloopers (*Plusia orichlacea fabricus*) are also prevalent in cabbage. Control is same as that of Diamond backmoth.

Tobacco Caterpillar (*Spodoptera litura*)

This insect has become a devastating pest on cabbage and cauliflower. Narayan *et al.* (2005) suggested IPM package for this pest. Cultural practices are collection and destroying the egg masses, ploughing the land before transplanting, installing pheromone traps (12/ha) and growing castor as trap crop along the borders. Chemical control measures are poison bait by mixing ricebran @ 10 kg with jaggery (1 kg) and methomyl 40 sp @ 150g to kill the hidden larvae. Spray of indocacarb 14.5 sc. @ 0.5g/l and methomyl 40sp @ 1.6g/l and also spray of fungal pathogen *Nomuraca rileyi* is effective.

Diseases

Fungal Diseases

Downy Mildew (Perenospara parasitica Persoon Fries)

Plant can be infected at any time during their growing period. Young plants infected early, shows as white mildew growth mostly on the under side or leaf, however, slight yellowing shows up the corresponding upper surface. The young leaf may drop off. Older leaves usually persist and the infected areas enlarge as they turn tan and papery. When the disease is severe, the whole leaf dies. It can be controlled by spray the seedling in the nursery bed with copper oxychloride (0.3 per cent).

Wire Stem (Rhizoctonia solani Kuhn)

It may attack different stage of growth. Young seedlings suffer the worst. Young plant shows soft water soaked spot on the skin just above the ground. The cotyledons wither and the plant eventually falls over and perishes. The fungus also causes head rot in cabbage. The first sign of infection is a reddish brown brown discolouration on the stem at the soil level. The area is constricted and the plant may be bent or twisted without breaking. Regulating moisture levels, seed treatment and soil drenching with dithane M-45 (0.2 per cent) or bavistin (0.1 per cent) can be used for protecion against the disease.

Black Spot (Alternaria brassicae (Berk)

The first symptom is small dark spot on the leaf surfaces. As the spot enlarges, concerntric rings develop. The diseased sopt enlarges progressively and defoliate plants. It is a destructive disease on seed crop affecting pods and seeds. Spray captan (0.2 per cent) or copper oxychloride (0.5 per cent) for the control of disease (Sridhar, 1982).

Cabbage yellow (Fusarium oxysporum f. Sp conglutinans (Wollenweber) Snyder and Hansen)

Affected plants have a sickly dwarfed yellow appearance, lower leaves drop one by one and growth of the plant becomes stunted. The fungus is seed borne and can also in soil persist many years. Disease development is promoted by warm weather conditions. Field sanitation, crop rotation and using disease free seeds are useful for controlling the disease up to some extent. Since, resistant cultivars are available, it is better to use them rather than to try other means of control. There are two types of resistant reported.

Type A

Single dominant gene controls the resistance, found first in cabbage cultivars. All Head Early. Copenhagen Market, Glory of Enkhuizen (Walker and Wellman, 1928). It is stable at different soil temperature. This apparently has been derived directly from *B. oleracea* var. *sylvestris.*

Type B

The cabbage cultivar Wisconsion Holander has this type of resistance. Both A and B types are found in Wisonsion All season.

Black Leg (Phoma lingum) (Tode ex Fr.) Desm

It occurs in most of the regions especially in area with rainfall during the growing period. The fungus is carried by the seed and hence it may occurs from the early stage. Stem of the affected plant when split vertically, shows severe black discoloration of sap stream. Whole root system decays from bottom up wards. Frequently, the affected plants fall over in the field. As the main infection is through seeds, hot water treatment can be prevented. Spraying of copper oxychloride or with organomercuric compound and use of resistant variety like Pusa Drumhead.

Bacterial Diseases

Black Rot of Curcifers (Xanthomonas campestris pv. campestris (Pammel Dowson)

The disease was first observed in the USA on cabbage in 1891, but the causal bacterium was identified during 1895. In India, disease was first described in 1928 on cabbage from pune, Maharashtra (Patwardhan, 1928). The infected plant become stunted, often one side of cotyledon turn yellow to black bend down and drop off prematurely. Many lower leaves are also shed off early. The remaining leaves may turn yellow with blackened veins. Development of 'V' shaped chlorotic to yellow lesions take place from the leaf margin. The vein and veinlet turn black and the leaf tissue necrotic and brittle. Lesion progress towards the midrib and from this systematic infection further spread into stem and root. Vascular tissues in affected parts become conspicuously black, followed by internal break down of fleshy tissue. Black sunken spot may develop on succulent seed stocks and siliqua. Considerable reduction in disease has been observed when seeds are treated with plantomycin (100 ppm) (Ksihun, 1981) or agrimycin 100 (100 ppm) (Rao and Kishun, 1986) or streptocyclin (100 ppm) (Shah *et al.*, 1985) or auromycin (1: 1000) (Trivedi and Patel, 1972).

Zonate Spot of Cabbage (Pseudomonas cichorii (Swingle Staff)

It was first reported from USA in 1956. In India, it was reported by Trivedi and Patel (1972) in 1961 from Solan. The spots on cabbage heads are circular to oblong, translucent, buff to brown, 7-12 mm in diameter, sunken and having typical zonations. Lesions on the inner leaves are irregular in outline, almost black in colour and localized near the margin of young leaves. The vascular bundles of stem become dark brown. Elongated and sunken lesion 5-7 cm in size also develops on petiole and mid ribs of the lower leaves. Control measures are same as given in black rot of cabbage.

Club Root (Plasmodiophora brassicae War)

This is a typical root parasite. Club root is particulary prevalent in soils with a pH above 7. Once affected, it causes irregular galls on the roots and these on the lateral roots take the shape of spindle. Leaves become yellowish and wilt.

The disease can be control in following ways : (i) At the planting time, 250 ml of 0.05 to 0.10 per cent solution of mercuric chloride may be applied for direct control. (ii) Roots are treated with 4 per cent calomel paste before transplanting and (iii) Use of resistant variety like Badger Shipper.

Physiological Disorders

Internal Tip-Burn

It is a very common disorder in cabbage. Interior of the head and margin of the foliage turn brown and die, leaving the head looking healthy outwardly. It is associated with the disturbance of water regime of the plant and coccurs due to calcium deficiency. Excessive nitrogen application, poorly drained soils, frequent watering, early planting, wider spacing and delayed harvests also favour tip-burn development. Selection for resistance to this disorder is the best method for control. Spraying of 200 ppm of calcium gluconate in water can also control it.

Head Bursting

Application of nitrogenous fertilizers in excess results in bursting of head in cabbage and knolkhol. The damage caused by insects are loosens the outer protective leaves and encourages bursting. Delayed harvesting of early verities prone to head bursting in cabbage. Sudden fluctuations in temperature also caused bursting in cabbage. Thus control of insects may reduce this damage up to 50 per cent. Use of nitrogenous fertilizer judiciously also decreases the damage of head burst.

HARVESTING

Cabbage is harvested when the heads is of suitable size, firm but tender. Harvesting is also done earlier, specially early in season when the prices are high. They have a tendencey to burst or loosen the leaves beyond the marketable stage. Irrigation to matured crop causes cracking of heads. Always harvest with few outer leaves to minimize the damage of head during transit. The early vearieties take 60 to 80 days, medium 80-100 days and late varieties take 100-130 days for harvesting after transplanting.

YIELD

The yield of early cabbage ranges between 200-300 q/ha, while that of late verieties between 320 and 400 q/ha. Hybrid crop yield upto 75 tonnes/ha.

POST HARVEST HANDLING

Grading

The grading should be done according to the size and quality of head. Grading can be done either according to weight per head or condition of the head before taking them to market. It will fetch more price. The Indian standard institution has laid down the principle for grading, based on solidity, trimming freedom from withering buffiness, bursting, subtract and seed stem and damage caused by discolouration, freezing, disease and insects. There are three grades i.e. Super, Fancy and Commercial for all type of Cabbage. It ensures higher price to the producers. Most markets prerfer a small to medium size Cabbage head, usually 1.3 kg to 1.8 kg is the most desirable (Singh and Singh, 2007).

Packing

The Cabbage should to uniformly trimmed to the same number of wrapper leaves and should be packed accordeing to size. Since many retail stores price Cabbage by the head, uniformity of size is very important. Cabbages are packed in shallow basket, and carried by head-loads or in carts to near by market or in trucks to distant markets. For local markets or *haat* cabbage heads are packed in gunny bags and upper portion of gunny bag close with dry grass and stitiched. Big baskets made with bamboo chips or Arahar sticks, are also used for packing of cabbage. Most of the Cabbage packed in open-mesh bags holding 35-40 kg Cabbage to send the distant market by trucks. Many markets require that the bags be labelled with count as well as weight of Cabbage in the bag (Singh and Singh, 2007). Bags contaning 16 heads are most popular size in the market. Some cabbage is packed in wooden containers and in crates for distant markets.

Storage

The marketable heads can be stored for 4-5 days under ordinary conditions, where as they can be stored for several weeks in the cold storage at 0°C to 1.7°C with 85% to 87% relative humidity. The best storage temp. is 0°C at 90-95% relative humidity and cabbage can be stored for about 2 to 8 month at this temperature and humidity (Mercontilia, 1989). A good air circulation will avoid condensation if a uniform 0°C temperature is maintained.

SEED PRODUCTION

Cabbage is cross pollinated crop and readily crosses with other member of the cole group. In India seeds are produced only in hills. Being a biennial the cabbage require two seasons to produce seeds. First season the heads are produced and in the following season seed production.

Method of Seed Production

There are three methods of cabbage seed production viz. seed to seed, head to seed and stump to seed method.

Seed to Seed Method

The plants are allowed to produce seed in their place of seeding stage (*in situ*). This is a commercial method of cabbage seed production. In this method labour is not needed for uprooting, storage and replanting of head. Early seed yield is obtained. However, selection cannot be done for trueness of head.

Head to Seed Method

Selected true to type heads are transplanted for the production of seeds. Outer whorls of the leaves surrounding the heads, are removed before transplanting. Cross cut/slit is given on the head and left as such. Cross cut help in proper aeration, initiation of growing point, bolting and production of seed branches (Fageria *et al.*, 1995).

Stump Method

Heads are romoved (and marketed) and stumps are allowed to produce flowering stalks from axillary buds. In fact. this method is not commonly followed but it is a economical.

Cultural Practices

Cultural requirement for all the three methods more or less are similar. As seed to seed method of cabbage seed production is followed comercially, the following cultural practices may be taken into careof.

Raising of Seedlings

Nursery land is well cultivated and divided into convenient size of beds and raised the level of 10-15 cm high above the ground. For plnating of one heatare area, about 60-80 sqm area is sufficient. For this much area application of about 2.5 to 3.0 kg ammonium sulphate, one kg each of single superphosphate and muriate of potash is needed.

The seeds are sown in shallow furrows (2.3 cm deep) spaced at 10-12 cm apart by dropping the seeds at 5-7 cm distance and at 1.5 to 2 cm depth. The seeds are covered gently either with the same soil of furrows and spreading mixiture of screened farmyard manure or compost that prevents formation of soil crust. Covering organic manure also helps in raising temperature around the seeds which enhances early germination. Mulch layer can also reduce direct effect of solar radiation. The nursery bed is watered with sprinkler. Always avoid wet conditions in the nursery. As soon as germination of seeeds is completed, the mulch layer should be completely removed and any delay in removal of it will cause poor growth of seedlings.

Direct Sowing

Cabbage seeds may be sown on ridges in main field and mulched simultaneously with perforated polythene film. Mulching increases soil temperature and seedling emergence. It has been found to advance emergence by 2 days and shorten the growing period by 21-24 days. Head weight at harvest is greatest from direct sowing+mulching (Mikhov *et al.*, 1995). However, direct sowing is not so easy and probably, therefore, it is not followed in India on commercial farming of cabbage crop.

Sowing Time and Seed Rate

As seed production of cabbage is done in hills in the month of march to June for heads. Whereas July-August for seed production. About 400-500 g of high quality seed is sufficient for raising seedling for planting of one hectare crop.

Effect of Shading on Cabbage Seedling

The rates of photosynthesis and root respiration decline in shaded plant. Shaded seedlings have less carbohydrate in the top and roots than those exposed to light. Rooting ability after transplanting to cabbage seedlings depends largely on carbohydrate content in the roots during the seedling stage. If nursery plants are subjected to low light intensity conditions, carbohydarate levels in the roots decrease because of lower photosynthetic activity. Decrease in carbohydrate content causes a decrease in activity of those enzymes which associated with respiration. Diminished photosynthates and lowered enzyme activities inhibit root respiration. Consequently a larger amount of energy is consumed for basal metabolism for survival which detracts from root formation after planting (Fukuoka *et al.*, 1996).

Transplanting

About 4 to 6 weeks old seedling are transplanted at the following distance.

Early Varieties	60×30 cm
	45×45 cm
	30×30 cm
Mid season var.	50×45 cm
Late Varieties	60×45 cm
	60×60 cm

The above mentioned planting distances may be adjusted according to local conditions and cultivation practices.

In situ method of cabbage seed production is better than replanting. Further *in situ* with large heads result higher seed plants yield and seeds quality in terms of seedling vigour and percentage of normal and abnormal seedling and dead seeds (Verma *et al.,* 1996).

Nutrition

Both for head and seed yields requirement of nutrients is high, particularly for nitrogen, phosphorus and potassium. The doses of nutrients vary according to season, soil types and management practices. In Table 1 doses for guide line are given, though they are general recommended doses in various states.

Table 1: Recommended doses of manure and fertilizers for Cabbage (Tandon, 1991)

State	Nitrogen kg/ha	Phosporus kg/ha	Potash kg/ha	FYM t/ha
Andhra Pradesh	60	25	60	40-50
Arunchal Pradesh	120	80	40	15
Assam	80	60	60	10
Gujarat	75	37.5	37.5	10-15
Haryana	125	50	50	50
Himachal Pradesh	125	75	30	-
Karnataka	150	100	125	25
Mdhya Pradesh	150	100	40	20
Mahrashtra	75	0	0	50-100
Mizoram	60	50	40	5-10
Orissa	125	40	60	-
Punjab	125	62	62	100
Rajasthan	150	45	45	25
Tamil Nadu	180	125	100	30
Uttar Pradesh	120	60	60	30
West Bengal	120	60	60	15

Method and Time of Application

Farmyard manure or compost is incorporated during field preparation that is before transplanting. One third dose of nitrogen and whole quantity of phosphorus and potassium is applied basally at transplanting. One third nitrogen is top dressed 3 weeks after transplanting and remaining one third dose of nitrogen during flowering and fruit setting. Both soil as well as foliar application of nitrogen through urea have better response (Baghel and Singh, 1996).

Irrigation

As soon as transplanting is completed light irrigation is given that will facilitate proper setting of seedlings in the field. Subsequent irrigations are done at 10-15 days intervals. Uneven moisture after the heads are formed may cause head splitting or head bursting.

Interculture and Weed Control

During early stage of crop growth, 2-3 light cultivations are needed. These cultivations will also keep down the most of the weeds.

There are some important weed speacies associated with cabbage crop crop. *Cyperus rotundus, Trianthema monogyna, Chenopodium album, Eleusine aegyptiacum, Trigonella. Polycerata, Melilotus indica* etc.

The pre-emergence application of herbicides like oxyflurofen 0.75 lit a.i./ha is very effective (Leela, 1982).

Promotion of seed stalks

The natural formation of seed stalks in cabbage is not cent per cent and that it is not synchronized, further it takes much time for emergence. In order to enhance this process, cabbage heads are usually incised at their fully developed stage. The selected true to type heads as per variety grown are incised by making the cross cut on the top of the head. While making cut care is taken that cut is not so deep that damage the heart or growing point.

Roguing

At least three roguings are required to maintain genetic purity of the variety grown. Roguings are performed at different stages of crop growth, these are :

1. *Planting out or before heading* : Roguing operation in cabbage is begun at planting out or before heading. At this stage, foliage is checked for its general characters. Plants showing different foliage characters are discarded.

2. *Most plants have formed heads* : This is the second roguing, done when heads have formed on most plants in the crop. At this stage, heads are checked based on shape, relative size, colour, firmness etc. Also plants should be removed which are either too early or too late in head forming habit.
3. *Maturity stage* : The third or final roguing is performed when seed stalks reach to maturity stage. At this stage, all plant are removed which are either too early or too late in maturity.

During all roguing operations, diseased plants should be removed invariably and they should be destroyed out of the seed plot. Such plot after completion of roguings should meet the field standards presented in Table 2.

Table 2 : Field standsards for cabbage seed production (maximum permitted in per cent)

Factors	Foundation	Certified	Ramarks
Off-type	1.0	0.5	*At and after
Other crop plants	-	-	flowering after
Objectionable weeds	-	-	maturity stage
Diseased Plant	0.1*	0.5*	

Table 3 : Seed standards for cabbage seed production (In percentage)

Standard	Foundation	Certified
Pure seed (Minimum)	98.0	98.0
Inert matter (maximum)	2.0	2.0
Other crop seed (maximum)	0.05	0.10
Total weed seed (maximum)	0.05	0.10
Objectionable weed seed (maximum)	None	None
Germination (minimum)	70.0	70.0
Moisture (maximum)	7.0	7.0

Harvesting

Generally, cent per cent pods may not attain matrity at a time, therefore, havesting is to be completed in 2-3 lots, that will also avoid shattering of the seeds. However, on commerical scale when it is not possible to complete harvesting then it should be done when most of the pods have attained maturity, (above 95%).

Harvesting is performed by cutting the seed stalks above the ground. Height of cutting depends on nature of the branching. In all cases, brachces should not be left without cuttings them.

Cut seed stalks are heaped on cemented floor for drying at least 4-5 days.

Threshing and winnowing

Threshing is down with help of wooden sticks on small scale and on commercial scale electrically operated thresher is used.

Seed material is then, cleaned through winnowing which is performed by blowing against natural winds or before fans.

Seed yield

Cabbage seed yield is higher when seed production is done by seed to seed method rather heads or seed or stump method. The seed yield also varies due to culitvar, climatic conditions and management practices.

On an average 500-600 kg/ha seed yield is assumed as good yield.

Drying

Seed lot is allowed to dry in open sun for about 4-5 days if desired moisture (7-8%) is not reached, the seeds should be dried in driers, if available.

Grading

Through screening small, light, ill developed shrunken, rotten and split seeds are taken out from healthy seeds.

Packing

As per market demand, seeds is packed in air-tight packet (polythene) or other glass as well metallic containers. Containers are labeled with information like name of the crop and variety, purity, germination percentage, source etc. Cabbage seed should meet the standards given in Table 3.

Storage

Packed seed of cabbage is stored under well ventilated dry and cool place. Under proper storage conditions, cabbage seeds remain viable for about 4 years.

Ratooning of Cabbage

Ratooning of Cabbage is highly profitable and its gives higher yield with lower investment.

To get a ratoon crop, one or two green leaves at the base should be left undisturbed while harvesting the main crop. During harvesting a sharp instrument (like sharp knife) should be used to remove the cabbage head from the stem so that it makes a sharp cut in the main stem without damaging it much. Immediately after the harvest the stem should be covered with the leaves or paper to protect it from direct sunlight. The field should be irrigated immediately after the harvest. After 15 to 20 days new cabbage buds will appear around the tip of the stem. The number of newly grown buds will vary from 4 to 20 depending upon the variety used. When these buds attain the size of 'button', only 2 to 3 buds should be retained and other removed. At this stage urea should be sprayed at a concentration of 0.2 per cent and the field should be irrigated as and when necessary. The ratoon crop will be ready to harvest within 60 to 70 days after havest of the main crop.

The response of ratooning is best in the early and mid season varieties. Varieties like 'Golden Acre', 'Copenhagen Market' 'Early Drumhead' and 'Pride of India' give 100 per cent response in head formation, ranging from 3 to 20 heads. In the late season varieties the ratoon crop does not grow well because of the rise in temperature in the later part of the season. Although the size of the cabbages in the ratoon crop is smaller than that of the first crop, double the yield can be expected in the ratoon crop owing to production of highest number of cabbage heads per plant.

Ratooning is profitable because it does not require any extra care for raising seedling and planting. There is no need to apply fertilizer to the soil also. Moreover, the total duration of the crop is almost half of the main crop. Therefore the expenditure for irrigation and other cultural practices for the crop is also reduced.

REFERENCES

Allen, C.L. (1914). Cabbage, Cauliflower and Allied Vegetable. Orange Judd Publishing, New York.

Anonymous (2000). AICVIP, Recommendations (1971-1998), IIVR, Varanasi, India.

Aykroyd, W.R. (1963). The nutritive value of Indian foods and the planning of satisfactory diets. ICMR special Rept., 42. New Delhi.

Baghel, M.S. and Singh, D.B. (1996). Response of soil and foliar application of urea on cabbage. *Recent Hort.*, 3(1): 108-110.

Bahadur, Singh, J. and Singh, K.P. (2004). Response of cabbage to organic manures and biofertilizers. *Indian J. Hort.*, 61 (3) : 278-279.

Bailey, L.W. (1930). The cultivated Brassicas, *Gentes herbarium*, 2: 211-67.

Bauer, A. (1932). Der naterstoffertuzug unscrer kulturplanzen. Ernahar pflanza, 28 : 241.

Bhagchandani, P.M., Verma, T.S. and Singh, N. (1977). Study on selection of cabbage cultivar (s) for summer cultivation in the hills. *Indian J. Hort.*, 34 (1) : 60-63.

Chatterjee, S.S. and Mukherjee, S.K. (1957). Sudies on vegetables in West Bangal II : Seed production of cabbage (*B. oleracea* var. *capitata*). *Indian J. Horti.*, 22 : 60-68.

Chhonkar, V. S. and Singh, R. (1964). Effect of gibberellic acid on growth and yield and quality of cabbage. *Indian J. Hort.*, 21 (1) : 50-63.

Chhonkar, V.S. and Jha, R.N. (1963). The use of starters and plant growth regulators in transplanting of cabbage (*Brassica oleracea* var. *capitata* L.) and their response on growth and yield. *Indian J. Hort.*, 20 (2) : 123-128.

Choudhury, B. (1976) Vegetables, National Book Trust, New Delhi.

Dave, S.K. and Bambhaniya, L.B. (1993). Performance of cabbage varieties under south Gujarat conditions. *Indian J. Hort.*, 50 (1) : 77-79.

Dixit, H.C. and Singh, Vinay (1997). Effect of boron and calcium on yield and nutrition of cabbage (*Brassica oleracea* convar *capitata* var. *capitata*). *Indian J. Agric. Sci.*, 67 (2) : 87-8.

Fageria, M.S., Arya, P.S. and Singh, A.K. (1995). New trend in cabbage seed production. *Sceintific Hort.*, 4: 101-108.

Fukuoka, N., Yoshioka, H., Shimizu, E. and Fuziwara, T. (1996). Effect of shading cabbage seedling on their physiological processes and rooting ability after transplanting to the nursery. *J. Japanese Soc. Hort. Sci.*, 65 (3) : 545-551.

Gates, R.R. (1953). Wild cabbages and their effect of cultivation. *J. Genet.*, 51 : 363-372.

Gawade, M.H., Patil, J. D. and Kakade, D.S. (2004). Studies on effect of intercrops in yield and monetary returns of cabbage. *Agric. Sci. Digest.*, 24 (1) : 69-70.

Gill, H. S., Singh, J.P. and Swarup, V. (1979) Inheritance of pod length, number of seeds per pod and seed yield in cabbage. *Vegetable Science*, 6 (2) : 98-105.

Gautam, K.C., Mani, V.S., Das, B. and Khakar, M.S. (1976). Weed killers in cabbage cultivation. *Indian Farming*, 26: 11-14.

Gupta, A. (1987). Effect of nitrogen and irrigation on cabbage production. *Indian J. Hort.*, 44 : 241-244.

Gutpa , O.P. and Singh, R.P. (1965). Preliminary observation on the effect of tenoran and eptam 6E herbicide in certain vegetable crops. *Indian. J. Hort. Sci.*, 355.

Helm, J. (1962). Die Lacimiaten Sipp-en Von *Brassica oleracea* L. *Kulfur planze* ; 11 : 92-210.

Jadhav, V.M. and Sreenivas, L. (1968). Influence of spacing-cum-irrigation on growth, yield and quality of cabbage (*Brassica oleracea* var *capitata* L.). *Poona Agric. Coll. Mag.*, 58: 18-25.

Khokhar, N.S., Singh, R.P. and Prasad M. (1970). Effect of different levels of nitrogen, phosphorus, potssium and FYM on the yield of cabbage. *Indian J. Agron.*, 15 : 9-12.

Kishun, R. (1981). Effect of different chemicals on black rot of cabbage. *Proc. 3rd Int. Simp. Plant Pathol.*, New Delhi, 250 (Abstr).

Kohle, D.S. (1985). To study the effect of Azotobacter and Azospirillium alone and in combination under graded levels nitrogen on growth, yield and quality of cabbage cv. Golden Acre. *M. Sc. Thesis*. Mahatma Phule Krishi Vidyapeeth Rahuri, India.

Lal, D. (1938). Cabbage cultivation. *Nagpur Agric. Coll. Mag.*, 12 (3) : 198.

Lal, G. (1993). Agrotechniques for cole crops. In : Advances in Horticulture, Vol. 5, Vegetable crops : Part 1. Eds : K.L. Chadha and G. Kalloo, (1993) Malhotra Publishing House, New Delhi- 110064 India. pp. 503-508.

Lal, S., Mangal, J.L. and Batra, B.R. (1986). Effect of canal water, underground brackish water and their combinations in different ratio on growth, yield and quality of cabbage and cauliflower. *International Seminar* on water management in Arid and Semi-arid Zones, pp. 11-18.

Lawson, H.W. and Wiseman, J.S. (1978). Competition between weeds and transplanted spring cabbage. *Hort. Res.*, 18 : 113-126.

Leela, D. (1982). A decade of weed research in horticultural crops. *Pesticides.* 15 (10) : 3-8.

Leela, D. (1982). Weed control by herbicides in cabbage, bottle gourd and water melon. *Indian Hort.*, 39 : 252-256.

Mangal, J.L., Jassin, A.H. and Mahmood, S.K. (1987). Effect of plant spacing and age of seedlings on yield and quality of Cabbage. *Haryana J. Hort. Sci.*, 16 : 121-124.

Mani, V.S., Gautam, K.C., Das, N., Das, B. and Singh, Y.R. (1973). Chemical weed control in cabbage, onion relay cropping system. *Proc. 3rd All India Weed control Seminar* Hisar, pp. 57.

Medhi, G. and Borthakur, U. (1999). Methods and rate of nitrogen application on growth and yield of cabbage (*Brassica oleracea* var. *capitata*) *Haryana J. Hort. Sci.*, 28 (1&2) : 130.

Mikhov. M.A., Antova, G.S. and Vuldev, S. (1995). Studies on seedlingless cultivation of early head cabbage by precise sowing. *Bulgarian J. Agric. Sci.*, 1 (3) : 259-264.

Mishra, S.G., Gupta, A.K., Varshney, M.L. and Sharma, K.N. (1984). Effect of naphthalene acetic acid and chelated zinc on vitamin C content and yield of cabbage. (*Brassica oleracea* var. *capitata*) *Prog. Hort.*, 16 (1&2): 92-94.

Mohanty, B.K. and Hussain, M.M. (1998). Effects of nitrogen and phosphorus on cabbage. *Orissa J. Hort.*, 26 (1) : 106-107.

Nath, Prem, Srivastava, V.K., Dutta, O.P. and Swamy, K.R.M. (2008). Vegetable Crops — Improvement and Production. PNASF, Bangalore, pp. 126-130.

Nieuwhof, M. (1969). Cole crops - Botany, Cultivation and utilization. World crops Book, Leonard Hill, Landon.

Patil, J.D., Rampise, S.A., Kakade, D.S. and Desai, U.T. (2002). Productivity and Profitability of cabbage intercropped with leafy vegetables. *Intl. Conf. Veg.*, Bangalor 11-14 November, 2002. Abstract : pp. 797.

Patwardhan, G.B. (1928). Bulletin, Department of Agriculture, Bombay, pp. 30.

Prabhakar, B.S., Srinivas, K., Mohan Doss, V., Subramanian, T.R., Rao, J.V., Shukla, V. and Sharma, C.B. (1985). Yield and price variability in the analysis of multiple cropping systems. *Singpore J. Pri. Ind.*, 13: 57-63.

Prakash, Ved and Bhardwaj, S.N. (1965). Influence of foliar application of molybdenum and cobalt on growth of cabbage. *Indian J. Hort.,* 22 (344) : 349-350.

Purseglove, J.W. (1975). Tropical crops - Dicotyledons. Longman, London.

Rajanna, K.M., Mathod, J.C. and Patil, V.S. (1995). Effect of different levels of NPK on yield of cabbage cv. Pride of India. *South Indian Hort.,* 43 (142) : 57-58.

Rao, A.N.S. and Kishun, R. (1986). A note on the control of bacterial black rot (*Xanthomonas campestris* (Pammel) Dowason) of cabbage (*Brassica oleracea* var. *capitata* L.) by seed treatment. *Presticides,* 20: 47.

Saini, G.S. (1996). A Text Book of Vegetable Production published by Aman Publishing House, Madhu Market Budhana Gate, Meerut (U.P.), India. pp. 96-97.

Seshadri, V.S. and Chatterjee, S.S. (1996). The history and adaptation of some introduced vegetable crops in India. *Vegetable Science,* 23 (2) : 114-140.

Shah, A., Srivastava, K.K., Roy, A.J. and Bora, S.S. (1985). Control of black rot disease of cauliflower by seed treatement. *Progressive Hortic.,* 17 (1) : 72-74.

Sharma, K.C. (2005). Foliar application of micronutrients on yield and quality of cabbage. *Crop Res.,* 8 (2): 6872.

Sharma, K.C. and Arya, P.S. (2001) Effect of nitrogen and farm yard manure on Cabbage (*Brassica oleracea* var. *capitata*) in dry temperate zone of Himachal Pradesh. *Indian J. Agric. Sci.,* 71 (1) : 60-61.

Sharma, K.C. and Verma, S. (2000). Performance of some cabbage cultivars under dry temperate high hills of Himachal Pradesh. *Indian J. Hort.,* 57 (3) : 248-249.

Sharma, S.K. and Kumar, Pradeep (2002). Effect of organic sources in cabbage production. *Intl. Conf Veg.,* Bangalore, 11-14 November 2002. *Abstract,* pp. 211-212.

Sharma, S.R., Singh, Ram and Singh, Narendra (2001). Pusa Ageti : a new heat tolerant cabbage. *Indian Hort.,* 46 (2) : 22-23.

Shoemaker, J.R. (1970). Response of direct seeded cabbage using different spacings and planting and thining methods. *Proc. Fla. Hort. Soc.,* 82 : 112-115.

Singh, D.N. and Nath, Vishal (2011). Varieties and Hybrids of Vegetables. Satish Serial Publishing House, Azadpur, Delhi, (India), pp. 13-28.

Singh, D.N. and Singh, S.P. (2007). Production Management of Underutilized Vegetables. Agrihortica Publications. Junagadh (Gujarat). India. pp. 79-80.

Singh, N.P., Bharadwaj, A.K., Kumar, A. and Singh, KM. (2004). Modern technology on Vegetable Production. International Book Distributing company, Lucknow.

Singh, P., Kumar, R. and Mangal, J.L. (1990). Performance of trickle, surface and sub-surface irrigation for cabbage (*Brassia oleracea* var. *capitata*) *11th Int. Congress* on the use of plastics in Agriculture, pp. 59-66.

Singh, R.V. and Naik, L.B. (1988). Response of cabbage to plant spacing, nitrogen and phosphorus levels. *Indian J. Hort.,* 45: 325-328.

Singh, S.D. (1978). Effect of planting configuration on water use and economics of drip irrigation systems. *Agron. J.*, 70 : 951-954.

Sinha, A.P. (1977). *Progressive Horticulture,* Vol. 6, pp. 37-40.

Snogerup, S. (1980). Brassica crops and wild Allies, Ed. S. Tsunoda; K. Himda & C. Gomezcampo, Japan Sceintific Societies Press Tokyo.

Som, M.G.; Jana, P.K. and Pal, S.S. (1976). Soil moisture regims and nitrogen levels on the yield and consumptive use of water by cabbage. *Vegetable Science,* 3 : 131-137.

Sridhar, T.S. (1982) Control of *Alternaria* disease of Vegatables. *Seed & Farms,* 8: 43.

Srinivas, K. (1984). Note on response of cabbage to plants density and fertilizers. *Indian J. Hort.,* 41 : 277-279.

Swarup V. and Gill, H.S. (1964). The use of marker gene in hybrid seed production in cabbage. *Curr. Sci.,* 33 (10) : 315.

Swarup, V. and Sharma, B.R. (1965). Inheritance of some quantitative character in cabbage. *Indian J. Genet.,* 25 (1) : 57-64.

Swarup, V. and Singh, D. (1963). Studies on hybrid vigour in cabbage. *Indian J. Genet.,* 23(1) : 90-100.

Swarup, V., Gill, H.S. and Singh D. (1963). Studies on hybrid vigour in cabbage. *Indian J. Genet. Plant Breed,* 5 : 30-35.

Swarup, V., Gill, H.S., Kalda, T.S. and Bhagchandani, P.M. (1968) Pusa Drum Head A Promising Variety of cabbage. *Indian Hort.,* 12: 6-8.

Swarup, Vishnu (2006). Vegetable Science and Technology in India. Kalyani Publishers. New Delhi. pp. 656.

Tandon, H.L.S. (1991). Fertilizer Recommendations for Horticultural Crops in India. A guide book pub Fertilizer Development Consultation Organization. New Delhi.

Tewari, R.N., Chattaerjee, S.S. and Swarup, V. (1979). Inheritance of resistance to black rot, *Xanthomonas campestris* (Pam) Dowson, in Cabbage. *Vegetable Science,* 6(1) : 27-36.

Tewari, R.N., Gill, H.S. and Singh, Ram (1977). Influence of different planting times on head shape in cabbage variety Golden Acre. *Progressive Hort.,* 8: 35-38.

Thakur, P.C. and Gill, H.S. (1976). Effect of nitrogen, phosphorus and potash fertilization on seed yield of cabbage (*Brassica olercea* var. *capitata*) . *Indian J. Hort.,* 33: 262-265.

Trivedi, B.M., Patel, P.N. and Verma, J.P. (1978). Zonate Spot of cabbage in India. *Sci. Cull.,* 44 : 182-183.

Vaidya, S.J. and Patil A.V. (1965). Indluence of date of transplanting and age of seedlings on growth, yield and quality of cabbage II. *Indian J. Agron.,* 10: 420-422.

Verma, T.S., Joshi, S. and Thakur, P.C. (1996). Effect of *in situ* and replanting in cabbage with different head sizes on seed productivity and its quality. *Recent Hort.,* 3. (1) : 147-150.

Walker, J.C. (1952). Diseases of Vegetable Crops, MC Graw- Hill Book Co. Inc; New York.

Walker, J.C. and Wellman, F.L. (1928). *Journal of Agricultural Research,* Vol. 37, pp. 233-41.

Wange, S.S., Patil, P.L., Meher, B.B. and Karkeli M.S. (1995). Response of cabbage to microbial inoculants and incremental levels of nitrogen *J. Maharashtra Agric. Univ.,* 20 (3) : 429-430.

Yawalkar, K.S., Kataka, P.N. and Srivastava, M.P. (1962). Commercial Fertilizers in India Agri-Hort. Publishing House. Nagpur.

Zimdahl, L. (1980). Crop weed competition - A Review. Published in the United States by the International Plant Protection, Centre, Oregon State University, Oregon, U.S.A.

□□□

3. KNOL-KHOL

Botanical name	:	*Brassica oleracea* var. *gongylodes* L. (Syn. *B. caulrapa* or *B. oleracea* var. *caulorapa*)
Family	:	Brassicaceae
Chromosome number	:	2n = 18

Knol khol or kohlrabi is a cool season, cole crop grown for its enlarged and fleshy stem. It is short duration winter vegetable, tubers are sowllen stems which develop above the ground may be due to secondary growth in thickness (Kalloo, 1993). The tuber is harvested before it becomes tough or woody. The word 'kohlrabi' is the German name for cabbage-turnip, resembles an above ground turnip (Choudhury, 1967).

ORIGIN

The knol khol has originated in Mediterranean centre of crop origin, specially in the coastal countries of western Europe. It was developed from the wild cabbage (*Brassica oleracea* var. *sylvestris*) through introgression, mutation and selection. Knol-khol in its present form appeared during 15th century in north Europe (Boswell, 1949). It was introduced to India long back (Nath *et al.*, 1976). According to Encyclopaedia Britannica, khol-rabi was first described in the eighteenth century as another cabbage of European origin (Shanmugavelu, 1989).

DISTRIBUTION

According to Garade, the knol khol was discussed in Century Herbals and stated that it was grown in Italy, Spain and Germany (Helm, 1962). It is grown in temperate, subtropical and tropical countries of the world where the mean temperature is 20ºC-25ºC for a period of four months. In India, it is cultivated in states like Kasmir, Uttar Pradesh, Himachal Pradesh, West Bengal, Madhya Pradesh and in some of the southern state it also cultivated as a minor crop.

BOTANY

Knol-khol is a biennial, herbaceous vegetable crop grown as an annual which produces a swollen, bulb like stem. The edible part is a swollen stem called tuber or knob which arise from the thickening of stem tissue just above the ground. Glaucous leaves have long slender petioles and grow out of this stem arranged in a compressed spiral on the bulbous part. It is a temperate crop which produces edible vegetable as an annual in the plains and in the hills. For seed production it is biennial in the hills. The edible portion is globular to slightly flattened stem (Nath *et al.*, 1976).

Plant breeding

Knol-khol is a biennial and cross-pollinated crop like cauliflower and cabbage with bisexual flowers. Since it fails to produce seeds in the lains, no effort was made either to maintain or to improve upon the available strains. It has CMS with sterile Ogura cytoplasm from radish similar to cabbage/ cauliflower. In India most of the varieties were introduced from European or American countries. The regional Research centre Katrain, Kullu Valley (H.P.) of IARI is the only centre, where knol-khol research is being carriedout (Nath *et al.*, 2008).

COMPOSITION

It is high in minerals like calcium, magnesium, potassium, phosphorus, sodium and sulphur. It also contains vitamin A and C. The nutritive value of knol khol (per 100 gm of edible portion) is given below (Choudhury, 1967):

Moisture	92.7g	Calcium	20 mg	Copper	0.09 mg
Protein	1.1g	Magnesium	18 mg	Sulphur	143 mg
Fat	0.2g	Oxalic acid	10 mg	Vit. 'A'	36 I.U.
Minerals	0.7g	Phosphorus	35 mg	Thiamin	0.05 mg
Fibre	1.5g	Iron	0.4 mg	Riboflavin	0.12 mg
Other Carbohydrates	3.8g	Sodium	0.12 mg	Vit. 'C'	85 mg
Calories	25	Potassium	37 mg	Nicotinic acid	0.5 mg

Uses

Knol khol is characterized by the formation of khob which arise from the thickening of stem tissue above the cotyledons. It is the portion which is used as food (Gill, 1993). It is primarily used as cooked vegetables. The small heads may be boiled in water after adding salt and salt water is drained. The heads are cut into four pieces, cooked and added with salt and vinegar and served as salad (Chauhan, 1965). The leaves are used as greens. In Europe too, leaves of early cultivars are used for human consumption (Gill, 1993).

VARIETIES AND HYBRIDS

Popular Varieties of Knol-khol

The popular varieties of knol-khol are Early White Vienna, White Vienna, Early Purple Vienna, Purple Vienna, Kyote No. 3 (Nath *et al.*, 1976), King of North and Large Green.

Szentesi Faber

It is a summer cultivar, has a thin, pale green skin and crisp flesh, early bearing, produce a higher marketable yield and show greater resistance to alternaria leaf spot (Bogdan, 1986).

Roggli's

Out door white and is medium type, resistant to bolting and tuber do not become fibrous soon.

Goliath

It is a late cultivar maturing between 70-100 days after planting. Tubers are white and blue type.

European varieties

Wiesmoor Forcing white, Gaugels Forcing white are quick growing (4-6 week of Planting) and resistant to bolting.

Short duration and resistant to low temperature varieties

Stuicka, Polni Ranar, Czekes, Blanr speak, Weiver blan, Prasake bele, Rogglis, Soroksari, Knaufs Frushweis, Prasfer weisser Treib, Trero, Primerera (Frydrych, 1971).

Varieties suitable for summer cultivation

Blaso, Pollux, Reliktess Bleu. These are most suitable cultivars for summer cultivation in temperate regions.

Moderate resistant to Fusarium wilt

Vienna White 1350 (Kardaeva and Katarzhine, 1979).

Resistant to Plasmodiophora brassicae

K 240 from Bulgaria.

Resistant to Xanthomonas campestris

Szentesi Tartos Kek (Pesti and Fodor, 1982).

Cultivars resistant to cold and free from bolting

Wiesmoor White Forcing, Kuauf's Early White and Moravia (Luzuy, 1965).

Varieties with short foliage

Wiesmoor White Forcing and purple variety Knauf's Ideal (Hartrath, 1977).

Some promising varieties available in India through seed dealers viz., 'Suttons' Earliest white (Suttons): Green Glory (Verma); Golith white (V.Sadashiv); Earliest Purpole, Early White, King of Market, Green Pearl, Petsri, Michihli (Pocha).

Hybrids

The first hybrid was developed by the firm Roggli. These hybrids were Rowel, Roblau and Rosta (Anonymous, 1984). These are early bearer. There are several hybrids developed by seed companies.

Rallis India : Lakshmi;

Tropica Seeds (Green CO.) : Champion, Dakini,

Annapurna Beej : Winner No.4 Imp., Raja, Kirtiman, Bishal;

Bejo Sheetal Seeds, Korist, Kossak and Kolibri.

Detail characteristics of varieties and hybrids described by Singh and Nath, (2011) are given below :

Early White Vienna

An early variety takes 50-55 days for the development of knobs after transplanting. Knobs are globular-round and green with fewer leaves. Flesh is cream-white, tender and crisp.

White Vienna

An early variety having dwarf plant. Stems (tubers) globular , light green and very smooth. The flesh creamy white, tender with delicate flavour. It takes 50 to 60 days for knob formation (Anonymous, 1998).

Early purple Vienna

An early variety which takes 55-60 days for knob formation. Foliage purple, knobs globular round, large with purple skin and light green flesh.

Purple Vienna

Purplish-blue leaves, knobs are globular round, large in size with bluish-purple, tinge, flesh light green. It takes 55 to 65 days from transplanting to knob-formation.

King of North

Plant height 20-30 cm, dark green foliage with broad leaf-blade. Knobs are flattish round and dark green in colour. Leaf-sheath is large and well-spread over the knob. Mature in 65-70 days.

Large Green

It is a late variety which takes 60-70 days for harvest. Plants are vigorous with dark green foliage, knobs, flattish round and dark green, adapted well to hilly regions.

Palam Tenderknob

It is an improved variety of knol khol was bred through mass selection at Department of Vegetable Science, HPAU, Palampur. It has small light green foliage. It forms turnip-shaped knobs, which remain split free. These are round, flat, stringless, light green, sweet and fleshy. These knobs attain marketable size of 5-7 cm diameter weighing 250-300 g in about 80 days. Its average yield potential is 250-300 q/ha (Kalia and Pathania, 2006).

Korist

It matures in 55-60 days after sowing, The approximate weight is 150 g and has good standing ability up to 30 days after maturity. It keeps growing even after 65 days and can grow up to 300 g. in size, without cracking and without fibre development.

Kolibri

It is attractive red coloured knol khol with more sweet taste. It matures in 55-60 days after sowing.Approximate weight is 150 g and has good standing ability up to 30 days after maturity. It is tolerant to cracking and there is no fibre.

Kossak

It is a big sized knol-khol, which matures in 65-70 days after sowing. The approximat weight of tuber is 1.0 kg.

Champion (F_1)

This hybrid variety grown in dry and cool climates. It gives high yield and very uniform, green bulb of globe shape, weigh about 600-700 g at maturity. Very good bulb quality, maturity 60-65 days. Erect growing habit and very good shelf life.

Dakini (F_1)

Plants are erect medium and tuber mature early in about 57 days after sowing. Heavy yielder, pale green bulb of semi-globe shape. Weight 500-600 g at maturity. Very good shelf life.

EWV Improved

It is developed by Tokita Seed, India. Pale skin, white flesh, crisp knobs are flat round to roundish. Ready for harvest after 55 days of transplanting. Average knob weight about 200-250 g. Knobs are with less foliage and smooth surface.

CLIMATIC INFLUENCE

Knol-khol is a cool season crop which thrives in a relatively cool moist climate. It can withstand extreme cold and frost better than other cool season crops. Nath *et al.* (1976) reported that the crop grows well for the tuber production at 10-20°C, though the quality is better under lower temperature. Early varieties are more prone to premature bolting and require above 15°C temperature for cultivation. The temperature of 20-25°C are considered optimum. It is reported that purple varieties often bolt more rapidly than white varieties (Lal, 1993). It has been reported, however, that the long oval tubers, which occasionally occur in cultivars with round or flat round tubers, are probably vernalized tubers, whose development has been arrested (Chatterjee, 1986). Being a hardy vegetable compared to cabbage and cauliflower, knol-khol can be grown in frost-prone areas.

SOIL REQUIREMENT

Knol-khol can be grown in all types of soil, though it make no high demands upon it. However good condition and fertility favour growth in an uniform manner. According to Sharma and Singh (2001) a rich garden soil will produce excellent knol-khol.

SOWING

Seeds are sown in lots from the end of August to the end of November for a continuous crop (Choudhury, 1967). Generally sowing is done in August, September and October for early, main and late varieties respectively. Sowing in north Indian plains is done from September-November and in the hills from March-August. In places where winter is mild, sowing may be done in October (Sharma and Singh, 2001).

Seed rate

About 800-1000g of seeds is sufficient to grow a crop for one-hectare land. Treat the seeds with 2.5g captan/thiram 3g/bavistin 1g/kg of seeds.

Nursery Management

- Prepare nursery beds of size 3m x 1m and 15cm high from ground level. A nursery area of 2.5 cent is required for raising seedlings to plant one hectare land. Drench the soil with 0.2% captan solution to avoid damping off. Add adequate quantity of farm yard manure. Sow seeds of one cm. depth. in a row and 10 days after germination spray on the seed bed 1.8 ml of thiodan in one litre of water.

PLANTING

The seedlings are planted out generally at 25 x 25 cm, 25 x 30 cm, 25 x 40 cm or 30 x 45 cm according to the available intercultural facilities (Chatterjee,

1986). Give pot watering to the seedling just after planting for quick establishment in the main field. Closer spacing decreases the size of the tubers, but will increases the yield per hectare. If planted to close some of the tubers may be harvested early to provide more space to the rest of the plants.

NUTRITIONAL REQUIREMENT

According to Laske (1962), knol-khol is sensitive to a relative increase of the N dressing but P treatment produced little effect. However with increased dose of potash the percentage production of the weight of tubers to the total yield increased. Backer (1956) reported that the nutrient removal from a hectare of knol-khol crop was 100 kg of nitrogen, 80 kg of phosphorus, 160 kg of potash and 60 kg of calcium. Chaudhury and Som (1969) found that best yield was obtained with 100.7 kg nitrogen and 60 kg phosphorus/ha in sandy loam soil with low available nitrogen and phosphorus. A yield of 20 tonnes per hectare removes about 100 kg of nitrogen, 85 kg of phosphate and 170 kg of potassium (Choudhury, 1967).

Application of NPK @ 75:50:50 kg/ha have been recommended for knol-khol in Orissa (Anonymous, 1988-89). The application of NPK @ 80:60:60 kg/ha in sandy loam soils of medium fertility will give good yield (Lal, 1993). However, a balance fertilizer dose of 100 kg N, 50 kg P_2O_5, 50 kg K_2O/ha is optimum to obtain maximum yield of knol-khol. Incorporate 25 tonnes of farm yard manure per hectare at final ploughing. Apply half nitrogen, full phosphorus and half potash per hectare at the time of planting as a basal dose. Give rest half nitrogen in 2 splits 25 days and 40 days after transplanting as first and second top dressing. Mix half of potash with second split of nitrogen at 40 days after transplanting.

Use of Micronutrients

(1) Application of dhanzyme granules @ 12-16 kg per acre in two splits first at 10-15 days after planting and second 20-30 days later increase tuber yield and tuber quality. (2) Application of dhanzyme liquid @ 1ml/lt of water (180 ml per acre) as a foliar spray, with high volume sprayer at 6-8 leaf stage and 15-20 days later improve tuber yield and tuber quality (3) Application of plantaid as foliar spray 3ml/lit. of water, twice at an interval of 3-4 weeks when plant shows symptom of deficiency. It improve the plant growth, good tuber formation as well as better knob/tuber size. (4) Incorporation of plantaid (S) @ 25 kg per hectare in soil in a single or split dose increase vegetative growth, tuber size, yield and quality.

Use of Biofertilizers

Chattoo *et al.* (1997) reported that biofertilizers had a beneficial effect on growth, yield and quality of knol-khol. In their study they used *Azospirillum* and *Azotobacter* as seed inoculant (500 g/ha), seedling treatment (2.0 kg/ha) and soil inoculant (2.5 kg/ha). They found that *Azospirillum* inoculation markedly increased growth, yield and quality of knol-khol and resulted in a nitrogen economy of 25%. *Azospirillum* also proved better than *Azotobacter* in respect of growth, yield and quality of knol-khol.

IRRIGATION

Knol khol plant has shallow root system and is very sensitive to soil moisture stress. When plants face shortage of water during the early period of growth, the start of stem swelling will be delayed and yield will be reduced (Hegde, 1993). At the same time plentiful water supply resulted in too rapid growth which may cause the tubers to crack, especially when initial growth is slow. Mangal *et al.* (1982) working on sandy loam soils at Hisar (Haryana) found that irrigation at pan evaporation coefficient of 1.25 (irrigation depth/ cumulative pan evaporation) was optimum for knol khol. This required 6 irrigations amounting to 300 mm water. Bhatnagar *et al.* (1986) reported minimum weight loss and maximum shelf life of knobs under 1.0 PEC.

INTERCULTURE

Knol-khol being a shallow-rooted crop, most of its roots are within 45-60 cm of the top soil. Deep cultivation should therefore, be avoided. Shallow and frequent cultivation should be given to provide soil mulch. One hoeing and weeding should be given 20 days after transplanting. Follow this with top dressing and earthingup, it promotes stability to the plants. Subsequent hoeing is done 40-45 days prior to application of second dose of fertilizers then earthing up is given in order to produce quality knobs/tubers.

Mulching

The plastic mulch is beneficial to increase the yield of knol-khol. During the night the mulched soil lost heat. During nights the slow release of heat was observed and during the day it has warmed up quickly (Klima, 1971). The knol-khol sown in boxes or in soil blocks had stronger stems and plants were more compact when the surface of the substrate was covered with granules of polysterene products, eg- styromull which reflects light (Bushmann, 1973).

Inter Cropping

Intercropping refers to growing two or more dissimilar crops simultaneously on the same piece of land, crop intensification is in both time and space dimensions. Intercropping enhance further production by efficient use of monetary and non-monetary inputs (Palaniappan, 1985). Higher yields in nutritional terms were obtained with intercropping, particularly when beet roots, peas or knol-khol were grown with okra or capsicum (Prabhakar and Shukla, 1990).

Crop Rotations

Saini (1996) suggested that following crop rotations can be taken profitably with knol-khol.

1. Knol Khol-Carrot-Amaranthus
2. Okra-Knol khol-Bittergourd
3. Bittergourd-Knol-khol-Tomato
4. Clusterbean-Knol-khol-Chilli
5. Amaranthus-Knol-khol-Brinjal
6. Bottlegourd-Knol-khol-Cowpea
7. Cowpea-Knol-khol-Muskmelon
8. Okra-Knol-khol-Chilli
9. Greengram-Knol-khol-Bittergourd
10. Blackgram-Knol-khol-Watermelon

WEED CONTROL

Knol-khol seedlings are highly sensitive to weeds especially during the first two weeks after transplantation. Weeding should be done at regular intervals to keep down the weeds. Manual weeding is helpful for small scale production of the crop. Indirect seeded crop, at least three to four weeks weed free period is required to avoid reduction in crop yield as observed by Zimdahl (1980). Bhan (1972) found nitrofen and BV 201 to be effective herbicides for controlling weeds and increased yields of knol-khol when compared with the weedy check. Singh *et al.* (1974) found nitrofen at 2.0 kg/ ha and alacholor 1.5 and 2.0 kg/ha to be promising for knol-khol. Leela (1987) got good weed control in transplanted knol-khol with butachlor, 1.5 kg, fluchloralin 1.0 kg and oxyfluorfen, 0.25 kg a.i./ha.

PLANT PROTECTION MEASURES

Some important insect pests and diseases alongwith their control measures are :

Insect Pests

Mustard saw fly. (Athalia sp)

The adult is minute black fly that lays eggs inside the leaf tissues. The caterpillar feeds on the leaves of young seedlings. It can be controlled by spray of malathion @ of 2ml/l. of water twice at 7 days intervals.

Mustard aphid (Lipaphis erysimi)

The greenish or black aphids damage the leaves by sucking the cell sap. The affected leaves get curled and plant wither away and die. Spray malathion @ 0.5% twice at 7 days interval to control the pest. Need based sprays of endosulfan 0.2% can easily control mustard aphid.

Cabbage butterfly (Pieris brassicae)

The young green caterpillars feed on the surface of the leaves and skeletonise them. In severe case leaves, tender shoots, flowers and fruits are completely destroyed resulting in the rugged appearance of the attacked plants. Spray nicotinic sulphate @ 0.05% to check the infestation. Weekly sprays of dipel @ 0.5 kg/ha were found very effective.

Semi-looper (Plusia sp)

The green caterpillars, which are loop while moving, attack the crop and cause serious damage. Spray nicotinic sulphate @ 0.05% or Thiodan @ 0.15% to control the semi looper.

Diamond back moth (Plutella xylostella)

This, causes damage to the plant and tubers. Larvae feed on the leaves, leaving intact the parchment like epidermis. Some times they produce holes in the leaves. The growth of young plant is greatly inhibited. Spray endosulfan @ 2ml/l. of water in pre tuber stage to control the infestation. Spray of cartap hydrochloride followed by neem-seed-kernel extract (5:1000 w/v) were found effective. Spray of malathion @ 2 ml/l of water to control the pest at harvesting stage and keep seven days waiting period.

Diseases

Damping-off

It is caused by *Pythium sp, Rizoctonia sp., Fusarium sp.* It is common in nursery seedlings. Rotting starts in the collar region of the seedlings. Drench the nursery bed with 0.1% solution of Brassicol/Captan/Fytolon. Treat the seeds with Agrosan GN or Ceresan @ 2g/kg or Thiram 3g/kg of seed at the time of seed sowing in nursery.

Black rot

Black rot is caused by *Xanthomonas or Pseudomonas* sp. The margin of leaves turn yellow. The veins becomes dark and vascular region of the main stem becomes discoloured. The spraying of streptocycline at 100- 200 ppm is also effective to check the spread of disease. The disease may be controlled by following crop rotation, the hot water treatment of seeds at 48.9°C for 30 minutes and use resistant varieties.

Leaf spot and Blight (Alternaria brassicae)

The small dark-coloured spots, appear on the leaves which spread rapidly to form circular lesions. In severe cases concentric dark rings and brown colour develop. Spray the crop with mancozeb @ 0.2% at 10-15 days interval. Treat the seed with bavistin @ 1g/kg and apply three sprays of indofil M-45 @ 0.2% at an interval of 15 days soon after appearance of disease.

Club root

This disease is caused by *Plasmodiophora brassicae.* In severe cases malformed roots present a clubbed appearance. Foliage wilts in sunny days and recovers towards the evening. It is prevalent in acidic soil. This disease can be minimized by liming the soil to raise its pH above 7. Nursery growing area can be treated with pentachloronitrobenzene, chloropicrin, cynamide and formalin approximately two weeks before sowings. Crop rotation and cultivation of resistant varieties are suitable control measure.

Black-leg (Phoma lingam)

The young plant in nursery bed are affected usually. The fungus attacks at the base of stem and roots and the plant wilts. Seed treatment with bavistin @ 1.5g/kg seeds, hot water treatments of seed at 50°C for 30 minutes and follow proper crop rotation for control of black-leg.

Physiological Disorders

Pre-mature bolting

Development of seed stalks before the formation of sufficient edible part is known as premature bolting. Knol-khol varieties are susceptible to bolting lack a juvenile phase and become generative, if exposed after germination to low temperatures (below 10°C) even for a week. The vernalizing effect of low night temperatures can be countered by high temperatures during the day. In slow bolting varieties, older plants are more sensitive to low temperatures than the younger plants. Purple varieties often bolt more rapidly compared with the white ones. Young plants if exposed to low temperature for a considerable time are more sensitive to bolting. Seedlings with 3-4 inches wide leaves at 10°C are considered critical for premature bolting. Therefore, high temperature is desirable for sensitive early varieties. Bolting may also be hereditary but is influenced by environmental conditions to a considerable extent.

HARVESTING

Tubers are harvested before they are full grown, because old tubers eventually become fibrous. Usually tubers may be harvested when they are of

6-8 cm diameter. In late maturing varieties sometimes bigger tubers of 10-12 cm diameter are also harvested. Early cultivars are ready for harvesting within 60-70 days after transplanting. However, late cultivars of Europe matures between 70-100 days after transplanting (Hazra and Som, 2006). The harvesting should be done twice or thrice a week. For its marketing the main root is cut-off and the enlarged stem alongwith the leaves are tied up.

YIELD

Individual tubers weight ranged from 200-250g while the yield per hectare may vary from 180 to 250 q/ha under Indian conditions. Late varieties are usually higher yielding than early ones. Hybrid varieties give more yield than early and late improved varieties. Frydrych (1971) reported that the yields increase with the increase in photosynthetic activity.Greater leaf area gave better yield.

POST HARVEST HANDLING

For better marketing and to get higher price from the produce, it is necessary to put them in proper grading and packing before sending to market for sale. Singh and Singh (2007) stated the suitable grading, packing and storage of knol-khol are given below :

Grading

The grading is done according to the size of tubers appearance and maturity of harvest of the crop. Big size matured, uniform and tender tubers with brightness and lustre are graded separately as "super" grade; medium size good tubers with uniform shape, texture and appearance are kept in "fancy" grade. Tubers with variable size, shape, uniform, free from bruiseness, pest, diseases and any spots. are graded in "commercial' grade.

Packing

For local market packing is done in bamboo basket/dalia/jhali. Tubers are packed in gunny bags for transport to distant market. Big baskets made of bamboo chips/arhar sticks are also used for packing produce to distant markets. For export purpose tubers are packed in plastic basket/shallow basket, plastic crates, corrugated fibre board boxes.

Storage

Under ordinary condition tubers can be stored for two to three days but under cold storage condition it can be kept for relatively longer period (Nath *et al.*, 1976). Storage life and quality of the wrapped produce are better than the unwrapped (Kopec, 1967).

SEED PRODUCTION

Knol-khol does not produce seeds in the plain of India. The seed is produced only in hills like Kullu, Katrain, Lahaul and Sipti and Kashmir valley. Plants do not bolt until they have grown for at least 6 to 8 weeks at temperature of 4°C to 7°C. Knol khol does not have any such variety that can produce seed in the plains. Temperatures even as low as –12° to –14°C are not reported to be harmful. The altitude is not a barrier and seed crop can be grown upto 10,000 feet above sea level or even more (Sharma, 1972).

Method of Seed Production

There are two methods for seed production i.e. head to seed method (transplanting method) and seed to seed (*in situ*) method. The seed to seed (*in situ*) method is followed in knol-khol seed production at Kullu and Kashmir valley (Singh *et al.*, 1959).

Seed to Seed Method

In seed-to-seed method severe winter condition may require some protection from freezing and thawing. The crop is allowed to over-winter and produce seed in the original position where they were first planted in the seedling stage (Agrawal, 1980). After roguing when the heads are fully matured in autumn, a shallow furrow is ploughed on each side of a row and a large ridge of soil thrown against the plants by potato ridger. The top of the plant is left exposed (Singh *et al.*, 1959). This method gives higher and early yields and cost of production is also less and commonly followed at Katrain and Kashmir valley.

Head to Seed Method

In this method, true to type, mature plants are selected and uprooted during autumn. The leaves all round the knobs and tops are removed, keeping the crown intact and plants are immediately replanted in such a way that the whole stem, up to the base of the knob, is pushed into the soil, so that the knob almost touches the ground and rests at the surface. At katrain and other places, this method is essentially followed for raising nucleus seed (Singh *et al.*, 1959; Agrawal, 1980).

The seeds of knol-khol is usually raised by in-situ method. Three types of plants are usually observed in the flowering stage as determined by the extent of branching of the flowering shoots (Sharma, 1972).

These types are : (a) non-branched type, (b) inter-mediate, and (c) fully branched. The higher proportion of a fully branched plants in a seed plot gives higher seed yield. There is, however, no appreciable difference in the quality of knobs produced by the seeds harvested from any of these three types. But it is important to rogue out inferior plants of knol khol and only well developed true-to-type knob, be selected.

Isolation Requirement

Knol-khol is a cross-pollinated crop and pollination is mainly done by bees.The seed field must be separated from fields of other varieties, fields of the same variety not confirming to varietal purity requirement of certification and from all other kinds of cole crops, at least by 1600 metres for foundation seed class, and 1000 metres for certified class (Anonymous, 1988).

Brief Cultural Practices

The cultural practices like nursery preparation, seed rate, seed sowing seedlings raising, planting, fertilizers application, intercultural operations and plant protection measures are same as described in crop production chapters.

Crop Inspection and Roguing

The crop should be inspected before the marketable stage. Off-type plants can be rogued out on the basis of knob shape, colour of leaves and maturity time of knobs/tubers. The plants are selected for well developed knobs, shape of the knobs, free from seed borne diseases and infestation of pests and for number of leaves around the knobs (Sharma, 1972). A minimum of three inspections shall be made, the first before the marketable stage of the knob; second when knobs have been formed and the third at the flowering stage (Anonymous, 1988).

The off type plants observed at the flowering stage, are usually determined by the extent of branching of the flowering shoots. The higher the production of fully branched plants, the greater is the seed yield. Therefore, care is necessary in nucleus seed production to select the right type of plants at flowering stage.

Subsequent roguing for off types, diseased plants affected by diseases such as phyllody, black leg, black rot, soft rot or leaf spot should be done from time to time as required (Agrawal, 1980).

Harvesting and Threshing

When the pods/siliques turn yellow and the seed becomes brown, harvesting can be commenced to avoid shattering, the seed crop may be harvested in 2 to 3 lots. Early plants are harvested first and the remaining crop is cut when about 75% of the pods turn yellowish brown. Harvested crop is collected on threshing floor in heaps for curing for 4 to 7 days. Every day heap is turned up and down. After curing, the seed is threshed and separated from straw. The seed is thoroughly dried in the sun and processed (Singh and Gill, 1993).

Seed Yield

Seed yields have been found to vary considerably from place to place depending upon climate, soil, crop management factors etc. However, the average seed yield has been found to be 1125 kg/ha (Singh and Gill, 1993).

REFERENCES

Agrawal, R.L. (1980). Seed Technology. Oxford and IBH Publishing Co., New Delhi.

Anonymous (1984). First hybrid kohlrabi virotis from the firm Roggli, *Gemuse.* 20(4): 152-154.

Anonymous (1988). Indian Minimum seed certification standards. The Central Seed Certification Board, New Delhi.

Anonymous (1988-89). Agriculture Guide Book. Directorate of Agriculture and Food Production, Government of Orissa, Bhubaneswar.

Anonymous (1998). Varieties and hybrids developed by Division of Vegetable Crops, IARI, New Delhi.

Backer Dillinger, J. (1956). Handbuch des gesantus Gemiisebanes. 6th edn. Paul. Parey, Berlin and Hamburg.

Bhan, V.M. (1972). Weed control in vegetable crops. *Third International Symp. Subtrop. Trop. Hortic.* Bangalore, pp.73.

Bhatnagar, D.K., Singh, B.P. and Mangal, J.L. (1986). Keeping quality of knol-khol as influenced by different irrigation intensities and nitrogen doses. *Haryana J. Hort. Sci.,* 15: 233-238.

Bogdan, I. (1986). Certified cultivar: Szentesi Feber, Kohlrabi, *Kerteszet es Szoleszet.* 36(6): 7.

Boswell, V.R. (1949). Our Vegetable travellers. *Natural Geog.,* 96: 145-217.

Bushmann, S. (1973). *Hort. Abstr.,* 44(8): 5612.

Chatterjee, S.S. (1986). Cole crops In.: Vegetable crops In India. (Eds.T.K. Bose and M.G. Som.) Naya Prokash, Calcutta.

Chattoo, M.A., Gandroo, M.Y. and Zargar, M.Y. (1997). Effect of *Azospirillum* and *Azotobacter* on growth, yield and quality of knol khol (*Brassica oleracea* var. *gongylodes* L.). *Veg. Sci.,* 24(1): 16-19.

Chaudhuri, B.B. and Som, M.G. (1969). Effect of nitrogen and phosphorus in Kohlrabi, *Hort. Sci.,* Calcutta, 2: 13-18.

Chauhan, D.V.S. (1965). Vegetable production in India. Ram Prasad and Sons: Agra-3.

Choudhury, B. (1967). Vegetables, National Book Trust India, New Delhi.

Frydrych, J. (1971). Photosynthetic activity of diploid and tetra ploid forms *of B. oleracea,* var. *gongylodes, Photosynthetische* 5: 38.

Gill, H.S. (1993). Improvement of Cole Crops. In Advances in Horticulturel, Vol.5. Vegetable Crops. Part I (Eds: K.L. Chadha and G. Kalloo) Malhotra Publishing House, New Delhi.

Hartrath (1977). Kohlrabi under glass; Varieties and spacing experiments, *Gemuse*, 13(1): 8-10.

Hazra, P. and Som, M.G. (2006). Vegetable Science.Kalyani Publishers, New Delhi.

Hegde, D.M. (1993). Water management in vegetable crops. In Advances in Horticulture vol.5-Vegetable crops. Part 1. (Eds: K.L. Chadha and G. Kalloo) Malhotra Publishing House, New Delhi.

Helm, J. (1962). Die Lacimiaten Sipp-en Von *Brassica oleracea* L. Kulturplanze; 11: 92-210.

Kalia, Pritam and Pathania, N.K. (2006). Palam Tender knob: a new early-maturing knol khol. *Indian Horticulture*, Vol.51(5): 26-27.

Kalloo, G. (1993). Kohlrabi, "*Brassica oleracea* var. *gongylodes*. In Genetic Improvement of Vegetable Crops (Eds. G. Kalloo and B.O. Bergh). Pergamon Press, New York.

Kardaeva, N.S. and Katarzhina, L.A. (1979). Resistance of *Brassica* varieties to *Fusarium* wilt, *Trudy Prikl. Bot. Genet. Sel.* 64:141.

Klima, T. (1971). *Acta Universitatis Agriculturae*, 10(1): 3-14.

Kopec. (1967). *Hort. Abstr.*, 37: 2452.

Lal, G. (1993). Agro-Techniques for Cole crops. In Advances in Horticulture Vol.5. Vegetable Crops: Part 1. (Eds: K.L. Chadha and G. Kalloo) Malhotra Publishing House, New Delhi.

Laske, P. (1962). Manuring of vegetables in field cultivation. Published by Verlagsgesellschaff for Ackerbau Hannover.

Leela, D. (1987). Weed control by herbicides in knol khol and radish. *Tropical Pest Management*, 31: 215-219.

Luzny, J. (1965). Testing the collection of kohlrabi and the F_1 generation for bolting, *Rostlinna Vyroba*, 2: 45.

Mangal, J.L., Pandita, M.L., Batra, B.R. and Singh, S.S. (1982). Effect of irrigation intensities and nitrogen on growth and yield of knol khol (*Brassica oleracea* var. *Caulorapa*) variety White Vienna. *Haryana J. Hort. Sci.*, 11:127-131.

Nath Prem, Velayudhan Sundari and Singh, D.P. (1976). Vegetables For The Tropical Region. ICAR., New Delhi.

Nath, Prem; Srivastava, V.K.; Dutta, O.P. and Swamy, K.R.M. (2008). Vegetable Crops - Improvement and Production. PNASF, Bangalore, 143-145.

Palaniappan, S.P. (1985). Cropping System in the Tropics – *Principles and Management.* Willey Eastern, New Delhi, pp. 6-9.

Pesti, M. and Fodor, L. (1982). Study of the susceptibility of *Brassica oleracea* crops to the pathogen *Xanthomonas campestris*, pv. *Campestris Novenyvedelem*, 22: 537.

Prabhakar, B.S. and Shukla, V. (1990). Response of vegetable intercropping system to fertility regimes. *Proc. Int. Symp. Natural Resource Management Sustainable Agric.*, New Delhi, 1, 2003.

Saini, G.S. (1996). A Text Book of Vegetable Production. Aman Publishing House, Meerut (U.P.), India.

Shanmugavelu, K.G. (1989). Production Technology of Vegetable Crops. Oxford and IBH Publishing Co. Pvt. Ltd., New Delhi.

Sharma, B.R. (1972). Seed Production of Cabbage and Knolkhol, paper presented at the *Summer Institute on Vegetable Seed production*, O.P.A.U., Ludhiana, 29 May - 24 June, 1977.

Sharma, S.R. and Singh, R. (2001). Knol khol (*Brassica oleracea* var. *gongylodes*). In Vegetables, Tubercrops and Spices. (Eds: S. Thamburaj and Narendra Singh) DIPA-ICAR, New Delhi.

Singh, B., Khurana, S.C. and Pandita, M.L. (1974). Chemical Weed Control in Cauliflower, Knolkhol and Turnip. *Haryana J. Hort. Sci.*, 3:182-189.

Singh, D.N. and Singh, S.P. (2007). Production Management of Underutilized Vegetables. Agrihortica publications. Junagadh, (Gujarat) India. pp. 158-170.

Singh, D.N. and Nath, Vishal (2011). Varieties and Hybrids of Vegetables. Satish Serial Publishing House, Azadpur, Delhi (India). pp. 44-47.

Singh, H. and Gill, S.S. (1993). Seed Production of Vegetable Crops. In Advances in Horticulture vol.6-Vegetable Crops: Part-2 (Eds: K.L.Chadha and G. Kalloo) Malhotra Publishing House, New Delhi.

Singh, H.B., Thakur, M.R. and Bhagchandani, P.M. (1959). Vegetable Seed Production in Kullu Valley - II Seed Prod. of Cabbage, Cauliflower, Knol khol and Chinese cabbage. *Indian J. Horti.*, 16: 153-61.

Zimdahl, L. (1980). Crop Weed Competition - A review published in the United States by the International Plant Protection Centre, Oregon State University, Oregon, USA.

□□□

4. BRUSSELS SPROUTS

Botanical Name	:	*Brassica oleracea* L. var. *gemmifera* DC.
Family	:	Brassicaceae
Chromosome Number	:	2n = 18

Brussels sprouts is important member of family brassicaceae after cauliflower and cabbage, a delicious cold climate winter vegetable that has been grown around Brussels in Belgium since 13th century (Shoemaker, 1953). According to Boswell (1971) brussels sprouts in its present form appeared around 1750 in Brussels (Belgium) as the name indicates. In 1800 it reached France and England. It is grown for the edible swollen axillary buds, known as sprouts or buttons or mini-cabbage (Nieuwof, 1969).

ORIGIN

The origin of this plant remains obscure and has been ascribed to both savoy, cabbage and various forms of kale e.g., thousand-headed (Ockendon and Smith, 1993). The earliest known reference is to spruyten, mentioned in the market regulations of Belgium towns as early as 1213 (Helm, 1963). Whether these were Brussels sprouts as we known them or just forms of decapitated cabbage is uncertain. In 1696 Zwinger described a many-headed cabbage which was probably Brussels sprouts (Toxopaeus, 1979). However, the crop only became widely known in Europe in the late eighteenth and early nineteenth centuries and the first reference to the breeding of Brussels sprouts was probably by Van Mons (1820).

DISTRIBUTION

Brussels sprouts is a very popular crop in North-West Europe and California state of USA and it tooks 500 years to reach English gardeners. It is widely grown in Australia, Netherlands with small scale in Belgium and France. In India, it is cultivated in Kodaikanal and Nilgiris of Tamil Nadu and Maharashtra. Recently it has become popular around big cities to meet the demand of big hotels catering to the foreign tourists. It is a delicacy with the foreigners and is slowly gaining ground with the vegetable growers in the vicinity of big cities like, Delhi, Bombay, Calcutta, Madras and Bangalore etc (Gill, 1993).

BOTANY

Plants are tall biennial erect with good leaf protection, stem unbranched, makes normal longitudinal growth 2-3 feet tall; axillary buds forming compact heads or sprouts, about 3 cm (Purseglove, 1974). The buds in the axils of usually petiolate leaves develop in to miniature heads of few centimetre in diameter and it is the buds which are called sprouts resemble the cabbage head but they are much smaller size . This is the reason that brussels sprouts is some time called miniature cabbage (Gill, 1993). Brussels sprouts are mainly cross-pollinated crops and have a sprophytic self-incompatibility system.

COMPOSITION

Brussels sprouts are rich source of vitamin A, B and C. They contain minerals like P, S, Ca, Fe, I and Cu and other nutrients. In addition, they have certain anticarcinogenic properties, whose effects attributed to the hydrolysed glucosinolate derived products such as isothiocynates and indoles.

Approximate composition per 100 g of edible portion

Moisture	85.5g	Fibre	1.2g
Riboflavin	0.16 mg	Protein	4.7g
Fat	0.5 g	Minerals	1.0g
Phosphorus	82.0 mg	Carbohydrate	7.1g
Potassium	4.77 mg	Chlorine	22 mg
Magnesium	26.0 mg	Vitamin C	72 mg
Vitamin A	210 I.U.	Calcium	43 mg
Sodium	7.9 mg	Sulphur	212 mg
Thiamin	0.05 mg	Nicotinic acid	0.4 mg
Iron	1.8 mg		

Uses

The sprouts seems to be increasing in popularity, especially as a frozen product. These can be used as salad or cooked as single or mixed vegetable with potato or other vegetables. Sprouts after cutting vertically into two halves are fried with basan (gram flour) and paste can be used for the preparation of pakoras. Boiled sprouts can also be consumed after adding salt, butter and black pepper powder depending upon test (Sharma and Singh, 2001). Usually sprouts are eaten as raw, boiled or steamed, cooked and pickled. It is also used as canned product.

VARIETIES AND HYBRIDS

There are two distinct type of cultivars, dwarf and tall (Sharma and Singh, 2001) and some hybrids have been developed and grown commercially in foreign countries.

Dwarf cultivars

They have short stem less than 50 cm in length and sprouts are small and crowded on the stem. These are mostly early cropper and most suitable for the areas where growing season is short. Common cultivars are Improved Long Island, Early Morn, Dwarf Improved, Frontier Zwerg and Kvik.

Tall cultivars

Tall cultivars are grown in areas where growing season is longer especially in England, other parts of Europe and north Indian hills. Strains of cultivar Eveshan Rapid (Sweden) and Wilhelmsburg (Germany) are early in maturity but do not withstand hard winters.

Late and hardier cultivars

Red Vein, Hilds Ideal, De Rosny Polarstjernen and Amager.

Other varieties

High yielding dwarf strains are Bollo, Cambridge, Cambridge No.1, Cambridge No.2, Cambridge No.3, Cambridge No.4, Cambridge No.5, Dalington Early, Favourite, Goliath, Fill basket, Ovmskirth Giant, Rouslench late, Wroxton.

Aphid resistant varieties

Uremo Inter, Irish Elegans. Roodnef.

Powdery mildew resistant varieties

Bin Dee, Engard 2077, Frigo Star, Groeningse late, Seven Hills, Solid line, Triumph.

Hybrids

The hybrids marketed by different seed companies of Japanese and Europe are Jade Cross, Avon Cross (first F_1 hybrids by National Vegetable Research Station at Wellesbourne). Doreman, Alkazar, Merlon, Poster, Rampart, Fortress, Predora, Rovoka, Sonara, Ladosa, Pearl Crystal, Oliver, Valiant, Rogor, Rider, Boxer, Richard Royal Marvel, Asgard, Kundry, Rasmunda and Stephen.

Special characteristics of some important hybrids

S. No.	Name of hybrid	Special character	Reported by
1.	King Arthur	Suited to mechanical harvesting, best cropper.	Wood (1970) Nuyten *et al.* (1973)
2.	Peer Gynt	Good yielder	Thompson *et al.* (1973)
3.	(i) Rola cross (ii) Uremo Post (iii) Top grade No.6977	Frost resistant	Meduldcramans (1971)
4.	(i) Topscore Dutch (ii) Trigomox Hybrid (iii) Berner market swiss hyrbid (iv) Jade Cross Japan hybrid	Best yielder	Woykee *et al.* (1972)
5.	Lanceolat, Orion	Outstanding culinary qualities	Berrera *et al.* (1975)
6.	(i) Craton (ii) Merton	Early to mid-early Tall, large, firm sprouts, mid-early, fine hard round sprouts	Anonymous (1976) Anonymous (1976)

Shanmugavelu (1989)

Table 1 : Performance of some varieties of brussels sprouts grown in Himachal Pradesh

Variety	Days from transplant to harvest	Plant height (cm)	Sprout size (cm)	Sprout type	Yield/ 3 m lines (kg)
Hilds Ideals	123	107.50	3.6 x 3.6	Uniform	1.113
Express	105	99.50	4.6 x 4.0	Firm	1.225
Fill Basket	120	100.00	3.8 x 3.7	Tightly packed	1.000
Rubine	170	150.00	3.4 x 3.5	Firm	0.600
Reger	103	53.50	3.2 x 3.3	Lightly packed	0.950
Titurel	100	36.30	3.7 x 3.4	Lightly packed	0.950
Carvalient	136	62.60	4.2 x 3.6	Loose	1.153
Oliver	120	80.41	4.6 x 3.7	Uniform	0.900
Robert	123	93.60	4.4 x 4.3	Firm	0.762
BBS -74	80	65.00	2.5 x 3.0	Uniform	0.500

Peshin (1998)

Detail characteristics of varieties and hybrids described by Singh and Nath (2011) are given below :

Hilds Ideal

This variety has been identified suitable for growing in the country. Its plant grows up to 60 cm and produces 45-55 sprouts, each weighing 7-8 g and the sprouts are compact and possess good flavour. A negligible percentage of loose sprouts at the base of the plants is also marketable. It takes about 115 days for first picking. The yield per plant is 250-400 g in four pickings at an interval of 10 days. Delay in picking will lower the sprout quality (Sharma and Gill, 1986).

Jade Cross (F_1 hybrid)

It is 90 days hybrid cultivar. Sprouts are firm, dark green, closely packed on longstems, can be grown under wide range of growing conditions (Gill, 1993).

Diablo

It mature in 120-150 days after transplanting, during winter season. In order to get uniform sprouts, cut the top of the plant as soon as sprouts at the bottom start to develop. It can be boiled eaten with cheese spread. Plant height is 90 cm, winter variety, gives round solid green sprouts, which are well spaced on the stem. It was developed by Bejo Sheetal Seeds.

CLIMATIC INFLUENCE

Brussels sprouts are best adapted to temperate regions of the world, and require a long growing sesason with cool summers and mild winters to achieve high productivity. In India, it is mainly grown as a winter crop (Choudhury, 1967). The best quality sprouts are produced when days are sunny and light frost at night. Hot weather results in soft and open sprouts which are undesirable and are reducing tenderness. It thrive best in relatively cool moist climate. The temperature for seed germination is around 12°-16°C. Plant take optimum growth when temperature is 16°-20°C and at higher temperature the growth is arrested. However temperature requirement may vary from one cultivar to other. This crop is frost resistant and late cultivar can withstand temperature as low as -10°C. It can continue to grow even at 0°C. Short transition period from summer to winter will give lower yield. During vegetative phase, plants develop vegetatively and form sprouts and during generative phase the plants after exposure to chilling temperature (5°-8°C) for a period of 6-8 weeks form flowers and seeds. Seed production requires for a period about two month during winter coupled with mild summer and low rainfall 40-55 cm well distributed over the months.

SOIL REQUIREMENT

Brussels sprouts can be grown in both light as well as in heavy soils. The sandy loam soils are preferred for early and clay loam for main season crop. The soil should be fertile, well supplied with organic matter and well drained. Crop grown best at a soil reaction of pH 6.0 to 6.8 but are not as sensitive to high acidity as cauliflower (Thompson and Kelly, 1972). Uniform growth of the plant is a pre-requisite, i.e., quick growth in peat soil, garden soil or highly fertile soil makes the sprout loose . On a heavy soil plants grow more slowly and quality is improved.

SOWING

Brussels sprouts, like other cole crops are generally sown in raised nursery beds. Some times sowing is also done in cold house or cold frame and the young seedlings are pricked off at a spacing of 5 x 5 cm.

Sowing Time

The sowing in Europe and California (USA) is done in August - September or April. Similar timings are also followed in the north Indian hills depending upon altitude. However, sowing is done in August - September in the plains of north India.

Seed Rate

About 300-400 g seed is required for F_1 hybrid cultivars to cover one hectare land (Chatterjee, 1986). However the optimum seed rate is 750 g per hectare. Shanmugavelu (1989) reported that 1.0 kg seeds is sufficient for raising 5 to 6 hectare crop. 450 g (350 seeds/g) of seed require to raise seedlings for one hectare of land (Peshin, 1998).

Seed Treatment

The seeds are treated with Thiram or Captan or Agrosan @ 3 g/kg or Bavistin 1 g/ kg of seeds before sowing.

Nursery Management

Nursery beds are prepared thoroughly by 2-3 ploughing or digging followed by planking. After the final preparation of nursery beds two days before sowing, the nursery beds are drenched with 0.3% captan or thiram solution @ 5 litre/m^2 to control the pathogens like *Pythium, Rhizoctonia, Phytophthora* and *Fusarium* which are mainly responsible for damping off disease in the nursery. When the soil is sterilized with formalin (1:48), it must be done 3-4 weeks before sowing. Treated beds should be covered with alkathene sheet for about 6-7 days. It must be aerated by turning the soil upside down to kept upon for a week before swing (Sharma and Singh, 2001).

Seed Sowing

Seed sowing is done in rows spaced at 8-10 cm apart and 1.5-2.5 cm deep. For planting crop in one hectare, about 60-80m^2 nursery area is required for raising the seedlings. After sowing, the seeds are covered with fine soil and surface of the bed is made uniform. The beds are covered with fine layer of dry grass. This helps in retaining moisture for longer time and avoid crust formation. Watering with rose cane will help in maintaining uniform moisture.

Seedling Management

It takes 7-12 days to germinate. When the emerging seedlings have come to the soil level, dry grass cover is removed, otherwise seedlings will become pale-and-lanky. Seedlings are thinned properly and drenched with 0.2% captan or thiram solution to save them from post-emergence damping-off disease. Proper moisture is maintained to get healthy seedlings.

PLANTING

Planting Time

In north Indian hills, planting is done from August to mid September and also in May depending upon the altitude. Under north Indian plains, planting is done in September-October to get better yield. The planting should be adjusted in such a way that at the time of sprout formation temperature is mild. For early crops plants are set out from February onwards which are harvested in July. The late crops are planted out in May or in early June (Chatterjee, 1986).

Planting Distance

It depends upon the type of cultivars and method of harvesting. For single mechanical harvesting, close spacing between the plants is followed. In European countries, spacing of 70 cm x 45 cm in hybrid variety Valliant gave optimum yield. Spacing of 60 cm x 45 cm and 60 cm x 60 cm are recommended (Sharma and Singh, 2001).

Spacing of 90 cm x 90 cm is recommended for tall growing cultivars and for places having longer growing periods. Spacing of 75 cm x 60 cm and 90 cm x 90 cm are advocated for mid hill and high hill crop (Pestin, 1998). The plants are spaced 80-90 cm apart in rows 90cm apart (Thompson and Kelly, 1972).

The plants are transplanted to the field when they are 6 to 8 weeks old in USA (Thompson and Kelly, 1972). But in Indian condition seedlings become ready for transplanting after 4-5 weeks with 3-4 leaves or 10-15 cm long. Straight seedlings are transplanted alongwith soil ball during evening hours. Older seedlings generally give poor yield.

NUTRITIONAL REQUIREMENT

Brussels sprouts require generous manuring but high rate of nitrogen must be accompanied by high rates of potash. Application of nitrogen makes a sprouts dark green, which is normally preferred by the consumers. Ammonium nitrate is better source of nitrogenous fertilizer for this crop. Several researchers have studied the response of fertilizers and manure application in brussels sprouts and gave varying recommendations.

Table 2 : Nutrient requirement of brussels sprouts

N (kg/ha)	P (kg/ha)	K (kg/ha)	Organic manure (q/ha)	Remarks	Authors/source
80	60	180	-	For moderately fertile soil	Shanmugavelu, 1989.
60	40	120	-		
300	180	180	-		Anon, 1968.
25-50	150-175	150-180	-	For all type of soils	Loetis *et al* ., 1970.
60-75	60-75	60-75	150		Sirohi and Behera, 2000.
100-150	50-150	100-200	150-200	Basal dose	Sharma and Singh, 2001.
100-200				Top dressing	

Total quantity of farm yard manure is added at final land preparation. Full dose of phosphorus and potassium and half of nitrogen is applied at planting. Remaining half dose of nitrogen is divided into 2-3 equal doses and applied after 30, 45 and 60 days after planting.

Nath *et al.* (2008) suggested that soil should be incorporated with 25-30 tons of farm yard manure/ha, 200 kg nitrogen per hectare should be applied in equal splits viz., at the time of planting, a few days after planting, after first picking and also after the second picking. Phorphorus and potassium are applied @ 50 kg and 20 kg per hectare respectively, at the time of planting.

Use of Growth Regulators and Micronutrients

(a) Application of benzimidazole solution @ 200 ppm at 21°C for six days to sprouts, retain good colour, high rate of respiration and photosynthesis and chlorophyll retention is more (Eares, 1968).

(b) Tompkins *et al.* (1972) reported that foliar spray of 3000-4500 ppm of succinic acid caused more uniform sprouts and higher yields.

(c) Maurer (1976) reported that application of SADH @ 8000 ppm increased the marketable yield by 13 to 31 per cent. Boron and molybdenum deficiency can be overcome by applying borax @ 10-12 kg/ha and sodium molybdate @ 4-6 kg/ha in the soil.

(d) Application of Dhanzyme liquid @ 1 ml/litre of water (400 ml/ha) as a foliar spray, with high volume sprayer at 6-8 leaf stage and 15 to 20 days later improve sprouts yield and quality.

(e) Application of plantaid as foliar spray 3 ml/litre of water, twice at an interval of 3-4 weeks when plant shows symptom of deficiency improve the plant growth, flowering as well as better sprouts are harvested.

(f) Incorporation of plantaid (S) @ 25 kg/ha in soil in a single or split dose increase vegetative growth, flowering, sprout yield and quality.

IRRIGATION

Brussels sprouts is a shallow rooted crop and respond well to irrigation. The crop is irrigated immediately after transplanting of seedlings and subsequent irrigation depends upon the climate and soil conditions. The field may be irrigated when the soil is seen deficient in moisture. Wolde (1956) reported that the irrigation water applied through 3 cm wide porous bitumen pipe laid 40 cm deep and 1.5 m apart, increased the yield by 300 to 400 per cent in dry season. Water logging kills the roots very rapidly and as such heavy irrigation should be avoided. Sprouts formation is very critical moisture sensitive stage in Brussels sprouts (Hegde, 1993). Sufficient moisture is maintained during growing period to get better yield. Generally, the crop requires irrigation at 8-10 days interval. Higher yields are obtained through irrigation.

INTERCULTURE

Intercultural operations are to be done regularly to keep the crop free of weeds and aeration of root system. Generally, 2-3 hoeing and weeding are required to remove weeds and to make the soil porous for better aeration. Earthing up operation in this crop is very important and after top dressing of urea at least one or two light earthing up should be done in rows of the plants. The lower leaves are often removed to get good and early sprouts (Sirohi and Behera, 2000). Direct field sowing of Brussels sprouts followed by thinning required slightly more labour hours than did raising of seedlings followed by machine planting.

Pruning and Topping

Metcalf (1954) in Montana studied the effect of removing the terminal bud and lower leaves after sprouts started to form. Removing the terminal bud increased the total yield and average size of sprouts but reduced the number produced. Pruning the lower leaves before the sprouts were ready for harvest greatly reduced the total yield. Carten (1962) reported that topping in mid-September increased the yield and it came earlier. According to Korenberg (1967) topping of the crop on the 20th September improved the size and unformity of the sprouts. By close plant spacings and removal of growing tips after the formation of sprouts in late summer or autumn it is possible to harvest the whole crop after 4-6 weeks of topping. Topping method is more suitable for mechanical harvesting. The period of harvest can be manipulated to a certain extent by removing the growing points.

Intercropping, Crop Rotation and Succession Cropping

Leek and strawberry either separately or together with brussels sprouts can be grown. Celery and brussel sprouts in sequence with or without leek are suitable. Brussels sprouts are commonly rotated with peas in order to check the soil borne pathogen. Benoit (1974) reported that rotation with other crops like peas and celery helps in checking club root *fungi (Plasmodiophora brassicae)*.

WEED CONTROL

A large number of weeds occur during transplanting and early phase of the crop growth. By use of herbicides weeds population can be controlled successfully. Application of weedicide like trifluralin (1 kg/ha), fluchloralin (0.5 kg/ha), nitrofen (2 kg/ha), oxyfluorien (1 kg/ha), alachlor (0.2 kg/ha), butachlor (2 kg/ha) 2-3 days before transplanting helps in to control broad leaves weeds.

Cassidy (1975) recorded consistently good and safe weed control with simazine @ 450 g/acre as post-emergence. Pathak (1972) reported that a combination of trifluralin @ 337 g/ha with alachlor @ 1.8 kg/acre applied immediately after sowing give complete control of weeds and pests effectively. Ivany and Cutcliffe (1973) reported that prynachlor @ 4.4 kg/ha and IMC 3950 @ 6.7 kg/ha gave excellent weed control.

PLANT PROTECTION MEASURES

The most common insect pests and diseases attacking to brussels sprouts and their management are given below:

Insect Pests

Important Pests

The important pests are diamond back moth (*Plutella xylostella*), cabbage head borer (*Hellula undalis*), cabbage butterfly (*Pieris brassicae*), bihar hairy caterpillar (*Spilosoma obliqua*), tobacco caterpillar (*Spodoptera litura*) and cut worms (*Agrotis* spps.). Caterpillars of these insects damage the crop. Conventional pesticides are ineffective against diamond back moth, gram caterpillar and tobacco caterpillar.

Spray of cartap hydrochloride followed by neem-seed kernel extract (5:1000 w/v) were found effective. Need based sprays of endosulfan 0.2% have also been recommended since these are not harmful to parasitoids complex. In case of cabbage butterfly, weekly spray of dipel @ 0.5 kg/ha were found very effective. Most of the insecticides are found effective against caterpillars of this insect.

Cabbage cut

Cabbage cut worm damage the crop in nursery stage or immediately after transplanting. It can be controlled by applying methylparathion (2% dust) @ 5 kg a.i./ha after 3-4 days of transplanting around each plant. Spraying of endosulfan is also effective.

Nymphs and Adults of Cabbage Aphids

Nymphs and adults of cabbage aphids (*Brevicoryne brassicae*), potato aphids (*Myzus persicae*), mustard aphid (*Lipaphis erysimi*) and cabbage thrips (*Thrips tabaci*) also attack the crops. Aphids are causing more damage to brussels sprouts. Aphids and thrips can be easily controlled by giving need based sprays of endosulfan 0.2%. Endosulfan is also considered safe for useful parasites.

Diseases

The most common diseases attacking the brussels sprouts crop and their management are mentioned herewith.

Damping Off

Damping off is primarily a disease of nursery and occurs in two phases. In pre-emergence phase, the young seedlings are killed before they reach the surface of soil and post-emergence is conspicuous by topping over the infected seedlings.The disease is incited by the species *Pythium, Phytophthora, Fusarium, Phoma, Olpidium* and *Rhizoctonia.* The control measures have been discussed under nursery management.

Black Spot

A seed-borne disease caused by *Alternaria brassicae, A. brassicicola* and *A. alternata*. Dark coloured spots appear on leaf, stem, sprouts, heads and pods which spread rapidly to form circular lesion. Under humid conditions, concentric ring appears. Optimum temperature for spread of disease is 20-30°C but its spores are able to germinate from 1-40°C provided sufficient moisture is available. Fungicides like mancozeb, copper oxychloride and chlorothalonil are effective against the disease. Seed treatment with metalyxal compound and spraying of the crop with mancozeb @ 0.2% at 10 to 15 days interval from the time of first week had given better control.

Downy Mildew

This disease is caused by an obligate pathogen *Pernospora parasitica*. This is more prevalent in the cooler region of the world. In India, its occurrence is common during winters and early summer in the northern plains and in the hills respectively. The optimum temperature for the spread of disease is 16°C, though range for infection and spread is 10-20°C. It can attack the crop at any stage of growth including storage. Small chlorotic, irregular, translucent, light green lesions appear on leaf lamina in the initial stage with pronounced downy growth of the fungus on lower surface during high humid conditions. Lesion dry out and becomes necrotic leading to defoliation and death of plant. In severe form, it may infect sprouts.

The control measures and seed treatments suggested for Alternaria spot disease are the same. But one protective spray of ridomil MZ 72 @ 0.025% before the onset of favourable conditions followed by dithane M-45 @ 0.2% at 10 days interval is effective.

Rhizoctonia

The disease is caused by the fungus *Rhizoctonia solani*. It can cause damping-off, wire stem, bottom rot and crown rot, depending upon the growth stage and plant at which it attacks the crop. Since it is a soil borne disease, integrated control approach like disinfection of nursery growing area, modification in cultural practices such as deep ploughing, use of compost and photodegradable mulch, light irrigation and use of resistant/tolerant varieties should be followed.

Sclerotinia Rot

Sclerotinia rot is caused by the fungus *Sclerotinia sclerotiorum*. This is a disease of seed crop, generally starts during generative phase. Wet soft lesions may appear on leaf scars in lower region of stem, left behind after defoliation of old leaves. The affected plant parts are covered with white silvery growth so it is also known as white rot. In the pith region of affected parts sclerotia of different sizes are formed. The affected plant wilt and die in due course of time. Crop rotation with monocots, flooding the field continuously for 2-3 months, application of saw dust after transplanting helps in reducing the inoculum.

Fungicides like benomyl, thiophanate methyl and iprodine have been reported effective against this disease. Four to five protective sprays alternatively of 0.05% bavistin combined with 0.2% dithane M 45 followed by dithane M 45 (0.2%) alone, at an interval of 15-20 days starting from sprout initiation stage are recommended to control the disease.

Yellows

The disease is incited by *Fusarium oxysporum. F. Conglutinas.* It can infect the crop at any stage of growth. Leaves turn yellow, die and their dropping starts from lowest leaves. The vascular tissues become yellow to dark brown and plants die slowly. It is a soil-borne disease which is difficult to control. The answer lies in development and cultivation of resistant varieties.

Club Root

The disease is caused by *Plasmodiophora brassicae* and is more common in soil having pH below 7.0 in the temperate regions. It is a soil-borne disease. Small spindle-like spherical club-shaped swellings on root and rootlets are characteristic signs of the disease. Yellowish foliage, flagging and wilting of plants are typical symptoms of this disease. It is not common in India disease can be minimized by liming the soil to raise the pH above 7.0. Nursery growing area can be treated with pentachloronitrobenzene, chloropicrin, cynamide and formalin approximately two weeks before sowings. Crop rotation and cultivation of resistant varieties are the only measures.

Black Rot

It is caused by *Xanthomonas campestris* pv. *campestris* and is destructive in subtropical and warm temperate area of the world. The disease does not spread below 5°C and above 36°C. The characteristic symptom is the production of V-shaped chlorosis on the margin of leaves, however, in susceptible cultivars blackening of veins and internal leaf chlorosis have been observed. The bacterium transmitted by seed though it can survive for at least a year in diseased plant debris and surviving plants of crucifers. Hot water seed treatment of 52° ± 2°C for 30 minutes is the only satisfactory method available to check the seed borne infection. Stray plants should be destroyed in order to avoid secondary spread. The spraying of streptocycline at 100-200 ppm is also effective to check the spread of disease.

Bacterial Soft Rot

It is incited by *Erwinia carotovora* and can attack the crop at any stage of growth. It can only invade the host through an injury which may be mechanical or by some fungal or other organisms. Spraying of the standing crop with 100-200 ppm streptocycline or plantomycin combined with coper oxychloride (0.3%) proved most effective.

Physiological Disorders

Internal browning

It is a very common disorder in brussels sprouts. Interior of heads and margin of the foliage turn brown and die, leaving the head lookng healthy outwardly. It is associated with the disturbance of water regime of the plant and occurs due to calcium deficiency. It mainly occurs in large sprouted varieties in late autumn and the winter season. An application of nitrogen in excess favour the formation of large sprouts which are more sensitive to this disorder. Growing resistant varieties without heavy nitrogen dose keeps internal browning under control. Spraying of 200 ppm of calcium gluconate in water can also control it.

Browning (Brown rot or red rot)

Browning is caused by boron deficiency which is influenced by soil pH. The availability of boron decreases at neutral soil reaction. Young leaves of boron-deficient plants become dark green and brittle, the older leaves puckered, chlorotic and often drop-off. Water-soaked, light brown to dark brown spots formed on the stem and branches may ultimately lead to the formation of cavities and a hollow stem. Plant is manifest in loose, wrinkled or thickened leaves. In seed crop, sometimes flowers and siliquas collapse.

A top dressing of 10-15 kg/ha of borax (sodium tetraborate) can control boron deficiency effectively. However, even smaller amounts will suffice in acid soils. Spraying of 0.2-0.4 per cent solution of borax @ 2 kg/ha depending upon growth, soil reaction and extent of deficiency can also control boron deficiency (Bergmann, 1992).

Frost injury

The leaves of young seedlings turn yellowish- white on both the surfaces. Petioles become flaccid and white, midrib along with adjacent parenchyma and stem may also be injured. Younger leaves are particularly sensitive to frost, so that the centre of the head turns brown, while outwardly the head appears healthy. Frost injury can be minimized by irrigating the fields on anticipating the danger of frost and by raising the field temperature by gearing smoke (Sharma and Thakur, 1995).

HARVESTING

Harvesting begins usually from 3 to 3 ½ month after setting the plants (Thompson and Kelly, 1972). The normal crop takes 120 days for sprout production in north India and as soon as lower sprouts on the stems attain proper size and firmness, they may be harvested. Delayed harvesting result in poor quality sprouts.

Loose sprouts at the base of the stem should be rejected otherwise the whole lot will be spoiled. Sprouts should be solid, upper leaves folded closely, attractive dark-green in colour and uniform in size. The freezing industries require a sprout less than 3 cm in diameter. At least 3-5 harvestings are required to be done.

In the north Indian plains, harvesting period is January-February and in hills from November-March or July-September depending upon cultivars and altitude. It is difficult to hand pick crowded sprouts on the stem, which is more common in short stemmed varieties and cultivars having sprout with broader base. Generally, leaves are removed from the plants before harvesting. After one or two harvestings, plants are cut from the base and after removing leaves, harvesting is done in shed (Sharma and Singh, 2001).

According to Chatterjee (1986) sprouts should satisfy the conditions for marketing: (i) should be solid and have their top leaves folded closely, (ii) should be dark or atleast light green and not yellow, (iii) sprouts with small loose leaves at the base are considered poor in quality and (iv) similarly, open top is also considered poor in quality. In developed countries brussels sprouts are harvested by machine to save labour requirement. Cutcliffe (1967) reported that the plants are cut off at ground level and the stalks are mechanically loaded with sprouts attached for delivery to the processing plant. The components of marketable yields are number of plants/m^2, number of buds per plant, acceptable size fraction, and mean weight per marketable bud (Fisher, 1973).

YIELD

The yield of crop depends on various factors like cultivars, maturity period and method of harvest. Large sprouted cultivars yield on an average 100-150 quintal per hectare but normally spaced crop when picked only once, may yield between 30-50 quintal per hectare.

POST HARVEST HANDLING

Sprouts are trimmed to pre-preparation for marketing and dirt, decayed, diseased and discoloured leaves are removed before grading the products. Pre-cooling is required to retard ripening and deterioration before marketing.

Grading

Singh and Singh (2007) stated that Sprouts should be graded properly to satisfy the condition for marketing. Grading the sprouts is becoming more and more popular. For grading different sizes like (a) big (above 4 cm), (b) medium (3-4 cm), (c) small (2-3 cm) and (d) very small (below 2cm) size in diameter are used. Loose sprouts are sometimes marketed as "blowers".

Packing

The sprouts are placed in baskets or other containers as picked and carried or hauled to the packing house. In New York, they are packed in quart berry boxes and are shipped in ordinary 16 or 24 quart berry crate. The individual boxes are some times overwrapped with a transparent film which delays wilting and seems to prolong their shelf life. In California, most of the crop is packed in plastic film bags holding 12 ounces of sprouts, and 12 bags are packed in fibre board boxes (Thompson and Kelly, 1972). In India, crop is packed in perforated film bags holding 500 g to 1.0 kg of sprouts and 10 to 20 bags are packed in baskets/fibre board boxes and also packed in polythene ceramic film (0.013 mm thickness) for consumer packaging (Peshin, 1998).

Storage

The sprouts keep well in storage at 32°F and high humidity (95-98 per cent) for 6-8 weeks. It was found that 10% CO_2 and 2.5% O_2 as best concentration for retarding yellowing without injury at 32°F. Sprouts packed in polythene ceramic film (0.013 mm thickness) can be stored well for 16 weeks at 0°C and only for 11 days at 20°C. Storage of sprouts in sealed polythene packs at 0°-1°C in modified atmosphere, 2-5% CO_2 and 14.5-16.8% O_2 reduces the yield losses and improves their quality in comparison to open storage at same temperature (Sharma and Singh, 2001). The optimum storage condition for sprouts are 0°C and 95-98% relative humidity to keep them marketable for 3-5 weeks. For home use in regions where severe freezes to occur, the whole plant may be stored in a cool cellar and the sprouts removed as needed. In mild climate the plants are left in the field or garden throughout the winter (Thompson and kelly, 1972).

SEED PRODUCTION

Brussels sprout is a cross pollinated crop. Its seed can only be produced in the dry temperature zones of H.P. and other high hill areas where winter is very cool.

Method of Seed Production

There are two methods of seed production i.e. Head to seed method and seed to seed method.

Head to Seed Method

This method is usually employed for the production of nucleus seed. The selected true to type fully mature heads/sprouts are taken and then transplanted at other locations after removing the outer leaves.

Seed to Seed Method

In this method seed crop is raised on the same location where it was originally transplanted to get head,. In this method the plants are allowed to overwinter in their original position.

Isolation

Since it is a cross pollinated and cross pollination is very high (60%) as reported by Watt (1968). The botanical varieties of *Brassica oleracea* group i.e. cabbage, cauliflower, knol khol, kale, collard and other allies freely cross with each other. Hence proper isolation among these crops also required to keep purity of seeds. Isolation of 3000 m for breeder's seed production is recommended. The isolation distance required is 1600 m for foundation and 1000 m for certified seed production (Anon., 1988). Honey bees and bumble bees are main pollinating agents.

Brief Cultural Practices

Brief cultural practices like nursery management, nutritional requirement and intercultural operations are same as given in crop production section.

Roguing and Inspection

A thorough roguing of off-type plants should be done immediately after sprouts formation. The true to type plants having genetic purity of the variety are retained and rest are pulled out. Second roguing is done at the time of bolting. Similarly plants with deformed inflorescence are pulled out. Final roguing is done during flowering.

The field inspections are carriedout atleast four times. First inspection is done to remove off-type, disease and blind plants during early vegetative growth. The morphological qualities of head are checked during second inspection. Third inspection should be made critically for horticultural characters. Fourth inspection is done at the time of flowering.

Harvesting and Threshing

Seed crop is generally harvested when 60-70% of pods turn yellow or light brown in colour as the whole crop do not mature at one time. Therefore, harvesting is done in two or three lots. The harvested crop is kept for curing in heaps for 7-10 days. These heaps are frequently stirred to facilitate proper aeration and prevent premature germination during this process.

Threshing is done by beating with sticks after proper drying. Seeds are separated manually or mini pettes machine is used for this purpose. Seeds are then thoroughly dried in the sun to bring moisture level to 6-8 per cent for their longer storage life.

Seed Yield

Average seed yield is varies from 200 to 500 kg/ha (Kohli and Vikram, 2001).

REFERENCES

Anonymous (1968). *Hort. Abstr.*, 38(3): 704.

Anonymous (1988). Indian Minimum Seed Certification Standards. The Central Seed Certification Board, New Delhi.

Benoit, F. (1974). *Hort. Abstr.* 44(11): 761-762.

Bergmann, W. (1992). Nutritional Disorders of Plants. Gustav Fischer Verlag Jena, Stuttgart, New York.

Boswell, V.K. (1971). *The encyclopaedia Britanica*, 4:315.

Carten, A.R.(1962). *Expt. Hort.* 7: 7-15.

Cassidy, J.C. (1975). *Hort. Abstr.*, 45(1): 26.

Chatterjee, S.S. (1986) . Cole Crops. In Vegetable Crops in India, (Eds.: T.K. Bose and M.G. Som), Naya Prokash, Calcutta -6.

Choudhury, B. (1967). Vegetables. National Book Trust, New Delhi.

Cutcliffe, J.A. (1967). *Can. Agric.*, 12(2): 5.

Eares, C.A. (1968). *J. Hort.*, 43: 317-322.

Fisher, N.M. (1973). *Hort. Abstr.*, 43(11): 747.

Gill, H.S. (1993). Improvement of Cole Crops. In Advances in Horticulture vol.5-Vegetable Crops. Part-1 (Eds. K.L. Chadha and G. Kalloo). Malhotra Publishing House, New Delhi, India.

Hegde, D.M. (1993). Water Management in Vegetable Crops. In Advances in Horticulture vol.5- Vegetable crops Part-I. (Eds.: K.L. Chadha and G. Kalloo), Malhotra Publishing House, New Delhi, India.

Helm, J. (1963). Morphologisch- taxonomische Gliederung der Kultursippen von *Brassica oleracea*, Kulturpflanze 11, 92.

Ivany, J.A. and Cutcliffe, J.A. (1973). *Hort. Abstr.*, 44(2): 841.

Kohli, U.K. and Vikram, A. (2001). Seed production of Exotic Vegetables. In Advances in Seed Production and Technology of Vegetable Crops. Centre of Advanced Studies Horticulture (Vegetable). Dr. YSPUHF, Nauni - Solan (H.P.) 102-104.

Korenberg, H.G. (1967). *Hort. Abstr.*, 38: 408.

Loetis, J.N.J. and Notenboon, L. (1970). *Hort. Abstr.* 40(4): 1003.

Maurer, R. (1976). *Can. J. Pl. Sci.*, 56(2): 325-330.

Metcalf, H.N. (1954). Effect of leaf and terminal bulb removal on yield of Brussels sprouts, *Amer. Soc. Hort. Sci. Proc.*, 64: 322-326.

Nath, Prem; Srivastava, V.K.; Dutta, O.P. and Swamy, K.R.M. (2008). Vegetable Crops – Improvement and Production. Dr. Prem Nath Agricultural Science Foundation (PNASF), Bangalore. pp. 133.

Nieuwof. M. (1969). Cole crops: Botany, cultivation and utilization. Leonard hill Books, London.

Ockendon, D.J. and Smith, B.M. (1993). Brussels sprouts *Brassica oleracea* var. *gemmifera* DC. In Genetic Improvement of Vegetable Crops. (Eds:G.Kalloo and B.O.Bergh), Pergamon Press, New York.

Pathak, S.C. (1972). *Hort. Abstr.*, 43(6): 352.

Peshin, S.N. (1998). Promotion of Rare Exotic Vegetables .In. production and Post Harvest Handling of Vegetables. Centre of Advanced Studies, Horticulture-Vegetable (ICAR), Dept. of Veg. Crop . Dr. Y.S.P.U. H.F., Nauni, Solan, HP. pp. 80-87.

Purseglove, J.W. (1974). Tropical crops: Dicotyledons. ELBS, Longman, London.

Shanmugavelu, K.G. (1989). Production Technology of Vegetable Crops. Oxford & IBH Publishing Co. Pvt. Ltd., 66, Janpath, New Delhi.

Sharma, B.K. and Thakur, J.C. (1995). How to cope with physiological disorders of cole crops. *Indian Horticulture*, 40(1): 51-55.

Sharma, S.R. and Gill, H.S. (1986). 'Hilds Ideal', a promising cultivar of brussels sprouts. *Indian Horticulture*, 31:17-18.

Sharma, S.R. and Singh, R. (2001). Brussels sprouts, sprouting Brocoli and Knol Khol. In Vegetables, Tubercrops and Spices (Eds. S. Thamburaj and Narendra Singh), DIPA-ICAR, New Delhi.

Shoemaker, J.S. (1953). Vegetable growing. John Wiley and Sons, Inc. New York.

Singh, D.N. and Nath, Vishal (2011). Varieties and Hybrids of Vegetables. Satish Serial Publishing House, Azadpur, Delhi (India). pp. 8-9.

Singh, D.N. and Singh, S.P. (2007). Production Management of Underutilized Vegetables. Agrihortica Publications, Junagadh (Gujarat) India. pp. 44-58.

Sirohi, P.S. and Behera, T.K. (2000). Unusual Exotic Vegetables for higher profit. Production Technology Bulletin. Unit of Vegetable Research, Production and Demonstration, ICAR, New Delhi.

Thompkins, D.R. (1972). J. *Am. Soc. Hort. Sci.* 97(6): 772-74.

Thompson, H. C. and Kelly, W.C. (1972). Vegetable Crops. (Edn. 5th). Tata Mc Graw- Hill Publishing Co. Ltd., New Delhi.

Toxopaeus, H. (1979). The domestication of *Brassica* crops in Europe. Proc. Eucarpia meeting - Cruciferous crops, (Eds.: N.P.A. Van Marrewijk and H. Toxopeus). Wagemingen, 29.

Van Mons, J.B. (1820). Substance of a memoir on the cultivation and variation of Brussels sprouts, *Trans. Hort. Soc.* London, 3: 197.

Watt, L.E. (1968). *Euphytica,* 17: 74-80.

Wolde, M. (1956). *Hort. Abstr.*, 26: 27-47.

□□□

5. BROCCOLI

Botanical name	:	*Brassica oleracea* L. var. *italica* Plenk
Family	:	Brassicaceae
Chromosome number	:	2n = 18

Broccoli (*Brassica oleracea* var. *italica* Plenk) is a popular vegetable in many parts of the world. It is an important cole vegetable from Brassicaceae. It can be easily distinguished from cauliflower by having a head composed of differentiated flower buds rather than a curd (Gray, 1982). It is a new crop in India which is nutritious among cole crops being rich source of sulphoraphane, a compound associated with reducing the risk of cancer (Kalia, 1995). Broccoli is also a rich source of dietary antioxidants and contains higher levels of nitrate (Heaton, 2001). Broccoli is an Italian word from Latin. 'Brachium' meaning an arm or branch (Choudhury, 1967). The green forms called as calabrese is mainly cultivated for frozen food. Now broccoli has become a special vegetable crop in northern plains and also in hills of the country.

ORIGIN

Broccoli is a native of eastern Mediterranean region, in common with cauliflower, derived from ancient forms of *Brassica oleracea* (Gates, 1953). Italy is a centre of diversification. The ancient Romans also knew this plant and called it 'Cyma'.

Citations of sixteenth century herbalists by Nieuwhof (1969) and a European writers of the same period by Boswell (1949), indicate that cauliflower and perhaps broccoli were well known in eastern mediterranean areas. Broccoli possibly originated about 2500 years ago and the ancestral forms of modern cultivars were probably selected about the time of Christ (Schery, 1972). Snogerup (1980) suggested that broccoli might have originated as an abnormality by crossing between early forms of branching shrub kale and other botanical varieties, followed by human selections. However, Song *et al.* (1990) supported monophyletic origin of cultivated morphotypes in *B. oleracea.*

DISTRIBUTION

Broccoli is widely spread along coasts of the mediterranean and north Atlantica. Cultivation spread over to U.K., Japan, USA and Europe. A large part of the produce in the USA goes to the freezing industry. It is becoming popular in other part of the world only gradually. In India its cultivation has been started recently around the big cities like, Delhi, Bombay, Calcutta, Chenai, Bangalore.

BOTANY

The plant consist of a short stem and a rosette of leaves in early developmental stage. However, sometime, thereafter, elongation of the internodes begins and at a certain point the terminal axis proliferates into numerous branches. The plant is generally regarded as an annual and a period of cold temperature, is not required to induce flower bud initiation. Although flowering is hastened by exposure to low temperatures.

In broccoli a main head is produced terminally as a fleshy branched, elongated stem and small heads on the axil of leaves (Purseglove, 1974). The plants form a kind of head consisting of green or purple buds and thick fleshy flower-stalks. The terminal bud is rather loose, green or purple in colour depending upon the varieties. Flower stalks are longer than cauliflower. The sprouts in the axils of leaves develop strongly, specially after the removal of the terminal head. Both terminal head and the sprouts with bud clusters are consumed as human food.

COMPOSITION

Broccoli is a highly nutritious, rich in vitamins and minerals. It is fairly high in vitamin 'A' and ascorbic acid and contains appreciable quantities of thiamin, riboflavin, niacin and minerals like calcium and iron. Besides anticarcinogenic properties, broccoli has about 130 times more vitamin A content than cauliflower and 22 times more than cabbage. It is also a rich source of sulphoraphane, a compound associated with reducing the risk of cancer (Kalia, 1995). According to Hazra and Som (2006), the curd or heads of broccoli contains following nutrients per 100 g of edible portion.

Moisture	89.9g	Protein	3.3 mg
Carbohydrates	5.5 g,	Riboflavin	0.12 mg
Fat	0.2 g,	Phosphorus	79 mg
Calories	37 g,	Calcium	80 mg
Vitamin A	3500 IU,	Iron	0.8 mg
Thiamine	0.05 mg	Ascorbic acid	137 mg

Uses

Broccoli heads are quite tender, fresh, full of flavour and laxative, which consumed raw as well as cooked (Kalia and Singh,2004). It is also used as half boiled vegetables, salad and mixed in soup with juice of other vegetables. Some times it is cooked and also pickled. In some country it is used for processing, terminal buds used for freezing (Thompson and Kelly, 1983). It is also used as frozen food and used as fresh vegetable in the early month of the year.

VARIETIES AND HYBRIDS

There are Several Types of Crop Varieties

Green broccoli, Purple sprouting broccoli, white broccoli, Pale broccoli and Dark green broccoli The main difference between the varieties are colour of the bud (Singh and Nath, 2011).

Green types

Green types are more commonly cultivated; these are early, medium and late type as per their maturity. The cultivars having solid, firm head with small green buds and few buds and few lateral shoots are preferred.

Early cultivars

Early cultivars are ready for harvest after 40-45 days of transplanting. De Cicco, Greenbud and Spartan Early are common early cultivars (Honma, 1960). Early cultivars do not perform well under low temperatures. Waltham 29, Green Mountain and Coastal Atlantic are suitable for early and late planting.

Some Mid and Late Planting Cultivars

Some mid and late planting cultivars are Green Sprouting Medium and Green Sprouting Late, biennial in nature, matured in 100 days.

Purpole types

Purpole types are more hardy and can withstand lower temperatures. Most of them are grown in European countries and harvested during winter. Colour development is adversely affected when grown under glass house or polythene house, hence, these types are suitable for protected cultivation.

Recently some varieties has been developed from IARI i.e. Pusa Broccoli 'KTS1', 'KTS 2', 'KTS 3', 'Palam Samridhi', and Palam Vichitra.

Gray (1993) reported several forms of broccoli within the Italic group, which details are given below:

Purple sprouting broccoli

Over wintering N.W. European Biennials, branched, purple spears, variable, early and late forms overlapping.

Purple Cape broccoli

Over wintering N.W. European biennials, single purple heads variable.

White sprouting broccoli

Over wintering N.W. European biennials, branched, white spears. Distinct early and late maturing forms of which the late type more strictly within the botrytis group (Gray, 1982).

Purple Sicilian broccoli (Cavolfiore violetto di Sicilia)

Single heading, pale purple heads, also known in horticulture as "purple cauliflower".

Couve broccoli Roxo de Cabeca

Tall purple heading/sprouting type from Portugal.

Calabrese (Cavolo broccolo ramosa verde di Calabria)

Green sprouting broccoli from calabrian region. Now a worldwide crop and virtually single heading.

Black broccoli (broccolo neri)

Highly branched annual with dark-green spears and sickle-shaped leaves; from Rome area.

Hybrids

Large number of F_1 hybrids cultivars are being marketed by the different seed companies of Japan, USA and Europe. These are:

Extra early

Southern Comet, Premium Crop, Clipper, Laser.

Mid season

Corsair, Excalibur, Cruiser, Emerald, Corona.

Late

Late Corona, stiff, Kayak, Green Surf belong to calabrese group. These are superior in yield, uniform in size and maturity have gained popularity and commonly grown in developed countries.

Some Japanese and European seed companies are also marketing hybrid cultivars in India; like-Packman, Pirate, Brigadier, Cap oven, Zues Hi Cream, Bacclus, Mars, Green Lofi, Tandon, Land Star, Green Mail, Green Dome and Shigmeri (Sharma and Singh, 2001).

Some hybrids have been developed and marketed by Indian Seeds Companies (Singh and Nath, 2011). These are :

Century Seeds; CBH-1, CBH-2
Bejo Sheetal Seeds; Lucky, Fiesta
Golden Seeds; Rapid, Green Baby
Namdhari Seeds; NS 50
Nath Sluis; Nath special off-1
Rallis India; Prativa
Sandoz seeds; Everest
Clause International; Citation
Indo American export; Packmen and Piret

Detailed characteristics of commercial varieties and hybrids are described by Singh and Nath (2011) are given below :

Pusa Broccoli KTS 1

Its plants are medium tall (65-70 cm long), foliage waxy and dark green with slightly wavy margins. Head solid green with small buds and slightly raised at the center. The main head size and weight are about 6cm and 15.4 cm (height x diameter) and 350-450g. It is ready 125-140 days after sowing and 90-105 days after transplanting in hills and reduced by 5-10 days in plains. It is medium maturing variety. It is resistant to mild frost and snowfall (Verma and Sharma, 1998).

Palam Samridhi

It has medium plant type with medium sized, dark green leaves . It is a green sprouting type with large, compact terminal, heads borne on long stalk. Terminal heads weigh 300-400 g. The subsequent tender sprouts arising from leaf axils. Its terminal heads become ready for harvesting in 85-90 days after transplanting. Heads remain free from yellow eyes and bracts. Its average yield is 150-200 q/ha. (Kalia, 2002).

Palam Kanchan

It has long, broad, bluish-green, upright entire margin leaves with prominent white midrib and veins. The head is large, compact, attractive and yellowish-green in colour. The heads attain marketable maturity 140-145 days after transplanting. The average yield potential is about 250-275 q/ha (Kalia and Singh, 2004).

Palam Vichitra

It has medium, open, dark green leaves with redish tinge. Purple streaks are prominent on stems at seedling as well as adult plant stage. The head is medium, compact, and purple coloured and attractive. It attains marketable maturity 115-125 days after transplanting with an average yield potential is 225-250 q/ha (Kalia and Jamwal, 2006).

Punjab Broccoli 1

It is developed through selection by Punjab Agricultural University, Ludhiana. Foliage dark green with smooth leaf surface and wavy margin with bluish-tinge. Plant branched, bears large number of sprouts in the axil of leaves. It takes about 65-70 days from transplanting to harvest. Heads and spears are dark-green with bluish-tinge. Average yield is 700 q/ha.

Lucky

It matures in approximately 60-65 days after transplanting. Head weight is 600-800 g and gives compact curd with fine buds. It can planted in winter season in the plains.

Fiesta

It matures in 65-70 days after transplanting. Head weight is 800-1000g. It is very compact and has fine buds and very high yielder, can be planted in winter season in the plains.

CLIMATIC INFLUENCE

Broccoli thrives best in a cool, moist climate. The climatic condition in high hills of dry temperate areas are quite conducive for growth and development of the crop. Though warm climate causes loose heads, increasing temperature upto 35°C accelerates leaf expansion and increases the amount of radiation observed by the plant leading to an increse in the plant size and yield potential (Klaring, 1998). Light frost causes considerable damage to the buds but some Indian varieties are resistant to mild frost and snowfall. Average day temperature of 20-25°C at vegetative stage and 15-20°C at heading are optimum (Verma and Sharma, 1998). Hot weather during the harvest period results in an undesirable leafiness in the heads, and the heads develop so rapidly that it is difficult to harvest at the proper time (Thompson and Kelly, 1983). It withstands hot and drier conditions better than the cauliflower (Purseglove, 1974).

SOIL REQUIREMENT

Broccoli require medium type of fertile soil with good moisture holding capacity for rapid development of seedlings after transplanting. For commercial cultivation of broccoli, fertile soil with good moisture supply would be very much necessary (Chauhan, 1965). Sandy loam or clay loam soils rich in organic matters are preferred and well drained soil of 5.5-6.5 pH with better water retention capacity gives higher yield. Stagnation of water causes injury to the root system of the crop (Sharma and Singh, 2001).

SOWING

The seed sowing should be adjusted according to area, its topography, height above mean sea level and climatic condition.

Time of Seed Sowing

The sowing of seeds should be completed by mid-October under north Indian plains and low hills (upto 900 m) conditions. Its sowing should be completed by mid-September under midhill condition. Where as under high hill and dry temperature areas sowing is also done in March-April.

Seed Rate

About 400-500 g seed produce enough seedlings to plant one hectare broccoli. In some countries direct seed sowing is also in practice and the plants are thinned to the desired spacing after seeds have germinated. In such cases higher seed rate i.e. 800 g/ha is required.

Nursery Preparation and Seedling Raising

A nursery bed of 3.0m long x 1.0 m wide x 15 cm high; supplemented with 20-25 kg well-decomposed farm yard manure, 250 g single super phosphate, 20 g indofil M-45 fungicide and heptachlore or methyl parathion dust 200 g a day prior to sowing is essential for raising healthy seedlings. Seed treatment with bavistin/thiram/captan 75 WP @ 2-3 g/kg of seed further helps in checking spread of damping-off disease in nursery-bed. The seed should be sown in well prepared nursery beds in lines 2-3 cm deep spaced 5 cm apart. Drenching nursery with 0.2% solution of thiram/captan at 7-10 days interval can check the post-emergence spread of damping-off.

PLANTING

Planting Distance

Planting distance of 40 cm x 30 cm, 45 cm x 45 cm, 50 cm x 45 cm, 50 cm x 50 cm and 60 cm x 45 cm is followed depending upon season and cultivars and mode of harvest (sharma and Singh, 2001). In USA, specings of 30 cm x 70 cm or 40 cm x 70 cm are followed to get only single harvest i.e., main head only. Closer spacings generally reduces the head size and delayed the maturity. Verma and Sharma (1998), reported that spacing of 45 cm between and within rows is optimum and it may be reduced to avoid hollowness in the stem, a physiological disorder due to very rampant growth. Spacing of 50 cm x 40 cm gave optimum yield in variety 'Pusa Broccoli Kt Sel.1' under Delhi condition.

Three to four weeks old seedlings are transplanted in well-prepared field at given spacing during evening hours. Pot watering is given just after transplanting the seedlings in main field for proper establishment of the seedlings. Transplanting of old seedlings result into poor growth of plant and development of small size heads.

NUTRITIONAL REQUIREMENT

Broccoli requires plenty of manure and fertilizers like cauliflower but it needs more nitrogen especially in late growing season. Sharma (2000) reported that application of N_{175} : P_{75} : K_{60} + FYM 12.5 t/ha recorded the maximum yield of 63.12 q/ha which was at par with N_{150} : P_{75} : K_{60} + FYM 12.5 t/ha with yield of 57.59 q/ha. Singh and Singh (2000) obtained highest net head weight and head yield of broccoli with application of N_{150} and K_{50} kg/ha alongwith uniform dose of phosphorus 120 kg/ha under low hills subtropical conditions. Sharma and Singh (2001) suggested that 100-125 kg N, 60-80 kg P_2O_5 and 25-50 kg K_2O/ha may be applied depending upon soil fertility status and cultivars. Sharma *et al.* (2002) obtained maximum yield and high cost benefit ratio of 1:2.35 when N was applied in combination with P @ 90 kg N and 60 kg P_2O_5. According to Sood and Sharma (2002) fertilizers doses of 125 kg N, 100 kg P_2O_5 and 50 kg K_2O/ha has been recommended for broccoli under temperate condition of Himachal Pradesh. Kalia (2002) and Kalia and Singh (2004) advocated the use of about 20-25 tonnes of farm yard manure at the time of field preparation with a basal dose of 50 kg N, 100 kg P_2O_5 and 50 kg K_2O/ha and another two split of 50 kg N/ha (each time) as a top dressing first at the time of first earthing up one month after transplanting and second at head initiation stage. Chatterjee *et al.* (2005) reported that fertilizer dose of 150:60:80 kg N: P_2O_5 : K_2O kg/ha had produced maximum curd yield (124.07 q/ha) as well as highest cost benefit ratio (1:6.49). Incorporate 20-25 tonnes/ ha farm yard manures during final land preparation. Apply full dose of phosphorus and potassium and half dose of nitrogen before transplanting. Another half of nitrogen is applied in two split doses as top dressing after 30 days and 45 days of transplanting.

Nath *et al.* (2008) suggested that a quantum of 20-25 tons of farm yard manure per hectare in addition to 100-200 kg N, 60-80 kg P_2O_5 and 40-60 kg K_2O per hectare should be applied. Nitrogen is to be given in two split doses, the first as basal and the next after the first harvest.

Use of Organic Nutrition

Organic manures and natural growth promoters offer an alternative to chemical inputs and are being used in vegetable production. Worthington (1998) found that organically grown broccoli had 40% higher vitamin C content. Organically grown broccoli contains higher levels of vitamin C, minerals and secondary nutrients like glucosinolates (Heaton, 2001). Organic farming is nutritionally superior with good taste, lusture, keeping qualities etc. Chatterjee *et al.* (2005) found that among the organic sources, application of mustard oil cake + bio-fertilizer-II (Azotobacter + VAM + Potash mobilizer) produced significantly higher yield (103.70 q/ha) and cost benefit ratio (1:4.46). Organic sources of nutrition had produced significantly better quality of curd than inorganic sources. Poultry manure + biofertilizer-I (Azotobacter + Phosphate solubilizer + Potash Mobilizer) had produced curds having significantly highest chlorophyll (32.80 mg/100 g), ascorbic acid (8.30 mg/ 100 g) and reducing sugar (20.20%) (Table 1, 2, 3).

Sanwal *et al.* (2005) reported that integrated use of orgnaic manures and natural growth promoters significantly improved the quality attributes within the same tissues but not change the chemical composition significantly between the tissues. The floral portion of the heads of broccoli showed higher values as compared to edible stem for dry matter, TSS, acidity, ascorbic acid, 6-carotine and proteins and lower values of fibres, pH, reducing and total sugars and nitrate. More than 90% of fibre content of the heads was found in the stem (Table 4, 5, 6).

Table 1 : Effect of organic sources of nutrition on growth and yield of sprouting broccoli

Treatments	Plant height (cm)	Spreading of plants (cm^2)	Leaves/ plant	Curd initiation (DAT)	Curd maturity (DAT)	Curd weight (g)	Curd yield (q/ha)
T_1- Cowdung+biofertilizer I - (Azotobacter+ Phosphate solubilizer + Potash mobilizer)	30.86	2045.80	22.60	72.67	87.64	101.67	37.78
T_2- Poultry manure + biofertilizer-I	34.67	1938.33	23.53	74.00	92.31	109.00	40.50
T_3 - Mustard oil cakes + biofertilizer-I	39.93	2714.13	23.40	68.67	88.36	151.67	56.05
T_4 - Cowdung + bio fertilizer-II (Azotobacter + VAM + Potash mobilizer)	39.00	2180.73	24.40	71.00	90.35	113.33	41.98
T_5 - Poultry manure + biofertilizer-II	38.27	2460.40	22.67	70.32	90.33	102.67	38.02
T_6 - Mustard oil cake + biofertilizer-II	39.00	2756.47	24.67	68.33	86.32	280.00	103.70
T_7 - Recommended dose of fertilizer @ 150:60:80 kg N: P_2O_5 : K_2O kg/ha	38.87	3224.07	26.40	65.68	81.68	335.00	124.07
CD (5%)	4.74	344.17	2.01	2.56	4.48	46.49	5.20

Table 2 : Effect of organic sources of nutrition on curd quality of sprouting broccoli.

Treatments	Chlorophyll content (mg/100 g)	Ascorbic acid (mg/100 g)	Reducing sugar (%)	Total sugar (%)
T_1 - Cowdung+biofertilizer -I (Azotobacter+Phosphate solubilizer + Potash mobilizer)	32.80	81.40	2.35	2.60
T_2- Poultry manure + biofertilizer -I	38.30	80.30	2.20	2.50
T_3 - Mustard oil cakes + biofertilizer-I	30.70	74.80	2.10	3.00
T_4 - Cowdung + bio-fertilizer-II (Azotobacter + VAM + Potash mobilizer)	36.90	80.50	1.90	2.40
T_5 - Poultry manure + biofertilizer-II	35.60	83.60	2.05	3.50
T_6 - Mustard oil cake + biofertilizer -II	39.10	72.80	1.70	2.20
T_7 - Recommended dose of fertilizer @ 150:60:80 kg N: P_2O_5 : K_2O kg/ha	27.60	76.00	1.80	2.30
CD (5 %)	1.44	3.17	0.24	0.92

Table 3 : Effect of organic sources of nutrition on economics of sprouting broccoli.

Treatments	Curd yield (q/ha)	Cost of inputs (Rs./ha)	Gross income (Rs./ha)	Net profit (Rs./ha)	Benefit: cost Ratio
T_1 - Cowdung+biofertilizer-I (Azotobacter+Phosphate solubilizer + Potash mobilizer)	37.78	87,382.72	118,062.50	30,679.78	0.35
T_2- Poultry manure + biofertilizer-I	40.50	42,111.00	126,562.50	84,451.50	2.00
T_3 - Mustard oil cakes + biofertilizer -I	56.05	59,292.50	175,156.25	115,863.75	1.95
T_4 - Cowdung + bio-fertilizer -II (Azotobacter + VAM + Potash mobilizer)	41.98	87,382.72	131,187.50	43,804.78	0.50
T_5 - Poultry manure + biofertilizer-II	38.02	42,111.00	118,812.50	76,701.50	1.82
T_6 - Mustard oil cake + biofertilizer-II	103.70	59,292.50	324,062.50	264,770.00	4.46
T_7 - Recommended dose of fertilizer @ 150:60:80 kg N: P_2O_5 : K_2O kg/ha	124.07	41,397.58	310,175.00	268,777.42	6.49
CD (5 %)	-	-	-	-	-

Chatterjee *et al.* (2005).

Note : (i) All the organic manures were applied 15 days before transplanting as cow dung manure @ 50 t/ha, poultry manure @ 8 t/ha and mustard oil cake @ 3 t/ha. (ii) All forms of bio-fertilizers were inoculated as root dip of seedlings for 15-20 minutes just before transplanting.

Table 4 : Effect of organic manure on dry matter, fibre and pH of stem and floral portion of broccoli.

Treatment	Dry matter (%)		Fibre (g/100 g)		pH	
	Floral	Stem	Floral	Stem	Floral	Stem
FYM (20 t ha^{-1})	12.16	6.73	1.29	2.78	7.9	8.1
Poultry manure (10 t ha^{1})	12.41	6.88	1.30	2.85	8.0	8.1
Pig manure (11 t ha^{-1})	12.28	6.89	1.32	2.95	7.7	7.9
Rabbit manure (11 t ha^{1})	12.81	6.94	1.31	2.46	7.3	7.2
Neem shield (4.5 t ha^{-1})	12.10	6.81	1.28	2.80	7.6	7.8
FYM + Panchakavya (16 t ha^{-1} + 4 spray)	12.55	7.55	1.30	2.72	7.4	7.5
FYM + Amritpani (16 t ha^{-1} + 4 spray)	12.89	7.61	1.27	2.52	7.4	7.6
Poultry manure + Amritpani (8 t ha^{-1} + 4 spray)	13.06	7.46	1.27	2.64	7.3	7.6
Poultry manure + panchakavya (8 t ha^{-1} + 4 spray)	12.66	7.55	1.26	2.71	7.3	7.5
NPK alone (100 % NPK 150::60:60 kg ha^{-1})	12.27	6.79	1.42	2.98	7.0	7.2
CD at P = 0.05	0.17	0.12	0.06	0.09	0.17	0.18

Table 5 : Effect of organic manure on chemical composition of stem and floral portion of broccoli.

Treatment	TSS (%)		Reducing sugars (%)		Total Sugars (%)	
	Floral	Stem	Floral	Stem	Floral	Stem
FYM (20 t ha^{-1})	6.9	6.0	18.73	19.23	34.86	36.15
Poultry manure (10 t ha^{1})	7.3	5.2	19.61	19.84	34.67	36.70
Pig manure (11 t ha^{-1})	6.6	4.8	19.46	19.76	34.66	38.10
Rabbit manure (11 t ha^{-1})	7.4	6.0	18.59	18.87	34.62	35.92
Neem shield (4.5 t ha^{-1})	6.6	4.8	18.87	19.29	39.56	41.87
FYM + Panchakavya (16 t ha^{-1} + 4 spray)	6.9	6.1	19.41	20.01	43.06	44.20
FYM + Amritpani (16 t ha^{-1} + 4 spray)	7.0	6.2	19.84	20.24	42.66	44.59
Poultry manure + Amritpani (8 t ha^{-1} + 4 spray)	7.1	5.4	19.38	19.69	37.79	40.24
Po ultry manure + panchakavya (8 t ha^{-1} + 4 spray)	7.0	5.6	19.30	18.54	35.04	37.98
NPK alone (100 % NPK 150:: 60:60 kg ha^{-1})	6.4	4.8	18.52	18.80	33.02	35.11
CD at P = 0.05	0.12	0.10	0.36	0.36	1.38	1.40

Table 6 : Effect of orgnaic manure on acidity and vitamins of broccoli.

Treatment	Acidity (%)		Ascorbic acid (mg/100 g)		β-Carotene (mg/100 g)	
	Floral	Stem	Floral	Stem	Floral	Stem
FYM (20 t ha^{-1})	0.26	0.09	52.46	38.24	6.85	4.25
Poultry manure (10 t ha^{-1})	0.31	0.06	52.89	37.98	7.00	4.30
Pig manure (11 t ha^{-1})	0.18	0.08	46.75	34.26	6.90	4.10
Rabbit manure (11 t ha^{-1})	0.24	0.08	48.64	35.54	7.25	4.35
Neem shield (4.5 t ha^{-1})	0.27	0.13	44.50	34.42	7.28	4.30
FYM + Panchakavya (16 t ha^{-1} + 4 spray)	0.23	0.08	55.76	39.16	8.30	4.75
FYM + Amritpani (16 t ha^{-1} + 4 spray)	0.20	0.09	56.24	40.15	7.80	4.50
Poultry manure + Amritpani (8 t ha^{-1} + 4 spray)	0.24	0.10	57.86	42.64	8.25	4.60
Poultry manure + panchakavya (8 t ha^{-1} + 4 spray)	0.26	0.10	56.88	40.80	8.80	4.85
NPK alone (100 % NPK 150::60:60 kg ha^{-1})	0.20	0.12	37.55	28.45	6.55	4.50
CD at P = 0.05	0.02	0.01	1.68	1.46	0.45	0.32

Sanwal *et al.*, (2005)

Note: Foliar spray of 10% solution of Panchakavya and amrit pani was done at 15, 30, 45 and 60 days after planting.

Use of Biofertilizers

Use of bio-fertilizers (Azotobacter and Phosphate solubilizers)is helpful in increasing the head quality and head yield of broccoli and it also reduces the requirement of nitrogenous fertilizer and efficiently increases the use of phosphorus to the plant. It can be applied in different way (Bhattacharyya *et al.,* 2000).

Seed Treatment

About 200 g of biofertilizer is required to treat 10-14 kg of seeds. Suspend one packet of 200 g in approximately 400 ml water and mix it thoroughly. Pour this mixture on seeds and mix with hands to obtain uniform coating on each and every seed. Spread the seeds in shade for drying for 10-15 minutes then sow them immediately.

Seedling Treatment

Prepare the suspension by mixing 1 kg of culture in 10-15 liters of water. Get seedlings required for 1 acre and make small bundles of seedlings and keep the root portion in the suspension for 15-20 minutes. Transplant these immediately. Generally the ratio of inoculant and water should be 1:10 approximately, i.e. 1 kg packet in 10 litre of water. Usually 1.5-2.0 kg of N bio fertilizer required for one hectare.

VAM Fungi

Vegetable crops are inoculated with VAM fungi in the nursery beds. The inoculum consisting of spores of fungus mixed with soil and infected root bits of the host plants on which they are raised in either spread uniformly in the nursery bed or placed in rows over which seeds are sown. The seedlings are colonized by the fungus 45 days after transplanting. Inoculated plants show better growth and improve nutrient content. Addition of inorganic phosphates should be reduced by 25-50% depending upon the soil P content. 200-500g of inoculum could be added depending upon the concentration of VAM spores alongwith farm yard manure and neem cake(Asokan *et al.*, 2000).

Use of Chemicals

Micronutrients like zinc, boron, molybdenum and iron are essential for good yield of broccoli and to check the whiptail and browning disorders caused by deficiency of molybdenum and boron. These disorders can be corrected in following ways :

(a) Soil application of borax @ 10-12 kg/ha and foliar spray @ 0.25% (2.5 g/litre. of water) at 10 days interval starting from 30 days after planting improve head quality.

(b) Tracel-2 or Multiplex @ 10-12 kg/ha or its 0.4% solution in form of spray at twice on an interval of 15 days improve the yield and head quality.

(c) Application of dhanzyme granules @ 8-16 kg/acre in two splits first at 10-15 days after planting and second 20-30 days later increase head yield and head quality.

(d) Application of dhanzyme liquid @ 1ml/litre of water (180 ml per acre) as a foliar spray, with high volume sprayer at 6-8 leaf stage and 15-20 days later improve head yield and head quality.

(e) Application of plantaid as foliar spray 3ml/litre of water, twice at an interval of 3-4 weeks, when plant shows symptom of deficiency. It improves the plant growth as better flowering as well as better florate formation.

(f) Incorporation of plantaid (S) @ 10 kg/ac. in soil in a single or split dose increase vegetative growth, flowering, head yield and quality.

IRRIGATION

Broccoli crop is most sensitive to moisture stress during head formation and enlargement stages although a moisture shortage at any period of growth reduces yield. Growth of broccoli plants gets reduced significantly when moisture is deficient during the early period of growth before heading, but as soon as a plentiful supply of water is available during head formation, the plants resume normal growth and the final yield of marketable produce will not be severely affected (Hegde, 1993). Frequent but light irrigation are essential throughout the crop growth span. Irrigations at an interval of 7-8 days are required depending upon weather conditions to get good harvest. Soil moisture deficit during head enlargement also reduces quality of harvested produce (Choudhury, 1967). Irrigation at 0.75 bar SMT during vegetative stage and 0.25 bar SMT during post-heading stage give optimum yield under Delhi condition.

INTERCULTURE

Broccoli is a shallow rooted crop which requires frequent shallow hoeing to provide proper soil mulch and to check weed growth. Two to three hoeings are required for proper aeration and root development. An earthing-up a month after transplanting is essential as it helps in plant anchoring, root aeration, weed check, as well as mixing of first split dose of nitrogen and second earthing is given at head initiation stage. Deep hoeing should not done to avoid injury to roots.

WEED CONTROL

Weeds cause a problem during early stages planted crops. If not checked these compete with the main crop for nutrients, water, light and aeration etc, affecting its crop growth adversely.At later stage, the crop develops its own canopy covering the ground and, thus, smoothers the weeds. Pre-planting application of herbicides like stomp 30 EC (Pendimethalin) @ 3.3 litre/ha or basalin (Fluchloralin) @ 2-2.5 litre/ha ensuring that sufficient moisture exists in the soil (Kalia, 2002). If labour is not a problem then manual weeding should be preferred, as chemicals pollute the soil environment (Kalia and Singh, 2004).

PLANT PROTECTION MEASURES

The important pest and diseases alongwith their management are mentioned below:

Insect Pests

The serious insect pests of broccoli are cabbage web worm (*Hellula undalis*), imported cabbage worm (*Pieris rapae*), cabbage looper (*Trichoplusia ni*) and aphids (*Brevicoryne brassicae*).

Cabbage web worm (Hellula undalis)

The caterpillar feeds on the growing point of the seedling. If a growing plant survives after this attack it forms several small unmarketable heads. Seedlings in nursery beds should be covered with nylon net cage to prevent the adult moths from laying of eggs on the foliage. Spraying insecticides like endosulfan or malathion @ 0.2% is useful to control this pest.

Imported cabbage worm (Pieris rapae)

The larvae feed on the leaves and may completely skeletonised the foliage, thus reducing quality and quantity of yield. This pest can be controlled by Bt (*Bacillus thuringiensis*) spray in different formulation.

Cabbage looper (Trichoplusia ni)

The larvae feed on the foliage and reduces both quality and quantity of yield . As it is surface feeder. it can be controlled by application of endosulphan @ 0.2%. Formulation of *Bacillus thuringiensis* can also check the population. This insect is also susceptible to nuclear polyhedrosis virus (NPV).

Aphids (Brevicoryne brassicae)

The nymphs and adults suck plant sap and transmit virus diseases. Heavily infected plants are stunted and yield is reduced. They also excrete sticky honey dew that coasts the plant. Shooty moulds grow in this honey dew, reducing photosynthesis and making produce unattractive. Pirimicarb a selective insecticide sprayed @ 0.5 kg a.i./ha controls aphids of this crop. Use of insecticidal soap innundative, release of *Chrysoperia carnea* to control aphid on broccoli is successful.

Diseases

Two bacterial diseases, black rot and soft rot and fungal disease like powdery and downy mildew are most serious in broccoli.

Black rot

It is caused by *Xanthomonas campestris* pv. *campestris.* Cotyledons of infected seedlings turn black, shrivel and drop-off. Initial infection on true leaves is a small wilted or yellowed 'V' shaped are at the margin. The pathogen survives on crop debris and is seed borne. Hot water treatment of seeds (soaking in hot water for 20 minutes at 50°C) eliminates the pathogen. Resistant varieties are used where black rot is common in a locality.

Soft rot

It is caused by *Erwinia carotovora* pv. carotovora. This disease is most prevalent from early to late in the season and develops rapidly under hot and humid conditions. Initial infection is usually found on the outer petioles that are in contact with root at the bud formation stage of the plants. The symptoms as water soaked areas develop in petioles, tissues become soft and then watery rot progresses. The infected tissue gives off an offensive sulfurous odour which in part may be due to the secondary bacterial growth. Avoid wounding plants during cultivation, harvest, packing and storing. Use of resistant cultivars is best solution of this problem.

Powdery mildew

This disease is caused mainly by *Sphaerotheca fuligenea* which becomes severe in slightly warm and humid conditions. Symptoms first appear as white spots on the surface of older leaves which enlarge and increase in number and cover both the surfaces of the leaves and stems. The seeds treated with hot water at 50°C for 30 minutes eliminate the bacterium from seed to grow into healthy seedlings. Spraying with karathane @ 2 g/litre of water control the disease.

Downy mildew

It is caused by *Peronospora parasitica*. The pathogen has a wide host range and also occurs in other cruciferous vegetables. The small yellow leaf spots of indefinite size and shape occur in leaf surface. The lesions are produced on the lower leaf surface and are covered with white downy mould of conidia and conidiophore. This disease is promoted by the cool and wet weather in winter months. It is common in young plants and is characterised by the appearance of purplish brown spots on the under side of the leaves and upper surface of the leaf. Downy mildew can be controlled by sowing treated seed with apron 35 WS (metalaxyl) @ 2 g/kilo of seeds or by spraying plants with dithane M-45 @ 2 g/litre at 8-10 days intervals. Sanitation and crop rotation help in reducing the infection of the disease.

Physiological Disorders

Sharma and Thakur (1995) observed a number of physiological disorders in broccoli and suggested their control measures which are given below:

Internal tip burn

This disorder is characterised by light brown tissue, which may later appear to dark brown or even black. The affected tissue losses moisture and takes on a papery appearance. This symptom is enhanced by high nitrogen application, high relative humidity, calcium deficiency and delayed harvests also favour tip burn development. Selection for resistance to this disorder is the best method control. Spraying of 200 ppm of calcium gluconate can also control it.

Browning

Browning caused by boron deficiency which influenced by soil pH. Young leaves become dark green and brittle. Water soaked areas are also seen on the stem that extend towards the curd and browning of the individual florets occur in the curd. A top dressing of 10-15 kg/ha of borax (sodium tetraborate) can control boron deficiency effectively. However, even smaller amounts will suffice in acid soils. Spraying of 0.2 – 0.4 per cent solution of borax @ 2 kg/ha depending upon growth, soil reaction and extent of deficiency can also control boron deficiency.

Whiptail

The leaf blades fail to develop properly and remain strap-like and extremely savoyed. The leaves are ruffled and distorted. In severe conditions only the midrib develops, and hence the name whiptail. Growing points are severely deformed and do not produce marketable curds. Sprouting near the base of the plants is also stimulated in extreme cases. Deficiency of molybdenum causes whiptail syndrome especially in high acidic soils.Because high manganese concentrations in such soils hinders the uptake of molybdenum which seldom occurs when the soil pH is 5.5 or higher. Whiptail can be controlled by the application of lime or dolomite lime stone to raise the soil pH upto 6.5 or higher. It can also be controlled by soil application of 1-2 kg of sodium or ammonium molybdate per hectare. The molybdenum may be mixed with fertilizer and applied in irrigation water. However it is not very effective in acid soils owing to a low fertilizer efficiency of these soils.Drenching nursery bed with 2 g of sodium/ammonium molybdate/m^2 two weeks before transplanting controls whiptail. Spraying of 0.01-0.1 per cent molybdate solution @ 250 g/ha is also effective.

Hollow stem

In heavily fertilized soils, particularly with nitrogen, rapidly growing plants broccoli develop hollow stems and curds. It is maximum at the time of first harvest, which declines with successive harvests. Stems do not show any apparent signs of disintegration. Hollow stem may be controlled by spacing the plants closer and with optimum use of nitrogenous fertilizers.

Frost injury

In sprouting broccoli, the clusters freeze, turn brown and ultimately rot even on experiencing a light forst. Forst injury can be minimized by irrigating the fields on anticipating the danger of frost and by raising the field temperature by creating smoke.

HARVESTING

Broccoli takes considerable time to mature. It is very important that broccoli is harvested at the correct time (before the buds open and when buds clusters are compact). Early type mature in 40-50 days, mid in 60-100 days and late after 100 days of transplanting. Usually the crop is ready for harvest from December-March in the plains, from November-April in the hills and August-October in higher hills of north India. Delayed harvesting cause opening of buds (heads) and loosening of head making it unfit for market. The central head is removed alongwith 15 cm long fleshy stem. The leaves are removed from the cut stem and head. In most of the cases, only main head is harvested and side speaks are not marketed (Sharma and Singh, 2001).Lateral shoots develop in the axils of the leaves and these develop into marketable heads from 1 to 3 or 4 inches in diameter. These lateral shoots produce a continuous harvest for several weeks (Thompson and Kelly, 1983). In developed countries, mechanical harvesting is done and the yielding component is only the central head. The heads may be 15-25 cm in diameter and weight 250-600 g.

YIELD

The average yield per hectare is about 80-150 q/ha.

POST HARVEST HANDLING

The shelf-life of broccoli heads and spears is significantly affected by temperature (Sharma and Singh, 2001).

The spraying of 2,4, 5-trichlorophenoxyacetic acid on the broccoli 3 or 4 days before harvest prevent the loss of chlorophyll from the heads for about one week. This material at very dilute concentrations also prevent dropping of the florets during subsequent handling operation (Marth, 1952).

Grading

Grading is done as per size and quality of heads or bunch of buds. Big size heads with compact cluster of buds are separated and kept in "super grade". Side or lateral shoots of medium size cluster with good appearance graded as "fancy grade". Small lateral and loose heads with open buds cluster are marketed as "commercial" grades (Singh and Singh, 2007).

Packing

In the packing house the broccoli heads are usually trimmed to a length of 6" and tied in a bunches weighing about 700g of uniform packing. The bunches are generally packed in flat crate which hold 28 bunches (Thompson and Kelly, 1983). In India bud clusters are usually packed in flat baskets which holds 20-25 bunches or 20 kg; but a variety of packages are used for marketing broccoli in other areas. Broccoli heads can be wrapped in transparent films for market. Packaging in low density polythene bags helped in keeping the broccoli heads more fresh and green under low temperature(Sharma and Singh, 2001).

Storage

Storage life of broccoli is not much influenced by relative humidity. It is stored in an oxygen-free atmosphere to prevent yellowing of the heads or buds (Lieberman and Hardenburg, 1954). Ethylene is one of the important volatile compounds emitted by broccoli in storage. Production of ethylene and the ensuing yellowing is reduce by refrigerating the packaged broccoli (Liberman and Spurr, 1955). Degradation of ascorbic acid and chlorophyll is much slower under cool storage. Broccoli can also be stored for 3 days at room temperature of 12.28°C-16.77°C with relative humidity 74.79%. But under cold storage it can be stored upto 21 days at temperature of 3°C with 88% relative humidity.

SEED PRODUCTION

Broccoli can freely seed at elevations of 1000 to 1200 m but can produce some seed in the plains of northern India as well (Choudhury, 1967). According to Sirohi and Behera (2000) broccoli varieties flower early (in February) and as such set seeds in the plains. It is a cross-pollinated vegetable crop which require 1000-1500 m isolation distance between two varieties of broccoli and also with one cole crop to another cole or crusciferous vegetable crops grown for seed production to check cross pollination.

For seed production the young seedlings are transferred to field at six true leaf stage. Seedlings are planted 30-60 cm apart with upto 90 cm between rows. The terminal growing points are usually removed at the end of winter (during the final roguing) which stimulates the production of inflorescences from the secondary shoots. For successful pollination bee-hives can be placed adjacent to this seed crops at the start of anthesis.

The plants are examined before the formation of normal heads or floret period for vegetative characters. When the plants are at market maturity they are checked for head colour, form and shape. Pods are harvested by hand as they reach maturity and placed on tarpaulins to finish ripening. Then threshing and cleaning of seed is done. The seeds are dried to 5 per cent moisture content for storage in vapour proof containers.

REFERENCES

Asokan, R., Mohandas, S.and Anand, L. (2000). Biofertilizers and biopesticides for horticultural crops. *Indian Horticulture,* 45(1): 44-47.

Bhattacharyya, P., Jain, R.K. and Paliwal, M.K. (2000). Biofertilizers for vegetables. *Indian Horticulture,* 45(2):12-13.

Boswell, V.R.(1949). Our vegetable travellers. *National Geographic magazine,* 96: 145-217.

Chatterjee, B., Ghanti, P., Thapa,U. and Tripathy, P. (2005). Effect of organic nutrition in sprouting broccoli (*Brassica oleracea* L. var. *italica* Plenck). *Vegetable Science,* 32(1): 51-54.

Chauhan, D.V.S. (1965). Vegetable production in India. Ram Prasad and Sons, Agra-3.

Choudhury, B. (1967). Vegetables. National Book Trust, India, New Delhi.

Gates, R.R. (1953). Wild cabbage and their effect of cultivation. *J. Genet.,* 51:363-372.

Gray, A.R. (1982). Taxonomy and evolution of broccoli (*Brassica oleracea* var. *italica). Econ. Bot.,* 36: 397.

Gray, A.R. (1993). Broccoli, *Brassica oleracea* L. (Italica group). In Genetic Improvement of Vegetable Crops (Eds: G.Kalloo and B.O. Bergh). Pergamon Press, New York.

Hazra, P. and Som, M.G. (2006). Vegetable Science. Kalyani Publishers, New Delhi.

Heaton, S.(2001). Organic Farming, Food Quality and Human Health. *Soil Association,* Bristol, U.K.

Hegde, D.M.(1993). Water Management in Vegetable Crops. In: Advances in Horticulture Vol.5-Vegetable Crops Part I. (Eds.: K.L. Chadha and G. Kalloo). Malhotra Publishing House, New Delhi.

Honma, S.(1960). Spartan Early- a new variety of broccoli, *Quart,. Bull. Mich. St. Univ. Agric. Expt. Stn.,* 42: 470

Kalia, P. (1995). Biochemical assays and yield performance of sprouting broccoli genotypes of Himachal Himalayas .*Cruciferae Newsletter,* 17: 90-91.

Kalia, P. and Jamwal, R.S. (2006). Palam Vichitra: a new broccoli. *Indian Horticulture,* 51(5): 16-17.

Kalia, Pritam and Singh, Yudhvir. (2004). Palam Kanchan: a new nutritious broccoli. *Indian Horticulture.* 49(2): 36.

Kalia, Pritam.(2002). Palam Samridhi: a new variety of sprouting broccoli, *Indian Horticulture,* 47(1): 24-25.

Klaring, H.P. (1998). Growth of broccoli as affected by global radiation and air temperature. *Gartenbauwissenschaft,* 63(3): 116-122.

Lieberman, Morris and Hardenburg, R.E.(1954). Effect of modified atmospheres on respiration and yellowing of broccoli at 75°F, *Amer, Soc. Hort. Sci. Proc.,* 63 : 409-414.

Lieberman, Morris, and Spurr, R.A. (1955). Oxygen tension in relation to volatile production in broccoli.. *Amer. Soc. Hort. Sci. Proc.*, 65: 381-386.

Marth, P.C. (1952). Effect of growth regulators on the retention of color in green-sprouting broccoli. *Amer. Soc. Hort. Sci. Proc.*, 60: 367-369.

Nath, Prem; Srivastava, V.K.; Dutta, O.P. and Swamy, K.R.M. (2008). Vegetable Crops - Improvement and Production. *PNASF*, Bangalore. pp. 142.

Nieuwhof, M. (1969). Cole crops-Botany, Cultivation and utilization, World Crops Book-Ser, Leonard Hill, London.

Purseglove, J.W. (1974). Tropical crops. Dicotyledons,. ELBS, London.

Sanwal, S.K., Rai, N., Yadav,D.S. and Yadav, R.K. (2005). Compositional differences of stem and floral portions of broccoli under different organic manures and natural growth promoters. *Vegetable Science*, 32(1): 73-75.

Schery, R.W. (1972). Plants for Man, pp.508-9.

Sharma, B.R.and Thakur, J.C. (1995). How to cope with physiological disorders of cole crops, *Indian Horticulture*, 40(1): 51-55.

Sharma, K.C. (2000). Influence of integrated nutrient management on yield and economics in broccoli (*Brassica oleracea* L. *italica* Plenk) under cold temperate conditions. *Vegetable Science*, 27(1): 62-63.

Sharma, S.K., Sharma, R. and Korla, B.N. (2002). Response of N and P on growth and yield of sprouting broccoli (*Brassica oleracea* var. *italica*). *Indian J. Hort.* 59(3): 313-315.

Sharma, S.R. and Singh, R. (2001). Cole crops: Brussels sprouts, Spruting Broccoli and Knol khol. In Vegetables, Tubercrops and Spices (Eds: S. Thamburaj and Narendra Singh). DIPA-ICAR, New Delhi.

Singh, A.K. and Singh, A. (2000). Influence of nitrogen and potassium on growth and head yield of broccoli (*Brassica oleracea,* L. var. *italica* L. Plenck) under low hills subtropical condition of H.P. *Vegetable Science*, 27(1): 99-100.

Singh, D.N. and Nath, Vishal (2011). Varieties and Hybrids of Vegetables. Satish Serial Publishing House, Azadpur, Delhi (India) : 5-9.

Singh, D.N. and Singh, S.P. (2007). Production Management of Underutilized Vegetables. Agrihortica Publications. Junagadh, (Gujarat) India. pp. 31-43.

Sirohi, P.S. and Behera, T.K. (2000). Unusual Exotic Vegetables for higher profit. *Production Technology Bulletin*, IARI, New Delhi.

Snogerup, S. (1980). Brassica crops and Wild Allies.(Ed: S. Tsunoda, K. Himda and C. Gomezcampo), Japan Scientific Societies Press, Tokyo.

Song, K., Osborn,T.C. and Williams, P.H. (1990). Brassica taxonomy based on nuclear restriction fragment length polymorphisms (RFLPs), 3. Genome relationships in Brassica and related genera and the origin of *B. oleracea* and B. rapa (Syn. *Campestris), Theor. Appl. Genet.*, 79: 497.

Sood, S. and Sharma, J.J. (2002). Production potential and economic returns of different harvest management systems in broccoli (*Brassica oleracea* L. var. *italica* Plenck) under temperate conditions. *Vegetable Science*, 29(1): 43-44.

Thompson, H.C. and Kelly, W.C. (1983). Vegetable Crops. Tata Mc Graw-Hill Publishing Company Ltd., New Delhi.

Verma, T.S. and Sharma, S.C. (1998). 'Pusa Broccoli KTS1': new Broccoli. *Indian Horticulture*, 43(3): 16-29.

Worthington, Virginia (1998). Effect of agricultural methods on nutritional quality: A comparison of organic with conventional crops. *Alternative Therapies*, 46: 58-69.

❑❑❑

6. CHINESE CABBAGE

Botanical name	:	*Brassica pekinensis, B. Chinenesis*
Family	:	Brassicaceae
Chromosome number	:	2n = 20

Chinese cabbage grow well in winter season, as a salad crop. It is a herbaceous annual and most important crop cultivated in China since the fifth century. It is grown in small scale in India (Singh and Singh, 2007).

ORIGIN

The native place of chinese cabbage is China where it has been in cultivation since fifth centry, chinese cabbage might have originated from hybridization between turnip (*B. campestris* ssp. *rapifera*) from north China and pak-choi [*B. campestris* L., ssp. *chinesis* (Rupr Olsson)] from south China where they were grown together in young Chou (Thompson and Kelly, 1983). Artificial crosses between these species have proved this hypothesis. It is indigenous to china and eastern Asia.

DISTRIBUTIONS

Chinese cabbage is an important crop of Taivan, Korea, Japan and Mongolia as well. It is little grown in countries like India and the USA and is considered as new vegetable (Swiader *et al.*, 1993). In India it is grown near big cities during winter and is supply to beig hotels and resturants and give higher returns.

BOTANY

The plant of Chinese cabbage is annual, sometimes a biennial, and has few characteristics same common as cabbage (Thamburaj and Singh, 2003). There are two distinct species of this vegetable namely:

Heading type Pet-sai (Brassica pekinensis)

Resemble cos-lettuce but produces a much larger heads which is elongated and compact.

Leafy type Pak-choi (Brassica chinensis)

This resemble swiss chard in growth habit. The leaves are long dark-green and oblong or oval and do not form a solid head. It is also called chinese mustard. The midrib is broad and light in colour.

COMPOSITION

Chinese cabbage is a good source of protein, vitamin, carbohydrate and minerals (Bose *et al.*, 1999). The 100 g of edible portion contain: water 95.0g, protein 1.2 g, fat 0.1 g carbohydrate 3.0g, phosphours 40 mg, potassium 253 mg, riboflavin 0.047 mg, vitamin C25 mg, calcium 43 mg, niacin 0.6 mg, energy 145 K cal and vitamin A 150 I.U.

Uses

Chinese cabbage is used as pot herb and consumed fresh in salads or cooked as vegetable like other green vegetables (Singh and Singh, 2007). It can also be used as a substitute for lettuce and celery.

VARIETIES AND HYBRIDS

There are Different Types of Chinese Cabbage

(a) Loose leaved variety (var. *dissoluta* Li). (b) Semi-heading variety (var *in fraction* Li). (c) Fluffy topped heading variety (var. *laxa Tse* et Lee). and (d) heading variety (var. *cephalata* Tse et Lee). This variety is further classified into three type: (i) Ovate type (*f. ovata* Li), (ii) Flat topped type, (*f. depressa* Li) and (iii) Cylindrical (*f. cylindrica* Li).

Chinese cabbage varieties have also been classified into three groups according to growing seasons (Singh and Nath, 2011).

(i) Spring varieties (January to April)

(ii) Tropical varieties (April to July)

(iii) Autumn or main season varieties (August to September)

Characteristics of some important varieties and hybrids (Kalloo and Bargh, 1993) are given below:

Tianjin Qingmaye

It is grown in Tianjin area of China since very long time. It has good quality, low cellulose content, disease resistance, yield storage and transport quality.

Naebyung 60 Days

It is a early cultivar, has significantly more vitamin C (36.8 mg/100g fresh weight) than other cultivars.

Raseosamgye and Olympic

These two cultivars have highest percentage of marketable heads (85%) at harvest in comparison to others.

Nepos, Nestor, Nerva and Spectrum

These cultivars perform well and do not produce bolter when planting is done in cold glass houses from second week of March and harvesting is done in third week of May.

Chiifu, Kaga, Hotoren, Aichi and Kenshin

These cultivars are popular in China having good morphological character and perform good in their growing region.

Peking

It is a fairly late Japanese variety, released in late 1979. It has a tight head weighing about 3.0 kg and few leaves.

Chihili

It produces a tall narrow and cylindrical head.

Wong Bok

It is short and very chunky in shape.

In India, there are not many varieties reocommended for cultivation under local conditions. Punjab Agricultural University has released Chinese Sarson No. in 1987 for general cultivation in the state (Singh and Nath, 2011).

Chinese sarson No. 1

It is non-heading type with light-green, broad leaves and semi-erect plant habiti. Its mid-rib is white, succulent and tender. It takes about 30 days for the first cutting after transplanting. It is quite rejuvenating, giving 6-8 cuttings. Its average yield is 4 tonnes/ha. It posseses field resistance to alternaria leaf spot.

Hybrid 59

It is an early maturing, higher yielding and more heat tolerant than others.

Yuan Pao 2

It has been developed for commercial cultivation.

65 Days

It produces more than 80 tonnes per hectare yield.

Michihli

It is heading type give yield of 54.2 t/ha. Cylindrical head and tightly folded dark green leaves, medium maturity and uniform head. White midribs very crispy and tender.

Manha (Cheongbang x Yungching 30 Days)

It is a good performer and has moderate resistance to *Erwinia carotovora* and turnip mosaic virus.

Sambok (Yungching 60 Days x Ta Feng Feng Lu)

This hybrid shows moderate resistance to *Erwinia carotovora* and turnip mosaic virus and also had good performance.

Taoyuna Yashu 2

This F_1 hybrid is responsive to nitrogen fertilizer and high temperature and it is high yielder.

Sumo

It has dark green foliage, leaves hairless and shiny. Mature in 55 to 60 days after sowing. Head oval shape, weighing 1.8 to 2.0 kg. The crisp head has a pleasant flavour and is ideal for salad and good substitute of head lettuce. Resistant to virus, soft rot and downy mildew.

Tropic Early

It produces marketable heads within 40 days after transplanting in tropical low land areas. The head is very firm and the leaves are hairless. thick and very juicy. It is suitable for fresh consumption. frying and processing. It is tolerant to bacterial soft rot and mosaic virus.

Rex (CC 24)

Rex is adapted and cultivated in the tropical highlands. It mature in 65 days after sowing. The head is oval shape and light green in colour with an internal orange yellow colour and very attractive. It is resistant to club root disease and shows good tolerance to soft rot and downy mildew.

CLIMATIC INFLUENCE

In northern India, Chinese Cabbage is cultivated during winter (October to February) season. Because of its sensitivity to high temperature. Chinese cabbage cultivation in low land area is confined to the winter months. During the summer, this crop is grown at high altitudes on limited scale (Choudhury, 1983). It thrives best at temperature from 15°-21°C. Below this temperatue seed stalk may emerge before heads are produced (Thamburaj and Singh, 2003).

SOIL REQUIREMENT

Chinese cabbage grow best in sandy loam soil. The soil should be well fertile, abundant in moisture but free from water logging (Bose *et al.*, 1999).

The optimum pH range is 6.2-6.8. A good muck soil rich in nitrogen and retentive of moisture is ideal for growing chinese cabbage.

SOWING

The crop of chinese cabbage can be raised by direct seeding as well as by transplanting the seedlings. Direct seeding is more commonly followed as it is believed to give a head start over transplanted crop. Secondly, the crop is saved from the soft rot incidence because there is no injury that sometimes occour due to transplanting. Hand planter may be used to save labour. To raise one hectare crop, 500g seed is required through transplanting method and 2.5 kg for direct seeding (Singh and Singh, 2007).

Spacing varies according to cropping types and cultivars used. Generally, 37,000 plants/ha (60 x 40 cm) is standard for early maturing cultivars white 28,000 plants (70 x 50 cm) for medium maturating cultivates and 24,000 plants (75 x 55 cm) for late maturating cultivars. Punjab Agricultural University (PAU), recommendes a spacing of 30 cm between plants as well as rows, Thinning is very important to maintain spacing between plants in the rows. Seedlings are thinned at 5-6 leaf stage, about three weeks after sowing. While thinning, diseased, off-type and leggy seedlings are removed (Bose *et al.*, 1999).

PLANTING

Transplanting can be done either on flat beds of appropriate sizes or on ridges each accommodating 1-2 rows. Seedlings of 8-10 cm long are transplanted in the main field. A light irrigation is given just after the planting of seedlings in the main field.

NUTRITIONAL REQUIREMENT

Chinese cabbage is a heavy feeder and requires adequate manuring for profitable yield. 15-20 tonnes/ha of well rotten farm yard manures/compost incorporated at final ploughing in the field (Singh, 1978). Chinese cabbage requires N,P,K fertilizers in relatively large quantities as compared with other leafy vegetables. Punjab Agricultural University has recommended 15-20 tonnes/ha farmyard manure, 125 kg N and 75 kg each of P_2O_5 and K_2O per ha (Singh and Singh, 2007). International Potash Institute Berne, Switzerland, has recommended 150-360 kg K_2O/ha for chinese cabbage. Whole P_2O_5 and K_2O and 1/3rd N is applied as basal dose before transplanting and remaining 2/3rd N as top dressing at 30 days and 45 days after transplanting. Band placement of nitrogen fertilizer is recommended in standing crop followed by irrigation. Singh *et al.* (1966), stated that it is advisable to give a dose of nitrogenous fertilizer of about 25g to each plant of chinese cabbage after every cutting for enhancing fresh growth of leaves.

IRRIGATION

Crop of chinese cabbage requires a continuous supply of moisture for uniform growth and development of heads. The optimum growth of plant occurs at moisture content of the soil varying from 60-100% of field capacity with average of 80% (Thamburaj and Singh, 2003).

First irrigation is given just after transplanting for quick establishment of seedlings in main field. There after watering is done depending on the weather and soil conditions. Light soils require frequent irrigation than heavy soils (Bose *et al.,* 1999). During dry weather conditions, the frequency of irrigation is more. Irrigation is given at 8-10 days interval if there is no rain in the season.

INTERCULTURE

In chinese cabbage, intercultural operations like hoeing and weeding are important to keep the field free from weeds and for proper utilization of nutrients by crop. The roots of chinese cabbage are shallow and surface feeders and therefore deep hoeing should not be done (Choudhury, 1983). Generally, 2-3 heing and hand weeding is given before top dressing (application of fertilizers).

Mulching

Mulching is useful for the conservation of mosture and control of weeds and also additional organic matter, especially when plant materials are used as mulch, helps in activating micro-organisms in soil, improvement in the quality of vegetables and control of certain diseases and pests. Leaves, straw, hey, crop wastes, different kinds of paper or polythene may be used as mulch (Thamburaj and Singh, 2003). Black polythene mulch is effective in controlling weeds, conserving moisture, including better growth and higher yield of Chinese cabbage. Plastic mulch increase the soil temperatue and the yield. Mulching with biomass viz. hay, leaves, coffee husks reduce day temperature and increase the moisture content. Mulching especially with coffee husks has beneficial effect on growth and yield (Swiader *et al.,* 1993).

WEED CONTROL

Removal of weeds eliminate competition for moisture, nutrients, light and air. Weeds can be controlled either manually or by the use of herbicides or a blend of both (Singh and Singh, 2007). Chemical weed control is economic because it saves time and labour. (a) Pre-plant application of trifluraline @ 0.51/ha or basalin @ 0.5 1/ha give excellent weed control of monocot as well as dicot weeds with higher yield. (b) Pre-plant application of stomp 30 EC (1 kg a.i./ha) followed by first hoeings/weeding at the critical crop weed competition period 35-40 days after planting and seocnd light hoeing/weeding 60 days after planting is effective incontrolling most annual and broad leave weeds.

PLANT PROTECTION MEASURES

Important insect pests and diseases of chinese cabbage and their control is reported by Singh and Singh (2007) are given below :

Insect Pests

The serious pests of chinese cabbage are web worm (*Hellula undalis*), imported cabbage worm (*Pieris rapae*), cabbage looper (*Trichoplusia ni*) and aphids (*Brevicoryne brassicae*).

Cabbage web worm (Hellula undalis)

The caterpillar feeds on the growing point of the seedling. If a growing plant survives after this attack it forms several small unmarketable heads. Seedling in nursery beds should be covered with nylon net cage to prevent the adult moths from laying of eggs on the foliage. Spraying insecticides like endosulfan or malathion @ 0.2% is useful to control this pest.

Imported cabbage wrom (Pieris rapae)

The larvae feed on the leave and may completely skeletonised the foliage, thus reducing quality and quantity of yield. This pest can be controlled by Bt (*Bacillus thuringiensis*) spray in different formulation.

Cabbage looper (Trichoplusia ni)

The larvae feed on the foliage and reduces both quality and quantity of yield. As it is surface feeder, it can be controlled by application of endosulphan @ 0.2%. Formulation of *Bacillus thuringiensis* can also check the population. This insect is also susceptible to nuclear polyhedrosis virus (NPV).

Aphids (Brevicoryne brassicae)

The nymphs and adults suck plant sap and transmit virus diseases. Heavily infected plants are stunted and yield is reduced. They also excrete stickly 'honey dew' that coats the plant. Shooty moulds grow in this 'honey dew' reducing photosynthesis and making produce unattractive. Pirimicarb a selective insecticiede sprayed @ 0.5 kg ai/ha controls aphids of this crop. Use of insecticidal soap innundative, release of *Chrysoperia carrnea* to conrol aphid on broccoli is successful.

Diseases

Two bacterial diseases, black rot and soft rot and fungal diseases like powdery and downy mildew are most serious in Chinese cabbage.

Black rot

It is caused by *Xanthomonas campestris* pv. *campestris*. Colyledons of infected seedlings turn black, shrivel and drop off. Initial infection on true leaves is a small wilted or yellowed 'V' shaped area at the margin. The pathogen survives on crop debris and is seed borne. Hot water treatment of seeds (soaking in hot water for 20 minutes at 50°C) eliminates the pathogen. Resistant varietics are used where black rot is common in a locality.

Soft rot

It is caused by *Erwinia carotovara* pv. *carotovara*. This disease is most prevalent from early to late in the season and develops repidly under hot and humid conditions. Initial infection is usually found on the outer petioles that are in contact with root at the wrapping stage of the plants. The symptoms as water soaked areas develop in petioles, tissues become soft and then watery rot progresses. The infected tissue gives off an offensive sulfurous adour which in part may be due to the secondary bacterial growth. Avoid wounding plants during cultivation, harvest, packing and storing. Use of resistant cultivars is best solution of this problem.

Powdery mildew

This disease is caused mainly by *Sphaerotheca fuligenea* which becomes severe in slightly warm and humid conditions. Symptoms first appear as white spots on the surface of older leaves which enlarge and increase in number and cover both the surfaces of the leaves and stems. Spraying with karathane @ 2g/litre of water control the disease.

Downy mildew

It is casued by *Pernospora parasitica* The pathogen has a wide host range and also occurs in other cruciferous vegetables. The small yellow leaf spots of indefinite size and shape occur in leaf surface. The lesions are produced on the lower leaf surface and are covered with white downy mould of conidia and conidiophore. This disease is promoted by the cool and wet weather in winter months. It is common in young plants and is characterised by the appearance of purplish brown spots on the under side of the leaves and upper surface of the leaf. Spraying with dithane M-45 @ 2 g/litre in early stage of infection helps to some extent in controlling this disease. Sanitation and crop rotation help in reducing the infection of the disease.

HARVESTING

Chinese cabbage is harvested when the heads are fully developed. It is generally become ready for harvest after 80-90 days of planting. The main criterion of judging when to harvest heading types is that the heads must be fully developed and firm and leafy types should develop proper size of leaves and they should be tender (Bose *et al.*, 1999). At maturity the whole plant is cut form the base. Heads are cut with *a sickle,* after removing diseased or injured outer leaves (Singh and Singh, 2007). The harvested plants should not be exposed to sunlight because of excessive wilting. Then the harvested heads or leaves are packed and send to market immediately.

YIELD

Yield depends upon cultivars and maturity period of the crop. Yield varies from 10-15 tonnes/ha.

POST HARVEST HANDLING

Grading

Grading makes orderly marketing and ensures high price for the well graded products. Before grading, the heads are uniformly trimmed to the same number of wrapper leaves to bring uniformity of the product. Grading of chinese cabbage is done on the basis of size and firmness. Generally, best quality full develop firm and tender heads are arranged in 'super' grade. The medium size heads with unifrom shape, good lustre and free from spots are kept in 'fancy' grade. Over sieze heads with dull colur and loose frame and free from diseases and pests are arranged in 'commerical' grade (Singh and Singh, 2007).

Packing

It is an important consideration in the process of marketing. Proper packing makes transportation and retail sale of the chinese cabbage more oncvenient. It is also increases the shelf life of the product because of the protection provided by the packages from external injuries and from being shrivelled (Singh and Singh, 2007). Generally , all kinds of packages are used for packing chinese cabbage. Heads are packed in plastic bags or corrugated boxes as required by market. Heads are usually laid in packages rather than placed upright. Non heading types leaves are packed in crates or baskets and send to local market. Sometimes flat baskets, boxes and various kinds of crates are also used to pack the produce (Thompson and Kelly, 1983).

Storage

Chinese cabbage can be stored in the field itself under extreme cold conditions. After harvesting plants are arranged closely in 3-4 rows length wise and is covered with straw matting. In warmer climates, produce is stored in a refrigerator or in a warehouse. The shelf-life of chinese cabbage is highly influenced by pre-harvest factors. Sunshine and high temperatue before harvest increase weight losses dring storage and adversely affects marketable quality. It can be stored for several weeks at 0ºC and of about 95% relative humidity in cold storage (Swaider *et al.,* 1993). The self life of chinese cabbage can be improved by low O_2 (2%) in combination with low CO_2 (2%).

SEED PRODUCTION

Chinese cabbage is a cross-pollinated crop. It produces good quality of seeds under north Indian plains. Both head to seed and seed to seed methods are used for seed production. The isolation distance of 1600m is given between varieties of chinese cabbage as well as the varieties of other cruciferous crop, to produce pure seeds. The corp is grown in rows of 60 cm apart and 30 cm between plants. In *petsai* for head to seed crop the enveloping leaves of each chinese cabbage head is removed to expose the undamaged interior growing point. Roguing during vegetative and flowering period is important for successful seed crop (Singh and Singh, 2007).

The mature seed pods are harvested when they turn into light brown colour. The individual seed stalks are harvested and placed on or over tarpaulins. When ripening is completed the seeds are threshed. The threshed seed is dried in the sun down to about 10 percent moisture content. Singh and Singh (2007) average seed yield of Chinese cabbage is 850 kg/ha.

REFERENCES

Bose, T.K., Som, M.G. and Kabir, J. (1999). Vegetable crops (eds.). Naya Prokash, Calcutta-6.

Choudhury, B. (1983). Vegetables. National Book Trust, New Delhi, India.

Kalloo, G. and Bergh, B.O. (1993). Genetic Improvement of Vegetable Crops. pergamon Press Oxford, UK.

Singh, D.N. and Nath, Vishal (2011). Varieties and Hybrids of Vegetables. Satish Serial Publishing House, Azadpur, Delhi, India. pp. 41-43.

Singh, D.N. and Singh, S.P. (2007). Production Management of Underutilized Vegetables. Agrihortica Publication, Junagadh (Gujarat), India. pp. 71-80.

Singh, Kirti (1978). Manurial Requirements of Vegetables. Indian Council of Agricultural Research, New Delhi.

Singh, Kirti; Thakur, M.S. and Singh, J. (1966). Do you Know these uncommon leafy vegetables. *Progve. Fmg.* Ludhiana, 3(2): 17.

Swiader, J.M.; Ware, G.W. and M.C. Collum, J.P. (1993). Producing Vegetable Crops. International Book Distributing Co., Charbagh, Lucknow, India.

Thamburaj, S. and Singh, Narendra, (2003). Vegetables, Tuber crops and species. (eds) Indian Council of Agricultural Research, New Delhi.

Thompson, H.C. and Kelly, W.C. (1983). Vegetable Crops. Tata Mc Graw-Hill Publishing Company Ltd, New Delhi.

□□□

7. KALE

Botanical Name	:	*Brassica oleracea* var. acephala DC.
Family	:	Brassicaceae
Chromosome number	:	2n = 18

Kale (Karam-sag) is an underutilized vegetable. It is not yet commercialized in any part of India. It is introduced in nineteenth century in India (Anon., 1980)

ORIGIN

The origin place of kale is probably southern and western Europe (Knott, 1988).

DISTRIBUTION

In India, kale is grown as small scale only in Assam, Kashmir, Maharashtra, Gujarat (Baroda area), Himachal Pradesh, Uttar Pradesh, Uttaranchal, Punjab, Haryana, Niigiri hills and near big cities (Singh and Singh, 2007).

BOTANY

Kale is a biennial but cultivated as an annual. Rosette of leaves and young shoots at the top of the stem which resemble cabbage prior to heading (Nick and Bradley, 1999). It does not form a head like cabbage.

COMPOSITION

Kale contains water (85%), energy (149 k cal), proteins (6.4%), fat (0.5%), carbohydrate (33.1 g), fibre (1.5%), calcium (141 mg), phosphorus (153 mg), iron (1.7 mg), sodium (17 mg), potassium (401 mg). It contains vitamin A (8,900 IU), thiamin (0.11 mg), riboflavin (0. 13 mg), niacin (1.0 mg), ascorbic acid (120 mg) and vitamin B_6 (0.27 mg). The composition is based on 100 g fresh edible portion (Anonymous, 1980).

Uses

The leaves of kale are cooked similar to cabbage greens or in stir fries. Kale is used as winter vegetable when other vegetables scare. The leaves are cooked as single or mixed with other vegetables (Thompson and Kelly, 1981).

VARIETIES AND HYBRIDS

Singh and Nath (2011) reported that following are important varieties of kale:

Dwarf Varieties

Dwarf Green Curled Scotch, Dwarf Moss Curled.

Medium Tall Varieties

Moss Curled, Fiamburgen Market.

Tall Varieties

These are comparatively less productive. Hybrids mature earlier and tend to be more productive. Scotish and Siberian varieties are more suited for temperate regions.

Karam Saag

It is grown in Jammu and Kashmir.

CLIMATIC INFLUENCE

Kale is temperate vegetable crop. Kale is the hardiest crop and can withstand temperature as low as 10° to 15°C (Thompson and Kelly, 1981). The quality produce is obtained under cool temperature. A few varieties are less hardy, can be grown in sub-tropical climate. It prefer full sun.

SOIL REQUIREMENT

Kale is grown in a well drained sandy loam soil and it is ideal soil for higher production. It is more tolerant to salt. Land is prepared upto good tilth (more pulverized). The soil pH ranges from 5.5 to 6.8 (Singh and Singh, 2007).

SOWING

Kale crop is raised by planting of seedlings. Therefore, seeds are sown in nursery beds. Sometimes, crop is also raised by sowing of seeds directly in the field (March-May, August-November on the hills and August-November in the plains).

The best time of sowing is in the months of August-September. About 350 to 400 g seeds is sufficient to raise seedling for growing one hectare crop (Singh and Singh, 2007). For direct sowing, 2.0 to 2.5 kg/ha seed is needed.

PLANTING

The seedlings become ready for planting 5-6 weeks after sowing seeds in the nursery beds. The best distance of planting in the main field is 45 cm x 30 cm or 60 cm x 50 cm (Sharma and Srivastav, 2004). Two to three seedlings are planted at a place as per given spacing and sometime wet land planting is also preferred to avoid mortality of the seedlings. So that maximum plant population can be maintained (Singh and Singh, 2007).

NUTRITIONAL REQUIREMENT

Kale is heavy feeder of nutrients. It respond better to organic manure and therefore, 15-20 t/ha farm yard manure or compost is incorporated during field preparation. Besides, 130-150 kg nitrogen, 60-80 kg phosphorus and 40-60 kg potassium per hectare are given, depending on soil test (Singh and Singh, 2007). Nitrogen is applied into two splits, phosphorus and potassium are applied as basal dose i.e., just before planting. First half dose of nitrogen is top-dressed 25-30 days after planting and the second half dose of N is applied 45-50 days after planting in form of foliar feeding.

IRRIGATION

As soon as transplanting of seedling is complete, light irrigation is given. The subsequent irrigations are given according to season, stage of growth, type of soil and other management practices. Accordingly, interval of irrigation is decided. Normally, Kale crop is given irrigation at 15-20 days interval (Singh and Singh, 2007). It can tolerate drought for short period, without affecting yield significantly.

INTERCULTURE

Generally, One or two shallow hoeing is given to keep the soil loose for proper growth and development of the plant and to keep the plot clean. First hoeing is given at 15-20 days interval before topdressing and second hoeing is given at 30-35 days just before application of second dose of fertilizers (Nick and Bradley, 1999).

WEED CONTROL

Weeds may pose problem and therefore, one or two hand weeding is done just after hoeing of the plot (Singh and Singh, 2007). Kale plants cover the soil later on, then weeds suppressed automatically.

PLANT PROTECTION MEASURES

Insect pests and diseases are not problem in kale cultivation.

HARVESTING

Kale become ready for harvesting 90-120 days after planting. The period of harvesting is November to January. Harvesting is done by removing whole plant (Singh and Singh, 200&0.

YIELD

Yield varies according to varieties, planting season and management practices etc. (Singh and Singh, 2007) reported that about 150 to 250 quintals of yield is obtained.

POST HARVEST HANDLING

Just after harvesting, it is desirable to sort out healthy and quality leaves. Leaves are bundled and packed in bamboo baskets. The produce is sent to market for sale without any delay because it is highly perishable in nature.The storage life of leaves is very short under ambient condition. However, leaves can be kept under refrigerated storage for longer period (Singh and Singh, 2007).

REFERENCES

Anonymous. (1980). The Wealth of India: Raw materials and industrial products. Vol. 3. *Council of Scientific and Industrial Research,* New Delhi.

Knott J.E. (1988). Handbook of Vegetable Growing . John Wiley and Sons, New York.

Nick, J.M.A. and Bradley, F.M. (1999). Growing fruits and vegetables organically, Rodale press, Emmaus, Pennsylvania.

Sharma, R.R. and Srivastav, M. (2004). Plant propagation and nursery Management. International Book Distributing Co., Lucknow.

Singh, D. N. and Nath, Vishal (2011). Varieties and Hybrids of Vegetables. Satish Serial Publishing House, Delhi. pp 43-44.

Singh, D. N. and Singh, S.P. (2007). Production management of Underutilized Vegetables. Agrihortical Publication, Junagadh, Gujarat.

Thompson, H.C. and Kelly, W.C. (1981). Vegetable crops. Tata Mc.Graw Hills Publishing Co. Ltd., New Delhi.

□□□

Chapter-3

Leafy Vegetables

1. SPINACH

Botanical name	:	*Spinacia oleracea* L.
Family	:	Chenopodiaceae
Chromosome number	:	2n = 12, 24

Spinach is known as vilayati palak, an important potherb or leafy vegetable, of which the leaves and tender shoots are consumed fresh or processed and grown in the USA and Europe. The shape of leaf is quite different from that of palak. The name "spinach" is derived from Spanish word *ospinaca* (Ryder, 1979). The edible part of spinach is a compact rosette of leaves . It is liked by the common people as a cooked vegetable (Nath *et al.*, 1987). The crop is more common in the hilly areas than other parts of India (Kohli, 1993).

ORIGIN

Spinach is native to central Asia (Southwest Asia), most probably persia (Iran). This was taken to spain by Moors from where it spread to other parts of the world (Choudhury, 1967). It was known to China as early as the seventh century as well as to the ancient Greeks and Romans and grown in Persia about 2000 years ago. It was introduced into North America by Arabs (Shoemaker, 1947).

DISTRIBUTION

The cultivation of spinach spread to the rest of the world from Spain (Pandita and Lal, 1986). It is grown in U.S.A, Europe, Rome, China, Greece, Persia and Arabian countries. In India it grows only in hilly areas of colder place. Now spinach is grown all over the plains. However, it is popular in the southern states also (Nath *et al.*, 1987).

BOTANY

Spinach is a herbaceous annual plant, for leaf production and biennial for seed production. It produces rosettes of fleshy leaves, which may be crinkled or smooth in the vegetative phase. Later the stem elongates and form flower stalks during the reproductive phase, with narrow pointed leaves (Pandey and Kalloo, 1993). Flowers vary from six to twelve per cluster but all of them do not develop at the same time. Male plants bolt and flower earlier than female plants and die soon after flowering (Rosa, 1925). Flowers may be staminate, pistillate, or hermaphrodite and remain receptive for a week or long (Mac Gillivray, 1961). Spinach is dioecious bearing male and female flowers on different plants, in addition to monoecious with various degrees of male and female flowers on the same plants. (i) Seeds are two types : rounded seeded (summer spinach) and (ii) prickly seeded (winter spinach).

Plant breeding

Spinach is a cross-pollinated and wind-pollinated. Polyploidy breeding has been reported in spinach. Heterosis upto 30-40% has been observed in F1 hybrids and utilized for development of downy mildew and mosaic resistant hybrids and cultivars. No significant effort has been made to improve upon the old local strains available in various parts of the country. Some of the varieties introduced from the western countries are being grown in certain parts where it is commercially produced. Bhagchandani *et al.* (1979) reported that additive gene effect was observed for leaf size and partial dominance for yield. The inheritance of leaf size depended on a minor gene. Nine gene groups were involved in the inheritance of yield.

Taxonomy

Spinach is quite different from spinach beet, but it is related to Swisschard, sugar beet, table beet, pig weed and salt beet. Two old and primitive forms of *Spinacia oleracea* were described by Girenko (1968) as *Spinacia tetranda* and

Spinacia turkestanica, they are diploid relative of Spinach and also have 2n = 18 chromosomes. The both species tetranda and turkestanica grown as wild in north west India and Nepal. However, Mehra and Malik (1963) have reported 2n = 12 chromosomes in *S. suaeda* species and concluded that the basic number of the chenopodiaceae family is X = 9.

Floral biology

Spinach plants produce small flowers. Pollens are small in size and light, produced in abundance and carried away by air currents. Pollination takes place by wind. Unpollinated pistillate flowers remain receptive for 2 to 3 weeks (Shoemaker, 1947).

There are 5 main sex classes observed by Pandita and lal (1986), described as below:

Extreme male

This type of plants, produce only male flowers and are practically useless because of their bolting. These plants do not produce seeds.

Vegetative male

These plants bear male flowers and bloom at the same time as the classes which follow but do not produce seeds.

Monoecious plant

The plants bear both female and male flowers on the same plant and produce seeds.

Hermaphrodite plant

Bisexual flowers having both male and female parts are produced by these plants. They may also bear male flowers and these plants produce seeds.

Female plants

These plants bear only female flowers and produce seeds.

Sex development in spinach is governed by many factors. Such as :

a) As increase in nitrogen level decreases the proportion of female plants. The female plants have more leaves than male plants (Quagliotti, 1972). However, femaleness can be enhanced, by cutting shoot of young plants and allowed to form adventitious roots.

b) The maleness is enhanced, by treating the plants with GA_3 25 mg/l after cutting the leaves. Hence, leaves play an essential role in sex expression and this role is related to gibberellins synthesis in the leaves

(Chailakhyan and Khryanin, 1979). It was observed that the treatment with gibberellin and CCC influenced the formation of reproductive organs through regulation of growth processes but sex was not determined directly (Cajlahjain *et al.*, 1969).

c) Photoperiod and plant growth regulators showed marked effects on the sex expression. Sex ratio (male to female) is near 1 in long days. The number of female plants flowering in long days can be, increased by gibberellins, whereas kinetin reduced flowering in male plants. Abscisic acid and Amo 1618 reduced flowering in female plants. In short days gibberellins stimulated flowering in female more than in male plants and at higher temperature flowering (maleness) is reduced (culafic and Neskovic, 1980).

COMPOSITION

Spinach is a very cheap source of vitamin A, B and C, iron and calcium and also contain appreciable quantity of ascorbic acid and riboflavin (Oomen and Grubben, 1978). It is also contains protein, carbohydrate and mineral (Nath, 1967). According to Aykroyd (1956) Spinach contains (Per 100 g of edible protein portion) following:

Moisture	92.1g	Carbohydrate	2.9g	Vitamin A	9300 IU
Fat	0.7g	Sodium	58.5 mg	Nicotinic acid	0.5 mg
Fibre	0.6g	Iron	10.9 mg	Riboflavin	0.07 mg
Calories	26	Copper	0.01 mg	Vitamin C	23.0 mg
Protein	2.0 g	Potassium	206.0 mg	Sulphur	30.0 mg
Minerals	1.7 g	Oxalic acid	658.0 mg	Magnesium	84.0 mg

Uses

Spinach is normally used as a cooked vegetables. It has got some medicinal value also. The edible part is a compact rosette of leaves. It is also used as salad in combination with lettuce and other vegetables. Spinach is also processed, primarily canned and fried (Singh and Dhaliwal, 2001). Usually leaves, petals, stalks are used for processing.

VARIETIES AND HYBRIDS

Spinach varieties have been classified on the basis of seeds and leaves as: (i) prickly- seeded and (ii) round seeded. According to Mac Gillivray (1961) spinach is of two types, prickly-seeded and (ii) smooth seeded. The prickly seeded spinach is an old one and was used in Europe in 1931, where as the smooth seeded was known in 1952.

Classification on the basis of leaves characters: (1) smooth leaves, (2) savoy-leaves.

Spinach cultivars have also been classified in different ways.

According to Kinney (1986) cultivars, classified into 4 groups : (i) Norfalk or Bloomsdale (ii) Round leaved (iii) Thick leaved (iv) Prickly seeded.

Geise and Farley (1929), classified spinach cultivars into 2 groups:

(a) Flat-leaved type and (b) savoy-leaved type. Each class was further sub-divided into 3 sub-classes: (i) short-season group (ii) long standing group (ii) slow-growing group.

Magruder *et al.* (1938) classified spinach as follow :

I) Leaves savoyed

A) Leaves heavily savoyed

- i) Plants late seeding
  - a) Leaves blue green
  - b) Leaves dark green
- ii) Plants not late seeding
  - a) Plants very early seeding
  - b) Plants medium-early seeding

B) Plants medium – savoyed

II. Leaves not savoyed (or slightly savoyed in cool weather)

A) Plants early seeding

- i) Seeds prickly
- ii) Seeds not prickly

B) Plants not early seeding

- i) Seeds prickly
- ii) Seeds not prickly
  - a) Leaves blue-green
  - b) Leaves medium green

According to Drewes (1943) spinach cultivars can be classified as follows:

I. Round-seeded

1. Smooth-leaved (or slightly savoyed in cool weather)

- i) Early bolting or not long –standing
- ii) Slow bolting or long-standing

2. Savoy leaved

i) Early bolting or not long-standing

(a) mosaic resistant (b) not mosaic resistant

ii) Slow bolting or long standing.

II. Prickly seeded

1. Smooth leaved (or slightly savoyed in cool weather)

i) Early bolting or not long-standing

(a) mosaic resistant (b) not mosaic resistant

ii) Slow bolting or longs- standing

2. Savoy-leaved

i) Early bolting or not long standing

(a) mosaic resistant (b) not mosaic resistant

ii) Slow bolting or long standing.

Ryder (1979) classified spinach cultivars on the basis of leaf type, into 3 groups (I) savoy (wrinkled leaf) (ii) semi-savoy (iii) smooth.

Detail characteristics of varieties and hybrids described by Singh and Nath (2011) are given below:

Virginia Savoy

This is prickly seeded cultivars having large dark green leaves with round tip. Plants are upright and have vigorous growth.

Early Smooth Leaf

Smooth seeded cultivar and produce small light green leaves with a pointed apex (anonymous, 1998).

Sporter

It is developed by Bejo Sheetal Seeds and high yielding F_1 hybrid. It has broad leaves without fibres and becomes very soft on cooking. It matures in 45 days and suitable for cultivation in both kharif and rabi season. It is tolerant to mildew races, 1, 2 and 3.

Spinach Early

It is developed by Tropica Green Co. and plant is upright with deep green leaves. It is early type and is harvested all the round year in tropical high lands. Spinach early is resistant to downy mildew races 1 and 3, heat and cold and also medium bolter.

Varieties/Genotypes/Hybrids resistant to Different Diseases

Downy mildew

Several genotypes including varieties, hybrids or species have developed as resistant against different races of downy mildew i.e. Early Giants, Hybrids 320, 308, Vikimum, Dorema, Matares, Vital, Fall green, Noord Brabant, West Mass, Triathlon, Triade, Hermes, Triangel, Trisolde, Solar, Tritan, Tristar, Correnta, Mazurka, Valta, Achilles, Atorasu, Aubade Hybrid, Capella Hybrid, Califlay Febris, Gurobarus, , G-one, Nores, Orai, Sevenade. Hybrid Soloman, Samson, Sanraito, EH7, EH-10, EH 424, EH 425 (Webb and Thomas, 1976).

Fusarium wilt (Moderate level of resistance)

The sources hybrids lines EH 7, EH 10, Hybrid 50, EH 424, EH 425. Ozarka, Green Valley and WRG 70-5 are sources of resistance to wilt (O'Brien and Winters, 1977).

White rust

Varieties/hybrids/lines exhibiting resistance to white rust caused by *Albugo occidentalis* Wilson, are Fall Green, Green Stand, Hybrid 178, WRG 70, Green Valley 11 x PH 1452, 4C x 119, Coho (Black and Dainello, 1986).

Mosaic (CMV-1)

Resistance source – Dixie Market and Fall Green Roga, Lavires and Mona Lisa.

Beet yellows

Transmitted through aphids (*Myzus persicae* Sulz), but may be transmitted by other aphids and mechanically also. Dutch varieties, least susceptible – Roga, and Ravires (Chod, 1985).

Beet mosaic

Least susceptible lines-Novera, Estivato and Mona Lisa. Roga and Lavires are resistant to beet yellow (Chod, 1985).

CLIMATIC INFLUENCE

Spinach is strictly winter-season crop and withstands freezing weather better than other crops. It tends to produce flower at the cost of leaves in long days (Chauhan, 1972). A low temperature or cool night makes plant more erect and savoying in leaves. Maximum yield is obtained under short day and moderate temperature. The crop grows best between temperature range of 15 and 26°C. The crop is both day-length and temperature sensitive. Hence, it's cultivation is best suited to the hills in temperate regions (Pandita and lal, 1986). High atmospheric humidity favours succulent and tenderness of foliage. Thus during early growth temperature of 15-21.1°C is required and seed germinate well at temperature of 10-15.5°C. Seed sowing at 1.5 cm deep at 25°C soil temperature give early emergence. According to Thakur *et al.* (1989), high temperatures about 12°C and long day conditions induce early bolting in the crop.

SOIL REQUIREMENT

Spinach can be grown on a wide range of soil. It should be fertile and well drained but produces good crop in lighter soils. It does very well in sandy loam soil. Spinach is susceptible to high acidity, hence, like spinach beet, it tolerates salinity pretty well. Thakur *et al.* (1989) reported that pH of 6.0 to 6.8 is good for successful crop production.

SOWING

Field Preparation

For raising spinach crop, soil is prepared by 3 or 4 ploughings followed by planking so that soil is well pulverized and levelled. After preparing soil thoroughly, beds of convenient size and irrigation channels are made. Sometimes before sowing, seeds are soaked in water overnight to hasten germination. Seed treatment with sulphuric acid also enhances the germination.

Sowing Time

Sowing time of spinach varies from place to place. In plains winter crop is sown in September-October (Nath, 1976) and in the hills, appropriate time of sowing is from July to September and March to June in the high hills (Anon., 1990). Prickly-seeded cultivars are best suited for autumn-winter crop in the hills, whereas smooth seeded cultivars are suitable for spring-summer in the hills and autumn sowing in the plains (Joshi and Joshi, 1960). Early sowing gave higher yield whereas the fresh and dry weight of plants were decreased with late planting but Vitamin C and Oxalic contents were increased.

Seed Rate

37-45 kg of seeds are sufficient to sow one hectare land. The seed rate is relatively higher because about 50% of the plants come out to be of male type with poor growth, which may be removed (Nath, 1976; Pandita and Lal, 1986).

Spacing

Spacing of 25 cm row to row and 10 cm within row is optimum (Anonymous, 1982).

Method of Sowing

There are two method of sowing (Saini, 1986), which are described below:

Broadcasting method

In this method seeds are broadcast in well levelled beds and mixed with soil thoroughly. This method is not so common and not desirable.

Line sowing method

It is more suitable method which facilitates weedings, hoeing, intercultural operation and cutting of leaves.

Seeds are sown, at a depth of 3-4 cm in the rows about 20-25 cm apart and covered with loose soil properly. If optimum soil moisture and temperature are available, seeds will germinate in about 10 days from sowing. In case of insufficient soil moisture, as light irrigation should follow in rows to ensure good germination.

NUTRITIONAL REQUIREMENT

Spinach requires more nitrogen and higher nitrogen level increases the yield and dry matter contents of the tops (Verma *et al.*, 1969; choudhary *et al.*, 1974). The spinach crop yielding 7.904 kg removes 60 kg nitrogen, 15 kg phosphorus and 50 kg potassium(Jacob and Uexkull, 1960). The yield is closely related to available nitrogen and it was increased by nitrogen in combination with phosphorus and potash; increased drymatter, crude protein and vitamin C content in spinach as reported by Cervato (1969) and Stanilova *et al.* (1972). Nitrogen application also influenced the yield of spinach under saline condition (Dhillon *et al.*, 1987).

Mehta (1959), is of the opinion that spinach requires 85-112 kg nitrogen per hectare to be given partly in the form of well-rotted manure and partly as ammonium sulphate in several doses after each cutting. According to Singh and Joshi (1960), a basal dose of 10 cartloads of farmyard manure and three top-dressings each of 112 kg of ammonium sulphate per hectare should be given to the crop. The maximum yield of spinach was obtained by Verma *et al.* (1969) by using 200 kg nitrogen/ha. Leaves having less oxalic-acid content which would provide enough availability of calcium in the body of human beings. The highest production efficiency of 462.10 kg green leaves per kg of nitrogen added with a net return of Rs. 30.81 per rupee invested on fertilizer was obtained by spraying 30 kg nitrogen/ha as urea alone (Mehrotra *et al.*, 1970).

Application of balance dose of 150 kg N, 88 kg P_2O_5 and 88 kg K_2O/ha give higher yield and quality spinach leaves. A basal dressing of 35-40 t/ha farm yard manure should be incorporated in the soil at the time of land preparation. Fertilizers should be applied in two splits i.e., half at sowing and half one month later, which are more effective than single application at sowing (Vergni and Huguet, 1971). A top dressing of about 20 kg N/ha after each cutting gives quicker growth of green, tender succulent leaves. 300 kg CAN, 310 kg superphosphate and 50 kg muriate of potash has been recommended alongwith 10 tonnes of FYM. CAN should be applied in two splits, half at the time of planting and other half after first cutting (Anon., 1990).

IRRIGATION

Spinach seeds require good moisture for germination therefore, it is necessary to give a light irrigation immediately after sowing if there is not sufficient soil moisture in the field. The second irrigation may also be given to soon after first irrigation to secure uniform and maximum germination. Spinach being a shallow rooted crop requires frequent irrigations. One irrigation at the time of sowing and subsequent irrigations at 10 to 15 days interval are needed for a good crop (Kohli, 1993).Ample soil moisture is important near harvest, which improves yield and quality of produce (Mac Gillivray, 1961). Deficit irrigation rates particularly, 50% ETC appeared to significantly reduce the leaf quality. Hence irrigation at 75% and 100% of ETC has highest water use efficiency (Daniel and Piccini, 2005).

INTERCULTURE

Regular light hoeing is necessary to keep soil loose for proper roots and shoot growth. The first light earthing up is given to keep plant in proper position. Then subsequent earthing up after one month at the time of top dressing is given.

Crop Rotations

According to Saini (1996), different crop rotation can be taken successfully for proper utilization of land and to get good profit. These rotation are given below:

1. Okra-Spinach-French bean
2. Cowpea-Spinach-Cluster bean
3. Bitter gourd - Spinach - Okra
4. Geengram-Spinach-Amaranthus
5. Amaranthus-Spinach-French bean

6. Maize-Spinach-Chilli
7. Cluster bean-Spinach-Brinjal
8. Cauliflower(Early)-Spianch-Muskmelon
9. Blackgram-Spinach-Bittergourd
10. Cowpea-Spinach-Watermelon

WEED CONTROL

In India generally weeds are controlled by hand weeding (Kohli, 1993). Two to three hand weeding is done to keep plot weed free. For large scale cultivation some weedicides have been recommended to control weeds which are economical and cheaper. Asulam @ 6 or 8 lit/ha give good weed control (Verlat, 1973). Application of alachlor and lenacil is more effective (Cantiliffe and Pathak, 1974). Pre emergence application of Hercules 22234 and benthiocarb give best weed control. Application of phenmedipham as post emergence @ 2.2 kg/ha give good control against broad leaved weeds.

Use of Chemical and Growth Substances

a) Seed treatments with sulphuric acid or bichromate mixture (NA + K) which destroys the endogenous inhibitors in the outer seed layers and enhances germination.

b) Application of micronutrients such as manganese and boron at 4.5 and 2.2 mg per one sq. m. enhance ascorbic acid carotene contents (Hulewicz and Mokrzecka, 1971).

c) Zinc (500 and 1000 ppm) in combination with GA_3 increases fresh and dry weight of Spinach (Omran *et al.*, 1973).

d) Application of GA_3 @ 200 ppm increase vegetative growth of plant.

e) Application of GA_3 10 ppm with tween 20 (surfactant) @ 500 ppm improved the petioe length, stem height and leaf blade area (Mini *et al.*, 2004).

f) Spray of GA_3 alone @ 20 ppm increased yield, facilitated harvesting and handling and improved keeping quality of spinach (Gonzalez, 1980).

g) Spray with $ZnSO_4$ @ 500 ppm + GA_3 @ 50 ppm increased seed yield (Omran *et al.*, 1973).

PLANT PROTECTION MEASURES

Insect Pests

Spinach crop is not damaged severely by insects-pest. Some of the important pests are aphids, beetles and caterpillars, damage the crop. They can be contolled by (I) spraying any insecticide like Endrin and Malathion. (ii) Spraying with Nicotine sulphate or dusting with 0.2-0.3 per cent Pyrethrum should be preferred. (iii) Crops which are kept for seed are usually attacked by aphids during January-February. This, however, can be controlled by spraying of malathion (2ml/litre).

Diseases

Spinach is damaged by several diseases such as damping-off, leaf spot and downy mildew.

Damping-off

It is caused by fungi *Pythium* sp. Seedlings are attacked by the fungus and killed. This disease can be controlled by treating seeds before sowing with captan, ceresan, or thiram @ 2.5 g/kg of seeds.

Leaf Spot

It is caused by fungus *Cercospora beticola*. Brown, dead circular spots are found on the diseased leaves. (i) This disease can be controlled with spray of 0.2 per cent dithane M-45 (0.2%) or benomyl (0.2%) at 15 days interval. (ii) Resistant cultivars of prickly-seeded type should be grown (Singh and Agnihotri, 1971).

Downy Mildew

This disease is caused by the fungus *Perenospora effusa*. Infested leaves show light yellow areas on the surface and young plants have pale green stunted and crinkled leaves. This disease can be controlled by the spray of 0.2 per cent dithane Z-78 thrice of the interval of 15 days.

White Rust (Albugo candida (Lev.) Kunze)

Rust like white raised pustules are formed on lower surface of the leaves and opposite each pustule on the upper surface a yellow patch is developed. More than 50 per cent yield is reduced due to this disease (Dainello *et al.*, 1981). This disease can be minimized by maintaining good sanitary conditions and distruction of infected materials. The control of this disease are spraying of difolatan (0.3%) , daconil (0.1%), dithane M-45 (0.2%), ridomil (0.1%) or aliette (0.1%) (Sharma and Sohi, 1982).

HARVESTING

The crop is ready for first cutting 3-4 weeks after sowing (Kohli, 1993). Harvesting can be manipulated as per requirements and varies from 5-6 leaf stage to just before seed stalk formation. A plant with seed stalk is considered unmarketable. Plant is cut just above the root. It is then washed, trimmed, graded and bunched or bagged.

Usually cutting is done manually and it gives higher yield than mowing. It gives 3-4 cutting in the season. Harvesting should be done in evening hours to protect leaves from dew, which cause crisp and break of leaves.

YIELD

The average yield of green leaves is about 50-60 q/ha.

POST HARVEST HANDLING

After harvesting all diseased, broken and yellow leaves should be removed from the produce.

Grading

The leaves are graded according to size, shape, and colour. The best quality leaves with uniform size and shape having well develops colour are arranged in "super grade". Medium type leaves with uniform shape and size having good texture, are kept in "fancy grade" and small leaves with free from deformities and spots alongwith with normal shape and fair texture are arranged in "commercial grade". Then the leaves are tied in small bundles according to grade for easy handling and marketing.

Packing

Spinach is packed for "shipment" in various types of containers including the round bushel basket, hamper and crates. For long hauls, crushed, ice is placed in the container and shipment is made in refrigerator cars or trucks. As a rule the quality of ice placed in the containers is about equal to the weight of the spinach (Thompson and Kelly, 1983).

A considerable portion of the spinach sold in retail stores is prepacked bags made of transparent films, which allow, gas ecahange but maintain high humidity by reducing evaporation of water (Thompson and Kelly, 1983). The above system of packing is done only in advanced country. In India, spinach is packed in gunny bags, bamboo baskets, cartoon, plastic tray, plastic crates and vegetables trays for market sale.

Storage

Storage of spinach is a problem and cannot be stored for a long time except for few hours under oridinary conditions. However, the crop after harvesting can be stored for 10 to 14 days at 0°C with 90-95% relative humidity (Nath, 1978). Lower temperature is best for spinach storage (Guerrero and

Valencia, 1972). The quality of spinach remained good for about 35 days when stored at 1°C (Ogato *et al.*, 1974). Storage of green with vaccuum pre-refrigeration and at 1°C temperature in a modified atmosphere (in plastic packing with a diffusion window) increases the storage life upto 40 days (Andre *et al.*, 1980). The recommended shipping and storage conditions are 0°C and 90-95 per cent relative humidity. Spinach can be stored for 10-14 days at above mentioned conditions (Singh and Dhaliwal, 2001).

Processing

For processing, spinach is to cut above the growing point so that only leaves are harvested and allows the plant to grow for subsequent harvsting. It gives 3-4 cuttings, then transported to the processing plant for processing and or canning. Spinach has a large leaf surface to weight ratio and has very high respiration rate. It is therefore, essential to cool the harvested product immediately and rapidly to prevent wilting and weight loss. Spinach may be cooled from 19.4-20.8°C in 10 minutes by vaccuum cooling (Singh and Dhaliwal, 2001).

SEED PRODUCTION

Spinach is a cross pollinated crop so it requires isolation distance of 1600m for foundation seed production (Anon, 1971) and 1000 m for certified seed production between varieties. For seed crop field preparation, manurial and other cultural requirements are the same as for green leaf crop.

Seed Sowing

Singh and Joshi (1960) observed that prickly seeded varieties are better suited for the autumn-winter crop in the hills. For the spring summer sowing in the hills and autumn sowing in the plains the smooth seeded varieties which have other obvious advantages easy threshing, grading and storage of seeds are more preferred.

Roguing

For pure seed production roguing is essential. First roguing is done before appearance of flower stalks and plants not true to type are removed. Second roguing is done at the time of blooming and early blooms are removed. Most of early bolters are male and have no effect on seed yield.

Harvesting and Seed Yield

Spinach produces seed stalks in about 65 days and seed in 150 days. Seeds mature slowly and unevenly. Seeds shatter badly from dried plant hence, the seed plants are harvested before they are fully dried. Plants are cut and dried in the field. After complete drying, the plants are threshed and seeds are cleaned by winnowing. Average seed yield is 7.5 to 10.0 q/ha (Choudhary, 1967).

According to Singh *et al.*, (1962) late cultivars in the hills produce less seed of about 100 kg/ha. Seed yield is also adversely affected by cuttings (Hussain *et al.*, 1970).

REFERENCES

Andre, P., Blance, R., Buret, M., Chambroy, Y., Dauple, P. ; Flanzy, C., Pelisse, C. and Vergniaud, P. (1980). *Revue Horticole.*, Suisse No.207 pp.37-42.

Anonymous (1971). Indian minimum seed certification standards. *Central Seed Committee, Ministry of Food and Agriculture, Community Development and Co-operation,* New Delhi.

Anonymous (1982). Package of practices for Rabi Crops. Haryana Agric. Univ. Hisar.

Anonymous (1990). "Sabji Utpadan" Directorate of Extension Education, Dr. Y.S. Parmar University of Horticulture and Forestry, Solan (H.P.).

Anonymous (1998). Varieties and Hybrids. Division of Vegetables Crops. IARI, New Delhi.

Aykroyd, W.R. (1956). Health Bulletin 23. Ministry of Health. Govt. of India.

Bhagchandani, P.M., Thakur, P.C. and Singh, N. (1979). Diallel analysis of leaf size and yield in Spinach. *Indian J. Agric. Sci.*, 49(5): 364-367.

Black, M. C. and Dainello, F. J. (1986). Comparison of percentage leaf area with white rust lesions and two other methods for evaluating partial resistance to *Albugo* spp. in spinach, *Phytopathology* 76: 1087.

Cajlahjain, M.H. , Martinvic, L. N. and Koeankov, V.G. (1969). Doklady Akademii Nauk, USSR, 189 pp.662-5.

Cantiliffe, D.J. and Pahatak, S. C. (1974). *Hort. Science*, 9: 470-72.

Cervato, A. (1969). Annali della Facolta di Agraria, Universita Catholica del S. Cuore, Milano, 10: 389-421.

Chailakhyan, M. Kh. and Khryanin, V.N. (1979). Planta 44: 205-207.

Chod, J. (1985). Susceptibility of some spinach cultivars and hybrids to beet mosaic virus, beet yellows virus and cucumber mosaic virus. *Zeszyty Problemowe* Postepov. Nauk., Rolnicyyeh, 291: 89.

Choudhary, A.R., Rashid, M.M. and Hussain, A. (1974). Horticulture, 2: 39-43.

Choudhury, B. (1967). Vegetables. National Book Trust, New Delhi, India.

Culafic, L. and Neskovic, M. (1980). Physiol, Plantarum, 48: 588-90.

Dainello, F.J., Heineman, R.R. and Jones, P.K. (1981). Texas Agric.Exp. Sta. No.PR-3878, pp. 7.

Daniel, L. and Piccini (2005). Yield and leaf quality of processing spinach under deficit irrigation. *Hort. Sci.*, 40(6): 868-870.

Dhillon, K.S., Dhillon, S.K., Singh B. and Kansal, B.D. (1987). Effect of different levels of nitrogen on yield and chemical composition of spinach . *P.A.U.J. Res.* 24:431-36.

Drewes, H. (1943). Spinach Varieties. *Mch. Spl. Bull.*, pp. 225.

Geise, F.W.and Farley, H.B. (1929) Variety Studies of Spinach, *Md. Bull.*, pp. 312.

Girenko, M.M. (1968). Sex changes in spinach varieties, *Trud. Prikda Bot. Genet Selek*, 40: 28.

Gonzalez, A.R. (1980). *Proc. Ann. Meeting Arkansas State Hort. Soc.*, pp. 68-70.

Guerrero, G.J and Valencia, M. A. (1972). Revista ICA. 7:121-38.

Hulewicz, D, .and Mokrzecka, E. (1971). Zeitschriftfur pflanzenernahrung and Bodenkunde, 130: 214-25.

Hussain, A., Safar, A. and Malukera, A.M. (1970). Effect of number of cuttings on seed production of Spinach (*Spinacia oleracea*). *Punjab Fruit J.* 32: 8-9.

Jacob, A. and Uexkull, H.V. (1960). Fertilizer use, Verlags Gesellschaft Fur Ackerban Mbh, Hannover.

Kinney, L.F. (1986). R.I. Bull., pp. 41.

Kohli, U.K. (1993). Agro-Techniques for Leafy Vegetables. In Advances in Horticulture Vol.5-Vegetable Crops: Part 1. (Eds: K.L. Chadha and G.Kalloo). Malhotra Publishing House, New Delhi, India.

Mac. Gillivray, J.H. (1961). Vegetable Production. Mc. Graw-Hill Book Co., Inc., New York.

Magruder, R. (1938). U.S. Dept. Agri. Misc. Pub., pp. 316.

Mehra, P.N. and Malik, C.P. (1963). Cytology of some Indian Chenopodiaceae, *Caryologia* 16: 67.

Mehta, Y.R. (1959). Vegetable Growing in Uttar Pradesh, Published by Bureau of Agri. Inf., U.P.

Mini and Krishna Murthy, K. (2004). Leafy Vegetables. Agritech Publishing Academy, New Delhi. pp. 221-232.

Nath Prem. (1967). Major green leafy vegetables. *Extn. Bull.* 9: Univ. Udaipur.

Nath, P., Velayudhan S. and Singh, D.P. (1987). Vegetables for the Tropical Region. ICAR, New Delhi.

Nath, Prem (1976). Vegetables for the Tropical Region . pp.109, ICAR, New Delhi.

Nath, Prem; Srivastava, V.K., Dutta, O.P. and Swamy, K.R.M. (2008). Vegetable Crops Improvement and Production, PNASF, Bangalore. pp. 162-163.

O'Brien, M.J. and Winter, H.F. (1977). Evaluation of spinach accessions and cultivars for resistance to *Fusarium oxysporum, wilt*. I. Greenhouse bench method, *J. Am. Soc. Hortic., Sci.* 102: 424.

Ogato, K., Ito, T. and Iwata, T. (1974). *J. Food. Sci. and Tech. Japan* 21:394-9.

Omran, A.F., Abo-Elsaod, I.A. and Ashour, N. T. (1973). *Gartenbauwissenschaft*, 38: 441-16.

Oomen, H.A.P. C. and Grubben G. J.H. (1978). Tropical leaf vegetables in Human Nutrition Communication 69. Department of Agriculture Research, Koninklijk Institute Voor de Tropen. Amesterdam.

Pandey, S.C. and Kalloo, G. (1993). Spinach, *Spinacia oleracea* L. In Genetic. Improvement of Vegetable Crops, (Eds. G.Kalloo and B.O. Bergh). Pergamon Press, Oxford.

Pandita, M.L. and Lal, S. (1988). Leafy vegetables. In vegetable crops in India. (Eds: T.K. Bose and M. G.Som). Naya Prokash, Calcutta.

Quagliotti, L. (1972). *Elette*, 18: 5-19.

Rosa, J.T. (1925). Sex expression in spinach. *Hilgardia* 1: 259-274.

Ryder, E.J. (1979). Leafy Salad Vegetables, A.V.I., Publishing Co., Inc., Westport Connecticut 195, USA.

Saini, G.S. (1996). A text Book of Vegetable Production. Aman Publishing House Meerut (UP), India.

Sharma, S.R. and Sohi, H.S. (1982). Effects of fungicides on the development of downy mildew and white rust of radish. *Indian J. Agric. Sci.*, 52: 521-524.

Shoemaker, J.S. (1947). Vegetable Growing, Wiley. New York.

Singh, D.N. and Nath Vishal (2011). Varieties and Hybrids of Vegetables. Satish Serial Publishing House, Delhi, India. pp. 371-374.

Singh, H.B. and Joshi, B.S. (1960). Our leafy vegetables *Farm Bull. 56*. Farm Infmn Unit, Directorate of Extension, Ministry of Food and Agriculture, New Delhi.

Singh, H.B., Thakur, M.R. and Bhagchandani, P.M. (1962). *Indian J. Hort.*, 20:148-54.

Singh, S. and Dhaliwal, M. S. (2001). Leafy and salad vegetables. In Vegetables, Tubercrops and spices, (Eds. S. Thamburaj and Narendra Singh) Indian Council of Agricultural Research, New Delhi.

Singh, S.B. and Agnihotri, J.P. (1971). *Indian J. Hort.* 28:70-79.

Stanilova, O., Boboshevska, D. and Vitanova, G. (1972). Nauchni Trudove Vissh Selskostopanski Institute 'Georgi' Dimitrov, Obshcho Zemledelie, 24: 35-46.

Thakur, P.C., Verma, T.S. and Joshi, B.S. (1989).Technology for production of more spinach seed. *Indian Horticulture*, July-September; 89: 29-31.

Thompson, H.C. and Kelly, W.C. (1983). Vegetable Crops Tata Mc Graw Hill Publishing Co. Ltd., New Delhi.

Vergni, P.and Huguet, C. (1971). Pepinieristes., Horticulturs Maraichers, No. 122. pp. 33-39.

Verlat, J.G. (1973). Groenten Fruit Netherlands, 28: 1483.

Verma, A.N., Srivastava, D.C. and Sharma, R.K.. (1969). Effect of nitrogen supply on growth, yield, ascorbic acid content in spinach (*Spinacea oleracea* L.). *Plant Sci.*, 1: 189-91.

Webb, R.E. and Thomas, C.E. (1976). Development of F_1 Spinach hybrids. *Hort. Science* 11: 546.

□□□

2. SPINACH BEET

Botanical name	:	*Beta vulgaris* var. *bengalensis* Hort.
Family	:	Chenopodiaceae
Chromosome number	:	2n = 2x = 18

Spinach beet is also known as Indian spinach, beet leaf or palak is very common leafy vegetable. It is one of the most popular winter leafy vegetable grown during winter throughout the country for green tender and succulent leaves used as vegetable.

ORIGIN

Spinach beet is probably the native of Indo-Chinese region (Nath, 1976). It was known in China as early as 647 AD. Aristotle mentioned red chards and Theophastus mentioned light green and dark green chards used in the 4th century BC. The Romans used spinach beet as feed for animals and man. It is closely related to beet root and swiss chard. By the 16th century, it was widely used for feeding animals during winter. Red beet featured in Roman receipes of the second and third centuries and was recorded in English receipes of 14th century. In Germany, it was first described in 1557, where referred as roman beet. The crop was introduced into USA in 1800, where it was known as garden beet (Campbell, 1976).

History and Adaptation

Most of the *Beta species* have their centre of diversity in the Mediterranean/ European regions. *Beta* spp. have been used as a pot herb since 2000 BC. There is a mentioned in the medical treatises "Charaka Samhita" and "Susruta Samhita" about garden spinach as "Palankya" in Sanskrit but its identification as *Spinacea oleracea* by Ray and Gupta (1980) and Ray, Gupta and Roy (1980) respectively, may need further checking and verification. There is one type generally found in Europe and known as "Perpetual Spinach" and this is different from Swiss Chard (*B. vulgaris* var. *cicla)* used as a pot herb. Probably the farmer is more close to Indian spinach or "Palak" of north Indian Plains. It is not known when it was introduced into India. Due to absence of many pot herbs for growing in the winter season of north India, early introductions might have obtained wide spread acceptance of the consumers.

The popularity of this leafy vegetable in Bengal in eastern India, is brought out by the fact that not only its leaves and stems in the young stage, are relished, even the seed stalks are consumed as a vegetable. The tropicalization of *Beta* which includes temperate sugar beet and garden (red) beet, may have started possibly in Bengal, which is the probable reason for the designation of "Palak" as *B. vulgaris* L. var. *bengalensis*. There is however, no authentic chronological record to testify this nomenclature.

In India, earlier improvement through natural selection by the farmers might have been in the direction of larger succulent leaves, late bolters and ability for regeneration after repeated cuttings. A hybrid selection by seedsmen between an early bolter of garden beet and palak, known as "Banerjee's Giant" is very popular since nearly four decades in Bengal. A biennial developed by the Regional Station of Indian Agricultural Research Institute at Katrain (Kullu Valley), called "Pusa Harit" is a derivative of a cross between palak and sugar beet. It may be pointed out that there is yet no monogerm variety in palak, either naturally occurring or consciously evolved.

The adaptation of palak to subtropical conditions of India, has not been studied in depth. It is a pot herb known from Caraka, Samhita (Ray and Gupta, 1980) days and has a long history of cultivation in India. The genus *Beta* is an essentially temperate/semi-temperate one consisting of sugarbeet, garden beet and Swiss chard, with palak also belonging to the same genus, having acclimatised to subtropical/tropical conditions. Interestingly, with such a long history of cultivation, the genetic variability observed in this species (*vulgaris* var. *bengalensis*) is not wide enough, commensurate with an outcrossing (wind pollinated) taxon. Further, palak is popular only in Indian subcontinent and no other country is known to cultivate this pot herb extensively (Seshadri and Chatterjee, 1996).

DISTRIBUTION

Spinach beet is widely grown in all parts of tropical and sub-tropical countries i.e., USA, Europe, China, Japan and Asia. It is grown throughout the year in many parts of India, particularly in states like Uttar Pradesh, West Bengal, Punjab, Haryana, Delhi, Madhya Pradesh, Bihar, Maharashtra and Gujarat. It is less common in southern states, but now this crop is becoming popular in south India.

BOTANY

The spinach beet (*Beta vulgaris* var. *bengalensis* Roxb) belongs to the family chenopodiaceae. It is closely related to beet root and Swiss chard. The plants are herbaceous biennial for seed production while it is annual for the edible leaf production. The leaf of spinach beet is edible, whereas the root of beet root is used as vegetable. The spinach beet is different from spinach (*Spinacia oleracea* L.), the former is common palak with succulent leaves having entire margin, whereas the spinach is Vilayati Palak with lobed leaf margin.The former is popular in northern states whereas the letter in southern states. The flowers are bisexual. The male and female parts are present in the same flowers in spinach beet, whereas in the spinach the male and female flowers are usually borne separately. Studies on floral biology and pollen morphology and physiology in spinach beet have been reported by Purohit and Nath, 1969; Nath and Purohit, 1969.

The leaves are rosette at early vegetative phase with long petioles. The leaf margin is entire. Plant has tap root system with number of fibrous roots. Inflorescence is racemose with bracts of cymose type (Nath, 1987; Pandita and Lal 1993).

Plant breeding

It is a wind cross pollinated crop. Efforts were made at IARI, New Delhi to evolve high yielding varieties, which resulted in the new varieties like "All Green" (Singh and Joshi, 1960) and "Pusa Jyoti" (Anon, 1971b). At the university of Udaipur, a high yielding variety "Jobner Green" suited to alkaline soil condition was evolved (Nath, 1967, 1969a). Similarly, H.S. 23 has been developed by HAU, Hisar. The improved varieties developed by selection from local types, hybridization between palak and sugarbeet, Swiss Chard or beet root and colchicine induced polyploidy are available (Nath *et al.*, 2008).

The basic methods employed for the development of spinach beet include introduction, selection especially mass selection and hybridization. Mass selection is also adapted for the maintenance of cultivars. The best individual plants are selected on the basis of their phenotypic performance and their seeds are composited for raising the following generation. Large number of plants are selected to avoid inbreeding depression.

Floral Biology

Palak is having small sessile, hermaphrodite perfect flowers borne in the axils of leaves in 2-3 clusters. Stamens are five, ovary is multilocular perigynous, pollens are light and wind pollination is effective i.e., anemophilous (Dark, 1971). Therefore, it is highly cross pollinated. The anthesis takes place between 7 a.m. and 5 p.m. Anther dehiscence takes place between 8.30 a.m. and 6.30 p.m. depending upon the temperature and humidity. The stigma become receptive 8 hours before anthesis and remain, receptive for 10 hours after anthesis (Purohit and Nath, 1970). The fruit is a seed ball containing two to three seeds.

COMPOSITION

Spinach beet is highly rich in vitamin A and C and also contains good amount of protein, calcium and iron. It has been regarded as a rich, yet cheap source of vitamins (Nath and Subramanyan, 1972). The details of composition is given below:

(per 100 g of edible portion)

Moisture 86.4 g	86.4 g	Minerals	2.2 g
Fat	0.8 g	Other carbohydrates	6.5 g
Fibre	0.7 g	Calcium	380 mg
Calories	46	Iron	16.2 mg
Phosphorus	30 mg	Thiamine	0.26 mg
Vitamin A	9770 IU	Nicotinic acid	3.3 mg
Riboflavin	0.56 mg	Vitamin C	70 mg
Protein	3.40 g		

(Aykroyd, 1950)

It is also a good source of folacin. The major carotenoids are lutein (1503 mg) and beta-carotene (1095 mg) per 100 g edible portion. Folate is in the range of 123-251.4 mg per 100 g edible portion, of which 51.0 mg is in the free form.

Uses

Spinach beet or palak is primarily used as a cooked vegetable. Tender, soft and succulent leaves and stems of palak are used as vegetable. It has some medicinal properties also. The herbaceous parts of spinach beet are mildly laxative.

VARIETIES AD HYBRIDS

Varieties/cultivars/cultures based on pigmentation of mid-rib and leaf veins are divided into two groups. One is with reddish mid-rib and leaf veins and the other with green mid-rib and leaf veins.

All Green, Pusa Palak, Pusa Jyoti, Pusa Harit, Jobner Green, Palak No. 51-16, Banerjee's Giant, Pusa Bharati, Arka Anupama, Ooty-1,HS-23, Pant Compsite are important commercial cultivars.

Sungro seeds

Sungro All Green.

Annapurna Beej

Red Kanka, White Kosla, Light red Kosla.

Tokita seeds

F_1 upright.

Singh and Nath (2011) reported the detail characteristics of varieties and hybrid developed by different organizations which are given below:

All Green

It is a pure cultivar and produces uniformly green tender leaves and gives about 6 to 7 cuttings at 15-20 days interval. It is a heavy yielder and yields about 125 q/ha of green leaves. During winter season it produces seed stalks in 75 days.

Pusa Palak

It is a selection from a cross between Swiss Chard and local palak. It produces uniform green leaves without any purple pigmentation. It is a late bolting cultivar.

Pusa Jyoti

This cultivar is evolved as a result of treating the seeds of cv. All Green with 2% colchicine solution for 24 hours and selection made from subsequent generations upto 10 years. It is a giant-leaved strain and produces large, green, thick tender, succulent and crisp leaves of good taste even when eaten raw as salad. When cooked they get easily minced and develop nice flavour and colour. Plants are very vigorous, quick growing and regenerate quickly after each cutting. It gives 6 to 8 cuttings and yields over 49 t/ha. It is rich in potash, calcium, sodium, iron and ascorbic acid and can be grown throughout the year.

Silver

The improved variety 'Silver' supplied 10-15 t/ha of leaves in two picking in July and September-October. Two additional picking may be taken if the root of the existing crops are sown in poly house.

Pusa Bharati

It is similar to Pusa Jyoti, the hitherto popular variety. Its cordate shaped leaves are 25 cm long and 14 cm in breadth, the petiole length being 15 cm smooth and tender, the leaves of Pusa Bharati are pure green without any red pigmentation. They may be picked up 30-40 days after sowing for 150 days duration. Its plants flower 175-180 days after sowing. On an average it gives 50 tonnes/ ha leaves in winter and early-spring seasons. It is ideally suited for both kharif and summer seasons.

Arka Anupama

It provides medium-large, dark-green, succulent, crinkled and attractive leaves. It is a late bolter with vigorous initial growth and regenerated at a faster rate. Its leaves become ready for first picking 35 days after sowing. After that they can be cut at early 10-12 days interval (up to 75 days). The total yield comes 41.08 tonnes/ha of fresh green from 4 picking. It has been developed by crossing IIHR 10 and IIHR-8 followed by pedigree method of selection. It is resistant to cercospora leaf spot under field condition. Leaves are rich in chlorophyll, carotene, vitamin E and iron contents.

Pusa Harit

It has been developed by hybridization between sugar beet and local palak and suitable for cultivation in the hills throughout the year mainly because of its chilling requirement, a character inherited from its sugar beet parent. Plants are upright, vigorous growing with uniformly thick, green, slightly crinkled giant sized leaves. It is a heavy yielder with remarkable ability for rejuvenation. It has late bolting habit and wide range of adaptability to varying climates. It can also tolerate alkaline soils.

Jobner Green

It was evolved by spontaneous mutation (point mutation) detected from the population of a local collection Sl.No. 5. It produces uniform green, large, thick, succulent, tender leaves with strong flavour having entire margin. Leaves are easily cooked and taste is comparable to cv. All Green. It is a prolific yielder and gives an average yield of about 296 q/ha.

HS23

This cultivar is released by HAU, Hisar. It produces dark green, large thick, tender and juicy leaves. It is a quick growing cultivar, gives first cutting about 30 days after sowing and a total of 6-8 cuttings at 15 days interval.

Palak No. 51-16

This cultivar has been developed by the department of Agiruclture, Maharashtra State. It produces green leaves, gives several cuttings and yields about 160 q/ha. It is a late bolting cultivar.

Banerjee's Giant

It is a very popular cultivar evolved by S.P. Banerjee in West Bengal. This cultivar has been developed from a cross between local palak and beet root. It produces large thick leaves with very succulent stem and fleshy root.

Ooty 1

A variety developed at the Horticultural Research Station, Udhagamandalam of TNAU by selection from a Himachal collection. First harvest is possible in 45 days followed by three more cuttings at 15 days intervals.

Pant composite

A cercospora leaf-spot resistant variety developed at GPUAT and T, Pantnagar, It produces tender, juicy uniform green colour leaves.

Red Kanka

It has red blood colour, grow very fast, soft and succulent leaves, generally cultivated in winter season. It is ready for harvest 25 days after seed sowing. One time harvest is common with this variety.

White Kosala and Light red Kosala

Its cultivation is done in rainy and winter season. Both tyeps are preferred by different locality and grown as a common leafy vegetable in states like Orissa, Bengal, Bihar, Jharkhand and Parts of Chattisgarh.

F_1 Upright

It has shiny, very dark green leaves and 18-20 leaves per plant, very tasty, easy for harvesting. Resistant to downy mildew, rich in minerals.

Sugro All Green

It has uniform green, smooth, tender leaves. Start first cutting 25-30 ays after sowing. Give 5-6 cuttings at 15-20 days interval. A late bolter, has very good market preference.

CLIMATIC INFLUENCE

Spinach beet can be grown successfully in tropical and subtropical climatic condition. However, demand for green leaves during winter is very high. Therefore, main crop is raised during winter season and it is known as cool season crop. It can withstand frost better than other vegetable crops. High temperatures, especially long days cause bolting, thus reducing its market value and making the plants unfit for producing edible leaves. It is a short-season crop. Higher yields are obtained under short day and mild temperature. Succulency and tenderness of the leaves are increased under high atmospheric humidity.

SOIL REQUIREMENT

Spinach beet/palak can be grown in a wide range of well drained fertile soils but it produces a good crop in lighter soils. It does very well in sandy loam soil. Crop can tolerate slightly alkaline and saline soils. It can be grown successfully up to a pH of 10.5 (Nath, 1987). Better results as regards quality and yield were obtained in the soil with pH 7, but the variety "Jobner Green" gave excellent performance in soil with pH 7 to 10.5 (Nath, 1969a). In general soil pH should be in the range 6 to 6.5. Low soil pH is harmful for growth and development of crop.

Soil is prepared by 3 to 4 ploughings or harrowings followed by plankings so that soil become loose, free from clods. Levelling of plot can be done when it is undulated.

SOWING / PLANTING

Spinach beet is raised through seeds. They are sown directly in the field.

Sowing time

Spinach beet is sown almost throughout the year under suitable climatic conditions. However, there are main sowing times in the plains are : (i) January-February, (ii) June-July , (iii) September-Novemebr. In hills, seeds are sown from April-June.

In the plains, Desi cultivars are sown in June to November and imported cultivars in October to December. In the hills, desi cultivars are sown in March to May and imported one in July to August (Rai and Yadav, 2005).

Seed rate

In order to raise one hectare area of spinach beet crop, about 25 to 30 kg seed is required. Proportionately the quantity of seed is increased as viability of seed is decreased. The seed germinates in about 10 days if proper soil moisture and temperature are available (Nath *et al.*, 2008).

Distance of sowing

Seed is sown at the distance of 22 to 25 cm between rows and 10 to 12cm between plant to plant depending on variety and season. For early crop sowing is also done in rows about 22 cm apart and later on thinning to 5 to 8 cm within the row. In both the cases, seeds should be placed at a depth of about 4 cm.

Method of sowing

Spinach beet, traditionally sown by broad cost method but it is unscientific and therefore, undesirable. Scientific and modern method is row sowing. Sowing in rows facilitates, weeding, watering, interculture, cutting of leaves etc. In this method, shallow furrow is opened at 2-3 cm deep and seeds are dropped followed by covering of the furrow. Under favourable soil moisture and temperature, the seeds germinates 8-10 days after sowing.

There are different factors affecting the germination (poor germination) of seeds (Singh, 1989). (1) viability of seeds is low, (2) Adequate soil moisture, is lacking at the time of sowing. (3) Optimum temperature is not available for seed, (4) Sowing of seeds when done, deep i.e., beyond 4 cm. (5) Formation of soil crust before emergence of seedlings (6) Attack of pests and or soil borne/ seed borne diseases.

NUTRITIONAL REQUIREMENT

As spinach beet is leafy vegetable, it requires more nitrogen for crown growth. The quantity of manures and fertilizers are variable according to soil, climate conditions, variety and purpose of the crop i.e., whether crop is grown for greens or for seeds. It is better to apply fertilizers based on the soil analysis. However, for good yield 200-250 q/ha farm yard manure would be enough during field preparation. Besides, application of 80 to 100 kg/ha nitrogen, 60 kg/ha phosphorus and 60 kg/ha potash should also be given in form of inorganic fertilizers (Rai and Yadav, 2005). In the absence of soil test, PAU recommends 25 tonnes of farm yard manure, along with 87.5 kg N and 30 kg P_2O_5/ha. The practices followed (Singh and Joshi, 1960; Nath, 1967; Yawalkar, 1992) are: 20 to 24 cart loads of FYM, 50 to 190 kg of nitrogen, 88 kg of phosphorus and 88 kg of potash per hectare. Pal *et al.* (2004) recommended 100-150 kg nitrogen, 80 kg phosphorus and 80 kg potassium per hectare. The FYM should be mixed thoroughly in the soil at the time of field preparation. The entire quantities of phosphorus and potash alongwith one fourth quantity of nitrogen are applied at the time of planting. The other three splits doses of equal amount of nitrogen are applied after each cuttings.

Foliar spray of urea @ 1.5 %, 15 days after germination and after each cuttings up to third cutting gave the highest yield of 438.80 q/ha in Jobner Green (Khandelwal and Nath, 1971). GA at 10 ml/l in combination with 1 per cent urea has been reported to give best yield (Rao *et al.*, 1972).

IRRIGATION

There should be sufficient soil moisture for proper germination of seeds. As far as possible, pre-irrigation should be done for rapid and uniform germination. Sometimes, due to pre-irrigation, sowing is delayed. Under such circumstances, it is advisable to irrigate crop after sowing. Irrigation depends on the soil condition, planting season, cultivar etc. Light soils require more water at frequent interval as compared to heavy soils. Spring-summer crops require more water at frequent interval as compared to winter crop. Irrigations in summer and in winter are done at 4 to 5 and 8 to 10 days interval, respectively. Generally, rainy season crop does not need artificial irrigation but drainage is most important. Sometimes, when there is long drought period occurred, during rainy season, irrigation is needed otherwise plants will bolt and edibility of leaves passes quickly and quality of leaves is reduce.

INTERCULTURE

Thinning should be done to provide space for accomodation and development of large uniform plants. Shallow hoeing is essential because the spinach beet plant cannot compete well with the weeds. About 2-3 hoeing are done at early stages of crop growth that will keep down most of the weeds and will loosen the soil for proper aeration. Sometimes light earthing is given for protection of tender leaves from soils during watering.

WEED CONTROL

Hand weeding is still a common practice to control weeds. Generally 2-3 hoeing cum hand weeing are required to control weeds. Chemical control of weeds may also be adopted with Pyrazone at 2.4 to 2.8 kg per hectare applied as pre emergence (Pandita and Lal, 1993).

PLANT PROTECTION MEASURES

Insects Pests

Aphids

These insects are small in size and found in large number on the tender portions of plant. They suck the sap from the plant tissues causing plant to become weak. They transmit the mosaic disease from one plant to another. The growth of the plants is checked. Quality of the leaves is reduced. Aphid can be controlled by the spray of 0.1 per cent solution of malathion 50 EC (1 litre/ha) dissolved in 625 litre of water). Harvesting of leaves should not be done for 6 to 8 days at least. Soil application of thimet 10 G granules at 18 kg/ha.

Leaf eating caterpillar (Laphygma exiqua)

They feed on leaves and make holes. The quality of leaves is deteriorated. Yield is reduced. Affected leaves are not acceptable to consumers. The caterpillar can be controlled by the spray of 0.1 per cent solution of malathion 50 EC at the rate of 1 litre/ha dissolved in 625 litres of water. Wait for harvesting 6 to 8 days after spraying the insecticide.

Fungal Diseases

Damping-off (Pythium sp.) : Seedlings are attacked by the fungus and killed. It can be controlled by treating seeds before sowing with any one of the fungicides like cersan, captan or thiram @ 2.5 g/kg of seeds.

Downy mildew (Peronospora spinaciae Laubert)

Symptoms appear as yellow to brown angular spots on the leaves with downy growth on the lower surface. In severe case it attacks the stems and petioles. The causal organism can survive in soil as oospores, in seed as dormant mycelium, on the seed as Contaminating oospores, or on perennial spinach beet (Singh, 1985). To control the disease (1) Use clean seeds. (ii) Remove crops debris and destroy. (iii) Follow a three year crop rotation. (iv) Spray dithane Z-78 at 0.2 per cent.

White Rust (Albugo occidentalis Wilson)

The fungus causes white blisters appear on the lower surface of the leaves and opposite each blister on the upper surface, a yellow patch is developed. It can be controlled by adapting crop rotation; keeping the field free from weeds, by growing smooth-leaved cultivar (Dainella *et al.*, 1981); by spraying copper fungicide like Bordeaux mixture at 1 per cent.

Leaf spot (Eterosporium variable Cke)

Old leaves show dirty white, water soaked, circular spots. Similarly, the other fungus *Phyllosticia chenopdii* Sacc, causes light yellow specks, which develops into spots, later affected leaves dry up. It can be controlled with spray of blitox at 0.2 per cent; by growing disease tolerant/resistant cultivars.

Anthracnose (Colletotrichum spinacie Ellis and Halsted)

Affected leaves show water-soaked spots , which later coalesce and cause drying of the leaves. On seed stalks, elongate spots are formed. Black dots are also seen on seeds. To control the disease (1) Use clean and healthy seeds (2) Remove crops debris and destroy (3) Follow crop rotation.

Bacterial Diseases

Bacterial soft rot (Erwinia carotovora)

This disease appears on leaves in transit i.e., it is a post-harvest disease. Water-soaked areas are developed, rotting occurred in the packed leaves due to lack of aeration an cleanliness. The control measure are (1) do not allow broken leaves, (2) after harvesting, do not allow for longer time to store, (3) always clean the leaves before packing, (4) packing materials and storing place should be perfectly cleaned.

Viral Diseases

Spinach mosaic

The disease is characterized by light and dark green patches and mosaic mottling of infected leaves. The leaves are reduced in size, distorted and invariably show chlorotic sectors which are almost white. The infected disease is caused by a strain of cucumber mosaic virus (CMV). The virus is easily transmitted by sap inoculation. The virus is also transmitted by insect vector

Myzus persicae in a non-persistant manner. It can be controlled by (i) Removal and destroying of affected plants. (ii) By controlling aphids. (3) By growing varieties like Virginia savoy.

HARVESTING

Spinach beet leaves become ready for first cutting in about 25-30 days after sowing. Only well grown, green succulent an tender leaves should be trimmed at the height of 5 to 7.5 cm above the ground. The leaves are cut with the help of a sharp knife. Subsequent cuttings can be done at the interval of 15 to 18 days depending on the variety. Nath (1967) reported that 'Jobner Green' was ready for subsequent cuttings in 13-15 days, whereas "All Green" took 15-18 days. If sown in the first week of September near Jaipur. 'All Green' allowed 6 to 7 cuttings, whereas "Jobner Green" produced 8 to 10 cuttings only. During summer, only one harvesting is recommended. Harvesting is not done early in the morning because there is dew on the crop.

YIELD

The average yield of green edible material (leaves) is about 100 to 150 q/ ha. The yield and quality of leaves are affected adversely if harvesting is not done at regular intervals.

POST HARVEST HANDLING

After harvesting , the produce is washed, trimmed, grade and bunched before marketing. The quality of leaves deteriorates, the attractive appearance of leaves and their turgidity are lost within 24 hours of harvesting. Rotting starts soon when leaves are lost within 24 hours of harvesting. So all the post-harvest operation should be attended quickly and timely.

Grading

The spinach beet leaves are graded according to their size, shape, tenderness and colour developments. Well developed leaves of uniform shape and size with bright colour and tenderness are kept in "Super grade". Medium size leaves with good texture and uniform shape alongwith freshness are kept in "Standard grade". Small leaves with good shape, light colour, fresh and free from any spot or blemishes are arranged in "Commercial grade" for market purposes.

Packing

After grading leaves are tied in bunches according to grades then packed. Spinach beet leaves are big in size (broad), tender and delicate so it should be packed properly to protect from any injury and to avoid breakage of leaves.

The bunches are packed in wet gunny bags/bamboo chips baskets or in arhar stick basket or in vegetable crates and cover with moist gunny bags/cloth to keep produce fresh. Packing is done according to grade along with bunches then stitched properly.

Storage

It is not possible to store the fresh green leaves for more than few hours under ordinary conditions. The green leaves can be stored at low temperature (0°C) and high relative humidity (90-95 per cent) for about 10-14 days (Nath *et al.*, 2008).

SEED PRODUCTION

Palak is a winter season crop, but has a much wider range of adaptability to climatic conditions. It produces seed freely in the north Indian plains. It is a wind cross pollinated crop and the two varieties should be kept about 1600 m apart to produce pure seeds (Anonymous, 1971). The spinach beet does not cross with the spinach either way. For certified seed production 1000 m of isolation distance is maintained.

Brief Cultural Practices

Field is prepared to fine tilth by ploughing, 3-4 harrowings, followed by levelling. Sowing should be done in October to November for seed crops. Nucleus/Breeders/foundation seeds are collected from approved agencies. About 25-30 kg seeds are sufficient to raise one hectare seed crop. Seeds are sown in rows 20-25 cm apart. Place the seed close to each other at the depth of about 2 cm. Sufficient soil moisture is required for good germination. Well rotten farm yard manure @ 350-400 q/ha should be incorporated in the soil at field preparation. A quantity of 60 kg/ha nitrogen is top-dressed after each cutting. Frequent irrigation is given as per requirement and according to season to get good yield. Crop field is to be kept weed free, especially at early stage and completely clean of pigweed (bathua). Usually 2-3 cuttings are taken from the seed crop then after plants are left to produce stalks.

Roguing

In order to maintain genetic purity of the variety, it is essential to do roguing three times viz., before flowering, flowering and fruit development and maturity stage. All off types, early bolters and pigweed plants are removed from the seed plot. Final roguing operation should meet the field standards as per Table 1.

Table 1 : Field standards for palak seed production (Max. permitted %)

Factor	Foundation	Certified
Off types	0.10	0.20
Other crop plants	-	-
Objectionable weed plants	-	-
Diseased plants	-	-

Table 2 : Seed standars for palak seed production (%)

Standard	Foundation	Certified
Pure seed (Minimum)	96.00	96.00
Inert matter (maximum	4.00	4.00
Other crop seed (Maximum)	0.10	0.20
Total wee seed (maximum	0.10	0.20
Objectionable weed seed (maximum)	0.10	0.20
Germination (minimum)	60.00	60.00
Moisture (maximum)	9.00	9.00

Harvesting

Palak seed crop is harvested 150-180 days after sowing depending on season, soil type and varieties grown. Whole plant can be harvested at a time when fruits/seeds are fully ripen. Harvested plants are heaped on cemented threshing floor for drying.

Threshing and winnowing

Dried plants are threshed by beating them with stick. Thresher can also be used on availability (Singh, 2001). Seeds are winnowed against wind and cleaned seeds are spread for drying till 9 per cent seed moisture is reached (Table 2).

Seed yield

The average seed yield is about 6 q/ha. When crop is raised under improved management, higher yields are obtained.

Storage

Spinach beet or palak seeds having their moisture percentage is about 9.0 can be stored for longer period of time (3-4 years) under low temperature and low humidity. The seeds are packed in polythene bags or in other air tight containers.

REFERENCES

Anonymous (1971a). Indian minimum seed certification standards. *Central Seed Committee, Munistry of Food and Agriculture, Community Development and Co-operation*, New Delhi.

Anonymous (1971b). New IARI vegetables varieties, IARI, *Extension Bull.* No.2. Indian Agricultural research Institute, New Delhi.

Aykroyd, W.R. (1950). Health bulletin No.23. Ministry of Health, Government of India.

Campbell, G.K. (1976). Sugarbeet. In Evaluation of crop plants (Sammond NW Ed.) Longman , London.

Dainella, F.J., Jeineman Prand Jones, R.K.(1981). Texas Agriculture Experiment Station, No.PR 3878, P.7.

Dash, S.O.S.(1971). Experiment on the cross-pollination of sugar beet in the field. *J. Natt. Inst. Agric. Bot.* 12: 242-246.

Khandelwal, R.C. and Nath, Prand (1971). Effects of foliar application of urea on growth and yield of spinach beet. *Farm J.*12 (11 and 12): 29-32.

Nath, P. (1987). Vegetables for the Tropical Region. Indian Council of Agricultural Research, New Delhi.

Nath, P.(1976). Vegetables for the Tropical Region. ICAR Low Priced Book Series No.2, ICAR, New Delhi.

Nath, Prem (1967). Major green leafy vegetables. *Extn. Bull.* 9 ; Univ. Udaipur.

Nath, Prem (1969a). A new mutant in *Beta vulgaris* var. *bengalensis. Indian J. Hort.* , 26 (3&4): 156-159.

Nath, Prem and Purohit, S.P. (1969b). Studies on pollen morphology and physiology in spinach beet. *J. Palynol.* 5(2): 111-121.

Nath, Prem and Subramanyam, S. (1972). Palak, a rich yet cheap source of vitamins. *Indian Hort.* 17(2): 23-24.

Nath, Prem; Srivastava, V.K., Dutta, O.P. and Swamy, K.R.M. (2008). Vegetable Crops. Improvement and Production. Published by *Dr. Prem Nath Agricultural Science Foundation*, Bangalore.

Pandita, M.L. and Lal, S. (1993). Leafy vegetables in Vegetable crops in India. In (Bose, Som and Kobir, (eds.), Naya Prokash, Calcutta, India.

Purohit, S.P. and Nath, P. (1970). Studies on floral biology in spinach beet (*Beta vulgaris* var. *bengalensis* Hort.) *Madras Agric. J.* 57: 98-105.

Rai, N. and Yadav, D.S. (2005). Advances in Vetetable Production. Research Co. Book Centre, New Delhi.

Rao, S.N., Panapathi, K., Mashuddin, S. and Narayana, Rao, M.N. (1972). Proceedings of 3rd *International Symposium on Sub-tropical Horticulture*, Bangalore, Vol. 2, pp. 191-198.

Ray, P., Gupta, H.N. and Roy, M. (1980). "Susruta Samhita". Scientific Synopsis. Indian Natl. Sci. Acad. New Delhi.

Ray, P. and Gupta, H.N. (1980). "Caraka Samhita". *Scientific Synopsis*, Indian National Sci. Acad., New Delhi.

Seshadri, V.S. and Chatterjee, S.S.(1996). The history and adaptation of some introduced vegetable crops in India. *Vegetable Science*, 23(2): 114-140.

Singh, D.N. and Nath, Vishal (2011). Varieties and Hybrids of Vegetables. Satish Serial Publishing House, Delhi, pp.348-351.

Singh, H.B. and Joshi, S.A. (1960). Our leafy Vegetables. ICAR *Farm Bull.* No.56, New Delhi.

Singh, R.S. (1985). Diseases of Vegetables. Oxford and IBH Publishing Co., New Delhi.

Singh, S.P. (1989). Production Technology of Vegetable Crops. A.R.C.C., Karnal, Haryana (India), pp. 298-303.

Singh, S.P. (2001). Seed production of commercial vegetables. Agrotech Co.Publishing Academy, Udaipur-313002.

Yawalkar, K.S. (1992). Vegetable Crops of India. Agril. Horticultural Publishing House, Nagpur.

□□□

3. FENUGREEK

Botanical name	:	*Trigonella foenum graecum* L. and *Trigonella foenum corniculata*
Family	:	Leguminoseae
Chromosome number	:	2n = 16

Fenugreek (*Trigenella foenum-graecum* L.), belongs to the family Leguminosae (Fabaceae) sub family Papilionaceae; also known as Methi, Menthi, Mentulu, Vendayam, Venthayam Menthya, Hulabaha, Shamlita, Fenugreek, Sag, Menthai, Uluva, Vendia, Keerai; is an important leaf vegetable cum condiment having several medicinal properties. Commonly grown throughout the country especilly in areas having cool and dry climate. Fenugreek (methi) is extensively cultivated in Rajasthan, Madhya Pradesh, Gujarat, Maharshtra and Punjab for its dry use of condiment. Being a leguminous crop it improve the soil by fixing the atmospheric nitrogen into the soil (Saini, 1996).

ORIGIN

Fenugreek is a native of South-Eastern Europe and West Asia. It is found growing wild in North-Western India (Bailey, 1950). It is also considered that Eastern Europe and Ethiopia to be the places of origin of Fenugreek (Choudhury, 1967).

DISTURIBUTION

Fenugreek is spread to India, Argentina and Mediterranean countries (South France, Morocco and Lebanon). It is grown all over India for leafy vegetable, spice and condiment, fodder and medicinal purposes. However, it is grown on commecial scale in some states like Rajasthan, Madhya Pradesh, Gujarat, Uttar Pradesh, Maharashtra and Punjab (Som and Maity, 1993). India is one of the major producers and exporters of methi seeds to Saudi Arabia, Japan, Sri Lanka, Korea and United Kingdom.

BOTANY

Botanically Fenugreek, methi is known as *Trigonella foenum graecum* L. (common methi *T. corniculata* (kasuri methi or champa methi) belongs to the family leguminosae sub family fabaceae is a herbaceous annual, quick growing and produces erect shoots to a height of 40-70 cm; leaves light to dark green with or without pink margin, produces flowers in axil of leaves in pinkish-white pods are slender and straw coloured ripe, beak-shaped and are about 8-10 cm long with 8-15 yellowish brown coloured and smooth surfaced seed. It is self-Pollinated crop, produces seeds freely. In contrast kasuri methi is slow growing, remaining in rosette condition during most of its vegetative growth period; flower bright yellow with orange brown on long stalk; pods 2-3 cm sickle-shaped; leaf blobbed; flowers and seeds smaller and scented. Seeds are small, hard, yellowish brown, smoth and oblong.

The cultivated froms are diploids with 2n=16. A sterile hexaploid (2n=44) has been recorded. A mixoploid plant with chromosome numbers between 24 and 44 has been detected among tetraploids. Abberrational frequency is higher in tetraploids. Environment plays significant role on the expression of morhological and physiological characters including grain yield.

Plant breeding

Fenugreek is a self pollinated crop. It is an herbaceous annual which produces seed freely. Very little efforts have been made in collecting, maintaining or utilization of different genotypes for the improvement of this crop (Nath *et al.*, 2008).

Floral biology

The flowers of fenugreek are hermaphrodite, They are self-compatible and self fertile. Therefore, self pollination is a rule. The flowers in common methi or desi methi are whitish where as bright orange to yellow flowers are in Kasuri or champa methi. Anthesis takes place between 10.30 A.M. and 5.30 P.M. with peak between 11.30 A.M. and 12.30 Noon. Stigma become receptive 12 hours after the opening and remains receptive for about 10 hours after the opening of flowers. The plants flower in about 30-37 days after sowing and the duration of flowering phase is 7-18 days. The spieces is typically self-pollinated and clestogamous/destogamous. Pollen fertility ranges from 95-98% in the unopened flower buds and 67-80% in the open flowers. The pods mature within 60-70 days of sowing (Thangaraj and Vijayakumaran, 2001).

Classification

There are two types of methi, grown in the country.

I. Common methi (Trigonella foenum-graecum L.)

This is quick growing herb producing upright shoots to a height of 40-70 cm. It produces dark green leaves with or without pink margins.

Two-to-three white flowers are produced in axils of each leaf. Pods are straw coloured and boat shaped. This is harvested either by clipping young plants from base or by uprooting. It is cultivativated almost in all parts of the country.

II. Kasuri methi (Trigonella corniculata)

This is also known as Marwari methi or Champa methi and it gives better peformence in cooler parts. This is generally secnted types. Growth of plant is slow initially and remains in rosette condition during vegetative growth period. It produces bright orange to yellow flowers pods are sickle-shaped and smaller than common methi. Its green leaves are used as vegetable while its seeds are used as spices and as pulse. Harvesting is done by clipping leaf growth several times. Harvesting by uprooting entrie plants is not practiced.

COMPOSITION

Fenugreek is rich in minerals, protein, Vitamin A&C. Detail nutrition value is given below (per 100 g of edible portion).

Moisture	81.6g	Chlorine	165 mg
Fat	0.9 g	Thiamine	0.05 mg
Protein	4.4 g	Calcium	360 mg
Fibre	1.1 g	Oxalic Acid	13 mg
Minerals	1.5 g	Sodium	76.1mg
Carbohydrates	6.0 g	Iron	17.2 mg
Calories	49	Potassium	51 mg
Magnesium	67 mg	Sulphur	167 mg
Phosphorus	51 mg	Vitamin A	6450 zo
Copper	0.26 mg	Niotinic Acid	0.7 mg
		Vitamin C	5.4 mg

Aykroyd (1963)

Uses

The green tender leaves of fenugreek are rich source of minerals (iron and calcium), protein, vitamin A and C, are used as vegetable alone or in combination with potatoes. The young plants especially the tender shoots and the aromatic green leaves are also used in the form of sag. The seeds and dry leaves of methi are used as a condiment and for medicinal purposes. The seeds are also used for dye making and extraction of alkaloids or steroidal substance diosgenin. Dried leaves and flowers are used for flavouring of foods (katyal, 1985). The seeds flour is mixed with wheat flour for making chapaties which are nutritious and delicious. The seeds are used in colic, enlargement of liver, spleen and rickets (Pruthi, 1979). Use of methi to diabetic patients is beneficial due to fibrous substance. Leaves and stems of methi make good animal fodder.

Fenugreek is lowering bloods cholestrol levels proprity. Seeds augment hunger and bring about contraction of the intestine, They ward off vomiting, cough and useful in piles. Methi seeds fried in ghee mixed with seeds of anise and salt and made into a paste, which is given to check diarrhoea. Seeds are used to induce child birth and to increase breast milk production. Application of paste of methi seeds is useful in mitigating burning sensation of the soles and the. palms. The medicinal properties of methi are slimming down the body, dissolution of oedema and swellings of the wounds, strengthening the body, aphordisiac or stimulative of the urge of sex and invigorative to the nervous tissue. For control of pimples on the face, grind methi leaves can be used. Fenugreek leaves are also useful in cases of feeble digestion and in rheumatism. Paste prepared from methi seeds when rubbed frequently on the face increases the lustre of the skin.

VARIETIES AND HYBRIDS

In certain parts of the country, fenugreek also grown as a green fodder, but distinct types of methi are used for fodder purpose and for leaf vegetable purpose. Both scented and non-scented cutivars are also grown traditionally.

Some Improved Varieties

Lam Sel-1, Pusa Early Bunching, Kasuri selection, Col, Rajendera Kranti and RMT-1, Methi No. 14, Metho No. 47, UM-112, Barbara and Kasuri Improved. Lam selection 1, CO1, Rajendra Kranti and RMT-1 are dual purposes varieties suitable for seed as well as for green purpose.

Singh and Nath (2011) reported detail characteristics of improved varieties and hybrids developed by different institutions which are given below:

Pusa early bunching

An early variety developed at IARI, New Delhi, is quick growing, producing upright shoots in bundhes and having white flowers, yellow and bold seeds. Its pods are 6 to 8 cm long, green in colour having 10-12 seeds per pod. First cutting can be made after 55-60 days of sowing It is a heavy yielder.

Lam selection-1

It is a selection from the Regional Agricultural Research Station at Lam, Guntur of N.G. Ranga Agricultural University plants are bushy in nature with medium height. It is a high yielding variety with better leaf quality. It is capable of yielding 740 kg of grain per hectare. The green yield is around 10-12 t/ha under irrigated conditions. It contains 5.3% of protein in seed (Rao *et al.,* 1981).

Prabha (NLM)

It is released by college of Agriculture, Jobner, Rajasthan, Suitable for cutivation in Rajasthan and Gujarat. It is a pure line selection from a local collection of Nagpur district. Semi erect growth, leaves soft, Sweet with dark green colour, can tolerate many pests and diseases. (Sharma and Bhati, 1984).

Rajendera Kranti (RMT-16)

This is a selection made at the Dholi Station of the Rajendera Agriculture University. The plants are medium tall, bushy and are suited for raising as a pure crop as an intercrop. It yields around 1,200-1,400 kg/ha in a crop duration of 120 days. Seeds contains around 9.50% protein.

RMT-1

It is a selection from the SKM college of Agriculture, Jobner, Rajasthan. The selection yields around 1500 Kg of seed from a hectare over a crop duration of 140-150 days. The variety is tolerant to root disease and is moderately resistant to powdry mildew disease. The seeds contain 21.0% protein.

CO1

It is a dual purpose variety released by the Tamil Nadu Agriculture University, Coimbatore. The crop matures in 85-90 days with a grain yield of 680 kg/hectare. When grown for leaf, the yield is around 9 tonnes/ ha. Because the plants are dwarf and has a short duration, it suits well for intercropping and relay cropping. The seeds contain 20-22% protein while, the leaf contains 15.9% protein (Ram Das *et al.,* 1982).

Pusa Kasuri (Kasuri selection)

It is a slow growing variety and remains in a rosette condition during most of its vegetative growth period. Its seeds are small and dark brown in colour and its flowers are yellow. Its pods are smaller in size than the former variety. It is a heavy yielder. Leaves of this variety have special fragrance. It becomes ready for first harvest after 60 to 65 days of sowing.

Methi No. 47

It is an improved variety of Maharshtra, high yielding and containing more vitamin C in the leaves.

Methi No. 14

It is developed by Maharashtra Agiculture Department. It is high yielding and contain more Vitamin C in the leaves.

HM-57

It is suitable variety for Haryana and it is capable of yielding 1000 kg of grain per hectare.

UM-50

It is superior cultivar for the greens and seed yield.

UM-112

The plants of this vatiety are very long and cultivated for both seed and leaf purpose. It cantains more amount of diosgenin, which is used for preparation of medicines related to embryo distortion.

IC-74

It is a local cultivar having erect greens, more branching type plants, good as greens as well as seed.

CO-2

Plants are medium sized, high yielding, less susceptible to root-rot, suitable to grow in coimbatore, Dharmapuri, Guddalore, Ramanathapuram, Karur, Tiruchirapalli, Perambalur and tirunelveli districts, maturity period is 95-100 days with an average yielding of 140.0q/ha.

HM-10

Plants are bushy, bold, yellow attractive seeds moderate tolerant to leaf spot disease, maturity period is about 140-150 days with an average yield of 20.1 q/ha.

RMT-143

Plants are medium sized seeds are typical yellow colour. Tolerant to powdery mildew. Maturity period is 140-150 days with an average yield of 16 q/ha, suitable to grow in heavier soils of Chittor, Bhilwara and Jodhpur area (Rajasthan). It was released in 1997.

Barbara

It is having erect plants, yellow seeds and high trigonellin content in seeds.

Lam selection

The plants are medium in height and bushy with better yield potential.

Lam Slelection 1

It is developed by APAU, Andhra Pradesh through selection. Plants are medium in hight and bushy with more branches having better yield potential of greens, 30 percent more than local cultivars. Seeds are golden yellow in colour. It matures in 90 days and produces an average yield of 7.40 q/ha. It is tolerant to root -rot, powdery mildew, caterpillars and aphids.

Hissar Sonali

Plants are bushy, semi erect with bold yellow attractive seed, moderate tolerant to leaf spot and rot diseases. Maturity period is 140-150 days with an average yield 19.0 q/ha.

Hissar Suvarna (HM 103)

A quick growing dual purpose high yielding varity suitable for cultivation throughout the country. The plants are tall and erect with more number of branches. It takes 130-140 days maturity period and yield 16 to 19 q/ha bold seeds with typical yellow colour. This variety is tolerant to cercospora leaf spot and powdery mildew diseases. it was developed through pureline selection from local germplasm collected from Haryana Agricultrual University, Hissar.

Hissar Madhavi (HM-350)

It was developed by HAU, Hissar through single plant selection from the line PME 46-1. It is a quick growing dual purpose, plant erect, tall with more branches. Pods are long, dark green leaves, typical yellow colour. It yield 19 to 20t/ha seeds. It is suitable to grown in Hrayana, Rajasthan and Gujarat. It is tolerant to powdery and downy mildews.

Hissar Mukta (HM-34)

This variety was developed by Haryana Agricultural University, Hissar through a pureline mutant from an indigenous collection from

U.P. This is a quick growing seed type suitable to grow in Haryana, Rajasthan and Gujarat under both irrigated and rainfed conditions. Seeds are bold with tan green seed coat colour. It yields 20-30 q/ha seeds. It is tolerant to both powdery and downy mildew.

CLIMATIC INFLUENCE

Fenugreek is a winter season crop and fairly tolerant to frost and freezing weather. It can also be grown as a hot weather crop. It is mainly cultivated in moderate or low rainfall areas while the areas with heavy rainfall are not suitable for its cultivation. Kasuri methi needs comparatively cooler climate. Fenugreek is both a tropical and temperate crop. The crop had a wider adaptability and is grown in a wide range of climatic conditions. It is grown from sea level upto an altitude of 2,000 metres. Places receiving moderate or low rainfull are best suited for the crop. Heavy ranifall areas are to be avoided. Dry weather is essential during crop maturity period.

SOIL REQUREMENT

This crop is grown in soils rich in organic matter and free from water stagnations. The crop is adapted to all types of soils, but it performs well in well drained loamy soil. Heavy loamy soils are used if adequate drainage facilities and rich organic matters are available. Heavy black cotton soils are the best suited ones for rainfed crop. The optimum soil pH for the crop lies between 6.0-7.0 (Som and Maity, 1986). The crop is fairly tolerant to salinity. However saline and aidic soils are to be avoided.

SOWING

Sowing time

The sowing time is so adjusted that the pod development and seed maturity phase coincide with a fairly dry period free from rains. Sowing of seed may be done from third week of October to second week of November for seed production, while for leafy vegetable the common fenugreek is sown from the middle September to middle of March. First fortnight of November is the best time for sowing of fenugreek (Anon, 1988). In Karnataka leaf crop is grown throughout the year. For grain crop, there are two distinct seasons for fenugreek, viz. June-July and September-December. In Rajasthan, Sowing is done in the first fortnight of November. It is grown from the middle of August to Middle of November in Karnataka.

According to Som and Maity (1986), common methi for leafy vegetable is sown from middle of September to middle of March. The sowing for both the species in the hills is March to April and October. When it is grown as a leafly vegetable, the duration of the crop is only 30-40 days and thus it can successfully be adopted in multiple cropping systems (Singh and Joshi, 1960).

Seed rate

The seed rate for common methi is 25 kg, while 20 kg of seed per hectare is enough for kasuri type. For seed crop 20-25 kg healthy seeds are required to sow one hectare land (Anon, 1984). For leafy vegetables 30-35 kg seeds per hectare is required (Rai and yadav, 2005). In Tamil Nadu about 20 kg of seeds are sown in an hectare. Punjab adopts a slightly higher seed rate of 30 kg per hectare (Thangaraj and Vijaya kumaran, 2001). For broad casting a heavy seed rate of 40 kg is used.

Sowing Method

Seeds are either sown in lines or broadcast. Broad casting of seed is done when there is sufficient moisture in the soil. The seed is generally broadcast in beds and the surface is raked to cover it. Improved seeds should be sown after inoculation with Rhizobium culture (Sharma and Bhati, 1984).

Spacing

A spacing of 10-15 cm is followed within the lines. A depth of 2-3 cm is normally adopted. Improved seed should be sown in rows 30. cm apart at 2 to 3 cm depths on well prepared friable soil (Sharma and Bhati 1984). Line Sowing in rows 20-25 an apart facilitate weeding and interculture operation during the initial stages of crop growth (Choudhury, 1976). A spacing of 30x30 cm has been found ideal for fenugreek (Anon, 1988). The common methi may take 5-6 days to germinate and kasuri 7-8 days. However in cool climate complete germination requires a few days more.

Pre-sowing seed treatments

Seeds are soaked in water for 2 days prior to sowing to enhance germination. Seeds are soaked in cycocel solutions at 50-100 ppm for improving germination and enhancing seedling growth. Soaking in a solution of 0.6% EMS is recommended for improved germination and survival of seedlings. Seeds are also treated with rhizobium culture prior to sowing (Thamburaj and Singh, 2001).

Land preparation

The soil should be brought to fine tilth by ploughing three or four times. Clods are broken and stones are removed. Farmyard manure or compost is applied at the last ploughing. Field is levelled thoroughly. Flat beds of convenient size are formed (3x2 m) with irrigation channels running around between every two rows of beds.

NUTRITIONAL REQUIREMENT

Fenugreek being a legume, does not require heavy quantity of nutrients. Fenugreek seeds removes N, P, and K in the ratio of 2:1:2 from the soil. The uptake of calcium, iron and zinc is also fairly high. The uptake per hectare of notrogen, phosphorus and potassium is of the order of 10.0, 3.5, 8.2 kg respectively. The uptake of calcium and magnesium is of the order of 1.4 and 1.8 kg. Other micronutrients are also absorbed considerably, viz. Iron (123g/ha); Manganese (14g/ha), zinc (39g/ha) and copper (12 g/ha). Therefore to maintain a steady state of productivity, adopt from the major nutrients, viz notrogen, phosphorus, potassium and magnesium particularly and zinc are to be included in the fertilizer schedule. The response to nitrogen is seen in the form of increased pod number and grain yield upto 45 kg N per hectare. Positive effects on yield are also seen with phosphorus. The biomass production is high with higher doses of organic matter and mussories rock phosphate in combination with lead and iron. For getting succulent foliage and early growth the soil should be fertile enough FYM (15 tonnes), 40 kg N and 20 kg of P_2O_5 are applied at the time of ploughing and before sowing. Two to three top dressing of nitrogen, 20 kg each time, should be done perferably after alternate cuttings in kasuri methi. In common methi one topdressing can be done if it is kept for cutting.

Fenugreek is a leguminous vegetable. It fixes nitrogen from the atmosphere upto 283 kgs/ha (Saxena and Ahmed, 1984) per annum and therefore the crop can be grown on residual fertility from the previous heavily manured and fertilized crop (Singh, 1989; Som and Maity, 1993). For guide line, it is to be known that the fenugreek crop needs 50-100 kg N, 50-60 kg P_2O_5 and 50-60 kg K_2O per hectare. Hardman and petropoulos (1975) studied the response of fenugreek to find inoculation with *Rhizobium mililotii* and obtained taller plants with a higher seed yield but the maturity was delayed. Ramkrishanan *et al.* (1976) were of the same view about the initial inoculation of seeds with *R. mililotii.* Recommended doses of NPK and Farmyard manure for some states of the country are given below :

State	Nitrogen (kg/ha)	Phosphorus (kg/ha)	Potassium (kg/ha)	FYM (kg/ha)
Andhra Pradesh	40	40	20	15
Arunachal Pradesh	45	30	60	15
Bihar	20	60	40	0
Himachal Pradesh	75	40	40	10
Gujarat	20	25	20	15
Karnataka	100	50	0	12
Madhya Pradesh	40	30	15	10
Rajasthan	20	50	0	0
Maharashtra	100	50	50	10
Tamil Nadu	50	25	0	0
Uttar Pradesh	40	40	20	25-30

(Tandon, 1991)

Parkeek and Gupta (1981) and Mahantesh (1993) also reported response of fenugreek to N and P application. It was found that the application of 60 kg N/ha along with the basal dose of 40 kg p/ha increases the yield of vegetable leaf and seed significantly in fenugreek. Banafar *et al.* (1995), reported the response of fenugreek to nitrogen and phosphorus in Madhya Pradesh. An increase in the level of N increased the yield of seed from 406 to 1200 kg/ha and yield of vegetable leaf from 290 to 490 kg/ha, but the effect was statistically significant only upto 60kg N/ha. Similarly P application @ 45 kg/ha increased the yield of seed and vegetable leaf to 1105 and 375 kg/ha compared with 776 and 270 kg/ha in the control, respectively. Basavaraja *et al.* (2002) reported the influence of nutrients on growth and yield of fenugreek. The results of two years experimentation indicate that the recommended dose of festilizers (N-100 kg, P_2O_5-50 kg/ha and no K) was better in achieving good plant height (51.95 cm), number of secondary branches/plants (5.75) and other yield characters. However, the treatment having 30 kg of additional K_2O gave significant and highest seed yield (8.08 t/ha) which might be due to the cumulative effect of increased pod length (8.04 cm) and higher number of seeds/pod (16.60). Nath *et al.*, 2008, concluded that higher yield of fenugreek (both foliage and seed) could be obtained by supplementing additional 30 kg of potassium. So, it is confirmed that the response of the crop to potassium is more significant than to nitrogen and phosphorus. An additional dose of 20-30 kg potassium increases the yield of foliage as well as seed. Narayana and Jain (1971) found that fenugreek seedlings grown in pots treated with nitrofen at 0.1-0.8 ppm gave higher numbers of nodules.

Use of plants growth Regulators

A spray of ascorbic acid solution @ 250-450 ppm improves the plants growth and larger number of flowers and fruits per plants with heavier seeds (chinoy *et al.*, 1975). According to Desai and Pathak (1965) the spraying of GA_3 (10 to 100 ppm) greatly enhanced internodal length and more number of leaves. Faul and Singh, (1967), reported that Sodium 2,3-Isobutyrate at 0.05 per cent applied before initiation floral buds produced higher yield with less pollen sterility.

INTERCULTURE

The fenugreek seed may take 5-8 days for germination after sowing. The crop should be thinned when three to four leaves start sprouting to retain the plant distance of about 10 cm within the rows (Sharma and Bhati; 1984). While, Chauhan (1972) recommended a spacing of 5.0 to 7.5 cm within the row. Shallow intercultural operations during early stages of crop only possible because close spacing is followed for leaf production. Two weedings and hoeings are necessary during the early stages of the crop. The crop is slow growing at the initial stage. The growth becomes faster after four to five weeks and the crop does not allow weeds to compete with it.

Tipping of the young shoots is done after 10-30 days of sowing. This is done at a heght of 4-5 cm from ground level. If the crop is grown for leaf, subsequent cuttings are taken at 15 days interval in the case of grain crop also, it is customary to take one or two cuttings for leaf before allowing the crop to flowering and fruiting under North Indian conditions. Normally two clippings are taken at 50-80 days of seeds sowing which do not affect the yield of seed crop. Cultivars yielding higher grain yield tend to produce low leaf yield.

WEED CONTROL

To keep the crop free from weeds first hoeing and weeding is done at the time of thinning and the second within 45-50 days after sowing. Occasional weeds are suppressed by this crop. Broad leaves weeds should be pulled out at the time of first cutting. At the time of thinning, weeding and earthing up is given. Top dressing with fertilizers is a also done at this time.

The chemical weed control can be done by pre-sowing application of fluchloralin at 1.12 kg/ha in methi crop that gives above 50 per cent more seed yield over the unweeded crop (Mali and Sawalik, 1987; Ramana *et al.*, 1994). Pendimethalin treatments cause severe phyto-toxicity to the fenugreek crop, showing scorching of leaves, rosetting and stunting of the plants, resulting in 68-70 per cent mortality of germinating seeds (Ramana *et al.*, 1994).

IRRIGATION

Sufficient soil moisture is necessary to obtain rapid and uniform and quick germination and constant growth of crop which can be provided by pre-sowing light irrigation. Subsequent irrigations are given at 10 to 15 days interval for better plant growth. Irrigations at flowering and fruit developing stages are most essential for seed yield. Higher seed yield is obtained when crop is irrigated at IW/CPE ratio of 1.0 to 0.8 (Anonymous, 1991). Fenugreek raised for leaf production requires light irrigation after each cutting of leaves (Singh, 1989).

Leaf cutting managment

Mehti is grown as leafy vegetable. The crop is harvested by clipping the leaves and young shoots and allowing the plants to regrow before harvesting again. In Gujarat, for seed production, the methi crop is allowed to mature without any leaf harvest. It is not necessary that the variety which gives high seed yield will also give high leaf production and vice versa (Mehta *et al.*, 1995).

PLANT PROTECTION MEASURES

Insect- pests and their control

Prophylactic spraying of Ekalux (0.05%) or Quinalphos (0.04%) or Malathion (0.1%) or Phasphamidon (0.03%) or Dimethoate (0.03%) or Methyl-demeton (0.03%) on seed crop of methi will prevent damages from a few pests such as aphids, leaf eating caterpillars and pod borers.

Diseases and their control

The crop of methi is suffered mainly by the following diseases.

Powdery mildew (Erysiphe polygoni DC)

Disease appears with white floury patches on both the sides of the leaves and other plants parts. Comparatively dry weather is favourable for the spread of disease. Usually disease appears late in the season. The disease can be controlled by dusting of sulphur 5 per cent at the rate of 15 kg per hectare.

Downy mildew (Peronospora trigonella. Gaum)

Affedcted leaves show yellow patches on upper surface and cottony greyish violet mycelium on lower surface. It can be controlled by spraying of Difolaton at 0.2 per cent. Cleaning of plant debris also help in checking spread of disease.

Damping off (Rhizoctonia solani Kuhn.)

Infected seeds and emerging seedings are having water soaked spots. Sometimes, they emit unpleasent odour. Seedlings may toppled down at the ground level. It can be checked by pre and post emergence of Brassicol or Bavistin. Drenching with Carbendazim at 0.1 per cent.

Leaf spot (Cercospora traversions 50 CC)

Round to semi circular spots on the leaves with brown margin and white at the centre appears. It can be controlled by taking crop rotation. Maintaining field sanitation. Selection of healthy seed for sowing. Spraying the standing crop with Zineb (co. 2%).

HARVESTING

The first cutting is made by sickle after 25-30 days of sowing when the plants are 15-20 cm high, leaving 2-3 cm stubs for the production of new stalks. The old leaves become bitter and unfit for consumption. The young plants are nipped at the ground level in common methi after 20 days of sowing. After another 15-20 days the plants are often uprooted, bunched and marketed. Kasuri methi can give 5-6 cuttings at an interval of 15-20 days. Leaves are very perishable in nature therefore, they are marketed soon after harvesting (Nath *et al.,* 2008).

YIELD

The common methi (Fenugreek) yields leaves about 70-80 q/ha and Kasuri 90-100 q of leaves per hectare. However a yield of 1.2-1.5 tonnes of seeds is normlly harvested under irrigated condition. In ratoon crop, a yield of 500-600 kg/ha is gathered.

POST HARVEST HANDLING

Grading

Leaves and root are washed properly to remove soil and dust particles from root zone and leaves. In green stage for leaf purpose whole plants are used for marketing.

Plants are graded according to size, tenderness and freshness and lusture. Grading is done in "super", "fancy" and "commercial" grades. Tender plants with uniform size and colour alongwith lusture are kept in "super" grade. Medium type plants with uniform size, good looks and colors are arranged in "fancy" grade. Plants with variable in size and uniform with normal colours and free from spots and blemish are kept in "commercial" grade for marketing. Multicut varieties are not graded and generally sale as loose with mix of all type and stage of leaves as a whole.

Packing

Plants are tied in small bundles according to grade and packed in gunny bags/bamboo baskets and in Arhar stick baskets. For distant market in big sities bundles are packed in vegetable crates and coverd with moist bags or cloth to keep the plants fresh for longer period. Some times small bundles of about 250g in weight are prepared for marketing.

Storage

The produce is kept in cool and moist condition under shade. It does not stand strong for more than a day. Fully dried leaves of methi can be stored for one year without any damage. The seeds can be stored for two years (Chauhan, 1972). Kasuri methi is available in dry form (dried leaves) in small packets of different size and this way dried leaves can be used allround the years.

SEED PRODUCTION

Both common and kasuri types seed freely can be produced in the plains of India. The practices normally followed for vegetable production are also for seed production. The crop can be grown as pure seed crop or after taking two to three cuttings, the plants are left for seed production. Singh and Joshi (1960) reported that in Kasuri Selection, it is advised to take two-three cuttings after that allowed to produce seed. But Saini (1975) was of the view that one cutting should be taken, it give the yield significantly more than pure seed crop. According to Choudhury(1976) common methi (cv Pusa Early Bunching) produced an average seed yield of 12-15 quintals per hectare, while the cutivar Kasuri Selection gave only 6-7 quintals, when cutting were not taken.

Roguing

Seed crop of methi requires thorough roguing of off-types including diseased plants at least thrice during early stage of crop growth i.e. before flowering, during flowering and fruiting and at maturity stages. Plants which are upright growing in spreading variety and spreading type in upright growing variety are removed. Similarly, late flowering and maturing plants noticed in early maturiting variety must be uprooted. Besides, the plants do not conform the characters of the grown plants of the varity should not be allowed in the seed crop. Simultaneously, diseased plants are up rooted and burnt out of the field. Finally the crop should met the field standards (Table 1).

Field Inspection

A minimum of two inspection shall be made, the first before flowering, the second during flowering and fruiting stages.

General Requirements

Isolation

Seed fields shall be isolated from the contamintants shown in column 1 of the table below by the distances specified 1 & 2 of the said Table.

Table 1 : Field standard for fenugreek seed production.

Contaminants	Minimum distance (meters)	
	Faoundation	Certified
Fields of other varieties	10	5
Fields of the same variety not conforming to varietal purity reqirements for certification.	10	5

Specific Requirements

Factors	Maximum permitted (per cent)*	
	Foundaion	Certified
Off types	0.10	0.20
**Objectionable weed plants	0.010	0.020

*Maximum permitted at the inspection.
**Objectionable weed shall be: senji (Melilotus spp.).

Table 2 : Seed standards for methi seed production (%).

Factors	Standards for each class	
	Foundation	Certified
Pure seed (minimum)	98.0 per cent	98.0 per cent
Inert matter (maximum)	2.0 Per cent	2.0 Per cent
Other crop seeds (maximum)	10/kg	20/kg
Total weed seeds (maximum)	10/kg	5/kg
*Objectionable weed seeds (maximum)	2/kg	20/kg
Other distinguishable weed seeds (maximum)	10/kg	20/kg
Germination including hard seeds (minimum)	70per cent	70 per cent
Moisture (maximum)	8.0 Per cent	8.0 per cent
For vapour-proof containers (maximum)	6.0 per cent	6.0 per cent

*Objectionable weeds are same as given above.

Harvesting

Seed crop of methi takes about 120 to 160 days for maturity. At proper stage of maturity, almost leaves become yellow and fall. Care should be taken under ripening stages of pods for harvesting. Shattering of seeds may cause when harvesting is delayed while early harvesting may cause shrinking of seeds. Such seeds may also have poor germination percentage (Singh and Singh, 1996).

Harvesting is done by cutting of plants at the ground level, and they are pilled up on threshing floor, for drying, away from the other variety or commercial crop. Some time the plant is pulled out entirely when the pods turn yellow, made into small bundles and sun dried seeds are separated by beating with stick. Separated seeds are winnowed and cleaned before sun drying.

Threshing and winnowing

Dried material is threshed with the help of thresher or by beating material on wooden plank. Threshed material is winnowed against wind. while winnowing all chops is taken out.

Drying

Cleaned seeds are spread on cemented floor or on tarpaulin for 5-6 days till seed moisture reaches to 8.0 per cent. Seeds can also be dried in electrically operated drier.

Seed yield

Seed yield of methi varies from 60 to 80 kg/ha depending upon type of variety and management practices. Improved varieties may yield up to, 100 to 120 kg of seeds per hectare.

Storage

Methi seeds which are properly cleaned and dried are packed in moisture proof polythene bags for next season use under ordinary conditions (dry and well ventilated).

Seeds are also stored in gunny bags lined with polythene paper. Vacuum gravity separator are used for cleaning fenugreek seeds. The seeds are to be stored with an initial moisture level of 7-8% and at an equilibrium relative humidinty of 40%. Specific quality standard; of seed should be as colour of seed-orange-yellow; volatile substance-0.02% moisture-9% and other character-high diosgenin. The ASTA has prescribed the standard for fenugreek i.e. Total ash maximum-5%, Acid insoluble ash-1.0%, moisture-8.5% and US Sieve size-95% should pass through US No. 25 sieve.

REFERENCES

Anonymous (1988). Status paper on spices, board, Ernaculam, Cochin, pp 1-118.

Anonymous (1991). Annual Report 1990-91. All India coordianted project on spices, Calicut, Kerala, India.

Anonymous (1998). Varieties and Hybrids. Divison of Vegetable Crops, IARI, New Delhi.

Aykroy, W.R. (1963). The nutritive value of Indian foods and the planning of satisfactory diets. *ICMR. Special Rept.*, 42, New Delhi.

Baboo, R. and Sharma, R.K. (1995). Nutrient uptake and yield of fenugreek (*Trigonella foenum-graecum* L.) as affected by nitrogen, phosphorus and cutting managament. *Vegetable Sci.* 22 (2) : 77-80.

Bailey, L.H. (1950). The standard Encyclopedia of Horticulture. The Mac Millan Co., New York.

Banafar, R.N.S., Tiwari, R.J. and Jain, R.C. (1995). Response of fenugreek (*Trinonella foenum graecum*) to nitrogen and phosphorus on Chromustert soil of Madhya Pradesh. *Indian J. Agric. Sci.*, 65 (11) : 821-822.

Basavaraja, N., Hosamani, R.M. and Nandi , V.R. (2002). Influence of nutrients on growth and yield of fenugreek. *Intl. Conf. Veg.*, Bangalore, 11-14 November, 2002. Abstract; pp. 207-208.

Chauhan, D.V.S. (1972). Vegetable Production in India (3rd edn.) Ram Prasad and Sons. Agra, India, pp. 256-60.

Chhibba, I.M., Kanwar, J.S. and Nayyar, V.K. (2000). Yield and nutritive value of diffrent varieties of fenugreek (*Trigonella* spp.). *Veg. Sci.*, 27(2): 176-179.

Choudhry, B. (1976). Vegetables. National book Trust India, New Delhi. pp.256-260.

Gill, H.S. and Singh, Hari (1988). Effect of planting dates and leaf cutting on seed yield of methi. *Punjab Agric. Univ. J. Res.;* 25 : 206-09.

Hardman, R. and Petropoulos, G.A. (1975). Planta Medica. Vol. 27 : pp. 53-57.

Katyal, S.L. (1985). Vegetable Growing in India. Oxford & IBH Publishing Co, new Delhi.

Mali, A.L. and Sawalik, S.N. (1987). Studies on weed control in fenugreek *(Trigoneralla foenum graecum L.). Indian J. Agron.* 32: 188-189.

Mavai, Deepak; Lal, S.; Singh, N. and Thakral, K.K. (1998). Response of Fenugreek seed crop to fertilizer application. ISVS: *Silver Jubliee National Symposium.* ICAR- PDVR, Varansi. 12-14 December. P. 155.

Metha, K.G., Patil, P.G., Kachhadia, B.T. and Edison, S. (1995). Varietal response to leaf yield of fenugreek (*Trigoneralla foenum graecum* L). *Indian Cocoa, Arecanut and spice J.* 19 (3) : 91-93.

Nath, Prem, Srivastava, V.K., Dutta, O.P. and Swamy, K.R.M. (2008). Vegetable Crops - Improvement and Production. *Dr. Prem Nath Agricultural Science Foundation,* Bangalore, 168-171.

Pruthi, J.S. (1979). Spices and Condiment National Book Trust of India, New Delhi pp. 118-123.

Rai, N. and Yadav, D.S. (2005) Advances in vegetable production. Researchco Book Centre New Delhi.

Ram, Das, S., Armugam, R., Sayed, S., Shah, H. and Muthuswami, S. (1982). CO-1. fenugreek-A high yielding grain-cum-green Fenugreek. *South Indian Hort.*, 30: 242-243.

Ramana, B.S., Singh, Hari; Saondhy, K.S., and Singh, Dalgit (1994). Effect of Herbicides for weed control in seed crop of methi (*Trigonella corniculata* L.). *J. Res Punjab Agric Univ.* 31(4): 401-411.

Rao, T.S. and Babu, M.K. (1981). Lam Sel., 1. A high yielding fenugreek type for Andhra Pradesh. *Ind. Cocoa Arecanut, Spices J.*, 4: 83-86.

Saini, S.S. (1975). Effect of cutting on the seed yield of Methi (*Trigonella foenum-graecum*). *Haryana J.Hort Sci.*, 4 : 82-85.

Saxena, P. and Ahmed, S.T. (1983). Intensive agriculture. 20:15-16.

Sharma, R.K. and Bhati, D.S. (1984). Prabha. A promising variety of fenugreek for Rajasthan and Gujarat. *Indian Cocoa, Arecanut and Spices J.*, 8:4-15.

Singh, D.N. and Nath, Vishal (2011). Varieties and Hybrids of Vegetables. Satish Serial Publishing House, Delhi. pp. 358-362.

Singh, S.P. (1989). Production Technology of Vegetable Crops. Agricultural Research Communication Centre, karnal (Haryana), India. pp. 309-313.

Singh, V.B. and Singh, Kriti (1996). Spices. New Age International Publishers, New Delhi, pp. 131-138.

Som, M.G. and Maity, T.K. (1986). Fenugreek. In Vegetable Crops in India. (eds) T.K. Bose and M.G. som. Naya Prokash, Clacutta-6 pp. 680-86.

Som. M.G. and Maity, S.P. (1993). Fenugreek In: Vegetable Crops (Ed. Bose, T.K., Som, M.G. and Kabir, J) Naya Prokash, culcutta.

Tandon, H.S.L. (1991). Fertillizer Recommendations for Horticultural crops. *A guide book. Fertilizer Development and Consultant Organization*, New Delhi.

Thakral, K.K.; Tehlan, S.K. and Lal, S. (1998). Effect of time of sowing and spacing on seed crop of fenugreek. ISVS : *Silver jubilee National Symposium*. ICAR.PDVR. Varanasi 12-14 December. p.160.

❑❑❑

4. CORIANDER

Botanical name	:	*Coriandrum sativum* L.
Family	:	Umbelliferae
Chromosome number	:	2x = 22, 24

Coriander (*Coriandrum sativum* L.); is also known as Dhania, Dhaniya, Dhaniyalu, Kothamalli, Kottambri Dhane, Kastumbri, Dhanyaka, Kusbara; belongs to the family Apiaceae (Umbelliferae) is a tropical plant and is capable of tolerating drought and heat. Its name is derived from the Greek words' Koris',meaning bed bug because of the unpleasant, fetid, bug-like odour of the green unripened friuts. The leaves are used for flavouing soups and other foods. The fruits are an importent ingredient of curry powder, usully contributing the greatest quantity of all the spices followed by turmeric. Coriander is grown throughout the year for leaf purpose.

ORIGIN

Coroander is a native of Mediterranean region speially southern Europe and Asia minor. Precisely, Italy is presumed as the native place of coriander. The origin of coriander (green leaf and seeds) is virtually lost in the antiquity.

History

Coriander is one of the fist seed spices to be used by mankind as early as 5000 B.C. The Coriandr seeds were found into the Egyptian tombs of 960 B.C. Further, Persia grew coriander 3000 years ago and it added fragnance to the hanging gardens of Babylon. Hebrews knew it also, since in the old Testament in the Book of Exodus, 'Manna' is dercibed as being like white coriander seed. In the third century BC, Eromans also found coriander seed as an excellent seasoning for popular foods.

BOTANY

Coriander is thin-stemmed, small, annual bushy herb, 30 to 90 cm in height, with much branches, tops or umbels. Leaves are alternate, compound and petiole has a pair of stipules which sheath the stem at the base. The compound leaves become segmented and linear as they reach the upper extremities. Both leaves and stems, may the whole plants have a strong pleasant aroma for which they are loved and used in cooking for flavouring and garnishing dishes. The inflorescence is a compound umbel and usally comprises about 5 smaller umbels. The corolla is made up of five petals. Flowers are both bisexual and unisexual the latter being mostly staminate. The flower possesses five stamens and the ovary is inferior and bilocular. The fruit is globular, round and small. Its surface has longitudinal fine groove and ribs alternately. When pressed, the fruit break into two locules, convex outside and concave inside, in each of which, there is one seed. The fruits are mild, delicately fragnant due presence of volatile oil. Seeds are pale white to light brown in colour.

Floral Biology

Coriander is an androemonoecious i.e. both staminate and perfect flower (hermaphrodite) are produced on the same plant and in the same umbellet (Pillai and Nambair, 1982). Generally, hermaphrodite flowers open early than male flowers. The stigma remains receptive for five days and pollen remains viable for a maximum of 24 hours. The flowers are protandrous, with the anthers maturing before the stigmas become receptive and an out crossing rate of 19 per cent is obtained, suggesting that the coriander is a facultative cross pollinator (Gladis *et al.*, 1996). Pollination is effectively brought about the bees. The higher fruit setting is recorded during the first three days (as the stigma remains receptive up to five days). Flower starts opening at 6 a.m. and reaches a maximum by 12 to 2 p.m. (Hill, Albert, 1992). Wind seems to play no role in pollination of coriander. Hymeropterans are most effective and efficient at carrying pollens. The visual impact of the compound umbel, presence of exposed nectar and availability of abundant pollens are the major attractants for insects visiting the umbels (Kaul *et al.*, 1989).

Under normal growing environment coriander flowers in 45 to 60 days after seed sowing and matures in 70-120 days. There is decrease in moisture content with advancing seed development. Natural detergent fibre, acid detergent fibre, hemicellulose, cellulose and lignin levels increase with maturity of seed. The presence of phylic acid may not vary at different stages of seed development (Gupta *et al.*, 1991). A variety possessing higher proportion of hermaphrodite flowers resulting higher seed set.

DISTRIBITION

Coriander is widely distributed in India. The main state in which coriander is grown on commercial scale, in India are: Rajasthan, Madhya Pradesh, Maharashtra, Bihar, Gujrat, Uttar Pradesh, Punjab, Andhra Pradesh, Orissa, Tamil Nadu and Karnataka. India is the largest producer of coriander in the world. Besides India, it is also cultivated in Myanmar, Pakistan, Turky, Morocco, Palestine, Romania, Czechoslovakia, Guatemala, Mexico etc. India exports coriander seeds to Beharain, Canada, Guyana, Indonesia, Jordan, Kuwait, Malaysia, Mauritius, Mozambique, Oman, Saudi Arabia, Singapore, South Africa, Srilanka, UAE, USA. The coriander powder is exported from India to Oman, Saudi Arabia, Singapore, South Africa, Srilanka, UAE, U.K., USA etc. (Singhal, 1996).

Crop improvement

Plants taller in nature, possessing more number of primary and secondary branches and umbel are considered ideotypes. More number of seeds in the umbel implies more yield in coriander. The number of hermaphrodite flowers and male flowers decide the seed number in umbels. Hence selection is carried for types possessing more number of hermaphrodite flowers in the umbels.

Exotic type introduced by the Central Indian Medicinal Plant Organization (CIMPO) from Bulgaria have adopted very well to Indian conditions. Most of the released varieties have been developed as a result of simple selections. Selection with multipinnate leaves possessing heavy yield potentialities have been indentified in CIMAP. Selection showing erect compact habit with purple stem having essential oil content between 1.6-1.7% have been also isolated.

CIMPO S 33 was released for its tall erect and basal branching habit. It is capable of yielding around 2.1 tonnes/ha in a duration of 143 days. This selection is small seeded with a higher essential oil contents of 1.3%.

The bold seeded varietes UD 1, UD 20 and UD 21 are selections made in Rajasthan. The variety UD 20 with a yield capacity of 1.2 tonnes/ha was released again as RCR 20 from Rajasthan Agricultural University, while GAU 1 and GAU 2 are selection from Gujarat. A small seeded variety Karan (UD 41) was also released from Rajasthan. In Madhya Pradesh, Gwalior 5385 and Katwari type coriander selection were made.

Tamil Nadu has so far made three selection, viz. CO 1, CO 2, and CO 3. Among these, CO3 is a dual propose coriander used as grain cum leaf type. Bihar has indentified a variety Rajendera Swati'. Guntur centre from Andhra Pradesh has also made selections like Swathi, Sadhana and Lam Sel. 2 (Thangaraj and Vijayakumar, 2001).

Hydridization

Coriander is basically a cross pollinated crop. Hybrids in coriander reveal heterosis for plant height and yield. The characters of either of the parents are dominant in F_1 progeny. The crop does not exhibit inbreeding depression. The flowers are minute and manual emasculation and crossing is cumbersom. The seeds of different country origin respond differently to gamma rays. Types from India and China respond very poorly to gamma rays. A doses up to 5 kr has a stimulating effect on various traits (Thangaraj and Vijay kumar, 2001).

COMPOSITION

According to Aykroyd (1941) Coriander contains 87.9% moisture, 3.5% protein, 0.6% fat, 6.5% carbohydrate, 1.7% mineral, 0.14% calcium, 0.06% phosphorus and Vitamin A 10460 to 12 600 I.U. per 100 g fresh leaves.

Coriander seeds contain the following nutrients as given by Pruthi (1979) per 100 g are mention below Table 1 :

Table 1 :

Constituents	Approximate quantity	Costituents	Approximinate quantiy
Moisture	6.3 g	Carbohydrates	24.0 g
Protein	1.4 g	Fat	19.6 g
Fibre	31.5 g	Mineral matter	5.3 g
Calcium	0.8 g	Phosphorus	0.44 g
Sodium	0.82 g	Iron	0.006 g
Potassium	1.2 g	Vitamin A	175 IU
Riboflavin	0.26 mg	Thiamine	0.23 mg
Niacin	3.2 mg	Vitamin C	12.0 mg
Volatile oil	0.3 g	Non- volatile	22.0 g

Uses

The entire Plant of young coriander is used in preparing chutneys, and the leaves are for flavouring curries and soup. The seeds are extensively employed as condiment and medicine. The oil of coriander is used as the valuable ingredient in perfumes and also for flavoring in food industries. The coriander seeds are as spices in the preparation of curry powder, pickling spices, sauces and seasonings. They are used for flovouring pastry, cookies, cakes and tobacco products. Criander is also employed for flavouring liquor, particularly gin. It is one of the important ingredients in the manufacture of food flavourings, bakery products, meat, fish and salads, soda and syrups, candy preseerves and liquors (Rai and Yadav, 2005).

The oil extracted from coriander seeds is used cheifly as a flavouring agent for spirituous liquors and in the cocoa and chocolate industries. It is also employed in medicine as a carminative or as flavouring agents to cover the taste or correct the gripping qualities of other medicines. Oil is also used as an ingredient in perfumes. It harmonises well with jasmine. Good quality oleoresin is extracted from coriander seeds which is used for flavouring beverages, Pickles and other delicacies. The residue from distillation can be used as a good cattle feed, as it is rich in protein and fat. Besides above, coriander fruit is highly aromatic, stimulant, carminative and antispasmodic. The fruits are called in infusion or decoction. The seeds are generally chewed to vanish bad smell breath created due to consumption of garlic. The roasted seeds are useful dyspepsia. The paste made from seeds is useful in pain in cephalagia. The juice of fresh plant is applied in erythema. The decoction with milk and sugar is beneficial during bleeding piles. The infusion or powder of fried seeds is very useful in colics of childern. Coriander reduces intoxicating effect of spirituous liquors (Rai and Yadav, 2005).

VARIETIES AND HYBRIDS

For Essential oil Production

Small seeded varieties are perferred whereas for spice use, large seeded varieties are mostly grown. Detail characteristics of varieties and hybirds reported by Singh and Nath (2011) are given below:

Sadhna (CS-4)

Plant semi- erect, tall, good bearing habit. It is suitable for rainfed condition. Early mturing, seed yield 10-12 q/ha in 115 days. Oil is 0.2 per cent. It is resistant to aphids and mites. Grown in Andhra Pradesh.

Swathi (Lam CS-6)

Plants grow to medium height. It is suitable for rainfed areas of Andhra Pradesh. It is early maturing which helps in escaping powdery mildew incidence. The seed yield is about 9-10 q/ha in 95-105 days. Essential oil content is about 0.3%. It is tolerant to powdery mildew.

CS-287

Earliest type with 78 days duration, small bold grain suitable for rainfed crop. Tolerant to wilt and grain mould. Yield 5.10 q/ha.

CO 1

Tall plant suitable for both leaf and seed production. Water requirement is not so high and cultivated as irrigated as well as rainfed. Crop duration is 110 days. Seed yield is 5.0q/ha.

CO 2

This is dual purpose variety. The yield potential for seed production in 90-100 days is 6-8 q/ha and the green leaf production is 10 q/ha in 40 days after sowing. Tolerant to drought, oil content is 0.3%.

CO 3

This variety is superior to both CO1 and CO2 in adaptability and seed production (8-10 q/ha). Suitable both for kharif and rabi seasons, lower incidence of fusarium wilt. Seed colour is attractive brownish yellow. All these three varities (Co1, Co2 and Co3) are released by Tamil Nadu Agril. University.

Pant Haritima

This variety is develped at G.B. Pant University of Agril. and Technolgy, Pantanagar. It yields better as green leaf as well as seed. Plant growth is heavy to medium, fine seeded, late maturing, high potential for seed production (15-20 q/ha). It is tolerant to stem gall disease (*Protomyces mecroporus*).

NP (D) 92

It is a selection from local material grown around Delhi area. Seeds are small, light green in colour having less fibre.

Small seeded

Seeds are small with less fibre and is most suited for growing in temperate regions. Volatile oil content is higher than bold seeded ones. The seed colour is light green.

Bold seeded

Seeds are bold having more fibre and is most suitable for growing in tropical and sub-tropical regions. Volatile oil (Coriandrol) is lesseer than small seeded ones and colour as light brown.

Caribe

It is developed by Bejo Sheetal Seeds, Jalna, Maharashtra. This hybrid has got broad leaves, nice aroma and is tolerant to bolting. Mature in 50 days after planting. It can be planted in kharif, rabi and in mild summer. It gives multiple cuttings, escaping powdery mildew incidence. The seed yield is about 9-10 q/ha in 95-105 days. Essential oil about 0.3%. It is tolerant to Powdery mildew.

Sindhu

Plant growth is medium, suitable for unirrigated condition of Andhra Pradesh. High yielding (10 q/ha seed yield), tolerent to disease and pests.

Gujarat Coriander-1

Plant growth medium, suitable for unirrigated conditions. The varity is bold seeded having higher yield potential and seed yield is about 11 q/ha bold in 110 days. Doing well in Gujarat.

Gujarat Coriander-2

Plant growth medium, suitable for unirrigated areas. It is bold seeded having higher yield potential of about 14 q/ha seed yield in 120 days duration.

Naraul selection

Plant growth medium, water requirement is less, bold seeded and high yielding (15-20 q/ha). Mostly grown in Haryana.

Gwalior No. 5365

This variety is resistant to fusarium wilt, medium seed size, produces 8 to 10 q/ha seed yield in 100-130 days. Generally grown in Madhya Pradesh.

Chimpa S-33

Plant grows tall,late maturing, high seed yield, 18-20 q/ha in 180-210 days. It is introduction from Bulgaria, Suitable for cultivation in Madhya Pradesh.

Merrocan

Seed yield 10-12 q/ha in 140-160 days, oil content is high. Suitable for growing in Madhya Pradesh.

Rajendra Swati

Plant grow medium in height, resistant to stem gall, disease and fruitfly. Variety is an early maturing. Seed yield is about 12-14 q/ha in 100 days. Seeds are rich in oil content. It is suitable for intercropping It is released from Rajendra Agricultural University, Dholi Campus, Bihar.

RCR-41

Tall erect plants, round and small seeds. Suitable for irrigated tracts, oil content 0.25%, resistant to stem gall and tolerant to wilt, late type with 130-140 days duration, seed yield 12 q/ha.

RCR-47

Plants grow in medium height, resistant to stem gall disease, seed yield is about 12-14 q/ha in 120-140 days. This variety is suitable for irrigated areas.

Kiran (UD 14)

This variety is suitable for irrigated areas, resistant to stem gall, seed yield 12 q/ha in 130-140 days, oil content is 0.25 percent.

UD-20

Plant growth is medium, bold grain, suitable for growing in unirrigated areas, seed yield is 12 q/ha in 120 days.

DH-5

It is devloped by CCSHAU, Hisar through selection from a local collection and indentified in 1993 for irrigated condition. The plants are medium-tall and bushy. It matures in 120-130 days and produces an average yield 18 to 20 q/ha under good management, seeds are round attractive and medium (13.8g/1000).

Katwan type

It is recommended for Madhya Pradesh. It is very popular among the farmers due to maximum green leaf cutting, late maturity and taller growth habit. Crop duration is 150-170 days. The average yield is 10-12 q seed/ha.

Lam CS-2

It is a medium tall (80 cm) variety with more branching and bushy in nature. Mericarp is straw coloured. Crop duration is about 105 days. The average yield is 1300-1400 kg/ha. It is reasonably tolerant to pest and diseases as compared to local varieties. It is suitable for cultivation in Andhra Pradesh, Rajasthan and Gujarat.

Lam CS-4

A fairly tall variety with good branching habit and more number of basal leaves. Crop duration is about 115 days. The variety is reasonably tolerant to pest and diseases as compared to the local. The average yield is 1400-1500 q/ha.

Lam CS-7

It is a selection from north India, acclimatized to conditon of Andhra Pradesh. It is dwarf variety with more leaves of good flavour. Average yield is 9 q/ha. The essential oil content of seed is also high (0.3-0.48 per cent).

RCR-435

The cultivar is developed by RAU, Rajasthan through half-sib recurrent selection from a local collection of Jalore district and identified for release in 1995 for irrigated condition. The Plants are bushy with quick early growth and medium-sized grains (140 g/1000). It matures in 110-130 days and produces average yield of 10.05 q/ha.

RCR-436

It is devloped by RAU, Rajasthan through half sib recurrent selection from a local collection of Kota and identified for release in 1995 for limited moisture conditions. Plants are bushy with quick early growth and bold grains (16.0g/1000). It matures in 90-100 days and produces average yield of 11.09 q/ha under limited conditions.

RCR- 446

It is also devloped by RAU, Rajasthan through half-sib selection from a local collection of Jaipur district and identified for release in 1997 for irrigated condition. The plants are leafy and erect with higher number of grains per umbel. Seeds of this cultivar are medium in size (13.5 g/ 1000). It matures in 110-130 days and produces an average yield of 12 q/ha.

CLIMATIC INFLUENCE

Coriander can be grown successfully during rabi season as it require cool climate. It is grown throughout the year for leaf purpose with the exception of March-May months. But for grain, the crop is adopted to specific seasons. The cultivated area should be free from severe frost during the months of February-March, when the crop is in flowering and seed setting stage. Cloudy weather and high atmospheric humidity during flowering and fruiting stage encourages pests (aphids) and disease incidence. The plants require relatively cool weather during the early stage for better vegetative growth, while a dry and relatively high temperature favours higher seed production. Lancttow Wing *et al.* (1984) reported that early bolting in summer is more common than winter and leads to low yield of seed. It is grown as irrigated and rainfed crop. For leaf purposes, it can be grown throughout the year, provided irrigation facilities are available.

SOIL REQUIRMENT

Coriander grows well under a variety of soils but thrives best on well drained medium to heavy soils. In India, it also grows well in black cotton soils. It can be cultivated in almost all types of soils provided sufficient organic manure applied under irrigated conditions. Unirrigated crop may be cultivated only in havy soil which has better water retention capacity. Black soil with high retention of moisture is best suited under rainfed condition of Andhra Pradesh, Rajasthan, Tamil Nadu etc. It can be grown successfully in soil with a pH 6-7. Bhati (1989a) recommended that coriander can be grown effectively in both light sandy to heavy loam soil by change of variety.

SOWING

Coriander is propagated (sown) by seeds. The seeds are directly sown in the main field by drilling method on commercial scale.

Sowing time

Sowing of coriander is undertaken from September to November for rabi crop. The best time for sowing of coriander is from 15 october to 15 november in most of the commercial growing areas for seed crops. The early sown crop may come to flowering stage earlier and become more vulnerable to frost damage, in north Indian conditions. Late sowing affects the crop growth and its yield adversely, but it is safe in areas vulnerable to frost incidence. In Malwa region of Madhya Pradesh, Kharif crop is sown during August- September and in June-July under Coimbatore condition of Tamil Nadu, but the rabi crop gives better seed yield than the kharif crop.

Seed treatment

Coriander fruit contains two seeds which are fully capable of germination. Therefrore, it is highly essential to divide fruit into halves (mericarps) by rubbing on rough floor with a wooden roller holding by hands at both the ends. This operation not only reduce 50 per cent seed requirement (both halves) but also enhance early and high percentage of germination.

In order to enhance germination, separated fruits are soaked in water for over night (12-24 hours). Water-soaking has resulted in 71.33 per cent germination followed by soacking seeds in GA_3 at 100 ppm 70.29 per cent and NAA-teated seeds 68.69 per cent. Untreated dry seed showed germination upto 55.13 per cent (Banafar, 1994).

Factors affecting high germination percentage in coriander seeds

1. Optimum temperature range 20-30°C.
2. Healty seeds free from any seed borne disease.
3. Current year seeds (not old).
4. Sowing depth is not beyond 2 cm.
5. Seeds from second order umbel give higher germination percentage (Ponnuswamy and Ramkrishan, 1985).
6. Treated seeds with fungicides like Brassicol, Captan, Dithane Z-78 (Prasad, 1988).
7. Soaked seed in water or in growth regulator before sowing.
8. Seeds stored in moisture proof containers at low temperature.
9. Seeds harvested at right stage of maturity.
10. Seeds obtained from winter season *(rabi)* crop have higher germination percentage than rainy *(kharif)* crop.
11. Varieties have difference in germination ability.

Seed rate

Method of sowing and spacing

Seed rate varying from 20-30 kg per hectare in Punjab and 12-19 kg per hectare in south India. The seeds are split into halves before sowing. Split seeds germinate little earlier than whole seeds. Soaking the seeds in water for 12 to 24 hours and drying them in the shade for 12 hours which causes quicker germination. Soaking coriander in warm water, 10ppm IAA, 20 ppm NAA or 50 ppm GA increased germination percentage and leaf yield. Before sowing, seed treatment is given with thiram at the rate of 2.5 g/kg seed. Sowing is done with gorru (seed drill) adopting a spacing if 20x15 cm in between and along the lines in Andhra Pradesh (Rao and Reddy, 1984). Banafer and Singh (1984) have reported that coriander seed is sown in rows 25 to 30 cm apart with a spacing of 5 to 10 pm cm between plants at a depth of 1.5 to 2 cm for irrigated crops and a depth of 4-6 cm for unirrigated crop in Madhya Pradesh. In irrigated soils, where fertility is high, 40 cm row spacing is desirable, otherwise, 30 cm row spacing is optimum for good yields (Bharti *et al.*, 1984). Under rainfed, 30 cm row spacing is otimum for good yields (Bhati *et al.*, 1984). Under rainfed condition, deeds are broadcast and covered with a country plough.

NUTRITIONAL REQUIREMENT

Coriander require very less nutrients for production of green leaves, so farmer generally do not apply fertilizer to crop. Incorporate 100 to 120 q/ha well rotten farmyard manure at the time of land preparation. For getting high yield, apply 20-30 kg of nitrogen, 40 kg phosphorus and 20 kg potassium per heetare as basal dose in last ploughing, have been recommemded under irrigated condition of Andhra Pradesh (Rao and Reddy, 1984). Fertilizer recommendation for irrigated crop is in the ratio of 40:30:20 kg nitrogen, phosphorous and potash/ha, respectively. Whereas, under unirrigated crop 10 kg/ha nitrogen and 20 kg/ha phosphorus have been recommended in Madhya Pradesh condition (Banafar and Singh, 1984). In Rajasthan condition, 20 kg of nitrogen, 30 kg of phosphorus and 20 kg of potassium per hactare have been recommended for good crop of coriender in unirrigated areas. For irrigated crop, nitrogen is increased upto 60 kg per hectore as reported by Bhati *et al.* (1984). Full quantity of phosphorus and potassium and half quantity of nitrogen may be applied at the sowing and remaining quantity of nitrogen four to five weeks after sowing.

Rahman *et al.* (1990) recommended application of 60:40:40 kg of NPK/ha for red sandy loam soil of A.P. However the National Research Centre for spices recommends application of 10-15 tonne FYM + 30:40:20 kg of NPK/ha as basal dose for irrigated crop and for rainfed crop basal application of 25T FYM + 15:20:0 kg of NPK/ha (Edison and Kallupurackal, 1991).

Cultivar such as Gujarat Coriander-1 has been reported to give higher yield at 60,30,30 kg/ha NPK, respactively (Ughreja and Chuandawat, 1992). Coriander sown in October produced higher yield. It may be because of the fact that K-accumulation became higher when doses of NPK were applied at 40:40:20 kg/ha (Venkateswarlu, *et al.*, 1994).Result revealed from the study conducted at NRC for Spices, Calicut indicate that coriander crop yielding seed of 631 kg per hectare remove 12 kg N, 4.5 kg P_2O_5, 11 kg K_2O, 5.1 kg Ca, 1.0 Kg Mg, 88g Fe, 16g Mn, 26g Zn and 11g Cu. It indicates that removal of nitrogen is highest and potassium is also more or less equal to nitrogen. Amongst micronutrients, removal of iron is highest followed by Zinc (Rethinam and Sadanandan, 1994).

Increasing application rates of N and P at 60 and 45 kg/ha, respectively increases N and P uptake and results in hegher yields (Tiwari and Banafar, 1995). Application of Nimin (neem seed extract) coated urea at 60 kg N/ha give highest return (Singh *et al.*, 1997).

Table 2 : The experiments conducted by various researchers revealed the best doses of NPK (kg/ha) for coriander in some states.

State 1	N 2	P_2O_5 3	K2 4	Reference 5
Andhra Pradesh	60	40	20-40	45
	-	60	-	27
	40	40	20	63
Bihar	60	40	20	3
Delhi	120	15	-	48
Gujarat	60	30	38	62
Haryana	60	-	-	59
Karnataka	40	60	30	8
Madhya Pradesh	40	30	20	35
	60	45	-	60
	80	40	30	61
Rajasthan	20	30	20	8
	-	93	-	53
	60	+Nimin	-	55
	50	50	60	4
Tamil Nadu	20	40	20	39
Uttar Pradesh	100	30	-	5
Weat Bangal	40	-	-	16

Micronutrient like Zn at 5 ppm has been found to increase coriander seed yield (27%) over control (Gupta and Raj, 1980) and so also spraying of Cu ($Cu_2SO_4.5H_2O$) at 0.25 per cent at 25, 37 and 49 days after sowing has been found to reduce stem gall disease, increase seed yield and essential oil content (Maurya, 1990).

IRRIGATION

During rainy season, coriander does not require irrigation expcet in long dry spell. However, winter season crop is given pre-sowing irrigation when soil moisture is lacking or a light irrigation is done as soon as sowing is completed, for early and better germination. The second irrigation is needed 7-8 days after sowing. Generally, it takes 8-10 days for proper germination. There after, the irrigation may be 6 or 7. Irrigation at pre-sowing and seed filling stages are essential.

No irrigation is necessary for coriander grown in black cotton soils, whereas, in light soils two three irrigation are required in south Indian conditions. For irrigated crop in addition to pre-tillage irrigation, three to four irrigation are riquired under north Indian conditions.

No doubt, sufficient soil moisture should be provided to the crop throughout but care should be taken for irrigating coriander crop at its critical stages (i.e. seed germination, branching, flowering, and fruit setting, seed filling stage). Higher yield is obtained when irrigations are given at critical stages (Tomar *et al.*, 1994). Irrigation with sodic water results in greater percentage reduction in seed yield (Yadav *et al.*, 1996).

Irrigation of coriander crop with sewage water at 3 times a week had been found to have a signifant reducation in the number of days to flowering, and significant increase in plant height at maturity and seed yield, compared with control (Devi 1991).

INTERCULURE

The germination is completed in a about 12-15 days after sowing. The first hoeing and weeding are done in 30-35 days after sowing. Thinning of plants is also done simultaneously to plant distance in line-sown crops, leaving two plants per hill and at this time the remaining half quantity of nitrogen may also be given. Second and third weedings are generally done at an interval of 30 days depending upon the weed growth (Rai and Yadav, 2005). Weeding and hoeings are done to remove the weeds, which provides better soil aeration to the crop and to preserve the soil moisture.

WEED CONTROL

In fact, hand weeding and hoeing check the weed growth in early stage of vegetative growth of coriander crop. However, Raghuvani *et al.* (1985) has recommended pre-emergence application of 0.9 kg/ha fluchloralin for higher yield, closely followed by 0.75 kg oxadiazon pre-emergence. Oxadiazon gave the highest net return and weed control efficiency (98 per cent). Malviya *et al.* (1989) observed best weed control with pre-emergence oxadiazon at 3.0 kg/ha. Kothari *et al.* (1989) observed 76.5 and 71.9 per cent weed control with pendimethalin at 0.75 kg/ha and fluchloralin at 1 kg/ha, respectively. Where as Thakral *et al.* (`1992) suggested pre-emergence application of pendimethalin and Fluchloralin each at 1.5 kg per hectare to control most of the weeds. Pre- emergence application of 1 kg Methazole or prometryn and 4 kg chloroxuron in spring season coriander whereas autumn application of 1 kg linuron or 0.5 kg prometryn and 4 kg chloroxuron have proved very effetive in controlling weeds with zero per cent damage to coriander crop (James *et al.*, 1991).

PLANT PROTECTION MEASURES

Insect Pests

Coriander crop is harmed anly by a few pests. The aphid is one of them.

Aphids (Hyadaphis coriandri Das)

Coriander is seriously affected by aphids particularly during flowering and fruit setting periods. Aphid incidence is generally greater on the late flowering varieties than early flowering ones. To control the aphid spraying the crop with Thiometon 25 EC at 12 ml in 10 liters of water or Endosulfan 30 EC at 20 ml in 10 liters of water or Phosphomidon or Monocrotophos before flowering, is recommended (James *et al.*, 1992).

The insecticides which are used for the control of aphids also kill the useful insects like honey bees which are pollen carrier help in pollination. The visiting hours of these useful insects are 10 a.m. to 4 p.m. therefore, It is advisable not to spray insecticide(s) for the control of aphids between 10 a.m. to 4 p.m. The spraying operation may be conducted before 10 a.m. an/or after 4 p.m.

Rao *et al.* (1982) reported that monocrotophos at the rate of 0.5 kg/ ha effectively controlled the *Thrips tabaci* and *Scirtothrips dorsalis* and chloropyrifos 0.5 kg/ha controlled *Spodoptera litena.* Bhutani (1984) observed efficacy of different insecticides on coriander plants and its stored seeds against various pests. He suggested efficient control of aphids (*Hyalopterus pruni and Bervicoryne coriandri*) by spraying of 0.05 per cent endosulfan and 0.03 per cent dimethoate or quinalphos for sap sucking bugs (*Nezara viridula and Agonoscelis nubilis*). Leaf feeding caterpillar, lapidoptera (*spodoptera litura and S. exignua*) can be controlled by flooding the field, collecting the egg masses and spraying of 4 per cent carbaryl or endosulfan. For the control of stored beetels (*Lasiederma serricare* and *Trogoderma eranarium*), heat treatment and fumigation have been recommended. Mital (1989) obsreved that control of aphids (*Hyadaphis coriandari*) by a spray of insecticides like ethylparathion, endosulfan and thiometon, which are safe to pollinator honeybees. It was concluded that for safe consumption of endosulfan treated coriander as a green, leafy spice, a waiting period of 11 days should be followed after application of the insecticide. No such waiting period was needed in case of mature fruits (Jain and Yadav, 1989).

Diseases

Powdery Mildew (Erysiphe polygoni)

The initial stage of this disease is marked by the appearanace of a whitish powdery mass on the leaves and stem. If the disease is unchecked at this stage, the whole plant becomes almost white with the powdery coating. Early infection does not permit seed formation, however, if infection occurs at a late stage, seed formation may take place but the seed will be small and shrivelled, thus affecting the yield and quality.

Control Measure

Spraying the plants with 0.2 per cent solution of wet sulfur or karathane can effectively control the disease. Spray should be repeated after 10 to 15 days of first spraying. Dusting the crop with 20 to 25 kg sulfur dust per hectare would also control the disease. Dusting may be repeated if necessary.

Stem Gall (Protomyces macrosporus)

This is a soil borne disease, which appears in the form of tumour like swellings on all aerial parts. The fungus is restricted to tumours-only. The affected leaves and fruits become hypertrophied and appear to lose flavour and other properties.

Control Measures

Bhardwaj and Shrestha (1985) applied four foliar sprays of thiram (0.2 per cent) captafol (0.2 per cent) and carboxin (0.1 per cent) at fortnightly intervals from the time of disease infestation and reported that they were more affective than seed treatment or combined seed and soil treatment in controlling stem gall and increasing yield. Kiran (UD-14) varity had been found highly resistant to this disease under field conditions.

Wilt (Fusarium oxysporum and F. Coriandrii)

The disease can attack the plants at any stage of growth. Severity of infestation increased with age. The infested plants show drooping of the terminal protion followed by withering and drying up to leaves eventually resulting in death. When the wilt infection occurs in the roots, it is very difficult to control this disease and it causes heavy losses.

Control Measures

(i) Use disease free seeds obtained the wilt free crop of the proceeding year. (ii) Treat seeds with suitable fungicides before sowing.(iii) Do not grow coriander continuously for more than three years in the same field. (iv) Provide deep summer ploughing to the land. (v) Avoid heavy saturation of soil moisture.

Use of Growth Substances

Amruthavalli (1978) studied sex-expression in coriander as affected by growth regulators. Coriander plants at the five leaf stage sprayed with GA, ethrel (ethephon) or CCC (chormequat) each at 100 ppm. The application of growth regulators give significantly higher nutrient uptake (Lal *et al.*, 1997). Male flowers numbers/umbels are increased by GA but reduced by ohter growth substances. The resulting ratio of bisexual and male flowers were the highest with CCC (1:0.79) followed by ethrel 91:0.96) as compared with 1:1.20 for untreated (control). Interaction of gibberellic acid (GA_3) and benzyladenine (BA) on flowering and sex expression in Bulgarian coriander was studied by Amrutavalli (1980) who concluded that GA_3 (25 ppm) improved flower, yield and strengthened the male tendency. Benzyladenine (BA) produced the opposite effects. Benzyladenine in combination with GA_3 did not alter GA_3 influence on the onset of flowering but strongly reduced its effects on flower, yield and sex ratio. Prakash and Kumar (1984) soaked seeds in various growth regulators at vairous concentrations. Treatments were continued until the flowering. The morphology of the umbels and sexual status of flowers were affected by the treatments. The highest number of flowers was produced by 50 ppm IAA, while the highest frequency of bisexual flowers (93.91 per cent) was produced by 50 ppm ascorbic acid.

Magnesium chloride 8 kg/ha applied at the waxy stage of ripening had good desicating properties and made harvesting easier. The treatment advanced fruit drying and ripening and increased essential oil accumulation but slightly reduced tinatal content and the germination of the seeds was also improved (sharygina, 1985). Early and higher germination have also been recorded by soaking coriander seeds in solution of GA_3 (50 ppm) or IAA (10 ppm) or warm water for 12 hours in comparison with no soaking. In addition to soaking treatment, when such soaked seeds are wrapped in polythene sheet or gunny bag cloth, the germination percentage increased, and it was earlier than unwarpped control. The germination percentage is maximum in 50 ppm GA3 + gunny bag cloth wrapping, followed by 20 ppm NAA + gunny bag cloth warpping.

HARVESTING

Coriander plants, can be nipped off when it attain the height of 5-6 cm so that the whole plant may not be uprooted. Later on, for garnishing purpose, the leaves and stems are nipped when suitable length of the plant is reached. Seed crop matured in about 90-150 days depending upon variety, soil and growing season. Crop exculusively grown for greens are harvested afters 30-35 days of sowing. Staggered sowing and harvesting make the fresh green leaf available over longer period.

The kharif season crop matures earlier than rabi season. Harvesting is done when the grain changes its colours to straw or light brown colour. Bhati (1989) indicated that when the crop was harvesting with 50 per cent or 100 per cent fruits truning yellow, the crop yield ws significantly higher than the 100 per cent green fruit stage harvest. The plants are cut or pulled and filled in to small stocks in the field to wither for two or three days. Threshing is done by beating with sticks or by rubbing with hands. The produced is winnowed, cleaned and dried in partial shade. For storage, seeds should not contain more than six per cent moisture. After drying, produced is stroed in gunny bags line with paper or in polythene bags to aviod entrance of moisture (Rai and Yadav, 2005).

YIELD

The green leaf yield varies from 30 to 40 quintals per hectare when crop a fully grown for green leaf purpose, while the seed yield varies from 15 to 20 quintal per hectare.

POST HARVEST HANDLING

Grading

The plants after harvest cleaned properly and make into bundle to represent the grades for sale. Grading is done in "super" "fancy"and "commercial" grades according to size, tenderness, leaf colour and freshness. Good quality with uniform sized tender leaves are kept in 'super grade' and medium size of leaves with good texture and freshness arrange in 'fancy' grade. The Plants with older age and dull in colour and free from spots and blemish are arranged in 'commercial' grade. There after plants make bundles of different grades and send to market for sale.

Packing

The bundles are packed in wet gunny bags/bamboo chips baskets or Arhar stick baskets or in vegetable crates and cover with moist gunny bags or cloth to keep produce fresh. Packing is done according to grade alongwith bundles then stitched properly and send to market.

Storage

The produce should be kept in cool and moist condition before disposal. It can be stored in well ventilated store with moist condition for 12-14 hours and time to time water sprinkle on the packets to keep fresh well dried leaves can be stored for one year. Coriander is available in this form in packets. Thus the dried leaves can be used throughout the year.

SEED PRODUCTION

Coriander seed can be produced freely in plains of the country as well as on the hills. It is a cross pollinated crop and requires 800 metres isolation distance between two vatieties. Better seed yield and quality is obtained when cutting or nipping of green leaves or stem is not done. Seed crop is sown from second week of October to second week of November in the plains and 15th February to 15th March on high hills. The recommended cultural practices have to be followed during its growth period.

Culture Practices

Roguing

A minimum of the three roguing are essential to maintion the crop as per field standerd (Table 3).

First roguing

The first roguing should be done before flowering preferably within 45 days of sowing to determine volunteer plants, off- types.

Table 3 : Field standerd for coriander seed production (maximun permitted in per cent)

Factor	Foundation	Certified
off-type	0.10	0.50
Other crop plants	-	-
Objestionable weeds	-	-
Diseased Plants	-	-

Second roguing

The second roguing should be done during the 50 per cent flowering. At this stage late flowering plants, diseased plants are completely uprooted.

Third rouguing

The third roguing should be done during maturity and prior to harvest. At this stage, the true nature of plant can be judged. All later maturing plants should be uprooted.

Table 4 : Seed standards for coriander seed production

Factor	Foundation	Certified
Pure seed (maximum%)	98.0	97.0
Inert matter (maximum %)	2.0	3.0
Other crop Seeds (maximum%)	10/kg	20/kg
Weeds seeds (maximum%)	10/kg	20/kg
Germination (minimum %)	65	65
Moisture (maximum %)	10	10
Moisture (maximum%) for Vapour proof Containers	8	8

Harvesting

When coriander fruits start turning from green to yellwish- brown colours (50 to 60% fruits) stage indicates for harvesting (100-120 days or so after sowing). One should not wait for fully mature umbels because there is fear of shattering of fruits. Not only this, delay in harvesting causes dull colour of the fruits which affects quality. Early harvesting before reaching desirable stage also causes poor quality as well as less yield.

Harvesting of coriander is performed by uprooting and/or cutting plants by sickle. The harvested plants are made into small bundles and they are staked under partial shade that will help in retaining greenish colour of the fruits which is rating as quality colour.

Drying and Threshing

Harvested produce is allowed to dry under partial shade for about 5-6 days or so. Completely dried plants are threshed by beating them with sticks. The seed free from chafs and other materials are taken out before winnowing.

Winnowing and Drying

After threshing, the material is seived, winnowed carefully by winding up almost all chafs, dust and dirt are eleaned. Clean seeds are then allowed to dry on tarpaulin or clean cemented floor for 4-5 days or so till moisture percentage of seed reaches to 10.0. Solar drier or electric drier may also be used for ealry drying, if available.

Storage

The dried seed is filled in plastic bags or in gunny bags lined with palstic sheet. The mouth of the each bag is seamed and stored under clean, dry ventilated place.

Seed Yield

On an average, 10 to 12 quintals or so seed yield of coriander is obtained from an area of one hectare. One leaf cutting at 30 days gives better seed yield (Babu and Ranot, 1995). The production vaires with the seasons, soil types, cultivar, incidence of pests and diseases.

The productivity of coriander in India is about 442 kg/ha which is very low that can be improved a lot with respect to seed yield potential and quality with respected to essantial oil content (0.40 to 0.8%). The European countries having coriander with very high essential oil content (1.4 to 1.7%).

The seed yield per plant recorded high heritability with moderate value of genetic advance as percentage of mean. The number of branches and fertile umbels per plant are positively and significantly associated with the seed yield per plant (Choubey, 1991).

Spraying of 5 ppm mixtalol (Aliphatic alcohol) enhances plant growth, umbels per plant, hermaphrodite flowers per umbel, improves sex-ratio (hermaphrodite : staminate) per umbel and chlorophyll content of leaves and produces more seed yield (Lal *et al.*, 1996).

REFERENCES

Amruthavalli, S.A. (1978). *Current Science*, Vol. 47: 129-136.

Amruthavalli, S.A. (1980). *Indian Journal of Plant Physiology*, Vol. 23: 14-20.

Aykroyd, W.R. (1941). Health Bulletin No. 23, Nutrition Research Laboratory, Koonoor.

Baboo, R. and Ranot, N.S. (1995). Effect of cutting management, nitrogen and phosphorus on growth and yield of coriander. *Indian J. Agron.*, 49(2) : 253-255.

Banafar, R.N. and Singh, K, (1984). *Indian Cocoa Arecanut and Spices Journal*, Vol. 8: 2.

Banafar, R.N.S. (1994). Influence of hormones and chemicals on the germination of coriander seed. *Coriandrum sativu* L). *Gujarat Agric. Univ. Res. J.*, 20 (1): 111-113.

Baswana, K.S., Pandita, M.L. and Sharma, S.S. (1989). Response of coriander to date of planting and row spacing. *Indian J. Agron.*, 34(3): 355-357.

Bhardwaj, L.N. and Shrestha, S.M. (1985). *Agriculture Eco. Systems and Environment*, Vol. 13: 314-319.

Bhat, V.R. and sulikeri, G.S. (1992). Effect of N, P. and K on seed yield and yield attributes of coriander (*C. sativum* Linn). *Karnataka J. Agric. Sci.*, 5 (1) : 26-30.

Bhati, D.S. (1989a) Abstract papers of First *National Saminar on seed Spices*, held at RAU, Bikaner (Jaipur) during. October. 24-25, 1989, pp: 20-21.

Bhati, D.S. (1989b). Abstract papers of First *National Seminar on Seed Spices* held at RAU, Bikaner (Jaipur) during October, 24-25, 1989, pp. 19-20.

Bhati, D.S. and Shakhawat, M.S. (1994). Effect of soaking date, row spacing and nitrogen on a qulity parameter of coriander. *Progr. Hort.*, 26(12): 14-18.

Bhati. D.S., Jain, M.P., Sharma, R.K., Bhatt, B.N. and Choudhary, G.R. (1984). *Indian Cocoa, Arecanut and Spices Journal,* Vol. 84 pp: 89.

Bhatta, G.K.M.G. and Hiremath, P.C. (1989). Chemical control of coriander blight. *Curr. Res.* 18 (12): 175-176.

Butani, D.K. (1984). Pesticides, 18: 15-17.

Choubey, P.C., Sharma, B.R., Verma. V.K. and Banafar, R.N.S. (1991). Variability studies in coriander (*Coriandrum sativum* L.) *Scientific Hort.*, 2: 137-144.

Devi, P. (1991). Growth estimates of sewage irrigated coriander and fenugreek. *Adv. Plant Sci.*, 4(42): 394-396.

Edison, S. and Kallupurackal, J.A. (1991). *Indian Cocoa, Arecanut* and *Spices Journal*, 13: 130-144.

Gladis,T., Diagne, O., Spahillari, M. and Hammer, K. (1996). Reproduction of medicinal and aromatic plants in the Gatersleben gene bank. *Proc. Int. Sym. Breed. Res. on Med & Aromatic Plants.* Quadin bury, Germany, 30 June, 4 July, (1996).

Gupta, Kaushalya; Thakral, K.K., Agora, S.K. and Wagle, D.S. (1991). Studies on growth, structural Carbohydrate and Phytate in coriander during seed development. *J. Sci. Food & Agric.*, 54(1): 43-46.

Gupta, V.K. and Raj, H. (1980). A note on effect of Zinc application on the yield and phosphorus nutrition of coriander. *Hayrana J. Hort. Sci.*, (1&2): 82-83.

Hill, Albert, F. (1992). Economic Botany. Tata McGraw-Hill Publishing Company, Ltd Bombay, New Delhi.

Jain, P.C. and Yadava, C.P.S. (1989a). Relative susceptiability of coriander varieties to coriander aphid (*Hyadaphis coriandrii* Das), *Indian J. Ent.*, 50(4): 535-536.

Jain, P.C. and Yadava, C.P.S. (1989b). Incidence of pests and their control on Coriander. *Indian Cocoa, Arecanut* and *Spices J.* 13(2): 62-65.

James, T.K., Rahman, A. and Douglas, J.A. (1991). Phosphorus requirement of coriander in black clay soil (Vertisol) of low available P_2O_5. *Indian Cocoa, Arecanut* and *Spices* J. 12(93): 112-113.

Katole, N.S., Sharma, O.L. and Dungarwal, H.S. (1989). Abstract papers of first *National Seminar in Seed Spices*, held at RAU. Bikaner (Jaipur) during October, 1989, pp:13.

Kothari, S.K., Singh, J.P. and Singh, K. (1989). Tropical Pest Management, 35: 2-5.

Koul, A.K., Hamal, L.A. and Gupta, S.K. (1989). Pollination mechanism in *Coriander Sativam* Linn (Apiaceae). *Proc. Indian Acad. Sci. Plant Sciences,* 99(5): 509-515.

Lal, H., Rathore, S.V.S. and Dhaliwal, K.S. (1997). Seed quality and nutrient uptake by coriander as influenced by irrigation and mixtalol spray. *J. Indian Soc. Soil Sci.*, 45(2): 230-234.

Lal, H., Rathore, S.V.S. and Kumar, P. (1996). Influence of irrigation and mixtalol spray on growth, flowering and yield of coriander. *Indian J. Soil Conservation,* 23(2): 141-146.

Lancttow, Wing, K.F., Bhusawan, M. A. and Rajkonar (1984). *Tachnical Bulletin,* Ministry of Agriculture, Fisheries and National Research, Mauritius, 4: 139-144.

Malviya, D.D., Kavani, H.D., Peter, J.C. and Raghvani, B.R. (1989). Abstract papers of first *National Seminar on Seed Spices held* at RAU, Bikaner (Jaipur) during October, Vol. 24-25, pp: 13.

Maurya, K.R. (1990). Effect of micronutrient on yield and essential oil content of coriander (*Coriandrum Sativum* Linn). *Indian Perfumer,* 34(4): 263-265.

Mital, V.P. (1989). Abstrct paper of first *National Seminar on Seed Spices,* held at RAU, Bikaner (Jaipur) during October, Vol. 24-25, pp.37.

Pillai, F.K.T. and Nambiar, M.C. (1982). Cultivation and utilization of aromatic plants. *Bull,* R.R.L. Jammu, pp. 167-189.

Ponnuswamy, A.S.and Ramkrishan, V. (1985). Germination and vigour of Coriander (*Coriandrum sativum* L.) seeds and relative to their position in umbel of diffrent orders. *South Indian Hort.,* 33: 30-34.

Prakash, G. and Kumar, N. (1987). *News letter,* 16: 7-8.

Pruthi, J.S. (1979) Spices and Condiments. National Book Trust of India. New Delhi.

Raghuvani, B.R., Malvavia, D.D. and Kavani (1985). *Hort. abstract,* 56: 2725.

Rahman, M.O., Hari Babu, R.S. and Rao, N.S. (1990). *Indian Cocoa, Arecanut* and *Spices Journal,* 13: 130-133.

Rai, N. and Yadav, D.S. (2005). Advances in Vegetable Production. Research Co Book Centre, New Delhi.

Rao, D.M.K., Ahemd , K. and Rao, T.S.R. (1982). *Indian Cocoa Arecanut* and *Spiecs Journal,* 7(1): 15-16.

Rao, M., Venkata and Venkata Reddy, P. (1984). *Indian Cocoa Areeanut* and *spices Journal,* 8: 1.

Rethinam, P. and D. Sadanandan, A.K. (1994).Nutrition and management of seed spices *Advances in Hort.,* 9(1): 499-513.

Sharygina, I.S. (1985). *Hort. Abstact,* 55: 1405.

Singh, D.N. and Nath, Vishal (2011). Varieties and Hybrids of Vegetables. Satish Serial Publising House Delhi. pp: 353-357.

Singh, R., Singh, S.D., Yadav, B.L. and R. (1997). Effect of nimin coated urea on nitrogen uptake and utilization efficiency by coriander (*Coriandrum sativum* L.). Crop Res., 13(3) : 653-656.

Singh, V.V. and Singh, Kirti (1996). Spices. New Age International Publishers, New Delhi.

Singhal, V. (1996). Coriander. In: Indian Agriculture. Indian Economic Data Centre. New Delhi, pp. 528-529.

Thakral, K.K., Singh, G.R., Pande, U.C. and Srivastava, V.K. (1992). Effect of nitrogen levels and cuttings on the production of green leaves and seed yield of coriander cv' Naraul' *J., Res* HAU, 22(1): 35-39.

Thakral, K.K.; Khurana, S.C. and Srivastava, V.K. (1989). Evaluation of herbicides in coriander seed crop. *Proc. First National Symposium on seed Spices*, Jaipur. Spices Boad, Ministry of Commerce, Govt. of India, Kochin, pp. 109-119.

Thangaraj, T. and Vijayakumar, M. (2001). Coriander (*Coriandrum sativum* L.). In: Vegetables, Tuber Crops and Spices. (Eds. Thamburaj an Singh, 2001). Indian Council of Agricutural Research. pp: 371-377.

Tiwari, R.J. and Banafar, R.N.S. (1995). Application of nitrogen and phosphorus increase seed yield and essenital oil of coriander. *Indian Cocoa, Arecanut* and *Spices J.*, 19(2): 51-55.

Tomar, S.S., Gupta, K.P., Abbas, M. and Nigam, K.B. (1994). Effect of irrigation and fertility levels on growth and yield of coriander. *Indian J. Agron.*, 39(3): 442-447.

Ughreja, P.P. and Chundawat, B.S. (1992). Nutritional studies in coriander (*Coriandrum sativum* L.) III. Levels of nutrient in various Plant parts. *Gujarat Agril. Res.J.* 18(1): 70-77.

Venkateswarlu, K., Reddy, S.E. and Reddy, Y.N. (1994). Potassium uptake efficiency of certain coriander cultivars sown at different dates. *Indian Cocoa, Arecaunt* and *Spices J.*, 18(3): 92-96.

Yadav, H.D., Singh, S. and Kumar, V. (1996). Response of winter spices to sodic water irrigation in light textured sodic soils. *Res. J. HAU.*, 26(1): 56-55.

□□□

Chapter-4

Salad Vegetables

1. CELERY

Botanical name	:	*Apium graveolens* L.
Family	:	Apiaceae (Umbelliferae)
Chromosome number	:	2n = 22

Celery belongs to the family apiaceae and rank second (after littuce) in importance amongst salad crops (Saini, 1996). Celery is a moisture loving, long-seasoned annual, usually grown under cool condition. It is mainly grown for its long fleshy leaf stalk and looks some what like leafy onion. It is widely grown in USA, France and in other European countries commercially and is available in the market throughout the year. In the high elevation of southern part of India, it is grown on limited scale for salad purpose (Yawalkar, 1969). In some pockets of northern India like Punjab and Uttar Pradesh celary is grown mainly as seed crop. The seeds are exported to European countries as condiment. It is also grown as fresh vegetable in kitchen gardens or home gardens.

ORIGIN

Celery is a native to marsh lands from Sweden to northern Africa and from southern Asia to the Caucasus. The centre of origin of the genus *Apium* is considered to be the mediterranean basin (Simmonds, 1976). The place of orign of celery extends from Sweden southward to Algeria, Egypt, Abyssinia, and in Asia even to the Caucasus Baluchistan and the mountains of India. It is likely that *A. graveolens* was first cultivated as a medicinal crop as early as 400 BC in Egypt and in the Roman Empire. Celery was domesticated as a true vegetable in Italy in the sixteenth century (Vilmorin, 1950).

DISTRIBUTION

The genus *Apium* is spread throughout the world, particularly in the mediterranean basin, Australia, New Zealand, South Africa and South America. . Celery is a popular crop widely cultivated throughout the world. (Carlos F. Quiros, 1993). Celery is derived from *A. graveolens* L., which occurs wild as a marsh plant throughout temperate Europe and Asia (Purseglove, 1965). In India it is grown in Punjab, Uttar Pradesh, and some cooler regions of the southern parts of the country.

BOTANY

Celery, *Apium graveolens* is a biennial plant, although grown as annual crop (Thompson and Kelly, 1983). The plant is characterised by enlarged, tender, edible, petioles or leaf stalks. It is a strong-smelling, glabrous herb to 1m tall; root leaves pinnately compound; flowering stem grooved and jointed; flowers white very small, in small compound umbels (Purseglove, 1965). Fruits are rigid, 1 to 1.5 mm long and 1mm in diameter contains a small seeds. Seeds are small brown, slow in germination, united or separated pericarp, and some with stalk ends. 10g seeds counts about 26450 and seeds are somewhat bitter. The dry ripe fruits are used for flavouring and also medicinally.

Taxonomy

According to Carlos F. Quiros (1993), *Apium graveolens* L, comprises three distinct cultivated forms or taxonomic varieties, which are grown for different economic purposes. As per the part of the plant consumed, these varieties are classified as follows:

Apium graveolens, var. dulce, common name celery

The plants of this type develop succulent, solid petioles which are consumed mostly raw in salads or as celery 'sticks' and are often cooked in soups and stews. It is widely grown in Europe and in the America.

Apium graveolens var. rapaceum, common name celeriac or root celery

It is characterized by the development of enlarged hypocotyl and root tissue. The resulting glob-like structure, which has a strong flavour, is consumed mostly cooked in stews and soup or grated on salads. This crop is grown mostly in northern and eastern Europe.

Apium graveolens var. secalinum, common name smallage or leaf celery

The plants of this group have slender, leafy and often hollow petioles, and are used mostly for their leaves as a condiment garnish and for medicinal purposes. Smallage is consumed mostly in Asia and in the Mediterranean countries. Annual forms of this varieties are also grown for seed which is used as a condiment and for beverages.

Floral biology

Celery flowers are protandrous, with pollen being released 3-6 days before the stigma is receptive. At this time the anthers will drop and the two stigmas will unfold in an upright position, becoming receptive (Honma, 1959). The degree of protandry varies, which makes reliable hybridizations difficult to perform because of the possibility of accidental selfing.

Celery has a very high degree of outcrossing, approximately 70%, due to wind and insect pollination (Orton and Arus, 1984). The pollinators observed at Davis visiting celery flowers included a large number of species, covering ten families of wasps and bees and eight familier of flies (Quiros *et al.*, 1986).

In order to induce seed-stalk development, plants require a period of vernalization 6-10 weeks at 5-8°C when they are at least 4 weeks old (Honma, 1959). Due to a wide range of responses to the cold treatment, it is often difficult to synchronize the crosses since plants will flower at a different times. However, pollen can be stored for 6-8 months at -10°C with a viability decline of approximately 20-40%, thus providing flexibility to perform crosses most of the time (D. Antonio and Quiros, 1987).

COMPOSITION

Celery is an important source of minerals, vitamins and aminoacids. Leaves are more nutritive than stalks especially in vitamin 'A', protein and calcium. Traces of copper and arsenic are also in tuberous root. The herb contains glucoside apiin (Anonymous, 1952). Nutritive value of celery stalk is given below in per 100g of edible portion (Choudhury, 1967).

Moisture	93.5g	Calcium	30.0 mg
Fat	0.1 g	Iron	4.8 mg
Fibre	1.2g	Vitamin'A'	867 I.U.
Carbohydrate	3.5g	Vitamin 'C'	6.0 mg
Protein	0.8 g	Riboflavin	0.05mg
Minerals	0.9g	Thiamin	0.12 mg
Calories	18	Nicotinic acid	0.3 mg

Uses

The leafstalks and petioles are eaten as salad, in soups, sauces, puree, fried and spiced. Its tough outer petioles are the basis of celery soup. The young swollen petioles are eaten after blanching, either raw or as flavouring (Ryder, 1979). The seeds of celery is used as condiment in european countries and spice in India. Celery can be preserved in cans after blanching. The dry ripe fruits are used for flavouring and also medicinally. They yield 2-3% of a volatile oil with a persistent characteristics odour of the plant. It is used for flavouring salt and as a fixative and ingredient of novel perfumes (Purseglove, 1965).

As a medicine, used to combat certain maladies and drunkenness, and also as an aphrodisiac. Iranians boil seed oil for inhalation of vapor to cure headaches. Seed is diuretic, a digestive aid. To induce menstruation wild celery seed is used in Egypt. In India it is mostly grown for seed to export for culinary sauces, oleoproteins and tonics. In the late 1800s and earlier this century, tonic, sodas and elixirs prepared from crushed celery seeds and were widely consumed in the USA (Russell, 1984, Garfield, 1985). It was claimed that these beverages were antiflatulent, diuretic, nerve builders and aphrodisiacs.

VARIETIES AND HYBRIDS

Vilmorin (1950) reported that two varieties of solid stem Italian celery were available in France, 'Solid celery' and Striped rose'. In 1875, Chemin selected a line known as 'Solid Golden White Celery' which became the variety 'Paris Golden' Yellow Self-blanching. In 1884 Pascal selected a green celery line from 'Paris Golden Yellow self blanching' which was known in the seed trade as Pascal celery. In 1887, both varieties were introduced in North America under the names of 'White Plume' for the self-blanching type and 'Giant Pascal' for the green type.

Celery cultivars are two types, viz., self-blanched and green cultivars.

Self blanched type

Self blanched or golden self blanch is tends to blanch easily and has less demand due to low vitamin ' A' content. They are Michigan Improved Golden,

Cornell 19 (thick petioles susceptible to brown spot), Cornell 619 (thick petioles, resistant to fusarium yellow), Supreme Golden, Golden plume or Wonderful, Golden self-blanching, Golden No 15 and Florida Golden.

Green cultivars

Green cultivars are gaining importance now a days. They are: Utah (dark green), Utah 52-70 (resistant to brown spot), Utah 10-B (have boron and magnesium deficiency), Giant Pascal, Emerson Pascal, Summer Pascal (resistant to brown spot) and Ford hook. All green cultivars are resistant to *Fusarium* yellows. Emerson Pascal is resistant to early and late blights also, but has a tendency to bolt.

Singh and Nath (2011), described the detail characteristics of cultivars, which are given below:

Cultivars from IARI

Some of the celery cultivars (Anonymous, 1998) acclimatized at IARI, sub station, Kulu valley, are given below (Richharia, 1957).

Standard Bearer

It is an early medium tall plant, produces medium pink stems with longitudinal streaks, stalk solid and of good size and flavour.

Wright's Grove Giant

This cultivar, is medium late, tall growing,high yielding produces large white stalks of fine quality.

Fordhook Emperor

It is late cultivar, plant dwarf and stocky, produces solid, white, thick, broad and tender stalks.

Triumen

It matures in 85 to 90 days, has dark green coloured attractive foliage and medium long petioles. It has good standing ability.

Celery Paris

It is very resistant to heat and rain. The foliage is tall and dark colours. Stalks are thick smooth and fleshy. Seeds are very small (more than 2500 seeds per gram). Plants can be harvesed 2-3 months after transplanting.

Celery France leaf

It has vigorous dark green and sturdy foliage with long, thick and smooth stalks.

Table 1 : Some of the varieties and their horticultural traits are given below

Variety	Duration (days)	Plant height (cm)	Colour	Leaf size (cm) L-B	Yield 3 m line (kg)
Utah-52-70	100	57	Dark Green	17.4-9.2	2.738
Golden Self Blanch	82	31	Medium yellow	14.6-14.6	1.533
White gem	100	35	Green	13.2-5.3	2.325
Grant Red	70	67	Light pink	12.0-6.0	2.000
Belder	67	18	Stem Dark green	12.0-5.3	1.700
White solid	68	32	Light green	14.7-5.8	1.300

Peshin, 1998
Utah-52-70 is best for winter and Golden self blanch for summer sowing.
L = Length; B = Breadth

CLIMATIC INFLUENCE

Celery grow best when the weather remains relatively cool (October to March) with a well-distributed moderate rainfall during the growing season (Thompson and Kelly, 1983). It needs plenty of sunshine for good growth. Celery is moisture loving plants. It can be grown in dry regions as an irrigated crop. The optimum temperature (2-10°C) range is 15-21°C. Low temperature causes bolting and shorten the time required for vernalization. It is favourably grown above 16°C temperature. Higher temperature causes bitterness in the leaves. According to Naik (1958), the crop comes up well in a places having low humidity and where plenty of sunshine is available.

SOIL REQUIREMENT

A good well drained sandy loam soil well supplied with organic matter is preferred for celery crop. It requires well pulverized fine soil for good crop. It does not thrive well on a soil having acidic more than pH 5.5 (Choudhury, 1967). The crop is slightly tolerant to acidic soil. It does not perform well at a pH above 6.7. Celery is moderately sensitive to salinity (Francois and West, 1980). Field should be well prepared by ploughing, harrowing , planking and compacting of loose soil.

SOWING

Sowing time

Mallik (1956) suggested the sowing time is (a) August and September for plain. (b) January-February for temperate region and (c) March for main crop for general purpose in temperate region.

Seed rate

About 250 g seeds is sufficient in order to raise seedling for planting one hectare land (Choudhury, 1967).

Seed treatment

Soaking of seeds prior to sowing hasten germination. BA+GA4 (10+50 ppm) give good germination. PEG (polyethylene glycol) treatment enhance germination rate and treatment for 14 days is more effective.

Nursry raising

Seeds are sown in well prepare nursery beds in rows or broadcast. Sowing in rows gives an advantage for intercultural operations. Seeds are sown in nursery bed at 1-2 cm deep. Seeds are slow in germination which takes about one month to germinate. Germination at high temperature will lead to weak growth of the seedlings at the beginning. Therefore, after germination in moderate hot bed, the seedlings are transferred to cooler place and provided with abundant ventilation. Germination is satisfactory at 10-20°C and good at 15° or 20°C, poor at 25°C and nil at 30°C (Nakamura *et al.*, 1982).

PLANTING

The seedlings of 45-60 days old within 10-15cm high are suitable for transplanting. Hardening of seedlings by withholding water for 10-14 days in nursery lower the tendency of bolting. Strong-stiff stemmed seedlings are transplanted in well prepared plots at spacing of 60 cm from row to row and 15 cm within rows. In the plains of India, transplanting is done in rows of 60 to 75cm apart during January. On the hills, transplanting is done in May.

For commercial cultivation, the seedlings are transplanted in well manured plots, 75-90 cm apart in rows and 15-20 cm between plants in rows. Saini (1996) reported that 10 cm spacing within plant gave better yield of marketable celery than in wider spacings.

NUTRITIONAL REQUIREMENT

Celery requires heavy manuring being a shallow-rooted crop. So the major portion of plant nutrients are drawn from the top soil. The yield of 20 tonnes of above ground green growth per hectare removes 140 kg of nitrogen, 55 kg of phosphorus and 22 kg of potassium (Choudhury, 1967). However, the nutrient requirement of celery is vary according to soil type and soil fertility. The fertilizers requirement for celery crop is N 85-114 kg /ha; P_2O_5 136-272 kg /ha and K_2O 136-340 kg /ha recommended for different type of soils. A balance fertilizers dose of 140 kg nitrogen, 55 kg phosphorus and 60 kg potassium per hectare is required for raising successful crop. Depending upon the nutrients availability and fertility status of soil, use of 200 kg nitrogen, 100 kg phosphorus and 100 kg potassium per hectare is beneficial.

The best yield and quality can be obtained by using sodium nitrate @ 225 kg/ha (in split application of 150 +75 kg) for the main crop or 150 kg/ha when celery is grown as a catch crop (Vulsteke, 1979). High rates of mineral forms of N adversely affected flavour in celery by increasing nitrate content.

The farm yard manure and fertilizers are applied in different manner: (a) Farm yard manure/compost @ 25 tonnes/ha is incorporated in soil at final land preparation. (b) Nitrogen is given in three splits i.e. one third at each stage i.e. 25, 45 and 60 days after transplanting. (c) Entire phosphorus is given as basal dose before planting. (d) Potash is given in three split doses i.e. 25, 45 and 60 days after transplanting alongwith nitrogen.

Use of micronutrients

Nutrients like calcium, magnesium and boron deficiency have been observed in different growing places of celery.

Calcium deficiency

Calcium deficiency seems to result in the trouble known as black heart. It first shows as a tip burn on the young leaves and spreads quickly to most of the heart tissues. This is followed by drying, blackening and in severe cases the killing of the entire heart. Deficiency can be check by foliar application of 0.10 molar solution of calcium nitrate or calcium chloride.

Magnesium deficiency

Magnesium deficiency causes chlorosis to the plants, which affect the yield and market quality. Spraying of plants with magnesium sulphate at 12 kg /ha will prevent the occurrence of chlorosis.

Boron deficiency

Boron deficiency appear due to heavy dose of nitrogen and potassium fertilization and results in cracked stem with lesions on the inner and outer surface of the petioles and over the vascular bundles. The adjoining epidermis curls outwards followed by a dark brown colour of the exposed tissue.

Application of commercial borax @ 12 kg/ha to be applied in the form of solution near the base of the plant is effective. It may also be applied in dry form or can be mixed with the fertilizers. Spraying the crop thoroughly with 1.5 % boric acid solution reduces the incidence of cracked stem. More amount of borax (12-15 kg/ha) would be required on a neutral or alkaline soil than on acid soil.

IRRIGATION

Celery is a moisture loving plant and unless soil is naturally moist, the application of water is necessary. It is wise to apply water before the first indication of wilting or any check of growth (Hegde, 1993). For succulent and tender stalks, high soil moisture is necessary. Water must be regularly supplied at frequent intervals (8-10 days in winter season). Prior to one month harvest more than adequate moisture is needed as the most rapid growth occurs at this time. Saini *et al.* (1989), in Ludhiana observed highest yield of seed crop of celery when irrigations were given at 50 mm CPE. Irrigation management in celery crop is very important to minimise the deep percolation of water and proper release of nitrogenous fertilizers for the plant. Celery requires about 750 mm water. Furrow irrigation and sprinklers are widely used for celery. Sprinkling in the evening is better.

INTERCULTURE

Good quality crop of celery depends on timely intercultural operations by which a continuous, fairly rapid growth of plants can be assured (Mallik, 1986). Celery responds more to the cultivation than many other crops. As the rootlets are very fine, brittle and spread near the soil surface, which are liable to be broken at every operation of soil, shallow cultivation should be practiced (Saini, 1996). Generally 3-4 hoeing and earthing up is given to the crop before and after each top-dressing. Soil should be hoed lightly but frequently and all lateral shoots should be removed as they appear. Soil mulch give beneficial effects to the celery crop. Radish or early lettuce can be grown with celery as cover crop and it can also be successfully grown after early potato.

Blanching

Exclusion of light from the leafstalks of growing plants make them devoid of chlorophyll and this process is known as blanching . Blanching is done either by wrapping paper around the leaf stalk or by earthing up soil as the plants grow. (Choudhury, 1967). Proper blanching makes the crop crisp, reduces acrid flavour, improves the flavour, keep the plant tender by causing inner tender stalks to elongate rapidly, which adds to its quality and ultimately forms a very wholesome salad (Naik, 1958). The earthing up is followed in gradual stages as the plant grows, once a week covering the basal portion of the plant higher and higher till only to the leaf tops are left exposed. Generally blanching is completed with three earthing up operation and some times in two.

WEED CONTROL

Since celery is a long duration crop, weeds are troublesome on most soils. Soil should be hoed lightly but frequently and hand weeding is done after each hoeing to keep crop free from weeds. But for large scale cultivation weedicide /herbicide spray is cheaper and economical for weed control. Some weedicides have been recommended.

Weedicides

(a) Maloran (chlorobromuron) at 5 kg or afalon (linuron) at 4.0 kg or gesagard (prometryne) at 5 kg/ha give adequate weed control on peat soil (Borowy and Kossowski,1979).

(b) Chloroxuron at 1.5 or 2.5 kg/ha alongwith ethofumesate at 600g/ha applied as pre-emergence treatment on seed bed give the best weed control and also improve the quality of celery seedlings (Lifshitz *et al.*, 1982).

(c) Bains *et al.* (1975) got excellent weed control and increased yield in celery with trifluralin, fluchloralin and nitrofen.

PLANT PROTECTION MEASURES

Insect pests

Leaf minor, aphids, thrips, celery worm, carrot weevil, celery hopper, celery looper, carrot rust fly, celery caterpillar and slug are some of the pests which attack celery crop.

Leaf miner (Pegomyia hyocyami)

The white maggot feeds on tissues between upper and lower layers of leaf causing serious injury to leaf. Spraying of malathion @ 2 ml per litre of water control the pest.

Aphids (Myzus sp.)

Aphids attack the foliage and reduce the plant growth and quality. To control the aphids spray malathion or endosulfan @ 2 ml/litre of water.

Carrot weevil, green celery worms, thrips, leaf hopper, celery looper, celery caterpillar, carrot rust fly, and slug.

They are controlled by:

(a) Spraying of dimethoate or methyldemeton @ 2 ml/litre of water at weekly interval check the pests incidence.

(b) Endosulfan or fenvalerate @ 2 ml/litre of water spraying at 10 days interval is effective to control the pest incidence.

Diseases

Some important diseases are early blight, late blight, damping-off, fusarium yellow, bacterial soft rot and bacterial leaf spot.

Early blight

It caused by *Cercospora apii*. It appears as scattered small circular brownish yellow spot with concentric rings and enlarge rapidly having a light brown central area. Spraying of blitox-50 @ 3g/ lit of water at 21 days interval thirce chek the infestation. Corp-rotation and field sanitation are also advisable to check this disease (Walker, 1952). Growing of resistant varieties i.e. Emerson and Pascal, resistant to early blight. Hot water treatment of seed at 48 to 50° C for 30 minute is effective to control the early blight.

Late blight

The disease is caused by *Septoria apii-graveolentis.* Small circular water soaked area develops on tips and margins of older leaves. Later on turn black and all plant parts are affected. Spraying with Dithan Z-78 @ 3g/litre of water, at 2 week intervals check the disease. Growing of resistant varieties and hot water treatment of seeds at 48-49°C for 30 minutes is recommended to check the disease.

Damping-off

It is caused by *Pythium* sp. Topling of infected seedlings at any time after emergence is generally observed. Infected tissue becomes soft and water-soaked, stem become constricted at the base and ultimately collapses even within a day (Walker, 1952). Application of 4 g brassicol/sq.m and soil treatment with formaldehyde at 1:50 upto a depth of 10 cm are recommended.

Fusarium wilt/fusarium yellows

It is effected by *Fusarium oxysporum.* The symptoms are stunted growth of plant, vascular discolourations, rotting of crowns, petioles and root. Cultivation of celery repeatedly in same land should be avoided. Growing of resistant varieties like All dark green and Pascal are suitable control.

Pink Rot

A soil-borne fungus (*Sclerotinia sclerotiorum*), pink rot is a destructive disease and prefer cool and moist conditions for development. The disease causes damping-off in the seed bed, a light pinkish rot of stalks in the field, and a watery soft rot in transit and storage. The early lesions are light brown with pinkish-brown borders. Eventually, the tissues become water-soaked and very soft. Soil fumigation or spraying with fungicides will help to control pink rot in the seed bed and in the field. Field sanitation and careful washing and packing are effective in preventing the disease in transit.

Bacterial leaf spot and soft rot

They are caused by *Pseudomonas* and *Erwinia carotovora.* Plant become small circular reddish spots with pale green border develop on leaves. In soft rot collar portion rot and plants fall down. Diseases develop in hot humid weather. Spraying with dithane Z-78 is effective. Spray of bordeaux mixture at an interval of one week control the diseases.

Viral Diseases

Two significant viral diseases of celery are aster yellows and western celery mosaic. Aster yellows is spread by leaf hoppers and causes plants to be stunted with yellow and brittle leaves. The primary method of control is to eradicate the leaf hopper vectors. Western mosaic is spread by aphids. Leaves of infected plants exhibit vein clearing followed by mottling and twisting. Plants become stunted. The most effective control of mosaic has been achieved in California by country-mandated 3 week celery-free periods during which all fields must be harvested and no new plantings allowed.

Physiological Disorders

Chlorosis

This is characterized by yellowing of older leaves due to a magnesium deficiency. It can be corrected by spraying magnesium sulfate ($MgSO_4$) at 23 kg per hectare on the foliage every 2 weeks (Swiader *et al.*, 1994); soil application of magnesium is usually not effective. Most "Summer Pascal" cultivars are fairly resistant.

Cracked stem

It also called "brown-checking". This disorder is first evident by a brownish mottling of the leaf, appearing along the margins. This is accompanied by a brittleness of the petiole, followed by crosswise cracks in the outer layers of the petiole. The tissues surrounding the cracks turn brown. Roots may also turn brown and the laterals die (Swiader *et al.*, 1994). The disorder is caused by a boron deficiency and is accentuated by high rates of nitrogen and potassium. Soil application of 1.14-2.28 kg of boron per hectare broadcast before planting or foliar application of 580g of boron per 378.5 litre of water per hectare is effective in preventing the disorder. 'Tall Utah 52-70' appears less affected by the disorder than other cultivars.

Black heart

This is occur due to calcium deficiency which causes the heart (inner leaves and growing point) to turn black. Bacterial soft-rot organisms invade the affected tissues, resulting in black, water-soaked centres. Occasionally, outer leaves are also affected and may turn yellow. High temperatures, rapid growth and water stress are associated with this disorder. Large plants are more susceptible, especially with high nitrogen rates, indicating that the disorder may be growth-related (Swiader *et al.*, 1994). It is more serious in regions where irrigation waters contain high levels of sodium. Calcium chloride 2.30 to 3.60 kg or calcium nitrate 2.30-6.80 kg per 378.5 litre of water per hectare, sprays directed at the growing point will help to control the disorder. Black heart may occur in the field as well as in storage (Bergmann, 1992).

Pithiness

This disorder is a breakdown of the thin-walled parenchyma cells that make up the major part of the leafstalk, resulting in soft tissue and a hollow cavity. It, usually develops in the outer petioles first and has been attributed to too rapid growth (Swiader, *et al.*, 1994), particularly with high nitrogen rates. Some cultivars are more susceptible than others.

HARVESTING

Celery has no definite stage for harvesting. The leaf stalks take about 9 months from seed to harvesting (Purseglove, 1965) and plants are ready for harvest within 75-90 days after transplanting (Yawalkar, 1969). Crop generally takes 100-120 days for harvesting after transplanting the seedlings according to growing period, season and place. Rapid growth is essential for good yield and better quality. Plants should not be allowed to become over mature as they will develop pithy petioles which will be unmarketable. When the plants are 40 cm high side suckers are removed and send to market. Plants become ready for harvest in plains by the middle of May and on hills in November (Thompson and Kelly, 1957). Plants are cut just below the soil surface. Cuts are given low then plants are trimmed and prepared for the market.

YIELD

Yield varies between 330 to 560 q /ha with average yield of 450 q /ha.

POST HARVEST HANDLING

Treatment

The treatment of BA at 10 ppm as post harvest dip extend freshness and market acceptability (Writter *et al.*, 1962).

Grading

After cleaning (stripping to remove suckers and diseased and damaged leaves then washing), grading is done into "super", "fancy" and "commercial" grade, according to quality, size and texture. Best quality with uniform shape, size and brightness materials kept in "super grade". Good materials with uniform size and similar shapes along with freshness and good looks are arrange in "fancy grade". Average, with good shape and size without any blemishes arrange in "commercial grade". Grade specifications are changed from time to time depending on marketing demands (Singh and Singh, 2007).

Packing

Pre-packaging celery is practiced to some extent in all producing regions. The number and the size of stalks depends to the kind of baskets or container used, 20-25 heads/stalks are usually packed in baskets. Film bags wraps and basket over wrapped with film are used to some extent. Celery packed flat is more susceptible to damage from pressure then packed upright.

Crate is most commonly used for shipment purpose, some are wire bound, some are nailed and some half-crates usually 10 or 11 inches in width, about 20 inch long and a mainly about 20 inch in depth (Carey, 1950). Celery for local markets is tied into bunches of 6 to 12 stalks and handled without packing in basket (Singh and Singh, 2007). Celery over wrapped with film must permit the movement of air and gases in basket in order to prevent anaerobic respiration.

Storage

Celery is generally stored in trenches and in cold storage. Stalks do not decay quickly in storage if grow in continous supply of water. The optimum temperature in the cold storage required for the crop is 0°C at 95-98 % relative humidity, where it remains in good condition for 2-3 months. Celery in storage absorbs foreign flavour. So it is advisable to keep away the crop during storage from odour of other crop or product (Choudhury, 1967).

SEED PRODUCTION

Celery is cross-pollinated crop so isolation distance of 1600 m for foundation seeds and 800 m for certified seed production is given to avoid cross pollination between any two varieties and also to check the mechanical mixture in the seeds.

The young seedlings are planted in the field at 5 leaf stage at 30 cm apart with 60 cm between rows.

The first roguing is required when the young plants are planted and next roguing is done in vegetative phase before starting of anthesis. Early bolted plants are discarded. Other cultural practices and agronomic techniques are same as given in crop production. Although celery flowers some time in March in the plains, its commercial seed production can be taken up only in the hills. The seed crop is mature and ready for harvest when the plants have become yellow and the seed is hard and grey brown colour. Seed should be dried upto 10 per cent moisture content. Average seed yield is 19 q/ha.

REFERENCES

Anonymous (1952). Wealth of India, *CSIR*, Vol.6, New Delhi.

Bains, D.S., Singh, A. and Singh, R. (1975). A note on the comparative efficiency of some herbicide for weed control in celery (*Apium graveolens*). *Indian J. Weed Sci.*, 7: 123-124.

Bergmann, Werner (1992). Nutritional disorders of Plants, Gustav Fisher Verlag Jena. Stuttgart. New York.

Borowy, A. and Kossowski, M. (1979). Biuletyn Warzywniczy, 23: 245-58.

Carey, L.C. (1950). Containers in common use for fresh fruits and vegetables, *U.S. Department of Agri. Farmers' Bull.* 2013.

Choudhury B. (1967). Vegetables. National Book Trust, India, New Delhi.

D'Antonio,V. and Quiros, C.F. (1987). Viability of celery pollen after collection and storage, *Hort. Science*. 22: 479.

Francois, L.E. and West, D.W. (1980). *J.Amer. Soc. Hort. Sci.*, 107: 952-4.

Garfield, E. (1985). From tonic to psoriasis: Stalking celery secrets. *Current. Contents* 18: 3.

Hegde, D.M. (1993). Water Management in Vegetable Crops. In Advances in Horticulture vol.5.Vegetable Crop Part.1: (Eds: K.L. Chadha and G.Kalloo) Malhotra Publishing House, New Delhi, India.

Honma, S. (1959). A method for celery hybridization., *Proc. Am. Soc., Hortic. Sci.* 73: 345.

Lifshitz, N.; Abayov, Y. and Aviram, H. (1982). Hassadeh, 62: 1570-73.

Mallik, P.C. (1956). Bagban. *Hort. J.*, 6:18-19.

Mallik, S.C. (1986). Celery. In Vegetable Crops in India (Eds.: T.K. Bose and M.G.Som). Naya Prokash, Calcutta-6.

Naik, K.C. (1958). Horticulture in South India. Min. Food and Agri., New Delhi, pp. 104.

Nakamura, S., Teranishi, T. and Aoki, M. (1982). *J. Jap. Soc. Hort. Sci.*, 50: 461-467.

Orton,T.J. and Arus, P. (1984). Out crossing in celery, *Euphytica*, 33:741.

Peshin, S.N. (1998). Promotion of Rare Exotic Vegetables. In. Production and Postharvest Handling of Vegetables. *Centre of Advanced Studies, Horticulture-Vegetable* (ICAR), Dept. of Vegetable Crops. Dr. Y.S. P.U.H.F. Nauni-Solan, H.P.

Purseglove, J.W. (1965). Tropical Crops: Dicotyledons, E.L.B.S. Longman. London.

Quiros, C.F., Hulbert, S.B. and Durgan, M.E. (1986). Cytological and genetical studies of a male sterile celery, *Euphytica*, 35:867.

Quiros, Carlos, F. (1993) Celery (*Apium graveolens* L.).In: Genetic Improvement of Vegetable Crops (Eds: G.Kalloo and B.O., Bergh).Pergamon Press, New York.

Richharia, R.H. (1957). Plant Breeding and Genetics in India, Scientific Book Co. Patna.

Russell, M. (1984). Heart of the matter: Stalking the roots of celery's history, *Wall Street Journal*, 9:20-84.

Ryder, E.J. (1979). Leafy salad vegetables. AVI Publishing Co, Westport, Connecticut.

Saini, G.S. (1996). A text Book of Vegetable Production, Aman Publishing House, Meerut (U.P.). India.

Saini, S.S., Cheema, S.S., Randhawa, G.S. and Mahey, R.K. (1989). Effect of irrigation and nitrogen on celery. *Indian J. Agron.*, 34: 341-345.

Simmonds, N.W. (1976). Evolution of crop plants. Longman, London and New York.

Singh, D.N. and Nath, Vishal (2011). Varieties and Hybrids of Vegetables. Satish Serial Publishing House, Azadpur, Delhi (India): pp. 351-353.

Singh, D.N. and Singh (2007). Production Management of cenderutilized Vegetables. Agrihortica Publications. Junagadh (India). pp. 59-70.

Swiader, J.M., Ware, G.W. and Mc Collum, J.P. (1994). Producing Vegetable Crops. International Book Distributing Co., Charbagh, Lucknow.

Thompson, H.C. and Kelly, W.C. (1957). Vegetables Crops. Tata Mc.Graw-Hills Publishing company Ltd., New Delhi.

Vilmorin, R.L. (1950). Pascal celery and its origin. *J.N.Y. Bot. Gard*, 51: 39.

Vulsteke, G. (1979). Land en Tuinbouw, No.196.

Walker J.C. (1952). Diseases of Vegetable Crops. Mc Graw-Hill Book Co., Inc; New York, pp. 230-55.

Writter, S.H., Dedolph, R.R., Tuli, V. and Gilbert, D. (1962). Proc. *Amer. Soc. Hort. Sci.* 80: 428-34.

Yawalkar, K.S. (1969).Vegetable crops of India. Agri-Hort. Publishing House, Nagpur.

□□□

2. PARSLEY

Botanical name	:	*Petroselinum crispum* Mill
Family	:	Apiaceae (Umbelliferea)
Chromosome number	:	2n = 22

Parsley is the most popular garden herb cool season corp. It is one of the most familiar and widely cultivated of the garden herbs. It is a biennial or short lived perennial and a cool season crop (Choudhury, 1987). It was used for flavouring by the Greeks and Romans. Parsley can be grown successfully through out the tropics, but tends to achieve out rather quickly near the equator. It is one of the most useful of all herbs.

ORIGIN

Parsley is a native to southern Europe and has been in cultivation for over two thousand years. It is now naturalized in most temperate countries (Purseglove, 1984). It was used by the ancient Greeks and Romans.

DISTRIBUTIONS

Parsley is distributed in all temperate part of the world. USA, Europe; Canada. UK, Italy and Asia are major growing area of parsley. Parsley is one of the most familiar and widely cultivated of the garden herb. It can be grown successfully throughout the tropics, but tends to die out rather quickly near the equator. It is not common in India (Swaider *et al.*, 1994).

BOTANY

The plant is erect glabrous biennial or short-lived perennial, 30-70 cm tall with a stout taproot; rosette leaves in a dense tuft, dark-green, shiny, 3-pinnate, leaflets 1-2 cm long, cuneate, lobed, often much crisped; umbels 2-5 cm in diameter, flat topped. Flowers yellowish. The seed, which is usually slow to germinate. As the flowering shoots develop they should be cut off at the base, it is cross pollinated (Thompson and Kelly, 1972).

COMPOSITION

Parsley is very rich in vitamin A, C, iron, minerals, protein and sugars. The plant yields a volatile oil of which the active principle is apiol. It contains (per 100g of edible portion): moisture 74.6 g, protein 2.2g, fat 0.3 g, carbohydrate 1.5 g, vitamin C 90 mg, vitamin A 5200 IU, iron 17.9 mg, thiamin 0.04 mg, riboflavin 0.18 mg, calcium 125 mg, niacin 0.5 mg and phosphorus 90 mg (Aykroyd, 1963).

Uses

The leaves are used as a garnish and for flavouring sauces, soups, omlets, and stuffing. It is minced and stew over potatoes, carrot or other boiled vegetables (Singh and Singh, 2007). The plant yield a volatile oil, of which the principal is apiol (Franz and Glasl, 1976). The small bushy plant and crinkled leaf type, moss curled parsely is also used as a ornamental plant in pots and in small beds in the gardens. Field parsely is easy to prepare an unusual accompaniment to meat. The parsely smell is supposed to absorb the intoxicating fumes of wine and thus prevent any usual effects of over indulgence (Yomaguchi, 1983). The swollen root of turnip-rooted parsely is eaten as a cooked vegetable.

VARIETIES AND HYBRIDS

Parsley cultivars can be grouped into different five type. They are (a) plain leaf, (b) celeray leaf, (c) curled leaf, (d) fern leaf and (e) turnip rooted. The most common and attractive type is curled leaf, which contain sub-type (i) double curled, (ii) ever green and (iii) triple curled (Kalloo and Berg, 1993).

The best known varieties are Moss Curled, Extra Double Curled, Fern Leaved, Curled Dwarf, "Extra Double Curled Paramount". Imperial Curled.

Detail characteristics of varieties and hybrids reported by Singh and Nath (2011) are given below :

Hamburg

It is a variety of turnip-rooted parsley. It has tough stem. Deeply cut plain leaves and moss curled. It is dwarf with finely cut serrated and deeply curled leaves and grown for its edible root. It is grown in the vicinity of the large cities, where it is utilized mainly by the foreign population.

Parsley Petra

It matures in 75 days after transplanting. It can be sown from August to January. It is vary uniform with very attractive curled leaves. High dry matter content and tolerant to bolting.

Hybrids

Festival 68, Hybrid. No. 41 and Hybrid No. 42. are used for cultivation in several parts of the parsley growing area to obtain higher yield (Madzharova, 1971).

CLIMATIC INFLUENCE

As parsley is a temperate crop; it requires cool climate with higher humidity and well distributed more rainfall during the vegetative growth period. In some regions where winters are relatively mild, the crop is grown in cold frames (Choudhury, 1976). The temperature of 15-20°C is suitable for its vegetative growth. In northern India crop can be taken during October to February (Singh and Singh, 2007).

SOIL REQUIREMENT

A good well-drained sandy loam. Well supplied with organic manures is preferred for Parsley cultivation. It requires well pulvarised fine soil for good crop. Fertile sandy soil is ideal for its cultivation (Swaider *et al.*, 1994).

SOWING

Sowing time

Sowing time in northern region is September–October and in southern area, December–January. About 500-700g seeds/ha is sufficient to grow seedlings for covering of one hectare land. The seeds are sown directly in the field or seedlings are raised in nursery beds. The spacing of 45 cm row to row and 15cm between plants is optimum to get maximum yield (Singh and Singh, 2007).

Nursery raising

The seeds of parsley are very small in size and germinate slowly (80-20 days after sowing), so seeds should be soaked in water for 24 hours to improve germination. Seeds are sown in well prepared nursery beds. The seeds should be sown in February-March in hills for summer supplies and August-September, in mid hill, for spring crops. The optimum temperature of germination is 45° to 60°F. Seeds also can be sown in boxes and seedlings transplanted when they have six leaves (Swaider *et al.*, 1994). Some times parsley seed is partially mixed with radish or cabbage seed. This helps in marking the parsley row so that hoeing and weeding may be done properly.

PLANTING

The seedlings of 45 to 60 days old are planted in well prepare field at given spacing in flat bed. Care is taken for proper establishment of plant. Parsley is also grown as a forcing crop in green houses, hot beds, and cold frame in winter and spring in some cases. In some regions where winter are relatively mild. The crop is grown in cold frames (Thompson and Kelly, 1972).

NUTRITIONAL REQUIREMENT

The farm yard manure or compost @ 15 tonnes per hectare are incorporated in the field at final land preparation. Fertilizers combination of

80:60:80 kg; N. P_2O_5, K_20/ha is optimum for higher yield of Parsley. Even 60 kg N, 40 kg P_2O_5 and 20 kg K_2O/ha is judicial fertilizers dose to obtain economical yield (Singh and Singh, 2007).

Fertilizers are given in split doses for better utilization by plants. It should be given as: (a) nitrogen in three equal split, i.e. 25, 45 and 60 days stage, (b) entire amount of phosphorus is applied as a basal dose before planting and (c) potash in 3 split along with nitrogen i.e. 25, 45 and 60 days stage.

Use of Chemicals

Auto tetraploids are produced on treatments with 0.1% colchicine. The medim containing 1 mg 2, 4-dichlorophenoxy acetic acid per litre of water give best results obtained from growth and petioles.

IRRIGATION

Parsley require frequent and light irrigations for proper growth and developments. Proper drainage is essential for the successful raising of crop. The field is irrigated immediately just after each fertilizer application.

INTERCULTURE

Frequent shallow cultivation is required to make soil loose and for proper development of roots. Generally 3-4 hoeings is given before top-dressing. Earthing up is given after each hoeing and application of fertilizers.

WEED CONTROL

Parsely plants ae very much bushy and quite slow to spread over the surface of the soil. So weed growth in the field is a major problem, hence weed control is very important in parsely crop. Usually 3-4 hand weeding is done after hoeing of the plot. Weed can also be controlled by a combination of cultivation and use of herbicides (Singh and Singh, 2007). Weeding is done after each hoeing before top-dressing of fertilizers.

PLANT PROTECTION MEASURES

Insect Pests

The important inserct pests of parsley are leaf miner, carrot weevil, green celery worms, thrips, leaf hopper, celery looper, celery caterpillar, carrot rust fly, aphid and slug. Thier Control measures are given below:

(a) Spraying of dimethoate or methyldemeton @ 2 ml/litre of water at weekly interval check the pests incidence.

(b) Endosulfan or fenvalerate @ 2 ml per litre of water spraying at 10 days interval check the pest incidence.

Diseases

The most common diseases are early bligh, late blight, damping-off, fusarium yellow, bacterial soft rot and bacterial leaf spot.

Early blight

It is caused by Cercospora apii It appears as scattered small circular brownish yellow spot with concentric rings which rapidly having a ligh brown central area. It is seed born disease, hence hot water treatment of seed at 48°-50°C. for 30 minutes is effective to control the disease. Spraying of Blitox-50 @ 3g/litre of water at 21 days interval thrice check the infection. Crop rotation and field sanitation are also important to check the disease.

Late blight

This disease is caused by *septoria apii graveolentis*. Small circular water soaked area develops on tips and margins of older leaves. Later on turn black and all plant parts are affected. Spraying with dithan Z-78 @ 3g/litre of water, at 2 week intervals check the disease. Growing of resistant varieties are also solution to control the late blight.

Damping off

It is caused by *Pythium* spp. Toppling of infected seedlings at any time after emergence is generally observed. Infected tissue becomes soft and water-soaked, stem become constricted at the base and ultimately collapses efe within a day. Application of 4 g brassicol per liter of water and soil treatment with formaldehyde at 1:50 upto a depth of 10 cm are suitable to control the disease.

Fusarium wilt

The fungus *Fusarium oxysporum* is caused damage to the parsley crop. Symptoms are stunted growth of plant, vascular discolourations, rotting of crowns, petioles and root. Cultivaion of parsley repeatedly in same land should be avoided. Growing of resistant varieties.

Bacterial leaf spot and Soft rot

They are caused by *Pseudomons* and *Erwinia carotovora*. Plant become small, circular reddish spots with pale green border. In soft rot, collar portion ror and plants fall down. Symptom develop in hot humid weather. Spraying with dithane Z-78 @ 3 g per litre of water is effective control. Spray of Bordeaux mixture at an interval of one week to control sofl rot disease is effective to check the disease (Kunovski, 1970).

HARVESTING

Parsley takes 90-100 days from planting to first harvesting. In harvesting of the parsley, the outer and larger leaves are harvested and tied in bunches for market. By this method the plant continues to produce a marketable product for many weeks. The plants continue to give leaf cuttings for several months. The harvested leaves are washed thoroughly which are packed loose or bunches for market (Singh and Singh, 2007).

YIELD

The average yield of green matter is 120 q/ha.

POST HARVESTING

Grading

The parsley leaves are graded according to their size and shape, tenderness and clour development. Well developed leaves of uniform shape and size with bright colour and tenderness are kept in "Super grade". Medium size leaves with good texture and uniform shape with freshness are kept in "Standard" grade. Small leaves with good shape, light colour, fresh and free from any spot or blemishes are arranged in 'commercial' grade (Singh and Singh, 2007) for market purposes.

Packing

After grading leaves are tied in bunches according to grades then packed. The bunches are packed for shipment in a small size vegetable crates, 4 dozen bunches are packed in each crates or in bushel baskets. Crushed ice usually is placed in the package, and shipment is made under refrigeration to distant markets (Thompson and Kelly, 1972).

Storage

Singh and Singh, (2007) reported that parsley leaves can be stored in cold storage for one week at 0°C temperature with 95% relative humidity.

SEED PRODUCTION

It is a biennial crop require temperate climate so seed production is possible only in hilly area, at an altitude of 1200 m or above. The crop is cross-pollinated so isolation distance for foundation sed production is 1600m and for certified seed production 800 m is require to produce pure seeds.

Generally, seed is multiplied by seed to seed method, although stock seed is produced by the root to seed method.

Seed to seed method

The seeds are sown @ 3 kg/ha in rows of 50cm apart and after germination the plants are thinned out at a distance of 30-35 cm between the plants.

Root to seed method

The seed is sown @ 6 kg/ ha in rows of 50 cm apart, the plants are lifted and selected for replanting at 50 cm apart in rows of 1m apart.

The young plants in both seed to seed and root to seed methods are checked for foliage characters. In root to seed method parsley is assessed for roots shape and quality before re-planting.

Other cultural operations and agronomic package of practices are same as given in crop production section. In the first year crop attain vegetative growth and in second year flowers and form seeds. The seeds in umbels are ready for harvesting when they turned brown. The seeds are separated from umbels from thresher and then cleaned properly. The seed is dried to 10% moisture content for better storage and germination. An average seed yield is 10 q/ha.

REFERENCES

Archakova, L.I. (1978). Biological and economic feature of parsley on the Kola Penninsula. *Byull. Vsesoxyuzenogo Ordena Lenina. Irdena Druzhby Narodov Inst. Omeni N.I. Vavilova* 82: 46.

Aykroyd, W.R. (1963). ICMR Special Rept. Series No. 42.

Ayonymous (1970). New Vegetable Varieties. Deutsch. Gartenbau. 7: 247.

Ayonymous (1998). Varieties and Hybrids. Division of Vegetable Crops. IARI, New Delhi.

Bejo Sheetal (2000). Hybrid Seeds bejo sheetal seeds, Pvt. Ltd. A-3, Old M.I.D.C. Jalna- 431203, Maharashtra, (India).

Choudhury, B. (1976). Vegetables, National Book Truest, India, New Delhi. pp. 182-183.

Franz, C. and Glasl, H. (1976). The essential oils of parsley, II. Comparative investigations of fruit, leaf and root oils of some parsley varieties. *Qualitas Plantarum* 25: 231.

Franz, C., Fritz, D. and Eisenmann, J.C. (1982). Parsley of industrial proccssing. 1. Var and sites, *Gemuse*, 18 (11): 380.

Indo American Exports (2001). Parsley Variety (M/s Petroseeds Company Inc: U.S.A.) P. B. No 7099, 17th cross. 2nd 'A' Main BSK 2nd stage, Bangalore-560070.

Kalloo, G. and Berg, B.O. (1993). Genetic Improvement of Vegetable Crops. Pergamon Press, New York.

Khlebnikov, V.F. and Rudenko, N.M. (1984). Study of the pathogen causing powdery mildew in parsley in relation to breeding for resistace, *Genet. Osnovy sel. Sel'sk rontenii i Zhivoinykj, Kishinev Moldavean SSR. stiinca*, 110.

Kunovski, Z. (1970). Resistance of several parsley cultivars to the pathogen of white leaf spot. *Rast. Zasit.* (*Pl. Protect*), 18 (3): 28.

Levandovskaja, L. (1965). Leafy froms of celery and parsley, *Kartofel i Ovosci* (Potato and Vegetables) 10(1): 30.

Madjarova, D. and Bubarrova, M.G. (1978). New froms obtained by hybrization of *Apium graveolens* and *Petroselinum hortense*, *Acta Hortic.* 73 : 65-72.

Madjarova, D. and Georgiev, G. (1971). New Plant forms obtained by hybridisation between parsley (*Petroselinum hortense* Hoffn) and celery (*Apium graveolens* L). *Comptes Rendus del Acad. des Sci. Agric. Bulg.* 4, 261.

Madzharova, D. (1971). The new parsley varieties "Festival 68. B". *lgarski Plodve. Zelenchustii i Konservi* 4: 13.

Moravec, J., Kvasnicka, S., Hondnota, C.S., Oorud Perrzeli, V.E. and Svetoven, S. (1963). The value of Czechoslovak varieties of parsley in the world collection. *Ovoncharstivi z zelinarstivi* 11: 149.

Oomen, H.A.P.C. and Grubben, G.J.H. (1978). Tropical Leaf Vegetable in Human Nutrition. Comm.69 Royal Tropical Institute, Amsterdam and Orphan Publishing Company, Willemsland, Curacao pp. 87&95.

Purseglove, J.W. (1984). Tropical Crops-Dicotyledons. Vol. 1&2. (edn. 1 reprinted 1984). Longman group Ltd. England pp. 652.

Singh, D.N. and Nath, Vishal (2011). Varieties and Hybrids of Vegetables. Satish Serial Publication House, Azadpur, Delhi (India). pp. 368-369.

Singh, D.N. and Singh, S.P. (2007). Production Management of Underutilized Vegetables. Agrihortica Publication Junagadh, Gujarat, India. pp. 191-198.

Swiader, J.M., Ware, G.W. and Mc Collum, J.P. (1994). Producing Vegetable Crops. International Book Distributing Co. Charbabh, Lucknow.

Thompson, H.C. and Kelly, W.C. (1972). Vegetable Crops. Tata Mc Graw-Hill Publishing Company Ltd., New Delhi. pp. 272-273.

Yomaguchi, M. (1983). World Vegetables, AVI Publishing Co., Westport, Connecticut.

Zelevein V.M. (1974). Bio-chemical characteristics of parsley in Prem province, Tr. Perms. Kh in-t 106, 92.

□□□

3. LETTUCE

Botanical name	:	*Lactuca sativa* L.
Family	:	Compositae
Chromosome number	:	2n = 18

Lettuce is most important salad crop, generally cultivated for its tender crisp leaves/heads, which are used as salad, containing appreciable amount of minerals and vitamins (Sharma *et al.*, 2001). The leafy type lettuce is a rich, source of protein and minerals and popular for its daily use in diet particularly in Europe. Recently it is gaining popularity in India too, owing to its nutritional qualities (Singh *et al.*, 2000). However, it occupies an important position for its cultivation in tropical and subtropical countries also. The first appearance of seed stalk varied from 59 (cv. Emperor Forcing) to 112 (cv. Italian Ice) days after sowing. It thrives best during the cooler parts of the growing season, and to be of the highest quality, their growth must be fairly rapid and continuous (Thompson and Kelly, 1972).

ORIGIN

Lettuce is originated in the Mediterranean area (Vavilov, 1935). The earliest indication of its existence came about 4500 B.C. in tomb paintings in Egypt, which showed plants with narrow pointed leaves of the type that is now sometimes called asparagus lettuce. It apparently was similar to the modern form found in Egypt and known as balady (meaning local). It is also believed that lettuce has been originated in Asia minor, the Trans-Caucasus, Iran and Turkistan (Mehta, 1959). Lettuce spread rapidly through the Mediterranean basin, particularly in the ancient Greek and Roman civilizations. It subsequently spread to western Europe and then to the New World, where it was mentioned as early as 1494 (Hedrick, 1972; Ryder and Whitaker, 1976). It is thought most likely to have originated from a progenitor similar to *L. serriola* L. (Durst, 1930; Ryder and Whitaker, 1976) which was cultivated for forage or oil. It was reported to be abundant in Haiti by 1565 and Brazil in 1647. The Portugese or the British most likely introduced it into India during 16th century.

DISTRIBUTION

Lettuce is grown in all continents, but largest consumers and producers are USA and Europe. Large areas of lettuce are also grown in south eastern Australia, Japan, China, Israel, Northern Mexico, Chile, Argentina, Brazil and Peru (Pink and Eavan, 1993). under cultivation. In India it is very much popular in northern part of the country and grown as winter season vegetable (Singh and Dhaliwal, 2002).

BOTANY

Lettuce is a highly self-pollinated crop, annual to biennial upto 120 cm tall leafy herb with milky juice (Wilfred, 1924). Stem erect, 30-100 cm, glabrous not setose with a slender tap-root. Basal leaves in rosette, undivided or runcinate-pinnatifid, shortly petiolate, cauline leaves simple, ovate to orbicular in outline, entire cordate-amplexicaul, sessile, not held vertically. Apex blunt. Inflorescence a dense, corymbose, flat-topped panicle. Heads with 7-35 florets, ligules longer than involucre (Ferakova, 1977). Pollination takes place from 9.0-9.5 am (Pal, 1975). Fruit is a typical achene and seed ex-albuminous. Achenes 6-8 mm, obovate, narrowed and often finely muricate at the apex, greyish. Beak white, filiform, as long as body. Pappus white.

Plant breeding

Since several varieties of lettuce introduced from outside performed well and the demand is not too high, no serious efforts have been made on studying variation and gene action for further improvement of the crop. There are certain known heat resistant varieties available which need to be introduced and evaluated in India. Such varieties are "Summerlong Buttercrunch" and 'Bibb (Butterhead) and 'Green Ice' (Leaf). Some small varieties like 'Tom Thumb' (Butterhead), 'Sweet Midget' (Cos) and 'Ruby' (Leaf) should also be tried for growing in containers (Nath *et al.*, 2008).

Taxonomy

Lettuce belongs to the family compositae and genus lactuca having 8, 9 and 17 pairs of chromosomes. There are several wild species of *Lactuca* which are mostly used as source of resistance to insect pests and diseases.

L.sativa

It is the most diverse species having several different forms that are nevertheless completely infertile. It is quite likely that this species has been derived through selection from hybrid between *L. serriola* and a third species (Lindquist, 1960).

L. serriola

This has usually small seeds, reflexed involucre; spines, shoot formation at the stem base, relatively narrow leaves that may have entire or toothed margins and a rosetting tendency.

L. salinga

It is similar to *L. serriola* in several respects but differ in having very narrow, toothed leaves and bearing sessile flowers on a spike compared to pedicelled flowers on panicles as in *L. serriola.*

L. virosa

It is morphologically distinct. It has large flattened seeds; strongly rosetting, blue-green leaves; includes biennial as well as annual forms.

Lactuca sativa, L. serriola, L. virosa, and *L. salinga* belongs to the group with 9 pairs of chromosomes and cross with varying degree of success with each other. The isolation of these species within the genus is possibly due to genetic barriers.

There are strong incompatibility barriers between the cultivated *Lactuca* species and the 17 chromosome species (Ryder and Whitaker, 1976). *L. sativa* and *L. serriola* are most closely related and cross readily with each other. Both cross with difficulty with other two species; *Lactuca sativa* should be used as the male in a cross with *L. salinga* (Lindqvist, 1960). According to Thompson and Ryder (1961), the *sativa-virosa* cross yields sterile hybrid.

Floral biology

Inflorescence is head or capitulum (composite flower) consisting of a large number of flowers. The elongation of seed stalk produces a single terminal capitulum and a series of branches with many capitula forming the panicle. Each capitulum is composed of an involucre of overlapping bracts surrounding several florets: as few as 10 to over 20. Each floret consists of a ligulate (strap-shaped) yellow corolla enclosing the sexual parts. The stamens are fused to form a tube, and anthers shed pollen at the inside surface as the flower open in the morning. The elongating style with two stigmatic lobes sweeps the pollen along as it emerges from the top of the stamen tube.

The capitulum begins to swell about 24 hours before anthers as the florets enlarge. The top circle of bracts opens. The top circle of brackets opens. The following morning the florets elongate and open, exposing the staminal column, through which the style and stigmas emerge. The stigmatic lobes separate, forming a V, and then curl backwards, signaling the end of receptiveness. Flowers open only once. The process takes place in 1-2 hours. Higher temperature and increased light speed up the process (Bassett, 1986). Anthesis starts at sunshine and flower fully opens by late morning. Pollination takes place from 9.0-9.15 am (Pal, 1975).

COMPOSITION

Lettuce is rich in Vitamin A, and minerals like calcium and iron. It also contains protein, carbohydrates and Vitamin C. According to Choudhury (1967) the 100 g of edible portion contain following:

Moisture	93.4 g	Sodium	500 mg
Protein	2.1 g	Sulphur	27 mg
Fat	0.3 g	Vitamin A	1650 I.U.
Fibre	0.5 g	Calories	22
Minerals	1.2 g	Magnesium	30 mg
Carbohydrate	2.5 g.	Iron	2.4 mg
Chlorine	23.0 mg	Thiamin	0.9 mg
Calcium	50.0 mg	Riboflavin	0.13 mg
Nicotinic acid	0.5 mg	Vitamin C	10 I.U.

Uses

Its tender leaves and heads are chopped and used as salad with salt and vinegar. However, vitamin C is almost entirely lost if it is cooked. Lettuce is also known to be sedative, diuretic and expectorant. Stem lettuce, eaten raw or cooked is widely used in Chinese cookery (Kalloo, 1986). There are some less common uses for lettuce. A cigarette containing no nicotine is made from lettuce leaves. Seeds of primitive form of lettuce in Egypt are a source of edible oil (Knowles, 1978). Unlike its close wild relatives, *L. sativa* does not contain the bitter sesquiterpene lactones, lactucin and lactupicrin. These have sedative properties and *L. virosa* L. has been cultivated for production of latex, which is used to manufacture the drug lactucarium for pharmaceutical use (Mabberley, 1987).

VARIETIES AND HYBRIDS

There are four types of lettuce: (a) Head type (var. *capitata*). (b) Leaf or cutting type (var. *crispa*). (c) Cos or romian type (var. *longifolia*). (d) Asparagus type (var. *asparagina*).

Thompson (1951) and Rodenburg *et al.* (1960) recognized six morphological types of lettuce; crisp head, butter head, cos or romaine, leaf or bunching, stem and latin.

Crisp head

The cultivars are heading types with wrinkled non-wrapper leaves, round large head (upto 1 kg), brittle-textured and tightly folded with green outer and white or yellowish inner leaves, having excellent shipping and handling abilities, var. New York, 515, Imperial 44, Great Lake and Alamo-1.

Butter head

It forms relatively small, loose heads, with broad, oily, crimpled, soft-textured leaves vary in colour, size and appearance. Var. Big Boston, White Boston, May King, Salamander and Waya head. These are most popular in Northern Europe and includes summer and winter type.

Cos or Romaine

The cultivars have elongated leaves forming a leaf shaped head; outer leaves are dark or light green with heavy ribs, inner leaves are finer and lighter in colour; coarse in appearance but eating quality is quite good. Var. Effel Tower, Pairs Island, Paris White and Dark Green Valmaine.

Leaf or Bunching

The cultivars are non-heading or leaf type; colour, heat sensitivity and quality vary but all produce a rosette or leaves. Var. Black Seeded Simpson, Prize Head, Australian and Salad Bowl for out door growing and Grand Rapids for greenhouse production.

Stem

It is also called 'celery lettuce' and is grown for its thick stem which is pealed and eaten either or cooked as vegetable. The leaves are edible but inferior in quality to the leaves of the other types var. Celtuce.

Latin

It resembles butterhead in forming loose head but the leaves are elongated and similar to Cos lettuce. It is mainly grown in the Mediterranean countries and in South America. Var. Gallega, Coriolla, Verde, Criolla Blenca and Madrilene (Ryder, 1979).

Varieties distributed by IARI, New Delhi; Chinese yellow (Loose leaf), Imperial 847 (Cabbage head), May king, Paris White (Cos), Slowbolt (Leaf lettuce).

Varieties from seed dealers

Singh and Nath (2011) have described the detail characteristics of varieties and hybrids developed by private seed dealers in India; mentioned below :

Suttons

Golden Boll, A.I. Unrivalled, Little Gem.

Pocha's

Big Boston, Burpeana, Waya-head, Great lakes, New York or Wonderful, Black seeded, Simpson

Sada Shiv's

Ice Berg, Imperial, Paris Cos.

Verma's

Big Boston, Wonderful.

Nunhems

Revolution, Bergamo, Dublin.

Tropical seeds

Minetto, X-01, Panama, Red sun,, Fast Fall.

Great Lakes

It is crisp head type and has large head with firm green leaves and outer leaves being blistered. It is fairly resistant to tip burn but susceptible to powdery mildew (Choudhury, 1967).

Chinese yellow

Early, loose leaf variety, light green crisp and tender leaves (Anonymous, 1998).

Revolution

It has sensational red colour, very slow bolting, resistant to Bremia NL-1-16 and good vigour are important features of this Lollo Rosa type variety. Good for outdoor and indoor cultivation.

Bergamo

This Lollo bionda type variety is bigger and slow in bolting. Fine serated leaves are very uniform in shape and have a high quality appearance. Resistant to bremia NL1-4,6,7,10,13-15 and suitable for glass house and outdoor cultivation.

Fast Fall

This black-seeded loose-leaf lettuce tolerant to diseases. This variety performs large and vigorous plant, produce attractive highly filled crisp, light green leaves, popular all around Asia for fresh leaves consumption and can be harvested after 40-45 days.

Dublin

This is a new medium sized variety with a fresh medium green colour. The head is filled quickly and it generates a firm fare sized ball, which is well closed at the base. Resistant to bremia NL 1-16.

Punjab Lettuce No. 1

It is a non-heading type. The leaves are light green shining and crisp and bears loose leaves. It takes 45 days from sowing to first harvest, of fully developed leaves. Average yield is 8.75 tonnes per hectare. It gives high yield under suitable package of practice and in favourable climatic condition.

Red Sun

A very attractive and fine tasting leaf lettuce. Loose leaf type, with maturity 45-50 days after sowing. An all-year round growing type, with distinct glossy red colour leaves. Good resistance to bolting. Displays excellent development at high temperatures.

Panama

Panama is a loose leaf type, very resistant to diseases, light green coloured leaves. The variety is slow to bolt.

Minetto

Crisp head, batavia type, medium-small size. Leaves are crispy, deeply notched and medium-green. Black seeded. Maturity 25 days nursery and 34 days from transplanting to harvesting. Minetto is resistant to heat and humidity (it can form heads even in tropical lowlands) and show very high tolerance to tip burn.

X-01

It is a butterhead type. The head is medium size; leaves are slightly wavy and medium-green colour. Black seeded. Maturity 55 to 60 days after sowing. It is tolerant to heat and is slow bolting.

Mansang

It is a new romaine type cultivar developed from a cross between Late Texagreen and Ithaca by the lettuce breeding team of National Institute of Highland Agriculture (NIHA), RDA in 2002. This new cultivar has brown seed colour and grey-green leaves. Bolting is a little later than 'Chaser green'. Average market yield is 33.9 t/ha. It shows a good taste, resistant to lettuce mosaic virus and tolerant to soft rot caused by *Erwinia carotovora* sub sp. *carotovora.*

Adam

It is crisp head lettuce cultivar of NIHA, Korea Republic. It is a cross of 'Montello' and 'Salad Crisp'. It has good growth, compactness, head weight and high yield of 45.2 t/ha.

CLIMATIC INFLUENCE

Lettuce does well in a relatively cool growing season with average monthly temperature of 12° to 15°C (Nath *et al.*, 1976). High temperature promotes seed stalk causes a bitter taste in the leaves and accelerate the development of tip-burn. Head lettuce rots in hot, rainy or humid weather. Seeds become dormant and do not germinate properly when the soil temperature is above 22°-30°C (nath *et al.*, 2008). Artificial light, fluorescent tube produce higher chlorophyl and ascorbic acid.

SOIL REQUIREMENT

Lettuce will tolerate a variety of soils but does not grow well on mineral soils. Hester *et al.* (1936) reported that in loam soil yield increased as the acidity decreased from pH 5.0 to 6.5.

Soil should be friable, rich in organic matter and have a high water holding capacity to ensure uniform moisture supply. Well drained sandy loam with rich in organic matter is best and pH 5.8 to 6.6 is optimum. It is sensitive to highly acidic soil (Nath *et al.*, 1976). Where earliness is important, sandy loam soil is preferred. Otherwise silt loam is more suitable.

SOWING

Sowing time

Since lettuce is a cool season crop it is sown during September-November in the plains. At high altitudes, the sowing is done from March to June (Singh and Dhaliwal, 2002).

Seed rate

To raise seedling so for one hectare 400 to 500 g seed is required. The nursery area of 50-70 m² is sufficient to raise crop for one hectare land. Seeds should be treated with 0.2 % brassicol or captan before sowing.

Nursery management

Seed beds are sterilized with formaline to protect from soil borne diseases or drench with 0.2% brassical or captan. Seeds are sown in line at 15 to 20 cm apart or broadcast. Soon after sowing, the seed bed is irrigated with rose can. Beds are irrigated on alternate days and seedlings are sprayed with 0.2% dithane M-45 or difoltan at 10-15 days interval. After 5-6 weeks of sowing seedlings are ready for transplanting (Saini, 1996).

Spacing

The spacing of 45 cm from row to row and 25 cm plant to plant should be given in flat bed to have optimum plant population for higher yield. Sharma *et al.* (2001) reported that plant spacing of 45 x 30 cm was found best for getting optimum yield.

PLANTING

The seedlings of five to six weeks old are transplanted at given spacing in the main field. Hardening before transplanting is to be done. Withholding water for about 6-8 days before transplanting helps hardening. Soon after, the field is watered for proper establishment of seedlings.

Nath *et al.* (2008) stated that seedlings are transplanted 4-6 weeks after sowing in well prepared plots at distance of 20-25 cm in the rows and the rows are kept 30 to 45 cm apart. The sowing is done in small flat beds but sometimes it is transplanted on ridges depending on the soil and irrigation facilities. Sometimes it is sown directly in the fine beds in rows at 15-20 cm apart. The plants are thinned out to 3 cm when crowded.

NUTRITIONAL REQUIREMENT

Lettuce plant has a small and shallow root system and the roots are concentrated in the top 60 cm, therefore surface soil should be rich in nutrients. According to Mehta (1959) and Yawalkar (1969) farm yard manure 18 to 36 tonnes, 50 to 86 kg of nitrogen, 24 to 50 kg phosphorus and 50 to 54 kg of potash per hectare are common for lettuce. Choudhury (1967) reported good results, with 100 kg nitrogen, 60 kg each of phosphorus and potassium per hectare. Application of about 10-15 tonnes of farm yard manure supplemented with about 50 kg each of N and K_2O and 90 kg P_2O_5/ha should meet the nutrient requirement of lettuce (Singh and Dhaliwal, 2001). On sandy and sandy loam soil apply 40-45 kg N and 75-100 kg each of P_2O_5 and K_2O in the absence of any farm yard manure application.

On silt loam and clay loam, apply 25 kg each of N and K_2O and 50-75 kg P_2O_5/ha. On Punjab soils 30-40 tonnes of farm yard manure, 62.5 kg N and 30 kg P_2O_5 has been recommended. Sharma *et al.* (2001) reported that 65 kg of nitrogen, 40 kg P_2O_5 and 40 kg K_2O per hectare has been recommended for sandy loam and slightly acidic soil of Himachal Pradesh. Whole of farm yard manure, $P_2O_5$and K_2O along with $1/3^{rd}$ of N are applied before transplanting at the time of field preparation. The balance of N is applied six weeks after transplanting and before the formation of heads (Singh and Dhaliwal, 2001).

Use of Chemicals

The effect of various chemicals and growth substances on growth of lettuce has been studied extensively. (a) Application of zinc sulphate/ molybdenum @ 0.5g/ m2 is beneficial for high yield. (b) Application of sulphur-coated urea give beneficial effect on plant growth (c) Application of nitrapyrin (a nitrification inhibitor) give high yield. (d) The fresh weight of plant increased with IBA and NAA @ 50 or 100 ppm (Singh and Saimbhi, 1967). (e) Use of GA_3 at 10 mg/litre of water can give more synchronous flowering and increase seed yields (Gray *et al.*, 1986). (e) Pre-harvest field sprays of BA at 5-10 ppm helped lettuce to retain its, fresh, green condition for an additional 3-5 days after packing. Post harvest treatment with BA also proved effective in delaying senescence.

Crop Rotations

Saini (1996) reported that following crop rotations are profitable with lettuce:

1. Amaranthus-Lettuce-Frenchbean
2. Maize-Lettuce-Chilli
3. Cauliflower(Early)-Lettuce-Cluster bean
4. Cowpea-Lettuce-Clusterbean
5. Cluster Bean-Lettuce-Brinjal

IRRIGATION

Lettuce requires a constant and abundant supply of moisture for optimum growth throughout its vegetative phase. There appears to be no particular stage of growth when the plant is especially sensitive to soil moisture conditions, although dry soils condition occurring just before harvesting appear to have a maximum effect on reducing yield (Hegde,1993).The roots of lettuce plants stop growing about 2 weeks before harvest and this may explain why lettuces respond well to irrigation at this time (Choudhury, 1967; Yawalkar, 1969). Irrigation interval varies with soil type, age of plant and climatic conditions. It is better to irrigate once in 5 to 6 days on sandy soils and 8 to 10 days on heavy soils.

Lettuce is generally irrigated either by furrow irrigation or sprinklers. Flooding is undesirable as the water would touch the plants. When sprinklers are used, it is better to irrigate in the early mornings. Great loss from tip burn and soft rot may result from careless irrigation of nearly ready leads by a sprinkler in the heat of the day (Hegde, 1993).

Trickle, sub-surface and sprinkler irrigations have been attributed to higher yield than furrow method. Trickle irrigation increased yield by 30 % over furrow irrigation. Fertilizers can also be applied with trickle irrigation. Surface irrigation maintained uniform moisture in the soil. High yield of good quality lettuce was produced from the plots irrigated during the main growing period with NPK irrigation water (Sagiv *et al.*, 1981).Drainage is as important as supplying irrigation, for excess of soil moisture may cause rotting.

INTERCULTURE

Since it is a shallow rooted crop, shallow cultivation is recommended. In the plains, irrigation at frequent interval is essential for rapid development of leaves. Excess irrigation in heavy soils causes the rot and burning of leaf edges (Nath *et al.*, 2008).

Hoeing is done intermittently for proper aeration to make soil loose and keep clean the plots. 2-3 hoeing is given before top dressing and then earthing is done. During summer the use of cheese cloth, burlap, palm leaf, shade reduce the heat and helpful in growing, crisp and sweet leaves (Shanmugavelu, 1989).

Mulching

Mulching is useful if there is rains in cropping season. It prevents soil erosion around the root zone and raises soil-temperature during the winter months, thereby, stimulating the plant growth. Mulching prevents the leaves from touching the soil and also from rotting. The locally and easily available rice straw can be used as mulching material (Singh and Dhaliwal,2001). Polythene mulch (black sheeted), polythene film use as a mulch and give good protection against infection with *sclerotinia* minor and also maintain moisture and suppress weeds.

Protected Cultivation

Lettuce is grown under glass or in plastic tunnel in temperate countries. Light temperature and CO_2 are artificially provided. Seeds are sown in plastic or wooden trays/boxes. Adequate moisture is maintained to improve germination and to enhance growth of seedlings. Seed boxes are kept at 12°-15°C. At the initiation of first true leaf (12-15 days after sowing), seedlings are picked out in small parts or soil balls. After 4-5 weeks, these plants are shifted in the greenhouse in beds at spacing of 45 x 45 cm, or 45 x 30 cm or 30 x 30 cm depending upon cultivar. Day/night temperature of 18°-20°C/14°-18°C and CO_2 at 1000-1500 ppm are maintained (Singh and Dhaliwal, 2001).Artificial light with fluorescent tube is provided. Nutrients are also provided in high quantity than field grown crop.

WEED CONTROL

Weeds cause heavy loss in yield. After 2-3 weeks of crop emergence, removal of weeds is essential for the control of weeds, hand weeding is a common practice and 3-4 hand weeding are adequate throughout the crop period. Lettuce requires at least 3 weeks of weed-free period after seeding to prevent crop yield reduction as reported by Zimdahl (1980).

Some herbicides were found effective for the control of weeds. Incorporation of benfluralin at 0.9 kg/ha in heavy soil before planting controlled most of the weeds. Several herbicides were tested in sandy loam soil and the pre-emergence application of proplam + diuron, chloropropham + sulfallate and propyzamide were performed well.Among these sulfallate was the best (Saini, 1996).

PLANT PROTECTION MEASURES

The Lettuce crop is attacked by numbers of insect pests and diseases which detail symptoms and control measures are given below:

Insect Pests

Aphids (Macrosiphum barri)

It is one of the most destructive insect attacking lettuce crop. It attacks on foliage and suck plant sap. Spraying of dimethoate or methyl parathion @ 2 ml/litre of water control the aphids in the crop. The spray of insecticides like malathion or carbaryl is effective for control of this pest. The use of 3-4 per cent nicotinic dust gives satisfactory control if applied when the temperature is around 18° to 21°C (Choudhury, 1967).

Cabbage looper (Trichoplusia ni)

It prefers lettuce as host for oviposition.Younger plant are more preferred for oviposition than older plants. Spraying of malathian @ 2 ml/litre of water control the pest.

Corn earworm (Heliothis zea) and tobacco bud worm (Heliothis virescens)

These two pests are also attacking lettuce. Spraying of malathion/ endosulphan @ 2 ml per litre of water is effective to check the infestation.

Diseases

Damping off

This is caused by *Rhizoctonia solani* and *Pythium ultimum*. Damping off is of two types, pre-emergence and post-emergence. In the first case seed germination is inhibited. Small seedlings are attacked at the soil surface in case of post emergence. There are shrinking and brown rotting at the stem of the seedling at the soil line and these seedlings thus fall down. In heavy wet soil at high temperature, there is more severity of the disease. Stem, leaves, petioles, mid rib and complete head are also affected by the pathogens. Pythium causes severe stunting and rosetting followed by wilting and death of plants. Lateral roots are severely rotted by this pathogen. To control the disease, seed should be treated with 2% ceresan and nursery soil should be drenched with 0.2% difolatan before sowing. Seedlings should be sprayed with 0.2% difolatan at an interval of 10-12 days.

Root rot

This is caused by *Pythium polymastum*. This disease is characterized by severe stunting and rosetting followed by wilting and death of the plant. Lateral roots and cortical tissues of the tap root are severely rotted. The disease can be controlled by removal of affected plants, and crop rotation.

Downy mildew

This disease is caused by *Bremia lactucae* Regel This is a serious disease of indoor and outdoor crops. Excess moisture enhances this disease. Light green to yellow areas develop on the upper surface of the leaves which later become brown. A downy white mass appears on underside of the leaves. To control the disease spray 0.2% dithane Z-78 (Walker, 1952). Grow resistant cultivar like Polo, Columbus, No.33, Soraya, Dolly and Imperial-17.

Lettuce drop

This is caused by *Scalarotinia sclarotiorum* and *S. minor*. Pathogen attack the stem near the ground. In advance stage, leaves dry and plant collapse. There is a soft and watery rot on stem and leaf bases. To control the disease soil sterilisation with steam air mixture at 70°-80°C is helpful and single spray of benomyl at high conc. (3-4 kg/ha) @ 440 litre/acre is effective to check the disease.

Botrytis rot

It is a disease in the green house plant of lettuce. Stem at soil line shows brown necrotic lesion and often the leaves touching the soil are also infected. To check the disease control the humidity and low temperature. Field soil should be dry as soon as possible.

Slimy soft rot (Erwinia sp.)

This is a sort of bacterial rot which first appears as water-soaked, greasy, soft lesions, which spread rapidly and later turn dark brown and slimy. The disease causes considerable damage in head lettuce and can be checked by the removal of the heads and also keeping the soil surface relatively dry. To control the disease, weekly spraying of copper oxychloride (Blitox-50) @ 3 g / litre of water is advised.

Lettuce mosaic

This is seed borne disease and the leaves exhibit mottled yellow area. In severe case plants fails to form head. Head become irregular in shape and show internal tip necrosis and stimulate rusty brown discolouration. Spraying of metasystox or rogor @ 1.5 ml/litre of water to check insect vector (aphids), is recommended and mosaic free seed should also be used.

Big Vein

This disease, caused by soil borne virus, is very wide-spread and results in serious losses in some areas, specially where lettuce is grown on the same land for several years. The first symptom of this disease, observed in young plants, is a yellowing along the veins. This is followed by thickening and crinkling of the entire leaf. The plant continues to grow, but produces, smaller, less firm heads than normal plants. The virus does not penetrate the above ground part of the plant, but is confined to the roots where it multiplies. It persists for a long time in the soil (Thompson and Kelly, 1972). Growing lettuce in virus free soil and rotation with crop that do not harbor the virus are control measures for field grown crops. Soil application of methyl bromide, dazomet, chloropicrin or 1,3 dichloropropene is effective to check the big vein.

Physiological Disorders

Tip burn

It is visible on the lateral margins of the inner leaves of mature head and more in glass house crop. Anderson (1946) indicate that water deficiency, as shown by high soil-moisture tension, is the primary cause of tip burn. Various reasons like -high temperature, light, maturity and duration, excess of nitrogen, calcium and boron deficiency and ontogeny age of plant. High Mn content, soil moisture content and high endogamous level of IAA. This disorder has been found to be associated with the stage of rapid growth and conditions favourable for rapid growth increase this malady. Close spacing resulted in lower incidence of tip burn because of slow growth. Raising of dark period, relative humidity, increasing ca content in the leaves. $CaCl_2$ application can also reduces the incidence.

Internal Rib Necrosis

Both rusty brown and internal rib necrosis are associated with lettuce mosaic. Both are controlled when lettuce mosaic is controlled through use of mosaic free seed (Shanmugavelu, 1989).

Ozone injury

Ozone at high concentration induced foliage injury to lettuce. This injury is characterized as non-parasitic troubles like crushing and brushing, russet spotting, and brown stain due to excessive CO_2 and pink and rib discolouration. To control this disorder ozone concentration should be corrected.

HARVESTING

The harvesting of lettuce is depends on the type and the purpose for which it is grown. Madariaga and Knott (1951) showed that the length of time from planting to harvest varied from 70 to 150 days. Leaf lettuce varieties are usually allowed to develop to full size, though they are often cut soon after they are half grown and for home use it is harvested at any time (Saini, 1996). Head lettuces are harvested when the heads are well formed, of good size and solid. A head harvested too early will be unfilled and soft. An over mature head will show deterioration like cracked or broken ribs, bitterness, yellowing of the green leaves and excess blanching of inner leaves. Crisphead lettuce is considered mature and ready for harvest when the head is firm hard and weighs about 1 kg although the standard size of crisp head for harvesting varies from region to region. Summer grown lettuce will be ready for harvest in 50-60 days while winter grown lettuce will take about 120 days (Singh and Dhaliwal, 2001). The time of harvest is directly related to temperature and may range from 50 to 120 days.

Lettuce is cut manually with a long handled sharp knife. Where the lettuce drop is serious, it is best to make the cut above the leaves resting on the ground so as to eliminate the disease spores and prevent their spread throughout the crate (Thompson and Kelly, 1972). It is better not to harvest immediately after a rain or while the dew is on it because the leaves being crisp and brittle break easily in handling.

YIELD

The average yield of lettuce per hectare varies from 100 to 140 quintal.

POST HARVEST HANDLING

Grading

Lettuce should be properly graded and plants free from diseases, insect and other injury should only graded and sent to the market. In case of head lettuce only solid head should be graded. Generally lettuce are graded in to 'super', 'standard' and 'commercial' grade. The best quality with uniform shape and size, brightness and lustre are kept in to super grade. Medium size heads with good look and texture are arrange into standard grade. Other type, free from blemish, diseases and other injury are put in to commercial grade (Singh and Singh, 2007).

Packing

Lettuces are generally packed in paperboard cartoons, wooden crate, plastic crate and splint baskets for different purposes. Commercially grown lettuce is packed for shipment in paper board cartoons. Standard vegetable crate is also used for packing and it hold 4-6 dozen heads packed in three layers with crushed ice between the layers and other the top. In some area a crate smaller than the standard vegetable crate is used for crisphead lettuce. Two dozen heads are packed in this crate. The heads are packed in two layers, the bottom layer placed with the stem end down and the top layer with the stem end up. Leaf lettuce grown in greenhouse, is packed in splint baskets, usually of 24 quarts capacity, which deliver about 10 pounds of lettuce (Thompson and Kelly, 1972).

Storage

Leaf type lettuce cannot be stored under ordinary conditions because they loose their moisture very soon. The head lettuce can be stored for 3 weeks at 0°C with 90-95% RH, whereas leaf type lettuce can be stored for 1 weeks at 0°C with 95% RH in cold storage (Pantastico *et al.*, 1975). Storage helps to prevent market gluts and makes available the produce during the period of slack production.

SEED PRODUCTION

Lettuce crop is biennial in habit. The seed stalks are produced just after full vegetative growth. Slightly high temperature is required for flowering. Early planted crop flowers in the plains and some quantity of seeds can be produced while for large scale seed production the crop should be raised in hills at the altitude of 1000m and above.

Lettuce is a self pollinated crop with cleistogamus flowers but, one to six per cent cross-pollinations has also been reported (Thompson *et al.*, 1958).

Land requirements

Select field on which lettuce was not grown within the previous two years, unless the crops field were inspected by the certification agency and were found free from seed-borne disease infection. In addition, the field should be well-drained, fertile with high organic matter.

Isolation distance

Seed field must be separated from fields of other varieties and fields of same variety not conforming to varietal purity requirement of certification at least by 50 metres for foundation seed production and 25 metres for certified seed production (Arya, 2003).

Brief Cultural Practices

Time of sowing

September to October.

Source of seed

Obtain nucleus/breeder's/foundation seed from source approved by a seed certification agencies.

Seed rate

Nursery sowing 0.5 to 0.75 kg per hectare

Direct sowing 2.0 to 2.75 kg per hectare

Sowing and transplanting

The seeds may be sown in raised nursery beds in rows 3 to 4 cm apart. Five to six weeks old seedlings later transplanted at a distance of 50 to 60 cm from row to row and 20-30 cm plant to plant.

In case of direct sowing, the seed crop is sown in line in well prepared field in rows 50-60 cm apart. The seedlings are thinned to approximately 20-30 cm apart.

Other agronomic practices for raising seed crop are same as for the crop grown for vegetable purpose.

Ideal true to the type plants are selected for seed production.

The leaf type produce seeds easily. In case of heading type, the plants of solid heads with crisp texture are selected and cut without damage to the internal growing point.

Tearing or making cut in head type

In the head type, a slight cut, or tearing the upper leaves by hand, helps the seed stalk to emerge easily (Singh *et al.*, 1962).

Roguing

Roguing should be completed before bolting starts (at the marketable stage) in loose head/leaf type varieties, or when heading is completed in heading type. Direct bolters, ill formed plants, off-type plants, diseased plants affected by mosaic, yellow mosaic and wild lettuce plants should invariably be removed from time to time as required. Another roguing is done at bolting stage.

Harvesting and threshing

Lettuce seeds ripen unevenly, and fully mature seed shatter readily. It is advisable to harvest under dry condition when 50 per cent of the inflorescences show fully developed white pappus or fluff. The plants are cut by hand and left to dry on the ground for a few days then threshed. Later, the seeds are collected by gently shaking them (Agrawal, 1985). The seeds are also dried up to 5.5 per cent moisture content and stored in vapour proof containers.

Seed yield

The leafy types usually produce 500 to 600 kg seed per hectare. However, the heading type produces only 100-125 kg seeds per hectare.

According to Agrawal (1985), the leafy type give seed yield about 500-600 kg/ha and the head type produce the seed only about 100-125 kg/ha.

REFERENCES

Agrawal, R.L. (1985). Seed Technology. Oxford & IBH Publishing Co., New Delhi.

Anderson, E.M. (1946). Tipburn of lettuce: Effect of maturity, air and soil temperature and soil moisture tension, *Cornell Bull.* 829.

Anonymous. (1998). Varieties and Hybrids Developed by Division of Vegetable Crops. I.A.R.I., New Delhi.

Arya, (2003). Vegetable breeding – Production and seed production, Kalyani Publishers, New Delhi. pp. 74.

Bassett, Mark, J. (1986). Breeding Vegetable Crops. The AVI Publishing Company, INC, West Port, Connecticut, USA.

Choudhury, B. (1967). Vegetables. National Book Trust, India, New Delhi.

Durst, C.E. (1930) . Inheritance in lettuce.III, Agric. Exp. Stn., Bull. 356.

Ferakova, V. (1977). The genus *Lactuca* L. In Europe P.68. Univerzita Komenskeho, Bratislave.

Gray, D., Steckel, J.R.A., Wurr, D.C.E. and Fellows, J.R. (1986). The effects of applications of gibberellins to the parent plant, harvest date and harvest method on seed yield and mean seed weight of crisp lettuce. *Ann. Appl. Biol.* 108: 125.

Hedrick, U.P. (1972). Sturtevants edible plants of the worlds. Dover Press, New York.

Hegde, D.M. (1993). Wter management in vegetable crops. In advances in Horticulture vol.5-Vegetable crops Part-1: (Eds.: K.L. Chadha and G. Kalloo), Malhotra Publishing House, New Delhi, India.

Hester, J.B., Parker, M.M. and Zimmerley (1936). Liming coastal plain soils. *Va.Truck Exp. Sta. Bull.* 91.

Kalloo. (1986). Salad Crops . In Vegetable Crops in India (Eds: T.K. Bose and M.G.Som). Naya Prokash, Calcutta-6.

Knowles, P.W. (1978). Personal communication. Universityof California, Davis.

Lindqvist, K. (1960). On the origin of cultivated lettuce. *Hereditas*, 46: 387.

Mabberley, D.J. (1987). The plant book. A portable dictionary of the higher plants. Cambridge University Press, Cambridge.

Madariaga, F.J. and Knott, J.E. (1951). Temperature summations in relation to lettuce growth, *Ammer. Soc. Hort. Sci. Proc.*, 58: 147-152.

Mehta, Y.R.(1959). Vegetable growing in Uttar Pradesh. Bureau of Agriculture Information, Uttar Pradesh.

Nath, P., Velayudhan, S. and Singh, D.P. (1976).Vegetables for the Tropical Region, ICAR, New Delhi.

Nath, Prem, Srivastava, V.K., Dutta, O.P. and Swamy, K.R.M. (2008). Vegetables Crops - Improvement and Production. PNASF, Bangalore. pp. 173-175.

Pal, A.B. (1975). *Progressive Hort.*, 7: 62-68.

Pantastico, Er. B., Chattopadhyay and Subramanyam, H. (1975) Post harvest Physiology, Handling and utilization of Tropical and Subtropical Fruits and Vegetables (Ed.: Er. B. Pantastico), AVI Publishing Co, West port, Connecticut, USA.

Pink, D.A.C. and Keane, M. Eavan (1993). Lettuce *Lactuca sativa* L. In Genetic Improvement of Vegetable Crops (Eds.G. Kalloo and B.O. Bergh). Pergamon Press, New York.

Rodenburg, C.M., Basse, H. ,Glaschke, E.B.,Drouzy, J., Trebouchet, G., Haigh, J.C., Watts, L.E. and Huyskes, J.A. (1960). Varieties of lettuce. Instituunt voor de veredeling van,Tuinbouwgewassen, Wageningen.

Ryder, E. J. and Whitaker, T.W. (1976). Lettuce. *Lactuca sativas* (Compositae). In Evolution of crop Plants (Ed. N.W. Simmonds). pp.39-41, Longman, London.

Ryder, Edward, J.(1979). Leaf salad Vegetables. AVI Publishing, Co. Inc. Westport, Connecticut.

Sagiv, B., Bar-Yosef, B. and Eliah, E. (1981). Preliminary Rept. Division of Scientific Publication, Bet, Dagon No.792, pp. 42.

Saini, G.S. (1996). A Text Book of Vegetable Production. Aman Publishing House, Meerut, U.P. (India).

Shanmugavelu, K.G. (1989). Production Technology of Vegetable Crops. Oxford and IBH Publishing Co. Pvt. Ltd., New Delhi.

Sharma, D.K., Chaudhary, DR and Pandey, DP (2001). Growth and yield of lettuce cv. Alamo-1 as influenced by dates of planting and plant density. *Vegetable Science*, 28(1): 38-39.

Singh, D., Atwal, A.K. and Bajaj, K.L. (2000). Evaluation of some promising type lettuce cultivars for leaf and quality characters. *Vegetable Science*, 27(1): 92-93.

Singh, D.N. and Nath, Vishal (2011). Varieties and Hybrids of Vegetables. Satish Serial Publishing House, Azadpur, Delhi (India). pp. 363-366.

Singh, D.N. and Singh, S.P. (2007). Production Management of Underutilized Vegetables. Agrihortica Publications, Junagadh, (Gujarat), India. pp. 180-190.

Singh, H.B., Thakur, MR and Bhagchandani, PM (1962). Vegetable seed production in Kullu Valley: Part IV-Seed production of Peas, Lettuce, Celery, Parsley, Onion, Leek and Spinach. *Indian J. Hort.* 20(3&4): 148-154.

Singh, K. and Saimbhi, M.S. (1967). Response of lettuce (Lactuca sativa L.) to plant growth regulator sprays. *Punjab Agric. Univ. J. Res.*, 41: 382-387.

Singh, Surjan and Dhaliwal, M.S. (2002). Leafy and Salad Vegetables. In Vegetables, Tubercrops and Spices (Eds. S. Thamburaj and Narendra Singh). DIPA – ICAR, New Delhi.

Thompson, H. C. and Kelly, W. C. (1972), Vegetable Crops. Tata Mc Graw -Hill Publishing Company Ltd., New Delhi.

Thompson, R.C. (1951). Lettuce varieties and culture, *U.S. Dept.Agr. Farmer's Bull.* 1953.

Thompson, R.C., Whitaker, T.W., Bohn, G.W. and Van Horn, C.W. (1958). Natural cross pollination in lettuce, *J. Am. Soc. Hortic. Hortic. Sci.* 72: 403.

Thompson, R.D. and Ryder,E.J. (1961). Description and pedigrees of nine varieties of lettuce. *USDA, Tech., Bull.* 1244.

Vavilov, N.I. (1935). The origin, variation, immunity and breeding of cultivated plants. *Chron. Bot.*, 13 (1/6): 1-366.

Walker, J.C. (1952). Diseases of Vegetable Crops. McGraw-Hill Book Co., USA.

Wilfred, M.R. (1924). The Botany of crop plants.

Yawalkar, K.S. (1969). Vegetable crops of India. Agri. Horticultural Publishing House, Nagpur.

Zimdahl, L. (1980).Crop Weed Competition – A Review, Published in the United States by the International Plant Protection Centre. Oregon State University, Oregon, USA.

❑❑❑

Chapter-5

Leguminous Vegetables

1. PEAS/GARDEN PEA

Botanical name	:	*Pisum sativum* L.
Family	:	Fabaceae
Chromosome number	:	24 =14

Pea is commonly known as garden pea (matar), is very popular because its greeen pods as well as dry seeds are in great demand for cooking as vegetable and as pulse, respectively. Pea crop is a short duration and easy to cultivate. It has its origin place probably in Ethiopia. It is with human beings since stone age. It is relatively hardy and grow best under cool climate. It is a very common crop during winter throughout the plains of India. In india, pea is being grown in various states. Peas are suitable for canning, freezing and dehydration and therefore they are available in the off-season. Both the green pods containing green immature seeds and dry seeds are rich in protein (25.6%) and are excellent sources of vitamins and minerals (Salunkhe and Desai, 1984).

ORIGIN

According to vavilov (1926) pea has different centres of origin. Area comprising central Asia, the near East, Abyssinia and the Mediterranean is the centre of orign based on genetic diversiy. Cultivated garden pea is not seed in the wild state and it might have been originated from wild field pea or other related species.

Pea or garden pra is a very popular winter season crop and is native of Europe and western Asia (Thompson and Kelly,1957). Pea (*Pisum sativum* L.) originated in the Near East and Mediterranean region and has been grown for food and feed since early Neolithic times (Zohary and Hopf, 1973).

Importance

Pea continues to serve an important role in modern agriculture as a nitrogen fixing rotational crop with cereals. The high protein concentration contributes favourably to diets of people in many developing countries. World production of dry pea is estimated at 10.6 million MT from 5.9 million hectares (FAO 2001). In addition, an estimated 7.2 milion MT of fresh peas for canning and freezing are produced on approximately 876,000ha worldwide.(FAO 2001). The United States accounts for less than 2% of the world's dry pea crop and only 17% of the fresh pea crop. Canada is the largest producer of dry peas at 25% of world production (FAO 2001). Dry pea production in the US is centered in the Pacific Northwest (PNW) states of Washington and Idaho, however, production in the midwest states of North Dakota and South Dakota, Nebraska and eastern Montana has increased steadily in recent years and promises to exceed production in the PNW. Fresh peas for canning are produced primarily in Washington,Oregon, Idaho, Wisconsin, Minnesota, and New York and several other states also have significant production.

Peas play important role in human diet as well in the national economy of India. It is an important vegetable among the fresh vegetable and is rich in protein. The crop is usually grown for its green pods which contain immature seeds and are cooked as a vegetable, alone or with potatoes. pea is a very common crop grown during winter thrughout the plains of north india and as summer vegetable in the hills. There is an appreciable acreage under this crop. Pea contributes to 4.81% of total area and 3.1% of total production of vegetables in India. The area under pea was 2.89 lakh ha with a production of 27.04 lakh tons in the year. 1998-99 (Swarup, 2006). Its consumption is very high when compared with other vegetable crops. It is rich in protein, carbohydrates, vitamin A and C, calcium and phosphorus. It also contains a small quantity of iron. The green pod in certain varities is edible and is usually cooked in various vegetable preparations in different regions of the country. It has a good market in the dehydrated and canned forms.

History and Adaptation

Archaeological studies have indicated that pea has been cutivated since ancient times. Janick *et al.* (1969) described the association of genus *Pisum* with man at least from stone age. The cultivation of field pea was a very old practice. Renfrew (1973) reported that in Austria its discovered from Bronze age, whereas in Germany they were grown in Iron Age. Evidence from carbonized remains indicate that pea has been cultivated with cereal crops such as wheat and barley since its domestication in 6-7000 B.C. Limited historical evidence makes it difficult to determine the exact location of domestication, however, soon after domestication pea and other legumes spread to other parts of the world with the movement and activities of man (Marx, 1977). The nutritional value and ability to store it for extended period has contributed to widespread production of the pea.

The pea plant is unknown in wild state but some primitive plants recorded in central Asis, near East and Northern Africa are found to be cross fertile with modern cultivated pea. The primary center of diversity is deemed to be Central Asia with secondary centre in the near East, in the Ethiopia and Mediterranean countries. The small seeded field pea referred to as *Pisum arvense,* might have given rise to cultivated garden pea through mutation to white flower and large seeds. Purseglove (1974) however regarded *P. elatius* a wild species of Russia as ancestor of cultivated pea.

In India, pea was found to exist in North India even before the arrival of Aryans (De Candolle 1896) Similarly this crop plant existed in Western Asia, before it was cultivated as per De Candolle's surmise. The record of peas has been made in Harappan civilization (2500 B.C.-2000 B.C.) from Harappa itself. Carbonised seeds have been discovered from chirand, Neolithic site in Bihar (Vishne Mittre, 1974). A few scattered seeds were found mixed with wheat seeds from Navadatoli-Maheshwar a Chacolithic site in Madhya pradesh and also in Osmanabad in Maharastra (Vishnu Mittre, 1974). There is a Sanskrit equivalent for pea, mentioned in "Susruta Smhita" (3rd to 4th cintury A.D.) as "satina" (Ray *et al.,* 1980). The existence of pea in India before the advent of Aryans is possibly true. But its cultivation as a pulse crop of round and non-sweet types has to be distinguished from vegetable (as garden pea), in respect of its domestication and its probable time.

It is admitted that small, round and non- sweet varities of peas represent and intermediate stage of domestication and sweet, round or wrinkled seeded types of temperate region have to be recognised as the advanced ones. In India, the former types are grown as a pulse crop entirely under rainfed condition in Uttar pradesh and Madhya pradesh and part of north Orissa because of its hardy nature. Small podded with less number of small seeds, these plants stand very well to the heavy rainfall in eastern Madhya pradesh where they are broadcast in August in low hilly tracts. Incidentally they carry a fair degree of resistance to powdery mildew. 'Kaparakheda' variety of Madhya pradesh may belong to this group.

But the sweet garden pea varieties now in cultivation might have been intoduced during British days. Most of the early garden pea varieties were developed in temperate regions of Europe probably in 16th to 17th century. Some of the hardy round seeded garden pea varieties like "Bonia", "Kaip", "Asauja" etc. may possibly be from early introduction of the British. They are less sensitive to wilt even under warmer conditions. Varieties of T-series developed by Dr. T.R. Mehta and other in Uttar Pradesh about 50 years ago, made a big impact as mid season vegetable varieties in U.P. and Punjab. Most of these are being used even now in breeding programmes (Seshadri and Chatterjee, 1996).

Another interesting feature of garden pea introduction is to be found in the late maturing variety "Kala Nagni" grown in Himalayas from Himanchal padesh to North Eastern Hills. It is an important variety in Darjeeling and Sikkim grown both during rains (July sowing) for harvest in October-November and again in autumn (November which over winters in higher altitudes to be harvested in April-May.) This variety 'Kala Nagni' (also known as Local Darjeeling or Local Sikkim) - so called because of black hilum- is adapted to heavy rains and resistant to *Fusarium* wilt.

The present day cultivated varieties especially early maturing, sweet and wrinkled seeded ones, like the popular "Arkel" a French introduction, are becoming popular and in fact are causing genetic erosion, displacing the hardy, drought tolerant and round seeded ones of the previous centuries. The adaptation of pea in India is significant because of its use as pulse crop and the farmers' selection went towards round and bold seeds and hardiness to adverse environment like drought or with minimum one ot two irrigations. In Uttar Pradesh, it is grown as a rainfed crop during winter season while in eastern Madhya pradesh and western Maharashtra, it is able to withstand heavy rains in the off season (August-November).

DISTRIBUTION

Pea is acool season crop best suited to production in the temperate regions. In the warmer areas of the tropics pea is restricted to production in the cooler high lands. It is grown in U.S., England, France and other European countries from mid-September to Fenruary-March. In Austria winter pea is grown and harvested at maturity and sold for seed. In India,pea cultivation is distribution mostly in Uttar Pradesh, Bihar, Madhya Pradesh and Himanchal Pradesh and some colder part of Orissa and Bengal.

BOTANY

Pea (*Pisum sativum* L.) is a self pollinated crop and belongs to the family Papilionaceae (Fabaceae). The fruit which is eaten is a 'pod' botanically. The plant is a harbaceous annual with vining habit. The leaves are pinnately compound with one to three pairs of leaflets. the terminal leaflet and occasionally the upper lateral are modified as tendrils. (Hawrhorn and Pollard, 1954). The tap root grows up to 120 cm depending on soil type. Root nodules form on roots by bacteria, *Rhizobium leguminosarum*. Legume bacteria fix nitrogen from atmosphere in the soil (root nodules).

Inflorescence is a raceme. The flowers borne singly or in pairs on elongated peduncles arising in the leaf axils. Flowers are white, pale cream or purplish in colour. The calyx consists of five sepals forming a tube united at the base. The corolla cosists of five petals. The large, broad upper petal (Standard) encloses the two lateral ones (wings); the lower ones are united to form the bud. Stamens are 10 in number, 9 form a cleft on the upper side surrounding the ovary and the tenth is free. Ovary is superior, one celled having two rows of ovules, arising an adjacent parallel placentae.

Plant breeding

Breeding objectives of garden pea are high pod yield, high shelling percentage, suitable for freezing, resistance/tolerance to diseases and pests. Breeding systems include pedigree mathod, backcross method, mutation breeding, bulk method, individual plant selection and single seed descent method (Nath *et al.*, 2008). Micro propagation of peas is done by tissue culture using apical meristems of seedlings and through somatic embryogenesis (Arya 2003).

Pea has been grown in India for several decades and is quite adapted to this part of the world. There are a good number of local strains and exotic varieties available. Most of the exotic varieties were introduced in India and used as commercial varieties and hence significant efforts were not made to breed varieties for various purposes. There has been an acute need for breeding varieties for high yield, better processing and cooking qualities and disease resistance (nath *et al.*, 2008). Some of the genetic information with regard to economic characters and male sterility have been reported (Joshi and Singh, 1959; Brahmappa and Singh, 1977; Narsinghani and Singh, 1979; Narasinghani *et al.*, 1979; Pal *et al.*, 1982). Rust and powdery mildew resistant varieties have released and variety Arka Ajit and Arka Karthik, possessing high yield and rust resistance characters also been released from Indian Institute of Horticultural Research, Bangalore.

Taxonomy

The pea belongs to the family Leguminosae and sub-family papilionaceae (Fabaceae). The chromosome number of pea is 2n =14 (Das and Kalloo, 1970). Based on crossability and cytogenetic evidence, the number of species within the genus *Pisum sativum* includes the following sub-species (Simmonds, 1979).

(1) *Pisum sativum* Linn. var. hortense: Garden pea.

(2) *Pisum sativum* Linn. var. arvense: field pea.

(3) *Pisum sativum* Linn. var. *macrocarpum*: Edible podded pea.

(4) *Pisum sativum elatius*: Wild form.

(5) *Pisum sativum sysiacum*: Wild form.

(6) Wild progenitor of *pisum sativum* is unknown.

The genus *pisum* is a member of the family papilionaceae tribe Viciae and is composed of two species, *P. sativum* L. and *P. fulvum* Subth and Sm. After a significant amount of study and discussion *Pisum sativum* has been further divided to include several subspicies, *P.s.* ssp. *elatius, P.s.* ssp. *humile, P.s.* ssp. *arvense* and *P.s.* ssp. *hortense*. The literature indicates that ssp. *elatius* and ssp. *humile* are the progenitors of the garden pea *P.s.* ssp. *sativum* (Zohary and Hopf 1973). *Pisum s.* ssp. *arvense,* includes the field pea and the 'Austrian' winter pea, both possessing colored flowers and variously pigmented seeds. Sub species *hortense,* the garden pea, is harvested at the succulent immature stage and consumed fresh or preseved through canning or freezing.

Genetic diversity within the genus *pisum* has resulted in a myriad of uses for the crop. Dry peas are used for animal feed and for human consumption in soups and processed snacks. Some minor uses of the dry pea are as a cover crop, green manure (ploughed prior to seed development) and bird feed.

Floral Biology

Pea is self pollinated with negligible natural crossing due to showy flowers. (0 to 3%). In early varieties, formation of flowers is at lower node number than the taller and late varities. The anthesis is maximum between 9 a.m. to 12 noon. The anther dehiscence starts before opening of flower. At 16°C the stigma remains receptive for 3 days after emasculation at 20 to 24°C retains good receptivity for only one day. Temperature affects the duration of stigma receptivity (Anon, 1970).

The fruit is a pod, green or yellowish green in colour, varying in size and shape with the variety. The pods are plumpy, the seeds are globular and either wrinkled, dimpled or smooth. The thin transparent seed coat encloses an embryo with two cotyledons and a well developed hypocotyl.

COMPOSITION

Pea or garden pea or vegetable pea is a highly nutritive vegetable containing high percentage of digestible protein (7.2%), alongwith carbohydrates (15.9%), vitamin A (405 I.U.) and vitamin C (52 mg/ 100g). It is also rich in minerals viz., calcium, potassium, phosphorus and iron. Fresh green peas are low in sodium. Pea protein is moderately deficient in the sulfur containing animo acids metheionine and cysteine.

According to Aykroyd (1941), it contains water 72.1%, carbohydrates 19.8%, protein 7.2%, fat 0.1% and mineral matters 0.8% (0.02% calcium, 0.08% phosphorus, 0.0015% iron). According to Mac Gillivray (1953). Shelled pea contain 74.3% water, and the food value per 100g of edible portion is energy, 101 calories, protein 6.7g, calcium 22 mg, vitamin A 700 I.U., ascorbic acid 28 mgs., thiamin 0.25 mgs., riboflavin 0.15 mgs and niacin 2.10 mgs. Rao and Purewal (1968), reported that pea has moisture 11g, protein 22.5g, Fat 1.8g, carbohydrates 62.1g, calcium 64 mg, iron 4.8 mg, riboflavin 0.15 mg, thamin 0.72 mg and niacin 2.4 mg. Pea is very rich in protein and is very valuable to those who cannot eat meat. On dry weight basis, garden pea (wrinkled seeded) contains 10-13% total sugar, 33% starch and 24-28% protein while fields pea (smooth seeded) contains 6.7% total sugar, 46% starch and 23-25% protein. Lower protein in smooth seeded may be due to higher starch content and greater proportion of amylopectin than wrinkled seeded. Flavour of the seed depends upon the sugar content.

Uses

The green pods in certain varieties is edible and is usually cooked in various vegetable preparations in different regions of the country. It has good market in the dehydrated and canned forms.

Fresh green immature shelled seeds of almost all varieties and the green pods (as whole) of only a few varieties are used as vegetable. The dry seeds of pea are used as pulse.

Peas are used in making soup, chat, dal, besan, concentrate. green fodder etc. Peas are rich source of digestible protein and good content of vitamins (A,B,C and G) and minerals (Phosphorus, potassium, calcium, sulphur, iron etc.). peas are valuable for vegetarians. Minor uses of the dry pea include extraction of starch and protein for food processing and industrial applications. The fresh canned or as a frozen product.

The most widely produced pea in the world is the dry edible yellow pea although many other types are produced for a variety of uses. Primary uses are in animal rations while a small fraction is used in human diets. It is difficult to estimate the quantity used for human food due to limited reporting from many develpong countries, but estimates approach 40-50%. The dry green pea is produces primarily for human consumption in soups and snacks. The fresh green pea can either be frozen, canned or consumed fresh. The Austrian winter pea is anothr type of dry pea which when harvested at maturity can be sold for seed or more commonly used as an ingredient in bird feed. Due to its excessive biomass production, the Austrian winter pea is commonly used as a green manure crop to improve soil tilth and fertility. A rather minor pea type produced for the snack industry in the Orient is the marrowfat pea. It is also harvested dry and is characterized by extremely large seed size and an irregularly shaped, dimpled seed surface.

VARIETIES AND HYBRIDS

There are number of varieties of peas, they are grown for different purpose. The green peas are easily canned, frozen and dehydrates. For convenience, some varieties are placed under different groups according to their special characteristics.

In pea, the flowering begins at lower nodes and continues upto the stem. Flowers may appear as early as sixth node in very early flowering cultivars and as late as 18th node in late flowering cultivars. In general, plant types that flower in 9th to 11th nodes are considered early cultivars.

Early Wrinkled Seeded

Arkel, Early Badger, Early December, Early Giant, Little Marvel, VL Ageti-Matar-7 (dimpled seeded), Hisar Harit, Jawahar Matar 4 (JM-4 or GC, 195).

Early Smooth-Seeded

Alaska, Asauji, Lucknow Boniya, Meteor.

Mid to late maturing wrinkled seeded

Alderman, Bonneville, Jawahar Matar-1, Jawahar Matar-2, Khapar Kheda, Lincoln, N.P. 29, Perfection New Line, IP-3 (pant uphar).

Mid to late maturing smooth seeded

Rachna

Green seeded varieties

Alaska, Arkel, Bonneville, Field pea-48, Jawahar Matar-1, Jawahar Matar-4, Thomas Laxton, VL Agenti Matar-7, Hisar Harit, IP-3(pant uphar).

Double-podded varieties

Bonneville, Kanawari, Khapar Kheda, T-19, VL Ageti Matar-7.

Edible podded varieties

Aparna, Matar Ageti-6, Mithi phali, Sylvia.

Varieties suitable for cool climate

Alderman, Early Giant, Arkel, VL Ageti Matar-7

For canning, important pea varieties are Early Badger, Lincoln, Alaska-81 etc; for dehydration, p-87, Arkel, N.P.-29 etc and for freezing, Thomas Laxton, Alderman, Taichung-12 etc.

Varieties developed by different institution/organizations

IARI,New Delhi

Arkel, Bonneville, Sylvia, N.P.29, Asauji, Meteor.

IARI, Regional Station, Katrain

Lincoln

IIHR, Banglore

Arka Ajit, UN 53-6, Arka Karthik.

IIVR, Varanasi

Kashi Nandini (VRP 3), Kashi Shakti (VRP7), Kashi Udai, Kasji Mukti.

TNAU, Coimbatore

Ooty 1.

NDAU &T Faizabad

NDVP 8, NDVP 10.

PAU, Ludhiana

Punjab 88, Matar Ageta 6.

CSAUA &T, Kanpur

Azad p-2(PRS 4), Azad P-3(PARS3).

JNKVV, Jabalpur

Jawahar Matar 1 (JM 1, GL 141), Jawahar Matar 2, Jawahar Matar-3 (Early December), Jawahar Matar 4(JM 4, GL195), Jawahar Matar 15, Jawahar Matar 54, Jawahar pea 71, Jawahar pea 83, Harbhajan.

GBPUA & T, Pantnagar

Pant Uphar (IP 3), pant Sabji Matar 3.

HAU, Hisar

Hisar Harit (PH 1).

VPKAS, Almora

VL Matar 3, VL Agethi Matar 7 (VL 7), VL 8, Vivek (VL Matar 6).

HPKV, Palampur

Palam Priya.

Hybrids

Pro Agro

PGS Hybrid Green Pearl.

Detail characteristics of varieties and hybrids described by Singh and Nath (2011) are mentioned herewith.

Arkel

It is an early wrinkled-seeded, most popular exotic variety introduced from England and tested by IARI, New Delhi, occupies large area almost in northern and central India (Choudhury and Ramphal, 1975). The plants are dwarf. Pods are dark green, sweet, sickle shaped, incurved towards the sutures having pointed distal end and with 7 to 8 green ovules per pod. The seeds are green, wrinkled and bold in size (100 seed weight of 17g). Harvesting starts in 50-55 days after sowing. The yield potential of this variety is 50q/ha and shelling percentages is 40. The variety is highly susceptible to collar rot if sown very early, especially under higher temperature conditions. It is suitable for fresh market and dehydration.

Bonneville

This is a wrinkled-seeded mid season most popular variety and introduced from USA. The plants are 60 cm tall. Pods are light green, 8 cm in length with 8 bigger green sweet ovules. The matured seeds are green and wrinkled bold (100 seed weight 17g). This variety flowers in 55-60 days and first blossom appears at 13-15th node. This variety is highly susceptible to powdery mildew. Its yield potential is 100q/ha with shelling percentage 45.

Sylvia

It is an edible-podded and tall growing variety. It takes 60 days to flowering and first blossom appears at 15-17th node. Pods are about 8.5 cm long, yellowish green and 5-6 seed/pod with a shelling percentage 40.

NP-29

This cultivar has been developed through selection at IARI, New Delhi. It is medium-tall, double-podded cultivar with dark green, foliage. It flowers in 75-80 days and first blossom appears at 14-16th node. Pods are green, straight, about 7.5 cm long and 6-7 seeded. It has shelling percentage of 50 and is highly suitable for dehydration (Jain and Choudhary, 1963).

Asauji

The selection of this variety has been made from the material collected from Amritsar and released from IARI New Delhi. It is a dwarf, early, green and smooth seeded cutivar suitable for early sowing. It flowers in 30-35 days and first blossom appears at 6-7th node. The pods are produced singly, about 8 cm long, curved, dark green, narrow and appear round when fully developed having 7 seeds/pod. The pods are ready for harvesting in about 60 days. It gives a shelling percentage of about 45 (Pal *et al.*, 1956).

Meteor

It is a round smooth seeded early variety and introduced from England. Plants are 35-40 cm tall, dark green and flowers borne generally singly. Pods are dark green, 8.7 cm long, well filled with 7 seeds, having shelling percentage 45. Pods mature in 58-60 days and suitable for early October sowing (Choudhury and Ramphal, 1975).

Lincoln

It is a dwarf to medium-tall, single podded cultivar. Flowering takes place 55-60 days after sowing and first blossom appears at 11-12th node. Pods are dark green 9.5-10 cm long, 6-7 seeded with shelling percentage 45. It is suitable for late sowing. Pods retain good colour after harvest and are good for canning.

Arka Ajit

It is resistant to powdery mildew and rust. Yield 10t/ha in 90 days.

UN53-6

A snap pea where the whole pod is edible. Yield 8-9t/ha in 90 days.

Kashi Nandini

This is earliest variety and can be sown from 25 October to 25 Novermber in northern plains. Plants are erect, 60 cm high, dark green with 7-8 pods/plant. Pods are 8-9 cm long, attractive with well filled 8-9 ovules, shelling quality being 47-48%. The pods are ready for picking 55 days after sowing. Total green pod yield being 110-120q/ha. It is tolerant to leaf miner and pod-borer.

Kashi Udai

An early maturing variety, its plants grow 60 cm high. It is sown from October end to 25 November in plains. Plants with dark green foliage have short internodes and 8-10 pods/plant. Pods are 9-10 cm long, attractive with 8-9 bold ovules/pod with shelling quality of 48%. The pods are picked up 60-65 days after sowing, total green pod yield being 100-110q/ha.

Kashi Mukti

An early maturing garden pea, it is resistant to powdery mildew. Last week of October to Novermber is ideal sowing in plains. Plant height is 50-53 cm. The plants are slender with shorter leaves, pods are 8.5-9.0 cm long, attractive filled with 8-9 bold and soft-textured seeds, with 48-49% shelling and yield of 110-120q/ha.

Kashi Shakti

This is high yielding, medium-maturing variety. It can be sown in November in plains. Plant height is 90-98 cm. Plants have dark green foliage with 11-12 pods/plant. Pods are 10-10.5 cm long, well filled with 7.5-8.5 bold seeds, shelling quality of 48-49% and yield of 140-160q/ha.

Arka Karthik

It is a mid-season variety growing to a height of 65 cm. It does not require any staking. It has combined resistance to powdery midew and russt. Its pods are ready for picking 60 days after sowing. They are 10.5-11.0 cm long, seeds are bold, green and sweet with 58% shelling. It has a yield potential of 11 tonnes/ha in 90 days.

Ooty 1

A dwarf variety with a yield of 11.9t/ha in 90 days, resistant to white fly.

NDVP-8

This mid season variety has been developed at NDUA&T, Faizabad. It gives an average yield of 100q/ha. It is recommended for release and cultivation in 1997 for punjab, Uttar Pradesh and Bihar.

NDVP-10

This mid season variey has been developed at NDUA&T, Faizabad. It gives an average yield of 117q/ha. It is recomended for release and cultivation in 1998 for Panjab, Uttar Pradesh and Bihar.

Punjab-87(87-1)

It is a derivative of the cross Pusa-2 x Morrasis-55 and released from punjab. Vine length is 55.5 cm. Leaves are dark green and consisting 1-2 flowers/axil. Days to flowering are 72 days. Pod is 9.3 cm in length, 1-2 pods/axil, green ovules, sweet, mature seed bold and wrinkled, 7-6 seeds/pod and 48.3 per cent shelling.

Punjab-(88-2-C)

It is developed by selection from the cross 'Pusa-2' and 'Morrasis 55'. Plants are tall with medium size pods and having 7 green ovules per pod. The seeds are green, wrinkled and bold. This variety is also susceptible to powdery mildew.

Matar Ageta 6

It is an improvement over 'Arkel' and 'Harabona' and released in 1989 by PAU, Ludhiana (Dhillon and Singh, 1993). It can plant as early as Harabona and yields more than either of them. It can withstand high temperature, prevelant at the time of planting in northern India. The germination rate is high, crop stand good and yield plenty. Plants are dwarf, quick growing, erect with green foliage and ready for first picking within 7 weeks after sowing. Pods are long, 12-15 in number, borne singly and in pairs, well filled and containing six grains on an average. The shelling percentage is 44. The grains are not as sweet as that of 'Arkel' but taste better than those of Harabona. The seed is smooth, dented and light green with high dry matter, chlorophyll and crude protein contents and a low value of per oxide activity (O.D./min/mg/protein) compared with Arkel and Harabona.

Hara bona

It is developed through pure line selection from the local field pea variety and released in 1980 from punjab. Vine length is 40 cm having dark green leaves. Flowering starts 21 days after sowing. Days to first picking is 45-50 days. Pods are 7.8 cm long, dark green and 5-6 seeds per pod with 40 per cent of shelling. Green seeds are less sweet. Mature seeds are bold, round, dimpled form both sides and green in colour. Seed yield is 20 q/ha. This variety escapes powdery mildew and pea rust.

Azad P-1

It is a derivative of the cross 6414 x 6405 and released in 1983 from Uttar Pradesh. Plant is 80-90 cm tall having dark green foliage. Days to flowering is 40-45 days. Pods are smooth, dark green, 8-10 cm long, narrow (1.2-1.4 cm), very tightly filled, 8-10 seeds/pod and 30-40 pods/plant. Shelling percentage is 50-55. Pods are slightly curved at the distal end. Mature seed is wrinkled. This variety escapes powdery mildew and rust. Seed yield is 17q/ha.

Azad P-2 (PRS-4)

It is powdery mildew resitant variety, developed at Vegetable Research Station, CSAU&T, Kanpur, derived from the cross Bonneville x 6587. Plants are tall (130-150 cm), erect, brached with light green foliage and white flowers, bears 20-25 pods/plant (double podded). Pods are medium with 6-7 ovules, nearly straight, light green, smooth and firm, Horticultural maturity duration is of 90-95 days with 120q/ha green pod yield. Physiological maturity duraion is of 125-130 days. Seeds are wrinkled and brown suitable for cultivation in late sown condition and powdery mildew prone areas. Rcommended for release and cultivation in 1990 for Punjab, Uttar Pradesh, Bihar, Rajasthan, Gujarat, Haryana, Delhi, Madhya Pradesh and Maharashtra.

Lucknow Boniya

It is early smooth-seeded cultivativar of in the plains (Singh and Joshi, 1970). It is dwarf white seeded cutivar flowers in 40 days and first blossom appears at 8-9th node. The pods borne singly, small, narrow, green and 4-5 seeded.

Madhu

It is derivation of the cross 6126 x Sylvia, released in 1973 in Uttar Pradesh. Plant height is 130-150 cm, foliage yellowish green. Days to flowering 50-55 days. Pods are 12-15 cm long, light green, smooth, membrane less, straight, 2-2.5 cm broad, 20-25 pods per plant and 5-6 seeds per pod. Mature seed smooth, round and of grey colour. This variety is susceptible to powdery mildew.Seed yield is 8-10 quintals per hectare.

T-19

This has been developed by Agriculture Department of Uttar Pradesh. It is a medium-tall and double-podded variety. The plants flower in 55-60 days and first blossom appears at 12-14th node. pods are yellowish green, slightly curved, 8.5 cm long and 6-7 seed/pod having shelling percentage 45.

Jawahar Matar-1 (JM-1or GC141)

It is a derivative of the cross T-19 x Greater Progress and released by Central Sub-Committee on Release of Varieties in 1977 for all pea growing areas of the country. Plant height is 82 cm at maturity. Leaves are normal shaped, green in colour, leaflet size of 5x3 cm. Flowers are white colour, two flowers per axil and flowering begins from the 11th nodes. Days to initiation of flower is 55 days. Days to 50 percent flowering is 62 days. Pod setting begins from 11th nodes and naturally 2 pods per axil. Pods are straight with beak like out growth at the lower end, green colour, medium thick pod wall and 8-9 seeds/pod. The shelling percentage is 52. The mature seeds are green in colour, wrinkled and containing 18.9g/100 seeds.

Jawahar Matar-2(JM-2or GC477)

This is a derivative of the cross Russian 2 and 'Greater Progress'. The plants are 60 cm tall. It has bigger pods having 9 green bigger sweet ovules. This green pod yield is around 100 q/ha. Shells are cmparatively thicker and have better keeping quality suitable for transportation. The matured seeds are green, wrinkled with 100 seed weight of 17g.

Jawahar Mater-3 (JM-3or Early December)

Wakankar and Mahadik (1961) have bred this cultivar in Madhya pradesh. A very early variety developed from a cross of 'T-19' and 'Early Badger'. It is a dwarf cultivar. This variety produces light pods, roundish oval, with a length of 5-7 cm and width of 1 cm having 4-5 ovules/pod. The seeds are green, wrinkled and bold (100 seed weight 18 g). The first picking can be done in 50 days after seed sowing. The yield potential of this variety is 40q/ha green pods.

Jawahar Matar-4 (JM-4or GC 195)

It has been developed from the cross 'T-19' x Little Marvel'. Plants are 55-60 cm tall and posses medium size green pods (7.0 cm), having 5 or 6 ovules/pod. The matured seeds are green, wrinkled and bigger in size (18g/100g seed). First picking starts after 75 days (Wakankar *et al.,* 1974). It yields about 60q/ha.

Jawahar Matar-5 (JM-5)

It is released in 1980 from Madhya Pradesh. Plants are 1.5-2.0 meters in length with large internodes, greenish yellow foliage, two pods/axil and 5-6 seeds per pod. Fruiting begins from 13th node. Pods are comparatively thin walled. Mature seeds are yellow and wrinkled. Seed index varies from 10-20 g/100 seeds. Seed yield is 10q/ha. This variety is immune to powdery mildew.

Jawahar Peas-4 (JP 4 and now JM6)

It is a cross of 'Local yellow Batri' x ('6588' x '46c'). The variety possesses medium size pods with 5-6 bigger size green ovules. It yields around 90 q/ha in plants. The seeds are yellow and medium sizes (100 seed weight 15g). This variety is most suitable to hilly areas and can be planted in August (Tiwari and Narsinghani, 1986), when the eprevailing temperature is low. It is planted after mid August as soon there is a break in rains. The sloppy lands help in good drainage, while the remaining monsoon showers keep the crop continuing. There are no irrigation facilities on these hillstocks. The pods are ready by October end or early November. The extra early green pods fetch very high price in the market.

Jawahar Peas-71 (JP 71)

It is a progeny of double cross (Arkel x JP 829) x (JP 501 x JM 1) having powdery mildew resistance belongs to mid season groups. The plants are 50 cm tall and pods are medium sized (7 cm) with 6-7 ovules. The seeds are green, wrinkled and bigger (100 seed weight 17 g). The yield of green pods is 120q/ha.

Jawahar Peas-83 (JP-83)

It is a mid- season powdery mildew resistant garden pea developed from a dounle cross (Arkel x JP-829) x (46-C x JP-501). The plant height is around 50 cm and the pods are bigger in size curved with 8 green sweet ovules. The cultivar possesses a high yield potential of 120-130q/ha green pods (Narsinghani and Bedi, 1990).

Knawari

It is a smooth-seeded main season variety. This is a tall-growing double-podded cultivar. Fowering takes place 65-70 days and first blossoms appear at 15-17th node. pods are about 8.5 cm long, yellowish green and 5-6 seeded with 40 percent shelling.

Khapar Kheda

It is tall growing double podded cultivar. Flowering takes place in 65-70 days and first blossom appears at 15-16th node. pods are 5.5 to 6 cm long and 4-5 seeded with 50 per cent shelling. It is very popular in Madhya Pradesh.

Harbhajan (EC33866)

It is an introduction and suitable for all dry areas of north India. Plants are dwarf. pods are green and long. Seeds are small (13g/100 seeds), round, wrinkled and yellow in colour. Fowering starts from 6th nodes. It is susceptible to powdery mildew.

Pant Uphar (IP-3)

It is a medium tall variety (70 cm), relatively thin stem, leaflet small in size and foliage light green. Flowers are white. Pods are medium size, 7.5 cm, long having 7 seeds/pod and the total yield is about 100q/ha. This variety is susceptible to powdery mildew.

Perfection New Line

It is a heavy yielding mid-season variety. The plant is medium-tall. Pod is about 8 cm, long dark green, sweet and well filled with wrinkled seeds. It is ready for first picking in 80 to 85 days (Nath, 1976).

PM-2

This variety has been produced from a cross of 'Early Badger' with 'IP-3'. The plants are 50-60 cm tall with green colour pods having

6 ovules. The seeds are wrinkled. The yield of green pods is about 52q/ha.

Pant Sabji Matar 3

Early season variety with long curved pods with 8-9 ovules. Picking starts 60-75 days after sowing. Yield 9t/ha.

Hisar Harit

It is a mid- early and high yielding. Bulk pedigree method of selection was followed in Bonneville x 023 for developing this variety. This variety is semi dwarf, early and takes 60 days for first picking; foliage green, single to double-podded; pod large green, well- filled; seed green dimpled after drying. The plants have small internodes the number of pods per plant is more than the other varieties and harvest index high.

Vivek-6

It is cross between Pant Uphar and VL Matar 3 and released by the Central Variety Release Committee in 1996. It is a medium-maturing variety takes 135-140 days and 80-85 days to maturity in mid hills and plains, respectively. The plants are dwarf and vigorous with dark green foliage. The pods are smooth, straight, medium sized (6-7 cm), light green and completely filled. The yield goes upto 123.68q/ha green pods, depending upon agroclimatic condition. It is tolerant to cold and moisture stress conditions. It is recommended for cultivation in Uttranchal, Himanchal Pradesh, Jammu & Kashmir, Haryana, Delhi and Rajasthan (Sri Dhar *et al.*, 1999).

Vivek Matar-10

Plants are quite distinct in both growth and yield attributes. The average green pod yield is 90-110 q/ha which mature in 120-130 days. Its plant height 50-60 cm, branched 3-4, first flower appear at 7-8 node, green foliage leaf normal, two white flowers/ peduncle, dark green slightly curved pod, mostly double pods/peduncle, 13-15 pods/plant, 143 pods/kg, 8-9 cm pod length, 45% shelling and greenish white wrinkled seeds. It shows 50% flowering in 80-85 days in hills and 60-65 days in plains. It is recommended for sowing from first week of November to last week of November in midhills as well as in plains. it is resistant to white rot, wilt, leaf blight and moderately resistant to powdery mildew under field conditions and tolerant to pod-borer.

VL-3

This line has been bred from a cross of 'Old Sugar' and 'Early Wrinkled Dwarf 2-2-a'. The variety has a height of about 65-70 cm and possesses light green pods of 6 cm in length having 6 green ovules. Yield potential of this variety is about 90 q/ha green pods. It is susceptible to powdery mildew.

VL-8

It is a mid season variety, developed at VPKAS, ALmora. It gives an average yield of 100 q/ha. It is recommended for release and cultivation in 1998 for Jammu & Kashmir, Himanchal Pradesh and Uttranchal.

VL Ageti Matar-7

It is an early maturing and high yielding variety and developed from VPKAS, Almora (Sri Dhar *et al.*, 1997). It matures in 120-125 days, a week earlier than Arkel (127-130 days) in Uttar Pradesh hills and in plains (60-65 days). The green grains are dimpled bold, very sweet with high T.S.S. (16.8 per cent). It yielded 93 and 103 q/ha green pods in hills and plains, respectively which is 19.6 and 14 per cent higher yield than Arkel. It is free from the incidence of powdery mildew.

Early Badger

It is a dwarf early wrinkled-seeded cultivar evolved at Wisconsin. It is suitable for sowing in early October. The plant flower in 40-45 days and first blossom appears at 10-11th node and ready for harvesting in 60 to 65 days. The yellowish green pods are borne singly, 7.5 cm long, well filled, 5-6 seeded sweet. It is a good canning cultivar having a shelling percentage 36. It is resistant to *fusarium* wilt and tolerate heat and drought (Choudhary, 1967).

Early Giant

It is a late variet. uited to hill of Himanchal Pradesh. Plants are tall and need staking. r ods are 9-10 cm long, dark green with 9-10 bold grains. Seeds are wrinkled.

Early Supperb

It is an English dwarf cultivar with yellowish green foliage. The plants are branched from base (Singh and Joshi, 1970). It is an early smooth-seeded variety. It flowers in about 45 days and first blossom appears at 8-10th node. The pods are borne singly, dark green and curved with 6-7 seeds. The shelling percentage is 40.

Swarna Rekha

It is an early maturing selection from local material of Bihar. It matures in about 120 days. Plants are semi-spreeding type with white flowers. Seeds are round, smooth and ceamy in colour with black hilum.

It is suitable for green pods and dry seeds as well. Its average yield is 80-100 q/ha of green pods and 15-20 q/ha of dry grains per hectare.

Palam Priya

It is a medium maturity (110-days), medium tall (60-67 cm), powdery mildew resistant garden pea. Its plants are vigorous with dark green foliage and branching habit. The pods are borne in double. These are smooth, straight, light green, well filled with average length of 8-9 cm. The average seeds/pod are 7-8 with about 60% shelling. Its green pods contain 15-16% TSS. The seeds become wrinkled upon maturity with light bluish green colour. On an average it gives 4-5 green pod pickings in hills with and average pod yield of 125-150q/ha.

Alaska

This is an early smooth seeded canning cultivar with bluish green seeds. plant is about 45cm tall and first blossom appears at 8-9th node after 38 days. Pods are borne singly, light green in colour, 7 x 1.25 cm, contain 5-6 small green seeds and shelling percentage 42 (Mac Gillivray, 1961).

Alderman

This is an excellent cultivar for home garden, shipment and freezing. Plants are dark green and 150 cm high. Pods are borne singly, 11.25 cm long with 8 to 10 seeds and excellent quality. This cultivar is suitable for hills (Yawalkar, 1969).

Little Marvel

This cultivar has been bred in England from the cross Chelsea Gem x Suttons Alaska. Plants are dwarf with dark green foliage. It flowers in 40 days and first blossom appears at 9-10th node. Pods are about 8 cm long, borne singly thick-skinned dark green, straight and broad containing 5-6 seeds. It gives shelling percentage of 40. The peas are very sweet and fine in quality (Singh and Joshi, 1970).

Kelvedon Wonder

It is having dwarf plant. Flowering takes place 40 days after sowing and first blossom appears at 8-9th node. Pods are curved, borne singly, green about 9 cm long and 6 seeded. The shelling percentage is about 40.

Thomas Laxton

It is a general-purpose cultivar with deep green foliage. Pods are of excellent quality, broad, and blunt, 8.25 cm long having cream and green coloured seed and suitable for freezing (Mac Gillivray, 1961).

Azad P-3

Plants are early, erect medium tall, dark green in colour. First flower appears in about 40 days after seed sowing. Pods are well filled, bold green and attractive. First picking can be ontained in 70 days after sowing, 3-4 picking are done. Average green pod yield is about 65-70 q/ha.

Ageta

Plants are small, erect and green in colour. It is suitable for early sowing. First flower appears in about 25 days after seed sowing and it takes about 6 weeks for first picking. Two picking are done. Average green pod yield is 50-55 q/ha.

Nath *et al.* (2008) revealed the research results obtained by several workers on performance of pea at different regions of the country, which details are given below :

Bhardwaj (1989) evaluated pea varieties against powdery mildew in Himanchal pradesh. Out of 99 varieties o fpea tested for their performance against powdery mildew for two successive seasons, only four varieties, viz., P185, T-10, P-6583 and BHU-456 were found resistant. Krishna Prasad *et al.* (1994) studied the interaction of genotype and environment in garden pea. On the basis of stability parameter, two varieties namely, VL-3 and KS-136 were found high yielders and suitable for unfavourable environments. Although the varieties, Sel-82 and KS-195 were high yielders, these two varieties were suitable for favourable environments only. Brar *et al.* (1995) reported the response of pea genotypes to leaf miner that IP3, KS136, JP71, JP83, ML21, ML22 and PH1 were moderately resistant to the leaf miner and these lines could be utilized for futher breeding purpose. Shridhar and Mani (1995) reported on the breeding of garden pea against stress condition for North Western Hills. The intensive breeding programme led to the development of VL-Ageti Matar 7, and early maturing high yielding with a faster vegetative growth habit derived from a cross between IP3 and Arkel. It possessed superiority over Arkel for pod yield by 13.3 per cent, about a week early in maturity, better germination, faster early vegetative growth and better horticultural traits with high degree of sweetness. The new early duration variety holds considerable promise due to its better ability in avoiding low temperature effects during normal and off seasons and better escape from powdery mildew disease in hills. Singh and Lal (1996) reported that the varieties of pea such as Rachna, DMR 11, Pant P5, HUP 2, HEP 4, JP 885, and KFP 103 have been released for cultivation in the powdery mildew prone areas of the country. Mehta *et al.* (1998) reported the resistance in pea genotypes to leaf miner. Study was conducted during 1993-94 and 1994-95 to evaluate 40 genotypes of pea for resistance to leaf miner (*Chromatomyia horticola* Goureau) at Palampur. 6 genotypes, viz 'NDVP 1', 'KS 136', 'KS 226', 'VGS', 'KS 225' and 'FC 1' were categorized as resistant (<1infestation index), 16 each as moderately resistant. Sharma *et al.* (2000) described the new high yielding powdery mildew resistant variety of vegetable pea, Pusa Pragathi developed by the Indian Agricultural Research Institute, New Delhi for commercial cultivation. This can be sown any time from late September to early December, ensuring high profits to the grower at a time when the pea harvest from the main crop is not available.

The production of snap pea is seriously limited by powdery mildew (*Erysiphe polygoni* D.C) and rust (*Uromyces fabae* [pers] de Bary) causing yield loss ranging from 50 to 100 per cent (Sindhan and Parashan 1984; Pal *et al.*, 1993 and Janila *et al.*, 2001). Mohan *et al.* (2002) evaluated the high yielding garden pea varieties resistant to rust and powdery mildew. With the objective of developing snap pea variety with combined resistance to powdery mildew and rust diseases, crosses were made among the parents involving Bonneville, a popular variety (susceptibal to both powdery mildew and rust), Freezer 656, resistant to powdery mildew from USA and IIHR 209 resistant to rust from Afghanistan. Pant *et al.* (2001) strdied the stability of yield components in garden pea. 'VL 6' performed the best regarding green pod yield (17.8t/ha)and seeds/pod (6.9) and was stable and suitable for favourable environments. 'Bonneville' was better performing and stable for 100 green seed weight (51.48g) and suited for favourable environments whereas, 'RPM34-2' was better performing and stable for shelling (51.8%) and showed average response to change in environmental conditions. A Promising line, IIHR 53-6, named as Arka Sampoorna with edible pod quality and combined resistance to rust and powdery mildew was selected in F_9 generation. Arka Sampoorna had maximum number of pods per plant (23.75) as compared to check and it was significantly higher. The pods were sweet, crisp and were devoid of perchment. Arka Sampoorna has given significantly high yield of 7.86t/ha, while in the check variety Oregon Sugar the yield was 5.45t/ha and the per cent yield increase in Arka Sampoorna over check variety was 44.22 percent. Disease intensity (PDI) for powdery mildew and rust in Arka Sampoorna showed combined resistance to powdery mildew (PDI 6.13) and rust (PDI 1.19) whereas; the check variety Oregan Sugar was susceptible to both the diseases. Aghora *et al.* (2002) evaluated the high yielding garden pea varieties resistant to powdery mildew and rust. Cv. Arka Karthik had significantly higher number of pods per plant (31.50), bigger pods , pod length (10.4 cm) and maximum shelling percentage of 62.17. Arka Karthik gave an average yield of 11.32 t/ha. The percent yield increase in Arka Karthik over check varieties Arka Ajit and Bonneville were 29.37 and 113.99 respectively. Arka Karthik was resistant to powdery mildew (PDI 3.25) and rust (PDI 0.76) whereas the check variety Bonneville was susceptible to both the diseases with PDI (39.97 for powdery mildew and 36.14 for rust). Arka Karthik was found to be suitable for processing (freezing) with high chlorophyll and sugar with low alcohol insoluble solids. Intensive work was undertaken in India on Breeding for desired qualities in garden pea. The multiple disease resistant lines are: JP-179, highly resistant to powdery mildew, resistant to *Fusarium* wilt, tolerant to rust, and resistant to leaf minor, JP-9 resistant to powdery mildew and *Bruchus*. JP Batri Brown 3 and JP Batri Brown 4 are resistant to rust and *Bruchus*. These are all multiple disease and pest resistant varieties, JP501 A-2 is resistant to powdery, mildew, *Fusarium* wilt and *Brouchus*. Arkel is suitable for dehydration. GC-195 and Kinnauri have high protein contents. The pea variety Leafless is tolerant to waterlogging. Hence different varieties are available with different quality attributes and suitability to various conditions.

CLIMATIC INFLUENCE

Pea is a cool-season crop best suited to production in the temperate regions. In the warmer areas of the tropics pea is restricted to production in the cooler highlands. Temperatures between 7 and 24°C are suitable for plant growth, but optimum yields are achieved between 13 and 21°C (Duke, 1981). Temperatures above 32°C during flowering often result in abortion and reduced seed set. It is relatively tolerant to frost when compared with Solanaceous and Cucurbitaceous crops. Blossoms and pods are susceptible to frost, whereas leaves and stems are relatively tolerant. Seed germinates better at the soil temperature of 10-18.3°C. The pea crop is produced primarily under rainfed conditions, but can also be irrigated. Irrigationis most common for the fresh market crop.

The seed can germinate at a minimum temperature of 5°C and the optimum temperature range is 18 to 22°C. Pods grow best in those regions, where there is a slow transition from cool to warm weather in spring.Temperature nearer to 30°C affects the quality of canning peas. For ripening this much temperature is helpful. Peas are very sensitive to drought.

SOIL REQUIREMENT

Pea can be grown on all types of soils but the early crop is expected in light soils and higher yields in loose, friable and heavy soils. It grows best at a pH of 5.5 to 6.7. It does not thrive in highly acidic or alkaline soils. If pH is lower than 6.0, liming is practiced to improve the soil conditions. It is advantageous, if soil is retaining adequate soil moisture thereby reducing number of irrigations.

As for as possible, soil is prepared up to good tilth. Two or three ploughings or harrowings followed by planking are sufficient to make soil pulverized that facilitaes seed gremination and further plant growth. The yield of pea decreases as salt level exceeds 4 ds/m,(Srimati and Ramaswamy, 1992).

SOWING

Sowing Time

Pea is grown as a rabi crop and sown from the beginning of October to middle of November in the plains and from the middle of March to end of May on the hills. In Darjeeling, this crop is sown from June to August, In Sothern states, pea is cultivated mainly in the hills, whereas in North India, three sowings are done during March-April, August-September and November-December.

Mid season varieties can be sown from the beginning of October to the middle of November in the plains. The early varieties can be sown in August-September. The cultivation of pea in Karnataka has been discussed (Pal and Nath, 1976).

In the plains of northern India seed pea is sown during mid-October to mid-November as winter crop (Gandhi *et al.*, 1991 and Dhaka *et al.*, 1992). In the hills. it is sown in May as an autumn crop. Pea can also be grown as early as in June-July in peninsular India.

Seed Rate

In order to sow one hectare area of early varieties of pea which have generally best growth, needs 100-120 kg/ha seed and for the mid-season and late cultivars, the seed requirement is 80 to 90 kg per hectare. When sowing is done in furrows, seed rate varies from 50-60 kg/ha. The pea seed of good quality with 95 per cent seed gremination is enough for sowing.

Spacing

Pea is sown at the row to row distance of 30 to 45 cm and plant to plant distance of 5 to 10 cm depending on type of cultivars, type of soil, time of sowing etc. On sandy loam soils sowing distance at 30 x 5 cm or 22.5 x 10 cm is advisable for higher seed yield of pea (Hooda *et al.*, 1994). Pea can also be sown at a row to row spacing of 22.5-30 cm and plant to plant spacing of 3.75-5 cm. Rai and Yadav (2005) stated that, depending upon maturity period and plant habit, pea is sown in rows 30 to 60 cm apart and within the rows 5 to 10 cm apart. In raised beds, it can be sown on both the sides of the bed which are 120 to 150 cm wide with furrows between them. Seeds are sown about 2.5 cm deep.

Seed Treatment

Pea seeds are treated with bacterium culture (*Rhizobium leguminosarm*). The jaggery or gur solution is mixed with the seeds thoroughly so that seeds are soaked well. There after, seeds are allowed to dry in shade prior to sowing (Katyal, 1985).

Soil inoculation with phosphobacterium is equally effective and sometimes better than seed inoculation with *Rhizobium* (Sarakanan and Nambisan, 1994). In the new fields where the pea crop has not been grown earlier the inoculation of *Rhizobium* sp. in the soil is recommended for better growth.

Method of Sowing

The pea is sown by drilling on commercial scale. The seeds are dropped in shallow furrows opened by a deshi plough and then after furrows are covered by light planking. Drilling can also be done with the help of Pora attached with plough. However, on small scale, sowing can be done with hand dibbling.

The depth of seed sowing is 2 to 3 cm. The wrinkled cultivars of pea are more tender than the round seeded ones; may be sown thicker in the rows (Shewell-Cooper, 1977). The duration of pod maturity reduces at higher plant densities but increases the difficulty of roguing(George, 1985).

Rai and Yadav (2005), suggested that pea seeds can be sown on flat or raised beds either by broadcasting or behind deshi plough in furrows, which are covered by usual planking. Some farmers drill the seed through pora (tube) attached to deshi plough and there is no need of planking. When the seed is sown too deep, it results in slow germination and poor stand. A depth of 5-7.5 cm is optimum according soil moisture and texture.

Singh *et al.* (1992) reported the effect of date of sowing, seed rate and chemical treatment on germination, flowering, maturity and yield of vagetable pea under rain fed condition. Pods per plant were highest when sowing was done on 10th October followed by sowing on 20th October. There was no significant difference of seed rate and chemical treatments on number of pods per plant. Pod weight per plant and yield per hectare were highest under 10th October sowing followed by sowing on 20th October. Pod weight per plant yield per hectare were significantly higher when seed rate was kept at 60.0kg/ha.

Dwivedi *et al.* (1998) studied the response of table pea cultivars to sowing dates and seed rate under Northern hill zone of Chattisgarh. Among the varieties. Bonneville produced significantly higher green pod yield (62.55 q/ha). Sowing of pes in mid October was found to be an appropriate sowing time and was recorded significantly superior pod yield (52.38 q/ha). Seed rate also influenced green pod yield significantly. It was maximum (43.85 q/ha) with 80 kg seeds per hectare followed by 60 kg (39.19) and lowest with 40 kg seeds/ha (31.48 q/ha).

Therefore, a seed rate of 70-80 kg/ha is required for garden pea. Planting can be done during middle of October with a spacing of 20-30 cm (row to row) and 5-10 cm (plant to plant). However, spacing and seed rate depends on the method of sowing (Nath *et al.,* 2008).

NUTRITIONAL REQUIREMENT

Being a legume crop, pea requires comparatively less nitrogen. High nitrogen dose has been found to increase vegetative growth on the cast of seed yield. Application of nitrogen along with phosphorus increses uptake of micronutrient by pea crop (Verma and Bhandari, 1996). In fact, application of phosphorus is much more important than that of nitrogen because phosphorus enhances symbiosis between plants and bacteria and they repay for it by fixing atmospheric nitrogen in the soil as well as in the plants (Verma and Bhandari, 1996) of more nodules. Potassium also makes plants to build up ability to produce more yield through nitrogen fixation. About 25 to 40 kg potassium per hectare usually fulfil the nutrition requirement of pea crop. Under rainfed conditions of Kashmir valley the seed yield of pea cv 'Rachna' was highest at 30 kg nitrogen and 50 kg phosphorus (Singh *et al.,* 1992). Whereas, in the hills of Kodaikanal (Tamil Nadu) pea cv 'Bonneville' resulted highest pod yield (6.55 t/ha) with 100kg of nitrogen + 150 kg of phosphorus + soil inoculation with phospho-bacterium (Sarakanan and Nambisan, 1994). Recommended doses of NPK and FYM are given in Table-1. Pea growers may reduce the doses when they intend to grow seed pea crop on comparatively fertile soil (Singh, 2001).

Besides, certain micronutrients also play important role in producing higher yield of pea. Among them 'B' plays vital role in mediating cell-surface interactions that lead to endocytosis of rhizobia by host cells and hence to the correct estabIighment of the symbiosis between peas and Rhizobium (Bolanos *et al.* 1996). On the basis of yield and quality and economical considerations, the optimum Mo rate has been found to be 1.0 kg per hectare for pea crop (Kovalavich, 1991). Application of Mn at 20 g/kg of seed has been proved best for increasing yield of pea cv 'Rachna' (Singh and Singh, 1995).

Table 1 : Recommeded doses of NPK and FYM are given for raising pea crop in some states of India (Tandon, 1991).

State	Nitrogen (N) Kg/ha	Phosphorous (P) Kg.ha	Potash (K) Kg/ha	Fym t/ha
Andhra Pradesh	15	50	20	20
Arunachal Pradesh	25	70	50	20
Gujarat	25	50	50	15
Haryana	30	50	0	20
Himanchal Pradesh	25	60	60	-
Jammu & Kashmir	30	50	50	-
Karnataka	37	60	50	12.5
Madhya Pradesh	30	75	40	20
Maharashtra	50	50	50	25-30
Orissa	25	50	75	-
Punjab	50	62	0	20
Rajasthan	25	40	0	20
Tamil Nadu Hills	180	125	100	15
Plains	50	60	30	12.5
Uttar Pradesh	25	50	50	10
West Bengal	20	60	50	15

Rai and Yadav (2005) reported that for getting good yield, farm yard manure should applied in quantity of 80-100q/ha at 15 days before sowing the seeds. The nitrogen, Phosphorous and potassium should be applied in ratio of 30:50:25 kg/ha, respectively. This nutrition should be given at the time of sowing of seeds as basal doses. Foliar application of 0.1 per cent ammonium moybdate can be used for increasing number of root nodules, yield, TSS and number of grains/pod in pea (Sharga and Jauhari, 1970).

Generally, fertilizers are applied by broad cast method which is not advisable. The best method of application is placement 3-5 cm below the seed. All the three fertilizers(NPK) are combined and can be applied as basal i.e. at the time of sowing. Organic manure is incorporated during field preparation.

Nath *et al.* (2008) summarized the results of nutrition on pea production, obtained by several researchers and stated that, depending on the fertility of the soil, manures and fertilizers have been recommended per hectare as 10 to 15 tons of FYM, 20 to 50 kg of nitrogen in the form of ammonium sulphate, 30 to 80 kg of potash in the form ofpotassium sulphate and 50 to 60 kg of phosphorus in the form of single super phosphat (Purewal, 1957; Yawalkar *et al.*, 1962). It is suggested that the complete dose of FYM may be mixed with the soil at the time of field preparation, whereas the complete dose of potash and phosphorus may be applied before sowing. The ammonium sulphate may be split into two doses with first dose at the early stage and the second at the flowering stage as top dressing. Sharga and Jauhari (1970) reported that the foliar application of 0.1 per cent ammonium molybdate, significantly increased number of root nodules, TSS and the length, diameter, fresh weight and number of grains per pod. Naik (1992) reported the response of early garden pea to plant spacing, nitrogen, phosphorus and potassium fertilization. The closest spacing (30 x 5 cm) recorded the highest yield. There was significant increase in yield up to 50 kg N/ha and 75 kg P_2O_5/ha. Response to potassium was not observed. Prasad and Prasad (1993) reported the effect of *Rhizobium*, phosphorus and GA_3 on yield of garden pea. *Rhizobium* inoculation, application of P_2O_5 at 90 kg/ha and GA_3 treatment at 50 ppm gave highest yield and significantly incresed the length of pods, number of pods, number of grains/pod and hundred pod weight in both the years of investigaion. Saini and Thakur (1996) reported the effect of nitrogen and phosphorus on vegetable pea in cold desert area. An application of 30 kg N/ha increased plant height, branches/plant, leaf area index, yield attributes (pods/plant, garins/pod and pod weight) and grains yield/ plant, which resulted in significantly higher green pod yield than without N. Response to phosphorous was observed up to 39.6 kg/ha (as the soil were deficient in p, bright sunshine and more ultraviolet rays led to luxuriant plant growth, and the zone was free of insects and diseases), showing significantly higher plant height, leaf-area index, green pod yield, yield attributes (pods/plant, grains/pod and pod weight) and grain yield/plant compared with that of 0,13.2 and 26.40 kg P/ha. However, when the green pod yield data were pooled, significant reponse was observed up to 52.8 kg P/ha.

Gupta *et al.* (1997)studied the effect of seed treatment and fertilizer application on qualitative performance of pea. Seed treatment with GA_3 at 100 ppm for 6 hrs and application of 60 kg P_2O_5 + 40 kg K_2O/ha increased the crop yield significantly. Kanaujia *et al.* (1997) studied the effect of phosphorus, potassium and *Rhizobium* inoculation on growth, yield and quality of pea. The growth and nodulation were incresed significantly with increasing levels of each of phosphorus and potassium (from 0, 30, 60, 90 kg/ha). On the other hand, yield and quality were incresed significantly up to 60 kg each phosphorus and potassium. *Rhizobium* inoculation also increased growth, yield and quality over uninoculated control.Arya *et al.* (1998) studied the effect of NPK nutrients on seed yield and yield attributes in pea. Maximum seed yield was recorded with N-45, P-40 and K-60 kg/ha. Parmar *et al.* (1999) studied the integrated nutrient supply system for DPP 68 vegetable pea in dry temperature zone of Himanchal Pradesh. The tratment receiving 75 per cent of the recommended dose of NPK (N15, P45, K22.5 kg/ha) both in the presence and absence of farm yard manure were found to be the most suitable treatment combination, giving higher green pod yield, besides saving 25 per cent of NPK fertilizers. Kanaujia *et al.* (1999) concluded that phosphorus and potassium @60 kg/ha each along with *Rhizobium* inoculation gave the highest pod yield and profit of Rs. 52657.00/ha.

Nath *et al.* (2008) concluded, that the fertilizers requirements of pea are 20-30 kg nitrogen, 60-70 kg phosphorus and 30-40kg potassium per hectare. In some cases, the crop responded well till 90 kg of phosphorus and 60 kg of potassium. *Rhizobium* inoculation increases the uptake of not only nitrogen but also uptake of the phosphorus and potassium. GA3 at 50 ppm has a favorable effect on growth and pod yield parameters (Pal, 2004).

Use of Plant Growth Regulators

Floliar spray of MH at 25 mg/l (Kumar and Sree Kumar, 1981) or CCC at 500 mg/l (Chandra *et al.,* 1987) before flowering has given best pod yield in pea (Kumar and Sree Kumar, 1981). Seed treatment with cytozyme at 1 per cent has also been reported to improve fresh pod yield in pea (Saimbhi and Gill, 1988).

IRRIGATION

The water requirement of pea crop is low. However, sufficient moisture is essential for germination, at the time of flowering fruit set and growth hence care should be taken to provide irrigation during this priod. For higher germination, pre-irrigation will provide proper soil moisture. In temperate regions, an adequate moisture through irrigation is maintained during and after flowering, where as in arid and semi arid areas and dry seasons sufficient water should also be given to ensure proper plant growth before flowering commences othe wise poor seed yield will be obtained (George, 1985). Pea cv. PH-1 gave highest pod yield (3.13 t/ha) with 4 irrigations i.e. at pre sowing, flowering fruiting and 20 days after fruiting (Basawana and Leoha, 1995). In pea cv 'Lincoln' produced highest yield per hectare with 4 cm of water applied at IW:CPE ratio of 0.8 (9350 kg/ha). This treatment also resulted in the highest water-use efficiency. The highest total water use occurred with 6cm of water applied at IW:CPE ratio to 1.0(Malik and Bhandari, 1994). In pea cv. Bonneville 3 to 4 irrigations after establishment were essential from flowering to first pod picking stage for maximum productivity under semi-arid condition.

INTERCULTURE

Hoeing once during early stage of crop growth will keep down germinating weeds and will provide good environment for plant growth. For higher yield, it is necessery to maintain crop weed free up to 45 days or so (Singh *et al.*, 1991).

Heoing is also done to keep down weeds and to provide good aeration for proper development of roots. In case of tall indeterminate varieties, it would be advisable to provide stakes to the plants to harvest better quality pods. Wire trellis supported by strings is generally used.

Inter cropping and cropping sequence

Singh and Nath (1995) reported the performance of pea varieties intercropped in ber orchard in semi-arid region. Arkel variety produced maximum pod yield (52.6 q/ha) as compared to other varieties without any adverse effect on the growth and fruit yield in ber followed by Azad pea. Intercropping of pea in young ber orchard was found more economical and suitable to increse per unit crop yield.

Saini (1996) observed that following crop rotations are highly profitable with pea :

1. Maize - Pea - Okra
2. Pearl millet - Pea - Okra
3. Pearl millet - Pea - Cluster bean
4. Cotton - Pea - Amaranthus
5. Paddy - pea - Bittergourd
6. Okra - Pea - Cowpea
7. Sponge gourd - Pea - Colocasia
8. Cauliflower (Early) - Pea - Brinjal
9. Cowpea - Pea - Chilli
10. Colocasia - Pea - Bottle gourd

WEED CONTROL

There are a few important weed species associated with peas depending on a region of where crop is grown. They are: *Amaranthus viridis, Anagalis arvensis, Asphodelus tenuifolius, Blumea lacera, B, trifoliate, B. weightiana, Carthamus oxycantha, Chenopodium album, C murale, Convolvulus arvensis, Coronopus didymus, Cynodon dactylon, Cyperus rotundus, Digera arvensis, Fumaria parviflora, Melilotus indica, Poa anrua, Polygonum* spp. *Trigonella polycerata, Viola* spp (Singh, 1997). The following herbicides are effective to control weeds in pea crop.

Table 2 : Chemical weed control in pea crop (pre-emergent)

Herbicide	Dose(kg/ha)	Reference
Prometryn	1.0-1.2	Bhalla and Singh (1981)
Propanil	1.0	Vishnoi *et al.*, 1982
Fluchloralin	0.9	Bhalla and Chourasia (1982)
Alachlor	1.0	Vishnoi *et al.* (1982), Singh *et al.*, (1984), Nanelal and Arya (1995)

Rai and Yadav (2005) reported that the useful method of weed control practiced in India is mechanical or manual, but manual, weeding is uneconomical and time consuming. Chamical weed control has been reported to be as effective as hand weeding and at the same time, it is cheaper and less time consuming. The various recommendations on use of weedicides in pea under All India Coordinated Vegetable Improvement Project are semmarized below (Anon, 2000).

- Lasso @ 0.75kg a.i./ha pre-emerged spray or tribunil @ 1.5 kg a.i./ha pre emergence or basalin @ 2kg a.i./ha pre plant incorporation along with one hand weeding at 45 days after sowing (DAS) have been found to be most effective for weed control under Pant Nagar conditions.
- Application of tribunil @ 1.87 kg a.i./ha pre-emergence spray has been recommended for controlling under agro-climatic conditions of Punjab.
- Application of stomp (pendimethalin) @ 0.5kg a.i./ha or lasso (alachlor) @ 1.5 kg a.i./ha as pre-emerged incorporation in the field followed by one hand weeding at 25 DAS, have been recommended under Sabour conditions.

Singh and Tripathi, (1997) studied the effect of various herbicides on weed control and yield of green peas. This investigation was undertaken during the winter season of 1983 and 1984 at the GBPUT, Pantnagar to evaluate the performance of various herbicides for weed control and pod yield of peas cv. Arkel under the sub tropical belt of

Tarai. Pre emergence spray of metribuzin 0.50 kg/ha with weeding at 45 days resulted in pod yield of 126.1 q/ha as compared to 110.9 q/ha with fluchloralin (0.96 kg/ha) and 109.5q/ha with methabenz thiazuron (1.4 kg/ha) both supplemented by one manual weeding. Post emergence spray of metribuzin 0.50 kg/ha failed to produce better yields as compared to its pre emergence application.

It was reported that basalin-45 EC (trade name of fluchloralin) @ 1.5 kg a.i./ha as pre planting application and lasso 50 EC (trade name of alachlor) @ 0.5 kg a.i./ha as pre emergence application and stomp 30EC (trade name of pendimethalin) @ 1.5 kg a.i./ha as pre emergence or post emergence appliction control the annual and broad leaved weeds (Arya, 2003).

PLANT PROTECTION MEASURES

Insect Pests

Pea stemfly (Ophiomyia phaseoli Tryon)

It attacks on early sown crop. Pupae and fully grown larvae can be seen in the stem just above the ground level where stem swells, cracks and becomes brownish, and weak. Such plants start drying after about 20-25 days of germination.

Control

Spray with 0.03 per cent Dimethoate or Monocrotophos at 0.04 per cent. Apply Furadan 30 kg per hectare in furrows with seed at sowing time. Seed dressing with Phrate or Disyston at the time of sowing provides effective protection against these flies for 2 to 3 weeks after germination (Jotwani and Butani, 1997).

Pea aphids (Macrosiphum pisi)

These small insects attack pea plants from January onwards. These aphids are green, long legged and soft-bodied. They suck the sap of younger parts of plants. There is a curling of leaves and pods, and plants are stunted in growth. Aphids are the carrier of pea mosaic.

Control

Apply Furadan 30 kg per hectare in furrows with seed at sowing time as a preventive concentration. Spray with Rogor at 0.03per cent of Metasystox or Malathion at 0.1 per cent concentration. Locate for varieties which are tolerant/resistant to aphids as 'Feltham' and Meteor.

Pea thrips

Pea thrips may be problem in very dry weather. They feed on the flowers, stems and pods. They lay their eggs on the pods. The orange grubs hatch out and feed first on the flowers and then on the pods and leaves causing slivery, mottled patches which later turn brown. Control measures of pea thrips are similar as for pea aphids.

Pea beetle

These beetles feed mainly on seeds. The adult beetles lay eggs on the dried seed, and the larvae hatch out and burrow into the seed. Leaving a small, round hole. infected seeds fail to germinate or germinate very poorly.

Control

Always obtain seeds from reliable sources. Sort out infected seeds before sowing.

Pea weevil (Bruchus pisorum)

The incidence of this pest is common in hills. The weevil lay eggs on small pods which hatch in 15-18 days. The larvae burrow through the pod into the seed and feed(Mac Gillivray, 1961).

Control

Hoe around the plant to stop the adults from laying eggs. Dust the crop with 1 per cent Rotenone at blossoming time.

Leaf miner (Phytomyza atricornis Meigen)

It is very common and severe pest of peas; found all over the pea growing areas in India. The pupae usually occur in the ground. As name indicates, it attack on leaves.

Control

Soil application of 0.5 kg ai/ha either Carbofuran or phorate at sowing and followed by another round of 0.5 kg a.i./ha 40 days after germination will control pea leaf miner very effectively (Srivasan and Shukla, 1980).

Besides the above insects, nematodes are serious pest. These can be controlled by raising crop in rotation, removal of affected plants from the plot and use of Aldicarb at the rate of 5 kg per hectare(Parvatha Reddy, 1985).

Diseases

Powdery mildew (Erysiphe polygoni DC)

White powdery patches form on leaves and pods. Dry weather favours the spread of disease. It is a seed borne disease.

Control

Clear out any old haulms left from previous plantings and burn them. Dust with sulphur or spray Sulfex (2.5 kg/ha), Thiovit (5 kg/ha) or spray Karathane or Dinocap 0.2% or Elosal 8 wp (0.5%) three times at 10 days interval. Avoid sowing in end Nov. and Dec (Pratibha Shukhal and Amin, 1991). Treat the seeds with hot water. Grow resistant varieties like Sugar Giant, Pant P-8, PMR-3, Rachana (Thakur *et al.,* 1996) and Field pea-48 (Sidhu *et al.,* 1994).

Downy mildew (Peronospora pisi sydow)

The spots of grey brown mould appear on the under surface of the leaves. This disease is not serious as powdery mildew. Incidence of downy mildew is only severe in very wet weather conditions. The disease is soil as well as seed borne.

Control

Uproot affected plants. Follow crop rotation. Seed treatment with hot water may be tried. Spray with Dithane Z-78 (Zineb) or Dithane M-45 at 0.2 per cent concentration.

Fusarium wilt (Fusarium oxysporum f. pisi)

Affected plants show yellowing and downward curling of leaves and incomplete filling of pods. The infection gets in through the roots. If the lower part of the stem is cut open, a discolouration is observed. In severe cases, the whole plant wilts and the stem shrivels.

Control

Follow long crop rotation. Treat the seed with Beniate (2.5 g/kg of seed). Use wilt resistant varieties e.g Alaska, Resistant Surprise, 'Rachana' is also comparatively wilt tolerant variety.

Bacterial blight (Pseudomonas pisi Sackett)

Water soaked spots on the stems, leaves and stipules appear. A white to cream coloured slimy ooze appears on the surface of the lesions. The infected area eventually turns brown and papery. Humid conditions favour the disease. It is seed as well as soilborne disease(Mac Gillivray, 1961).

Control

Remove affected plants and weeds. Use disease-free seed. Plant the crop at wider spacing. Avoid excess moisture with restricted irrigation. Treat the seed with Streptomycin sulphate (2.5 g/kg of seed).

Rust (Uromyces pisi de Bary and U. fabae de Bary)

Under humid conditions, this becomes serious in peas. The plants dry up Rust is mainly soilborne as teliospores in crop debris. Rust appears to survive on weed hosts belonging to Lathyrus, Vicia, etc. and the spores are wind blown on the main crop (Singh, 1985).

Control

Remove affected plants and crop debris and destroy them. Use high quality seed from reliable sources. Follow long crop rotation excluding broad beans, Vicia and Lathyrus etc. As for as possible, locate and destroy the weed hosts. Spray Dithane M-45 (0.2%) and Calixin (0.2%) and repeate it at 10-15 day interval. Total 3-4 sprays should be given.

Root rots (Rhizoctonia solani, Fusarium solani, pythium sp.)

This disorder is common. Mainly soil micro-origanism may cause it. Due to higher temperatures these micro-organism become active result in rotting of the roots.

Control

Do seed-dressing with Thiram (2 g/kg seed) before sowing. Grow tolerant varieties. Prefer smooth-seeded varieties for early sowing. Provide good soil drainage. Follow crop rotation.

Mosaic virus

Affected plants show mottled pale green or yellow patches on the leaves and pods. It can causes distorted pods. *Pisum Virus* 2 A is sap transmissible and in the field by vectors such as. *Aphis* spp. *Myzus persicae* and *Rhopalosiphum* spp. The virus is seed borne. Another is Enation Mosaic pea is caused by virus 1. Small outgrowth (enations) on the lower surface of some leaves is seen. Terminal growth of plant is checked, curling and rosetting of plants etc. are seen. It is sap transmissible with difficulty. It is not seed-borne. Aphis are vectors (Singh, 1985).

Control

Control insects-vectors carefully. Pull out affected plants and burn them. Locate resistant cultivars. Some cultivars viz., Canners perfection, Hundredfold and Little Marvel are resistant (Singh, 1985).

Factors Responsible for Low Production of Peas

Singh (2001) reported that following factors are responsible for low yield of pea :

1. Low yield potential of the varieties.
2. Susceptibility of diseases.
3. Low due to insect - pests.
4. Late - maturity.
5. Poor responses to inputs like fertilizers and irrigation.
6. Cultivation with poor production technology.
7. Raising of the crop under marginal and poos soils.
8. Physiologically ineffcient plant type.
9. Instability in the performance of the varieties.
10. Farmers are relucant to use costly inputs because of the risk.

Integrated Pest Management

The integrated pest management in peas includes cultural, mechanical and biological control. Use of tolerant or resistant varieties, crop rotation with non-leguminous crops, intercropping with linseed or mustard or barley reduces the pest population. Intercropping with mustard or faba bean reduces powdery mildew and rust. Mechanical control measures are seed cleaning, rouging of diseased plants and use of bird scarers (Nath *et al.*, 2008).

Biological control

Inundative release of *Coccinella septumpunctata* @ 1000 adults/4000 m^2 for aphids, at 10 days interval released twice. Soil amendment with pearlmillet residues, mustard care, or mustard pod straw reduces root rots. Seed treatment with *Bacillus subtlis, Pseudomonas* spp *Gliocladium virens* controls the root rot and damping off. Soil application of *Trichoderma harzianum* or spray of *B. subtlis* is effective against white rot disease. Nematodes are controlled by *Asperigillus* niger @ 10^8 spores per kg of seed or *paecilomyces lilacinus*. Tolerant varieties like DDR16, HFP4, AVT-2(D), HFP9510 and LFP227 are available for nematodes (Trehan, 2003).

HARVESTING

The green pods of early varieties are ready for harvest in 45-60 days while mid and late season varieties, which are generally indeterminate, take 70 to 100 days. As the pod attains marketing stage it turns dark to light green and the grains are well filled in the pods. In the processing industry, the maturity of pea is tested with the help of 'tenderometer.' The high quality of pea is tested by its tenderness and high sugar content.

Usually 3 to 4 picking of pods are made during the season, which is spread over 2 to 4 weeks at 7 to 10 days interval, mid and late season varieties require more number of pickings (Rai and Yadav, 2005).

Temperature has great effect on the rate of ripening of pea and at higher temperature there should be no delay in harvesting. For fresh market the pods are hand picked. The green pods are to be picked before the webbing starts. In case of mature pods seeds should be removed from pods by threshing, immediately or else, the pods will shatter.

YIELD

The early varities produce 25 to 40 q/ha of edible pods, medium varieties 65 to 75 quintals and dthe late varieties 85 to 115 q/ha, where the shelling percentage is 40 to 45.

POST HARVEST HANDLING

Grading

Harvested pod are separated into different maturity grades. Peas for processing are graded into four grades based on the size of shelled peas, and the smaller sized are considered to be the best quality and bring the highest prices from canners. The sieves used for separating the sizes have mesh ranging from 18/64 to 24/64 inch in diameter. While size is important in determining maturity of peas, many processors buy peas on the basis of tenderometer grades rather than sieve size. The tenderometer reading or some similar measure of quality is much better than size alone. Small peas may be over mature and of poor quality, while large peas may be tender and not yet mature.

Packing

When peas are prepared for market the over matured yellow pods, the flat pods, the diseased and insect injured pods and the trash should be removed before the peas are packed. In India pods are generally packed in gunny bags, crates and in special baskets. Baskets of Bamboo or sticks are also used in some areas. Now a days several type of packing materials are in practice for proper transportation of the green pea to distant markets. These packing materials are corrugated fibre boxes, fibre board boxes, and perforated boxes. Cartoon card board boxes, which keep the material fresh and in proper quality upto its last destination.

Peas are peacked for shipment in baskets, hampers, and boxes of various types. It is important to have the containers well filled so that it will not be stack when it reaches market. In western countries, peas shipped to long distance are pre-cooled with ice cold water and transpoted in refrigerated trucks in order to retain their sweetness and quality (Thompson and Kelly, 1983).

Storage

Fresh peas can be kept for over two weeks in cold storage at 0°C and 85 to 90% relative humidity, where as at 21.50°C they become unfit for sale at the end of 5 days. It can be stored for more than 2-3 days under ordinary room conditions.

SEED PRODUCTION

Pea is a self pollinated crop hence two cultivars can be grown in adjacent plots without any danger of out crossing. However, it is suggested that isolation distance should be given for production of pure seeds. Select field on which same kind was not planted in the previous season. In addition, the land should be well drained and with a neutral pH.

Isolation Distance

Seed pea plot is isolated from other plot of the same varieties or other grown for commercial purpose to prevent mechanical mixture at planting or harvest time. An isolation distance of about 50m is advisable for foundation seed production and that of 25m for certified seed.

The agronomic practices which are followed for the cultivation for green pods are the same for the crop for seed production.

Mehta *et al.* (1994) reported on the effect of studies carried out on the effect of spacing and seed rate on seed production of early variety of pea. Variety PH-1 has given significantly highest seed yield with closer spacings. Seed rates of 75 and 100 kg/ha being at par but proved significantly better than 50 kg/ha.

Mehta *et al.* (1997) reported the effect of sowing time and spacing on seed crop of pea. Sown on October 30th with the closest spacing (20cm) gave the highest seed yield.

Roguing

Seed pea crop is maintained genetically pure by proper roguing. It is conducted for off-types and diseased plants before flowering, during flowering and during the edible pod stage. At least, three roguing are essentially done. More roguings can be done if needed. Rouging is done at different stages.

Before flowering

The off-type plants which are taller, having variation in size and shape of leaves, particularly attention is given to checking the foliage, including stipples (George Raymond, 1985).

Flowering

This is the second critical roguing stage in which rogue plants are identified on the basis of blossom variations in colour, growth habit of stem. Early flowering plants from late flowering cultivar, and late flowering plants from early flowering cultivar are removed.

Pod formation

The off-type plants are removed on the basis of size, shape and curvature of the pod, late flowering plants should be removed along with weak, low-yield plants. Besides, all diseased plants perticularly those suffering from seed - borne pathogens like powdery and downy mildew, wilt bacterial blight rust, mosaic virus etc.

In fact, foundation seed plot requires more critical roguing than certified seed plot. This opertion should be repeated till uniformity in plant type conformed, and plot met the field standards (Singh, 1989) as per table 3.

Table 3 : Field standards for peas seed production (maximum permitted limit).

Factor	Fondation	Certified
Off-types	1.0	0.50
Other crop plants	Nil	Nil
Objectionable weed plant	Nil	Nil
Diseased plants	0.5	1.0

Harvesting

Harvesting of seed pea crop is done when about 25 to 30 per cent pods are ripe and almost all pods are mature. Plants are piled up for curing for 3-5 days. Delay in harvesting may result in shattering of pods.

Seed Extraction

Dried plants of pea are subjected to threshing operation followed by cleaning of the seeds.

Drying

Cleaned pea seeds are followed to dry in sun for about 5-days or so, drying of seeds can be also be done electrically operated driers. The moisture of seed should be nearer to 9.0 percent both for foundation and certified seed class (Table 4).

Seed yield

On an average 15 to 20 q/ha seed yield is obtained under good management. The 1000 grains weight for large seeded varieties is about 330 g and 150 g for the small seeded varieties.

Table 4 : Seed standards for peas seed production (in percentage)

Standards	Fondation	Certified
Pure seed (minimum)	98.0	98.0
Inert matter (maximum)	2.0	2.0
Other crop seed (maximum)	None	0.05
Total weed seeds (maximum)	None	None
Objectional weed seed (max)	None	None
Germination (minimum)	75.0	75.0
Moisture (maximum)	9.0	9.0

Storage

Well dried seed of pea having 9.0 per cent moisture is stored in gunny bags lined with polythene. The bags are staked in well vantilated dry room for next season use.

REFERENCES

Anonymous (1970). Pulse crops of India, Indian council of Agricultural Research Publication, New Delhi, India, pp. 259-261.

Anonymous, (2000). AICVIP Recommendations (1971-98) IIVR, Varanasi, India.

Arya, P.S. (2003). Vegetable Breeding-production and seed production : Kalyani Publishers, New Delhi. pp.74.

Arya, P.S., Sagar, V. and Singh, S.R. (1998). Studies on effect of NPK nutrition on seed yield and yield attributes in pea (*Pisum sativum* L.) *Haryana J. Hort, Sci,* 27(1).

Basawana, K.S. and Leoha, P.K. (1995). Response of pea (*Pisum sativum*) to irrigations. *India J. Agron.* 40(1): 139-140.

Bhalla, P.L. and Singh, S.P. 91981). Comparative efficiency and selectivity of herbicides on seed yield of pea (*Pisum sativum*). In: *Abstr. Papers, Conf., ISWS,* Khandwa, India.

Bhalla, P.L. and Chourasia, K.K. (1982). Influence of herbicidal treatments on the growth and seed yield of garden pea (*Pisum sativum*). In : *Abstr. Papers. Conf. ISWS*, Khandwa, India.

Bhardwaj, S.S, Shyam, K.R. and Dohroo, N.P. (1987). Performance of pea variety against powdery mildew in Himanchal Pradesh. *Veg. Sci.*, 14 (1) : 55-57.

Bolanos, L.Brewin, N.S. and Bonilia, I. (1996). Effect of boron on Rhizobium-Legume cell-surface interactions and nodule development. *Plant Physiol.*, 110 (4): 1249-1256.

Brahmappa and Singh, R.D. (1977). Combining ability for quantitative Characters in pea (*Pisum sativum* L.) *Indian J. Hort*, 3(1) :44-48.

Brar, K.S., Dhillon, G.S., Singh, Mohan and Dhillon, T.S. (1995). A note on the resistance in pea genotype to leaf minor, *Chromatamyia horticola* (*Goreau*) : *Haryana J. Hort. Sci.*, 24(2) : 156-158.

Chandra, R., Pollsetty, R. and Panwar, J.P.S. (1987). *Annals of plant physiology*, 3: 65-73.

Choudhary, B. (1967).Vegetable. National Book Trust of India, New Delhi.

Choudhary, B. and Rampal (1975). *Indian Horticulture*. 19: 25-27.

Das, K. and Kalloo, G. (1970). A technique of squashing stipule tips on pea. *Curr. Sci.*, 39: 494-495.

De Candolle, A. (1896). Origin of cultivated plants. Hafner publishing Co., New York.

Dhillon, G.S. and Singh, M. (1993). *Indian Horticulture*, 37: 16-17.

Duke, J.A. (1981). Handbook of legumes of world economic importance. Plenum Press, New York, pp. 200-205.

Dwivedi, R.K., Singh, V.K. Choudhury, S.K., Sangar, R.B.S. and Bajpai, R.P. (1998). Response of table pea (*Pisum sativum* L.) cultivars to sowing dates and seed rate under northern hill zone of Chhatisgarh, M.P. *Orrisa J. Hort.*, 26 (1): 17-19.

F.A.O. (2001). http://apps. fao. org.

Gajendra, Singh and Singh, P.O. (1995). Performance of pea varieties at different seeding times. *Annals of Agric. Res.* 16 (3): 384-386.

George, Raymond A.T. (1985). Vegetable seed Production Longman, London and New York. pp. 186-193.

Gupta, Arun; K., Samnotra, R.K. and Sudan, S.K. (1997). Influence of seed treatment and fertilizer application on quantitative performance of pea (*Pisum sativum* L.) cv. Bonneville. *Haryana J. Hort, Sci.*, 26 (1-2) : 151-153.

Hawrhorn, L.R. and Pollard, L.H. (1954). Vegetable and Flower seed production. The Blakistan Company, INC New York. Toronto.

Hooda, J.S., Singh, B.R. and Singh, V.P. (1994). Effect of sowing time and plant population on the yield and yield attributing characters of field pea genotypes. *Crop Research*, 7 (2): 299-302.

Jain, N.L. and Choudhury, B. (1963). Some studies on the dehydration of peas. *Indian J. Hort.*, 20 (2) : 129-134.

Janick, J., Schety, R.W., Woods, F.W. and Ruttan, V.W. (1969). *Plant Science*. Freeman, San Francisco.

Joshi, B.S. and Singh, H.B. (1958). A bloom mutant in garden pea. *Indian J. Genet.*, 10(2): 236-243.

Jotwani, M.G. and Butani, D.K. (1977). Insect-pests of leguminous vegetables and their control. *Pesticides*, 11(19): 33-38.

Kalia, P.(2005). Palam Priya-a garden pea. Indian Hort., 50 (3): 18-19.

Kanaujia, S.P., Rastogi. K.B. and Sharma, S.K. (1977). Effect of Phosphorus, potassium and rhizobium inoculation on growth, yield and quality of pea cv Lincoln. *Veg. Sci.*, 24 (2): 91-94.

Kanujia, S.P., Rastogi, K.B., Sharma, S.K. and Narayan, Raj (1999). Response of phosphorus, potassium and rhizobium inoculation on pod yield and economics of pea (*pisum sativum*) cv. Lincoln. *Haryana J. Hort, Sci.* 28 (112): 117-118.

Katyal, S.L. (1985). Vegetable Growing in India, Oxford & IBH Publishing Co., New Delhi.

Kovalavich, Z.S. (1991). Agro ecological basis of molybdenum dose for peas. *Agrokhimiya*, No; 10: 94-98.

Kumar, K.V. and Sree Kumar, V. (1981). *South Indian Horticulture*, 52: 168-176.

Mac Gillivray, J.H. (1961). Vegetable production, McGraw Hill Book Co., New York.

Malik, R.S. and Bhandari, A.R. (1994). Water requirement of pea (*pisum sativum*) in Inceptisol of mid-Himalayas. *Indian J.Agric. Sci.*, 64(12): 847-849.

Malik, R.S. and Jaiswal, R.C. (1977). Effect of salinity on germination, growth, yield and quality of pea. *Haryana J. Hort. Sci.*, 6: 181-185.

Marx, G.A. (1977). Classification, genetics and breeding. In: J.F. Suteliffe and J.S, Pate (Eds) Physiology of the garden pea. Academic Press, New York. pp. 21-44.

Mehta, C.P., Jaya, J.S., Malik, Y.S. and Ram.Mangal (1994) Effect of spacing and seed rate on seed production of early varieties of pea (*Pisum sativum* L.). *Haryana J. Hort. Sci*, 23(4) : 310-312.

Mehta, C.P.; Mangal, Narwal, A.K. and Malik, Y.S. (1997). Effect of sowing time and spacing on seed crop of pea (*Pisum sativum* L.) *Haryana J. Hort. Sci;* 26 (3-4) :263-266.

Mehta, P.K., Sharma, T.N. and Chandel, R.S. (1998). Resistance in pea (*Pisum sativum* var. arvense) genotypes to leaf miner (*Chromatomyia horticola*). *Indian J. Agric. Sci.*, 68(5): 271-273.

Mehta, Y.R. (1959). Vegetable growing in Uttar Pradesh. Bureau of Agriculture information, Uttar Pradesh.

Mohan, N., Aghora, T.S., Somkumar, R.G., Pal, A.B., Vani, Akella, Ganeshan, Girija and Rawal, R.D. (2002). Evaluation of snap pea resistant to powdery mildew and rust. *Intl. Conf. Veg.*, Bangalore, 11-14 November, 2002 Abstract: pp. 213.

Naik, L.B. (1992). Response of early garden pea (*Pisum sativum* L.) to plant spacing, nitrogen, phosphorus and potassium fertilization. *Indian J. Agric. Res.* 26 (4) : 210-214.

Nanelal, T.R. and Arya, P.S. (1995). Efficacy of various herbicides to control weeds in peas (*Pisum sativum* L.) *Crop Research*, 10(2): 179-183.

Narsinghani, V.G. and Bedi, Raman (1990). Genetic inheritance of powdery mildew resistance and its breakage in peas. *Proc. Natt. Acad. Sci.* (India). 60 (B) : 73-75.

Narsinghani, V.G. and Singh, S.P. (1979). Heterosis in mutant pea types. *Indian J.Agric, Sci.*, 49 (1) : 39-42.

Nath, P. (1976). Vegetables for the tropical region. ICAR low priced Book series No. 2. ICAR, New Delhi.

Nath, Prem; Srivastava, V.K., Dutta, O.P. and Swamy, K.R.M. (2008). Vegetable Crops – Improvement and Production. PNASF, Bangalore. pp. 102-108.

Pal, A.B., Vani, A. and Rawal, R.D. (1982). *Annual Report* Indian Institute of Horticulture Research. Bangalore, pp. 32-34.

Pal, A.B., and Nath, Prem (1976). Growing pea in Karnataka. *Intensive Agric.*, 14(4): 24-25.

Pal, B.P., Sikka, M.S. and Singh, H.B. (1956). Indian Journal of Horticulture, 13: 64-73.

Pal, Netra; Pachauri, D.C., Bhagchandani, P.M. and Sharma, R.K. (1993). Indian Horticulture, 38: 35.

Parvatha, Reddy, P. (1985). Chemical control of root-knot nematodes infecting peas. *Indian J. Nematod*, 15 : 120-121.

Pratibha Shukhal and Amin, K.S. (1991). Integrated disease management of powdery mildew of field peas (*Pisum sativum* L) caused by *Erysiphe polygoni* DC. *Legume Research*, 14(2).

Purewal, S.S. (1957). Vegetable cultivation in northern India. *Tech. Bull.*, 39. India Council of Agriculture Research, New Delhi.

Purseglove, J.W. (1974). Tropical Crops-Dicotyledons, Longmans, London.

Ray, P., Gupta, H.N. and Roy, M. (1980). Sansruta Samhita Scientific Synopsis. *Indian Natl. Sci. Acad;* New Delhi.

Renfrew, J.M. (1973). *Paleoethnobotany*-Metheum London.

Saimbhi, M.S. and Gill, B.S. (1988). *Haryana Journal of Horticultural Science*, pp. 255-260.

Saini, G.S. (1996). A Text Book of Vegetable Production. Aman Publishing House, Meerut (U.P.) India. 175-177.

Salunkhe, D.K. and Desai, B.B. (1984). Peas and Beans. In : Postharvest Biotechnology of Vegetables (CH.6), Volume. CRC press, Boca Raton, pp. 125.

Sarakanan, A. and Nambisan, K.M.P. (1994). Intergrated nutrient management in peas under hill soils of Kodaikanal Tamil Nadu. *Madras Agric. J.*, 81(10): 573-574.

Seshadri, V.S. and Chatterjee, S.S. (1996). The history and adaptation of some introduced vegetable crops in India. *Vegetable Science*. 23(2): 122-142.

Sharga, A.N. and Jauhari, O.S. (1970). Studies on the effects of foliar application of molybdenum on nodulation and quality of garden pea. (*Pisum sativum* L.) *Madras Agric. J.*, 57 (4): 216-221.

Sharma, B., Mishra, S.K. and Tyagi, M.C. (2000). Pusa pragati a new high yielding powdery mildew resistant variety of vegetable pea. *Indian Fmg.*, 50 (9): 22-23.

Shewell- cooper, W.E. (1977). Basic Book of Vegetable Growing. Barrie and Jenkins (2nd Ed), London.

Shridhar and Mani, V.P. (1995). Breeding garden pea against stress conditions for north-western hills. *Veg. Sci.*, 22 (1) : 48-51.

Sidhu, P.S., Berma, M.M. and Gill. A.S. (1994). Field pea-48-a green seeded variety of dry peas. *J. Research. PAU*, 31 (4):506.

Simmonds, N.W. (1979). Evolution of crop plants. Longman, New York, pp. 339.

Singh, D.N. and Nath, Vishal (2011). Varieties and Hybrids of Vegetables. Satish Serial Publishing House, Azadpur, Delhi (India). pp. 244-250.

Singh, G., Bhan, V.M. and Tripathi, S.S. 91984). Efficiency of herbicides in vegetable pea. *Indian J. Weed Sci.*, 16: 17-21.

Singh, G.P. and Singh, V. (1995). Effect of molybdenum and manganese on their uptake and yield in pea (*Pisum sativum* L.) *Indian J. Plant Physiol.* 38(1): 85-87.

Singh, H.B. and Joshi, B.S. (1970). Pulse crop of India. Indian council of Agric. Res. pp. 183-200.

Singh, H.B., Thakur, M.R. and Bhagchandani, P.M. (1962). Vegetable Seed Production in Kullu Valley. IV Seed production of peas, lettiuce, Calery, Parsley, Onion leek and spinach. *Indian, J. Hort*, 20(3&4) : 148-154.

Singh, J., Kumar, S., Bhan, V.M. and Balyan, R.S. (1991). Crop weed competition studies in pea (*Pisum sativum*) *Indian J.Weed Sci.*, 23(1-2): 67-68.

Singh, P.K.,Dasgupta, S.K. and Tripathi, S.K. (2004). Hybrid vegetable Development, Food Products Press, NY. USA. pp. 278-280.

Singh, P.P. and Tripathi, S.S. (1987). Effect of various herbicides on weed control and yield of green peas. (*Pisum sativum* L.) *South Indian Hort.*, 35(4): 328-331.

Singh, R.A. and Lal, S. (1996). Grow powdery mildew resistant varieties of pea. *Indian Fmg.* 45(12): 15-16.

Singh, R.C., Arya, M.C. and Kumar, Narendra (2000). Off-season pea cultivation a boon to farmers of hills. *Indian Fmg.* 49(10): 5-6.

Singh, R.P., Dubey, S.K., Prasad, M. and Khokhar, N.S. (1969). Studies on nitrogen, phosphorus, potash and FYM on yield of pea under irrigated conditions. *Indian J. Agron.* 14 : 112-117.

Singh, R.S. (1985). Diseases of Vegetables. Oxford and IBH Publishing Co., New Delhi.

Singh, R.S. and Nath, Vishwa. (1995). Performance of pea varieties as inter crop in ber orchard in semi-arid region. *Indian J. Hort,* 52 (2) : 137-140.

Singh, S., Mangal, J.L. and Srivastava, V.K. (1992). A note on the effect of date of sowing, seed rate and chemical treatment on germination, flowering, maturity and yield of vegetable pea under rainfed condition. *Haryana J. Hort. Sci.* 21 (1-2) : 96-98.

Singh, S.P. (1989). Production Technology of Vegetable Crops. A.R.C.C. Karnal, Haryana, India.

Singh, S.P. (1997). Principles of Vegetable Production. Agrotech Publishing Academy. Udaipur, Rajasthan, India, pp. 139-166.

Singh, S.P. and Singh, H.N. (1979). Genetics of seed number per pod and its inheritance in pea. *Indian J. Agric. Sci.* 49 (6) : 401-403.

Singh, T., Singh, B.V. and Panwar, B.S. (1972). Effect of pre-sowing treatments with phytohormones on the yield of pea. *Indian J. Agric. Res.* 6 : 279-284.

Singh, T.P. and Singh, K.B. (1970). Genotypic and phenotypic variability of yield and other quantitative characters in a collection of field pea. *Madras Agric. J.* 57 (2) : 723-726.

Sri Dhar, Mani, V.P., Ved Prakash and Lal, Shankar (1999). *Indian Horticulture,* 44: 19.

Sri Dhar, Mani, V.P., Dubey, S.B., Lal, Shankar and Ved Prakash (1997). *Indian Horticulture,* 41: 47-49.

Srivasan, K and Shukla, R.P. (1980). Control of pea leaf miner (*Phytomyza atricornis* Meigen) and aphids (*Acyrthousiphon pisum* Harris) with certain granular insecticides. *Pesticides,* 141 (12) : 20-22.

Srivastava, B.K. (1991). Morpho physiological response of garden pea (*Pisum sativum* L) cultivars to sowing dates. *Res & Dev. Reporter,* 8 (2) : 137-142.

Swarup, Vishnu (2006). Vegetable science and technology in India. Kalyani Publishers. New Delhi. pp : 656.

Tandon, H.L.S. (1991) Fertilizer recommendations for Horticultural crops. A guidebook. Fertilizer Development and consultation organisation, New Delhi, India.

Thakur, B.R., Kapoor, A.S. and Jamwal, R.S. (1996). Varietal resistance of pea to powdery mildew in dry temperate zone of Himanchal Pradesh. *Indian Phytopathol,* 46(1): 92-93.

Thompson, H.C. and Kelly, W.C.(1957). Vegetable crops MCGraw- Hills Book Co., Inc, New York, Toronto London.

Tiwari, Anamika and Narsinghani, V.G. (1986). Promising varieties of kharif peas with powdery mildew resistance. *Veg. Sci.*, 13 : 137-139.

Vavilov, N.I. (1926). *Turkey Prikl Bot Genet Sel,* 20: 1.

Vavilove, N.I.(1949). The original and variation, immunity and breeding of cultivated plants. *Chron. Bot.* 13 : 364.

Verma, V.L. and Bhandari, A.R. (1996). Study on the uptake of macronutrients by pea as affected by nitrogen and phosphorus levels. *Hort. J.* 9(1): 77-80.

Vishnoi, A.K., Joshi, S. and Joshi, M.C. (1982). Chemical weed control in (*Pisum sativum*). *Indian J. Weed Sci.,* 15: 72-73.

Vishnu Mittre (1974). Paleobotinical evidence in India- In "Evolutionary Studies in World Crops". Ed J. Hutchison Cambridge University Press.

Wakankar, S.M. and Mahadik, C.N. (1961). *Indian Farming,* 11: 8-12.

Wakankar, S.M., Yadav, L.N. and Kelkar, G.M. (1974). *Research Journal of JNKVV,* Jabalpur, 8: 192-195.

Yawalkar, K.S. (1969). Vegetable crops of India. Agril. Horticulture Pub. House, Nagpur.

Yawalkar, K.S., Kakota, P.N. and Srivastava, M.P. (1962). Commercial Fertilizers in India, Agri-Hort. Publishing House, Nagpur.

Zohary, D. and Hopf, M. (1973). Domestication of pulses in the old world. Science, 182 : 887-894.

□□□

2. FRENCH BEAN

Botanical name	:	*Phaseolus vulgaris* L.
Family	:	Fabaceae
Chromosome number	:	2n = 2x = 22

French Bean is one of the most important leguminous vegetable, also known as kidney bean, haricot bean, snapbean, navy bean, salad bean, string bean, common bean, dwarf bean or garden bean. It is grown for the tender immature, edible pods and for the dry ripe seeds and to a lesser extent for green shelled beans. The dry seed type are called as 'Rajmash' in India. Snap bean seed can also be used in dry state like the dry bean types (Sharma and Joshi, 1993).

ORIGIN

French bean originated in the new world, principally Central and South America (Kaplan, 1956). Southern Mexico and Central America are considered to be the primary centre of Origin, while secondary centre lies in Peru-Bolivia-Ecuador region of American continent (Vavilov, 1950; Smartt, 1976).

The domestication of *P. vulgaris* might have occurred in Brazil and Northern Argentina from a wild form of *P. arborigineus,* which is progenitor of *P. vulgaris* (Yarnell, 1965). The red Indian of America cultivated French bean along with maize, from them Whitemen started its cultivation (Sturtevant,1919). Spanish explorers introduced it into Europe (Purseglove, 1968). From Europe, it spread to Africa, the East Indies, India and other parts of the world (Simmonds, 1976).

DISTRIBUTION

French bean was cultivated by the Indians in both north and south America from ancient time. In Europe, the Frenchbean spread rapidly in the 16th and 17th centuries and reached England by 1594 (Pursoglove, 1968). From Europe, it was introduced in India during the 17th century (Simmonds, 1976). In India French bean is popular in Punjab, Haryana, Jammu and Kashmir, Himachal Pradesh, Western Utter Pradesh, Maharashtra (near Poona), Karnataka (near Bangalore), Andhra Pradesh, (near Hyderabad), Tamil Nadu (near Salim), Nilgiris, (Kodaikanal areas). In other states also, its popularity is getting momentum. In fact, French bean is in great demand in cities (Singh, 2001). *P. vulgaris* is now widely cultivated in many parts of the tropics and subtropics and throughout the temperate regions. In India, French bean is cultivated in the area of 1.50 lakh hectare with production of 4.10 lakh tonnes during 2000-01 (Swarup, 2006).

BOTANY

There are two type of french beans, namely bush beans (dwarf bush bean) and pole beans. Semi-pole types are also not uncommon. The leaves pinnately three-lobed and the leaflets are either broad ovate or long and narrow. Tap root grows up to 120 cm, and large number of adventitious roots arise from the base of the stem and tap root. Roots may or may not have nodules of the nitrogen-fixing bacteria. Dwarf bush types are popular. The dwarf bush type has small plant stature, height upto 40-60 cm, short, erect stem, with the main axis consisting of 4 to 8 shortened internodes, terminating in an inflorescence leaves alternate trifoliate, petiole long. Flowers borne an auxiliary, lax racemes. Pods slender narrow 4-6 seeded, glabrous, straight, or slightly curved. Light to dark green colour. It is self-pollinated, pollination takes place at the time the flower opens.

Plant breeding

French bean is a self-pollinated crop. The breeding objectives in French bean are for the improvement of vegetable types and dry bean types. The breeding procedures for the crop include hybridization, pedigree selection, single plant selection, bulk method of selection etc. Several exotic varieties were introduced from USA, Sweden, Mexico, Australia and Russia such as contender, Giant Stringless, Jampa, Kentucky Wonder, Tender green, Premier and others. In India, several improved cultivars are developed. They are pole types and bushy types which are early maturing either round or flat and stringless. Pant Bean 2 and YCDI are dual purpose types for both green pods and dry seeds (Nath *et al.*, 2008).

Local strains and some of the introduced varieties of French bean are available. Most of the genetic variations exist among the introduced varieties in the grower's field where no effort was made to maintain the genetic purity of a particular variety. Some 600 collections have been made in this crop (Bhagat *et al.*, 1979). Some of the introduced varieties which performed well under large scale production were recommended (Pal *et al.*, 1956; Singh and Prasad, 1967). A breakthrough was made by evolving a new variety 'Pusa Parvati' with the aid of mutation breeding (Swarup and Gill, 1968; Anon., 1971b). Some genetic studies with regard to economic characters have also been made (Patel and Chavan, 1961).

Taxonomy

French bean (*Phaseolus vulgaris* L.) belongs to the family Fabaceae. Its chromosome no. 2n= 2x= 22. Besides, there are other three cultivated species viz., *P. coccineus*: (Runner bean or scarlet bean). *P. lunatus*: (Lima bean, large-seeded), Sieva (small-seeded), butter or Madagascar bean *P. acutifolius*

(Simmonds, 1976) var. latifolius. (Tepary bean). *P. coccineus* is cross-pollinated whereas *P vulgaris* and other two are self-pollinated with only a negligible amount of cross pollination. The most important changes which have taken place in the fruiting characteristics of these species are in pod and seed size, although there has also been an appreciable increase in flower size in *P. vulgaris* and *P. coccineus*. Seed number per pod has changed in *P. vulgaris* and *P. acutifolius*, where upto nine seeds are found in wild forms but rarely more than 5 in most of the cultivars. In all species, an increase in permeability of the seeds to water has taken place under domestication, which is an important property for uniform germination and easy cooking (Miranda, 1974).

Floral Biology

Flowers are axillary racemose. They are white, rose, violet or yellow in colour. The corolla extends beyond the calyx and the keel is prolonged below the stigma. The calyx is five parted. Flowers may be white, rose, violet or yellow in colour (Alim, 1989). Anthesis occurs between 7 and 8 a.m. flowers do not close, but corolla is shed after a few days. Dehiscence of the anthers takes place in the evening before anthesis (Jones and Rosa, 1928). It is self-pollinated but natural crossing has been reported from 8.26 per cent at a distance of one metre and 2.63 per cent at a distance of 9 metres (Barrons, 1939). The fruit or pod (7 to 20 cm x 1.5 cm) contains 4 to 8 seeds, in some varieties upto 12. Pod is glabrous, straight or slightly curved, with prominent beak, yellowish to dark green in colour (Barrons, 1939). Seeds are borne alternatively on the margin of the two placentae located dorsally (Barrons, 1939). They are white green, yellow, pink, red, purple, brown or black sometimes mottled or striped. The seeds are oblong kidney shaped or globular. Hilum usually white.The average weight of 100 seeds ranges from 25 to 45 g. The germination of seed is epigial (Hawthorn and Pollard, 1954 and Salunkhe *et al.*, 1987).

Morphology of Pods

Pod structure has been altered in cultivars with a reduction in dehiscence and fibre content. Three distinct pod textures are found in the common bean; the parchmented which are very fibrous and dehisce strongly at maturity, the leathery types which are less dehiscent but split readily along the sutures and the fleshy or stringless pods which are indehiscent and do not splits readily. Varieties with parchmented pods are used only for dry seed production; leathery podded varieties can be used for green pod production when young; or haricot production when fully matured. Fleshy podded or stringless types are usually grown for the production of green pods (Smartt, 1976).

A remarkable range of testa colours occurs in the seeds of *phaseolus* species; five main colour groups have been identified, namely, white, black, red , ochre and brown.Superimposed on these ground colours may be various patterns of spots, fleeks and strips (Sharma and Joshi, 1993).

A further change which has been taken place with the disposal of *Phaseolus beans* to temperate regions of the world is in response to photoperiod. Since the genus is native to the tropics, many cultivars from the region are photoperiod sensitive and responsive to short days. There are also day-neutral types and all the temperate cultivars have been selected for day-neutrality or tolerance of long days (Simmonds, 1976).

COMPOSITION

Green french bean rank high as a cheap source of nourshing food. They are a valuable source of protein, calcium, iron and vitamins. Composition of french bean (green pod) per 100 g of edible pods (Gopalan *et al.*, 1982) is given below:

Moisture	91.4 g	Catrotene	132.0 mg
Fat	0.1 g	Thiamine	0.08 mg
Fibre	1.8 g	Sulphur	37.0 mg
Other carbohydrates	4.5 g	Vitamin "C"	24.0 mg
Protein	1.7 g	Nicotinic acid	0.30 mg
Minerals	0.5 g	Calcium	50.0 mg
Vitamin 'A'	221 I.U.	Phosphorus	28.0 mg
Riboflavin	0.06 mg	Iron	1.70 mg

Uses

In India it is grown for tender pod. It is an excellent vegetable both in form of young pods and dried seeds. It is used for culinary purpose, and form a good vegetable curry. Sometime mixed with other vegetable and give same taste after cooking. Generally its green pods are cut into pieces and fried sole vegetable and dry grains (rajmash) cooked as pulses. Pods contain pectic substances as calcium pectic 9 to 15 per cent (on dry weight basis). Beans are also said to be antidiabetic and good for bladder burns, cardiac, carminative, depurative, diarrhea, diuretic, dropsy, dysentry, eczema, emollient, hiccups, itch, kidney resolvent, rheumatism, sciatica and tenesmus (Duke, 1981). The dried beans are rich in protein and closely compare with meat. In India, it is mostly grown for tender vegetable, while in USA it is grown for processing in large quantities.

VARIETIES AND HYBRIDS

French bean varieties are of two types (1) dwarf or bush types and (ii) climbing or pole types. Each type bears green pods or wax pods, and varieties either string or stringless. Detail characteristics of varieties and hybrids described by Singh and Nath (2011) are given below :

Dwarf or Bush Type Varieties

Contender

This is an introduction from USA.This is an early variety, developed by Indian Agricultural Research Institute, New Delhi. The pods are stringless and dark green in colour. The seeds are oval in shape. This variety has been found to be resistant to mosaic and powdery mildew.

Giant Stringless

This variety was developed at IARI, New Delhi. Variety is early, pods are green in colour, medium size slightly curved and stringless. The seeds are glossy and yellowish brown. The yield potential is high.

Jampa

This is Mexican variety, suitable to grow in Maharashtra. This is very early variety, pods are available for first picking 50 days after sowing. The pods are flat, smooth, non-stringy and pale green in colour. The seeds are black, smooth, small in size and flat in shape. It is resistant to wilt and can grow successfully in warmer climate.

Pusa Parvati

This is an outcome of mutation breeding (cv. Wax pod, and American Variety) done at IARI, New Delhi. The variety is early . The pods are light green, attractive round. This is a high yielding, and most suitable for hills. The seeds are oval type.

Bountiful

This variety is high yielding, suitable to grow from September to February in south India. On main stem, pods form in cluster. Pods are tender green, available for harvesting within a week of flowering. The pods remain for long period even when left unharvested (Shanmugavelu, 1989). The seeds are flat type.

Premier

This variety is suitable in the plains of Tamil Nadu. This is less susceptible to wilt and mosaic. The seeds are black in colour and oval in shape.

Plant Anupama

This variety was developed at Pantnagar. This is an early maturing variety, suitable for both hills as well as plains of Uttar Pradesh and elsewhere too. The pods of this variety, are round, smooth tender, nonstring, translucent and attractive in appearance. This is resistant to angular leaf splt, tolerant to blight, common mosaic and sclerotinia mould (Ram and Singh, 1984).

Arka Komal (Selection-9)

This variety was developed at Indian Horticultural Research Institute, Bangalore (Karnataka). The plants are erect and bushing. The pods are straight and flat. The pods can tolerate long distance transportation (Rajendran and Satyanarayana, 1988). It has better yield performance than Pusa Parvati (Singh, *et al.*, 1993).

Arka Subidha (IIHR 909)

Plants bushy and photosensitive. Pods straight and oval, light green, stringless and fleshy. Yields 9 t/ha in 70 days. Developed at IIHR, Bangalore.

Arka Anoop

Its plants are bushy and photo-insensitive. It can be grown during all the three seasons. Its pods are ready for picking 45 days after sowing. Pods are flat, smooth, long (17-18 cm) and crisp with less parchment, the yield potential being 20.0 t/ha in 70 days. It is resistant to both rust and bacterial blight diseases.

TKD1

It is a hybrid derivative of a cross between two pole types, viz., Sel.1 and PV 118. The green tender pod yield ranged from 50-60 q/ha and matures in 90-100 days.

Selection-5

This is also developed at IIH, Bangalore. The pods are round stringless. It is suitable to grow in kharif season.

SVM-1

Pods green round stringless, 13-14 cm long. Ready for picking in 65-75 days. Average pod yield is 105-125 q/ha. Resistant to angular leaf spot both in leaves and pods. A dual purpose variety and thus can be used as vegetable and seeds as pulse. Recommended for mid hill areas.

Laxmi

Pods are formed in cluster of three, 13-14 cm long, stringless, green round and attractive. Pods ready for picking in 65-70 days. Seeds are white in colour with light yellow scar. Average green pod yield is 120-140 q/ha. Tolerant to angular leaf spot disease.

VL-Boni-1

This variety has been developed at VPKAS, Almora, plants are dwarf (30-40 cm) having white flower with purple tinged. Pods are found, light green, attractive, stringless and fleshy; first picking starts 45-60 days; moderately susceptible to foliar diseases; gives 100-120 q/ha pod yield. Suitable for multiple cropping and can be planted in the hills from March to July.

YED 1

It can give a green pod yield of 97.5 q/ha. Seeds are bold, attractive and dark-purple in colour.

Ooty 1

It has yield potential of 340 q/ha of green pods in a duration of 80 days. The variety is moderately resistant to leaf spot, anthracnose and pod borer.

Phule Surekha

Phule surekha have been developed from Jampa improved variety by employing pure line selection. It is a semi-determinate variety. Its pods are straight flat with average length of 9.25 cm and breath 1.07 cm. It is suitable for kharif, rabi and summer seasons (Photo-thermo-insensitive). It is specifically bread for small straight the edible pods and resistant to anthracnose, leaf crinkle, bean yellow-mosaic virus and wilt (caused by stem fly attack). It gives highest yield of 153.39 q/ha which was 53.68% higher than contender.

Kashi Param

Its plant determinate in height, grow 70 cm high and are bush type, with dark green leaves. Its pods are attractive, fleshy, average length 14.7 cm, round and dark green. Its green pods yield is 120-140 q/ha and suitable for cultivation in Jammu and Kashmir, Himachal Pradesh, Uttaranchal, Madhya Pradesh and Maharashtra, (Kalloo *et al.*, 2006).

Pole Type or Twining Type Varieties

Kentuky Wonder

This is a late maturing variety, developed at IARI, New Delhi. The pods are green in colour, large curved, round and stringed at later stage. The seeds are light brown in colour. This variety has high yield potential. Average pod yield is 100-125 q/ha.

KKL 1

It is a pole type best suited for elevation of 800-2,400 m. It has a potential yield of 70 t/ha of pods or 30 q of grains per hectare.

Pusa Himlata

It is a climbing or pole-type , french bean (*Phaseolus vulgaris* Linn.). Vine is medium in height (2.5 m long). The internode distance is small, each nodes bears 3-4 flowers. Pods are medium -sized (14 cm), round, meaty, stringless and light green the early maturing ready for picking in 60 days. Seeds are white and suitable for processing as green-shelled bean. Mature seeds also have good cooking quality and are suitable for use as dry bean (Rajmash).

Other varieties and genotypes

Swarn Priya, Black Prince, VL Bean 2, VL Bean 3, Aparna, IVFB-1, IVFB-2, IVRFB-1, VL 5, Sel 2 and Sel-4.

CLIMATIC INFLUENCE

French bean is a cool weather crop but can thrive in higher temperature than pea. It cannot withstand drought as well as very heavy rain fall and frost. It can be grouped into longday, short day and day neutral plants. Most of the french bean varieties are day neutral, (Choudhury, 1967) and hence the day length does not affect the seedling habit except some semi-vine varieties which are shortday type. The favourable temperature for crop growth, flowering and fruiting is 18-23°C and if temperature drops below 15°C, crop growth and development are adversely affected. This crop requires plenty of sunshine especially in first phase of its development.

It is a heat loving crop but not resistant to high temperatures. Below 8°C germination does not occur. At a temperature of 14°C, the seed set is reduced and that is mainly due to a reduction in ovule fertility rather than pollen germination. Above 30°C at flowering time, increased abscission of flower buds, flowers, young pods and poor fertilization and seed development in pods may occur.For best growth and yield the optimum soil temperature should be is 25°-30°C. Temperature for seed germination of pole type is ranged 20°-25°C. Best pod set with temperature range of 15° to 25°C for four hour after pollination (Chauhan and Mangal, 1993). French bean does well at places between, 1,000-2,000 m above the sea level with comparatively low rainfall of about 75-100 cm (Yawalkar, 1969). French bean prefers, shady places and, therefore, can be grown under shade also.

During hot weather, blossoms and pod may drop. Similarly, heavy and continuous rains result in dropping of blossoms and pods.

SOIL REQUIREMENT

French bean is a leguminous crop can be grown on wide range of soil. However, sandy loam fertile, well drained soils are best. High moisture and high nitrogen delay maturity. Soil is given pre-irrigation before preparation that help in making soil loose and friable. Sandy and sandy loam soils (Choudhury, 1967) are preferred for an early crop but heavier soils are desired for mid-season crop. It does not grow well in extremely acidic soils and is sensitive to high concentrations of aluminium and manganese. The most suitable soil is high concentrations of aluminium and manganese. The highest yield is obtained in soil with a pH between 5.3 and 6.0 (Prabhakar *et al.*, 2003).

SOWING

Sowing time

French bean can be sown twice in a year, in January-February and July-September in the plains (Choudhury, 1970) and March-June in the hills. Singh *et al.* (1975) have reported that French bean can be planted from April to July but best time of planting is 3rd week of May for obtaining maximum yield in hill areas.. In plains this crop can be sown twice but delayed sowing reduces the yield. Singh and Sharma (1970) reported that there was no significant interaction between variety and date of sowing.

Field preparation

According to Chauhan (1968) beans need a thorough preparation of land because of their nitrogen fixing capacity. Field should be thoroughly ploughed 2-3 times with soil turning and desi ploughs, followed by 1 to 2 harrowings to remove the weeds.

Distance of sowing

Distance of sowing varies with the type of French bean grown. Bush types are planted at 45-60 cm x 10-15 cm (row x plant) whereas pole types at 100 cm x 22-30 cm. Planting in double row is also done with 60-70 cm distance between two set of rows and 30 cm between two rows. Variety 'Uday' of French bean has been found to produce seed yield of 233 kg/ha at plant density of 2.2 lakh plants/ha (Alim, 1989).

High plant density (80,000 plants/ha) gave the highest yield in case of *phaseolus* beans (Thimmegowda *et al.*, 1974). But contrary to this Pandey *et al.* (1974) reported that french bean a crop of *phaseolus* group, produced more number of green pods per plant and maximum total yield under widest inter row spacing (30-60 cm). Singh and Behera (1999) recorded significantly higher green pod yield of 114.33 q/ha with closer spacing of 35cm x 25 cm.

Seed rate

The seed rate of French bean for the bush type of cultivars is about 70 to 90 kg per hectare whereas for pole type of cultivars it is 20 to 30 kg per hectare.

Seed germination of French bean is lost within 9 months in storage if moisture content is more than 6-7% at a storage temperature of 30°C. Therefore, it is essential to obtain freshly harvested seed of good quality (Chauhan and Mangal, 1993).

Seed treatment

Treat the seed with thiram @ 3 g/kg of seed 4-5 days before sowing. After that take up seed treatment with carbosulfan 25 St (marshal 25 st) @ 60 g/kg of seed. Moist the seed slightly before treatment for better attachment of the chemical.

Inoculation

When French bean is grown on newly reclaimed soils. Then the seeds should be inoculated with nitrogen-fixing bacteria before sowing. Pre-sowing inoculation of seeds with rhizobium culture can very well compensate for nitrogen and maximum yield, early seed emergence and better seed quality was recorded from inoculated seeds with lower application of NPK and compared to higher levels of N P K without inoculation (Kulkarni, 1979).

It is advisable to sow two seeds per hill and remove weak seedling after germination is completed. Generally sowing of two seeds per hill increased the yield as compared to one seed (Prasad *et al.*, 1978).

Two to three seeds are sown at a place in furrow or in the side of ridges. Seeds are dibbled by human labour or drilled mechanically at a depth of 2 to 3 cm (Chauhan, 1968). In certain places, broadcasting of seeds is practised but it may interfere with the cultural operation and to maintain optimum plant population. Generally line sowing is done for easiness in intercultural operation. Seeds should covered properly after sowing. On large scale. sowing can be done with seed drill or by opening of furrow up to 4-5 cm deep and followed by dropping of seeds. Though this method of seed sowing is efficient but desired or correct distance between plant to plant is difficult to maintain. Dibbling method takes time but distance is maintained properly. Seeds germinate 8-10 days after sowing under favourable conditions.

NUTRITIONAL REQUIREMENT

The nutritional requirement of French bean depend on the condition of soil. It is estimated that a crop with yield of 800 kg of grain and 700 kg of haulm removes 73.5 kg nitrogen, 31 kg phosphate and 61.2 kg of potash from the soil (Prabhakara *et al.*, 2003).

Table 1: Optimum doses of NPK and FYM found for French bean crop in some states.

State	Nitrogen (N) kg/ha	Phosphorus (P) kg/ha	Potash (K) kg/ha	FYM (t/ha)
Andhra Pradesh	20	50	50	10-15
Arunachal Pradesh	25	70	50	20
Assam	30	40	20	0.2
Haryana	30	40	0	10
Himachal Pradesh	50	100	50	10
Jammu and Kashmir	50	100	50	10
Karnataka	62	100	75	25
Madhya Pradesh	50	70	60	20
Maharashtra	50	20	50	10
Mizoram	30	40	20	5
Orissa	50	80	80	10
Punjab	30	40	40	10
Rajasthan	30	40	40	10
Tamil Nadu (Hills)	180	125	100	30
(Plains)	100	100	50	12.5
Uttar Pradesh	50	60	60	20

Source : Tandon (1991)

French bean responds better to application of organic matter, because it helps in improving physical, chemical and biological conditions of soil (Anonymous, 1953). Such conditions favour better plant growth and ultimately higher pod and seed yield. Nitrogen at higher level decreases nodulation. But nitrogen at low level enhances microbial activity in the soil and plants while phosphorus at double the level than nitrogen required for better uptake of nutrients and nodulation. Similarly , potassium plays important role for higher seed yields. Potassium has also been found to induce early flowering. Beans like other legumes, have a relatively high nitrogen demand during pod fill and senescence and abscission may result from competition for N. Singh *et al.* (1981) reported significant increase in French bean yield with increasing levels of nitrogen, i.e. (0-20 kg/ha) and 0-90 kg/ha of phosphorus. Significantly higher yield was recorded with 50 kg/ha nitrogen and 80.0 kg/ha phosphorus (Chandra *et al.*, 1987). Singh and Tripathi (1994) reported that application of fertilizers @ 62.5: 100:100 kg N, P_2O_5, K_2O/ha with closer spacing (35 x 25 cm) produced maximum green pod yield. Kanwar and Thakur (1972) reported that foliar spray of 15 ppm phosphorus (Dehydrogen phosphate) hastened the flowering. The fertilizer should be placed in bands about 70 to 90 mm to the side of seeds and slightly deeper than seeds.

Nath *et al.* (2008) reported that French bean responds well to phosphorus upto 100 kg though its recommended dose is 80 kg/ha. Being a leguminous vegetable it does not require much nitrogen. But for better nodule formation nitrogen application is necessary. Generally, nitrogen @ 60 kg, phosphorus @ 80 kg and potash @ 40-50 kg per hectare are applied to the crop. Also seed treatment with *Rhizobium phaseoli* culture (30 g/kg of seed) is beneficial for better nodule formation and nitrogen fixation. Under acidic soil conditions, application of NPK alongwith the basic application of lime 4.8 q/ha increase the crop yield.

Method and time of application

Organic manure is applied 10-15 days prior to sowing during field preparation. Fertilizers like nitrogen (half dose) and whole dose of phosphorus and potassium are applied as basal at the time of sowing.Remaining half dose of nitrogen is top dressed 3 to 4 weeks after sowing.

Use of micronutrients

Foliar application of various micro-nutrients produced a superior quality of pods and increased yield . Application of zinc and magnesium produced higher quality, increased yield and ascorbic acid. Beans are sensitive to higher concentration of aluminium and manganese and toxicity symptoms were noticed on plants (Thomson and Kelly, 1972). Yield of french bean was markedly increased when plants were treated with Zn + B + Mo + Mn at 0.2 % concentrations (Jana and Kabir, 1987). Besides, foliar application of micronutrient at 0.1 per cent can be done immediately when symptoms of any deficiency of them are noticed (Singh, 2001).

Application of growth substances

The plant growth, flowering, fruit set and pod yield are improved by use of growth substances in french bean. Growth regulators like PCPA @ 2 ppm, L-naphthlyacetamide or B-naphthal acetic acid at 5-25 ppm have shown good effect on fruit set when sprayed at prevailing temperature when pod set is impaired. Medhi (2000) found that application of NAA @ 15 ppm improved physiological as well as the yield attributes and recorded highest in comparison of other treatments. Spray of GA_3 @ 50-200 ppm proved beneficial in improving the vegetative growth of the crop (Prabhakara *et al.*, 2003). Spray of tracel once at flowering is effective in good pod setting and higher yield.

IRRIGATION

First irrigation is given immediately just after sowing of seeds if pre-sowing irrigation has not been given, for proper germination of the seeds. French beans are shallow rooted crops and sensitive to excessive moisture. Chauhan (1968) reported that the crop of french bean is irrigated at an interval of 7-10 days after the first irrigation. Irrigations are required in the early phases of crop growth just prior to bloom and pod development period. A good yield is produce even in soils with low moisture content (Choudhury, 1970). However, optimum soil moisture should be made available at the time of fruit set and pod development. The sprinkler irrigation wetting upto 60 cm depth of soil is beneficial to increase the seed yield. An additional irrigation is given as and when necessary. About 6-7 irrigation during the growing season would be required at regular interval for higher pod yield. Lack of adequate soil moisture may result in flower drop, reduce percentage of pod setting, length of pod and number of seeds per pod. Spring summer crop of French bean produces higher seed yield with IW:CPE ratio at 0.6 to 0.9 (Chakor and Rana, 1992).Stagnation of water causes the death of plants.

INTERCULTURE

The inter-space should be kept clean by frequent shallow cultivation especially during the first 4 weeks of the crop. The surface soil is loosen by hoeing as soon as it is dry after every irrigation. First light earthing up is given just after 10 days of germination when plant started growing. Earthing up provided stability and keep straight and in proper position to the plants in the field. Subsequent earthing up after first top-dressing of fertilizer i.e., an 25 days and next may be taken up 15 days after first top-dressing.

Mulching

Grass mulching was useful to prevent moisture loss from the soil and increase the root growth, nodulation, shoot and plant growth and also increased water use efficiency of legumes (Gupta and Gupta, 1983).

Staking

Pole type (Twining type) cultivars of French bean grow well on support made up of cane frames. They are also supported by erecting wooden poles connected with strings. If the plants are grown in rows, single stick of about 2m length should be fixed near each plant. Cross canes may be fixed on the vertical poles at spacing of about 40 cm. If the plants are grown in hills, the twigs and branches should be given good support.

Crop rotations

Saini (1996) suggested the following crop rotation with French bean for higher returns.

1. Maize-Pea-French bean
2. Clusterbean-Cabbage-French bean

3. Okra-Cauliflower (Mid-season)-French bean
4. Spongegourd-Pea-French bean
5. Bittergourd-Radish-French bean
6. Bottlegourd-Pea-French bean
7. Maize-French bean-Okra
8. Cluster bean-French bean-Cowpea
9. Cowpea-French bean-Bittergourd
10. Amaranthus-French bean-Cowpea
11. Okra-French bean-Tomato

WEED CONTROL

French bean requires one to one and half weeks of weed free period after emergence to prevent crop yield reduction (Zimdahl, 1980). On an average uncontrolled weeds caused 69.7%, reduction in the green yield of french bean (Singh *et al.*, 1984). Manual weeding is the common practice adopted to control weeds in French bean. About 2-3 weedings may be needed till the plants can cover well and to prevent weed competition with crops. When the canopy fully develops is smoothers the weeds.

French bean crop is harmed by some weeds like *Amaranthus viridis, Chenopodium album, portulaca oleracea., Raphanus raphanistrum, Cynodon dactylon, Cyperus rotundus.* Weedicides like Toke-E-25 at 1.0 kg per hectare or Alachlor (Lasso) at 3.0 kg per hectare may be applied as pre emergence (Leela, 1982).

Gupta and Singh (1965) found Eptam 6E to be promising herbicide for this crop. Dhuria and Leela (1970) got excellent control of weeds and enhanced yield in french bean var. Contender with alachlor 1.5 kg a.i./ha. Singh *et al.* (1984) found alachlor at 1.0 kg/ha very effective followed by nitrofen 1.5 kg/ha. Ali and Kushwaha (1987) got effective weed control and yield at par with treatments like pendimethalin 0.75 kg or oxyfluorfen 0.75 kg/ha when compared with one hand weeding or no weeding. Nath *et al.* (2008) suggested that weeds should be controlled either by hand hoeing or by the use of chemical herbicides like dinitro materials at the rate of 5 to 8 kg/ha.

PLANT PROTECTION MEASURES

Some important insect pests, diseases and their control measures are as follows :

Insect Pests

Bean beetle (Epilachana varivestis)

This is also known as Bean lady bird Beetle. Both the adults and larvae feed on all parts of the plant particularly on the lower surface of leaves. This pest can be controlled by procuring good seeds, removing and burning of seed and pods which are holed and spraying malathion or Carbaryl at 0.1 per cent, will control it effectively.

Bean aphids (Aphis and Macrosiphum sp.)

These are very small green to black insects which such the cell sap of the tender parts of the plants, mostly the leaves. Affected plants stop growing. The developing pods are covered with a black, sticky substance. There is curling of leaves, twisting of twigs and shedding of flowers. It may be controlled by spraying nicotine sulphate (1:600), lindane (0.1%) or parathion (0.025%). If the fruit formation has taken place, spraying with 50 per cent malathion at 3 ml in 5 litres of water may be done (Reddy, 1962, 1968). The edible pods may be harvested before spraying and the next harvest should be done after a week of spray. In case of parathion spray, harvesting after two weeks of spray is preferred.

Jassids (Emposaca facae E. Kraemeri)

They harm the crop very much. The control of aphids are also applicable to control these jassids.

Red spider mite (Tetranychus urticae)

During hot, dry conditions, mites attack French bean crop. They suck the sap of the leaves. It is advisable to uproot all affected plants and destroy them, and spray the crop with Malathion at 0.1 per cent 10-12 days interval, twice or thrice depending on degree of infestation.

Bean thrips

The adult is small and greyish in colour. Infested leaves turn pale and drop-off. Pods after infestation become silvery white in appearance. Uprooting of severely affected plants and spraying of Parathion at 0.05 per cent will control bean thrips effectively.

Hairy caterpillar (Ascotis imparata and Spilosoma obliqua)

The larvae feed on leaves and cause characteristic skeletonization of leaves. This can be controlled effectively by collection of larvae and spraying of endosulfan at 0.07 per cent.

Root-knot nematode (Meloidogyne sp.)

It becomes a serious pest in certain beans, in some regions. The attacked plants are pale, yellowish and some what dwarfed and have a tendency to wilt. The roots show typical galls which should not be confused with the bacterial nodules. It may be controlled by following long crop rotation or by soil fumigation with nematicides like Nemagon.

Diseases

Anthracnose (Colletotrichum lindemuthianum (Sacc and Magn) Briosi and Cav.)

Dark brown or black sunken spots with reddish or yellowish margins appear on veins, petioles, stems and pods and seeds. Through seeds fungus carry from season to season. Cool humid, rainy weather is more favourable. Fungus infect other legumes also tlike cowpea. Indian bean, Lima bean, Scarlet bean etc. To control use healthy seeds. Follow clean cultivation. Follow crop rotation, Avoid sprinkler irrigation. Treat the seed with fungicides like ceresan or thiram or dithane Z-78 or dithane M-45 (2 g/kg seed). Spray the crop with benlate (0.2%) or bavistin (0.2%).

Powdery mildew (Erysiphe polygoni DC)

White powdery floury spots are developed on both sides of the leaves and other parts of the plant except roots. In severe infection defoliation occurs. It is a seed-borne as well as soil-borne. Dry weather favours spread of disease. It can be controlled by collection and destruction of diseased plant debris. Dusting of crop with sulphur dust (200 mesh) at 25 to 30 kg per hectare at 10 to 15 days interval or spraying of crop with wettable sulphur or kelthane or karathane at 0.2 per cent or benlate or bavistin at 0.1 per cent.

Rust (Uromyces phasioli typica or U. appendiculatus Unger)

Small white, slightly raised spots are developed on the lower surface of leaves, and these gradually become dark brown to black. In severe, infection all above ground parts are affected. High humidity favours rust spread. Other bean crops (Lima bean, Indian bean , Scarlet bean etc.) are also affected with rust. Dusting of fine ground (200 mesh) sulphur at 25 to 30 kg per hectare, is effective to check the disease. Adjustment of sowing time, and finding out rust tolerant varieties, will help in controlling rust disease effectively.

Leaf spot (Cercospora cruenta)

Radish brown to brown water-soaked spots develop on lower leaves and pods, later on these spots appear ashy-grey with purplish borders. Spraying of copper oxychloride at 0.1 per cent will control this disease.

Ashy stem blight (Macrophomina phaseolina Goid)

Affected plants show black sunken cankers on seedlings. A small dark, sunken lesion on the stem is the first evidence of the disease. The lesion spreads very rapidly and the entire stem may be attacked, often killing the growing point. The disease is seed borne. It is severe in a warm humid areas. It has very wide host range. It is advisable to sow disease free seed, seed treatment with ceresan prior to sowing and to follow phyto sanitary measures.

Bacterial blight (Xanthomonas phaseoli Dowson)

Irregular water soaked, translucent, sunken, red to brown spots surrounded by a yellowish zone appear on leaves. Severe infection may cause falling off leaves. On stems, reddish brown to black stripes are appeared. The pods are also affected. Infected seeds show yellowish spots which are shrivelled and fail to germinate (Yawalkar, 1969). Affected plant or parts should be collected and burnt in the field. Seed should be procured from disease-free areas. Seed should be treated with organic mercurial fungicides (emisan, mercuric chloride) and antibiotics like streptocycline, plantomycin or agrimycin 100 ppm are benefical to control this disease.

Bean mosaic

This disease is caused by three different strains of the virus. The leaves become ruffled, crinkled and have mottled yellow areas on the green leaf surface. The pods from infected plants may be rough and deformed. The common bean mosaic and its variants are carried in seed from diseased plants and may be transmitted from diseased to healthy plants by aphids. This disease can be controlled by growing resistant varieties like Contender, Kentucky Wonder, Rival and Logan. It may be avoided by roughing some of the infected plants in the beginning itself.

Maturity standards

The different maturity parameters like seed size, per cent seed, dry-matter content, AIS (Alcohol insoluble solids) and distribution of pods according to sieve size are found to be reliable maturity standards. The processing industry use this practice in canning and freezing to judge quality by size distribution of seeds.

HARVESTING

The french bean is ready for harvest from 40 days onwards after germination depending on cultivation. It takes about 2-3 weeks after flowering for the pod to be ready for picking (Choudhury, 1967). The green pod for vegetable purpose are harvested when they are immature and fully grown but still tender. As the harvest is delayed, the total yield increases but the quality falls rapidly due to over maturity of pods, fibre development and rough surface.

About 2-3 pickings are made for bush bean, french bean while it may be 3-5 for pole beans. Picking is usually done by hand. In advanced countries, where manual labour is costly machine harvesting is done by mechanical pickers under once-over harvest system. Dry beans are harvested when a large percentage of pods are fully ripe and most of the remainder have turned yellow. Harvesting is done either by hand or by machine. Beans are required 1 and 2 weeks to cure, after which they are threshed either by bullocks or by machine.

YIELD

French bean yield vary considerably with the area, fertility of soil and the variety. The yield of green pods vary from 180-250 q/ha in bush varieties and 100-120 q/ha in pole type. The yield of dry beans or bean seed varies from 20 to 25 q/ha.

POST HARVEST HANDLING

After harvesting, pods are sorted out immediately then damaged and broken pods are rejected from the lots.

Grading

Green pods of french bean is arranged in three grades like 'super', 'fancy' and commercial grade. Good size pods of uniform shape with tenderness before development of seeds are kept in 'super grade' and fetch more price in the market. The medium size with good texture and colour before bulging of pods or before full development of seeds inside the pods, are arranged in 'fancy grade', which is preferred by processors for the purpose of processing i.e., canning. The large pods with developed seeds inside and dull colour and textures of different shapes are kept in 'commercial grade'.

But the seeds of dry bean (pole beans) are graded as per their seed size in following grades. i.e., 'extra fancy', 'fancy', 'extra standard'. 'Top standard' and 'standard'. Fancy grade and extra standard are preferred for processing industry (Silbernagel and Drake, 1978).

Packing

The french bean pods are packed in basket, cardboard boxes, corrugated fibre board boxes and wire bound boxes for distant marketing. For local market pods are packed in gunny bags or bamboo chips round baskets or sticks baskets and top is provided with cushening materials like paddy straw or green grasses and stitched properly.

Storage

The french bean can be kept in storage for several days at about 40°F and at humidities above 90 per cent. Even at 40°F it can be kept better. When beans are transported long distances during warm weather they are shipped in refrigerated cars. Hydrocooling the beans immediately after harvest (Thompson and Kelly, 1972).

SEED PRODUCTION

The seed production of french bean is akin to the production of dry bean (Rajmash). For seed crop it should be grown in such season so that they become ready for harvest during rain free period.

Land requirement

Select field on which the same crop was not taken in the previous season, unless the crop in the previous season was of the same variety and was certified. The land should be well drained.

Isolation distance

French bean flowers are cleistogamous, but they are selfcompatible and self pollinated, although some chance of cross-pollination (1.1%) have also been recorded (Qualiotti *et al.*, 1980) particularly in the tropics where the activity of insects (including thrips) is greater (Drifhout, 1981). Isolation distance of 50 metres for foundation seed between two cultivars, is maintained. It is 25 metres for certified seed crop.

Brief cultural practices

The cultural practices like sowing, nutritional requirement, interculture and plant protection measures are same as described in crop production section.

Roguing

Roguing is performed at four stages. They are :

Before flowering

The first roguing is done before flowering. This is based on plant habit, vigour, height according to type (bush type or pole type), leaf shape and colour. Besides, severely affected plants by diseases (Barrons, 1939) particularly seed borne diseases.

At flowering

At this stage, plants are removed on the basis of vigour and flower colour, and plants affected with seed borne diseases.

At pod developing stage

At this stage, plants are removed based on pod characters such as pod shape, colour and plants affected with seed-borne diseases.

At maturity

At this stage, late flowering and late maturing off-types can easily be detected, which are removed.

Harvesting, Threshing and Storage

During seed ripening stage, irrigation is stopped. Harvesting is done when pods are fully ripe and have turned yellow, i.e., about to shatter. Crop is harvested manually or by machine. The harvested plants are staked for 7-10 days for drying . Fully dried matter is threshed either by bullock or by threshing machine. All possible care should be taken that seeds are not injured because french bean seeds are susceptible to cracking. Seeds are cleaned against wind and spread for drying. Well dried seeds are packed according to market demand in polythene containers. Packed seed is stored under dry clean and well ventilated place. Culinary quality, colour, viability and longevity of beans are affected by storage environment. Seeds remain viable two to three years.

Seed yield

The average seed yield of french bean is about 12 to 18 quintals per hectare.

REFERENCES

Ali, M. and Kushwaha, B.L. (1987). Cultivation of rabi rajmash in plains. *Indian Farming*, 37 (2): 20-23.

Alim, M. (1989). Response of french bean genotypes to population densities during winter season. *Indian J. Pulse Res.*, 2(2): 125-128.

Anonymous (1953). Fertilizer for flower and vegetable production. ICAR, New Delhi.

Anonymous (1971). New IARI Vegetables varieties. *IARI Exten. Bull. No. 2,* Indian Agricultural Research Institute, New Delhi.

Barrons, K.C. (1939). Natural crossing in beans of different degrees of isolation. *Proc. Am. Soc. Hort. Sci.*, 36:637.

Bhagat, N.R., Kawalkar, T.G., Bhatia, H.P. and Hardas, M.W. (1979). Plant Introduction experiments with tender-podded French bean. *Veg. Sci.*, 6(2): 92-97.

Chakor, I.S. and Rana, K.S. (1992). Effect of irrigation and sowing date on yield of spring planted kidney bean (*P. vulgaris* L.) *Indian J. Agron.*, 37(3):590-591.

Chandra, R., Rajput, C.B.S., Singh, K.P. and Singh, S.J.P. (1987). A note on the effect of nitrogen, phosphorus and rhizobium culture on growth and yield of french bean cv. Contender. *Haryana, J. Hort. Sci.*, 16: 145-147.

Chauhan, D.V.S. (1968). Vegetable Production in India. Ram Prasad and Sons, Agra.

Chauhan, K.S. and Mangal, J.L. (1993). Agro-Techniques for legume vegetables. In Advances in Horticulture, Vol.5-Vegetable Crops,: Part-I (Eds: K.L. Chadha and G. Kalloo). Malhotra Publishing House, New Delhi.

Choudhury, B. (1967). Vegetables. National Book Trust, India, New Delhi..

Choudhury, B. (1970). Beans - Pulse Crops of India. Indian Council of Agricultural Research, New Delhi.

Dhuria, H.S. and Leela, D. (1970). *Annu. Rep.*, pp.69-70. Indian Institute of Horticultural Research, Bangalore.

Drifhout, E. (1981). Maintenance breeding of beans. *Proc. of FAO/SIDA. Technical conference on Improved seed production*. Nairobi, Kenya.

Duke, J.A. (1981). Handbook of Legumes of World Economic Importance, USDA, Bellsville, Maryland, Plenum, New York and London.

Gopalan, C., Ranashastri, B.V. and Balasubramanian, S.C. (1982). Nutritive Value of Indian Food. ICMR, Nat, Inst. of Nutrition, Hyderabad.

Gupta, J.P. and Gupta, J.N. (1983). Effect of grass mulching on growth and yield of legumes. *Agric. Water Management*, 6: 375-383 .

Gupta, O.P. and Singh, R.P. (1965). Preliminary observation on the effect of tenoran and epitam 6E herbicide in certain vegetable crops . *Indian J. Hort. Sci.*, 355.

Hawthorn, L.R. and Pollard, L.H. (1954). Vegetable and flower seed production. The Blakiston Company. Inc. New York. Toronto.

Jana, K.B. and Kabir, J. (1987). Influence of micronutrients on growth and yield of french bean cv. contender under poly house condition. *Veg. Sci.*, 14: 124-127.

Jones, H.A. and Rosa, J.T. (1928). Truck Crop Plants. Mc.Graw Hill Book Co. New York.

Kalloo, G., Rai, Mathura; Kumar, R., Prasanna, H.C., Singh, M., Kumar, S., Singh, B., Ram, D., Pandey, S., Lal, H., Pandey, P.K., Rai, S., Pandey, K.K. and Satpathy, S. (2006). New Vegetable varieties from IIVR, Varanasi, *Indian Horticulture*, 51(3): 22.

Kanwar, J.S. and Thakur, M.R. (1972). Effect of phosphorus and Zn on growth and flower initiation in four varieties of kidney bean. *Punjab. Hortic.*, 12: 162-170.

Kaplan, L. (1956). The cultivated beans of pre-historic South West. *Ann. Missouri Bot.Garden*, 43: 189-251.

Kulkarni, M.V. (1979). Yield and Quality of bean (*Phaseolus vulgaris* L.). Seeds produced in relation to rhizobium inoculation, nutrition and removal of early produced pods. *Thesis Abstr.*, 5: 45-46.

Leela, D. (1982). A decade of weed research in horticultural crops. *Pesticides*, 15(10): 3-8.

Medhi, A.K. (2000). Growth and Yield response of French bean (*Phaseolus vulgaris* L.) plant to foliar application of some plant growth hormones. *Orissa J. Hort.*, 20(2): 29-32.

Miranda, C.S. (1974). Evolutionary genetics of wild and cultivated *Phaseolus vulgaris* and *P. coccineus*, Ph.D. Theis, Univ. of Cambridge.

Nath, Prem; Srivastava, V.K.; Dutta, O.P. and Swamy, K.R.M. (2008). Vegetable Crops – Improvement and Production. PNASF, Bangalore. 110-113.

Pal, B.P., Sikka, S.M. and Singh, H.B. (1956). Improved Pusa Vegetables. *Indian J. Hort.*, 13(2): 64-73.

Pandey, G.K., Seth, J.N., Phogat, K.P.S. and Tiwari, S.N. (1974). Effect of spacing and P_2O_5 on the green pod yield of dwarf french bean variety Black Prince, *Prog. Hort.*, 6:41-47.

Patel, G.D. and Chavan, V.M. (1961). Inheritance of some characters in the field bean. *Indian J. Genet.* 21(2): 142-145.

Prabhakara, B.S., Naik, L.B., Mohan, N. and Varalakshmi, B. (2003). French bean (*Phaseolus vulgaris* L.). In Vegetables, Tuber Crops and Spices, (Eds:. S. Thamburaj and Narendra Singh). Indian Council of Agricultural Research, New Delhi.

Prabhakara, B.S., Naik, L.B., Mohan, N. and Varalakshmi (2003). Indian bean (*Dolichos purpureus* L.). In Vegetables, Tuber Crops and Spices (Eds.: S. Thamburaj and Narendra Singh). DIPA-ICAR, New Delhi.

Prasad, R.D., Singh, K.N. and Bondal, K.V. (1978). Effect of inter and intra row spacing and number of seeds per hill on yield of french bean in Kodai hills, *Ghana, J. Sci.* 16: 25-33.

Purseglove, J.W. (1968). Beans In.: Evolution of crop plants (Ed: N.W. Simmonds)p. 169.

Qualiotti, L., Lepori, G. and Raldi, A. (1980). Seed production (Eds. P.D. Hebblethwaite). Butter worths, London, Boston, pp. 569-584.

Rajendran, R. and Satyanarayana, A.(1988). Grow new French beans. *Indian Hort.* 33(2): 6-7.

Ram, Harihar and Singh, B.P. (1984). Pant Anupama: a French bean variety for Indian bean variety for U.P. hills. *Indian Farmers' Digest*, 17(4):19-20.

Saini, G.S. (1996). A Text Book of Vegetable Production.Aman Publishing House, Meerut (U.P.) India.

Salunkhe, D.K., Desai, B.B. and Bhat, N.R. (1987). Vegetable and Flower Seed Production. Agricole Publishing Academy, New Delhi (India).

Shanmugavelu, K.G. (1989). Production Technology of Vegetable Crops. Oxford and IBH Publishing Co. Pvt. Ltd. New Delhi.

Sharma, P.P. and Joshi, A.K. (1993). Improvement of French bean. Advances in Horticulture vol.5 - Vegetable Crops. Part-1 (Eds: K.L. Chadha and G. Kalloo), Malhotra Publishing House, New Delhi.

Silbernagel, M.J. and Drake, S.R. (1978). *J. Amer. Soc. Hort. Sci.*, 103:257-60.

Simmonds, N.W. (1976). Evolution of crop plants. Longman, London and New York.

Singh, D. N. and Tripathi, P. (1994). Effect of NPK fertilizers and Spacing on growth and yield of French bean. *Veg. Sci.* 21(1): 7-11.

Singh, D.N. and Behera, A.K. (1999). Response of French bean (*Phoseolus vulgaris*) to fertilizers and spacings . *Indian Journal of Agronomy* 44(3) : 597-600.

Singh, D.N. and Nath, Vishal (2011). Varieties and Hybrids of Vegetables. Satish Serial Publishing House, Azadpur, Delhi (India). 231-234.

Singh, D.N., Nandi, A. and Tripathi, P. (1993). Performance of French bean cultivars in north central plateau zone of Orissa. *Indian J. Agric. Sci.*, 63 (10): 658-659.

Singh, G., Bhan, V.M. and Tripathi, S.S. (1984). Weed control in French beans. *Indian J. Weed. Sci.*, 16: 124-126.

Singh, H.B. and Prasad, R. (1967). French bean – choice of varieties for north plains. *Indian Hort.*, 21(1): 21-22.

Singh, J.P. and Sharma, R.K. (1970). Effect of different sowing dates on yield of french bean . *Prog. Hortic.*, 2:39-44.

Singh, K.N., Prasad, R.D. and Tomar, V.P.S. (1981). Response of french bean to different levels of Nitrogen and phosphorus in Nilgiri Hills under rainfed conditions. *Indian J. Agron.*, 26:101-102.

Singh, M.P., Seth, J.N., Pandey, G.K. and Kuksal, R.P. (1975). French bean for the U.P. *Hills. Indian Hort.* 20: 17-19.

Singh, S.P. (2001). Seed Production of Commercial Vegetables. Agrotech Publishing Academy Udaipur (India).

Smartt, J. (1976). Tropical Pulses. Longman, London.

Sturtevant, E.L. (1919). Sturtevant's Notes on Edible Plants (Ed: U.P.Hedrick). New York State Agril. Exp. Sta. 27th *Annu. Rep.*, 2: 1-86.

Swarup, Vishnu (2006). Vegetable Science and Technology in India. Kalyani Publishers. New Delhi, pp. 656.

Swarup, V. and Gill, H.S. (1968). Mutations in French bean. *Indian J. Genet.* 28(1): 44-58.

Tandon, H.S.L. (1991). Fertilizer Recommendations for Horticultural crops. A guide book. Fertilizer Development and Consultation organization, New Delhi.

Thimmegowda, S., Fernades, I.J. and Krishnamurthy, K. (1974). Closer inter row spacing produced high yield in beans. *Curr. Res.*, 3: 156-157.

Thompson, H.C. and kelly, W.C. (1972). Vegetable Crops, Mc. Graw Hill Book Company, Inc. New York.

Vavilov, N.I. (1950). The origin, variation, immunity and breeding of cultivated plants. *Chron. Bot.*, 13:1-365.

Yarnell, S.H. (1965).Cytogenetics of Vegetable Crops - IV. *Legumes. Bot. Rev.*, 31: 247-303.

Yawalkar, K.S. (1969). Vegetable Crops of India. Agri. Hortic. Publishing House, Nagpur-1, India.

Zimdahl, L. (1980). Crop Weed competition. *A review*, Published in the United States by the International Plant Protection Centre, Oregon State University, Oregon, U.SA.

□□□

3. INDIAN BEAN

Botanical name	:	*Lablab purpureus* L. (Syn. *Dolichos lablab* L.)
Family	:	Fabaceae
Chromosome Number	:	2n = 2x =22, 24

Indian bean or Dolichos bean or lablab bean or hyacinth bean, also known as Common bean, field bean, Australian pea, Indian butter bean, Sem and wavy salad bean belongs to the family Leguminosae (Fabaceae). It is called in Western countries as Bonavist bean, Lablab bean, Sem bean and Egyptian bean. Dolichos is a Greek name meaning long pod and lablab is an Arabic or Egyptian name meaning dull rattle of the seed inside dry pod. It is an important leguminous pod vegetable of India and elsewhere in the tropics and in a few countries in temperates. It is also grown as a green manure crop and cover crop.

ORIGIN

Indian bean is supposed to have originated in India as wild forms of this bean are found in this country (Choudhury, 1967). From India it might have been introduce in China, Wsestern Asia and Egypt. According to Rao (1977), it is a native of India. Another view in that it is originated from South East Asia and Egypt.

DISTRIBUTION

Purseglove (1968) believed that the crop was taken to Africa and has subsequently been distributed to many tropical countries. It is widely distributed in China, Western Asia, Egypt and Sudan. In India it is grown throughout the country. It is grown as a field crop in Tamil Nadu, Andhra Pradesh, Gujarat, Madhya Pradesh and Maharashtra. The garden varieties are raise in Northern India but common in Southern, Western, Central and Eastern India.

It is commercially grown for pods used as vegetable and for dry seeds used as pulse . Indian bean is one of the popular protein rich vegetable legume crops in India. However, regional preferences are existed in the cultivation of Indian bean. For instance, green shelled seeds are mostly preferred in south India, whereas white pods are liked in eastern India. In north Indian Plains, people like green fleshy pods of Indian bean.

BOTANY

Indian bean (*Lablab purpureus* (L)) Sweet is perennial in bahit but cultivated as an annual or biennial. The plants are of two types, determinate and indeterminate. The determinates are bushy and indeterminates are vining or twining pole types.

It is a herbaceous perennial herb, often grown as an annual, 1.5-6 m tall; usually twining, but bushy forms also occur; very variable in colour of stems, foliage, flowers, pods and seeds and in shape of pods. Leaves alternate, trifoliate; petiole slender, laterally compressed, narrowly grooved above, swollen pulvinus at base. Inflorescences axillary, erect, long-stalked, stiff racemes. Flowers arising 1-5 together from prominent tubercular thickenings on peduncle; pedicels short, calyx 4 lobed, corolla white or purple; stamens 10, style curved with soft hair, stigma terminal, glabrous, capitate. Pods variable, obliquely oblong, often, curved and flattened, 5-15 x 1-5 cm, curved beak with persistent style; tuberculate along margin, 3-6 seeded. Seeds variable in size and colour, rounded, slightly compressed, white, cream, buff, reddish, brown or black or variously speckled; hilum white, prominent, raised, extending one-third distance around seeds. Weight of 100 seeds 25-50 g. The flowers are visited by insects so some cross-pollination probably occurs. Pollination and seed setting is reduced in cold weather (Purseglove, 1968).

Plant breeding

Though it is grown all over the plains of India, not much research work has been carriedout either to maintain the local useful strains or to improve upon the local strains for pod quality and yield (Nath *et al.*, 2008). The genetics of some of the economic characters have been studied (Cruz and Ponnaiya, 1969; Joshi, 1971; Rajendran and Sathyanarayana, 1981). A vine variety, 'Pusa Early Prolific' was recommended by IARI. It bears early long thin pods in bunches and is suitable for sowing both as early spring and autumn crop. Some of the promising strains have also been reported from coimbatore (Kunjamma and Rajashekaran, 1969). In Karnataka 'Hebbal Avare' of UAS, Bangalore is popular and grown commercially.

At IIHR Bangalore two line viz., 'IHR 93' and 'IHR 140' were reported to be rust resistant and high yielders with good pod quality. Breeding photo insensitive varieties both as vegetable and seed is gaining popularity and two varieties Arka Jay and Arka Vijay have been developed and recommended for commercial cultivation by IIHR, Bangalore (Nath *et al.*, 2008).

Floral biology

Indian bean flowers open between 9 to 17 hours, the peak period being 14 to 16 hours. The anther dehiscence occurs about two days earlier to flower opening (Pokle and Deshmukh, 1971). The secretion of the exudates starts at tip of the stigma four days before anthesis. It is maximum at the time of pollination after which it decreases gradually. Indian bean is self-pollinated but chances for natural crossing vary from 2 to 8 % (Agrawal, 1989; Kumar and Bhatanagar, 1989). Emasculation and pollination result in fruit set up to 33 % and is most successful when emasculation is carried out before noon and pollination is performed on the same or following days. Bagging after emasculation and pollination increase the percentage fruit set to 63; due to the protection of pollen and stigmatic exudates from the drying effects of the sun (Chowdhary *et al.*, 1989).

COMPOSITION

Indian bean is richer than French beans in its nutritive value. Its food value per 100 g of edible portion is given (Choudhury, 1967) below :

Protein	1.7 gm	Moisture	86.1 g
Minerals	0.5 gm	Fat	0.7 g
Other carbohydrates	4.5 gm	Fibre	1.8 gm
Calcium	50. mg	Calories	48
Iron	1.7 mg	Magnesium	34 mg
Potassium	129 mg	Sodium	55.4 mg
Sulphur	37.mg	Copper	0.13 mg
Thiamine	0.3 mg	Vitamin A	312 IU
Nicotinic acid	0.3 mg	Riboflovin	0.06 mg

Uses

The young pods and tender beans are popular vegetable in India and are also used elsewhere in the tropics for this purpose. The ripe and dried seeds are consumed as a split pulse in India. The beans are also sprouted, soaked in water, shelled, boiled and smashed into a paste, which is fried with spices etc. The haulms used for fodder as hay, silage and green manure. It is a soil improver crop (Thompson and Kelly, 1972). It also have medicinal used. The mature seeds, especially dark coloured ones, contain a 'trypsin' inhibitor and must be boiled before eating. The trypsin inhibitor is broken down by heat and a toxic cyanogenetic glucoside, soluble in cooking water is removed (Anonymous, 1979).

VARIETIES AND HYBRIDS

Indian bean has great variation in terms of plant habit, pod characters grown all over the country. Indian bean is very popular in north eastern India where large endemic forms and promising varieties have been identified (Parathasarathy *et al.*, 1987; Singh, 1989). Similarly in south and north India, certain varieties are grown.

In Indian bean two botanical varieties are recognized. They are *Lablab niger* var. lablab (var. typicus) (Syn. *Dolichos lablab* var. typicus) in which the long axis of the seed is parallel to the suture and *Lablab niger* var. *lignosus* (syn. *Dolichos lablab* var. *lignosus, D. lignosus* (in which seeds are at right angle to the suture). *D. lablab* var. typicus is a garden type and cultivated for its soft and edible pods. The variety lignosus is popularly known as carpet legume or field bean is a perennial, semi erect, hardy and drought resistant legume and mainly grown for dry seed as pulse and capable of growing in diverse soils. On the basis of growth habits, cultivars can be classified in to two groups, viz., the bushy field cultivars and twining pole garden cultivars.

Different cultivars respond differently to temperature and day length. The rainy season variety, if sown in summer, may give only vegetative growth.

The important cultivars grown in India are given below:

Pusa Early prolific	Babbal Avare-A	CO 10 Avarai
JDL 37	Deepaliwal	CO 11 Avarai
Kalianpur T-1	Val. Konkan-1	Rewa
Arka Jay	Val-125-36	Rajni
Arka Vijay	Pusa Sem-2	DB 1
Konkan Bhusan	Pusa Sem-3	DB-18
CO-1	CO1 Mochai	
CO-2		

The detail characteristics of varietires and hybrids described by Singh and Nath (2011) are given below :

Pusa Early Prolific

It is an improved vine variety. It bears early. The pods are thin and form in bunches. It is suitable for sowing in early spring and early autumn in the northern Indian Plains. It is recommended by IARI, New Delhi (Anonymous, 1998).

Arka Jay (IIHR Sel.1)

It is a dwarf bushy Dolichos bean, early in flowering (44 days) and pod maturity (another 19 days) are profitable for fresh supply and for processing industry. Average pod yield is 125 q/ha. It attain a height of 83 cm and do not reqire support or staking.

Arka Vijay (IIHR Sel 2)

It is also dwarf bushy Dolichos bean, take nearly 46 days for flowering and another 22 days for pod maturity. The average pod yield is 150 q/ha. It attains a height of 74 cm and do not require support.

Konkan Bhusan

It is a bushy type (60-75 cm height) bean that matures in 55-60 days after sowing and yielding 8-10 tonnes/ha. Pods are tender, green and stringless. A plant produces 125-180 pods, each weighing 3-4 g the pods is 10 cm long and 15 cm wide. The line DPL -D1 named as Konkan Bhusan. It was developed through cross between "Hebbal Avare 3" x Wal 1".

JDL-37

This variety is developed by Jawaharlal Nehru Krishi Viswavidyalaya at Jabalpur, Madhya Pradesh. It is higher yielder.

Kalianpur-T-1

This variety is pole type, developed by Chandra Shekhar Azad University of Agriculture and Technology, Kanpur, Uttar Pradesh.

CO-1

This variety developed from a Pure Line Selection from Coimbatore local. This was released in 1980 in Tamil Nadu. Plant height 60-70 cm, stem green, leaves trifoliate and valentines. Pods are flat, green when tender, turn into tan colour at maturity. Seeds are bold and black in colour. The duration of variety is 140 days.

CO-2

This variety was developed from the cross between Co. 8 x CO 1 and released in 1984 in Tamil nadu. This is photo insensitive, plants grow 60 cm tall, erect, bushy type. Flowers are pinkish purple and turn to bluish purple on fading. Pods are flat green when immature and turn to tan colour at maturity. Plant height 65-75 cm, determinate type. Flowers are white green in colour. Pods are green containing 2 to 3 seeds. The seeds are round, small sized and brown in colour.

Hebbal Avare-3

A derivative of the cross Hebbal Avare-1 x US-67-13, released in Karnatak in 1978; plant height 65-75 cm, erect and determinate growth habit, photo-insensitive, flowers of white colour, pods green, 2-3 seeded and seeds brown, round and small. Duration is 90-110 days.

Hebbal Avare-4

This variety is a determinate in habit. Variety is very early, start picking in 60 days and it yields about 60 q/ha of green pods.

Deepaliwal

This variety is a high yielding, pole type. Duration of maturity 90 to 95 days. This was developed at Punjabrao Krishi Vidyapeeth, Maharashtra.

Wal Konkon–1

This was developed from the cross between Wal 2-K_2 x Wal 125-36, released in Maharashtra for Konkan region. This variety is bushy type. The plant is having long tenderils. The leaves are smooth dark green in colour, days to flowering 50-60 days. Duration of the crop is 110-115 days. Seeds are brown in colour. The protein content is 14.68 per cent and resistant to yellow mosaic virus.

Wal-125-36

This variety is selection from local germplasm, released in Gujarat state. Plant height 60-75 cm, foliage green. Pods are borne in cluster. Seeds are milky white. The crop duration is 110-120 days.

Pusa sem-2

This variety has been evolved at IARI, New Delhi for cultivation in northern India. Pods are semi-flat, dark green, non-stringy, 15-17 cm long. The yield of green pods in field is about 14 t/ha and in trellies it is about 22 t/ha (Sharma *et al.*, 1997).

Pusa sem-3

This variety has also been evolved at IARI, New Delhi. This is recommended for cultivation in northern India. Pods are flat, green very tender, meaty, stringless, 15-16 cm long. About 17 t/ha green pods are obtained in field and about 28 t/ha in trellies. (Sharma *et al.*, 1997).

CO1 Mochai

It is high yielding, early culture with edible nature of green and dry bean. It is a selection from DL 2539 through single plant selection (Annappan *et al.*, 1989).

CO 10 Avarai

This culture has been evolved from a gamma irradiated seeds (24 kr) of a bushy variety 'CO6' (derivative of a cross between 'DL 244' (garden bean) and 'DL 3196' (field bean) and its progenies screened for green pod yield and quality. The culture is higher yielder than CO 9 and possess better values for sensory tests and biochemical analysis (Sivasubramanian *et al.*, 1989).

CO 11 Avarai

It is derived from cross between CO 9 (bush type) and white Yanaikathu (Pandal type), followed by selection during F_6 generation as Colt 22, After different yield trial finally named as CO 11 Avarai (Ramasamy *et al.*, 1990).

Rewa

It has a protein content of 25.11 per cent.

Some other superior cultures are CO 3 (DL 269), CO 4 (DL 453) and CO 5 (DL 692).

Hybrids

Some promising hybrids are CO 57 x DL 3196 (Higher yield, seed size and better pod quality), 6023A x 6010, 16019 x 6010 and 6014B x 6015 (higher ascorbic content). The hybrids are produced through hand-emasculation and pollination, making use of protruded stigma and use of selective differential gametocide like FW 450 (Peter, 1998).

Culture resistant to aphid

PLS 224, PLS 126, and PLS 121 (Rajsekharan *et al.*, 1983).

Resistant to pod borer

Cultivars PLS 24, PLS 16-1 (Chakravarthy, 1977).

Resistant to pod and seed damage

JDL 60, JDL 145 and JDL 18 (Parsai, *et al.*, 1989).

Resistant to root-knot nematode

Line IL 028, 29, 33 and 1949 (Azami and Patil, 1989).

CLIMATIC INFLUENCE

Indian bean is relatively a cool season crop. It is indeterminate in response to photo periods (Sheriff *et al.*, 1969), there being some short day and some long day types. Indian bean can be grown in both sub tropical and tropical climatic conditions. It is susceptible to frost and extreme hot weather. Though, there are varieties which are drought resistant. However, most varieties are adapted to temperature ranging between 18° and 30°C. Indian bean can be grown as dryland crop in regions with 630 to 890 mm rainfall (Chakraborty, 1993). It is observed that variety lignosus, popularly known as carpet legume or field bean, is a perennial, semi erect, hardy and drought resistant legume.

SOIL REQUIREMENT

Indian bean can be grown on a wide range of soils of average fertility (Singh, 1989). Soils should be rich in organic matter with good water holding capacity. The medium soils are best suitable. It grown luxuriantly in fertile soil on the cost of pod yield. The suitable soil pH is between 5.0 to 7.8. It cannot stand water logging conditions (Chakraborty, 1993).

SOWING

Indian bean is grown in the month of July-August. It can be grown early also in the areas where rains come early. Indian bean is raised as monoculture and also mixed with ragi or sorghum. Sometimes, it is grown with caster crop, in this way vine type of Indian beans also get support.

20-30 kg/ha seed sufficient to cover one hectare land. Seed should be treated with thiram @ 3 g or bavistin @ 1 g/kg of seeds.

Spacing of row to row 1.0 m and plant to plant 0.75 m is optimum for higher yield (Chakraborty, 1993). It is also sown at the spacing of 100 x 80 cm for seed production as a pure crop (Singh, 2001).

Land should be prepared thoroughly by ploughing 3-4 times and planking should be done to get fine tilth. There are two methods for sowing of dolichos bean *viz.*, Ridge sowing, by drilling or by dibbling and pit sowing. The pit size should be 45 x 45 x 45 cm.

2-3 seed are sown on ridges in hills or drilled in row at given spacing. In pit system, pits are dug at given spacing and filled properly then 3-4 seeds are sown in each pits and later thinning is done to allow two plant to grow per pit.

When it is raised as a mixed crop it is drilled between two rows of ragi or sorghum at 100 cm distance. Usually 2-3 seeds are sown per hill which later thinned to two or one healthy plant. When it is grown as mixed crop with sorghum, it twines on the sorghum stalk. The earheads of sorghum are harvested and then stalks are cut for animal fodder. Indian been has high fodder value.

In kitchen garden plants can be retained for 2-3 years. It is kept in summer to produce pod during the next season. Crop flowers in Nov-Dec. and are harvested in Jan and Feb. in frost free areas.

NUTRITIONAL REQUIREMENT

Although Indian bean is a legume crop it responds well to the fertilizers, about 25 tonnes of well composted organic manure is applied to the soil at the time of final preparation of the land. Application of 20 kg N, 60 kg P_2O_5 and 60 kg K_2O/ha is recommended (Prabhakara *et al.*, 2003).

Vijay (1972) reported that depending upon the fertility status of soil it needs 25-30 t of farm yard manure, 15 to 40 kg N, 50 to 75 kg P_2O_5 and 50 kg K_2O/ha. Complete dose of farm yard manure has to be incorporated at the time of field preparation. Half dose of nitrogen and whole dose of phosphorus and potassium is given as basal dose at the time of sowing. Remaining half dose of nitrogen is top-dressed 30-35 days or so after sowing.

Use of Chemical and Growth Substances

(i) Seed treatment with systematic chemicals zinc chloride solution gave the highest germination.

(ii) Seed soaking in water at 60°C for 10 to 30 minutes give maximum germination. Inoculation of rhizobium give higher yield about 11.5%.

(iii) A high rate of flower shedding a major factor limiting productivity of Indian bean. In some variety of Indian bean, it is also reported that only 10-20 per cent flowers develop in to mature pods. Fruit setting and final yield can be improved upto the great extent with the application of calcium chloride (0.1%) or NAA (100 ppm at flower initiation stage (Sharvashakar and Kulkarni, 1989).

IRRIGATION

Indian bean is commonly cultivated as a rainfed crop. The irrigation will be given sparingly in the case of irrigated crop. It cannot stand water logging. Hence, frequent irrigation should be avoided. Only occasional light irrigation may be given. It can tolerate comparatively dry conditions. However, during flowering optimum moisture may reduce flower drop which is reported to be more in this crop. Varietal character may also be one of the important reasons of flower drop and pod set. It is essential to maintain available moisture during flowering and pod development stage (Agrawal, 1989).

INTERCULTURE

Frequent intercultivation is done during early stage of crop to keep field free from weeds and to make soil loose for proper aeration of the root zone. Once the vine is spread in the field the interculture operation is not possible. Before vining, once or twice hoeing is done. It will control weeds and will enhance crop growth.

Staking

As soon as vining starts, plants should be trained by providing stakes with the help of thin bamboo/forest wood so that they can grow properly. Staked plants give more fruits and higher yield.

Intercropping and crop rotations

Choudhury (1967) reported that in central and south India bean is grown either as pure or mixed with sorghum. According to Saini (1996) the following crop rotations are common with Dolichos bean (Indian bean).

1. Dolichos bean - tomato
2. Dolichos bean - brinjal
3. Dolichos bean - chilli
4. Dolichos bean - okra
5. Dolichos bean - cowpea
6. Dolichos bean - cluster bean
7. Dolichos bean - bitter gourd
8. Dolichos bean - Watermelon
9. Dolichos bean - Muskmelon
10. Dolichos bean - maize.

WEED CONTROL

Shallow cultivation during the early stages of crop is necessary to check the weeds and to facilitate earthingup. Usually hand weeding is done to control the weeds of bean crop. Weeding is taken up after each hoeing before application of fertilizers. Weeds uprooted carefully by hands so the root of the plants will not disturb during early stage. A pre sowing application of fluchloralin @ 2 litre/ha checks the weed growth for 20-25 days. Afterwards the basin and pits are kept clean and weed free. At the later stages of crop growth, the weeds are kept under check due to the thick canopy of the crop (Prabhakara, *et al.*, 2003).

PLANT PROTECTION MEASURES

Insect Pests

Important insect pests are bean aphids, pod borers and jassids. Details of symptoms and control measures are given below:

Bean aphids (Aphis crassivora)

This pest causes characteristics damage to the tender aerial parts of the plants and produces symptoms like curling of leaves, twisting of twigs and developing fruits and sometime shedding of flowers. The infestation may assume serious portion at any stage of crop growth. It can be controlled by application of systemic organophosphatic insecticides viz., spraying with 0.05% phosphomidon or 0.02% methyl demeton.

Pod borer (Heliothis armigera)

It is serious pest of the pods in some parts of the country. The young caterpillars first-feed on the surface of the pods, bore into them and feed on the seeds. It can be effectively controlled if the egg masses, caterpillar and pupae are hand picked in the beginning of the infestation. Application of granules of aldicarb, carbaryl and carbofuran are also effective to control the pod borers. Application of quinalphos or phorate each @ 1.25 kg a.i./ha is also beneficial to check the pod borer infestation.

Bean weevil (Brochus sp.)

This beetle infests the pods when the crop is in the field and later feeds on the seed in the stored condition. The weevils can be killed by fumigation of/ with carbon disulphide at the rate of 1.5 to 4.0 kg to each 28.31 m^2 of space. Heating the beans upto 54.4°C for 30 minutes also destroys the beetle. If the bean seeds are to be used for seed purpose only, they can be mixed with 3 per cent DDT and then stored.

Jassids (Empoareca fabae and E. kraemeri)

They are also been found to be serious pests which produce typical chopper burn symptoms on crop when severely infested. It can be controlled by spraying with 0.05% phosphomidon or 0.025% methyldemeton.

Diseases

Anthracnose

It is caused by *Collectotrichum lindemuthianum*: This is most common diseases of beans and most severe in high rain fall subtropical to temperate areas than in tropical areas. The fungus is seed borne. The most characteristics symptoms of the disease are dark brown or black sunken spots like cankers on the pods. The lesions remain isolated by yellow-orange margins and under high humidity dull, salmon coloured ooze is formed in the centre. Similar spots are also found on the cotyledons and stem of young seedlings and when severe can cause seedlings mortality.

It can be controlled by (1) Use of healthy seeds, clean cultivation, avoidance of overhead irrigation. (2) Use of sulphur fungicides like thiram, dithane Z-78 and the systemic fungicides like benlate or bavistin give satisfactory control.

Powdery mildew

It is caused by *Erysiphi polygoni*, a common disease of beans and other legume vegetable and pulses. White powdery growth occurs on leaves, spreading to cover the stem and other plant parts. In severe cases, the entire plant dries up. Dusting with powdered sulphur spraying of wettable sulphur preparations (3g/l) or kelthane or dinacap (1 ml/l) are recommended for control.

Leaf spot

It is caused by *Corcospora cruenta* is a fairly wide spread disease. Lower leaves are first-affected . Water soaked lesions develop on the leaf lamina which soon turn reddish brown to brown. The spots are somewhat irregular being bound by veins. At later stage the spots appear ashy-grey with purplish borders. Such lesions may also develop on the pods. It can be controlled by spraying of copper oxichloride or organic sulphur fungicides.

Bacterial blight

It is caused by *Xanthomonas phaseoli*, which is most important bacterial disease of Indian bean and other beans. The disease is characterized by irregular, sunken, red to brown leaf spots surrounded by a somewhat narrow yellowish halo. Several spots may coalesce to form irregular patches. Premature leaf fall occurs consequent to infection. Seed treatment with organomercurial fungicides (emisan, mercuric chloride) and antibiotics like streptocycline, plantomycin or agrimycin 100 ppm is effective to check the disease.

HARVESTING

Harvesting is done after two and half month to three months of sowing and the plants continues to give pods throughout winter and spring months. Pods are harvested when they are green and succulent and have not become fibrous.

YIELD

The yield ranged between 150 to 250 quintals per hectare of green pods.

POST HARVEST HANDLING

Grading

The grading of green pods is done according to size, shape, texture, lusture, colour and tenderness to fetch high price and easy sale in the market. The superior type of pods with uniform size, shape and colour kept in 'super grade'. The medium size of pods with good texture, freshness and colour are arranged in 'fancy grade'. The big size pods with dull colour and developed seed inside the pods are arranged in 'commercial grade' and kept separately.

Packing

The green pods are packed in gunny bags/big round bamboo chips baskets and tops is stitched properly and send to local market for sale. Pods are also packed in fibre boxes, cardboard boxes, wire bound boxes and plastic crate and in plastic bags for transport to distant markets for sale.

Storage

To keep the pods fresh they should be stored in cool place in shade or in well ventilated room. Pods are cover with wet gunny bags or water may be sprinkled occasionally over them to keep fresh. Pods also can be stored in cold storage at temperature of 0.6°-1.7°C with 97% relative humidity for three days in good condition.

SEED PRODUCTION

Indian beans are self pollinated and are partially cross pollinated by insects. It is usually grown with cereals or other supporting crop. For seed production it should be grown as a pure crop with staking of the vines.

Isolation distance

Indian bean is mainly self-pollinated. Natural crossing, however is up to the extent of about eight per cent. The seed field is isolated from fields of other varieties and from same variety not conforming to varietal purity by at least 50 m for foundation seed production and 25 m for certified seed production (Agrawal, 1989).

Brief cultural practices

The cultural practices like, sowing, manurial requirement, interculture and plant protection measures are same as described in crop production chapters.

Roguing

It is important to rogue seed crop carefully to maintain genetic purity. Such characters as foliage colour, plant type, flower and pod characteristics, etc. should be carefully checked and off-types removed from the field at pre-flowering, flowering and maturity stage. In addition to off types, the plants affected by disease namely bacterial blight, anthracnose, ascochyta blight and bean mosaic should also be invariably removed time to time from the field (Agrawal, 1989).

Harvesting and threshing

Harvesting is done when a large percentage of pods are fully ripe and most of the remaining have turned yellow. Harvesting should be started before lower pod become dry enough to shatter. Harvesting is done usually by hand.

The crop after harvesting, is left in the field to dry for about seven to ten days. Later, it is threshed by bullock or by thresher. While threshing, care should be taken to keep the mechanical injury to minimum. After threshing and cleaning the seed should be dried to moisture content below 9 per cent before storage.

Seed yield

On an average, 12-15 quintals of dry seed yield of Indian bean from one hectare is obtained. Seed yield is improved by a foliar spray of 0.5 ppm triacontanol at flowering which decreases flower drop to a great extent (Jadhav *et al.*, 1987). High seed yield is expected from an ideal high yielding plant type having 3-4 inflorescences. Each inflorescence is with 4 nodes and the time interval between development of successive inflorescence is short (Prasad, 1984).

Seed storage

Indian bean with 9.0 per cent moisture can be stored under ordinary conditions for longer period of time. Germination of seeds stored for 40 months in polythene bags have approximately double that of those stored in cloth bags. None of the seed treatment prevented loss of germinability during storage, but treated seeds produced seedlings with increased root length, low values for electrical conductivity and free amino acids (Vanagamudi *et al.*, 1986).

REFERENCES

Agrawal, R.L. (1989). Seed Technology, Oxford and IBH publishing Co., New Delhi.

Annappan, R.S., Sivasubramanian, P., Rangaswamy, P., Sundrapandian, G., Ramanathan, T. and Iyemperumal, S. (1989). Col Mochai – A high yielding field lab lab. *Madras Agricultural Journal*, 76: 640-2.

Anonymous (1998). Varieties and Hybrids developed by Division of Vegetable Crops, IARI, New Delhi.

Anonymous, (1979). Tropical Legumes – Resources for future. National Academy of Sciences, Washington D.C., pp.279.

Azami, M.I. and Patil, B.D. (1989). Effect of root-knot nematode on carpet legume, *Dolichos lablab* var. *lignosus* with a note on screening of resistant lines, *Indian J. Nematol.* 18: 104.

Chakraborty, A.K. (1993). Lablab bean. In: Vegetable Crops (Eds: T.K. Bose, M.G. Som,. and J. Kabir) Naya Prokash, Calcutta.

Chakravarthy, A.K. (1977). Pod borer resistance in *Lablab niger* (L.) cultivars with special reference to the pod borer., *Adisura atkinsoni*. (Moore) [(*Lepidoptera noctuidae*)], M.Sc. (Ag.) thesis submitted to University of Agricultural Sciences, Bangalore.

Choudhury, B. (1967). Vegetables. National Book Trust, India, New Delhi.

Chowdhary, A.R., Ali, M. and Quadir, M.A (1989). Aspects of pollination and floral biology of lablab beans (*Lablab purpureus* (L) sweet). *J. Japanese Soc. Hort. Sci.* 58(3): 665-671.

Cruz, R.D. and Ponnaiya, V.T.S. (1969). Inheritance of pod and seed colour and pod shape in garden bean. *Indian J. Genet.*, 29(1): 139-141.

Jadhav, B.B., Patil, B.A. and Patil, V.H. (1987). Effect of tri-contanol on lablab bean and Indian mustard. *Indian J.Agric. Sci.* 57(1): 56-58.

Joshi, S.N. (1971). Studies on genetic variability for yield and its components in Indian beans (*Dolichos lablab* var. *lignosus*). *Madras Agric. J.*, 58(3): 367-371.

Kumar, M. and Bhatanagar, S.P. (1989). Stigmatic surface and exudate in Lablab purpureas. *Phytomorphology*, 39(1): 97-102.

Kunjamma, Harishi, V.K. and Rajashekaran, V.P.R. (1969) CO-2, CO-3, CO-4 and CO-5 the four economic strains of garden lablab (*Lablab* niger var. *typicus*). *Madras Agric. J.*, 56(7): 475-479.

Nath, Prem; Srivastava, V.K., Dutta, O.P. and Swamy, K.R.M. (2008). Vegetable Crops – Improvement and Production, PNASF, Bangalore, 114-115.

Parathasarthy, V.A., Medhi, R.P. and Singh, B. (1987). Varietal evaluation of Dolichos bean. Annual report ICAR, Research Complex for North Eastern Hill Region, Shillong, pp.73-75.

Parsai, S.K., Rawat, R.R. and Gupta, M.P. (1989). Relative susceptibility of field bean varieties to bean beetle (*C. phaseoli* Gyllihal), *Agric. Sci. Digest* (*Karnal*) 9(3): 136.

Peter, K.V. (1998). Genetics and Breeding of Vegetables. Indian Council of Agricultural Research, New Delhi.

Pokle, Y.S. and Deshmukh, N.N. (1971). Anthesis and pollination in garden bean (*Dolichos lablab* L.) Nagpur College of *Agriculture Magazine* 44:17.

Prabhakara, B.S., Naik, L.B., Mohan, N. and Varalakshmi (2003). Indian bean (*Dolichos purpureus* L.). In Vegetables, Tubercrops and Spices. (Eds: S. Thamburaj and Narendra Singh). DIPA - ICAR, New Delhi.

Prasad, T.G. (1984). Physiological investigations into causes of low productivity in field bean (*Lablab purpureus* (L) Sweet). *Mysore J.Agric. Sci.*, 18(1): 75-76.

Purseglove, J.W. (1968). Tropical Crops: Dicoty ledons. John Wiley, New York.

Rajendran, R. and Satyanarayana, A. (1981). Annual Report. Indian Institute of Horticultural Research, Bangalore.

Rajsekharan, B. and Babu, P.C.S. (1983). Screening of lablab varieties for resistance to aphid (*Aphis craccivora* Koch), *South Indian Hortic.* 31:50.

Ramasamy, P., Balasubramanian, M., Gnanam, R. and Rangasami, P (1990). Co11 Avarai, (*Lablab purpureus* var. typicus (L.) Sweet. *Madras Agricultural Journal*, 77: 121-4.

Rao, M. (1977). Genetic studies in lablab [(*Lablab purpureus* (L)Sweet] for green pod yield and its components. M.Sc. (Ag.) thesis, Tamil Nadu Agricultural University, Coimbatore, India.

Saini, G.S. (1996).Vegetable Production, Aman Publishing House, Meerut (UP), India.

Sharma, R.K., Tewari, R.N. and Pachauri, D.C. (1997). Pusa Sem 2, and Pusa Sem 3: Promising lablab beans. *Indian Hort.* 41(4): 46.

Sharvashakar, G. and Kulkarni, R.S.C. (1989). *Indian Hort.* 33 and 34 Jan-March and April-June, p. 24-27.

Sheriff, M.N., Rajagopalan, C.K. and Venugopal, K. (1969). Pod setting studies in *Dolichos lablab* L. var. typicus. *Madras Agric. J.*, 56(10): 661-662.

Singh, B. (1989). Performance of pole type Dolichos bean. Annual Report ICAR Research Complex for North Eastern Hill Region, Shillong, p.78.

Singh, D.N. and Nath, Vishal (2011). Varieties and Hybrids of Vegetables. Satish Serial Publishing House, Azadpur, Delhi (India). 235-238.

Singh, S.P. (1989). Production Technology of Vegetable Crops, A.R.C.S., Karnal (Haryana), India.

Singh, S.P. (2001). Seed Production of Commercial Vegetables. Agrotech Publishing Academy Udaipur, India.

Sivasubramanian, P., Annappan, R.S., Iyemperumal, S., Sreerangaswamy, S.R. and Pandiyan, M. (1989). CO10-A dual purpose lab lab. *Madras Agricultural Journal* 48: 144-154.

Thompson, H.C. and Kelly, W.C. (1972). Vegetable Crops, Tata Mc. Graw, Hill Publishing Company Ltd., New Delhi.

Vanagamudi, K., Karivara tharaju, T.V. and Ramakrishan, V. (1986). Seed storage studies in field bean (Lablab purpureus). *News Letters* of the *Association of Official Seed Analysts,* 60(1): 78-85.

Vijay, O.P. (1972). *South Indian Hort.*, 22 (2): 12-13.

□□□

4. BROAD BEAN

Botanical name	:	*Vicia faba* L.
Family	:	Fabaceae
Chromosome number	:	2n=12

Broad bean is an important winter legume vegetable. It is also known as faba bean (Chauhan and Mangal, 1993), horse bean, field bean, tick bean and windsor bean which is fourth most important legume crop in the world after beans, peas and chick pea (Filippetti and Ricciardi, 1993). It is the only bean and winter legume which can withstand sufficiently cold temperature.

ORIGIN

Broad bean is one of the oldest cultivated plant. The native place is mediterranean region or south-western Asia, and is closely, related to *V. pliniana* Trabut, which grows wild in Algeria. However, De Candolle (1967) indicated, either north Africa or the south Caspian Sea to be the place of origin of broad bean, while Egypt was referred to by Abdalla (1979). It was reported by Cubero (1974) and Ladizinsky (1975) that the mediterranean basin and the near and middle eastern regions are the probable site of origin of broad bean.

DISTRIBUTION

Broad bean is widely cultivated since pre-historic times and have been found in the remains of lake-dwellings in England and Switzerland. Broad bean was grown by the ancient Egyptians, Hebrews, Greeks and Romans. It was an early introduction into western Asia and reached China in remote times (Thompson and Kelly, 1983). Broad bean was the only edible bean known in Europe in pre-Columbian times and was introduced into new world after 1492 (Purseglove, 1965). The major countries are China, Ethiopia, Latin America, Egypt, Mexico and Brazil where broad bean is widely cultivated. Broad bean (*Vicia faba* L.) is an important rabi season pod vegetable of northern plains of India (Baboo and Kumar, 1996). However it is minor legume crop of the country and also grown in localized area of Kashamir, Rajasthan, Uttar Pradesh, Punjab, Haryana, Karnataka, Madhya Pradesh and Bihar.

BOTANY

Broad bean is glabrous, annual herb.The plants are erect and fairly tall (30 to 180 cm) with heavy, stiff-stems. Well developed tap-root to 100 cm or more with strong laterals; small roots with large clusters of small lobed nodules.

Stems stout, hollow, square, winged at angles, with 1-7 branches. Leaves pinnate, leaflets 2-6, sub-opposite or alternate. Inflorescences short axillary racemes, 1-6 flowered. Flowers fragrant, 2.5-3.7 cm long, stamens short diadelphous: style short bent, with tuft of hairs near tip; stigma terminal. Pods are glossy green, stout sub-cylindrical or flattened, beaked 5-10 cm long in field cvs. up to 30 cm long in garden cvs, fleshy with white velvety lining when young, tough and hard at maturity and contain 5 or 6 large seeds. Seeds very variable in shape and size, strongly compressed to nearly globular, white, green, buff, brown, purple or black, 1-2.6 cm long; hilum prominent, seeds are used as green shell beans or dried for storage. Weight of 100 seeds 40-180 g and 15 cm or more in length (Purseglove, 1968).

Taxonomy

Muratova (1931) classified *Vicia faba* L. in two subspecies with different maximum number of leaflets per leaf: faba (eufaba in her nomenclature) with more than four leaflets per leaf, and paucijuga, with fewer than four . The subspecies faba, on the basis of seed size, included three botanic varieties: *major, equina and minor*. In addition, botanic varieties were sub-divided into sub-varieties on the basis of geographical distinction. The species is taxonomically isolated from the genus, and although there is evidence of successful intercrosses in the genus, this is not the case for *Vicia faba* L. (Yamamoto, 1980).

According to Gunn (1978), *faba bean* differs from the other *Vicia* species on the basis of lack of tendrils and hilum aspect. The same section also includes *Vicia narbonensis, V. galilea* and *V. hyaeniscumus,* but following cytogenetic and electrophoresis analysis any close relation between these species and *Vicia faba* L. was excluded (Ladizinsky, 1975).

According to Purseglove (1968) some authorities divide it into 2 subspecies:

1) Sub sp. *paucijuga* Mur., comprising several Indian cvs.
2) Sub-sp. *eu-fava,* which occurs both in Europe and Asia, is sub-divided into 3 varieties:

a) Var. *Minor* Beck, tick peas or pigeon beans, with small rounded bolster-shaped seeds;
b) Var. *equina* Pers., horse beans, with medium-sized, bolster-shaped seeds, with an average length of 1.5 cm are also known as field bean.
c) Var. *major* Harz., the broad bean used for culinary purposes, with large broad flat seeds, with average length of 2.5 cm, and with longer pods.

COMPOSITION

Broad bean is rich in carbohydrate, protein, minerals and vitamins. The edible parts contain moisture 85.4 g, carbohydrate 7.2g, protein 4.5 g, fat 0.1g, calories 48 ,Vit.A14 I.U., thiamin 0.08mg, ascorbic acid 12.0 mg, calcium 50 mg and iron 1.4 mg (per 100g of edible portion) as reported by Hazra and Som (2006) . The dried beans contain about water 14.3%; protein 25.4%, fat 1.5%, carbohydrate 48.5%; fibre 7.1% and ash 3.2% (Purseglove, 1968).

Uses

The pods of broad bean are used as green, generally as cooked vegetable either singly or mixed with potatoes in curry form. It is also used in the preparation of mixed pickle with other vegetables. Boiled seeds of broad bean can also be consumed after applying salt and chilli or black pepper powder. Its pollen cause allergic and also green pods to some people. An illness some times fatal and known as "favism" is caused (Choudhury, 1967).

VARIETIES AND HYBRIDS

Germplasm Resources

Filippetti and Ricciardi (1993) reported that the International Board for Plant Genetic Resources (IBPGR) has developed a network of plant genetic resources centres, each with the responsibility of maintaining the world collection of a particular crop. The International Centre for Agricultural Research in the Dry Areas (ICARDA) has a world mandate for the improvement of faba bean. Since its inception, ICARDA has one of the largest faba bean germplasm collections for the continuous improvement of this crop. Faba bean germplasm is maintained at ICARDA in two collections:

1. International legume Faba Bean (ILB) collection (3305 accessions) containing the original germplasm accessions as received, usually heterogenous and heterozygous to a great extent.

2. Faba bean pure line (BPL) collection which has been derived from the ILB collection by a pre breeding process of cyclic single plant progeny rows.

The collection expeditions have also taken place in Cyprus, Egypt and Afghanistan. Collection of Vicia faba have also been made in the Mediterranean region by the Germplasm Institute, CNR, Bari, Italy and in the Ethiopia by Ethiopian Genetic Resources Centre.

The collection stored in Bari includes material from Algeria, Greece, Spain, Tunisia, Ethiopia, Italy, Egypt and Libya (Perrino, 1985). In Cyprus, the Agricultural Research Institute in collaboration with FAO and IBPGR collected

98 accessions of faba bean (Della, 1981). Afghanistan, Ethiopia, Germany, Turkey, Ciprus, Spain, Canada, UK, Morocco, Egypt, Iraq, Italy, Syria, and the former USSR are the main countries of origin of the major collections of broad bean germplasm.

There is no improved variety available in India, but several exotic varieties introduced from abroad (Singh and Nath, 2011) are :

Long pod type

Aquadule Claudin, Imperial White long pod, Masterpiece green long pod, Imperial green long pod, Red Epicure.

Windsor type

Imperial White Windsor, Gaint Four seeded Green Windsor, Imperial Green Windsor.

Local types

Bakla local it is a selection from local population grown in North Uttar Pradesh.

Synthetic varieties

A synthetic is a population of plants artificially produced by the breeder. A synthetic variety is an advanced generation of a seed mixture of selected genotypes (inbred lines, clones, hybrids etc.), maintained for a limited number of generations by hybridization in all combination among them.

Throw's MS

It was the first synthetic faba bean to be grown commercially in the U.K.

Maris Beagle

It is another English synthetic variety composed of five inbred lines and released in 1974.

Soravi and Avrissot

These synthetic varieties were released in France; the first is an example of international cooperation between the Plant Breeding Institute of Cambridge (UK) and the National Institute for Agricultural Research (INRA-France). Now most of the North European faba beans are synthetic varieties (Filippetti and Ricciardi, 1993).

CLIMATIC INFLUENCE

Broad bean is a temperate crop. It can be grown towards the edge of tropics as a winter crop. It is not suited to the low hot tropics where it may flower well but usually produces no pods (Purseglove, 1968). It is grown in summer and rainy season in different parts of the country. It has hardy plant and is the only bean which is grown as an autumn and a winter crop as it can withstand cold temperatures as low as 4°C. It is mainly grown at higher elevation. Seeds can not germinate in cold. The best temperature for pod setting are 15-25°C and at very high temperature during rainy season, plant drop their flowers and pods.

SOIL REQUIREMENT

The crop of broad bean can be taken on all type of soil from light sandy to heavy clay. It does not grow well on extremely acid soil. It prefers rich, well drained loamy soil with pH range of 6.5-7.5 for high yield (Choudhury, 1970). It tolerates salinity to some extent.

SOWING

Sowing time

The sowing time of broad bean is July- September for rainy crop and January-February for summer crop. In hilly area best time of sowing is from March to first week of May.

Land preparation

Crop needs a thorough preparation of land because of their nitrogen fixing capacity,(Choudhury, 1968). Field should be thoroughly ploughed 2-3 times with soil turning and desi ploughs, followed by one to two harrowings to remove the weeds.

Seed rate and spacing

About 70-100 kg seed of broad bean is required to cover one hectare land (Choudhury, 1967). Spacing of row to row is given 40-60cm and plant to plant 15-20cm for higher yield. Spacing of 45 cm x 15 cm is also advisable for higher yield. Soaking of seeds in water for 24 hours before sowing not only increases the germination also make faster to the rate of germination.

Seeds are sown in hills by dibbling, using 2-3 seeds /hill. Sowing is done in lines by dibbling at given spacing (Chauhan and Mangal, 1993). The seeds are also sown in shallow furrows of 15cm width formed at a spacing of 75 cm. In each furrow two rows of seeds are sown at a spacing of 25 cm in a zig zag manner along the furrow. It can also be sown in a single row system with a spacing of 45 x 15 cm (Prabhakara *et al.*, 2003).

Table 1: Effect of nitrogen (with or without seed inoculation) and phosphorus rates on pod and green shoot yield (q/ha), fibre content and net returns of broad bean.

Treatment	Pod yield (q/ha)		Green shoot yield (q/ha)		Crude fibre content (%) in green pods		Net return (Rs./ha)	
	1993-94	1994-95	1993-94	1994-95	1993-94	1994-95	1993-94	1994-95
N kg/ha								
0	28.4	30.7	59.6	64.5	1.0	1.1	13,864.50	13,244.50
Inoculation	38.8	40.4	81.5	84.8	1.5	1.6	18,959.75	15,709.25
35	33.3	35.2	69.9	73.9	1.4	1.5	16,694.34	14,565.34
Inoculation + 35	50.2	52.2	105.4	109.6	2.1	2.2	24,806.84	20,687.84
70	43.8	42.2	93.0	98.9	1.7	1.8	20,140.68	17,626.43
Inoculation +70	48.5	50.1	100.8	101.2	1.9	2.0	22,754.18	19,413.93
CD at 5 %	4.8	5.2	8.6	9.6	0.3	0.4	-	-
P_2O_5 kg/ha								
0	30.2	32.6	63.3	68.4	1.1	1.1	13,768.70	13,564.25
30	39.4	40.7	827	85.5	1.5	1.7	16,666.67	14,450.50
60	48.8	50.8	102.5	106.7	2.0	2.1	23,768.55	20,605.75
90	43.6	45.1	91.6	94.7	1.8	1.9	20,824.44	16,517.00
CD at 5 %	3.2	3.5	5.7	6.4	0.2	0.3	-	-

NUTRITIONAL REQUIREMENT

Broad bean is a legume, and as such only phosphatic and potassic fertilizers are generally applied. But it is a vigorous legume and makes such rapid growth that often a nitrogenous fertilizer gives a good response. The farm yard manure is applied @ 10 t/ha alongwith NPK at the rate of 20:50:40 kg/ha (Prabhakara *et al.*, 2003).

Baboo and Kumar (1996) reported that nitrogen is essential for vegetative growth of plants thereby the yield of green pods of broad bean and the shosphorus affects the maturity of the crop. Rhizobium inoculation accelerate the activity of nitrogen fixing bacteria and helps to fix abundant quantity of atmospheric nitrogen. The seed inoculation by rhizobium culture helps in nitrogen economy for the broad bean. In their study they found that application of seed enoculation plus 35 kg N/ha and 60 kg P_2O_5/ha alongwith uniform dose of 35 kg K_2O/ha significantly increased pod yield, shoot yield, fibre protein and phosphorus content in pod. Further, more return was obtained (Table 1).

The fertilizers should be placed in bands about 7 to 8 cm to the side of the seeds and slightly deeper than the seeds. One third of nitrogen with full dose of phosphorus and potash is applied as a basal dose before sowing. Though remaining nitrogen is given in two equal split at 25 and 50 days as top dressing after hoeing and weeding, then earthing is done to keep plant stable.

IRRIGATION

Broad bean is a deep rooting crop extending root system beyond 100 cm depth. This crop connot withstand water logging. Hence need well drained soil. Plant should be supplied with sufficient water to maintain a uniform growth.Either an excess or deficiency of water should be avoided (Hegde, 1993). First irrigation is given just after sowing of the seeds for better and quick germination. This is followed by light irrigation on the third day. Full bloom stage is the most responsive to water although in dry seasons, very low moisture from planting time until flowering and during early stages of growth can reduce vegetative growth (Choudhury, 1967). Pod set is dependent on the maintenance of favourable moisture supply to the stigmatic surface. It may be desirable to irrigate at 10-15 days interval during cool season.

INTERCULTURE

Broad bean require regular interculture operations to keep the weeds under check. and to provide a good condition for a crop growth and proper utilization of nutrients by crop. Generally, 2-3 intercultural operation is given i.e., at 20-25 and 40-45 days stage. It should be done before application of fertilizer and after that earthing is given to keep the plant in proper position and for better root and shoot growth (Singh and Singh, 2007).

Staking

The plants of tall varieties need staking and it is given by bamboo twigs or wooden sticks at an distance of one metre down both sides of the double row close to the plants. The plants are tied around the stakes with *sutuli* or twine 30-60cm above the ground. The side shoots arising from the base of the stem are removed to encourage straight erect growth of the plants (Prabhakara *et al.,* 2003).

WEED CONTROL

Inadequate weed control reduces yield so it is necessary to check the weeds during the early stage of crop growth. Shallow cultivation is given to control the weeds and to facilitate earthing. Usually, 2-3 hand weeding is needed till the plants can cover well and to prevent weed competition with the crop. In commercially cultivated crop weedicide application is economical. A pre-sowing application of fluchloralin @ 2 litre/ha check the weed growth up to 20-25 days during vegetative phase of crop.

PLANT PROTECTION MEASURES

Some of the important pests are pod borer, aphids, agromyzid fly and plant hoppers and diseases are powdery mildew, phytophthora, rust, septoria leaf spot and anthracnose. Possible control measure are :

(a) Spray the crop with phasphomidon (0.5 ml/litre) or monocrotophos (1.25 ml/litre) on appearance of aphids.

(b) Spray endosulfan or malathion (2 ml/litre)or carbaryl (2g/litre) at flowering to control pod borer.

(c) Spray the crop with wettable sulphur (3g/litre) or dinacap (1 ml/litre) when powdery mildow and rust is noticed.

(d) Spray indofil M-45 (2g/litre) to control fungal diseases like anthracnose and septoria leaf spot.

HARVESTING

The broad bean pods should be harvested before thay become tough and stingy and seeds become large enough to cause the pods buldge around the seeds.Generally, pods are harvested about 3-4 month after sowing at green shell stage before they are full grown and while seeds are small. At market maturity, the green pods are slightly rounded about 2 cm in width and 15 cm or more in length. As harvest is delayed, the total yield increases but the quality falls rapidly due to over maturity of pods, fibre development and rough surface. Picking is usually done by hand.

YIELD

The yield of green pods ranges from 70-100 quintals/ha and green bean yield, varies from 18–20 quintals/hectare.

POST HARVEST HANDLING

Just after harvesting, green pods are sorted out, after that diseased, borer affected, damaged and deformed pods are rejected from the lots. So that produce can give better price in the market.

Grading

The grading is an important aspect of beans marketing. The produce is generally sorted into different grades and attractive forms to fetch additional returns. Green pods are arranged in three grades i.e. 'super', 'fancy' and 'commercial'. Good size, tender pods with inform shape and glossiness are arranged in 'super grade' to get more return in the market. The medium size pod with good texture and colour before, bulging and full development of seeds in side the pods, are arranged, in 'fancy grade'. The dull colour pods with uneven size and shape, free from any deformity are kept in commercial grade to facilitate marketing (Singh and Singh, 2007).

Packing

It is an important consideration for transporting to produce to reduce their losses and maintain their freshness, succulence, and quality for longer period. The edible pods are packed in basket card board boxes, corrugated fibre board boxes and wire bound boxes for distant marketing. Green pods for vegetable purpose are also packed in gunny bag or bamboo chips round baskets or in Arhar sticks baskets for local market and top is provided with cushening materials like paddy straw (Singh and Singh, 2007) or dry grasses or paper cuttings and stitched properly so that produce can reach in good condition to the market.

Storage

The green pods for vegetable purpose can be kept in ordinary room condition for 2-3 days. In storage it can be store for 1.5-2 weeks at 4.4°-7.2°C temperature with 90-95% relative humidity.

SEED PRODUCTION

Broad bean is a self-pollinated crop, and as such, seed production is not a problem. For foundation seed production the varieties should be grown apart at a distance of 50 m and for certified seed production they should be 25 m apart.

The seed crop of broad bean is cultivated in the same way as grown for green pods. In many countries the crop in also grown for its seeds. The pods on drying shatter. Hence, care should be taken to see that dry pods are harvested when they turn brown. Unless the area under the crop increases sizeably the demand for seed would be less. As it is still grown mostly as a backyard crop, the home gardeners can leave a few pods from their crop for seed (Singh and Singh, 2007).

Seed yield

Haq (1982) suggested that seed yield improvement could be achieved by pure-line selection or mass selection of indigenous and exotic genetic resources and by hybridizing with selection within hybrid lines and the success would depend on the selection of parent and their adaptation to any given environment. About 1 to 1.5 tonnes of seeds can be produced in one hectare.

REFERENCES

Abdella, M.M.F. (1979). The origin and evolution of *Vicia faba* L., in *Proc. Ist Mediterranean Conf. Genet.* 713.

Baboo, R. and Kumar, S. (1996). Quality and pod yield of Broad Bean (*Vicia Faba,* L.) as influenced by nitrogen (with and without inoculation) and phosphorus applications. *Vegetable Science,* 23(1): 16-20.

Candolle, A.De. (1967). Origin of cultivated plants, *2nd edn. Report, Hafner,* New york, 1986.

Chauhan, D.V.S. (1968). Vegetable production in India. Ram Prasad and Sons, Agra.

Chauhan, K.S. and Mangal, J.L. (1993). Agro. techniques for legume vegetables. In Advances in Horticulture Vol.5 - Vegetable Crops: Part 1. (Eds. : K.L. Chadha and G. Kalloo). Malahotra Publishing House, New Delhi-110 064, India.

Choudhury B. (1967). Vegetables National Book Trust, India New Delhi.

Choudhury, B. (1970). Beans-Pulse crops of India. Indian Council of Agricultural Research, New Delhi.

Cubero, J.I. (1974). On the evolution of *Vicia faba* L., *Theor, appl. Genet.* 45-47.

Della, A. (1981). Report on the exploration and collection of broad beans (*Vicia faba*) in Cyprus, *Plant Genetic Resources Newsl.* FAO, Rome.

Filippetti, A. and Ricciardi, L. (1993).Faba bean (*Vicia faba* L.). In Genetic Improvement of Vegetable Crops (Eds.: G. Kalloo and B.O. Bergh). Pergamon Press, Oxford, U.K.

Gunn, C.R. (1978). Genno *Vicia* with notes about Tribe Vicieae (Fabaceae) in Mexico and Central America, USDA Tech. Bull., No. 1601, USDA, Washington.

Haq, N. (1982). *Pflanzenzuchtung,* 88: 1-12.

Hazra, P. and Som, M.G. (2006). Vegetable Science. Kalyani Publishers, New Delhi.

Ladizinsky, G. (1975). On the origin of the broad bean *Vicia faba* L., *Isra, J. Bot.*, 24: 80.

Muratova, V. (1931). Common beans (*Vicia faba* L.) *Bull. Appl. Bot. Genet.Plant Breed.*, Suppl. 50.

Perrino, P. (1985).Conservazione delle risorse genetiche vegetali in Italia: *Realizzazioni e programme, in Risorse Genetiche Vegetali: Autonomia di* Sviluppo, Centro. Internazionale CROCEVIA-Roma.

Prabhakara, B.S., Naik, L.B., Mohan,N. and Varalakshmi (2003). Broad bean (*Vicia faba* L.). In Vegetables, Tuber Crops and Spices. (Eds.: S. Thamburaj and Narendra Singh). DIPA-ICAR, New Delhi.

Purseglove, J.W.(1968). Tropical crops : Dicotyledons. Vol. 1, ELBS, Longman, London.

Singh, D.N. and Nath, Vishal (2011). Varieties and Hybrids of Vegetables. Satish Serial Publishing House, Azadpur, Delhi, India. 220-221.

Singh, D.N. and Singh, S.P. (2007). Production Management of Underutilizred Vegetables. Agrihortica Publications; "Shipra" I.V. Nagar Road, Girivihar Society, Gandhidham, Junagadh-362001 (Gujarat) India. 122-127.

Thompson, H. C. and Kelly, W.C.(1983). Vegetable Crops. Tata Mc. Graw-Hill Publishing Co. Ltd., New Delhi.

Yamamoto, K. (1980). On the interspecific hybrid progenies between *Vicia pilosa* M.B.and *V. amphicarpa* Dorth, Jpn. *J. Breed.* 30: 351.

□□□

Chapter-6

Root Crops

1. RADISH

Botamical name	:	*Raphanus sativus* L.
Family	:	Crucifereae
Chromosome number	:	2n=18

Radish is one of the important root vegetable which is cultivated for its enlarged edible root. It is a quick growing and short duration vegetable crop. Being a quick growing crop, it is easily planted as a companion crop or inter crop between the rows of other vegetables. It may also be planted on mends or ridges, separating one plot from another. It is cutivated all over India, especially near the city markets. Cultivation of radish both for root and seed is very easy. It is most suitable for raising in kitchen garden or home garden.

Due to introduction of several varieties which require dilfferent type of climatic conditions, it is possible to grow radish all the year round under north Indian plains. Radish is grown as off season crop during summer in mid as high hills.

ORIGIN

The recent study indicates that east of mediterranean, most probably china was the centre of origin of this crop. Werth (1973) reported that European type likely to be original in the area of east mediterranean sea to Caspian sea. Indian group of radish including rat tail type is believed to have originated in the area of their distribution though their ancestors have been lost (Bailey and Bailey, 1956; Mansfield,1959). The small European or table type radishes originated later than the large ones (Helm, 1957). White globular form were developed in 18th century and later on the red ones were also developed.

Importance

Radish is important root vegetable mostly consumed as a salad, mainly for its pungent taste. Young leaves are also cooked and eaten as vegetable. The rat-tail or mougri Radish (*R. sativus* var. *caudatus*) does not produce the characteristics fleshy root, but is grown for its long slender pods which are eaten raw or cooked. The Indian cultivars are usually white with conical shape of about 30cm length and tapering tip and the roots are pungent to mildely pungent depending upon the season of growing. Roots produces in early autumn season of Oct-Nov are usually pungent, while those of winter season Dec.-Feb. are mildly pungent to sweet. Radish is a common vegetable during winter season consumed by all sections of population and as short duration crop, it is extensively grown all over the country.

The acclimatisation of introduced verietis, mainly sub-toprical ones, has taken place in north india the prefence has always been to flavoured and pungent varieties. Some popular introductions are the Japanese white and Mino Early and the table varieties like white Icicle, Rapid Red, White tipped and Scarlet Globe are grown in small areas, while some native varieties like Pusa Desi, Pusa Rashmi, Pusa Chetki etc. are grown in large areas around the cities. There are now the varieties available for round the year production in north India. In eastern India, pink coloured roots are preferred by consumers, like Chinese Pink variety. There is a popular variety "Jaunpuri"/Newar" in eastern U.P., which produces white long and thick roots weighing about 5-10 kg. Even though China and Japan have a large range of radish varieties suiting to sub-tropical to temperate conditions, the tropical radish with heat tolerance, capable of forming roots even under day temp. of 35°C, is a significant feature of adaptation of asiatic races in India. Black radish a semi-domesticate, is found in north Indian hills and has been used in breeding spring radish variety called Pusa Himani (Seshadri and Chatterjee, 1996).

Radish contains glucose as major sugar and small amount of frutose and sucrose. Pectin and pentosans are also reported to be present. The characteristic pungent flavour is due to volatile isothiocyanates (trans - 4-methyl-thiobutenyl isothiocyanate), and the colour of the pink cultivars is due to the anthocyanin pigments.

History and Adaptation

It is one of the ancient vegetable, known in Egypt from early times (around 2000B.C.) and inscriptions of the inner walls of pyramids show that radish was cultivated as an important vegetable. It later spread to china around 500 B.C. and to Japan in 700 A.D. (Sirks, 1957). In china and Korea there exist the oldest records on domestication dated to be 400 and 100 B.C. respectively.

However radish is not known in wild state. It was earlier assumed to have been evolved from *R. raphanistrum* L. a widely distrubured weed in Europe. The area of maximum diversity is the region from the eastern Mediterranean to Caspian sea. The origin of the cultivated radish is not clear. It is surmised that it would have originated by natural or artificial crossing between *R. Laldra* and *R. maritimus* (Kitamura, 1958 and Tsunoda, 1991).

The introduction of Radish in India has not been Chronicled. 'Mulaka' is the Sanskrit equivalent for radish in ancient India. Radish was considered as a medicinal plant in Puranas (Sensarma, 1989). In the Sutra period (800 B.C.-300B.C.), the mention of Radish roots is made as aiding digestion when munched after meals. In 'Susruta Samhita' the medicinal treatise of 3rd-4th century A.D. radish has been recorded having medicinal Properties (Seshadri and Chatterjee, 1996).

DISTRIBUTION

It has spread to china about 500 B.C. and to Japan A.D. 700. The crop was cultivated from the thirteenth century B.C. in Egypt. This crop was cultivated in Europe in the fifteenth or Sixteenth century and was introduced to America at this time. It is also popular in Greeks, where it was highly prized and its cultivation was practiced at the time of the pharaohs. Radish is one of the most ancient vegetables. Inscriptions on the inner wall of the pyramid show that Radish was an important vegetable in Egypt. Also cultivated in Syria, Greece and Roman. It has spread throughout the world. It is widely cultivated in northern and southern plains as well as in the hills and almost all states of india. It is most suitable for raising in kitchen garden or home garden.

BOTANY

Radish, *Raphanus sativus* L. belongs to the family Crucifereae. It is a quick growing herbaceous annual. The enlarged edible roots are fusiform botanically which may differ in colour from white to red. There are two distinct genetical groups in radish. The Asiatic varieties which are primarily for tropical climates, produce edible roots in the first season and seed in the second season as a biennial crop, whereas the exotic or European. varieties produces roots in the plains of tropical and subtropical climate and seeds in the hills of temperature climate.

The leaves are rosette covered with stiff bristles. The fleshy root of radish is modified root (Fusiform) develop from both the primary root and the hypocotyl (Choudhary, 1983). Roots vary greatly in shape, size,length, time of root formation, length of time they remain suitable for making salad. The skin colour of radish is white, pink, red, black and various other shades. The flesh of radish is always white in all varieties.

Radish is having two distinct genetical groups:

(i) European or Temperate Types

They are early (20 to 30 days or so), roots are small in size less pungent, cannot tolerate high temperature. This type of radish are not grown in tropical and sub-tropical climate but for seed production, they are grown in the hills. The root are most suitable for salad making.

(ii) Asiatic or Tropical and Subtropical Types

These types of radish are grown for one season for root production and for one more season for producition that is the biennial for seed production. The roots are pungent in taste. They can tolerate high temperatures. These types are suitable for both vegetable and salad purpose.

Plant breeding

Radish is a cross-pollinated crop. The commonly adopted procedures for breeding are mass selection, recurrent selection, hybridization and heterosis. The F_1 hybrids are produced by single crosses, or double crosses. Transgression of useful traits is possible by wide interspecific crosses between *Raphanus* and *Brassica* species by protoplast fusion and genetic engineering (Nath *et al.*, 2008). These aspects are important for developing commercial cultivated species of the crop. Some studies on Pollen morphology and physiology in radish have been reported (Nath and Mehta, 1969). Different useful strains of Asiatic types have been reported from different regions of the country. A good number of temperate and Asiatic varieties were also introduced and some of them have been grown as commercial varieties for decades, but their seed production has been possible only in Kashmir, Kulu Valley and Nilgiri Hills. Usually the Asiatic types are more pungent than the temperate ones. Attempts have been made to improve upon the Asiatic strains by crossing with the exotic varieties to obtain new varieties with roots qualities of temperate types and ability to produce seeds in the plains(Singh *et al.,* 1971; Dutta and Vijay, 1970). An important achievement in this direction has been the release of the new radish variety 'Pusa Himani' (Singh *et al.,* 1971). Emphasis has also been laid on developing varieties resistant to diseases. Dutta and Vijay (1970) reported radish strain 'IIHR-1' to be highly resistant to white rust (*Albugo candida*). 'Arka Nishant', was the variety developed at IIHR, Bangalore in 1972 which produces high yield with the better quality roots in the southern region. Male sterile line has also been used for hybrid seed production in other countries (Ogura, 1968). In radish Ogura sterile cytoplasm is widely distributed in wild Japanese radish. Muthukrishnan and Arumugam (1977) reported that the root length is the most important contributory factor for yield.

Taxonomy

According to Banga(1976) there are four types of radish commonly cultivated in various region of world :

(i) Small cool season radish; (ii) Large radish with wider range of temperature adaptation; (iii) Rat-tail or mongri radish forming no fleshy roots but forming long slender (20-60 cm) pods. (iv) Fodder radish also producing no fleshy roots. All types of radish belong to the species *Raphanus sativus* L. with 2n=18, and botanically they are known as varieties radicula, niger, mongri (or Caudatus) and oleifera, respectively. The first two of which are grown for their tuberous roots, where as oleifera is grown primarily for the oil in its seeds (Kay, 1987). All above varieties intercross freely and also hybridize with wild *Raphanus* spp. The taproot (except in var. mongri) is swollen and varies from almost globular, about 1-2 cm in diameter in the salad types to as much as 1m long and 15 cm in diameter, cylindrical or conical in shape, in the oriental types and weighting upto 15 kg. The flesh is normally white, though in some, may be pink or red. In the salad radish, the skin is usually red (occasionally white); in the oriental radish it is normally white. Cultivated radishes are put under 3 different species based on plant morphology and ecological requirements.

(i) The indian types which are heat tolerant, pungent and mostly used for pickle making are called *R. indicus*.

(ii) The European types which are frost tolerant, rich in Carotene, long-day plants and strictly annuals are called *R. sativus* L.

(iii) The Japanese types which are intermediate to Indian and European types, growing in the coastal regions of Japan and neighbouring regions, and the turnip shaped giant radishes cultivated in the sakur-juna Island of Japan are called *R. raphanistroides*. *R. caudatus* L., the non tuber forming rat-trail radish is closely related to Indian types of radishes. The indian type of radishes got evolved in the Near Eastern and Indo-Burma centre of Origin. Its presumed ancestor has been lost.

The European types originated directly or through hybridization from a few wild species like *R. maritimus*, *R. landra* and *R. rostratus*. The Japanese types are derived from *R. sativus* f. *raphanistrodis* (syn. *R. raphanistroides*).

COMPOSITION

Radish green is very rich in vitamin A,B and C with little of carbohydrates, iron and protein. It is eaten both raw as salad or cooked in various ways. It has a cooling effect and it increases the appetite. It has some medicinal values too. 100g of edible roots contain 93.7% water, 4.2% carbohydrates, 1.1% protein, 0.1% fat, 0.7% fiber, 30IU vitamin A, 24 mg vitamin C, 37 mg calcium, 31 mg phosphorus, 260 mg potassium, 15 mg magnesium. 37 mg sulphur and chlorine (Aykroyd 1941). Trace element in raidsh include aluminium, barium, lithium, manganese, silicon, titanium, fluorine and iodine (upto 18 mg/100g). Pink skinned raidsh is generally richer in ascorbic acid than the white skinned one. Vitamin C content of raidsh root is greatly influenced by light condition. Radish contains glucose as the major sugar and smaller quantities of fructose and sucrose. Pectin and pentosans are also reported to be present. The leaves of radish are good source of extraction of protein on a commercial scale (Joshi, 1971 ; Pirie, 1971 and Bagchi *et al.*, 1977) and the radish seeds are a potential source of a non-drying fatty oil suitable for soap making, illuminating the edible purposes. The pods of the rat-tail radish contain moisture 92.3%, Protein 1.3%, fat 0.3%, fibre 1.1%, carbohydrate 4.3% and minerals 0.7%. They also contain 7.8 mg calcium and 2.4 mg phosphorus per 100g of edible portion.

Uses

Fleshy roots are eaten raw or as salad or cooked. The roots of radish are also cooked as vegetables. Radish top is cooked as leafy vegetable and is very rich in minerals and vitamin A and C. The pickle morabba and raita is also prepared from raidsh roots. Radish preparations are useful in liver and gall bladder troubles. The seeds are said to be peptic, expectorant, diuretic and carminative (Purwal, 1957). Radish is considered to be useful for patient, suffering from piles, liver trouble, enlarged spleen and jaundice. It is relished for its pungent flavour and is considered as an appetizer. Radish has refreshing and depurative properties. In homeopathy, radish roots and leaves are used for neurolagic headaches, sleeplessness and chronic diarrhoea. Leaves, roots, flower and pods are active against fram-positive bacteria. Radish also been used as a fodder crop in some countries like U.K. and South Africa. The rat tail or mongri radish, which is very similar of common radish, but does not produces the characteristic fleshy root is grown for its long slender pods, which are eaten raw as salad or cooked as vegetables.

VARIETIES AND HYBRIDS

Singh and Nath (2011) reported detail description of varieties and hybrids of radishes which are given below :

Arka Nishant

It is an improvement over a collection (IIHR-2) from Singapore, developed after 10 cycles selection and released for commercial cultivation in 1980 by IIHR, Bangalore. It is Asiatic variety and matures within 45-55 days. Roots are medium-sized (25cm x 3-4 cm), marble white, crisp texture, pleasant aroma, free from early bolting, pithiness, splitting and forking and each root weighing 300-400 g. Root and shoot leagth ratio is 1:1 with mild pungency. It yields 200-300 q/ha.

CO-1

It is a selection from germplasm type (RS-44-2-1) and released from TNAU, coimbatore in 1981. The roots are milky white, long (22cm), thick (12.5 cm girth) and each weight 220g. The roots contain 10.9 mg/100g of vitamin C per 100g and 1.20 per cent crude fibre. It bolts and sets seeds in the plains and hence very easy for the farmers to produce seeds for themselves. Roots have better consumer's preference. Top to root ratio is 0.90. It is susceptible to rust and aphids. This variety is suitable for growing round the year under different cropping systems. The crop duration range from 40 to 45 days. It yields 200-250q/ha.

Chinese Pink

It is an introduction identified by Dr. YS Parmar University of Horticulture and Forestry, Nauni (Solan). Roots are 12-15 cm long, semi stumpy to stumpy roots, pink with white colour towards the tip and takes about 50-55 days from sowing to root formation. It is a good cultivar for the hills but does well in plains with mild climate. It requires low temperature for seed production.

IIHR-1-1

Roots are thin long (30cm), each root weighing about 300 g and free from pre-mature bolting, pithiness, splitting and forking and mildly pemgent. Root surface is smooth and shining white. It is highly resistant to white rust and takes about 45 days to reach harvest maturity.

IIVR 1 (Kashi Sweta)

It is developed through selection from Pusa Chetki material, suitable for growing in spring, summer, autumn and winter season. It is ready for harvesting in 30-35 days after sowing. Roots are 25-30 cm long, 3.5-4 cm in diameter, straight tapering with blunt tip. This variety yielded about 500q/ha edible roots.

IIVR-2 (Kashi Ganga)

It is suitable for planting from September to February and harvesting within 40-50 days after sowing. Due to late pithiness, it can stand the field upto 10-15 days after ready for harvesting. Its foliage is soft and smooth. Roots are stright and tapering types, 35-40 cm long and 3.5-4.2 cm diameter, This variety yielded 450q/ha.

Japanese White

It was original named as Shiroaguri-kyo and released by IARI, New Delhi under the name Japanese white. Roots are 25-30cm long, 5cm in diameter, cylindrical and blunt at the tip. Skin is pure white and smooth. Flesh is snow white, crisp, solid and mild pungent. Tops are medium large with deeply cut leaves. It matures in 45 days. It is suitable for growing between October-December in the plains and July-September in the hills.

Kalianpur No. 1

It is a selection from local material released in 1982. Stem and foliage are green, smooth, top heavy, leaves long, broad and less lobed. Roots are 22-23 cm long, smooth, crispy, and white with green shoulder, thick and tapering. It is free from maustard sawfly, aphids and white rust.

Nandauni

This is developed from a local material and very popular in Himanchal Pradesh. The roots are long, tapering, smooth and light pink in colour. The tops are dark green with cut leaves.

Punjab Ageti

It is suitable for April to August sowing. Roots are long, thin with lower half-white and upper red. This variety gives an average yield 315q/ha.

Punjab Pasand

It is suitable for main season. Roots are long and white semi-stumped. The average yield of this variety is 535q/ha.

Punjab Safed

It is a derivative of cross white 5 x Japanese white. Roots are white, tapering smooth, mild in taste, free of forking, 30-40cm in length and 3.5 cm thick. It is a quick growing type with roots remaining edible for 10 days after attaining full size. It takes 50-60 days for root formation.

Pusa Chetki

It is a selection made from a material collected from Denmark (Choudhary and Sirohi, 1975). Roots are medium large and stumpy, pure white, tender and smooth and mildly pungent. It becomes ready for harvesting 40-50 days after sowing. Since it can tolerate high temperature, therefore, it has been found suitable for sowing from the middle of March to the middle of August. The cultivar sets seeds profusely in the plains because it bolts very early in the month of late October or early November. It may be grown as a summer and monsoon crop from April- Septeamber. During summer sowing, though the yield will be comparatively low, but it is compensated by the high price, it fetches at that time.

Pusa Desi

It is a tropical and subtropical cultivar, suitable for sowing from middle of August to early October in the northern plains. It is selection from local material. Roots are pure white, 30-35cm long, tapering with green stem end, pungent and heavy yielder, tolearnt to slightly higher temperature, medium light green top with cut leaves. It matures 50-55 days after sowing.

Pusa Himani

It is developed through hybridization between a temperate type (Black) and a popular Asiatic Type (Japanese White) and released from IARI, Regional Station Katrian (Singh *et al.*, 1971). It is suitable for December to February sowing in the plains, when no other variety can form such good roots. It is the only variety, which can be grown throughout the year in the hills farming. The roots are 30-35cm in length and 10-12 cm in girth with green stem and they are semi-stump to tapering with short tops. The skin is pure white. The flesh is crisp and sweet-flavoured with mild pungency. It matures in 55 days.

Pusa Reshmi

It is a main season variety in Asiatic group and suitable for mid September to early October sowing. Roots are 30-35 cm long, tapering white with green shoulder, mildly pungent, tolerant to slightly higher temperatures medium light green top with cut leaves. Roots are ready for harvesting in about 55-60 days.

Rapid Red White Tipped

It is an extra early European variety, which matures in 25-30 days. It is bunching and table type. Roots are small, globular, bright red with white tip. Flesh is pure white, crisp and snappy. Tops are short, mildly pungent in flavour and suitable for November-December sowing in the plains.

Scarlet Globe

Foliage is 10-15 cm tall. Root is round small, 2cm in diameter and bright red. Flesh is crisp and white. It takes 25-30 days from sowing to root formation and belongs harvesting will result in pithiness.

Scarlet Long

It is an early variety, having light green foliage with 15-20cm length Roots are long and tapering to a point. Flesh is white and crisp.

White Icicle

It is a medium- short European variety that matures in 30 days. the skin is pure white, thin and tender, whereas the flesh is white icy, crisp,juicy, mild and sweet flavour with juicy enough pungency to appeal the appetite. The root is solid Icile shaped straight and tapered. it has given a good harvest in the plains. Delay in harvesting leads to pighiness, suitable for October and November sowing in the plains.

Hisar Muli No. 1

It is developed at HAU, Hisar and is more popular in Haryana. Its roots are straight, white and ready to harvest after 50-55 days of sowing. It is sown from September to October.

Long White (F_1)

It can be grown both in high land and lowlands. Excellent root quality, uniform appearance with white skin with upright and dark green foliage. Root 600-800g each, highly uniform and 22-26 cm long. Tolerant to heat and drought. Resistant to Downy mildew, Fusarium yellow and other diseases. Mature in 45-50 days and developed by Tropica Green Co. seed company.

Longo (F_1)

Slow bolting hybrid with excellent root enlargement at low temperatures. Highly uniform roots of 35-40 cm long, 7-8 cm in diameter and 1.0-1.2 kg in weight. It is developed by Tropica Green Co. seed company. Resistant to fusarium wilt. Suitable for fresh market.

Shogunmino (F_1)

This F_1 hybrid developed by unicorn seeds, roots have beautyful white skin, very crisp and jusy texture. Roots uniform, long (30-40 cm), very slow in pithiness development and can grow well und high temperatures.

Summer Queen (F_1)

It is developed by Sungrow Seeds. Roots straight and tapering, white coloured 30 cm long, late in pitiness. Matured in 55 days performed best in summer and rainy season.

Hill Queen (F_1)

It matures in 55-65 days after sowing, roots straight, slightly tapering, lustrous white, 40-45 cm long very late in pithiness. Suitable for planting in Oct.-Nov. Roots have very little fibre, fetches premium price in the market and developed by sangro seeds.

Komal (F_1)

Roots cylindrical and stump, white, upper part some times has greenish tinge, length 20-25 cm, very late in pithiness. It is developed by Sungro seeds specially for areas where leaves are eaten, roots are less payment, leaves soft and palatable, suitable for planting from February to May.

In India there are many local varieties such as Jaunpuri giant in Jaunpur, Uttar Pradesh and Newari, Bombay red, Nadauni, Baramasi in Himanchal Pradesh. Though radish requires relatively low temperatures during growth period, it is possible to grow throughout the year. It has varieties suitable for different regions with different agroclimatic conditions. In southern and western parts of the country Asiatic radish is sown almost throughout the year. i.e. January-February and June-September. The European types can be sown during October-December. Like wise, in northern parts, starting with April season (with Pusa Chetki), raidsh can be sown in mid August to October (Pusa Desi), in mid-September to November (Pusa Reshmi) and in mid December-February (Pusa Himani). Singh *et al.* (2002) reported a new high yielding early Asiatic veriety Kashi Sweta. The roots are 25-30 cm long, 3.5cm diameter, straight, tapering and blunt at tip; shoot length: root ratio (3:3.2) with reduced pithiness as compared to Pusa Chetki. Even after 5-10 days of the harvesting stage the roots do not develop pithiness, thereby prolonging the marketablility without deterioration of root quality. It is ready for harvesting 30 days after sowing. It can be grown round from August to February in northern plains of India (Nath *et al.*, 2008).

CLIMATIC INFLUENCE

Radish is found growing all the year round due to different types viz. temperate type (European type) and tropical type (Asiatic type). Varieties having wide range with respect to temperature tolerance due to their ability to acclimatization. However, temperature range from 10 and 15°C is desirable to produce goood flavoured, textured and good sized roots. Small rooted varieties particularly temperate type when grown in hot weather result in tough and extremely pungent roots. Long rooted varieties perticulary of tropical type (Asiatic type) are hardy and therefore, they can be grown under high temperature and during rainy season, these varieties produce profuse flowering and fruiting and ultimately higher seed yields.

It is a cool season crop, but the Asiatic varieties can resist more heat than the European of temperate varities. It attains best flavour, texture and size at 10 to 15.6°C (Choudhury, 1967 and Nath *et al.*, 1987). Increase in soil temp. from 5 to 20°C favoured leaf development. At temp. 20-25°C uptake of nutrient is greater and dry matter content is higher but at a still higher temperature (25-30°C) the water requirement of the plants increases considerably and root deformation occurs. Higher leaves and dry matter production at 20 and 23°C, maximum root growths initially at 20-23°C but later at lower temperature (10-14°C). Whole plant weight increases with increasing light intensity. A long day as well as high temperatures lead to bolting without adequate root formation. During the hot weather the roots become tough and pungent before reaching the edible size and therefore should be harvested while young and small in size. Radish is more pungent at higher temperature. Pungency decreases with the advent of cooler temperature. The temperate types are generally small, mild pungent in taste and mostly used as a salad. Photoperiod also influence root development in radish. The root development under 12 hour photoperiod is more as compared to 8 hours and 16 hours (Angell and Hillyer, 1962). Low temperature is a critical factor causing flowering, which is accelerated by long photoperiod.

SOIL REQUIREMENT

Radish can be grown in all kinds of soils but the best results are obtained on sandy loam soils which are friable and contain high amount of humus.

Heavy soils produce roots of poor quality i.e. roots become, rough, tough, mis-shaped with more numbr of finrous roots (Sadhu, 1993). Radish is moderately tolerant to acid soils (pH 6.8 to 5.5). Soils should be well prepared by digging deeply and making friable. It is better if there is no clod, stone, weed plant in the fiels. In case, if it is essential to grow radish on heavy soils, then an adequate amount of sandy soil should be added during the land preparation. In order to provide equal distribution of irrigation water and to avoid water stagnation then the field leveling seems to be an essential operation.

Radish generally shows good tolerance to salinity and can be grown up to 6.0ds/m ECe without a significant loss in root yield and quality. It has more tolerance for Na_2SO_4 than NaCl. The uptake of K, Ca, or Mg is decreased with rising salt concentration and this effect is more pronounced in plant receiving Na_2SO_4 (Maliwal and Paliwal, 1979; Malik *et al.*, 1983 and Shimose and Hayashi, 1983). Similarly, seed yield of radish increased at about a rate of 10 per unit increase in soil salinity, and salt tolerance limit of radish seed crop has been found to be associated with electrical conductivity value of 5.30 ds/m ECe (Mangal *et al.*, 1991).

SOWING

Since, Radish is a cool season crop its cultivation is preferred during winter season in the plains. It can be sown any time between September to January in the northern plains as it is not affected either by frost or by extreme cold weather conditions. It is grown from March to August in hills. In the regions where summer is mild, it can be grown throughout the year except during few months of summer. In Bangalore, the radish roots are available practically for 8 to 10 months of the year but the best edible roots are available during November-December only. The temperate types are generally not planted till October.

But radish can also be grown throughout the year by using suitable varities. Details of period of cultivation for diffrent prominent varieties of different groups are give below:

Varieties	Period of Sowing of seeds	Period of harvesting
Pusa Chatki	Early April to mid October	Last week of September.
Pusa Desi	Mid August to Mid October	Last week of September to early December.
Pura Reshmi	Mid september to mid November	Late October to early January.
Japanese white	Mid October to mid December	Mid December to early March.
White Icicle	Late October to february end	Late November to March.
Pusa Himani	Mid December to Mid February	Mid February to Mid April.

Sarkar *et al.* (1978) reported that under short photoperiod carbohy drate levels continued to increase till maturity while under long photoperiod a decline in carbohydrate level was recorded after 60 days. Perecntage of total and reducing sugars as well as starch in both leaf and root showed gradual rise with the incerease in light intensity and maximum amount of carbohydrate was found in plants receiving full light.

Rao and Manohar (1990) Studied the effect of sowing time on yield and quality for different varieties of radish at Udaipur. Higher yield of good quality radish were obtined from October sowing than early or late sowings. Among the cultivars Nerima Long cultivar Produced maximum yield followed by Pusa Reshmi. The time of sowing in radish varies from region to region. As it is non-tolerant corp for high temperature, August-October is the suitable season for sowing. It requeries about 5.5 to 11.0 kg of seed for one hectare of area. It is usually grown. on ridges to facilitate good root production. It is grown as a single crop or as a companion crop. The plant rows or ridges about 22 cm high are kept about 45 cm apart, whereas the plants within the rows are about 8 cm apart. Rastogi *et al.* (1985) found that 60 x 45 cm is optimum spacing with 60 kg N/ha for getting the maximum seed yield. Sarnaik *et al.* (1987) concluded that half sized steckling are best for planting for seed in case of radiesh var. Pusa Rashmi. The temperate types can be sown in row about 20-30 cm apart. Generally, the seed sowing is done 1.25 cm deep in a staggered way to give continuous supply of the produced. It is aown strictly in the field and not in nursery beds. The seeds are mixed with fine soil or sand and placed in rows by hand and covered with soil.

NUTRITIONAL REQUIREMENT

General doses of manures and fertilizer which are applied in radish crop in various states of India are given in Table 1. In fact these are not exact doses required for radish but help to those who intend to venture firstly in radish cultivation. Optimum doses are location spacific and can only be decided accordingly.

Table 1 : Recommended doses of nitrogen, phosphoruse and potassium and organic manure (FYM) for radish crop are applied in some states of India (Tandon, 1991).

States	N(kg/ ha)	P_2O_5(kg/ha)	K_2O (kg/ha)	Fym (t/ha)
Arunachal Pradesh	100	50	50	20
Assam	50	50	100	20
Gujarat	40	20	20	10
Haryana	60	30	0	30
Himanchal Pardesh	50	40	40	0
Karnataka	75	37	37	20
Mizoram	60	50	40	5-10
Orissa	62	50	100	0
Punjab	62	30	0	37
Rajasthan	43	48	48	25
Tamil Nadu	50	100	50	25
Uttar Pardesh	40	40	20	20
Wast Bengal	75	40	50	15

A radish crop yield of 200 quintals removes 120kg nitrogen, 65 kg phosphorus and 100 kg potassium (choudhry 1976). Diffreant recommendation made by varius workers are FYM 25-40 tons, nitrogen 18-50 kg in the from of ammonium sulphate and phosphours 50 kg in the form of SSP potash 50 kg in form of muriate of potash. (Desai, 1948; Purewal, 1957; Yawalkar *et al.*, 1962). Nitorgen 50 kg/ha produces significant effect on the number, length size and fresh weight of radish leaf besides, enhancing the fresh weight and dry matter content of the root. Phosphorus at 40 to 80 kg/ha had significant increase in root yield of radish cv. Meghalaya Selection (Singh *et al.*, 1995). The FYM Should be mixed thoroughly at the time of field preparation, whereas the complete dose of potassic, phosphatic and one-half dose of nitrogenous fertilizers may be applied in rows before sowing. The second half dose of nitrogenous fertilizers should be applied as a top dressing along with the irrigation when the plant starts growing vigorously. Mehta *et al.* (1997) reported the response of radish to varying levels of irrigation and nitrogen. Highest seed yield of radish was recorded with ID/CPE ratio

1.6 which differed significantly during second year. Application of 60 kg N/ha resulted in significant increase in the seed yield during both the years that difference between 60 and 90 kg N/ha was non significant during second year. Krishnappa and Puttaraju (2002) studied the effect of nitogen and phosphorus levels on growth and yield of radish. Root yield and size increased significantly to varying levels of 'N' and interaction of 'N' and 'P_2O_5' levels. Significantly, the highest yield (355 q/ha) was in 150 kg N/ha over without application of 'N' and 50 kg N/ha. The increase in yield was mainly due to increase in length and diameter of roots.

Nath *et al.* (2008) concluded that the application of increased levels of nitrogen may improve the root and yield parameters, but irrigation levels should be optimum in order to avoid forking of roots. Irrigation at 10-15 days interval in winter and in summer at 7-8 days interval is sufficient. The soil should be highly fertilized to make nutrients available to the plants. Therefore FYM @ 20-25 t/ha, 30 kg nitrogen, 50 kg phosphorus and 50 kg potassium are the ideal doses of mojar nutrients in radish. But it has been observed that radish responds well to nitorgen @ 100 kg and phosphorus @ 100 kg hectare.

Asok Kumar *et al.* (2002) reported the effect of *Azotobacter* on growth and yield of radish of different varities. The two season's data indicated that the radish crop inoculated with 5 kg *Azotobacter* through soil at the time of sowing alongwith 75 percent of recommended 'N' recorded significant higher value for plant height, root length,root diameter, root yield, leaf yield and dry matter of roots and leaves compared to all other treatments. This treatment also recorded the maximum net returns of Rs. 62,204/ha and Rs. 58,604/ha with cost benefit ratio of 1:3.52 and 1:3.30 in the variety Pusa Chetki during rabi and kharif, respectively. Similar trends were observed in the varity Arka Nishanth but lesser cost benefit ratio.

Therefore, radish responds well to biofertillizer application along with the recommended doses of major nutrients. This could also help in improving the soil fertility status over a period of time. *Azotobacter* at 2.5-5.0 kg per hectare as soil application to be beneficial along with 75 per cent of the recommended nitogen (Nath *et al.*, 2008).

Use of Chemicals and growth subatances

(a) Soil application of boron @ 3.0kg/ha give higher yield. (b) Application of 1 or 2 ppm boron give quality roots and maximum top yield. (c) Singh and Dohare (1964) found that soaking in NAA at 10 and 20 ppm for sowing was not effective in stimulating germination of

Radish seeds. But the root yield was considerably increased by NAA 20 ppm. (d) Kumar (1973) obtined 100 per cent germination following treatment with GA_3 at a concentration of 200 ppm.

IRRIGATION

The irrigation requiremant is depends upon season, type of soil and amount of organic matter present. For rapid germination of seeds and production of tender and attractive roots, the soil should be contain plenty of moisture. If the soil is not moist enough at the time of sowing, the first irrigation is given immediately after sowing. During, summer month ferquent irrigation is necessary, otherwise the growth will be checked and roots will be pungent and tough making them unfit for market. The irrigation requirement of radish are found to be about 210-250 mm on sandy loam soil. Irrigation should be given st 5-7 days interval. The drip irrigation is useful in the production of radish in plastic tunnels; moist condition was found better in increasing the yield. Moisture stress increases the NO_3 contents in roots (Hegde, 1987) which is not conducive for health. In India, radish is raised both basin and furrow system of irrigation. However for higher yield and good quality roots, its cultivation may be taken under sub-surface drip system of irrigation.

INTERCULTURE

It is necessry to check the weed growth and provide soil aeration for better growth and yield. In radish, the growing roots tend to push out the soil surface. Therefore, proper earthing up is essential to get well-developed quality roots. Since, all the parts of radish is consumed, hence, chemical weed control should be avoided. Under intercultural operation, weak, lanky and crowded plants are removed to provide sufficient space, moisture and solar energy to remaining plants. Shallow hoeing is done at early stage of plants to facilitate root growth. When the root has started growing, earthing up should be done for enlarged root production.

WEED CONTROL

Weed pose much problem in radish crop during early stage of vegetative growth so weeding should be done at reguler entervals to keep down the weeds. There are few weed plants which can harm the radish crop. They are *Chenopodium album, Convolvulus arvensis, Cynodon dactylon, Cyprus rotundus, Poa annua* atc. (Singh, 1997). Perennial weeds can be controlled effectively by deep ploughing during field preparation and several other annual weeds by presowing/planting application of herbicides like Dalapon 1.0 kg, Fluchloralin 1.0 to 1.5 kg, Nitorfen 2.0 to 2.5 kg per hactare ane day after planting of radish steckling for higher seed yield (Gambhire *et al.*, 1983).

PLANT PRODUCTION MEASURES

Insecet Pests

Aphids (Myzus persicae, Brevicoryne brassicae, Lipaphis arysimi)

These are the most important that harm the crop. These suck the sap of the leaves, inflorescence and young pods. The cloudy and humid atmospheric conditions are favourable for the rapid increase in aphid population (Nath, 1976). It can be controlled by removal of severally affected plants.

The crop is sprayed with Nicotine sulphate or Malathion (0.01%) or Monocrotophos (0.05%). Seed crop planted early than the season is less affected than late sown crop.

Flea beetle (Phyllotreta sp.)

Flea beetles attack redish leaves and make small circular holes until they are skeletinized and lace-like They are active in sunny weather during spring. It can be checked by adoption of phytosnitary mesures including weed hosts and spraing of Malathion (0.1%) or Sevin (0.4%) at 10 to 15 days interval control flea beetle effectively.

Mustard sawfly (Arhalia proxima)

The affected leaves and fruits show holes as grubs feed on these. Hand piciking can be done on small scale and spraying of carbaryl (0.2%) or Malathion or Dichlorvos (0.05%) are effective control measures.

Diseases

White rust (Albugo candida)

Affected leaves show patches of white powdery substance on the under surface of the leaves. Most cool weather favours the disease development (Singh, 1985). Removal of weed plants and spraying of Dithane M-45 at 0.2 per cent or Bordeaux mixture at 0.8 per cent will control white rust infection.

Alternaria Blight (Alternaria raphani)

Lesions appear on leaves, stems, pods and seeds. The styler- ends of the pods become black and shrivelled. Infected seeds have very poor germination. Use of disease free seed, seed treament with hot water at 50°C for 30 minutes, spraying the crop with Blilox 50 or Copper

Oxychloride (0.5%) and adoption of proper crop rotation are effective control measures.

Radish Mosaic Virus (RMV)

Small chlorotic lesions appear on the leaves. It is transmitted through seeds. Removal of weed plants, spraying of Rogor 30 EC at 0.06 per cent at regular intervals to conrol aphids which are vector of RMV.

Root rot (Erwinia rhapontici)

It results in rotting of pith tissues and cavity formation and the plant wilts and die. Root dip in Agrimycin -100 @ 100 ppm is recommended.

Radish phyllody

It is another disease of the seed crop, which appears at flowering. The sepals, petals and carpels of affected flowers showed phyllody and the stamens become sepaloid. The degree of phyllody increased in the direction of carpels.

HARVESTING

Harvesting depends on cultivars, the edible roots get ready within 25 to 60 days. The temperate varieties take lesser time than the Asitaic variaties. The maurity of the roots for marketing will depend on the size of the roots as well as on the variety. It is necessary that the roots are harvested at the right stage because they tend to become fluffy and unmarketable thereafter. Roots also become bitter and pithy if the harvesting is delayed. In India, harvesting is done manually. A light irrigation may be given before harvesting to facilitate lifting of roots. The roots are washed, graded and tied in bunches.

YIELD

The Asitic improved varieties produce 150 to 250 q/ha of roots in 40-60 days, whereas the temperate varieties produce 50 to 80 q/ha in 25-30 days. Singh *et al.* (1971) reported that average yield of Pusa Himani may be as high as 390-471 qmintals per hectare both in spring-summer and winter sowings.

POST HARVEST HANDLING

Grading

After harvest edible roots are properly washed and cleaned the tops then graded and tied in bunches. The grading is done as per shape, size and tenderness of the roots in 'super', 'fancy' and 'commercial' grades. The top quality tender, uniform and straight roots are placed in 'super grade' and kept separately. The medium size roots with good texture and bright colour, having clean surface are arranged in 'fancy grade'. The roots with bigger size, over matured and without any blemish place in 'commercial grade' and arranged separately with labelled and send to the market in deffrent packing for sale.

Packing

The graded edible roots with shoots are tied in small bunches and packed in polythene bags, plastic bags, wire round basket, and C.F.B (Corrugated Fiber Board boxes) and sent to distant market for sale. Radish for processing purposes, cuts the roots from tops, then place in burlap bags for transportation to canning centre. For local market radish bundles are kept in gunny bags or big round bamboo basket and coushening materials are provide on top and stitched properly and transported to the market for sale. Some times tops are removed only roots are packed in basket, hampers or crates or in transparent film bags for marketing in advanced country (Thompson and Kelly, 1983).

Storage

Radish roots cannot be stored for more than 2-3 days under room temperature without impairing its quality. However it can be stored for about 2 months in cold storage at 0°C and 90-95% relative humidity. (Pantastico, 1975).

SEED PRODUCTION

Radish is a cross pollinated crop and hence requires isolations from other verieties to avoid any out-crossing with regard to purity of the seeds. It is recommended that the two varieties may be planted 1,600 m apart to produce pure seeds (Anon., 1971). When roots are fully developed, they are uprooted and true to type roots are selected. The selected roots are properly cut by removing one-third to one-half in length and one half to one-fourth portion to top leaving one fourth portion. These stecklings are transplanted in the well prepared field at the distance of 60 x 30 cm or 90 x 20 cm (Singh *et al.*, 1971). Some growers leave the plants with the roots in the field and allow it to produce seed stalks. It is not recommended, as the selection of good roots cannot be made and off-type plants which are not true to varity can not be removed. Also the seed quality and yield are affected if transplanting method is not followed (Nath *et al.*, 2008).

Roguing

This is very essential operatial for the production of high quality seed of radish. Almost all off -types, diseased plants and weed plants like wild radish, wild turnip, mustard (these weed plants may have the chances of cross pollination with radish) are removed from the crop.

The first roguing is done 20-30 days after sowing, the second when roots are lifted and replanted. At this stage all off-type roots based on colour shape and size and disease infection can effectively sorted out. The third roguing is performed during flowering and fruiting stages based on inflorescence colour maturity period etc. Fourth roguing can be done if needed, at maturity stage for late maturing plants (Table 2).

Table 2 : Field standard for seed production (maximum permitted in per cent)

Factors	Foundation	Certifid	Remarks
Oft -type	0.10	0.20	In root to seed method at the time of
Other Crop plants	-	-	replanting of roots, at and after
Objectionable weed plants	-	-	flowering, strict rouing for seed borne disease-Black rot, Black leg and Leaf
Diseased plants	0.10	0.05	spot

Kalvi and Nath (1970a) reported that the plants with two thirds top left and one half root cut or one fourth root cut proved to be superior to control with full top and full root in respect of new vegetative growth and seed stalk emergence. Kalvi and Nath (1970b) confirmed that the plants with two thirds trop left and one third root cut produced the highest seed yield of 7.4 q/ha. But the highest germinaion of 95.5 and 80 per cent was recorded in plants with two thirds top left and one fourth root cut in Kalyanpur and Marwari, respectively, than the control which gave 3.4 q/ha. Further investigations on the perormance of these seeds (obtained through various shoot and root cuts) on shoot growth and root and again on its subsequent seed production revealed establishing the point that better seeds could be obtained through these simple techniques (Nath and Mehta, 1971). Singh *et al.* (1960) reported 4.9 to 5.9 q/ha of seed from Japanese white.

Harvesting

Fully ripe pods are harvested. There is no fear of shattering of seeds becouse pods are indehiscent type. The seed stalks are cut and kept in small piles to dry.

Threshing and Winnowing

Completely dried pods are used for seed extraction. Threshed material is winnowed and seeds are separated, and again allowed for drying.

Grading

Radish seeds are passed through screen where small seeds are taken out.

Packing

Radish seeds packed in polythene bags, glass or plastic containers or some other suitable containers, glass jars (air-tight). High seed moisture leads to poor germination. Radish seed should meet the seed standaeds given in Table 3.

Table 3 : Radish seed standards (%).

Standard	Foundation	Certified
Pure seed (minimum)	98.0	98.0
Inertmatter (maximum)	2.0	2.0
Other crop seed (maximum)	0.05	0.10
total weed seed (maximum)	0.10	0.20
Objectionable (maximum)		
Germination (minimum)	70.0	70
Moisure (maximum)	6.0	6.0

Storage

Packed radish seeds are kept under well ventilated and dry places away from direct sun. Radish seeds remain viable up to 4 to 5 years, if stored well.

Seed Yield

The quantity and quality of radish seed vary due to type of varieties, climatic conditions, methods of seed production, soil type, management practices etc. Because of these factors, the seed yield varies from 250 to 400 kg or so per haectare; further seed yield can be improved when pollinating agents like honey bees are in large numbers. Promotion of bolting and flowering is enhanced by the application of GA_3 at 100 ppm to the seed or at the end of flowering on standing crop (Singh, 2001).

REFERENCES

Anonymous (1971). Indian minimum seed certification standerds. Central Seed Committee, Ministry of Food and Agriculture, Community Development and Co-oparation, New Delhi.

Ashok Kumar, R., Lingaiah, H.B., Raju, B. and Rajanna, K.M. (2002). Effect of azotobactor on growth and yield of radish (*Raphanus sativus* L.) cultivars, *Intl. Conf. Veg., Bangalore,* 11-14 November, 2002. Abstract: pp 196 -197.

Aykroyd (1941). Health Bulletin No. 23. Nutrition Research Lab., Koonoor.

Bailey, L.H. and Bailey, E.Z. (1956). Hortus second- A Concise Dictionary of Gardening and Gerneral Horticulture. Macmillon Co. New york.

Banga, O. (1976). Evolution of crop plants Ed. N.W. Simmonds, Longman, London & New York.

Choudhury, B. (1967). Vegetable. National Book Trust, Insia. New Delhi.

Desai, B.L. (1948). Pocha's Garden Guide. Bombay.

Dutta. O.P. and Vijay, O.P. (1970). Improvement of Asiatic root crops through selection and hybridization. *Annual Report*. IIHR, Bangalore.

Gambhir, O.P., Malik, Y.S. and Pandita, M.L. (1983). Chemical weed control in seed crop of radish. *Indian J. Weed Sci.* 15: 74-76.

Hegde, D.M. (1987). Effect of soil water potential, method of irrigation and nitrogen on yield, N up-take and water use of radish. *Ind. J. Agron.* 32: 245-249.

Helm, J. (1957). Ueberden typus der art *Raphanus sativus deran Gliedevberg and Synonynica. Kulture pfianze*; 5:41-54.

Kalvi, T.S. and Nath, Prem. (1970a). Vegetative growth and flowering as influenced by various root and shoot cuts in radish stecklings. *Farm. J.*, 11 (8): 17-22.

Kalvi, T.S. and Nath, Prem (1970b). Seed quality and yield as influenced by various root and soot cuts in radish steckling. *Punjab Hort., J.* 10 (1&2):137-148.

Kay, E. Daisy (1987). Root Crop,pp. 174 -186. Tropical Development and Rasearch Institute, Londan.

Kitamura, S. (1958). Varieties and transitions of Radish In Japanese radish, (ed) I. Nishiyama, *Jpn. Soc. fron. Sci.* Takyo: 1, 1958.

Krishnappa, K.S. and Puttaraju, T.B. (2002). Effect of nitrogen and phosphorus levels on the growth and yield of radish (*Raphanus sativus* L.) *Intl. Conf. Veg.*, Bangalore, 11 - 14 November, 2002 Abstract pp. 190.

Malik, Y.S., Mehrotra, N., Pandita, M.L. and Jaiswal, R.C. (1983). Effect of soil salinity levels on the yield and qulity of radish carrot and turnip *Indian. J. Agric. Sci.* 53(9): 861-862.

Maliwal, C.L. and Paliwal, K.V. (1979). effect of salts on the growth and chemical composition of carrot and radish. *J. Root Crops.*, 5: 50-50.

Mangal, J.L., Lal, S. and Hooda, P.S. (1991). Salt tolerance of radish seed crop. *Haryana Agric. Univ. J. Res.*

Mansfield, R. (1959). Die Kulturpfianzen baikeft. pp. 1-659. Akademic verlog. Berlin.

Mehta, C.P., Lal, S. and Singh, Harbir (1987). Response of radish (*Raphanus sativus* L.) to varying levels of irrigation and nitrogen. *Haryana J. Hort. Sci.*, 16(3-4): 287-292.

Muthu Krishnan, C.R. and Arumugam, R. (1977). Correlation and regression studies in radish. *Indian J. Hort.* 34: 163 -165.

Nath Prem and Mehta, S.S. (1969). Some Studies on pollen morphology and physiology in radish (*Raphanus Sativus* L.), *J. Palynol.*, 6:78-81.

Nath, P. (1976). vegetables for the Tropical Region. Indian Council of Agricultural Research, New Delhi, Low- Priced Book Series x 10.2.

Nath, Prem and Mehta, S.S. (1971). Shoot growth and root production in genetically uniform radish varieties as influnenced by seed obtained through diffrent shoot and root cut techniques. *Indian J. Hort.*, 29(2) :212-218.

Nath, Prem; Badra, T., Denton, Lanre (1987). Germplasm resources and genetic improvement of vegetable crops in the tropics, National Horticaltural Research Institute. Ibadan, Nigeria, pp. 169.

Nath, Prem; Srivastava, V.K., Dutta, O.P. and Swamy, K.R.M. (2008). Vegetable Crops – Improvement and Production. PNASF, Bangalore, pp. 182-187.

Nath, Prem; Velayudhan, Sundari and Singh, D.P. (1987). Vegetable for the Tropical Region. ICAR, New Delhi.

Ogura, H. (1968). Studies on the new male sterility in Japanese radish with special reference to the utilization of this sterility towards the practical raising of hybrid seeds. *Mem. Fac. Agric. Kagoshima Univ.*, pp. 39-79.

Pantastico, Er.B. (1975). Post Harvest physiology, handling and utilization of Tropical and Subtropical Fruits and Vegetables (Eds), AVI Publising Co. West Port, Connecticut, USA.

Purewal, S.S. (1957). Vegetable Cultivation in northern India. *Tech. Bull.*, 39, Indian Council of Agricultural Research; New Delhi.

Rao , B.N. and Manohar, M.S. (1990). Effect of sowing time on yield and quality of Radish (*Raphanus sativus* L.) cultivars. *Haryana J. Hort. Sci.*, 19 (3-4) : 333 -336.

Sadhu, M.K. (1993). Root Crops In.: Vegetable Crops (Ed. Bose *et al.*) Naya Prakash, Calcutta.

Sarkar, B., Mandal, R.K., Sandhu, M.K. and Bose, T.K. (1978). Effect of day length and light intensity on growth and development in radish. *Vegetable Science*, 5: 57-65.

Sarnaik, D.A., Baghel, B.S. and Singh, K. (1987). Seed yield of radish (*Raphanus sativus*) by different root cuts. *Journal Farming System*, 3: 25-27.

Seshadri, V.S. and Chatterjee, S.S. (1996). The history and adaptation of some introduced vegetable crops in India. *Vegetable Science*. 32(2): 122-142.

Shimose, N. and Hayashi, N. (1983). Salt tolerance of parsely, welsh onion, radish and cabbage. *Sci*: *Rept*. Faculty of agric, Oka Univ., 62 : 25-30.

Singh, A.R., Singh, R.and Shankar, K. (1983). Effect of nitogen, sulphur and potash fertilizers on mineral composition and metobolites of radish (*Raphanus sativas* L.). *Haryana J. Hort. Sci.*, 12 :71.

Singh, D.N. and Nath, Vishal (2008). Varieties and Hybrids of Vegetables. Satish Serial Publishing House, Azadpur, Delhi (India). pp. 61-65.

Singh, G.P., Pandita, M.L. and Singh, G.R. (1979). Effect of date of sowing on growth, yield and quality of diffrent cultivars of radish. *Haryana J. Hort. Sci.,* 8: 134-136.

Singh, H.B., Thakur, M.R. and Bhagchandani, P.M. (1960). Vegetable seed production in kulu valley. II *Indian J. Hort.,* 16(3): 153-161.

Singh, H.B., Thakur, M.R. and Bhagchandani, P.M. (1960). Vegetable seed production in kulu vally. III. *Indian J. Hort.,* 17(1) : 38-47.

Singh, J.P., Gill, H.S., Swarup, V. and Arora, R.S. (1971). Pusa Himani radish. *Indian Hort.,* 16 (1) : 13-14.

Singh, K., Saimbhi, M.S. and Singh, I.S. (1971). Effect of size of stecklings, spacing and nitrogen fertilization on seed production of radish. *J. Res.* HAU (Hisar). 1: 10-12.

Singh, Makhan; Batra, V.K., Singh, Jitender, Bhatia, A.K., Singh Virendra and Arora, S.K. (2002). Response of foliar application. *Intl. Conf. Veg., Bangalore,* 11-14 November, 2002 Abstract: pp 176.

Singh, R.S. (1985). Diseases of Vegetables. Oxford and IBH Pub. Co., New Delhi.

Singh, S.P. (2001). Seed Production of Commercial Vegetables. Agrotech Publishing Academy, Udaipur (India)., pp. 238-245.

Singh, V.B., Kar, P.L. and Tatung, T. (1995). Effefct of nitrogen and phosphorus on growth, yield and nutrient uptake of radish cv. Meghalaya Selection: *Adv. Hort. & Forst.* 4: 127-132.

Singh. S.P. (1997). Principles of Vegetable Production. Agrotech Publishing Academy Udaipur -31001 (India).

Sirks, M.J. (1957). Genen en phaenen, 2:2-10.

Srinivas, K. and Naik, L.B. (1990). Growth and yield of radish in relation to nitogen and potash fertilization. *Indian J. Hort.* 47: 114-119.

Tandon, H.L.S. (1991). Fertilizer Recommendations for Horticaltural Crops. A guide book Fertilizer Development and Consulation Orgnization, New Delhi.

Thompson, H.C. and Kelly, W.C. (1983). Vegetable Crops. Tata McGraw Hill Book Company Inc. New York. pp. 343-344.

Tsumoda, S. (1991). The history of cultivated plants in the tribe Brassicae. *Agric. Horti.,* 66: 283.

□□□

2. CARROT

Botanical name	:	*Daucus carota*
Family	:	Umbelliferae
Chromosome number	:	2n = 18

Corrot is derived from the wild carrot, Subsp, Carota, which occurs natural in Europe, Asia and Africa. It is of ancient cultivation in the Mediterranean region and has now spread throughout the world. It was a favourite vegetable in England in the 16th century and was taken to Virginia in 1609. The roots are used as vegetables and for putting in soups steams, curries and other dishes; the grated roots is used in salads; tender roots are pickled. Carrots are canned and may also be deliberated. The roots are a rich source of carotene and carotene concentrates are prepared from them. They may be pressed for juice, which is sometimes blended with orange juice. The roots and tops may be fed to live stock. The seeds contain an essential oil which is used for flavouring and perfumery. They contian about water 88.3% proteing 0.8 percent, fat 0.4 percent; carbohydrate 8.9 percent, fibre 0.8 percent and ash 0.8 percent.

The cultivated carrot has a swollen tap root, which may be short and stumpy or a long tapering cone. There with a smooth, tender, bright red or orange flesh, with minimum of core are preferred. It is an erect biennial; 30-100 cm. In height with solid stem, cvs 3-pinnate segments pinnatified, lobes lanceolate, petiole usually long; umbels 3-7 cm in diameter, becoming convave in fruit; bracts 7-13; fils usually white. Fruit oblong- avoid, 3-4 mm. Long primary ridges ciliate; secondary rigdes with hooked spines. The roots should be harvested while still young, carrot can be stored for several months without loss in quality.

Carrot is an important vegetable crop world wide both economically and as a source of Pro-vitamin A carotene. It is a cool season crop, grown all over the world in spring, summer and autumn in temperate countries and during winter in tropical and subtropical climate.

ORIGIN

Carrot is the most common root vegetable all over the country. Bailey (1948) reported that corrot is a native of Europe and was found as a weed in America. They were introduced into England during the fifteenth century in the Royal Gardens. Asia and North Africa and possibly North and South America (Thompson and Kelly. 1957). It was probably cultivated by the ancients but was not a common food plant. Asiatic types probably originated from central Asia.

Carrot originated in the area of Afghanistan. Carrot is derived from the wild carrot. Sub species *carota,* which occurnaturally in Europe, Asia and Africa.

History and Adaptation

Written records in Europe indicated that carrots, were cultivated prior to 10th century. The species *D. carota* is native of western Asia (Afghanistan). The largest diversity in morphologial characters of this species have been found to occur in that region. The roots vary in degree of remification, fleshiness and colour which ranges from white to anthocyaninsed. Wild types are found in South west Asia, southern Europe, north Africa and north America. Carrot was first introduced into China in 13th-14th century and Japan around 17th century. This crop was introduced in to India early in its history. There is the Sanskrit equivalent called "Girijana" for carrot which propably indicates the derivation of Hindi name 'Gajar'. The first setlers to Virginia brought seeds and soon carrots were grown by American Indians, carrot probably were first grown in salem gardens, about 1620 A.D. (Shoemaker, 1947). In India the carrot is said to have been introduced from Persia.

Even now the primitive black and red types are popular in India. The orange European types having higher carotenoid content, were believed to have been introduced later during the visit of European traders. Obviously, the black and red types might have come earlier from the cantral Asian region of its origin-probably Afghanistan, Persia and southern Russia. Ain-i-Akbari (Anon) records carrot in Mughal cuisine which explains the spread from central Asia (Seshadri and Chatterjee, 1996).

Punjab and kashmir are also considered a primary centre of origin with secondary centres in Ethiopia and north America. Himachal hills and Kashamir Valley still abound in its wild form where wild animals especially brown bear feed on its roots. In some parts of Kashmir people still eat the wild carrots (Yawalkar, 1980). In noth India highly coloured types of carrot are met which are not available in Europe. Types varying from absolute colour less to light lemon, light orange, orange and deep orange, light purple, deep purple and almost black are available. This is another indication of its being a primary centre of origin.

DISTRIBUTION

Carrot spread to the Mediterranean rigion by the 10th to 12th century, to Europe and China in the fourteenth century and to Japan in the fifteenth century. Orange carrots were first known in the seventeenth century in nothern Euorpe. Before that time only yellow and purple rooted types were cultivated. Now carrot is distributed throughout the world like temperate, tropical and subtropical countries. In India the carrot is said to have been introduced from Persia, and now grown all over country.

Carrot is an important vegetable in European countries, USA, China, Japan, Korea and India. It is grown in Punjab, U.P. Andhra Predesh, Karnatak, Assam, Haryana. The Asiatic carrots are generally red coloured because of anthocyanin pigment. The European types are orange because of carotene, a precursor of vitamin A. In India, mostly Asiatic carrots are grown probably due to appeal for red pigment. The carrots are grown throughtout country.

COMPOSITION

Carrot is a rich source of carotene (α and β carotene). In root sucrose is most abundant with endegenous sugar contents. The seeds contain an essential oil, which is used for flavour and in perfumery. Sugar and volatile terpenoid are responsible for carrot flavour. Cooked carrot had flovour due to reducing sugar content. For colour anthocyanin red is responsible.

According to Aykroyd (1941), Chatfield (1949 and 1954), Watt and Merrill (1964), fresh carrot contains per 100g. edibale portions : Water 88.6g, Protein 1.1g, fat 0.2g, Cabohydrate 9.1g, Fibre 1.0g, Vit. A 12,000 I.U. Vit. B_1 0.06 mg/m; B_2 0.06 mg/m. B_6 0.12 mgm; Nicotimic acid 0.5 mgm; Vit. C 2-10 mg; Vit. E 0.45 mg; Biotin 0.003 mg; FA 0.008 mgm; K. 331 mgm; Ca 39 mgm; Mg 21 mgm; Fe 0.8 mgm; P37, S21 and Cl 40 mg. According to Jain and Mukherjee (1966), Carotene content in the flesh and core portions has been found to vary from 5.86 to 9.10mgm. and 3.07 to 6.90 mgm per 100g.

Uses

The roots are used as raw as salad and as a cooked vegetables and putting in soups, stews, curries and other dishes, the grated root is used in salads; tender roots pickled. Carrots are canned and may also be dehydrated. Gajar halwa is a delicious dish, carrot jam is also popular. The roots may pressed for juice, which some times blended with orange juice. The roots may pressed for juice, which some times blended with orange juice. The carrot leaves are eaten in many countries, its tops can be used as good source for extraction of leaf proteins. Due to high nutritive value and cheap, highly acceptable processed products can be prepared from it, which are (a) canned carrot shreds in sugar syrup. (b) Carrot shreds for used as a break fast food. (c) Strained carrot pulp powder for use as a Baby food. Carrot tops are used as fodder and also preparation of poultry feed. Its fruits are recommened in chronic diarrhoea. A decoction of carrot is a popular remedy for jaundice in Europe and is also considered as source of sugar and as a substitute for coffee. An infusion of carrot has long been used as a folk medicine for thread worms. Carrot increases the quantity of urine and helps the elimination of uric acid. Carrot seeds are aromatic. Stimulant, and carminative. They are reported to be useful in disease of the kidney and in cropsy. A black type carrot is used to prepare kanji, a beverage very common in few states.

BOTANY

The carrot is an annual or biennial herb, with and erect to much branched stem, 30-120 cm high arising from thick, fleshy taproot, 5-30 cm long. Leaves are pinnately compound with long petioles expanded into stalk at the base. The edible carrot root is actually an enlarged fleshy taproot, of phloem, cortex and core or xylem. Good quality carrots have a maximum of cortex and a minimum of core of xylem. Inflorescence compound umbel. It is the king umbel or umbel of the first order (or primary umbel) that flowers first. The flowers are perfect with small petals usually white or yellowish. Anthers in a single umbel are completed in 7-9 days. Carrot seed is actually a fruit, an indeciscent mericarp, containing a single seed. The fruit is oblong about 3 mm long.

The carrot, *Daucus carota* L. belonging to the family Umbelliferae. has got fleshy edible root portion which is designated as conical root botanically. As in radish, the carrot also has two groups-Asiatic and Eruopean types. The asiatic type carrots produce roots and seeds freely in the plains, where as the European types produce good root but fail to produce seeds in the plains. The Asiatic type has its root ends predominantly conical whereas it is blunt in European type carrots. The other contrasting features of Asiatic types are: tropical and subtropical, annual, no chilling requirements for seed production more anthocyanin, more core heavy top, core is not coloured, ealry maturing and high sugar content.

Taxonomy

It is an annual herb for root production and a biennial for flowering and seed set. The stem is erect branched 30-120 cm high arising from the thick fleshy top root which is 5-30 cm long. The colour flesh may be white yellow, orange red, purple or dark purple.

Breeding methods, include mass selection and pedigree selections within different population. The other is hybridization and development of hybrids. This is generally based on two systems of cytoplasmic male sterility with different genetic backgrounds and origin. They have brown anthers (ba) and petaloid stamen. The brown anther type (ba) is present in all cultivated orange coloured open pollinated varieties. The cytoplasm of petaloid type (pt) is derived from a wild form of *Daucus carota* L. Bhagchandani and Choudhury (1971) noted the hetorosis and inbreeding depression in Asiatic carrot. It is reported that selection based on phenotypic values may prove quite effective because of high heritability estimates. Chandel *et al.* (1988) boserved that both additive and dominance variances were significant for others characteristics. Gill and Kataria (1974) found mark variation in dry matter, T.S.S., and carotene content among European and Asiatic varieties. European types of carrot were superior to Asiatic types in respect of all above characters. The bigger size of core and high percentage of moisture was mainly responsible for poor nutirtive value of Asiatic type of carrot.

Plant breeding

Carrot is a cross-pollinated crop with protandrous bisexual flowers. In Asiatic carrots, a good number of local useful strains are available in various regions of the country, and some of them are very high yielding. Because of the pleasant flavour liked by the Indian it has larger production and consumption than the European types which have a different flavour but excellent texture. In the plains at some of the research stations, it has been possible to maintain and evaluate some of the useful strains for utilization in breeding programmes (Nath *et al.*, 2008). Also efforts were made at IARI, New Delhi and its sub-station in Kulu Valley to cross the usefull strains of the Asiatic and exotic strains, with an intention to incorporate good root qualities of European types into high yielding Asiatic types. A carrot variety Pusa Kesar which has good qualities of both Asiatic and European types with an ability to produce seeds in the plains has been developed (Singh, 1963). In 1985 another Asiatic carrot Pusa Meghali was evolved from a cross between Pusa Kesar and Nantes. Also information has been gathered on the genetics of various economic characters in Asiatic carrots. Bienz (1968) reported that splitting was a varietal characteristic and was governed partially by genetic factors in addition ot the effect of spacing and nitrogen application. Dickson (1966) concluded that cracking susceptibility may be governed by a single dominant gene. Scibrjo and Zemoja (1996) were of the opinion that the carotene content was higher in varieties with orange colour. Efforts have been made to breed hybrid corrots by using male sterile lines. The new cytoplasmic male sterile system in an alloplasmic form of orange carrot is a corss between the wild carrot *Daucus gummifer* and the cultivated carrot *D. carota*. Recently the heat tolerant carrots like Kuroda, and New Kuroda, having deep oragne coloured and stumped roots, are imported from Japan and South Korea by private companies. They are becoming popular in the southern states (Nath *et al.*, 2008).

VARIETIES AND HYBRIDS

A large number of carrot varieties are growing at various places. They are locally suitable. However, a few varieties which are widely adopted to climatic and edaphic conditions and are succesfuly in various regious (Singh, 1989 and Yawalkar, 1985).

On the basis of temperature requirement to flowering varieties are classified into two groups.

1. Temperate or European or Bieannial types

For flowering. low temperature treatment (4.8-10°C) is needed. Such varieties, do not produce seeds in the plains of India. The varieties are Nantes, Chantenay, Imperator, Zeno. Pusa Yamdagni, Early Nantes and Ooty 1.

2. Tropical or Asiatic or Annual types

For flowering, low temperature is not needed. Such varieties produce seeds in the plains of India. The varieties are Pusa Kesar, Pusa Meghali, Sel-233 No. 29, Desi Red, NS 820, Hisar Gairic, Hybrid No. -1, Gajar No. 29, etc.

The detail characteristics of varieties and hybrids of different groups deseribed by Singh and Nath (2011) are given below :

Nantes

This variety is temperate type (European type). the is not suitable for seed production in plains but most suitable for root production. The roots are well shaped, scarlet orange in colour, cylindrical in shape with stumped end forming a small thin tail. The skin is thin and flesh is sweet and with pleasant flavour. It is ready in 120 days.

Chantenay

Chantaney is temperate type (European type). It does not produce seed in the plains. Roots are attractive deep reddish orange, smooth, root shoulders thick somewhat square, gradually tapering towards a distinctly stem end, flesh tender, sweet and fine textured. It takes about 120 days. This variety is excellent for canning and storage.

Nantes Half Long

The variety commonly known as Nantes is evolved at IARI Regional Station, Katrain; roots are small, slim, rough, sweet, cylindrical and stumpy with abrupt tail; the core is small and self coloured; yield is 15-20 t/ha in 90-110 days.

Early Nantes

Roots almost cylindrical terminating abruptly in small thin tail, 12-15 cm long, orange flesh with self coloured core; duration 90-110 days.

Chaman

Developed at SKUA&T, Srinagar, roots long, cylindrical and semi blunt; tolerant to cracking; yields 25-27 t/ha.

Ooty-1

Developed at Horticultural Research Station (TNAU), Udhagamandalam; roots are 25 cm long with deep orange colour, yield 45-50 t/ha, in 100-110 days.

Pusa Yamdagni

It was developed from a corss between EC 9981 × Nates Half long and released by IARI, Regional Station Katrain (HP). Roots are long, orange with self coloured core (character of Nantes Half long). This is high yielder, more carotene and minerals compared to Nates Half long. Roots are ready withing 120 days.

Zeno

This cultivar developed at Ooty (Tamil Nadu). This is a temperate type, high yielding having good quality roots.

Imperator

This variety is a cross between Nates and Chantenay. The foliage is large. It is a mid to late maturing variety. Roots are long and thick with short tapered end. The colour of the cortex is orange with slightly less pigmented core.

(II) Asiatic or Tropical types

Pusa kesar

This is an improved cultivar. It was developed by recombination breeding from a cross between 'Local Red' and 'Nantes Half Long'. This cultivar is high and quick growing. The central core is self coloured less branching of roots. Top leaves are short. Roots remain in field for longer period without bolting. It can tolerate high temperature.

Pusa Meghali

This variety was developed and recommended by IARI, from a corss between Nantes and Pusa Kesar. The roots are long teapering, orange with self-coloured core. The foliage growth is medium. The branching and forking of roots is very less.

Sel. 233

This cultivar was developed by Punjab Agricultrual University in 1978 from a cross between Nantes and No. 29. The leaves are dark green in colour, roots 15-18 cm long. smooth orange with light orange core. Disorders like cracking and forking are less. The roots are ready for harvesting in 90-100 days.

No. 29

This is selection from local material, developed by Punjab Agricultral University. The root are long tapering light red in colour.

Danvers

This is mid-season variety, with large foliage. Roots are medium to long in size with tapering slightly rounded end. The cortex is orange and yellowish core.

Desi Red

It has medium foliage, mature in 80-90 days after sowing. Roots tapering, 15-20 cm long, appealing red, uniform red, internal colour, core small and self coloured. High yielder.

Corumba

It has very uniform, medium long cylindrical roots. Excellent internal and external colour with good shelf life. Good for cold storage.

NS 820 (OP)

It is high yielding, early vigorous plants very uniform long, straight and cylindrical roots. Very small self coloured core. Crisp flesh with good sweetness and excellent quality. It has very good storage and transport qualities.

Namdhari Kuroda (OP)

An early vigorous, high yielder with uniform long cylindrical roots. Roots are smooth, deep orange with very small self coloured core. Flesh is very sweet, crisp and endowed with good storage and transport qualities.

Samurai Red

Medium sized vigorous plant with green leaves. Roots are 30-35 cm long, dark red with self coloured core, ready for harvest after 90 days of sowing. Roots are very uniform, smooth, sweet and juicy.

Kuroda Improved

A very uniform foliage and root variety. Deep range colour, self coloured core, 18-20 cm long, round shoulder with stump roots, ready for harvest after 80-90 days of sowing. Ideal for summer sowing to harvest in fall and winter.

Hisar Gairic (H C1)

It has been developed through mass selection from the germplasm collected from various parts of the country. It matures in 120 days, the leaves are light-green, roots long, tapering, attractive, light brick red thin self core, less fibre and forking. It gives an average yuield of 28.8 tonnes/ ha and carotene content is 96.2 mg/100g of fresh weight. The roots are 18.1 cm long and 3.9 cm across. The high yielding 'Hisar Gairic' is for early sowing.

Hybrid No. 1

It has been developed at Maharashtra hybrid seed Co. Ltd., Jalna. Roots are deep orange, uniform, length 7-9 inches, matures early; semi table for processing and fresh marketing.

Samsoon 196

It is Nantes type, very high yielding, having soft edible core. It gives long carrots with vigorous upright growing leaves, having good taste. Maturity mid early and tolerant to powdery mildew.

Bangor

It matures in approximately in 100 days. It is Borlikum type carrot, tolerant to powdery mildew and very high yielding.

Oranza

It is having edible core and is good for slicing, having good yield, dry matter content is 9 to 12%. Maturity mid early and Nantes types.

Carson

It matures in approximately 110 days from sowing. Shape is Chantenay type, having healthy foliage and gives attractive coloured carrots.

Premia (F_1)

Early hybrid and Nantes type for early fresh market. Short foliage and produces long (15-18 cm), slender cylindrical carrots with rounded shoulders. Its external and internal colour is an attractive orange red and its core is fine. Resistance to cracking is very good. Normal yield ranges betwen 45-49 mt/ha under optimum growing conditions.

CLIMATIC INFLUENCE

The corrot is cool season crop. It is relatively sensitve to temperature. It can tolerate a wide range of rainfall but water stagnation becomes a probelm. Carrot seeds germinate in the temperature range of 7 to 24°C. Carrot roots develop a good colour under the temperature range of 15°C to 21°C. Lower temperature (below 15°C) may result in formation of seed stalks which is suitable for carrot seed production. High (33/38°C) day/ night temperatures during polination, fertilization or the erly stages of seed development can greatly reduce carrot seed yield and seed quality (Elballa and Cantiliffe, 1996). The European or temperate types of corrot produce seeds only in the hills where as Asiatic or topical types produce seeds in plains of India. European types are more sensitive to temperature. Barnes (1936) reported that greenhouse grown carrots produced longest root between 10 and 15.6°C with poor colour and between 15.6 and 21.1°C with good colour. However, 23.9 to 29.4°C temperature produced shortes root, but the shape was not affected by the termperature. During the development of root the colour changes from white to yellowish white, whitish yellow, light yellow, dark yellow orange and then dark-orange gradually. The carotene content is reduced below 15.6°C and above 21.1°C.

SOIL REQUIREMENT

Being a root crop carrot, rquires deep, frieble loamy soil. For early crop, sandy loam soil is preferred. Carrot grown on heavy soils tend to become rough and coarse because the roots fail to penetrate the hard soil evenly. The optimum soil pH is 6.0 to 7.0. Mn toxicity may be noticed at higher pH through chlorosis of leaves. Plough pan restricts the growth and development of roots that may also cause root deformity. Idealy, carrots are grown on land with a high water table where crop never suffers form drought. However, water logging should always be avoided (Singh, 2001).

It is always advisable to grow carrot on soil which was manured well for a previous crop. The application of organic manures which are not properly rotten, should be avoided in the current growing season of the carrot crop because many roots will tend to be malformed, the roots being split into two or more forks. Stones can also result in crooked, deformed roots (Marshall, 1990). Carrot can be grown at soil salinity 4 to 6 ds/m Ece without significant loss in the yield (Malik *et al.,* 1983) However yield begins to decline at 1ds/m Ece (Mangal *et al.,* 1990).

It does not grow well on highly acidic soil. Heavy soil may check the development of root and cause forked lateral roots. Soil should be prepared by deep ploughing followed by cross harrowing. If necessary, planking may also be done to make soil clodless. It is desirable to prepare the field up to fine tilth to ensure maximum and early germination because seed is small and slow in germination.

SOWING

Carrot is grown in different places at different times according to temperature requirement.

Sowing time

Carrot can be grown almost all the year round where climatic conditions are mild. Several successive sowing may be done every fort night to obtain a continuous supply of roots. In the northern plains of India carrot is sown from the middle of August to the beginning of December (Choudhary, 1967 and Nath *et al.,* 1987). It is grown from March to August in hills. In northern plains, seeds of temperate types are sown druing October-December and that of Asiatic types are sown in August to early October. In soutern and central parts it can be sown in January-February, June-July and October-November (Nath *et al.,* 2008).

Seed rate

The seed rate varies from 8.0 to 10.0 kg. per hectare. For better yield, seed soaked for 12 hours in 50 ppm IBA solution.

Method of Sowing

Carrot is usually sown directly on ridges (10-15 cm high) or on flat beds (1.5 cm) in light soils. Commonly, the growers broadcast the seeds but the proper methods to sow them in rows, spaced 30 cm apart. The distances between ridges are 45 cm for Asiatic types and 30 cm for temperate types. For sowing carrot on level surface, the fields are divided in small plots, convenient for irrigation and the seed mixed with fine soil or coarse sand is broadcast and the soil surface is raked to cover the seed lightly. Depending on the moisture available, the seeds germinate in 5 to 10 days. Initially thick sowing of seed should be done. The thinning may be required after germination to maintain 5-7 cm distance between the plants within the rows.

NUTRITIONAL REQUIREMENT

There is immense scope for improving carrot yield potentiality and quality with balance application of N, P. and K. fertizers. Nitrogen is responsible for rapid elaboration of leaves in early stage of growth cycle in carrot, thereby increasing photosynthetic activity. On the other hand, potassium is responsible mainly for the translocation of assimilates to storage organ.

Carrot is a heavy feeder on fertilizers and manures. According to Thompson and Kelly (1957) a carrot crop yielding 106 to 114 quintals removes about 14 kg of nitrogen, 8 kg of phosphorus and 45 kg of potash. Depending on the fertility of the soil recommendations have been made in various regions as 25-60 kg nitrogen, phosphorus 25-50 kg, potassium 90-110 kg (Yawalkar, 1962).

The fertilizer comprising 80 kg nitrogen and 50 kg potassium with adequate manurial support may be recommended for obtaining higher growth, yield and quality of carrot in the plains of West Bengal (Sharangi and Paria, 1995). General dose of NPK at 50,40 and 30 kg/ ha presented in Table given below for Himachal Pradesh. However for obtaining higher yield and quality 50 kg nitrogen and 80 kg Potassium per hectate are applied in carrot cv. 'Pusa Kesar' (Singh and Singh, 1996).

Tandon (1991) reported the recommended doses of nitrogen, phosphorus and potassium and farm yard manure are applied in different states of India, which details given below :

State	N Kg/ha	P_2O_5 kg/ha	K_2O kg/ha	FYM t/ha
Andhra Pradesh	50	40	50	25
Arunachal Pradesh	100	50	75	20
Assam	60	30	120	25
Gujarat	60	40	30	10
Haryana	60	30	30	30
Himachal Pradesh	50	40	30	-
Karnataka	75	60	50	25
Maharashtra	100	80	120	25
Mizoram	60	40	50	-
Orissa	60	45	50	50
Punjab	60	30	75	30
Rajasthan	60	40	120	25
Tamil Nadu	180	125	100	30
Uttar Pradesh	40	50	50	10
West Bengal	60	40	30	10

The complete dose of FYM may be mixed with the soil at the time of field preparation and the complete dose of and potash phosphorus may be applied to the soil before sowing. It is suggested that the application of nitrogen in the form of ammonium sulphate may be given as a top dressing to increase the yield.

Table 1 : Root yield carrot at influenced by irritation and nitrogen regimes (Prabhakar *et al.*, 1993).

Treatments	Root yield (kg/ha)
Evaporation replenishment (%)	
25	80
50	15385
75	22220
100	26100
LSD (p=0.05)	
N, level (kg/ha)	
0	8610
40	17480
80	20820
120	23910
LSD (P=0.05)	2390

Rai and Yadav (2005) suggested that for getting good yield 200 quintals/ha well rotten farm yard manure should thoroughly mixed in the soil 15-20 days before sowing the seeds. Application of 80 kg nitrogen, 60 kg phosphorus and 60 kg per hectare potash are beneficial as inorganic fertilizers. Half quantity of nitrogen and full quantity of each phosphorus and potash should apply at the time of sowing. The remaining quantity of nitrogen may be given 30-45 days after seed sowing.

Prabhakar *et al.* (1993) studied physiological analaysis of growth and yield in carrot in relation to irrigation and nitrogen (Table 1). Results indicated that frequent irrigations given to replenish 100 per cent evaporation loss resulted in maximum dry matter production and distribution to both above and below ground parts, leaf area, leaf area duration (LAD), relative growth rate (RGR), net assimilation rate (NAR) and crop growth rate (CGR) leading to higher root yield as compared to irrigation schedule to replenish 25 to 70 per cent evaporation loss. Increasing levels of N fertilization produced higher dry matter and root yield through increased leaf area. LAD, RGR, NAR and CGR. Shanmugasundaram *et al.* (2005) conducted an experiment at Horticulatural Research station, Nilgiris on effect of graded levels of nitrogen, FYM, lime and biofertilizers. Results indicated that nitrogen treatment @ 200 kg/ha has shown maximum nitrogen content in the foliage (2.4%) and in roots (1.52%). A combination of FYM (10 t/ha), lime (5t/ha) and *Aztobacter* and Phosphobacterium (2kg/ha) with 120 kg nitrogen per hectare was found to be effective in improving the over all performance of the crop.

Nath *et al.* (2008) concluded that application of major nutrients in combination with biofertilizers and organic manures would benefit the crop. The range of major nutrients varies from region to region. But ideal doses of NPK for carot crop are found to be 40-50kg. 40-50 kg and 80-100 kg per hectare respectively.

IRRIGATION

It is advisable to irrigate the field lightly before seed sowing to ensure sufficient soil mosture for rapid and uniform germination. Irrigation should not be done immediately after sowing the seeds as it may result in the formation of soil crust which hinders proper emergence of seedlings. Carrot is a medium rooted crop and roots are mostly confined in the depth of 120 cm. It is very sensitive to soil moisture. Insufficient moisture for carrot increases roughness, decreases root size and yield. Low moisture grown carrot has strong and pungent flavour. Final stages of root development are very sensitive to moisture stress. Dryness of soil followed by rain or irrigation is likely to cause growth cracks in carrot roots.

Under in sufficient soil moisture roots do not develop properly, hence regular irrigation should be done at 7-10 days intervals depending on type of soil, climate condition etc. (Singh, 2001). Carrot should be irrigated before any wilting of leaves appears, since they may be able to obtain enough water to keep the leaves erect without having enough moisture to make the desired growth. Furrow system of irrigation is common but basin irrigation is also followed occasionally in India.

INTERCULTURE

Thinning is the most important operation in carrot crop. Within a week of seed sowing, the seedling start emerging through the ground. In line sown corp, these are thinned to space 4 to 5 cm apart in early stage (3-4 leaves). In broad cast sown crop, thinning is done twice, once earlier to remove surplus plants from the clump leaving only one seedling at a hill. The second thinning is done at a later stage by removing the larger roots for table use, leaving the smaller one to develop.

In the early stage the crop should be kept free of weeds which may be done by fequent hoeings. It facilitates in proper soil aeration and root development. Earthing up should be done to cover the developing roots with soil and thus to prevent discolouration of roots.

WEED CONTROL

Weeds do harm more to seed crop raised by seed than they do to seed crop on raised by root. At ealry stage of crop growth, shallow cultivation once will suffice the purpose. Pre emergence herbicides may be applied for effective control.

Weedicides if selected for seed production then pre-plant application can be to ensure weed free seed crop. Linuron 0.5-1.0 kg/ha (Randhawa and Saimbhi, 1979) or Oxyfluorfen 0.25 kg/ha (Singh and Malik, 1983) or Fluchloralin 1.5 kg/ha (Choudhary and Pathak, 1992). The common weeds of carrot are *Cyperus rotundus, Oxalis corymbosa, Portulaka oleracea, Trianthema monogyna, Chrysanthemum segetum.*

So to check the growth of above weeds, maloran 50 WP pre-emergence was observed to be superior to other treatments, whereas the highest root yield was recorded with both maloran 50 WP and gesagar 50 WP (Sharma and Bhagchandani, 1979).

Mangal *et al.* (1991) reported the effect of different concentrations of chloromequate (cycocel) under different moisture regimes on productivity of carrot. The best treatement combination was irrigation at 0.8PEC and two cycocel sprays of 500 ppm strength at 30 and 45 days after planting of stecklings. However, under 0.2 PEC irrigation treatment where minimum irrigations were done, application of 1000 ppm cycocel was more effective as compared to 500 ppm of this salt. Similar beneficial effect of 1000 ppm cycocel spray over 500 ppm cycocel was recorded under 0.4 PEC irrigation intensity. Carrot was tolerant to weeds with a pre-emergent and post-emergent application of imazamox (3.6 g a.i/ha) and triflusulfuron (7.3 g a.i./ha).

In general. TOKE 25, EC @ 2.0-2.5 kg a.i/ha maloran (3 g a.i/ha) pendimethalin (1 g a.i/ha) as pre emergent application are effective against weed in carrot crop and triflusulfuron (7-7.5 g a.i/ha) and imazamox (3.5 g a.i/ha) as post emergent sprays are recommended for weed control (Nath *et al.,* 2008).

PLANT PROTECTION MEASURE

Insect Pests

Carrot fly (Psila rosae)

This is serious pest of carrot. Flies are attracted by the pungent smell of the carrot foliage, and find small areas of disturbed soil near the rows burrow into the roots making them malformed decayed and unfit for consumption. Affected roots cause wilting of foliage. The foliage becomes rusty colour. For control follow minimum, cultural operations in standing crop. It is also advisable to plant a few rows of onion in the carrot field becuase smell of onion leaves acts as repellent to carrot fly. Spraying of Dimethoate at 0.03% twice to thrice will control carrot fly effectively. It will be better to locate resistant varieties like fly effectively. It will be better to locate resistant varieties like Nantes, which is resistant to carrot fly.

Wire Worm

The larvae which are shinny and gold-yellow in colour. They live in the soil. They bore into the roots and resulting the wilting of leaves. To control the wireworm there is need to cultivate soil properly. Weeding can be done at an early stage of crop growth that will not only control weeds but also ruduce the population of wireworm as they become exposed on the surface of the soil and are picked up by insectivorous birds.

Aphids

They suck the sap from the foliage, and also someitimes damage the carrot roots. The foliage of carrot become stunted, wilted and turned to greyish-green in colour. The infestation of aphids is more during dry weather conditions. Under such period, regular irrigations should be given. Aphids do harm more to seed crop compared to root crops, and therefore spraying of Diemthoate (0.03%) will be necessary to do as and when required.

Nematodes (Pratylenchush Sp; Heterodera carotal, Meloid-ogyne sp.)

They do serious damge to roots and making them unfit for human consumption. To control namatodes it is better to avoid growing carrot on plot which is infected with Nematodes. Digging out ou affected roots from the field is necessary. Soil can be treated with Nemagon. Growing of Tagetes in carrot field help in reducing nematode population ot considerable level. Application of Carbofuran at 3 kg/ha reduces significantly population of *M. incognita* (Devi and Das, 1994).

Diseases

Leaf spot

It is also known as cercospora leaf spot. (*Cercospora carotae* Solheim). The disease incidence is first noticed by the formation of lesions on the edge of the leaf segment, causing lateral curling. In severe cases leaf turn black, leaf petioles are also affected. However roots are not affected. To check leaf spot make provision for proper drainage, long crop duration, seed treatement with hot water at 50°C for 15 minutes, and spraying crop with Dithan M-45 (0.2%) or Copper oxychloride (0.2%) are effective control measures.

Leaf blight

It is also known as *Alternaria blight* (*Alternaria dauci* (Kuhn) Groves and Skolko). Brown to black spots appear on along the leaf margin of partcularly older leaves than younger leaves. The carrot roots are not affected by this disease. Humid weather is very favourable for the quick spread of disease. It is a seed borne disease. To control this disease selection of well-drained soil or provision of providing efficient drainage system, adoption of crop rotation, seed treatment with Captan or Thiram at 3g per kg of seed. Spraying of Maneb (0.2%) at 10 days interval and finding out leaf blight resistant/tolerant culitvar like imperial Long Scarlet.

Powdery mildew (Erysiphe umberllifercerum De Barry)

White powdery spots appear on the leaves, stems, flowers and fruits. Below the spots area become brown or purplish. Dusting of sulphur powder, spraying of Maneb 80 per cent at 2.5 kg/ha at least twice at an interval of 10 days are effective to check this disease.

Aster Yellow or carrot Yellow

Leaves become yellow with vein clearing, apical buds grow out into small chlorotic shoots resembel to witches' broom, older leaves become reddish, twisted and ultimately break off. Roots do not develop proper in shape and size. Storage roots produce more number of fibrous roots on them, quality of roots is reduced as they become bitter in taste and astringent flavour. Affected seed plants become stunted in growth. There is early formation of seed stalks on infected plants compared to healthy plants. There is poor fruit setting.

This disease is transmitted by six-spotted leaf hopper (*Macrosteles divisus*) Besides other crops are also infected with Aster yellow are celery, lettuce, potato, tomato etc. The Aster yellow disease can easily be controlled by the control of six spotted leaf hopper with cetain insecticides. Severely affected plants should be uprooted. Weed plants act as host plant for leaf hoppers and therefore such weeds should be removed.

Bacterial blight or Root rot : (Xanthomonas carotae Dowson)

Affected leaves show small, irregular yellow areas which turn dark brown and water soaked. On petioles and stems dark brown streaks are formed (Singh, 1985). Brown to maroon lesions are formed on the roots. The appearance is more common at the time of flowering and seed setting. To control this disease it is advisable that seed crop of carrot should be raised on the plot where carrot was not grown for the last at least three years before sowing. Carrot seeds should be treated with hot water (50°C) for 15 minutes. Planting of crop relatively at wider spacing proves beneficial.

White rust (Albugo candida (Lev.)) Kunze

Affected leaves develop white powdery patches on the under surface severe infection causes malformation of leaves, stem and foral parts. Moist and cool weather favour the disease. To control this disease make weed free plots and spray. Dithane M-45 at 0.2% solution.

Physiological Disorder

Carrot Spliting

Splitting or cracking of carrot is a major problems in carrot growing areas, splitting seems to be controlled by genetic factors and a number of other factors be involved. (a) Due to increase in amounts of N in the soil and that high soil concentration of ammonium compound cause more serious splitting.

(b) Early cultivars tend to split more than late cultivars. (c) Heavy side dressing with NH_4NO_3 at the root stage of 6.5 mm in diameter with spaced root increase spliting. (d) Wider the spacing, the greater amount of splitting and large roots are more likely to split than small ones.

Cavity Spot

This disorder appears as a cavity in the cortex, in most cases the subtending epidermis collapses to form a pitted lesion. It is of non pathogenic origin. (a) This disorder is induced primarily deficiency of Ca and associated with increase accumulation of K and a decrease accumulation of Ca. (b) It is observed that cavity spot disorder of carrot roots is manifestation of Ca deficiency, which may be induced excess K uptake during the outogeny of carrot plants. Increase in Ca levels in the growing medium result in increased Ca accumulation in the plants, which have significant affect on controlling of disorder.

HARVESTING

The carrot may be adjudged for harvesting at the marketable stage depending on the size of the root. Care should be taken that they should not grow beyond the marketable stage as they become fluffy and unfit for consumption. The common Asiatic varieties attain the marketable stage at 2.5 to 4 cm diameter at the upper end. The roots which are grown on ridges are easy to be pulled out than those in flat beds. In both the cases it is advisable to give light irrigation before harvesting which facilitates publling out the roots from the soil without any damage.

YIELD

The Asiatic varieties yield higher than the European varieties. The Asiatic varieties yield 20-25 t/ha whereas European varieties yield 10-15 t/ha. The roots are washed, cleaned and bunched in a fixed number, say half-a-dozen together.

POST HARVEST HANDLING

Grading

The carrots are graded according to shape, size, tenderness and brightness of the edilbe roots. Generally roots are separated in 'super', 'fancy' and 'commercial' grade. The top quality tender root with uniform size, colour are kept in 'super grade'. The medium size roots with good texture and colour arranged in 'fancy grade'. The big roots with clean surface without any belmish and splitting are arranged in 'commercial grade' for better marketing.

Packing

Graded roots are tied in bunches (6-12 roots in each) and packed in polythene bags, plastic bags, and baskets. Sometimes pre-packed in perforated polythene bags for storage. Most of the carrot roots packed in transparent film bags and send to market. For distant market roots are packed in plastic crates, CFB, boxes, cardboard boxes and in plastic tray/wire bound baskets. For local markets roots packed in gunny bags or round bamboo chips baskets and stitiched properly. Packing carrots in polythene bags reduced storage losses by 15.5 and 12 percent in cold stores and ventilated stores, respectively. Carrot packed in cotton nets lined with perforated polythene showed 2-4 percent weight loss and were substaintially fresher in appearance and contain no infected roots than those and in unlined nets. Lining bulk bins with polythene also reduced dehydration to an acceptable extent.

Storage

Fresh carrots cannot be stored for more than 3-4 days under ordinary condition. However long-term storage in cold store is possible without any appreciable change in quality. At termperatures of 0°C with 90 to 95 per cent relative humidity they can be stored for seven weeks (Pantastico, 1975).

SEED PRODUCTION

Carrot has two races for seed production point of view.

1. Biennial or Temperate or European type-remain vagetative in the first year and become reproductive in the following growing season. The exposure of the plant to temperatures below a critical level (vernalization) is essential for the induction of the reproductive stage, i.e. the formation of the seed stalk and flower development (bolting).
2. Annual or Oriental or tropical or Asiatic type. In the plains of Northern India the Asiatic or oriental type produce seeds freely.

The seed production of temperate type is limited to the hills only. Since it is a cross-pollinated corp, every care should be taken to keep the two varieties away from each other and the isolation distance given to produce foundation seed is about 1000 and for certified seed it is 800 m (Arya, 2003).

Table 2 : Field standards for carrot seed production (Maximum permitted in per cent)

Factors	Foundation	Certified
Off-types	0.01	0.20
Other crop plants	-	-
Objectionable weed plants	-	-
Diseased Plants	-	-

Table 3 : Seed Standards for carrot seed prodcution (In percentage)

Factors	Foundation	Certified
Pure seed (Minimum)	95.0	95.0
Inert Matter (Maximum)	5.0	5.0
Other crop seed (Maximum)	0.05	0.1
Total weed seed (Maximum)	0.1	0.2
Objectionable weed seed (Maximum)	-	-
Germination (Minimum)	60.0	60.0
Moisture content (Maximum)	8.0	8.0

Usually good quality and high yielding roots of variety are selected at the root marketable stage and transplanted in well prepared field after giving proper root and shoot cuts to ensure better qualtity and higher seed yield. The selected roots are planted at a distance of 75cm x 20 cm. Nath and Kalvi (1969a) reported the maximum of 82.8 per cent plant survival with two third top cut and no root cut, or one half root cut treatment and the minimum of 62.5 per cent with no top cut and three fourth root cut treatment in local Rajasthan variety Sharbati. The seed stalk growth was delayed with the increase in the size of the root cut. The size of the plants in terms of height and width was not affected by any of the shoot and root cut treatements. In another report (Nath and Kalvi, 1969b) the maximum seed yield per hectare was recorded from the plants which received the treatment of one third top cut and one half root cut in the two varieties. The highest seed germination of 72.2 per cent was obtained in the first order umbel, 59.7 per cent in the second order umbel and there was no germination within a month of harvest in the third order umbel seeds. These boservations suggested that the stecklings should be given the treatment of one third shoot cut and one fourth to one-half root cut to obtain better quality and higher seed yield in carrot. If possible the seed growers may not harvest the seeds of third order umbel, which has poor seed germination, thereby ensuring better germination percentage from the seeds collected from the first two orders of umbel (Nath *et al.,* 2008).

The off-type plants including premature bolters should be rogued out before flowering of the stalks. The harvesting is done when the secondary umbels are fully ripe and the third order umbels have begun to turn brown. Shattering is not a serious problem in carrots. The plants can be pulled and stacked for 330 to 550 kg of seed per hectare was obtained from the European variety Nantes in the Kulu Valley (Singh *et al.*, 1960), Whereas an average seed yield of 2,054 kg/ha with proper root and shoot cut treatements was botained from the Asiatic variety, but the plants with no treatment produced 1,461 kg of seed/ha only (Nath and Kalvi, 1969b). Malik and Kanwar (1969) recommended that all seeds smaller than 1.66 mm gathered from the late harvest should be eliminated from the seed stock because a good response to seed size was obtained by shoot growth, root development and yield. At the time of seed maturity a dry and hot weather is required.

Drying and seed extraction

Carrot umbels are dried within 5-6 days on open sun. While drying, turning of the material from bottom to up helps in proper drying. The seeds are extracted by beating the dried umbels with sticks. Thresher may also be used on commercial scale.

Seeds are winnowed well by blowing against the wind if thresher is not used/available. The seeds should so cleaned that there is no chaff of other inert matter. Cleaned seed should be dried to 8.0 per cent moisture (Table 3).

Packing and Storage

As per demand, the packing of the seed should be done in polythene bags, glass or plastic containers or some other suitable containers.

Packed seed should always be stored away from humid places.

Seed yield

On an average, carrot seed yield is obtained about 1200 kg to 2000 kg per hectare. One can expect more than this range depending on methods, climatic conditions, soil types, cultivars and management practices.

Treatments for higher seed yield

A single spray of 750 ppm GA_3 at the 8 to 10 leaf stage increase seed yields via an increase in seed number per umbel but seed germination has not been affected (Galmarini *et al.*, 1995). Higher seed yields and seed quality are obtained with 4000 gamma irradiation treatment of carrot seeds before sowing. Besides this treatment also increases nutritive values of roots and reduces the storage losses by 6.5 per cent (Singh and Singh, 1996).

REFERENCES

Anonymous () 'Ain-i-Akbari, English Translation by H. Bolckmann, Ed. SL. Goomer, Aadiesh Book Depot Delhi.

Arya, (2003). Vegetable breeding- Production and seed production : Kalyami Publishers, New Delhi- pp 74.

Aykroyd (1941). Health Bulletin No. 23. Nutrition Research Lab., Koonoor.

Bailey, L.H. (1948). Manual of cultivated plants. The Mac millan Co. New Yark.

Barnes, W.C. (1936) Effects of some environmental factors on growth and colours of carrots. N.Y. (Cornell) state memoir, 186.

Beiley L.H. (1928). The Standard Encyclopedia of Horticulture 1:674.

Benz, D.R. (1968). Evidence for carrot splitting as an inherite tendency. *Proc Amer. Soc. Hort Sic.*, 93:429-433.

Bhagchandani, P.M. and Choudhuri, B. (1971). Heterosis and inbreeding depression in carrot (*Dacucus carota* L.), *Progressive Horticulture*, 2 : 65-75.

Bhardwaj, L.N. and Shreshta, S.M. (1985). Agricultural Ecosystems and Environment, Vol. 13, pp. 314-319.

Caht field, C. (1954). Food composition minerals and Vitamins, F.A.O. Nutrition Studies No. 11.

Chandel, Kamaljit, Singh and Rattan, R.S. (1988). Epistasis, additive and dominance variation in carrot. *Veg. Sci.*, 15 : 31-37.

Chatfield C. (1949). Food composition Tables for International use, F.A.O. Studies No. 3

Choudhary, J.K. and Pathak, A.K. (1992). Chemical control of wood sorrel (*Oxalis maritiana*) in carrot crop and non-crop situation. *Indian J. Agron*, 37(3) : 537-540.

Choudhury, B. (1967). Vegetable, Natinal Book Trust, India.

Devi, G. and Das, P. (1994). Estimation of losses in carrot due to root-knot nematode, *Meloidogyne incognita* under field conditions. *J.Agric, Sci. Soc.*, North East India. 7(1): 109-110.

Dickson, M.H. (1966). The inheritance of longitudinal cracking in carrots. *Euphytica.*, 15: 99-101.

Elballa, M.M.A. and Cantiliffe, D.J. (1996). Alteration of seed stalk development, seed yield and seed quality in carrot by varying temperature during seed growth and development, *J.Am. Soc. Hort. Sci.*, 121 (6) : 107-108.

Galmarini, C., Borgo, R., Gavida, J.C. and Tizio, R. (1995). Effect of gibberellic acid on seed production in carrot (*Daucus carota* L.). Effect of different concentrations and application dates on vegetative cycle length and seed yield and quality. *Horticulture Argentina*, 14(37): 74-86.

Gill, H.S. and Kataria, A.S. (1974). Some biochemical studies in European and Asiatic varieties of carrot (*Daucus carota*) *Curr. Sci.*, 43 : 184-185.

Jain, N.L. and Mukherjee, S.K. (1966). Vegetable varieties for Processing and Preservation, *Ind. Hort.* X, No. 2.

Malik, B.S. and Kanwar, J.S. (1969). Effect of seed size and stage of harvest of carrot seed on germination, growth and yield of carrot. *Indian J. Agric. Sci.*, 38(6): 603-610.

Malik, Y.S., Mehrotra, X.I., Pandita, M.L. and Jaiswal, R.C. (1983). Effect of soil salinity levels on the yield and quality of radish, carrot and turnip. *Indian J. Agric. Sci.*, 53(9): 861-862.

Mangal, J.L., Srivastava, V.K. and Karwasra, S.P. (1990). Soil tolerance in vegetable crops. *Tech, Bull*, 1, Haryana Agric, Univ., Hisar (India).

Mangal, J.L.; Lal, S., Yadav, A. and Kohli, V.P. (1991). A note on the effect of different concentrations of chloromequate (cycocel) under different moisture regimes on productivity of carrot. *Haryana J. Hort Sci.*, 20 (1-2): 144-146.

Marshall, Cavendish (1981). The Fruit and vegetable. Garderner's Handbook, Pub. Marshall Cavendish Book Ltd. London.

Nath, Prem and Kalvi, T.S. (1969a). Vegetative growth and flowering as influenced by root and shoot cuts in transplanted carrot stecklings. *Farm J.*, 11(2) : 9-12.

Nath, Prem and Kalvi, T.S. (1969b). Carrot seed quality and yield as influenced by root and different order umbels and root and shoot cut treatements. *Punjab Hort. J.*, 9 (1&2) : 81-89.

Nath, Prem; Srivastava, V.K. Dutta, O.P. and Swamy, K.R.M. (2008). Vegetable Crops. Improvement and Production, PNASF, Bangalore. 177-181.

Nath, Prem; Velayudhan, S. and Singh, D.P. (1987). Vegetables for the Tropical Region, ICAR, New Delhi. pp 98-104.

Pantastico, Er. B. (1975). Post harvest physiology, Handling and utilization of Tropical and subtropical Fruits and Vegetables (Eds.) AVI, Publishing Co. West Port, connecticut, USA.

Poierre, M.D. St. and Bayer, R.T. (1991). The impact of domestication on the genetic variability in the orange carrot cultivated *Daucus carota* ssp. *sativus* and the genetic homogeneity of various cultivars. *Theor. Appl. Genet.*, 82 : 249-253.

Prabhakar, M., Srinivas, K. and Hegde, D.M. (1993). Physiological analysis of growth and yield of carrot (*Daucus carota* L.) in relation to irrigation and nitrogen. *Indian J. Hort.*, 50 (3) 251 : 260.

Rai, N. and Yadav, D.S. (2005). Advances in Vegetable Production. Researcho Book Centre, New Delhi.

Randhawa, K.S. and Saimbhi, M.S. (1979). Performance of some pre-and post-emergence herbicides in carrot (*Daucus carota* L.). *Haryana J. Hort. Sci.*, 5: 210-215.

Scibrja, G. and Zemoja, A. (1969). Agricultural techniques for carrot growing for carotene production. 131-139. *Abstr.* 21, 55, 335. *Plant Breeding Abstr.*, 39 (1) : pp 516.

Seshadri, V.S. and Chatterjee, S.S. (1996). The history and adaptation of some introduced vegetable crops in India. *Vegetable Science* : 23 (2) 114-140.

Shanmugasundaram, R. and Savithri, P. (2004). Effect of nitrogen levels, organics and amendments on nitrogen nutrition of carrot. *Agri. Sci. Digest*, 24 (1) : 5-8.

Sharangi, A.B. and paria, N.C. (1995). Growth, yield and qualitative responses by carrot to verying levels of nutrition and potassium. *Hort. J.* 8(2) : 161-164.

Sharma, R.K. and Bhagchandani, P.M. (1979). Chemical weed control in Carrot, *Daucus carota* L. *Veg. Sci.*, 6 (1) : 1-4.

Shoemaker, J.S. (1947). Vegetable growing. John wiley and sons Inc. New York.

Singh H.B., Thakur, M.R. and Bhagchandani, P.M. (1960). Vegetable Seed Production in Kulu Valley. III. *Indian J. Hort.* 17 (1): 38-47.

Singh, A.K. and Singh A.K. (1996). Effect of nitrogen and potash on seed yield of carrot (*Daucus carota* L.) Pusa Kesar. *Crop Res.* (Hisar), 12 (2) : 182-184.

Singh, D.N. and Nath, Vishal (2011). Varieties and Hybrids of Vegetables. Satish Serial Publishing House, Azadpur, Delhi (India), 54-57.

Singh, H.B. (1963). Pusa Kesar and Pusa Kanchan, Carrot and turnip varieties. *Indian Hort.*, 7 (4) : 21.

Singh, H.B., Thakur, M.R. and Bhagchandani, P.M. (1960). Vegetable seed production in Kulu Valley II. *Indian J. Hort.*, 16 (3) 153-161.

Singh, K.P. and Malik, Y.S. (1983). Chemical weed control in carrot seed crop. *Indian J. Weed Sci.*, 15: 21-16.

Singh, R.S. (1985). Diseases of Vegetables. Oxford & IBH Publishing Co. New Delhi.

Tandon, H.L.S. (1991). Fertilizer Recommendations for Horticultural Crops. A guide-book, Fertilizer Development and consultation organization, New Delhi.

Thompson, H.C. and Kelly, W.C. (1957). Vegetable crops. MC Graw Hill Book Co, Inc., USA.

Yawalkar K.S. (1980). Vegetable crops of India. Second Edition, Agri - Horticultural Publishing House, Nagpur.

Yawalkar, K.S. (1985) Vegetable Crops of India. Agri-Horticulatral Publishing House. Nagpur. 440010 (India).

□□□

3. TURNIP

Botanical name	:	*Brassica rapa* L. (Syn.*Brassica campestris*)
Family	:	Cruciferae
Chromosome number	:	2n = 20

Turnip is a cool season root vegetable grown mainly in temperate and subtropic regions. It is a frost-hardy, cool-season biennial that is grown as annual for their enlarged fleshy taproot. It has been in cultivation in Europe for over 4000 years. Turnip was introduced in Canada in 1540 by Cartier during his voyage of exploration, and was taken to Mexico in 1586 and Virginia in 1610 (Peter, 1998). It is grown as a winter vegetable in the north western states of India and as a summer crop in temperate climate. It can be successfully grown at elevation of 1500m or above (Brar and Nandpuri, 1972).

ORIGIN

Two main centre of origin have been indicated. The Mediterranean area is thought to be the primary centre of origin of European types (Bailey, 1963), while Eastern Afghanistan with adjoining area of Pakistan is considered to be another primary centre with Asia minor, transcaucasus and Iran, as secondary centres (Sinskaia, 1928). The parents of cultivated turnip are found wild in Russia, Siberia, and Scandinavia (Thompson and Kelly, 1957).

DISTRIBUTION

Turnip was cultivated by the ancients and ranked next to grapes and cereals in Italy. It is widely distributed in Europe, Germany, U.K., Russia and Asian countries i.e. Afghanistan, Pakistan, Iran, China, and India (Purewal, 1957). Turnip is now cultivated throughout the world. In India, turnip is grown on small scale in Delhi, Jammu and Kashmir, Himachal Pradesh, Haryana, Punjab and Uttar Pradesh (Nath *et al.*, 1976).

BOTANY

Turnip belongs to the family Cruciferceae and genus *Brassica, B. rapa,* L. emend. Metzger (syn. *B. campestris* Hooks f. & Thomas var. *rapifera* Metzger). The crop is highly cross-pollinated due to self incompatibility and male sterility (Peter, 1998). It is a herbaceous annual for the root production, whereas biennial for the seed production. Leaves and petioles of turnip are hairy and course and yellowish green in colour. Leaves are oblong to oval, serrate or pinnate. The inflorescence is a terminal raceme on the main stem. Flowers are hermaphrodite and are highly cross-pollinated mainly due to self-incompatibility in nature (Yamaguchi, 1983; Nath *et al.*, 1987). The root is napiform consisting of the hypocotyle, which swells and become spherical. A distinct taproot and secondary root arise from the lower part of the swollen hypocotyle. Thickening begins in the central part of the hypocotyle, followed by the upper and lower parts. Asiatic type sets seeds freely in the plains whereas the European type do not produce seeds in the plains.

Plant breeding

Turnip is cross-pollinated hence inbreeding is done in this crop. The important uses of inbreeding in cross-pollinated crops are to obtain uniformity in plant characters, to improve yield by individual plant selection and to combine suitable inbred lines in the production of hybrids and synthetics. Self fertilization is the most intense form of inbreeding. Until 3rd or 4th generations, inbreeding depression is not much. Mass selection, progeny selection and hybridization are the common breeding systems in turnip. Hybridization between Asiatic and temperate types has been effective in this crop (Nath *et al.*, 2008).

A number of good genetic stocks of Asiatic and European types are available in India. Some of them are very good varieties. Efforts were made in the past to maintain some of the Asiatic strains, but no intensive breeding work was taken up. However, an interesting achievement has been the release of the new varieties Pusa Kanchan (Singh, 1963), Pusa Sweti (Asiatic type), Pusa Chandrima (European type), (Singh *et al.*, 1971) from IARI, New Delhi. Nath *et al.* (2008) reported that Pusa Kanchan has been selected from cross between an Asiatic variety Local Red Round and the European variety Golden Ball. It has good qualities of both Asiatic and European type of varieties with ability to produce seed freely in the plains.

COMPOSITION

The roots contain vitamin B, and vitamin C in appreciable quantities and also have ascorbic acid, calcium, sugar, and protein, in fair amounts. Turnip greens are good source of calcium, iron, and vitamin A, C, and B (Choudhury, 1967). Details composition in per 100g of edible portion. (Aykroyd, 1963) is given below:

Moisture	41.6%	Ascorbic acid	43.0 mg
Carbohydrate	6.2 g	Calcium	30.0 mg
Protein	0.5 g	Phosphorus	40.0 mg
Fat	0.2 g	Fe	0.4mg
Thiamin	0.04mg	Vitamin A	15669 I.U.
Riboflavin	0.04mg (in green)	Vitamin C	43 mg
Energy	29 Kcal		

Uses

Turnip roots are used as vegetable and as salad. The roots are also cooked alongwith the tender top. European types of turnip are relatively sweet and more palatable and may be eaten sometimes raw. The Asiatic turnips use to prepare good pickles and to a limited extent in salads (Choudhury, 1967 and Nath, *et al.*, 1987). Some places it is used in several delicious dishes including with meat. Turnip leaves are some times used as vegetable but are extensively utilized as green fodder for feeding cattle (Gopalan *et al.*, 1971).

VARIETIES AND HYBRIDS

There are two distinct groups of turnip (i) biennial or temperate or European type; (ii) annual, tropical or Asiatic type. The seeds of former group can be produced in the hills and plains (Gill, 1993).

Turnip cultivars have also been classified into several groups on the basis of morphological characteristic i.e. shape and colour of the root and top (Shoemaker, 1949).

Classification

(1) White Fleshed, (2) Yellow Fleshed and (3) Foliage Cultivars.

White Fleshed

It has three different types : (a) Purple topped, (b) Green topped, (c) White topped.

(a) *Purple topped* : It has two types –

(i) Flat type, var. purple top Milan,

(ii) Globe type, var. Purple Top White Globe.

(b) *Green topped* : It has two types

(i) Globe type, var. Green Top White, Green Globe,

(ii) Long type: var. Cow Horn.

(c) *White topped* : It has four types :

(i) Flat type, var. White Milan, White Flat Dutch, Norfolk,

(ii) Globe type, var. White Stone, Quick Silver,

(iii) *Half-Long type*: var. White Egg, White Gem

(iv) Long type, var. Lily White.

Yellow Fleshed

It has three different types

(a) Purple topped, globe type, var. Aberdeen Purple Top.

(b) Brodge or green topped, globe type: var. Aberdeen Green top, Amber Globe.

(c) Yellow topped, globe type: var. Yellow Globe, Golden Ball, All Seasons.

Foliage cultivars

Flat Japan, Shogoin, Seven Top.

Variety suitable for saline and sodic soil

Sel 66, Snow ball, and purple top white globe.

Cultivars of both temperate or European and Asiatic types are now available in India. Detail characteristics of cultivars/varieties and hybrids described by Singh and Nath (2011) mentioned herewith :

Pusa Chandrima (Asiatic x European)

It is early maturing with high yield and possess all the good qualities of the European type. Roots are medium to large in size, 8-9 cm in length and 9-10 cm in diameter. The tops are medium but not so deeply cut. The skin is smooth, pure white, fine grained with sweet and tender flesh. The root attains maturity in 55-60 days after sowing. It is temperate or biennial cultivar and does not produce seeds in the plain of India (Gill *et al.*, 1971).

Pusa Swarnima (Japanese white x golden ball)

It matures 15 days earlier than Golden Ball and yields 40% more. The roots are flatish round, 6-7 cm long and 7-8 cm diameter with creamy yellow skin, flesh amber coloured, fine textured and mild flavoured. Tops are medium and not very deeply cut. This gives more seed yield than Golden Ball. It matures in 65-70 days.

Pusa Kanchan (local red round x golden ball)

It has good qualities of both the types. The skin is red, but the flesh is creamy yellow with excellent flavour and taste. The leaf top is shorter than the Local Red Round (Singh, 1963). It is earlier by 10 days than Local Red Round, but the roots can be kept longer in the field without being spongy. It sets seeds in the plain.

Pusa Sweti

It is an Asiatic cultivar suited for sowing in October. It has attractive white roots. It produces seeds in the plains of India. It matures in 45-50 days after sowing and yields 20-30 tonnes/ha (Anonymous, 1998).

Purple Top White Globe

It is a high yielding large rooted cultivar. The roots are nearly round, upper part purple, lower portion creamy, flesh is white firm, crisp and mildly sweet-flavoured. The top is small, erect with cut leaves . It is adapted to cooler months.

Golden Ball

The roots are perfectly globe-shaped, medium sized and smooth. It has bright creamy yellow skin and pale, amber coloured flesh of fine texture and flavour. The top is small erect and with cut leaves.

Early Milan Red Top

It is an extra early and high yielding cultivar reaching maturity in 45 days. The roots are deep flat with purplish-red tops and white underneath. The flesh is pure white, well-grained, crisp and mildly pungent. The tops are very small with 4-6 sessile leaves.

Snowball

This variety is an early temperate type, with medium sized small top. It has a duration of 55 days. Leaves are small and light-green. Roots are round and medium sized, flesh light-yellow.

Pubjab Safed-4

This variety is early maturing, commonly grown in Punjab and Haryana. The roots are attractive, medium size round and white. The taste is good.

L1

The root is round, pure white, smooth, rattailed and crisp with mild flavour. It can be harvested within 45 to 60 days after sowing. It yields about 262 quintals/ha.

4-White

It is an early maturing deshi type. Its roots are white, round and medium and developed in about 60 days. The average yield is about 200 q/ha.

4-Red

It is also an early maturing deshi type. Its roots are red and medium. It yields about 200 q/ha (Saini, 1996).

CLIMATIC INFLUENCE

Turnip is grown throughout temperate regions, subtropics and tropics where the temperature for a period of 3 months is 15.6°-21.1°C. The moderate climate is best suited for its cultivation. It is a hardy crop and it can tolerate frost and mildly freezing temperature. Higher temperature adversely affects quality; the root become woody, tough and bitter in taste in hot weather. Low temperature below 10°C are likely to cause flowering. The most favourable temperature for the development of the root and the ratio of root/greens are 10°-13°C air temperature, and 18°-23°C root temperature. Higher ascorbic acid content is obtained at greater light intensity (Nath *et al.*, 1987).

SOIL REQUIREMENT

Turnip can be grown on a wide range of soils in the presence of adequate organic matter. A medium deep, friable, well drained soil with 5.5-7.0 pH is recommended. High yields of best quality roots can be obtained in sandy loam soils (Kale and Masalkar, 1993). Extremely light clay soils or very sandy soils should be avoided and except these extremes a satisfactory crop can be grown on most of the soils. Turnip is moderately tolerant to acid soil (pH 6.8 to 5.5) to alkaline soil (pH 7.5 to 8.0).

SOWING

The sowing time of turnip depends on type of varieties and climate. In plains of India, Asiatic types (tropical types) are sown from end of July to September and the European types (temperate types) from September-October to December (Choudhury, 1967; Yawalkar, 1980). In the hills, sowing is done from March to May. The optimum sowing time is first week of October in East-Coast region, whereas, September is the suitable sowing time to get higher yield.

The seed rate of about 3 to 4 kg is required to raise crop in one hectare land. Seed treatment should be done with thiram or captan 3 g/kg of seeds. Spacing of row to row 30 - 40 cm and plant to plant 10-15 cm is optimum for high yield (Nath *et al.*, 1987). There are different methods for sowing/planting : (a) Ridge sowing. (b) Flat sowing and (c) Broadcasting.

Turnip does not transplant well, so it is raised by seed. The seeds are some times mixed with fine sand or ash to facilitate uniform sowing. Seeds are sown in rows 30-40 cm and 1.25 cm deep in furrow either on ridges or on flat beds. Some times broadcasting is done to save labour expenses (Bailey, 1965).

Sowing is also done on ridges by opening shallow furrows on the ridges, and seeds are sown 1.25 cm deep in furrows and covered properly.Ridge sowing is generally done in rainy season or in low lying area to facilitate drainage. In normal cases, seeds, germinate after 4 to 6 days of sowing.

NUTRITIONAL REQUIREMENT

Turnip requires high amount of nitrogen and phosphorus similar to other root-crops. The fertilizer doses recommended for turnip are 23-25 tonnes FYM/ha supplemented with 60-70 kg nitrogen and 40-50 kg each of phosphorus and potash (Choudhury, 1967; Yawalkar, 1980). According to Nath *et al.* (1987) 20-25 t FYM, 70-100 kg of nitrogen and 44 kg each of phosphorus and potash per hectare are recommended for higher yield. Fertilizer dose of 62:40:50 kg N:P:K/ha is recommended for red lateritic soil of Orissa (Anonymous, 1988).

Depending on the fertility status of the soils various fertilizer recommendations have been made (Swarup, 2006) and the practices being followed are (kg/ha): 20 to 25 tonnes FYM, 70 to 100 kg of nitrogen, 50 kg of phorphorus and 50 kg of potash.

The entire dose of farm yard manure, phosphatic and potassic fertilizers should be applied to the soil at the time of field preparation. Half of nitrogen is to be applied in the form of ammonium sulphate or calcium ammonium nitrate at the time seed sowing. Fertilizers are mixed properly and placed 7-8 cm below the seeds in the soil alongwith few quantity of farm yard manure and cover with soil for proper utilization. Another half of nitrogen should be applied as topdressing at the time of root formation.

IRRIGATION

Turnip is a shallow rooted crop and prefers liberal moisture supply for rapid growth, yield and quality (Hegde, 1993). Irrigation at 0.2 bar soil moisture tension at 15 to 22.5 cm soil depth was optimum for turnip on sandy loam soils (Devenport, 1962). This required 9 irrigations totaling 230 mm water. However, Pandey (1966) recorded that only about 3 irrigations totalling 135 mm water were enough when there were winter rains. Lal and De (1972) at Delhi observed high yield of turnips irrigated at 0.3 bar soil moisture tension at 22.5 cm depth which required 5 to 6 irrigations with about 310 mm water.

Both basin and furrow system of irrigation are adopted to irrigate turnips in India. The first irrigation is given after sowing of seeds if the seed bed has no conserved moisture. If there is adequate moisture in the field at the time of planting, the first irrigation should be given after thinning. Subsequent irrigations are scheduled according to the season at an interval of 3 to 6 days in summer and 10 to 12 days during cold weather (Hegde, 1993).

High moisture at the beginning of root swelling depresses the yield which is attributed to changes in nitrogen utilization within the plant. The crop should be supplied with plenty of water from root enlargement to harvest as it is the most sensitive critical stage and any stress will drastically lower the yields (Hegde, 1993). In general, the crop needs irrigation at an interval of 10-15 days (Sadhu, 1986).

INTERCULTURE

Shallow hoeing is given just after emergence of the plant to make soil loose for proper root growth. Next hoeing is given 25-30 days after sowing to keep plot clean and to make soil loose for proper root development The top dressing of fertilizer should be done after next hoeing (25-30 days after sowing) then plant is earthed up.

Thinning

Thinning is an important in turnip to maintain optimum plant spacing and required population. Plants are thinned out keeping a distance of 10-15 cm from plant to plant within the row , when germination is completed and plants are well established (Kale and Masalkar, 1993).

Earthing

In order to provide good conditions for root development, earthing up is done 25-30 days after sowing and after the second dose of application of nitrogen by putting some soil around the base of the turnip plant so that roots are covered. Some times, side shoots are developed which should be removed periodically. Earthing up and top-dressing of nitrogenous fertilizers are done simultaneously.

Crop Rotations

According to Saini (1996) following crop rotations can be taken successfully to increase the economy of vegetable growers.

1. Okra-Turnip - Cowpea
2. Cluster bean - Turnip - Okra
3. Bottle gourd - Turnip - Long Melon
4. Bitter gourd - Turnip - Chilli
5. Ridge gourd - Turnip - Tomato
6. Amaranthus - Turnip - Brinjal
7. Cauliflower (Early) - Turnip - Maize
8. Cowpea - Turnip - Chilli
9. Maize - Turnip - Okra
10. Greengram - Turnip - Bittergourd
11. Blackgram - Turnip - Clusterbean
12. Soybean - Turnip - Cauliflower
13. Garlic - Turnip - Cauliflower
14. Tomato - Turnip - Cabbage

WEED CONTROL

The growth of turnips being rapid in a short period, fertile soil is preferred and even requirement of fertilizers is also more. This condition encourages the growth of weeds to a great extent. Weeds, if uncontrolled, leads to a very poor yield of turnips (Leela, 1993). Some weedicides have been recommended for controlling weeds in turnip.

i) Nitrofen and alachlor were effective herbicides to check weeds (Singh and Sharma, 1968; Singh *et al.*, 1974).

ii) Pre-emergence application of alachlor 1.87 and 2.50 kg/ha and nitrofen 0.93 and 1.25 kg/ha gave excellent control of all weeds except. *Cyperus rotundus* and resulted in root yields higher than the hand weeded control (Saimbhi and Randhawa, 1976).

iii) Fluchloralin (0.9 kg/ha) applied as pre plant incorporation only in flat sown crop also gave very good weed control and highest root yield (Leela, 1993).

iv) Increased per cent of protein in thrips was recorded by Sharma *et al.* (1978) with alachlor 2.0 kg, nitrofen 1.87 kg, diuron 0.25 kg and linuron 0.25 kg/ha as compared to the weeded control.

v) Application of trifluralin (2.0 kg/ha) gave good control of broad leaf weeds with enhanced yield of turnips (Pandey and Gupta, 1979).

vi) Singh and Reddy (1990) reported that application of basalin @ 0.75 l/ha as pre plus post planting coupled with 100 kg N/ha is suitable for commercial seed production of turnip crop under hill condition.

PLANT PROTECTION MEASURES

The important insect pests and diseases of turnip and their control measures are given below:

Insect Pests

Flea beetle (Phyllotreta vittata Fabr.)

The turnip flea beetle is the most harmful pest, particularly during dry weather. They make small holes in the leaves ; in severe cases leaf show skeletonized appearance. Besides, they transmit turnip yellow mosaic virus. Hoeing and spraying carbaryl (0.2%) reduce the population of flea beetle considerably.

Wire worms

These are harmful in the soil which are cleared off for cultivation. They bore roots and make tunnels. To control wire worms, apply chloropyriphos dust before sowing and do regular shallow cultivation.

Gall weevil

They attack roots and make them swollen and disfigured. When cut the root galls, maggots of white in colour are found. Uprooting infested plants and application of chloropyriphos powder in the soil before sowing is effective.

Aphids (Brevicoryne brassicae L.)

Radish and cabbage aphids are seen in clusters on ventral side of leaves.They suck the plant sap. Affected plants curl and show yellowing symptoms. Spraying of 1-2% ziram, zineb, captan,check the aphids.

Mustard saw fly

Flies are most active in September to December. They feed on the leaves and make small holes. Females have a saw like ovipositor with which they cut the leaf tissues and make unfit for consumption. Spray of carbaryl, phosphamidon, endosulfan @ 0.2% is effective to control the pest.

Diseases

Powdery mildew (Erysiphe cichoracearum DC.)

Affected leaves show powdery patches of varying size. Dry atmosphere is favourable. Irrigation of crop regularly, removal of leaves as soon as disease appear and spray the crop with dithane M-45 @ 0.2% is effective to control the powdery mildew (Mittal *et al.*, 1987).

Downy mildew (Perenospora destructor (Berk) Casp.)

Affected leaves show grayish white patches on the under surfaces of the leaves. Leaves turn yellow and letter die. In order to control, remove affected leaves and destroy and spray with zineb at 0.2% and ridomil MZ @ 0.1 (Mir, 1987).

Bacterial soft rot (Erwinia carotovora)

This disease was noticed in Pusa, Bihar, during February, 1931 (Prasad, 1931). Affected roots become soft inside while outer skin remains hard and firm. Infection usually enters the root through any cut and injury occurred in the roots. For control uproot affected plants along with whole root and destroy. Avoid hoeing and liming. Crop rotation with non-host plants also reduce the disease.

Dry root

Affected Plants show dry brown sunken, canker-like areas on the roots. Control measures are ; avoiding the attack of slugs and snails. These cause the wounds on turnip roots which facilitate bacterium entry, avoiding use of fresh manures which are likely to carry infested roots and leaves. Uprooting and destroying affected plants is effective to check the dry root disease.

Alternaria blight (Alternaria dauci (Kuhn) Groves Skolko)

It is wide spread and when the foliage of turnip is infected, the roots may also become infected and show symptoms. The spots are nearly circular, often zonate and show various shades of brown or some times black. White mould or dark spores may cover the spot.To control the disease spray the crop with indofil M-45 or bavistin (200g) or blitox or any other copper fungicide 1200g, dissolved in 200 litres of water, for 3 times at an interval of 10 days commencing from 3 weeks age of the plant. Alternatively, foltaf or captaf (0.25%) or cuman L may also be sprayed.

Phyllody

It is common and causes severe damages to seed crops in the hills. It shows malformation of flowering shoots and the infected plants usually do not produce any normal fruits. To check the phyllody use disease free crops seeds. Sanitation of field crop is also effective.

Turnip yellow mosaic virus

This diseases is transmitted by flea beetle. Affected plants do not grow properly, root development slows down, quality deteriorates. To control the disease remove affected plants, control flea beetle and remove cabbage plants as they serve as an alternate host of turnip-yellow mosaic virus.

Turnip crinkle

This disease is characterized by crinkling of leaves. The infected leaves show rugged appearance develop yellow spots which soon coalesce to give chlorotic appearance to the leaves with necrosis. Later the leaves crinkle and wither away. Leaves emerging after inoculation, are diseased and the plants become rosette in appearance with stunted growth (Verma and Verma, 1961). Roguing of virus-infected plants and removal of weed hosts may help in minimizing disease incidence. Soil application of furadan @ 1.5 kg a.i./ha at the time of sowing seeds. 2-3 foliar sprays of either demecron (0.05%) or monocrotophos (0.05%) or rogor (0.05%) at 10 days interval check the disease.

Physiological Disorders

Brown heart

The deficiency of boron is said to be the cause of brown heart. Inner portion of the affected roots become gray or brown in colour. The roots become unfit for consumption. To control brown heart maintain adequate soil moisture i.e. do not allow turnip crop to grow under dry condition and avoid excessive liming. Apply well rotten farm yard manure and work in properly before sowing turnip seed, it will make the soil moisture retentive. Apply borax @ 20-25 kg/ha depending on soil test.

HARVESTING

Turnips are harvested when they reach to marketable size after 5-10 cm in diameter, large roots are often coarse in texture and bitter in taste. If turnips are harvested late, the roots become hard and fibrous and deteriorate rapidly in quality (Kale and Masalkar, 1993). Early varieties becomes ready for harvesting in 60-80 days whereas the late varieties 90-120 days. A light irrigation is necessary just before harvesting to facilitate lifting. The entire plants are uprooted, roots separated, cleaned and tops are cut. (Natarajan and Veeraragavathatham, 2003).

YIELD

The yield of turnip depends on variety, climatic condition, soil type, management practices adopted etc. However, yield varies from 190 to 250 quintals per hectare (Sadhu, 1986).

POST HARVEST HANDLING

Just after harvesting, the tops are completely removed. The roots are washed and cleaned in the water (better if running water is used). All the side roots are removed. This will give attractive appearance to the finished roots.

Grading

Before grading the tops of the turnips are cut near the surface of the crown. The roots are graded according to size in different grades like 'super', 'fancy' and 'commercial'. The good size of tender roots with uniform shape, brightness, lustre and smooth surface are kept in 'super grade'. The medium size roots with good texture and uniform shape with clean surface are arranged in fancy grade. The oversize roots without any blemish, dull colour with freshness are arranged in commercial grade and kept separately for marketing (Singh and Singh, 2007).

Packing

Generally roots are packed in bamboo chips baskets or Arhar sticks baskets or in gunny bags and tops are provided with some cushioning materials like grasses or paddy straw and stitched properly and send to local market (Singh and Singh, 2007). The quality roots for distance markets are packed in transparent film bag; C.F.B Boxes, Polythene bags, plastic bags, plastic crates, cardboard boxes, cartoons and wire round baskets to keep in good and fresh condition for higher prices in cosmopolitan cities like Bombay, Calcutta, Madras and Delhi. Some times roots are packed in perforated polythene bag to store in cold storage.

Storage

Turnip could be stored for 2-3 days under room conditions. However, at low temperature (0°C) and 95% relative humidity roots store well for several weeks (Yamaguchi, 1983). The washing of turnip roots improved the appearance and reduced decay (Haller, 1947). Under ordinary conditions, away from sun but under well-ventilated shed/room, turnip roots can be stored for about 4 to 6 days. Dipping the roots in hot paraffin retarded shrinkage and greatly improved the appearance because paraffin film on roots lowered the rate of respiration of the roots and tended to reduce the occurrence of internal break down (Thompson and Kelly, 1957).

SEED PRODUCTION

This root vegetable has two distinct races, namely

Biennial or temperate or European type

This type produces seeds only in temperate climate, which is prevalent in the hills of India.

Annual or tropical or Asiatic types

It can be sown for seed both in the plains as well as in the hills. It is not economical to produce seed in the hills of this type hence seeds are produced only in the plains.

Land requirement

Field should be free from same kind of crop since two previous years. Soil should be deep, light and friable and well drained.

Isolation Requirement

Turnip is cross-pollinated by insects. Field should be isolated from the varieties and other cruciferous like chinese cabbage, rape, mustard and rutabaga by at least 1600m for foundation seed production and at least by 1000 m for certified seed production.

Method of seed production

There are two method for seed production in turnip (Singh *et al.*, 1960).

Seed to seed method or in situ method

In general, this methods give better yield than other. However this method can be practiced only if quality nucleus seed is used to raise the seed crop.

Root to seed or transplanting method

It is desirable because it provides an opportunity to select desirable types of roots, thus enabling maintenance of high standard of seeds.

Annual cultivars are sown from July to September in plains and biennial cultivars are sown from last week of August to first week of September in hills. The seed rate is 3.75-5.0 kg/ha. Obtain nucleus/breeders/foundation seeds from approved agencies. The seeds are sown in ridges or flat bed. Spacing for in situ crop 45 cm x 10 cm row to row and plant to plant and for transplanting of roots 60cm x 45 cm row to row and plant to plant.

Sowing of tropical or Asiatic type

Seed to seed method

Seeds are sown in July in well prepared field on ridges 45 cm apart than flat sowing, because ridges facilitates drainage and better root development. Sowing is done in line or either on ridges or on flat bed, and the plants are thinned to 10 cm apart in the lines.

Manures and fertilizers

According to Agrawal (1980) a light dose of 20-25 tonnes farm yard manures per hectare should be applied before preparation of land. This should be supplemented by 35 kg nitrogen, 50 kg phosphorus and 50 kg potash initially, before the sowing. Another dose of 35 kg nitrogen, should be top-dressed when the roots just start forming. In second season (for transplanting of roots), apply 10 to 15 tonnes farm yard manure per hectare before preparing the field for transplanting. Supplement it with 50 kg phosphorus per hectare at the time of transplanting. Top dress 25 kg nitrogen per hectare in the pre-bolting stage, and another dose of 25 kg nitrogen before flowering starts. Chakrabarti (1983), Sandhu *et al.* (1966)and Singh and Reddy (1990) advocated the use of N at 175, 75 and 100 kg/ha respectively for getting optimum seed yield of turnip.

Weeding and earthing up

One weeding and earthing during the early stage of growth is necessary to promote root development. Another hoeing and earthing up is given before bolting so as to provide support to the seed plants, otherwise they may lodge during flowering and fruiting.

Irrigation

Irrigation may be given according to the need of the crop and the field should not be allowed to dry out during root development and flowering and fruiting.

Diseases and pests

Disease and insect pests should be kept under control by regular use of insecticides and fungicides as discussed in crop production section.

Root to seed methods

In this method, during November-December when the roots are fully matured, the plants are uprooted and true to type roots are selected. All under-developed, deformed, diseased and off types roots are discarded. After pruning the tap root and clipping the tops, leaving the crown intact, the selected roots are transplanted in a freshly prepared soil at distance of 60 cm x 45cm in such a way that the whole root is covered under the soil, leaving the crown exposed.

In the plains, the seed of the Asiatic types of root crops are generally raised by planting cut roots. The usual practise is to remove half to three fourth of the lower root before planting (Two third top left and half the root cut; one third top left and half of the root cut). This process reduces the yield considerably. It is therefore, advisable to cause as little injury to roots as possible before planting (Singh *et al.*, 1989).

The treated (roots which given cutting for planting) roots are either transplanted immediately in the field or stored at a temperature of 0°C and relative humidity of 90-95%. In kalpa valley of Himachal Pradesh, turnip roots are stored satisfactorily in trenches covered with a wooden plank and a thin layer of soil is spread over the plank. Sometimes the roots are furrowed down at the original place of sowing and removed during spring, then selected and planted. Tropical or Asiatic or annual type do not need any storage (Agarwal, 1980).

Flowering and fruiting

Bolting in European cultivars of turnip starts in the month of January and continues till March and the crop come in full bloom in the month of March-April depending on cultivar. The inflorescence of turnip is a typical terminal raceme. Honey bees are the chief pollinating agents. To ensure good seed set, pollination of all the flowers on inflorescence is necessary and to ensure it, a bee hive adjacent to the seed field should be provided (Singh *et al.*, 1960).

Roguing

Usually, root to seed method is followed for producing seed. Rogue out undesirable roots, namely off-type, diseased and malformed, at the time of transplanting. Subsequent roguing for off -types and other crop plants may be done at the bloom stage. Rogue out diseased (seed borne diseases) and severely insect damaged plants from the field as required from time to time.

Harvesting and threshing

Turnip fruits dehisce easily when fully dried. Hence shattering is a serious problem in fully dried seed crops of turnip. It is, therefore, advisable to cut the whole crop when 60-70% of it turns yellow brown in colour (Singh and Gill, 1993), at a stage when most of the pods turn yellow but not fully dried. Harvesting is done by sickle. Cut seed stalks near base of crown, preferably in the early morning. The crop can be piled in a heap, covered with a tarpaulin and allowed to cure for 4-5 days. The pile is turned upside down and allowed to cure for another 4-5 days. The period of curing, of course, depends on weather condition.

The crop is threshed by beating with a stick. The seeds immediately after threshing and winnowing should be thoroughly dried in sun, otherwise it will quickly lose viability. The seeds are then cleaned so as to make them free from small seeds and other inert materials. Seeds are store in polythene bags under cool, dry and well ventilated place for next planting season. Turnip seeds remain viable for about 4 years under good storage condition (Sharma, 1977).

Seed yield

Average seed yield is 600 to 800 kg/ha.

REFERENCES

Agrawal, R.L. (1980). Seed Technology. Oxford & IBH Publishing Co., New Delhi.

Anonymous (1988). Agriculture Guide Book. Directorate of Agriculture and Food Production, Government of Orissa, Bhubaneswar.

Anonymous (1998). Varieties and hybrids. Division of Vegetable Crops . IARI, New Delhi.

Aykroyd, W.R. (1963). The nutritive value of Indian foods and the planning of satisfactory diets. ICMR special Rept. 42: New Delhi.

Bailey, L.H. (1965). The Standard Encyclopaedia of Horticulture, 3: 3403-3404.

Brar, J.S. and Nandpuri, K.S. (1972). Cultivation of Root crops, Punjab Agricultural University, Ludhiana.

Chakrabarti, A.K. (1983). Effect of nitrogen and phosphorus on turnip seed crop. *Seed Res.*, 11:87-90.

Choudhury, B. (1967). Vegetables. National Book Trust India, New Delhi.

Choudhury, B., Sirohi, P.S. and Sharma, J.C. (1976). Pusa Sweti - an early maturing tropical turnip. *Indian Hort.* 21:7.

Devenport, D.C. (1962). Irrigation requirement of radish, turnip and beet root at two levels of nitrogen. M.Sc. thesis, Agronomy Division, IARI, New Delhi.

Gill, H.S. (1980). 'Pusa Swarnima'- A turnip variety with great promise. *Indian Hort.*, 24(4).

Gill, H.S. (1993). Improvement of Root Crops - Turnip. In. Advances in Horticulture, Vol.5- Vegetable Crops: Part-1 (Eds. K.L. Chadha and G. Kalloo) Malhotra, Publishing, House, New Delhi.

Gill, H.S., Singh, J.P. and Bhullar, B.S. (1971). 'Pusa Chandrima' - A high yielding turnip. *Indian Hort.*, 16:19-20.

Gopalan, C., Ram, Sastri, B.V. and Balasubramanium, S.C. (1971). Nutritive value of Indian foods. National Institute of Nutrition, Hyderabad.

Haller, M.H. (1947). Effect of root- trimming, washing and waxing on the storage of turnips, *Amer.* Soc. *Hort. Sci. Proc.* , 50: 325-329.

Hegde, D.M. (1993). Water Management in Vegetable Crops. In Advances in Horticulture vol.5-Vegetable Crop Part I (Eds.: K.L. Chadha and G. Kalloo) Malhotra Publishing House, New Delhi.

Kale, P.N.and Masalkar, S.D.(1993) . Agro-techniques for Root Crops: 5 turnip. In Advances in Horticulture vol.5 - Vegetable Crop Part I. (Eds: K.L. Chadha and G. Kalloo) Malhotra, Publishing, House, New Delhi.

Lal, B. and De. R. (1972). Fertilizer and moisture requirements of turnip (*Brassica rapa* L.). *Indian J.Agric. Sci.*, 42: 956-960.

Leela, D. (1993). Weed control in Vegetable Crops. In Advances in Horticulture vol.5 - Vegetable Crop Part 1 (Eds: K.L. Chadha and G. Kalloo) Malhotra Publishing House, New Delhi.

Mir, N.M. (1987). Screening of fungicides for field control of downy mildew (*Perenospora destructor*) on onions. *Indian J. Mycol. Plant Pathol.* 17: 321-322.

Mittal, S., Akram, M. and Saxena, S.K. (1987). Identity of powdery mildew of *Daucus carota* in Aligarh. *Natl. Acad. Sci. Lett.*, India, 10:383-384.

Natarajan, S. and Veeraragavathatham, D. (2003). Turnip (*Brassica rapa* L.). In Vegetables, Tubercrops and Spices (Eds. S. Thamburaj and N.Singh), DIPA- ICAR, New Delhi.

Nath, P., Velayudhan, S. and Singh, D.P. (1987). Vegetables for the tropical region. ICAR, New Delhi.

Nath, Prem; Srivastava, V.K.; Dutta, O.P. and Swamy, K.R.M. (2008). Vegetable Crops — Improvement and Production. PNASF, Bangalore, pp. 188-189.

Pandey, R.K. and Gupta, S.L. (1979). Comparative evaluation of different herbicides for weed control in turnip and residual assessment for fodder oat. *Curr. Agric.* 3: 147-150.

Pandey, S.L. (1966). Effect of different levels of Potassium and moisture on the chemical composition and yield of radish, turnip, and beet root, Ph.D. thesis. Agronomy Division, IARI, New Delhi.

Peter, K.V. (1998). Genetics and Breeding of Vegetables. Indian Council of Agricultural Research, New Delhi.

Prasad, H.H. (1931). A bacterial white soft rot of turnip. *Indian J. Agric. Sci.*, 1: 534-537.

Purewal, S.S. (1957). Vegetable cultivation in northern India. Tech. Bull. 39, ICAR, New Delhi.

Purseglove, J.W. (1977). Tropical crops : Dicotyledons, vol.I, English Language Book Societies, Longman, London.

Sadhu, M.K. (1986). Root crops. In Vegetable Crops in India (Eds: T.K. Bose and M.G. Som.). Naya, Prokash, Calcutta.

Saimbhi, M.S. and Randhawa, K.S. (1976). Effect of pre-emergence application of some herbicides on the control of weeds in turnip and radish. *Punjab Agric. Univ., J. Res.*, 13: 282-285.

Saini, G.S. (1996). A text Book of Vegetable Production. Aman Publishing House, Meerut, (UP), India.

Sandhu, K.S., Dhesi, N.S. and Kang, U.S. (1966). Effect of fertilization and spacing on turnip seed crop. *Indian J. Agron.* 11:45-50.

Sharma, B.R. (1972). Seed production of cabbage and knol khol. Paper presented at the Summer Institute on Vegetable Seed Production, PAU, Ludhiana, 29 May- 24 June, 1972.

Sharma, P.B.; Saimbhi, P.B. and Sharma, M.S. (1978).The chemical composition of turnip as influenced by some pre-emergence herbicides. *Indian J. Weed Sci.*, 10:19-22.

Shoemaker J.S. (1949). Vegetable Growing, John Wiley and Sons., Inc. New York.

Singh, A.K. and Reddy, G.R. (1990). Effect of herbicides on weed population and levels of nitrogen on seed yield and its components in turnip. *Vegetable Science*, 17(2): 119-126.

Singh, B., Khurana, S.C. and Pandita, M.L. (1974). Chemical weed control in cauliflower, knol khol and turnip. *Haryana J. Hort. Sci.*, 3: 182-189.

Singh, C.G., Pandita, M.L. and Khurana, S.C. (1989). Studies on effect of root age, size and spacing on seed yield of turnip(*Brassica campestris* var. *rapa* L.) , *Veg. Sci.* 16:119-124.

Singh, D.N. and Nath, Vishal (2011). Varieties and Hybrids of Vegetables. Satish Serial Publishing House, Azadpur, Delhi (India).

Singh, D.N. and Singh, S.P. (2007). Production Management of Underutilized Vegetables. Agrihortica Publications, Junagadh (Guj.), India.

Singh, H. and Gill, S.S. (1993). Seed Production of Vegetable Crops. In Advances in Horticulture, Vol.6 - Vegetable Crops: Part-2 (Eds: K.L.Chadha and G. Kalloo) Malhotra Publishing House, New Delhi.

Singh, H.B. (1963). Carrot and turnip varieties 'Pusa Kesar' and 'Pusa Kanchan'. *Indian Hort.* 7:24.

Singh, H.B., Thakur, M.R. and Bhagchandani, P.M. (1960). Vegetable seed production in Kullu valley III. Seed production of root crops. *Indian J. Hort.* 17: 38-47.

Singh, K. and Sharma, R.N. (1968). Weed control studies in vegetable crops. I-Preliminary studies with new herbicides. *Punjab Hort. J.*, 8: 245.

Sinskaia, E.N. (1928). The oleiferous plants and root crops of the family cruciferae. *Bull. Appl. Bot. Genet. Plant Breed.*, 19(3): 1-648.

Swarup, Vishnu (2006). Vegetable Science and Technology in India. Kalyani Publishers. New Delhi. 656.

Thompson, H. C. and Kelly, W. C. (1957). Vegetable Crops. Tata Mc Graw-Hill Publishing Company Ltd., New Delhi.

Verma, G.S. and Verma, J.P. (1961). Occurrence of a strain of turnip crinkle virus in India. *Indian J. Microbiol*, 1:37-41.

Yamaguchi, M. (1983). World Vegetables. AVI, publishing company. Inc. West Port, USA.

Yawalkar, K. S.(1980). Vegetable crops of India. Agri-Horticultural Publishing House, Nagpur, India.

□□□

4. BEET ROOT

Botanical name	:	*Beta vulgaris* L. sub sp. *vulgaris*
Family	:	Chenopodiaceae
Chromosome number	:	2 n = 18.

Beet root commonly known as beet, a very popular vegetable in eastern and central Europe, but is much less important in Western Europe and the USA, where it is known as garden beet (Pink, 1993). It is grown for its swollen root and is the horticultural form of *Beta vulgaris* subsp. *vulgaris* which also includes the agricultural types sugar beet , mangold and fodder beet. The species also includes chard (subsp. *cicla*) and sea beet (subsp. *maritima*), which is thought to be the ancestor of most cultivars (Campbell, 1976), which grows wild on sea shores in Britain and through Europe and Asia to the East Indies. The earliest form of domesticated beet was leaf beet. The leaves were eaten and the roots were used for medicinal purposes only (Debock, 1965).

ORIGIN

Beet root is probably a native of Europe (Thompson and Kelly, 1957) but some believe that it might have originated in Mediterranean region (Europe, North Africa and West Asia). It was introduced into India in remote time (Burkill, 1935). The first appearance of the improved beet in Germany is recorded about 1558 and in England about 1576. It was mentioned by Mc Mohan in 1806 as being grown in the gardens of America at that time. (Bailey, 1963).

DISTRIBUTION

Beet root was taken to northern Europe by invading armies and has been widely used throughout Europe, the middle East and India for food and fodder, particularly during the winter (Yamaguchi, 1983). Some of the important growing countries are Germany, England, America, China, Arab, Greece , Rome and India. It is grown in almost all the states of India but not as common as other root crops (Nath *et al.*, 1987).

BOTANY

The beet root, garden beet (*Beta vulgaris* L.) is a member of the family Chenopodiaceae. It is a wind cross pollinated crop having protand rous flowers. The garden beet is different from the sugar beet where emphasis is laid more on the sugar content of the root, where as the garden beet is used for salad and cooked vegetable purpose (Nath *et al.*, 1987).

Plants are usually biennial glabrous herbs with conspicuously swollen roots at junction with stem. For root production it is herbaceous annual. Leaves, often ovate and cordate, dark green or reddish, frequently forming rosette habit. Inflorescence in small 3-4 flowered cymes arranged on a spike, each subtended by a small narrow leaf; perianth of 5-segments, becoming thicker towards base as fruits ripen; stamens 5; ovary sunk in disc; stigma 3. Although the ovary is three carpellate, there is only one ovule per ovary. Fruits mostly an aggregate formed by the cohesion of 2 or more fruits held together by swollen perianth (calyx) bases and forming an irregular dry body, known as seed ball or the so called seed. However, if there is a single flower, a single germ seed will develop. The true seed is small, about 3 mm long, kidney-shaped and shiny to reddish brown in colour. A gram of beet seed ball counts about 50 seeds, which retain viability for 5-6 years under ordinary storage conditions. The roots are usually a deep red colour and may be globular or cylindrical (Purseglove, 1977).

Plant breeding

Garden beet is a cross pollinated crop. Selfing causes inbreeding depression. The method adopted to improve beet root are mass selection, family selection and hybridization. It produces seeds in the temperate climate. Hence no attempt was made to breed varieties in the plains. All varieties that are under cultivation are from abroad. Some of the imported varieties were found to be suitable for commercial production in the plains. Recently efforts were made to select suitable sugar beet varieties for extracting sugar rather than beet varieties for vegetable (Nath *et al.*, 2008). Some of the male sterile lines and monogerm seed varieties have also been reported for commercial utilization in the foreign countries (Stewart, 1961).

Morphology

The roots are thick and fleshy, and store large amount of reserve food. The upper portion of root develops from the hypocotyle and the lower portion from the tap root, from which the secondary roots arise. The shape and colour of root vary with the genotypes and are also influenced by temperature, soil, nutrition etc. During the development of fleshy roots, vascular tissues and storage parenchiyma tissues are formed alternately, which results in concentric rings and are visible in a median cross section of mature beet root (Yamaguchi, 1983). The exterior root colour varies from orangered to dark purple red depending on the cultivar, while the interior colour is influenced by cultivar, temperature, season, soil, nutrition etc. The colour of root is due to the presence of red-violet pigment, β-cyanins and yellow pigment, β-xanthins. The root system is extensive with taproot penetrating to a depth of 30 cm. The strong lateral roots grow diagonally for a distance and then downward to the length of tap root which contributes to the drought resistance nature of the crop (Natarajan and Veeraragavathatham, 2003).

COMPOSITION

Beet root is rich in protein, carbohydrate, calcium, phosphorus and vitamin C. The food value is given below (Aykroyed, 1963).

Per 100 g of edible portion

Moisture	87.7 g	Riboflavin	0.09 mg
Carbohydrate	8.8 g	Vitamin C	88.0 mg
Protein	1.7 g	Calcium	200mg
Fat	0.1 g	Phosphorus	55.0 mg
Minerals	0.8 g	Iron	1.0g
Thiamin	0.04 mg	Potassium	43.0 mg
Calories	43	Oxalic acid	40.0 mg

The red colour of table beets is due to betacyanin, a nitrogen containing compound, with chemical properties similar to anthocyanins. Beets also contain a yellow pigment, betaxanthin. The ratio of these two pigments varies with cultivation and changes during growth and with environmental conditions (Nilsson, 1973).

The beet tops is also rich in iron, vitamin A (Fe 3.1 mg, vitamin A 21000 I.U., Thiamine 110 mg and ascorbic acid 50 mg per 100 g edible portion), vitamin C and protein.

Uses

Beet root is a useful vegetable in a number of ways. The tender leaves and the young beet plants are used as greens (pot herbs). The swollen roots are eaten boiled or as vegetables and salad. They are also pickled and canned. The large sized beets are used for canning. Beets are cooked with other vegetables and with meat and is also grown for processing (Choudhury, 1967). The tops may be used much like spinach as leafy green for salads or boiled. The ancient Greeks and Romans used both the leaves and the roots as feed for animals and human.

VARIETIES AND HYBRIDS

Beets are generally classified into two major groups on the basis of the edible parts.

Beets grown mainly for their fleshy

The beets grown mainly for their fleshy roots belong to the crassa, or garden group and include yellow beets and the common table, or red beets.

Group is grown as leafy vegetables

This group is grown as leafy vegetables belong to the cicla group and include leafy beets and Swiss Chard.

Varieties of beet root has been grouped on the basis of shape of roots

Flat

Flat Egyptian.

Short top

Flattened at top and bottom with rounded sides and conical or tapered base. var. Crosby Egyptian, Early Wonder, Asgrow Wonder.

Round or globular

Roots are round or globular in shape, var. Detroit dark red, Crimson globe.

Half long

Length is shorter than long types, var. Half long blood, Winter keeper.

Long

Roots are long, may grow as much as 40 cm, quite popular in Europe, var. Long dark blood.

Hybrids

Mahyco-Mahyco Lal; Ralish India-Dayana; Clause International-Warrior.

Varieties resistant to physiological disorder like cracking and canker-Long dark blood.

Variety/Hybrids	Source
Crimson Globe	IARI, New Delhi
Detroit Dark red	IARI, New Delhi
Father Ball	USA
Egyptian Blood	USA
Edmond	USA
Eclipse	USA
Crosby Egyptian	USA
Early Wonder	USA
Ooty-1	ICAR
Mahyco Lal (F_1)	MAHYCO

Action (F_1)	BSS
Pablo (F_1)	BSS
Lalima (F_1)	Golden Seeds
Bolivar (F_1)	Nunhems
Ruby Queen	Peto Seeds
Ruby Queen Improved	Tokita Seeds
Bassano	-
Egyptian Blood	-
Corumba	-
Dayana	Ralish India
Warrior	Clause International

Detail characterstics of varieties and hybrids described by Singh and Nath (2011) are mentioned below :

Crimson globe

The roots are globular to flattened globe, medium red with little shoulders. The flesh is medium dark red with indistinct zones. The top is medium to tall with large, bright green leaves having maroon shade. It is heavy yielder (Anonymous, 1998).

Detroit Dark Red

Roots are perfectly round with smooth, uniform, deep red skin. Flesh dark blood red with light red zoning, tender and fine grained, top small, leaves dark green, fined with maroon. It is a heavy yielding cultivar maturing in 80-100 days (Anonymous, 1998).

Crosby Egyptian

Roots are flat globe with a small top root and a smooth exterior. The top is medium tall, green with red veins. This cultivar reaches maturity in 55-60 days after sowing and shows pronounced white zoning when grown in warm weather.

Early Wonder

The roots are flattened globe with rounded shoulders with a smooth, dark red skin. The top is heavy, green with red veins. This cultivar take 55-60 days after sowing to reach edible maturity.

Ooty 1

It is a selection from the local type. It is a high yielding, variety with potential of 31-45 tonnes per hectare. The skin is thin and the flesh colour is blood-red. The duration from sowing to harvest is 120-130 days. It is suitable for growing in south Indian hills (Natarajan and Veeraragavathaham, 2003).

Edmond

Moderate size, handsome, rounded, smooth, deep, red, good grain and flavour.

Eclipse

Uniformly globular, bright red, fine grained and sweet. One of the best quick growing early beets.

Egyptian Blood

Top quite small, roots fair size, rich deep red, a standard early variety.

Mahyco Lal

Top is medium with erect leaf crown, maturity 65-70 days. Root globe with close tail, deep red , crisp flesh. Uniform with smooth texture, retain orginal quality without ringing. Root 125-130 g each. Yield potential 40-45 t/ha and resistant to premature bolting.

Action

This is an early maturing hybrid, which become ready in 50-55 days from sowing. The average beet weights 100-150 g. It can stand in the field for 30 days after maturity, without cracking or losing colour and taste. The size keeps increasing upto 400 g.

Pablo

It matures in 55-60 days from sowing and weight 120-150 g each. On maturity, if not harvested, keeps growing upto 450 g, without cracking or losing taste and colour and does not develop fibre. It can be grown very well in all three seasons.

Lalima

Vigorous growing early maturing hybrid. Uniform roots can be harvested in about 60 days after planting. Tops are erect with bright green and glossy foliage. Roots are top to globe shaped, smooth, dark red, very sweet and free from zones. The hybrid has good tolerance to cercospora leaf spot, rhizoctonia root rot, downy and powdery mildews and drought. Yield potential is 45-50 t/ha, suitable for both fresh market and processing industry, especially for pigment extraction.

Ruby Queen

It is a vigorous growing early maturing variety. Uniform smooth shouldered beet with small tap root which holds its shape well even when crowded. The interior colour is deep red. Roots can be harvested in about 50-55 days after planting. Roots are crisp, sweet, round and dark red. Yield potential is 40-45 t/ha.

Bolivar

Very strong plant type produces round uniform beet, with smooth skin and deep red internal colour. Upright leaves facilitate early harvest. Very slow bolting. Good for fresh market and processing. Roots dark red, 200-250 g each, diameter 12-14 cm with round shape. First harvest within 45 days with good keeping quality.

Ruby Queen Improved

Plant vigorous, light green leaves with red vines petiole. Roots crisp, sweet, round, dark red and ready for harvest after 55 days of sowing. Slices are very dark red and free from white rings.

Corumba

It has very uniform, medium long cylindrical roots. Excellent internal and external colour with good shelf life.

Other favourite varieties

Golden Tankard, Golden Yellow, Mammoth, Mammoth Long Red, Seneca Detroit, King Red, Detroit Red, Detroit Perfect, Rubia, Formanova, Little Ball, Kogel and Red Ball, which are common in temperate countries.

CLIMATIC INFLUENCE

Beet root performs well under cooler conditions. It grows best in winter in the plains of India, also grow well in warm weather, but it attains best colour, texture and quality in a cool weather conditions (Choudhury, 1976). Beet plants are very sensitive to low temperature. Exposure to lower temperature of 4.5°-10°C for 15 days or more, bolting is likely to occur before the roots reach marketable size (Sadhu, 1986; Nath *et al.*, 1987). Beets require abundant sunshine for development of storage roots. Good quality roots rich in sugar and intense red colour are obtained always in cool (18.3 to 21.1°C) weather. In excess hot weather alternate light and dark red concentric circle appear in the roots known as zoning. High temperature causes poor root development (Lorenz, 1947).

SOIL REQUIREMENT

Beets can be grown on a wide range of soil , but deep well drained loams or sandy loams are considered best. Deep rich alluvial soil such as silt loam is also suitable for high yield. Heavy soils are not satisfactory for beets since the roots are likely to be asymmetrical in shape. It is very sensitive to acidic soil but thrives very well in alkaline soils with pH as high as 9 to 10. A soil pH of 6-7 is considered best for beet. If grown on alkaline soils, manganese sulphate should be included and in soil with a pH as low as 4.6, copper sulphate should be included in the fertilizer schedule (Sadhu, 1986).

SOWING

Land preparation : Soil should be thoroughly prepared by 4-5 ploughing, 15-20 cm deep followed by hoeing to get fine tilth. The soil surface should be smooth and loose and free from clode and trashes.

Sowing time

Sowing time for hill condition is March -July. In northern plains sowing is done in September-November and in southern plains sowing is done during July to November. Main crop is sown during October-November. If the temperature is high during sowing, the roots produced would be coarse with woody flesh and dull colour. Staggered sowing with regular intervals of one or two weeks would provide steady supply throughout the season.

Seed treatment

Treat the seeds with thiram @ 3 g or captan @ 2 g per kg of seeds. Seed treatment with PCNB, thiram, captan or dexon gave better seedling emergence and controlled pre-emergence damping-off (Natti, 1966). The seeds are soaked in water for about 12 hours before sowing to facilitate better germination in the field.

Spacing

Spacing of row to row 45-60 cm and plant to plant 10-15 cm is given for higher yield (Anonymous, 1988). Cultivars with small tops like Flat Egyptian are given closer spacing than those with heavy tops like Detroit Dark Red (Sadhu, 1986). As other root crops it is also preferred to be sown on ridges with routs 30 to 45 cm apart and with plants about 15 to 22 cm apart (Nath *et al.*, 2008).

Seed rate

About 5.0 to 6.0 kg seeds are sufficient to plant one hectare land. The seed usually contain one or more seeds and are called seed balls.

The seed balls are sown 2.5 cm deep in furrow either in flat bed or on the ridges as case may be. The use of seed drill for sowing is beneficial for better emergence. The seed ball should be placed in close contact with soil by compacting. To facilitate uniform and early germination, the seed bed should not be too wet or too dry. Seed germination is depressed by excess moisture.

NUTRITIONAL REQUIREMENT

Beet root responds well to fertilizers and plants must take uninterrupted rapid growth to produce best quality roots. A crop of beet root producing 25 t/ha yield removes 60 kg of nitrogen, 40 kg of phosphorus and 112 kg of potassium (Jacob and Unexkull, 1960). The amount and kind of different nutrients will vary with soil type, inherent fertility of the soil, previous fertilization programme, planting season and variety. High nitrogen fertilization tend to produce roots of poor quality (Nilssion, 1973). On the other hand, the quantity of nitrogen and potash should be increased in sandy soil.

Beet roots have a relatively high boron requirement. Its deficiency causes internal breakdowns black spot or dry rot (Tehrani and Moayed, 1971; Choudhury, 1967; Nath *et al.*, 1987). However, Hemphill *et al.* (1982) reported that boron and phosphorus did not affect yield but boron application reduced incidence of root canker.

Several researchers have studied the response of fertilizers and manure application in beetroot and gave varying recommendations (Table 1).

Table 1 : Fertilizers recommendation for beet root.

N (kg/ha)	P (kg/ha)	K (kg/ha)	Organic manure (t/ha)	Other nutrients (kg)	Authors/source
60-70	100-120	60-70	15-20	-	Choudhury, 1967
30	60	-	-	-	Paria, 1973
60	90	120	30	-	Perednov and Stepuro, 1978
180	180	240	-	-	Perednov and Stepuro, 1978
70	110	70	-	-	Natarajan and Veeraragav athatham, 2003
224	296	188			Hemphill (1983)
50	50	100			Anonymous, 1988
200	-	-			Hemphill, 1983
30	60	90	-	20	(Borax) Jana, 1987
62	40	50	-	-	Anonymous, 1988
70	100	70	25	-	Natarajan and Veeraragavathatham, 2003

The entire dose of farm yard manure should be incorporated in the soil at the time of final land preparation. Half the dose of nitrogen and full dose of phosphorus and potash can be applied as basal dressing and the remaining half nitrogen should be given as top dressing 25 days after sowing.

IRRIGATION

Beet root is medium rooted and responds well to irrigation. A constant water supply is essential for seed germination and for high yield of good quality beet. In light soil first irrigation should be given immediately after sowing. A deficiency of water may leads to reduction in size. Irrigation application at 0.5 bar soil moisture tension at 15-22 cm soil depth is optimum for beet root grown in sandy loam soil in Delhi condition (Devenport, 1962). The water requirement of beet is about 300 mm in 5-6 irrigations (Sadhu, 1986). However, when there is winter rains, only about 3 irrigation are sufficient for this crop (Pandey, 1966).

The crop is generally planted on ridges with furrow irrigation in India. Sivanappan and Padmakumari (1980) compared drip system with conventional furrow irrigation system for beet root at Coimbatore and noticed substantial saving in water by the drip system (177 mm water) as compared to furrow irrigation (858 mm water) apart from increased productivity.

Beet root is very sensitive to soil moisture at all stages of growth. Stress near harvest increases sugar percentage and reduces fresh weight of roots. With adequate soil moisture supply during growth, the tissues of the beet root remain tender. Overwatering may result in too large and succulent tops. When crop is grown for seed, an adequate supply of moisture must be provided during flowering and seed development (Hegde, 1993).

INTERCULTURE

Clean and shallow hoeing should be given to keep soil loose and to provide aeration for proper growth of plant. Deep cultivation is avoided in beet root because it may damage the crop. One or two earthings are also given to prevent the exposure of roots to sunlight which causes greening and lower the quality of produce.

Thinning

It is an essential operation in beet crop, because the seed ball is actually a fruit consisting of 2-6 seeds, each of which may germinate and produce a plant. Generally the plant emerge in groups unless segmented seeds or monogerm seed is used. Hand thinning is usually done. Sometimes thinning is delayed until some of the beets are large enough to use. When the large ones are removed the small ones are left to develop (Sadhu, 1986). Thinning at 3 cm produced roots of best quality (Shanmugavelu, 1989). Thinning is seldom practised with the crop grown for processing because of high cost of labour involved and small beets are preferred by processors (Thompson and Kelly, 1957).

Mulching

Sometimes mulching is given to the beet root crop for maintaining the moisture and to supress the weed growth. The proper mulching increase yield of beet at different spacing but closer spacing give better yield (Carbussina, 1953).

Crop rotation

Beet root considered is one of the important green house crop. It is not usually a main crop but is grown in between other crop such as lettuce, beans or tomatoes. In crop rotation, it should not be cropped after potato since the potato scab disease (which is common in potato) will find it easy to infect beet. Following crop rotation can be taken with beet root to fetch more return.

1. Beet root-Cabbage-Cluster bean
2. Turnip-Beet root-Knol khol (for ridges or bunds)
3. Beet root-Capsicum-Cowpea.
4. Pea-Beet root-Okra.

WEED CONTROL

Weed drastically decrease yield of beet root, so it should be controlled at a proper stage. Beet require 2-4 weed free weeks after 50% emergence to prevent yield losses (Zimdahl, 1980). Season long competition can prodeuce yield loss of 35-95% (Singh *et al.*, 1974; Upadhyay *et al.*, 1980). Singh and Parashar (1977) got significant increase in root and sugar yield, with pre-emergence sprays of chloridazon at 1.6-3.2 kg/ha. Bhan *et al.* (1977) recorded maximum root yield with pyrazon + phenmedipham (3.0 + 2.0 kg/ ha). Bhale Rao *et al.* (1977) controlled weeds successfully in sugar beet with a pre-emergent application of nitrofen at 1.8-2.5 kg/ha. Singh *et al.* (1978) got excellent weed control of problem weeds like, *Polygonum plebejum* and *Chenopodium album* with a pre-emergent application of chloridazon 3.0 kg/ ha on silty clay loam followed by a post-emergent application of phenmedipham 1.5 kg/ha. Singh *et al.* (1981) got effective control of weeds with a pre-emergent application of pyrazon 3.0 kg/ha consistently. Leela (1987) found one pre-emergent application of alachlor or butachlor at 1.5 kg a.i./ha or fluchloralin 1.0 kg a.i./ha to be effective herbicides for controlling weeds in beet-root for a period of 60 days. Chauhan and Motiwala (1985) found chemical weeding to be an effective approach for controlling weeds in sugar beet. The root yield increased by 35-54 per cent.

PLANT PROTECTION MEASURES

Beet roots generally attacked by certain insect pests like Beet leaf minor, Webworms, semilooper and few diseases such as leaf spot, downy mildew and virus. Their control measures are given below :

Insect Pests

Beet leaf minor (Pegomyia sp.)

Larva feed on the tissue between the upper and lower layers of the leaf, causing serious injury to the leaf, making it unfit for manufacturing of food, consequently the plant growth checked. Infested leaves present a blistered appearance. It can be controlled by destruction of all fallen leaves and other plant refuse after harvesting of root. Spraying of parathion @ 2ml/litre of water, in the under surface of the leaves control the pest.

Webworm (Hymenia sp or Loxostage sp.)

The adult may lay eggs on the leaves and the larvae attack the foliage, either by spinning small webs among the tender leaves or feeding on the underside, protected by small web. Spraying of rogor @ 1 ml/litre of water effectively cotnrol the pest.

Semi looper (Plusia sp.)

The green caterpillar feed on the foliage damaging it badly. Dusting the crop with 5% malathion or 4% endosulfan or 2% lindane dust control the semilooper. The application of 50 per cent wettable powder at the rate of 0.5 kg in 75 to 100 litres of water is also effective to check the green caterpillars.

Diseases

Leaf spot (Cercospora beticola Scce)

Small, circular spots of about 2 mm in diameter appear in great numbers. These spots have definite margins usually darker in colour than the rest of the lesions. The spot may drop out and the leaf may assume a shot-hole appearance. Petioles are also affected with spots tending to elongate. When numerous spots develop whole leaf becomes senescent, dies prematurely and drops. Older leaves are mostly affected (Walker, 1952). Defoliation occurs throughout the growing season resulting in reduction in root size and yield. High humidity usually favours the spread of this disease. To control leaf spot disease, copper oxychloride can be sprayed @ 0.1% thrice at 15 days interval. Crop rotation with other crops is also effective to control the disease.

Downy mildew (Peronospora schachtii)

Some times it become a serious disease of beet root which may occur at any growth stage. The disease is mostly prevalent during the cooler months. All above ground parts are affected. On the leaves, spots of various sizes appear, become lighter green in the upper surface, while on the under surface, the mildew noticed. Flower shoots on infected plants become stunted and distorted. The entire inflorescence has a compact appearance and excessive leaf development may give witches broom appearance. The fungus survives on the crop residues in the soil and is also carried by the seeds. For control of downy mildew take up following measures.

Adapt field sanitation, crop rotation, resistant cultivars and seed treatment with fungicide. Spray indofil Z-78 @ 0.3% thrice at an interval of 15 days to control the disease.

Viral Diseases

Beet roots are affected by a number of viral diseases likes, mosaic, curly top and yellow, purple leaf of beet.

Mosaic

Conspicuous mottling with chlorotic zonate ring spots are common . These may become necrotic with age. When these ring spots develop, their centres are usually green. Affected plants remain stunted and may lose some leaves. The disease is normally transmitted and spread by aphid. Spray metasystox or rogor @ 2 ml/litre of water to control aphids.

Curly top

The leaves are under sized and curled and the veins are transparent. It is transmitted by beet leaf hoppers. Control leaf hopper through spraying of endosulfan or fenvalerate @ 2 ml/litre of water at 10 days intervals twice to control the vector population.

Yellow

Older leaves of infected plants become chlorotic, thickened, leathery and brittle. The foliage becomes abnormally red or yellow and often dies. It is transmitted through aphids. Spraying of Malathion @ 2 ml/litre of water at 7 days interval twice to control the yellow is effective. Keep isolation from related plants, which may act as secondary hosts. Control of insect pests and destruction of infected plants prevent spread of this disease.

Purple leaf of beet

This viral disease was first recorded around Pune during 1958 (Capoor and Sharma, 1965). The infected plants are stunted and leaves have a tendency to stand erect and come closer, unlike the healthy plants where the leaves are broad, long and profuse. Leaves of infected plants show an unusual intense purple colour, while the young emerging leaves show it prominently. Few leaves develop minute necrotic lesions all over the lamina. The virus is readily transmitted by mechanical sap inoculation. The virus was identified as a strain of tobacco mosaic virus (TMV). Roguing of virus-infected plants and removal of weed hosts may help in minimizing disease incidence.

Physiological Disorder

Internal black spot or brown heart or heart rot

It is caused due to boron deficiency and deficient plants usually remain dwarf and stunted. The leaves are smaller in size and fail to develop normally and eventually turn brown or black and die. The leaves may assume a variegated appearance due to development of mixture of yellow and purplish-red blotches over parts or whole while the stalk of such leaves shows longitudinal splitting (Purvis and Carolus, 1964). Frequently, the affected plant has twisted leaves and exhibits a slight shortening and distorting of its leaf-stalk in the centre of its crown. The growing point may die and decay. The roots do not grow to full size and under severe deficiency they remain very small and distorted and have a rough unhealthy greyish appearance instead of being clean smooth. Their surfaces are often wrinkled and cracked within the fleshy roots, but more numerous on the light coloured zones or cambium layers (Lorenz, 1942).

The boron deficiency is common in alkaline soils or on soils relatively high in calcium. Overliming often brings about boron deficiency. Soil application of borax may control the disorder. The quality of borax vary from 5-50 kg/ha with nature of soil, soil-moisture and soil reaction (Natarajan and Veeraragavathatham, 2003).

HARVESTING

Beet roots are harvested within 60-75 days of sowing at marketable maturity, which is judged depending on the size ranging from 3 to 5 cm diameter. Beet root should be deep red , firm and free of cracks, fibrous roots or corky tissue. Elongated roots indicate crowding during growth and toughness (Maini, 1998). Usually pulled by hand, the tops are cut and after washing roots are stored in cool place.

YIELD

The average yield of beet varies from 250 q to 300 q/ha.

POST HARVEST HANDLING

After harvesting beet roots are de-topped and washed properly. Oversized beets are not in great demand and since they can be used only for diced beets or baby food rather than as whole beet or whole slices. So they are separated from the lot.

Grading

Beets are arranged in 'super', 'fancy' and commercial grade according to size, shape, lustre and brightness. The tender, small size beets with uniform shape, bright colour and lustre are arranged in 'super grade'. The medium size with bright colour and good texture beets are kept in 'fancy grade'. The large sized beets with clean surface without any blemish and spots are arranged in 'commercial grade' and packed separately for transportation of distant market for sale (Singh and Singh, 2007).

Packing

Usually beet roots are packed in big round bamboo chips basket, Arhar stick basket and sometimes in gunny bag (Roy, 1998) and top portion is provided with cushening materials then stitched properly and send to local market. The quality beets of different grades are packed in 4-6 numbers in polythene bags; transparent film bags; CFB boxes, pllybags, plastic bags, cardboard boxes, bushel baskets or similar containers (Thompson and Kelly, 1957). The packed materials are sent to distant market for big cities to fetch higher prices in developed countries.

Storage

Generally, ventilated storage is used for beet-roots. Beet roots are often stored in outdoor pits or banks or in common storage house in colder regions. The best temperature for beets is near the freezing point, but they should not be allowed to freeze. The air in the storage house or cellar should be kept rather humid to prevent wilting and withering of the roots. Beets keep well for a longer period in a cold storage at a temperature of 32°F with humidity of about 90 per cent . Beet can be stored for 2-3 days under ordinary condition. The storage life of beet root can be enhanced from 12-20 weeks at temperature of 1-4°C. (Kaul, 1998).

SEED PRODUCTION

Beet root is a herbaceous, long day biennial and temperate type and require two season to produce seed. There is no any tropical type of beets hence their seeds can be produced only in the hill of India. Beet requires exposure to low temperature of 4.4-7.7°C for 6-8 weeks for inducing the reproductive phase i.e. initiation of flower stalks (Madsen,1974). However, the period may extend beyond sixty days, depending upon the variety (Rekhi, 1976). Further the requirement may also vary considerably and is less if the average temperature is below 7.5°C (Agrawal, 1980).

Land requirement

Land should be free from beta species crop at least last 2-3 years and also should be fertile and well drained.

Isolation distance

Beet root is cross-pollinated by wind or insects,. Pollen normally sheds over a period of weeks, therefore, chances of contamination due to outcrossing are prolonged. In beetroots an isolation distance of 2000 m for foundation seed production and 1000 m for certified seed production, is require (Arya, 2003).

Method of seed production

There are two methods of seed production. They are: (1) Seed to seed or in situ method (over wintering method). (2) Root to seed or trnsplanting method (steckling method).

In seed to seed method the crop is over wintered in the field and the crop flowered in situ after the necessary thermal induction.

In root to seed method, the selected stecklings (roots) are lifted at the end of the first season and stored in shallow trenches for over wintering. These stecklings are replanted to produce seed in the following spring season. In India the seed production by root to seed method is done in Kulu valley (Agrawal,1980). Before re-planting one-third of the root tip is chopped and the top is also removed so as to retain about 10 cm top leaves. But planting of whole roots is recommended to get maximum seed yield.

Brief cultural practices

Good land preparation and adequae soil moisture is necessary.

The sowing of seed is done in end of June to mid July and during November-December. Nucleus/breeder seeds for foundation seeds crop and foundation seed for certified seed crop are necessary. Ridge sowing and flat bed sowing for dry climate is to be adapted. Spacing for in situ crop is 60 x 15 cm row to row and plant to plant. Seedlings (Steckling) are spaced 45 x 10 cm, row to row and plant to plant for tuber production.

Tubers are planted at spacing of 60 x 45 cm, row to row and plant to plant for seed production.

Fertilizer dose of 100-150:50:150 kg N, P_2O_5; K_2O/ha is optimum for high seed production. 60 % dose given to first crop for raising stackling. 40 % dose given to seed crop. In addition to that give 20-30 kg of Borax per ha to prevent cracking and canker of the roots. Borax also helps in good root development.

Irrigation at flowering is very useful. Afterwards irrigation is given as per requirements.

Roguing

The diseased and off type plants must be rogued out from the field. Rogue out early bolters, off type plants from time to time. In steckling method, before transplanting sort out all undesirable stecklings such as diseased, malformed and those not conforming to varietal, characteristic and reject them.

Flowering and seed setting

Seed stalk, elongation in beet starts in early part of April in kulu valley and the crop is in full bloom from mid May to Mid June. As stated already the inflorescence is a large panicle and seed maturity begins at the base of the panicle.

Harvesting

The beet seed crop can be harvested when two thirds of the seed are ripe. Generally,it is ripen in July. At this stage, seed at the base of the branches mature and turn noticeably brown. Ripe seeds some time shatter, but if plants are pulled early in morning when moist, shattering is prevented. When the summer is hot and dry, harvesting may start as early as the last week of June, but usually it is done from the first week of July to the end of July. Harvesting is usually done by hand as mowing is not satisfactory. Because ripening is uneven, two to three hand harvesting are often necessary (Singh *et al.*, 1960). The crop is stacked after harvesting for curing and then dried under sun.

Threshing

Seeds can be threshed with an ordinary grain thresher, combines can also be used for threshing. Rough cleaning with fanning will removes all shunken seeds, weeds and stalks, etc.

Seed yield

The average seed yield varies at 8 to 10 q/ha.

Storage

To remove excess moisture, maximum drying temperature should not exceed 40°C. The seed should be dried to 9% to 10% moisture for safe storage.

After cleaning, seeds may be rubbed and graded with the help of suitable machine and packed for storage. Beet roots seeds remain viable for a fairly long period under room temperature. For prolonging storage life and quality, seed moisture can be reduced by one percent (Doijode, 2004). At 0°C and 90-95% relative humidity, seeds can be stored up to 2-8 weeks.

REFERENCES

Agrawal, R.L. (1980). Seed Technology. Oxford and IBH Publishing Co., New Delhi.

Anonymous (1988-89). Agriculture Guide Book. Directorate of Agriculture and Food Production, Government of Orissa. Agriculture Information Press, Bhubaneswar, Orissa.

Anonymous. (1998). Varieties and hybrids. Division of Vegetable Crops, IARI, New Delhi.

Arya, P.S. (2003). Vegetable breeding – Production and seed production : Kalyani Publishers, New Delhi.

Aykroyed, W.R. (1963). The nutritive value of Indian foods and the planning of satisfactory diets, ICMR Report, series 43; New Delhi.

Bailey, L.H. (1963). Manual of Cultivated Plants. The Mac-Millan Company, New York.

Bhale Rao, A.C., Singh, O.P., Ghanekar, A.R. and Karve, A.D. (1977). Use of nitrofen for controlling weeds in sugar beet. In *Proc. Sixth International Convention*, STAI, DSTA and SISTA, Madras, pp.103-106.

Bhan, V.M., Maurya, R.A. and Negi, P.S. (1977). The influence of weeds on sugar beet yield and their control using herbicides. *Proc. ISWS/APAU Weed Sci. Conf.*, Hyderabad, pp.207.

Burkill, I.H. (1935). A dictionary of the Economic Products of the Malaya Peninsula. Crown Agents, London.

Campbell, G.K.G. (1976). Sugar beet in; Evaluation of Crop Plants. (Ed: N.W. Simmonds). Longman, London, 25.

Capoor, S.P. and Sharma, D.C. (1965). Purple leaf disease of beet. *Indian Phytopathol.*, 18:88-89.

Carbussina, N.M. (1959). *Hort. Abstr.*, 30: 506-507.

Chauhan, R.S. and Motiwala, M.P. (1985). Chemical weeding - An effective approach for controlling weeds in sugar beet. *Pesticides*, 19(6): 51.

Choudhury, B. (1976). Vegetables . National Book Trust India, New Delhi.

De Boke, Th. S.M.(1965).The genus *Beta*: domestication, taxonomy and interspecific hybridization for plant breeding. *Acta Hort.*182, 335.

Devenport, D.C. (1962). Irrigation requirements of radish, turnip and beet root at a two levels of nitrogen, M.Sc. thesis, Agronomy Division, IARI, New Delhi.

Doijode, S. (2004). Seed storage of vegetable crops. CBS publication, New Delhi.

Hegde, D.M. (1993). Water Management in Vegetable Crops. In Advances in Horticulture Vol.5 – Vegetable Crops Part-1: (Eds.: K.L. Chadha and G. Kalloo). Malhotra Publishing House, New Delhi – 110064, India.

Hemphill, D.D. Jr., Weber, M.S. and Jackson, T. L. (1982). Table beet yield and boron deficiency as influenced by lime, nitrogen and boron. *Soil Sci. Soc. of America J.* 46:1990-92.

Jacob, A. and Unexkull, H.V. (1960). Fertilizer Use, Varlags Gesellschaft Fur Ackerban Mbh., Hannover.

Jana, B.K. (1987). Response of micronutrients on yield and quality of beet root cv. Sutton's Globe. *Vegetable Science*, 14(2): 115-119.

Kaul, J.L. (1998). Post harvest Deterioration of Vegetables. . In Production and post harvest handling of vegetables. Centre of Advanced Studies , Horticulture-Vegetable (ICAR) Department of Vegetable Crops, Dr. Y.S. Parmar University of Horticulture and Forestry, Nauni – Solan, Himachal Pradesh – 173 230, India.

Leela, D. (1987). Weed control by herbicides in beet root (*Beta vulgaris* L.) *Proc. 11th Conf.* APWSS, Teipei, Taiwan, Nov. 29-Dec., 5, 239-249.

Lorenz, O.A. (1942). Internal breakdown of table beet, *Cornell Memoir* 246.

Lorenz, O.A. (1947). The effect of certain planting and harvest dates on the quality of table beets, *Amer. Soc. Hort. Sci. proc.*, 49: 270-274.

Madsen, A.(1974). Sugar beet in India. *Seed Tech. News*, 4(3): 1-2.

Maini, S.B. (1998). Maturity Indices of commercially important vegetable crops. In Production and post harvest handling of vegetables. Centre of Advanced Studies , Horticulture-Vegetable (ICAR) Department of Vegetable Crops, Dr. Y.S. Parmar University of Horticulture and Forestry, Nauni – Solan, Himachal Pradesh – 173 230, India.

Natarajan, S. and Veeraragavathatham, D. (2003). Beet root (*Beta vulgaris* L.) . In Vegetables, Tubercrops and spices, (Eds. S Thamburaj and Narendra Singh) Indian Council of Agriculturaal Research, New Delhi.

Nath, P., Velayudhan, S. and Singh, D.P. (1987). Vegetables for the tropical region. ICAR New Delhi.

Nath, Prem; Srivastava, V.K.; Dutta, O.P. and Swamy, K.R.M. (2008). Vegetable Crops - Improvement & Production. PNASF, Bangalore.

Natti, J.J. (1966). *Plant Dis. Reptr.*, 50:614-617.

Nilsson, T. (1973). The pigment content in beet root with regard to cultivar, growth development and growing conditions. *Swedish J. Agric.Res.*, 3: 187-200.

Pandey, S.L. (1966). Effect of different levels of potassium and moisture on the chemical compostion and yield of radish, turnip and beet root. Ph.D. thesis , Agronomy Division, IARI, New Delhi.

Paria, N.C. (1973). Indian Agriculturist, 17(2): 22-23.

Pink, D.A.C. (1993). Beet root (*Beta vulgaris* sub. sp. *Vulgaris*), In: Genetic Improvement of Vegetable Crops (Eds. G. Kalloo, and B.O., Bergh), Pergamon Press, New York.

Purseglove, J.W. (1977). Tropical Crops.-Monocotyledons, Longman, London.

Purvis, E.R. and Carolus, R.S. (1964). Hunger Signs in Crops. (Ed. H.B. Sprague). David Mackay Co., New York.

Rekhi, S.S. (1976). Sugar beet Seed Production in India, *Seeds and Farm*,2(1): 25-27.

Roy, S.K. (1998). Packaging Technology of Vegetables. In. Production and Post harvest Handling of Vegetables. Centre of Advanced Studies; Horticulture-Vegetable (ICAR), Dept. of Veg. Crops. Dr. Y.S. P.U.H.F., Nauni, Solan, H.P.

Sadhu, M.K. (1986). Root crops. In Vegetable Crops in India (Eds.: T.K. Bose and M.G. Som), Naya Prokash, Calcutta.

Shanmugavelu, K.G. (1989). Production Technology of Vegetables Crops. Oxford and IBH Publishing Co. Pvt. Ltd., New Delhi.

Singh, D.N. and Nath, Vishal (2011). Varieties and Hybrids of Vegetables. Satish Serial Publishing House, Azadpur, Delhi (India).

Singh, D.N. and Singh, S.P. (2007). Production Management of Underutilized Vegetables. Agrihortica Publications, Junagadh, (Gujarat) India.

Singh, G., Singh D. and Singh, B. (1981). Weed control studies in sugar beet. Proc. 8th *Asian Pacific Weed Sci. Sco. Conf.*, Bangalore. India. 22-29 November.

Singh, G., Singh, Y and Pant, P.C. (1978). Weed control studies in sugarbeet with phenmedipham and pyrazon herbicides. *Indian J. Weed Sci.* 10: 9-14.

Singh, G.B., Maheshwari, B.K, Prasad, J. and Singh R.N. (1974). Studies on weed competition in sugar beet (*Beta vulgaris* L.) . *Indian J. Weed Sci.*, 6: 115-120.

Singh, H.B., Thakur, M.R. and Bhagchandani, P.M. (1960). Vegetable seed production in Kullu Valley :III- Seed productin of root vegetables. *Ind. J. Hort.* 17 38-47.

Singh, Raghabir and Parashar, K.S. (1977). Weed control in sugar beet with herbicides. *Indian J. Agron.*, 22: 108-109.

Sivanappan, R.K. and Padmakumari, O. (1980). Drip Irrigatin. College of Agricultural Engineeirng. Tamil Nadu Agricultural University, Coimbatore, Tamil Nadu.

Tehrani and Moayed, G. (1968). *Diss. Abst.* No. 28.

Thompson, H.C. and Kelly, W.C. (1957). Vegetable Crops. Mc Graw Hill Book Co., New York.

Upadhyay, U.C., Khan N. and Chakrawar, V.R. (1980). Weed management studies in onion (*Allium cepa* L.) In abstr. Papers, *Annus. Conf., ISWS,* Bhubaneswar.

Walker, J.C. (1952). Diseases of Vegetable Crops. Mc Graw – Hill Book Co., New York.

Yamaguchi, M. (1983). World vegetables.AVI Publishing Company, Inc.,West Port, USA.

Zimdahl, L. (1980). Crop weed competition – A review Published in the United States by the International Plant protection Centre, Oregon State University, Oregon, USA.

❑❑❑

5. CHINESE RADISH

Botanical name	:	*Raphanus sativus* Linn var niger
Family	:	Brassicaceae
Chromosome number	:	2n = 18

The chinese radish is a favourite crop of the home gardeners because it is easily gown and is ready for use in vary short period (3-6 weeks) after seed sowing. It is a fast growing erect annual herb for root production (Kaul and Kanwar, 1997).

ORIGIN

Chinese radish is probably originated from the old world, cultivated long before Greeks and Romans came to know of it as 'rhaphanis' and 'radix respectively (Thompson and Kelly, 1978).

DISTRIBUTION

It is widely grown in USA, Japan, China, Europs and in others temperate part of the world. In india it is cultivated in Kinnaur (Himachal Pradesh), Leh and Kargil (Jammu and Kashmir) and scattered in other hill region. It is less known underutilized vegetable. It is found growing in kitchen gardens (Singh and Singh, 2007).

BOTANY

Chinese radish is biennial but cultivated as annual for root production. Seed production take anther season. The top of the plant is heavy. It is a large and heavy rooted, broad shouldered, thick skinned. The roots have an attractive pink colour and weighing on an average 1-2 kg each (Anonymous, 1972).

COMPOSITION

On the basis of 100g fresh edible portion, chinese radish contains: protein 1g, calcium 32 mg, iron, 1.4 mg, vitamin C26 mg, potassium 400 mg and energy 30 Kcal (Knott, 1988).

Uses

Chinese radish roots mildly pungent and an excellent vegetable for cooking. During severe winters of Ladakh, when no fresh vegetables are available (Singh and Singh, 2007) chinese radish is vailable in the market.

IMPROVED VARIETIES

Only local varieties are available. This root vegetable required attention to the vegetable breeder.

CLIMATIC INFLUENCE

Chinese radish require temperte climate. The roots are thick skinned therefore, they can adopt wider range of temperature adoption (Ware and Collum, 1980).

SOIL REQUIREMENT

Chinese radish is a root vegetable. For better growth and development of roots, loose textured soild are ideal because roots go to deeper depth for their growth. Heavy soild are not suitable rather harmful to root yield and quality. Heavy soil cause rough, mis-shaped and under-developed roots.

SOWING

Chinese radish seeds are sown directly in the field in the month of March to June. The sowing is done on ridges so that bulky roots get favourable environment. The destance of sowings is 45 cm x 30 cm (row x plant). While planting care should be taken that seeds are not sown deeper as precentage of germination is decreased.

NUTRITIONAL REQUIREMENT

In order to provide favourable conditions for untrient absorption, farm yard manure 10-15 tonnes/has is applied before seed sowing i.e. during land preparation. Application of organic matter help making soil light, friable and loose. Such soils have also better weter holding capacities. Besides, for obtaining further higher yield, application of nitrogen 100 kg, phophorus 60 kg and potassium 40 kg/ha is needed. Nitrogen half dose alongwith whole dose of phosphorus and potassium are applied just before sowing as basal dose. The remaining half dose of nitrogen is applied 3 weeks after sowing (Shoemaker, 1953).

IRRIGATION

For higher percentage of seed germination adequate soil moistrue is most important. Seeds should be sown either following pre-sowing irrigation or post sowing irrigation. For root development, there should always be sufficient soil moisture that can be maintained by light and frequent intervals (Marshall, 1981). Failure of regular irrigations may cause tough root, which are unfit for consumption and markets.

INTERCULTURE

Shallow cultivation is very helpful to provide suitable environment for getting up proper plant growth. This also keeps down the weeds. As soon as roots start bulzing upward, it is essential to do earthing so that they are covered. This operation not only increases yield but also quality of roots. Exposed roots direct to sun cause greening of roots that is not desirable (Singh and Singh, 2007). Green roots have not market value.

WEED CONTROL

Generally, 2-3 hand weeding is given during vegetative growth period at 15 days and 25 interval, respectively. It takes care of crops during the root development stage (Koul and Kanwar, 1997).

PLANT PROTECTION MEASURES

The most common insect pests and diseases alongwith their control are given below :

Insect Pests

Aphid (Myzus persicae, Brevicoryne brassicae)

They suck the sap of the leaves. Favourable conditons for aphid increase is cloudy and humid atmospheric conditions. The removal or destroying of affected plant parts, spraying with malathion (0.01%) or monocrotophos (0.05%) will control aphid effectively.

Flea beetle (Phyllotreta sp.)

Flea beetle attack leaves and make small circular holes until they are skeletonized and lace-like. Flea beetle become active in sunny weather during spring. These beetles can be control effectively by following phytosanitary measures and spraying of malathion (0.1%) or sevin (0.4%) at 10-15 days interval.

Diseases

Dapming-off (Rhizoctonia solani)

The fungus attacks particularly on seedlings which collapse at ground level. Sowing of seed thinly, thinning of crowded seedlings, avoiding sowing in wet conditions and seed treatment are effective in controlling the disease.

White rust (Albugo candida)

The causal fungus causes white powdery substance on the under surface of the leaves. Incidence of disease is more in wet conditions and are crowd. Spraying of dithane M-45 at 0.2 per cent is very effective.

Alternaria blight (Alternaria raphani)

Symptoms start with appearance of lesions on leaves, stems, pods and seeds. The styler ends of the pods become black and shrivelled. Infected seeds have very poor germination. Rains favour incidence of disease. It is advisable to use disease free seed. Treat the seed in hot water at 50°C for 30 minutes, spray copper oxychloride (0.5%) and follow crop rotation for control of disease.

Radish mosaic virus (R.M.V)

Small, chlorotic lesions, appear on the leaves. It is transmitted through aphids. One can control the disease by removing weed plants, spraying of malathion (0.01%) or monocrotophos (0.05%).

HARVESTING

The roots of chinese radish become ready for harvest in 65-70 days. The crop growing at higher altitute of about 7500 feet if harvesting is delayed, remain unaffected from any pithyness, pungency etc, it is due to its thick skin.

YIELD

About 250 to 350 quintals of roots are available per hectare (Singh and Singh, 2007).

POST HARVEST HANDLING

Harvested roots are cleaned for foliage and feeding roots. The unwanted roots are sortedout. the healthy root is grounded according to market demand (Singh and Singh, 2007). It is advisable to send produce in the market as soon as it can.

REFERENCES

Anonymous (1972). The Wealth of India : Raw Material and Industrial products. Vol. VIII. CSIR, New Delhi.

Knott, J.E. (1988). Handbook of Vegetable Growers. John Wiley and Sons, New York.

Koul, B.L. and Kanwar, H.S. (1997). Chinese radish is a promising vegetable for hills. Indian Horticulture, 42(1): 16-17.

Marshall Cavendish. (1981). The Fruit and Vegetable Gardeners' Handbook. Marshall Cavendish Book Ltd, London.

Singh, D.N. and Singh, S.P. (2007). Production Management of Underutilized Vegetables. Agrihortica Publications, Junagadh (Guj.), India.

Singh, Umrao; Wadhwani, A.M. and Johri, B.M. (1983). Dictionary of Economic Plants in India. ICAR, Publications, New Delhi.

Shoemaker, J.S. (1953). Vegetable Growing. Chapman and Hall Ltd., London.

Thompson, H.C. and Kelly, W.C. (1978). Vegetable crops. Tata Mc. Grow-Hill Publishing Company Ltd., New Delhi.

Ware G.W. and Mc Collum, J.P. (1980). Producing Vegetable Crops. Interstate printers and Publishers. INC, Illionis, Danvilla.

□□□

6. PARSNIP

Botanical name	:	*Pastinaca sativa* Linn.
Family	:	Apiaceae
Chromosome number	:	2n = 22

Parsnip is an underutilized vegetables. Its roots are used for the preparations of multiple items. It is considered as a winter vegetable because its flavour is not fully developed until the roots have been exposed to near-freezing temeratures for 2-4 weeks in the fall and early winter (Kalloo and Berg, 1993). The starch in the parsnip root changes into sugar, resulting in a strong, sweet, unique taste.

ORIGIN

Parsnip is a native to the eastern mediterranean region. The plant was known to the ancient Romans and Greeks, grown for its large, edible roots and for medicinal purpose. It was introduced to the West Indies in 1564 and brought to the U. S. by the colonists and was cultivated in the earyl 17th century in Virginea and Massachusetts. It is also cultivated in western Asia for its large fleshy tap root (Knott, 1988).

DISTRIBUTION

It is widely cultivated in Europe and America. In india, it is occasionally grown as an annual in kitchen garden in temperate regions (Anonymous, 1996).

BOTANY

Parsnp is biennial in habit but it is cultivated as annual. It is similar to carrot in anatomy and development. It is hardy, white roots store well in the ground all winter. The plant develops rosette of leaves and large fleshy tap root which is economic parts. Flowers in terminal compound umbels; fruit (commoly called head) broad, elliptic and ridged, make of two maricarps.

COMPOSITION

Parsnip contains moisture (72.4%), protein (1.3%), fat (0.3%), carbohydrates (23.2%), fibre (1.7%) and mineral matter (1.1%), calcium (50 mg), phosphorus (40 mg), iron (0.4 mg), vitamin A (30IU), thiamine (0.31 mg), niacin (0.4 mg), ascorbic acid 16 mg, riboflavin (0.06-12 mg) per 100 g edible portion (Singh and Singh, 2007).

Uses

Nick and Bradley (1994) reported that parship can be used in following ways :

1. The roots are consumed raw and after boiling as vegetable and salad.
2. They are also used for Jam, cakes, wine and in fritters or pan cakes.
3. They make an excellent soup and wines.
4. Roots contain more sugar than sugar beet which reduce to about three quarters when cooked.
5. The root are also fed to livestock not more than 20 per cent of the total fodder.
6. The parsnip possesses diuretic and carminative properties. The fruit is based as a folk medicine for stomach, bladder and stone troubles.

VARIETIES AND HYBRIDS

Only a few varieties are available. These ae Cobham Improved Marrow, Tender Harris Model, Andover, Lancer, Alba, Unicorn, White Gem, All American, Fullback and Hollow Crown etc. (Thompson and Kelly, 1981).

CLIMATIC INFLUENCE

Parsnip requires a long cool season which improves its quality. It is the crop of temperate climate. North Indian hilly conditions provide sufficient low temperature for over-wintering in the field itself. However, site that receives full sun is suitable (Thambugraj and Singh, 2001).

SOIL REQUIREMENT

Parsnip requires deep, loose, fertile, well-drained sandy loam soil having a pH of 6.0 to 7.0. Preparation of land is done thoroughly through deep ploughing followed by cross harrowing, Removal of all stones, pebbles, crocks, weed plants, stumps etc from the field is most desirable (Singh and Singh, 2007).

SOWING

Parsnip is raised by seed. The seeds are sown directly in the main field. The viability of parsnip seed is last for one year and therefore, it is advisable to use fresh seeds every year. Before sowing, seeds are soaked in water over night that helps in germination. Seeds germinate in 2-3 weeks at the temperature of 7°C.

The best time of sowing is the spring (April-May in high hills, March-August in mid-hills) season. The crop can also be sown in late winter. In plains, parsnip is sown in October-November. The temperature ranges from 60 to 65° F is good for growth. Higher temperature deteriorate the quality of roots (Thamburaj and Singh, 2001).

The optimum planting distance is 45 cm x 30 cm or 60 cm x 45 cm. The depth of sowing of seed is about 1.5 cm. About 7.0 kg seed is required for raising one hectar crop of parsnip.

NUTRITIONAL REQUIREMENT

Application of 10-15 t/ha well rotten farm yard manure is done during the land preparation. Application of partial rotten farm yard manure may cause forking of roots. Besides, 50 kg nitrogen, 50 kg of phosphorus and 25 kg/ha potassium are applied in the soil 2 weeks before planting. Higher doeses of nitrogen may also induce forking in roots (Singh and Singh, 2007).

IRRIGATION

Adequate soil moisture is good for the growth and development of tap root. Water stress condition may cause reduction in yield and quality of roots.

INTERCULTURE

Shallow intercultural operations i.e., first at 20-25 days and second after 45-50 days of sowing are done to keep soil loose for proper root development. Thinning is done when seedlings attain 10-12 cm height. This operation is essential to maintain proper distance between plant to plant (Knott, 1988).

WEED CONTROL

Generally, two hand weedings is given before topdressing of fertilizers for proper utilization of nutrients by crop. There is no need of use any herbicides.

PLANT PROTECTION

Insect pests and disseases have not been found serious.

HARVESTING

The parsnip roots are ready for harvesting in 3-5 months. Harvesting of parsnip roots starts when the foliage begins to die down. It takes about 26-30 weeks from sowing to harvest. It harvested in late autumn or early winter. In high hills, digging of roots is done in October-November and March-April in mid hills (Thamburaj and Singh, 2001).

The roots are dug carefully with the help of Kudali or spade because roots growth deep considerably.

YIELD

About 250-300 q/ha roots are obtained from well managed crop. The average size of root is from 12-20 cm.

POST HARVEST HANDLING

Just after digging, roots are kept under shade so that the soil adhered is removed after a couple of days. All undesirable roots including cut roots are separated from the lot. The extra ordinary large as well as small roots are also sorted out (Singh and Singh, 2007). The healthy lot is kept in layers of sand in cold dry shed. the roots can be stored for about 2-4 months at the temperature of 0°C and 90-95 per cent relative humidity.

SEED PRODUCTION

Singh and Singh, (2007) has given following guidelines for seed production of parship.

1. Select the larger and longer healthy roots.
2. Cut the root at the botton end of about 1/3 of the total length
3. Replant the root in well prepared land.
4. Raise the crop following all improved management practices to root production.
5. Plant will flower and set seeds.
6. Parsnip is cross-pollinated and therefore, the isolation distance of at least 1000 metres is essential.
7. Harvest seeds heads at full mature stage. Dry the seed heads, extract the seeds.
8. The seed yield is ranged 300-500 kg/ha.
9. Store dried seeds in air-tight containers.

REFERENCES

Anonymous (1996). The Wealth of India : Raw Material and Industrial products. Vol. 7:280-281. Council of Scientific and Industrial Research, New Delhi.

Kallo G and Bergh B.O. (1993). Genetic Improvement of Vegetable Crops. Pergamon Press, Oxford, U.K.

Knott, K.E. (1988). Handbook of Vegetable Crowing. John Wiley and Sons, New York.

Nick, J.M.A. and Bradley, F.M. (1994). Growing Fruits and Vegetables organically. Rodale Press, Emmaus, Pennsylvania.

Singh, D.N. and Singh, S. P. (2007). Production Management of Underutilized Vegetables. Agrihortica Publication, Junagadh, Gujarat.

Thompson, H.C. and Kelly, W.C. (1981). Vegetable crops. Tata Mc Graw Hill Publishing Co., New Delhi.

Thamburaj S. and Singh, Narendra. (2001). Text Book of Vegetables, Tuber crops and Spices. ICAR, New Delhi.

□□□

Chapter-7

Bulb Crops

1. ONION

Botanical name	:	*Allium cepa* L.
Family	:	Alliaceae
Chromosome number	:	2n = 2x = 16

Onion is one of the oldest known and most important commercial vegetable crop grown in India. It become so important bulb crop because it is short duration and quick growing having various uses, long storage life of bulbs and less losses during transportation. India ranks second in the world's production of onion, producing about 12 per cent of the total production, the first being China (Currah and Proctor, 1990).

Onion rank second in area of the vegetables and third in production in the world among major vegetable crops namely onion, cauliflower, green peas, Cabbage, tomato and green beans. About 40.02 million tonnes of onion is produces in the world from about 2.33 million has of land. India ranks second in area (0.42 million ha of onion), the first being China (0.45 million ha of onion). In production, India ranks second in onion (4.43 million tonnes) and the first being China (10.04 million tonnes), (Singh, 2003).

The production and demand for onion are relatively high. India is one of the exporters of onion. The important onion producing states are maharashtra, Tamil Nadu, Karnataka, Andhra Pradesh, Bihar and Assam. Maharshtra occupies first place with 19 per cent of area followed by Gujarat 17 percent (Anon., 2004-2005e). It is exported to Malaysia, Japan, Burma, Ceylon, Hong Kong, Iran and East African countries. Onion exports during 2006-07 touched 9.43 lakh tonnes against 7.7 tonnes during 2005-06, Maharashtra accounts 75 per cent of total onion exports. The onion production of India is 55 Lakh tonnes out of which about 28 per cent is from Maharashtra.

ORIGIN

Origin place of onion is reported to be central Asia and meditrranean region (Mc Collum 1976). The russian scientist vavilov (1951) reported that onion orginated in Pakistan. The onion originated from the region camprising north-west India, Afghanistan, the Soviet Republic of Tazik and Uzbek and western Tiensen (Jones and Mnn, 1963). Western Asia and the area around mediterranean sea are its secondary centre of origin. It was known in the earliest times, piectured on Egyptian monuments and is mentioned in the Bible (Chadha, 1990). It is now cultivated throughout the world.

Onion is cool-season vegetable. Its history can be traced back to at least 5000 years. Onion, as a food, medicine and religious object were known during the Ist Egyptian dynasty (3200 B.C.) as reviewed by Tackholm and Drar (1954). Egyptian gave them divine honour and fed them to the workers building pyramids for strenght. References of onion as food were also found in Bible and Koran. In India onion has been grown from ancient times as mentioned in a famous early medical treatise 'Chark Samhita', which inferred the teaching of an earlier authority of about sixth century B.C. many virtues were attributed to the onion by Charaka (Ray and Gupta, 1980).

DISTRIBUTION

Onion is one of the most important vegetable crops grown all over the world. The notable onion growing countries are China, India, USA, USSR, Netheriands, Japan, Sapin and Turkey. Of these India and China are the leading onion growing countries. China, India, USA and Turkey are the major onion growing countries in the world contributing about 50% of the total production. Onion productivity is highest in Japan (45.93 tonnes/ha) followed by USA (44.50 tonnes/ha), UK (42.65 tonnes/ha) and Netherlands (41.67 tonnes/ha). India inspite of having second position in production in onion has very low productivity 10.67 tonnes/ ha. The per capita availability of onion is highest in Netherland (41.46 kg/year) followed by Turkey (35.67 kg/year), Spain (24.86 kg/year) and Korea Republic (20.30 kg/year). In India per Capita availability is 4.51 kg/year which is quite low.

Onion is being grown in various states of India, the most important onion growing states are maharashtra, Tamil Nadu, Andhra pradesh, Bihar and Punjab. The other major onion growing states are Orissa, Uttar Pradesh, Karnataka, Rajastan and Gujarat. Regions where onion cultivation is concentrating are Nasik-Niphad and Jalgao in Mharashtra. Bellary and dharwar in karnataka, Madurai in Tamil Nadu, Mahua in Gujarat and Bolangir and Bhawanipatna in Orissa. The average yield of onion in India is poor (112q/ha) than the world average production of 140 q/ha.

BOTANY

Onion is herbaceous annual for the edible bulb production and biennial for seed prodcution having bisexual flower and is highly cross-pollinated crop. The edible protion is a modified stem known as bulb, which develops underground. Typical flavour due to a volatile oil known as allyl propyl disulphide and red colour is because of the pigments anthocyanin and yellow colour due to "Quercetin". Primary roots produces by the leading, after which all other roots are adventitious, arising from the short stem. Root stem superficial with most of the roots in the tops 30 cm of soil within a radius of 30 mm of stem. A way short flattened stem is produced at the base of the plant which increases in diameter as growth centres to give an increased core; the oldest frant later decays. The onion leaves form from a shortened crown stem (Stemplate). The attachment of swollen leaf bases encircle the stem, forms the bulb. The shape of the bulb may be flat, globe, cylindrical. Outer colour is white, yellow, red, brown, pink and green. Elongation of stem forms the flower stalk (Howthorn and Pollard (1954). Onion has fibrous root system.

Plant breeding

Nath *et al.* (2008) reported that onion has been grown in India for a long time and some of its derivatives are also available. Various useful strains are available in different growing areas of the country. At some places efforts were made to maintain and grow the useful indigenous strains. Some varieties were also introduced from other countries and few of them have proved to be promising and have been used for commercial production. Genetic studies on onion have reported for better keeping quality and for suitability to dehydration (Singh and Joshi, 1975, 1978; Singh *et al.*, 1976). Efforts have been made to study the genetics of various economic characters and to utilize the male sterile lines for the commercial hybrid seed production (Singh *et al.*, 1979).

In onion, breeding methods are mass selection, pedigree selection, back cross method, hybridization. In the development of hybrid, there are male sterile lines, maintainer lines and pollinator lines. A hybrid cultivar will result from a male sterile line and a pollinator line. Although India is one of the leading producers, very few workers attempted to test different hybrid combinations for heterosis and combining ability. Male sterile lines are required for heterosis.

Male sterility form indigenous source was found in Nasik White Globe cultivar at IIHR. It was transferred to six different genotypes which are now being used for exploiting heterosis. The two promising hybrids are hybrid 1 (MS 65 x SI 13-1-1-1) & hybrid-5 (MS 48 x S14-1-1-1) which gave good bulb yield of 45.5 t/ha (Pathak and Gowda, 2003).

At present National Research Centre for Onion and Garlic at Rajagurur (Pune) is maintaining 437 lines of onion. Several Indian varieties, indigenous germplasm, and exotic varieties were tested for resistance to pests and diseases. The line IHR-56-1 was resistant to Purple blotch; Red Creole, Pusa Red, Pusa Ratnakar. Arka Kalyan, N-2-4-1 and a few are moderately resistant to purple blotch. Pusa red, Rampur local, Patna Red are tolerant and Telagi red, White large, Poona Red, Bellary Red, Patna Red, N-257-9-1 are found resistant to basal rot.

Micropropagation of male sterile (CMS) plants of female parent can be useful in hybrid seed production. Haploids have been successfully produced from cross pollinated ovules and their diploids will be useful for the development of inbred lines within a shorter time. Interspecific hybrids can be obtained by protoplast fusion. In monocotyledenous onion plant, transformation by *Agrobacterium rhizogenes* has been successful (Arya, 2003).

Taxonomy

Onion belongs to the family Alliaceae, genus *Allium*, *A. cepa* L., chromosome no. is 2n = 16. Natural tetraploid plant with 2n = 32 have been observed. The species is divided in to different groups which are sometimes given the rank of botanical varieties as follows :

Allium cepa var. cepa

Common onion with large bulbs borne singly and propagated through seeds. It is mainly used as salad or in curries

Allium cepa var. aggregatum

Multiplier onion or potato onion : Producting small bulbs borne in clusters and generally propagated through small bulbs. Mainly used for seasoning curries.

Allium cepa var. ascalonicum

Produces bulbs in clusters on surface of soil. This perennial onion rarely produces seeds and is propagated through bulbs.

Allium cepa var. viviparum /proliferum

This viviparous plant produces a perennial underground bulb and is not widely cultivated. The bulbs produced in place of umbel are used for pickling and seasoning. Bulb-lets are also used for propagation.

Allium schoenoprasum (2n = 16, 24, 32) Chive

This is a hardy perennial bunching herb grown for their hollow green leaves. It is propagated by root division and is tolerant to extreme cold and drought.

Allum fistulosum

It is a perennial bunching species. It is very common in China and Japan and is grown for its edible tops and leaf bases.

The genus *Allium* was for long time included in the family Lilliaceae for its superior overy. It was later shifted to Amaryllidaceae as flowers are borne in a bracted umbel on top of a scape. *Allium* being the typical genus in the family amaryllidaceae, a new separate family status has been given to all the species under *Allium* as Alliaceae. Onion is highly cross pollinated and the pollinators are bees and other insects. The reason for cross pollination are heteromorphism and male sterility. Flowers are protandrous, the inner three stamens shed their pollen first followed by the outer satmens. The whole process takes 2-3 days after which the style reaches its full length and becomes receptive.

Floral Biology

The onion flowers are borne in the form of umbel terminally at flower stalk. Flower stalk is hollow and round (at maturity) in cross section and somewhat swellen at the middle or near the base. The umbel is enclosed with a papery spathe, consisting of 2-3 bracts, which split open by the pressure of the developing flower buds. The number of flowers may vary from 50 to 2000 depending upon the cultivars, time of planting, size and storage condition of mother bulbs. The flowers are white to bluish in colour. The stamens are 6 in 2 whorls, anthers bilocular; ovary supperior, 3 loculus with nectory at base. There are 2 ovules in each locule. Anthesis occus in early morning (6-7 hr). Anther dehiscence is between 7.00 and 17.00 hr and on next day also with peak between 9.30 and 17.00 hr. Pollen fertility is maximum on the day of anthesis. Stigma receptivity is also high on the day of anthesis (Jones 1923). Onion plants remain in bloom for over 30 days. Cross pollination is common by honey bees. Selfing also takes place as the insects visit many flowers in a single umbel (Singh and Singh, 1996). The fruit is a three lobed, three celled capsule, each locule containing 1 or 2 block seeds at maturity. Embaryo, which is mostly made of the cotyledon is spirally twisted, cylindrical and is embeded in the endosperm, which forms the major portion of the seed (Hawthorn and Pollared, 1954) ratio and pollinating agent significantly affect seed production. However the 1:4 ratio (Male fertile to male sterile plants) is recommended for higher productivity (Larenzon *et al.*, 1993).

Male Sterility

Male sterility formed by interaction of nuclear gene and cytoplasmic factor (Jones and clarke, 1993). It is prsumed that there are two types of cytoplasm - normal (N) and sterile (S). Plants with sterile cytoplasm 'S' in the presence of homozygous recessive nuclear genes result in male sterility (Sms ms). The plants with 'N' type of cytoplasm are always male fertile. Similarly, plants having dominant nuclear gene either in homozygous or heterozygous condition are male fertile. Another form of cytopasmic factor 'T' has been reported to control male sterility in onion (Schweiguth 1973).

Male sterile plants are usually designated as 'A' line. They can be maintained by crossing with male fertile 'B' line having fertile cytoplasm N (ms ms). B line can be identified by crossing 'A' line with several other lines and if the progeny turns out to be completely male sterile in a particular case, that particular parent will be considered as 'B' line with N (ms ms). Fertility of 'A' line can be maintained by crossing with 'C' line having a constitution of N (Ms ms) as pollen parent. Progeny segregates in the ratio of 1 male sterile : 1 male fertile. 'C' line is generally disignated as fertility restorer line.

Male sterility reported in the indigenous germplasm 'Nasik White Globe' was controlled by strong cytoplasmic factor, which is evident from the fact that the crosses obtained between male sterile plants and the 400 germplasm lines including exotic accessions did not produce even a single fertile progeny (Pathak *et al.*, 1987).

COMPOSITION

Onion is rich in minerals like phosphorus, calcium and carbohydrates. It contains protein and vitamin C., has some medicinal properties also. Detail is given below :

(Per 100g of edible portion) Moisture - 86.8g; Calories - 49; Carbohydrates - 11.6 mg; Vitamin C-11 mg; Fat - 0.1 mg; Protein - 1.2g; Calcium 10 mg; Thiamin - 0.08 mg; Nicotinic acid - 0.4 mg; Ribaflavin - 10 mg (Rao and Purewal, 1957).

According to Manjunath, 1948; Brahmachari and Augusti, 1962; fresh onion contains about 86.8% moisture, 11.6% carbohydrates including 6 to 9% soluble sugars, 1.2% protein, 0.1% fat, 0.2 to 0.5% calcium, 0.05% phosphorus and traces of Fe, Al, Cu, Zn, Mn and I. Vitamin A, B_1, B_2, C and nicotinic acid are present as 2.6 IU, 64 mg; 79 mg; 11 mg; 0.77 mg; per 100g onion respectively. Onion to be richest in Vanadium (680 to 750 μg 100g fresh weight) among variety of food stuffs examined (Roshania and Agrawal, 1981).

The pungency in onion odour is formed enzyme reaction only when tissues are damaged. The enzyme allinase hydrolyses the Salk(en)yl cysteine sulphoxides to produce pyruvate, ammonia and many volatile sulphur compounds associated with flavour and odour. Both beneficial and unpleasant effects of onions and their extracts are due to sulphur compounds. They contain a large number of sulphur amino acid peptides and their derived products. Augusti (1976) explained that onion contains an acrid volatile oil (0.05%) with a pungent smell. The oil is rich in sulphur and contains a variety of aliphatic disulphides including allyl or propanyl propyl disulphide, dispropyl disulphide, methyl propyl disulphide and their trisulphides. However the chief component is propenyl propyl disulphide, an isomer of allyl propyl disulphide.

Flavour is influenced by varieties, stage of the bulb, storage period, nutrition application etc. Large doses of sulphur application increases flavour strength. However, it depends on the sulphur content of the soil itself (Kumar and sahay, 1954; Singh and Batra, 1972). Steady increase in the volatile sulphur towards the maturity stage and maximum just before the tops begin to fall over and reduction in volatile sulphur delay in harvesting and lifting from the field was noticed by Purewal (1957).

Uses

The green leaves, immature and mature bulbs are used as vegetable preferably in combination with potatoes, and other vegetable forms. Onions are high in food energy, medium in protein content and rich in phosphorus, calcium and riboflavin. Bulbs are used in salad, soup, sauces and for seasoning of vegetables and foods and also for making pickles. Dehydrated bulbs are used for making flakes and powder that can be stored and transported for longer duration and distance.

Onions have several therapeutic and medicinal uses such as its juice is used as smelling on hysterial fits in faintness condition and incuring flatulence, dysentery and cholera. It is useful in reducing high cholesterol, weakness, lethargy, tiresomeness and increases vitality. The onions are a vital rejuvenator and revitaliser. It increases appetite. It suppresses the formation of gas. Onion is very useful to ward off sunstrokes in summer. Onions are used to allay irritation of insect bites. The onion features as a home remedy for colds and fever in some parts of western India. It has been described as having invaluable diuretic, expectorant and digestive properties. Onion contain anti-fungal factor (catechol) and therefore their use is very benefincial (Nadkarni, 1954).

The pungency in onion is due to volatile oil, allyl propyl disulfide which varies due to cultivar, growing season (temperature, humidity etc.), type of soil and soil moisture, stage of maturity, storage conditions and duration, management practices. Onions are most pungent at the end of full maturity stage when foliage begins to fall over. The outer skin of onion bulb is red due to quercitin. The sulphurous pungency of the bulb while drawing tears in preparation, losses its strong taste in cooking and has no substitute for its flavour or as a thickener, in curries and gravies (Indian Express 15th March, 1992).

Onion has blood sugar lowering effect. According to Augusti (1976) S-methyl cysteins sulfoxide and S-propyl cysteins sulphoxide, the precursors of the disulphide and their oxides (allicins) are found. All these sulphur compounds can remove thiols, which inactivate insulin by oxidation or thiol disulphide exchange reactions.

Onion has lipid lowering effect. Augusti (1974) reported that regular use of onion by a diabetic patient lowered his insulin doses considerably. Sainani *et al.* (1979) reported that the people who consume 50 to 80g onion and 7 to 10g gartic daily, they have significantly lower body fat and their serum levels of cholesterol, triglycerides, β-lephotein and phospholipids are significantly lower than those who abstain from onion use.

VARIETIES AND HYBRIDS

Commercially two groups of onion

There are commercially two groups of onion of various colour, shape, size, keeping qulaity, pungency etc. Big sized (5 to 10 cm diameter) bulbs and small sized (1.2 to 2.5 cm diameter) bulbs. Red and yellow varieties are most commonly used as vegetable and spice where as white varieties are suitable for dehydration and processing purposes. The name of varieties given with the name of place where it was found successfully growing viz. Patna Red, white Patna, Poona Red, Nasik Red, Yadgiri or Bellary Red etc.

Big Sized bulb varieties

Red skin, Sliver or white skin, Yellow skin, Big sized have a mild flavour sweet in taste. less pungent. Big sized are Red Globe, White Globe, Yellow Globe, White Patna, large Red, Bombay Red, Patna Red, Poona Red, Nasik Red, Bellary Red, Early Grand yellow scaled Onion and Pusa Red, N-53, VL-1, VL-67, VL-3, Udaipur 103, 101, 102, Pusa white flat, Arka Niketan, Arka Kalyan, Pusa Ratnar, N-2-4-1. Gujarat white, N-257-91.

Varieties with medium sized bulb

Hisar-2, Panjab Red Round, Co-4.

Varieties with small sized bulb

Prize Taber, Market-Gardner, Yellow Globe, Queen and paris, Siver skinned, Co1, Co2, Co3, Co4, MDU1, Pukekohe (for dehydration); White Imperial, Spanish and Riverina, Late brown; Early lockyar brown; Mulitplier (With 3-4 bulbs of small size).

Hybrids

The hybrid varieties are higher yielding and more uniform in bulb size than the improved old varieties.

Tropica Seeds (Green Co.)

Granex, Texas Yellow Grano.

MAHYCO Seeds

MOH-2.

Bejo Sheetal Seeds

Lucifer. Flare, Orient, BSS 262.

Nunhems Seeds

Rojo, 9539, 9730.

Malav Seeds

Malav-21.

Detail Characteristics of varieties and hybrids described by Singh and Nath (2011) are given below :

Red -Type

Punjab selection

Plant height 55-65 with 6-9 leaves per plant. Bulb globular, red, bulb diameter 5-6 cm and average wieight 50-70 gm. Bulbs are firm with good keeping qualities. Bulb yield 300q/ha. Developed at PAU, Ludhiana in 1995 by mass selection from indigenous material. TSS about 14%. The variety is suitable for growing in U.P., Bihar, Punjab, M.P., Maharashtra and Karnataka (Anan. 1988).

Pusa Ratnar

Bulb large, round, a deep red, keeping quality average with higher yield potential 300 q/ha, duration 125 days after transplanting. Bulb obvate to flat, globular, less pungent, neck drooping, T.S.S. 11-12%. Developed by NBPGR, New Delhi in 1975.

Pusa Red

It rquire short to intermediate day length, Plant height 30 cm, Bulb medium sized average bulb weight 70-90 g, bronze in colour, flat to globular shaped, less pungent. Less bolting, keeping quality good, duration 125-140/days after transplanting. Average yield-250 q/ha. TSS 12-13%, suitable for growing late kharif and Rabi season. Suitable for seed production under the hot and dry climate (Patil *et al.*, 1990).

N-2-4-1

Average plant height 46 cm, bulb light red, globe shaped average length and breadth 6.1 and 55 cm, pungent in taste. TSS 11.3%. High yield potenitality with good keeping quiality. Tolerant to alternaria blight and thirps. It was developed at MPKVK, Rahuri, in 1985 (Anon., 1988).

Pusa Madhvi (Line-102)

Bulb medium to large in size, light red in colour and flatish round in shape. Maturity 130-135 days after transplanting. The average yield potential 300 q/ha. Keeping quality good suitable for rabi cultivation.

Early Grano

Introduction from U.S.A., large globular bulbs, yellow coloured, storage quality poor mild pungent, good for salad purpose. Attain full size in 95 dyas after transplanting and matures in 115-120 days, heavy yielder (around 500 q/ha). T.S.S 6-7%, suitable for both kharif and rabi season.

Brown Spanish

Bulbs solid, reddish brown, oblong, mild pungent with high T.S.S. and good storage ability. A long day type suitable for cultivation at an altitude of 1000 m and above.

Arka Niketan

Bulbs globose with thin neck, light red, attractive colour, average weight 100-180g T.S.S. 12-14%, pungency and dry matter content also high. Excellent storage quality (5 month without much loss). Yield 34 tonnes of bulb/ha in 100-115 days after transplanting, can be grown both in rabi and Kharif Season.

Arka Kalyan

Plant height 70-90 cm. Bulbs are dark red, globular, 4-6 cm in size and moderately pungent. Average bulb weight 130g and T.S.S. 11-12%. Moderately resistant to the purple blotch disease, can be grown in kharif and Rabi (Pathak *et al.*, 1988). Mature in 100-110 days after planting and produce an average yield of 35 tonnes/ha.

Agrifound Dark Red

It is a seclection from AADF, New Delhi. Bulbs are dark red, globular in shape with tight skin, moderately pungent. T.S.S. 12-13%. Mature in 150-160 days from date of Trasplanting. Suitable for kharif and late kharif. Best storing capacity. Bulb yield ranged from 300-400 q/ha.

Kalyanpur Red Round

It is a popular variety of Uttar Pradesh, developed at C.S.A.U. & T, Kalyanpur. Bulbs are bronze-red coloured, globular in shape, moderately sweet, moderately pungent. Mature in 130-150 days from sowing. T.S.S. 13-14.9% and bulb yield 250-300q/ha.

Arka Pragati

It was developed at IIHR, Bangalore, from a local collection (IIHR-149 from Nasik. Plant height 50-60cm. Bulb globe shaped with thin neck and deep pink outer skin with high pungency. An early variety with good keeping quality. Yield 450 q/ha in 130 days. Can be successfully grown both in Kharif and Rabi season.

Baswant 780

Bulb attractive red, globular, TSS. 11-12%, keeping quality good, average bulb yield 200-250q/ha. Crop duration 100-120 days. Suitable for Kharif.

BL Piaz 3

It is developed at MPKV, Rahuri, has wide adaptability. Medium sized closed neck, round shape and tight skinned, light red bulbs weighing 80-150g and measuring 5 cm across. Maturity 160-170 days after planting. Resistant to purple blotch. Low incidence of thirps. It is a cross from Line BYG 2207 A and local collection.

N-53

Bulb bright scarlet red, attain full size in about 95 days after planting and average keeping quality. Yield 200-250 q/ha. Suitable of plating in Kharif T.S.S. 11-12%.

Udaipur 101

Bulbs are deep red, flattish-globlular and sweet with less pungency. Good for salad. Yield 200-300 q/ha. Matures in 150-160 days after sowing T.S.S. 12-14%.

Udaipur 103

Bulbs are less red than Udaipur 101, oblate-globular in shape, sweet but slightly more pungent. Yield 250-300 q/ha, maturity in 150-165 days after sowing T.S.S. 10.5-13%.

Hisar 2

Bulbs are bronze-red in colour, top sided flattish-globular, sweet but pungent. Maturity in 165 days after sowing T.S.S. 11.5-13.9%. Bulb yield 300 q/ha.

Agrifound light Red

Bulbs are light red and globular in shape. Bolting is less. It has good keeping quality. Maturity 160-165 dyas. After sowing. T.S.S. about 13%. Yield ranged from 300-385 q/ha.

Bangalore Rose

Small onion, bulb uniform in colour, shape, size and neck thickness, developed by IIHR. Most suitable for export. Three crops may be harvested in irrigated condition. However best quality bulbs for export are otained from the rabi crop.

Arka Bindu

It is a selection from rose onion has small bulbs (2.5-3.5 cm dia) of uniform size and shape, deep red in colour, with high TSS (18-20%), free from splits and premature bolters. Bulb yield 250q/ha in 100 days and developed for export market.

Arka Pitamber

It is a cross of UD102 and IHR396, Bulb uiform medium sized, yellow colour, globe shape, high TSS (10-12%), wt. 100-110g firm bulb, good keeping quality free from splits high stable bulb yield (30-35 tonnes/ha). Highly suitable for export to European countries, Japan, USA (meet 10% export requires of Yellow onion). Short day types.

White-Type

Pusa white Round

Bulbs are white and roundish flat in shape. T.S.S. 12-13 per cent. Average yield 300 q/ha. Suitable for dehydration. Maturity 125-130 days and very good keeping quality in storage. Developed from a local collection (Line 106) at IARI, New Delhi.

Pusa White Flat

Fulbs medium to large in size, flattish round with attractive white colour, T.S.S. 12-14%, maturity 120-130 days. Yield ranged from 335-350q/ha. Good keeping quality.

S-48

Developed at PAU, Ludhiana. Plant height 40-60cm. Bulb flattish round and white in colour, good texture and flavour. Suitable for dehydration. Good keeping and storage quality. T.S.S. 12-13%, Bulb yield 250-300q/ha in 140 days.

Gujarat White

Flattish round, medium in size and good keeping quality.

N-257-9-1

This variety developed at Mahatma phula Agril Univ. Rahuri (Ms). Globe shaped bulbs with attractive white colour high yield potenital with good keeping quality. Suitable for cultivation in rabi season.

Udaipur 102

Bulb white, round to flat, size medium to large. TSS 12%, average bulb yield 300-350q/ha, crop duration 120 days. Developed by RAU, Udaipur, Rajasthan.

Onion Hybrids

Arka kirthiman (Ms 65×sel-13-1-1)

Bulbs are red with firm texture; light red flesh with 10% TSS; takes 120-125 days to mature, shelf 5 months, good transport quality; free from split and bolters; resistant to biotic stresses under field condition. Bulb shape globe, weight 100-120g, TSS 8-10%, crop duration 130 days, yield 40-45 tonnes/ha, suitable for kharif and Rabi. Recommended for Andhra Pradesh, East Madhya Pradesh, Orissa.

Arka Lalima (MS0-48-sel-14-1-1)

Bulbs are deep red, globe, weight firm texture; red flesh with 12% TSS; shelf life of 5 months at ambient condition; good transport quality; resistant to biotic stresses under field condition; free from split and bolter; crop duration 140 days, bulb weight 110-130g, TSS 10-12%, yield 45-50 tonnes/ha; can be grown in kharif and Rabi season. Recommended for Andhra Pradesh. Madhya Pradesh. Orissa.

VL hybrid Piaz 67

Uniform colour and medium size, high yielder hybrid. Recommended for cultivation in U.P. hills and plains and other places.

Branching Onion (Parade)

The leaves are consumed when tender. No bulbs are formed and it is grown only for leaf purpose. Used as salad and in curries. Can be planted round the year.

Sallot Tropix

It takes 80 days for maturity after tranplanting. It is red coloured and very productive. It gives multiple bulbs of small size and medium pungency. It can be used for sambhar and can be picked at early stage for salad purpose. Recommended for Rabi Season.

Onion Liberty

Size : 200 to 250 gm shape round, maturity 100-105 days after transplanting. Colour: Yellow, Suitable for growing: Rabi Winter: It is a grano type onion, having thick rings, uniform round and very high yeild. It is good for processing industry.

Onion Lucifer

Size-100 gm shape round maturity 100-110 days after trasplanting. Colour attractive dark red good for Kharif and Summer. It is attracitve dark red in colour good for fresh market and very high yielding.

Onion Flare

Size-100 gm shape round maturity 110-115 days after transplanting. Colour red, suitable for growing in all seasons except in cold winter. It is auniform sized hybrid. good for storage, as well as for fresh market.

Onion Orient

Size 100g shape-round-maturity 100-120 days after transplanting. Colour red, suitable for all seasons except in very cold winters. It is very productive hybrid. which gives uniform shape, size and colour, good for storage.

Onion BSS 262

Size 100 gm shape-round maturity 105 days after transplanting. Colour-milky white suitable for Kharif & Rabi. It gives uniform sized round bulbs, with TSS 12-14%. Good for dehydration.

Rojo

This red hybrid produced large thick-flat to flattened globe bulbs with excellent exterior colour. It has medium storage and good resistance to pink root. Higher yield superior colour and wide adpatability makes this hybrid an excellent choice for farmers and the customers.

Rio Santiago

An early maturing hybrid 160-170 days, prodces round bulbs thus makes this hybrid unique in the short day red onion market. Globe shape bulbs are medium large in size and medium in pungency.

CLIMATIC INFLUENCE

Onion is a cool season vegetable, very sensitive to photoperiodic rquirements and can be grown under a wide range of clomate, best suited as a tropical crop. It does best where the season is mild whitout the extremes of heat or cold or excessive rainfall. The ideal temperature for vegetative growth is 12.8-23.0°C. For bulb formation it requires long days and still higher temperature (20-25°C). Even though Onion is treated as a long days plant, for bulb formation and its development, varieties differ in their response to length of day. Most cultivars grown in plains of north India are short day cultivars. Long day varieties will not produce bulbs under short day conditions and short day cultivars if planted under long day conditions will develop early bulbs. At least 10hr light for growth, development and bulbing with favourable temperature is essential. In India mostly short days onions are grown except in hills. Rainfall exceed 75-100 cm in area is not suitable for onion cultivation (Rao and Purewal, 1957). In such places only summer crop can be grown and 70% relative humidity reuires for good growth.

Extremes of climate are harmful for seed stalk formation and seed setting. Temperature between 4.5°C and 14°C is favourable for seed stalk formation and longer these tempertures prevail, the more flowering stalks are produced and also more flowers are produced in each umbel. Temperature play more important role than the day length in seed stalk development. Temperature throughout the year and day lengh during growing season are impotant factors for seed production.

SOIL REQUIREMENT

The onion is not exacting to its soil rquirements. Best soils are deep friable loams, alluvium, and where there is free drainage, absence of persistent weeds and presence of organic matter which favour production of ecellent crops (Yawalkar, 1969). The optimum pH range between 5.8 and 6.5.

Onion gives high yield on heavy soils like clay provided adequate quantity of humus is supplied to lighten the soil. In the light sondy soils, the maturity is earlier but the yield is low (Deshi *et al.*, 1965 and Pandey, 1989). The most desirable soil is the one that retains enough moisture and at the same time be favourable enough to be easily cultivated and to allow proper development of bulbs.

SOWING

The seed sowing in the nussery and seedling transplauting in the field vary from region to region in different parts of the country as given below (Rao and Purewal, 1957):

Sowing and Trasplanting time of Onion nursery in different regions

S. No	Region	Sowing time of nursery	Transplanting time of seedlings
1.	North India	October to the middle of November	Middle of December to January
2.	Poona and Nifad area	October	November
3.	Andhra Pradesh	October	November
4.	M.P. and Vidarbha	Middle of December	End of January
5.	West Bangal	Sept. to Nov.	October to December
6.	Orissa	Ist week of October	2nd week of November to 2nd week of December

Sowing time for Kharif onion in North India is May to June. Highest yield was obtained when nursery was raised in the late June or early July for harvesting in December or January (Sharma & Kumar, 1983).

Seed rate of 10-12 kg for rabi crop and 12-15 kg for Kharif corp is rquired to raise seedlings for planting in one hectare land (Anon., 1996; Sharma and Kumar, 1982-83). However in Kharif more onion seeds are rquired when there is incidence of damping off.

The spacing between rows 15 cm and between plants 8 to 10 cm has been recommended (Jyotishi and Pandy, 1969 and Dastane and Joshi 1964 and Rao *et al.*, 1967). Care should be taken to see that the seedlings are not set deeper than 2 to 3 cm.

Land Preparation

Medium land should be ploughed after harvesting of paddy for 4-5 times till a fine tilth is obtained. The ploughing may be shallow because the roots do not penetrate deep in the soil.

Seed treatment

Thiram 3g per kg of seeds or captan 2.5g per kg of seeds.

Nursery management

Seedlings are produced in the nursery bed for transplanting. About 25 raised beds 7.5 m long, 1.0 m wide and 15 cm high provide enough seedlings for plating of one ha land.

Apply 3-4 baskets full of FYM and 500g NPK (15:15:15) mixtures to each bed and mix it well in the soil, sow the seeds thinly in rows spaced at 5-7 cm distace and cover the rows with powdered manure then irrigate immediately. Subsequent irrigation is given at regular intervals. Protects seedlings by drenching regularly with captan (0.2%) or copper oxychloride (Blitox-50) 0.2%. Seedlings will ready for transplanting in 6-8week after sowing when small swellings are form.

PLANTING

Prepare 30m × 3m size beds and transplant 6 to 8 week old seedlings in 15 cm rows with 10 cm spacing within the plants. Patel *et al.* (1958) and Singh (1974) showed that the optimum age of seedlings for transplanting is 6 to 8 weeks, however, in case of rabi onions in Maharashtra, 8-10 weeks old seedlings should be transplated. The healthy seedlings 12-14 cm length are transplaned in the well-prepared field in rows about 15-20 cm apart and plant-to-plant distance of 7-10 cm depending upon the variety. In some areas, growing onion by sets is popular especially for kharif season. It gives better plant stand and yield but takes longer duration of the crop.

Seeds of big onions are sown directly in the field in Bellary and Dharwad area of Karnataka. For obtaining on early crop in kharif season, small sets/bulbs are used for planting. Seeds are sown @ 25g/ m^2 in nursery beds in mid January to the beginning of February. Selected sets or small bulbs (1.5^{-2} cm or 2-2.5 cm) are uprooted along with tops and stored in a cool place. They are transplanted in August. The sets are planted on ridges which are 30-45 cm apart and the distance between plants is 10-12cm. The multiplier onions are vegetatively propagated by bulblets. About 1.5 tones of bulblets are required for planting in one hectare (Nath *et al.,* 2008).

Singh *et al.* (1995) studied the effect of sizes of onion sets and planting distance on yield and economics of kharif onion. The highest bulb yield (170.63 q/ha) and net profit (Rs. 42269/ha) with benefit cost ratio (3.71) was recorded when 2 cm size onion sets were planted at 45 cm X 10 cm planting distance.

Shrivastava *et al.* (1996) reported response of onion varieties to plant spacing in relation to yield attributes of bulb productivity. N-53, N-2-4-1 and Agrifound Light Red had shown significantly better performance with the spacing of 10 cm x 10 cm in terms of number of leaves, leaf bio mass and marketable bulb yield. Kumar *et al.* (1998) reported the maximum yield of bulbs was obtained when the plants were spaced at 20 cm x 10 cm.

Sharma *et al.* (2002) reported the effect of methods and dates of planting on onion sets on bulb yield on kharif onion in Rajasthan. Mean of two years data revealed theat 15th August planting increased bulb yield by 20.05, 14.85 and 71.49 percent over 31st July 31st August and 15th September respectively.

Singh *et al.* (2004) evaluated the onion varieties and recommended a planting space of 20-25 cm between the rows and 6-8 cm between the plants.

Nath *et al.* (2008) coneluded that planting can be done at a pacing of 20 cm x 10 cm distance in onion. In case of small sets it is 30-45 cm x 10-12 cm wheras in case of multiplier onion it is 45 cm x 15 cm.

NUTRITIONAL REQUIREMENT

The onion crop appears to be a heavy feeder of nitrogen and potash. The onion crop yielding 274q/ha removes 81 kg of nitrogen, 40kg of phosphorus and 35.5kg of potash (Yawalkar *et al.,* 1962). Different recommendations have been made regarding the application of manures and fertilizers. According to Purewal (1962), Nath (1967) and Pande *et al.* (1969) the doses are: FYM 25-50 t/ha, nitrogen 50-110 kg, phosphorus 25-135 kg and potassium 50-110kg/ha.

Singh and Dhankar (1989) studied the effect of nitrogen, potash and zinc on growth, yield and quality of onion Cv. Pusa Red and Hisar-2. Different levels of nitrogen, potash and zinc significantly affected the growth, yield and quality of onion. High levels of nitrogen (N 160, 120 kg/ha) reduced bolting, increased plant growth, ascorbic acid content and yield of bulbs. Under 80 kg N/ha reduction in neck thickness, and increase in total soluble solids, dry matter, reducing, non-reducing and total sugar contents of bulbs were boserved. Application of potash (K_2O 100 kg/ha) alone sugar contents of bulbs were observed. Application of potash (K_2O 100 kg/ha) alone and in combination with zinc (K_2O 100 kg + $ZnSO_4$ 25 kg/ha) reduced bolting, neck thickness, increased plant growth, yield, ascorbic acid, TSS, dry matter, sugar and sulphur content of bulbs. Variety Hisar-2 indicated significantly low bolting, lesser neck thickness, higher growth of the plants. yield and ascorbic acid content of bulb as compared to Pusa Red.

Selvaraj *et al.* (1993) studied the effect of irrigation and nitrogen fertilization on bulb yield of small onion and concluded that onion at 1.2 IW/CPE ratio to a depth of 5 cm with 90 kg N/ha was found optimum to get increased yield of onion bulb. Summer season crop recorded higher yield than kharif crop due to better utilization of available soil moisture and nutrients, especially nitrogen.

Katwale and Saraf (1994) reported the response of onion to varying levels of fertilizer doses during monsoon season in Satpura plateau. The fertilizer dose of 125,60 and 100kg/ha of nitrogen, phosphorus and potash was found best to get highest returns. Warade *et al.* (1995) reported response of organic, inorganic and bio-fertilizers on yield of onion bulbs of cv. B-780. They concluded that the increased onion yield can be obtained by appliying FYM @ 40 t/ha along with mineral fertilizer @75:50:50 kg/ha NPK and inoculation with *A. brasilence*, thereby saving 25 per cent in the nitrogen. Kumar *et al.* (1998) reported influence of nitrogen on growth and yield of cv. Patna Red. 150kg N/ha was found better for maximum diameter of the thickest stem, maximum number of leaves per plant, maximum spread of the plant, minimum maturity period of bulbs, maximum diameter of the bulb, maximum fresh and dry weight of bulb, maximum length and maximum number of roots per plant, maximum thickness and maximum length and maximum number of roots per plant, maximum thckness and maximum number of rings per bulb and maximum yield of bulbs (q/ha) in comparison to lower dose i.e. 75kg N/ha and control. Singh *et al.* (2000) reported the response of kharif onion to nitrogen and phosphorus and potash application in eastern plains of Rajasthan. The crop responded significantly and almost linearly to potash fertilization upto the highest level of 83 kg/ha. Application of potash @ 83kg/ha also recoreded the highest weight per bulb (49.38g) and as a result enhanced the bulb productivity by 14.19 and 6.91 per cent over lower levels of 41.5 and 62.2 kg/ha respectively. It thus recorded the highest net returns of Rs. 11,943/ha with B:C ratio of 1.376 and incremental B:C ratio of 33.04. Singh *et al.* (2001) studied the integrated nutrient management of onion in rainy season, and found that average weight of bulb and bulb yield increased significantly up to N 120 kg/ha and farm yard manure upto 10 tons/ha. However the maximum net returns and benefit : cost ratio were recorded with N application @ 150 kg/ha and farm yard manure application @ 10 tons/ha. According to Yadav *et al.* (2002) varieity N-53 responds well to 100kg nitrogen and 150 kg potassium under irrigated coarse textured soils. It is iedeal for obtaining higher bulb production in kharif onion.

Arvaidya *et al.* (2002) reported the effect of planting dates and spacing on yield of onion bulbs in cv. N-53 under heavey rain fall zone of south Gujarat. The 15th November planting gave significantly highest yield during both the years but it was at par with 7th December planting in pooled result. Closer spacing. i.e. 10 cm x 10 cm gave highest yield during both the years and pooled results with bulb diameter and bulb weight were maximum in wider spacing i.e. 20cm X 10 cm. The yield was increased in linear order with increasing levels of nitrogen. Treatment with 150 kg N/ha gave significantly higher yield in pooled results but it was at par with 125 kg N/ha in both the years.

Dixit and Sharma (2002) reported the effect of phosphorus and lime application on productivity, puptake and build up of soil phosphorus taking onion as test crop on mountain acidic soils of western Himalayas. The productivity and uptake of P in onion crop increased significantly due to the residual effect of lime upto full lime requirement of 3.08 t/ha. The increase over no lime being to the tune of 41 per cent in the productivity of onion, whereas 40 per cent in the uptake of P. Likewise P application upto highest dose of 112.5 kg P_2O_5/ha significantly increased the following parameters: the increase being 52 per cent in productivity of marketable bulbs of onion and 91 per cent in P uptake over no phosphorus. Thamburaj (1991) reported the use of biofertilizers (*Azospirillum*) in onion. The application of nitrogen and *Azospiillum* had significant effect on yield of onion bulbs independently. On the basis of two years data it was obeserved that application of 100kg N/ha gave highest bulb yield (334.6q/ha) which was at par with 75kg N/ha

(223.2q/ha) and the increase in yield was 14.5 and 10.6 per cent, respectively over control (292.3 q/ha). Application of *Azospirillum* also gave significantly higher bulb yield of onion (320.9 q/ha) as compared to without *Azospirillum* (306.1 q/ha).

Hebbar and Prabhakar (2002) reported on integrated nutrient management and moisture conservation in rainfed onion. The experiment was carried out to study the effect of three leaves of FYM, two levels of fertilizers and neem cake on growth and yield of onion variety Arka Kalyan. Highest marketable bulb yield of 28.73 t/ha was recorded when 40 t/ha FYM plus 125:50:125 kg/ha of N, P_2O_5 and K_2O (recommended dose) was applied. In another experiment onion variety Arka Kalyan under rianfed condition, yields in rainfed bed were marginally higher as compared to flat bed under types of mulching materials. Mulching with black polyethylene sheet recorded significantly higher marketable bulb yield (28.63 t/ha) over grass mulch (25.16 t/ha) in flat bed system. Patil *et al.* (2002) reported on fertilizer use economy in onion. Among the different treatment combination, the highest bulb yield of 16.26 t/ha was recorded in the treatment of 100 per cent recommended dose (125:50:125 NPK kg/ha) of fertilizer (RDF) and FYM (30 t/ha) followed by the treatment of vermicompost (1 t/ha) plus 50 per cent RDF (15.70 t/ha) and 50 per cent RDF plus ACD-20 strain of *Azsphirillum* (15.38 t/ha). Three years experimental data indicated that the same bulb yield of onion could be produced by applying bioagents or vermicompost (1 t/ha) with 50 per cent reduction in RDF. This has an additional ecofriendly advantage over the chemical fertilizers.

Nath *et al.* (2008) on the basis of above finding concluded that the manurial and fertilizer recommendation for onion is : FYM 25 to 50 tons, nitrogen 80 to 100 kg, phosphorus 50 to 60 kg and potash 50 to 60 kg/ha. The FYM should be thoroughly mixed in the soil at the time of field preparation. Half of the nitrogen and full P_2O_5 and K_2O are applied at the time of transplanting. The remaining 50 per cent N is applied as top dressing 30 to 40 days after transplanting. Sulphur @ 50 kg/ha and 5 ppm zinc as foliar spray increases the bulb yield and their quality. A soil application of zinc sulphate at the rate of 25 kg/ha is also recommended for increasing the TSS and sulphur contents of bulbs.

Optimum Range of Nutrient Requirement in Onion

For nutrient (N,P,K) analysis, leaf portion of "top no white potion, one third to half grown" should be taken. Optimum range of N(%) = 5.00-6.00; P(%) = 0.3-1.2; K (%) = 3.5-5.00.

Deficiency Syndromes

Nitrogen

Deficiency causes interveinal or marginal chlorosis progress from light green to yellow. Entire plant becomes yellow under prolonged stress.

Phosphorus

Leaf appears dull, dark green blue green or red-purple, on the underside at the midrib, and veins.

Potassium

Leaf margins tanned, scorched, or have necrotic spots.

Calcium

Growing point are first to turn brown or necrotic. Tip burn of young leavs.

Magnesium

Marginal chlorosis or chlorotic interveinal on younger leaves. Chlorotic area may become necrotic and brittle.

Sulphur

Deformed leaves in Onion are most common symptom of its deficiency.

Zinc

Drying of leaf tips and whole leaves in advanced stage are typical symptoms.

Optimum Range of Micronutrient (ppm)

Fe = 60-300 ppm, Zn = 25-100ppm, B = 22-60ppm, Mn= 50-250ppm, Cu=15-35ppm.

Biofertilizers

Biofertilizers are nitrients derived from microbial sources such as *Azospirillum* (a nitrogen fixing bacterium) and vesicular-arbuscular mycorrhiza (VAM) that increase the uptake of phosphorus. The inoculation of biofertilizers increases vegetative growth; bulb wet and bulb yield of onion and net profit per hectare. In addition application of biofertilizers improve the production of plant nutrients like N,P,K, Ca,

Mg, Zn, Cu and Mu. It also improve the production of gibberellic acid, indole acetic acid, and dehydrozeatim which have positive influence on the physiological activity of the plants. Bulb quality is also superior with enhance level of TSS and sulphur.

Method of Application

Add sufficient carboxy methyl cellulose solution to the inoculation of *azospirrillum brasilens* sp. *nov.* for converting, it into sticky paste, to which onion seeds should be added in the proportion of 1 kg seed in 100g of paste and left to dry for 30 minutes in shade before sowing. In small forrows of the nursery bed, apply VAM culture @ 1 kg/m^2 (a mixture of *Gigaspora margarita* Becker and Hall and *Glomus fasciculatum* (Thax.) Gerdeman and Trappe of each, 225 spores per 50 ml of soil and make the soil compact. Sow the seeds in the furrow and cover with sand. Water the beds immediately. Mulch the nursery beds with paddy straw to improve germination of seeds. The twin problems can be appreciably taken care of by the use of biofertilizers.

Micronutrients

Application of micronutrients improve the quality of bulbs :

Soil Application

Soil application of Zn @ 6-10 kg/ha or spraying of Zn.SO4 (0.3%) 2-4 times is effective to increase the yield, TSS, sugar and ascorbic acid.

Copper and Boron

Copper and Boron @ 13.4 and 1.8 kg per hectare respectively is helpful to increase the yield.

Dhanzyme Granules

Application of Dhanzyme granules @ 8-16kg/acre; first at 10-15 days after planting and 2nd 20-30 days later to first dose.

Danzyme Liquid

Spray @ 1ml/ltr of water after 25-40 days of plating. It increases crop yield, improve crop qaulity and increase resistant to pests, disease and climatic stress.

Plantaid

For foliar spray use 1ltr of plantaid in 300ltr of wate in one acre. Two spray at an interval of 3-4 weeks should be given. Plantaid (S) @ 10kg/acre in single or split dose apply in the soil directly. It improve bulb yield and bulb quality.

IRRIGATION

Give light irrigation once in four to six days depending upon soil and weather condition. Water less frequently during the growing period and more frequently during the bulbing period. Over irrigation is harmful, the leave turn a poor green colour. Sandy soil needs more irrigation and clayey soil lesser, usually in rabi 15 to 18 irrigations are required for a good yield. During kharif 8-10 irrigations are enough if rainfall is normal. Normallly 10-12 ha inch of water is enough for the entire crop season.

The onion has narrow rather sparse root system. Irregular irrigation may increase the proportion of splits to almost double and over watering may favour thickness.

The irrigation should be stopped about 15 days before harvesting for proper maturity and curing of bulbs. This practice also reduces the losses in storage.

Depending on the climate and soil conditions the number of irrigations given will vary. The most important stage of irrigation is the bulb formation stage and the negligence at this stage results in cracking of the bulbs and low yield. Narang and Dasthane (1969) reported that the onion crop in all rquired 18 irrigations from transplanting ot harvest.

INTERCULTURE

Two to three hoeing are required for the onion crop, these helps in better growth of the plants. It is a shallow rooted crop, deep tillage is likely to injure the roots and thus decrease the yield. Moisture stress also decreases yield. Weeding, hoeing should be done at the time fertilizer application and topdressing i.e. 21 days and 45 days after transplanting.

WEED CONTROL

Since onion is a shallow rooted crop, keep the field free from weeds. Onion rquires almost perfect weed control throughout the growing period. Being a closely spaced planted corp, the onion rquires manual weeding which is quite costly. Sometimes labour, availability is also a problems. Specific weed control practices including intercultivation and use of herbicides should be planned prior to transplanting. Integrated weed control measure with a combination of cultural and chemical weedicides can be employed very beneficially.

In onion, normally seen weeds are; *Chenopodium album, Amaranthus viridis, Cyperus rotundus, Poa annua, Ecehinochloa crusgalli, Fumaria parviflora, Trigonella, polycerate* etc. To keep down the weeds very shallow cultural operations should be followed. Weedicides like cholro-IPC at the rate of 5 to 6 kg/ha control the weeds (Choudhury. 1967). Stomp (2.5 litres/ha) or tenoran at 2.5 kg/ha in 800 litres of water three weeks after transplanting can also check broad-leaved weeds. Nitrofen (TOKE-25) @ 6 litres/ha along with one hand-weeding is useful in controlling weeds in kharif. Singh (1996) reported on economic weed control in onion. Average of three year field data revealed that fluchloralin and oxyfluorfen at higher rates (i.e., 2.0 kg and 0.25 kg/ha. respectively) reduced the weed population and weed dry weight to a great extent, but the size index of onion bulb was recorded highest at the lower rates of fluchloralin (i.e., 1.0kg/ha + one hand weeding 45 days after trasnplanting). Fluchloralin 1.0 kg/ha PP1 + one hand weeding 45 days after transplant (DAT) treatment gave the highest bulb yield (178.89 q/ha) but it was on par with the treatment oxyfluorfen 0.25 kg/ha + one hand weeding 45 DAT, which gave 163.88 q/ha bulb yield. A preplanting application of oxadiazon (1kg a.i./ha) fluchloralin (2kg a.i./ha) and oxyflourfen (0.2 kg a.i./ha) along with one hand weeding after 45 days of transplanting is recommended to control the weeds in onion. Pre emergence application of nitrofen (1kg a.i./ha) and chloroxuran (1kg a:i/ha) one week after transplanting is also effective.

Thakral *et al.* (2002) recported that to obtain maximum number of total and healthy seedlings as well as highest dry weight of seedlings, trifluralin at 1.00 kg/ha as pre-sowing soil incorporation or pendimethalin at 1.00 kg/ha or alachlor at 2.00 kg/ha applied as pre-emergence were the most effective in controlling weeds.

Basalin (1kg active ingredient in 100 litre of water) can be sprayed before transplanting to make the crop weed free. With the help of chemical, only one weeding will be rquired and after 55-60 days of transplanting (Pal, 2004). Singh *et al.* (1997) investigated the effect of pendimethalin and fluchloralin alone and in combination with one hand weeding on control of weeds, productivity and economics in onion. The control of weeds through chemicals and mulches were superior to manual hand weeding in onion crop to get maximum profit per hectare.

Cropping System and Crop Rotations

Saini (1996) suggested that following crop rotation can be taken successfully with onion :

1. Capsicum - Onion - Longmelon
2. Fenugreek - Frenchbean - Onion

3. Maize - Potato - Onion
4. Cowpea - Potato - Onion
5. Okra - Pea - Onion
6. Greengram - Pea - Onion
7. Black gram - Potato - Onion
8. Brinjal - Onion
9. Bottle gourd - Onion
10. Sponge gourd - Onion
11. Soybean - Onion
12. Chilli - Onion

Shankar *et al.* (2002) reported the studies on onion based cropping system. In recent year soil fertility-fertilizer use research is focused on cropping sequences. Due to escalated chemical fertilizer production cost, impaired quality of produce, degraded soil and polluted surroundings necessitated the practice of cropping systems for vegetable production. Onion is a short duration bulb vegetable crop commercially grown through different cropping patterns. The present investigation was carried out during kharif, late kharif and rabi season to find out most economically viable. environmentally suitable and socially acceptable onion based cropping systems like aster-onion, marigold-onion. groundnut-onion, potato-onion, onion-wheat, soybean-onion and also with traditional and popular cropping systems of bajra-onion and bajra-wheat. Among the various cropping sequences evaluated, aster in kharif season followed by onion in rabi season recorded the highest C:B ratio of 1:2.2 followed by bajra-onion sequence (1:2.08) but legume based cropping patterns like preceding crop of groundnut in kharif season followed by onion in rabi season and soybean (kharif)-onion (rabi) sequences noticed the highest onion yield of 27.2 and 26.2 t/ha respectively. Based on the results, it is concluded that the most profitable onion based system was aster (kharif)-onion (rabi) on the same piece of land in one year whch is more economical and suitable for western Maharashtra conditions (Nath *et al.*, 2008).

PLANT PROTECTION MEASURES

Important insect pests and diseases of onion and their control measures are given below :

Insect Pests

Onion thrip (Thrips sp.)

It is a small, yellow sucking insect which causes white blotches on the leaves resulting in the browning of the tips. It is most injurious during dry weather and is seldom very destructive during rainy season. Dusts containing 3/4 of 1 per cent rotenone have proved satisfactory for use on bunching onions where the fresh green tops bring a marked preference (Thompson and Kelly, 1957). Wherever possible preventive measures should be used to restrict the thrips population. The cutting and burning of grass and weeds in waste places will destroy alternate hosts. Parathion, toxophene, heptachlor or dieldrin should not be applied to onions which are to be eaten green. In this case malathion (malathion 50 EC@ 750 ml/ha in 650 litre of water) should be used and it should not be applied within three days of harvest. The above chemicals may be used in case of the crop being grown for the bulbs. Jones *et al.* (1934) found some varieties like White Persian, Grano, Sweet Spanish, and Crystal Wax to be fairly resistant as compared to others.

Chemical control include; spraying with monocrotophos (0.15%), dimethoate (0.2%), endosulfan (0.2%) or malathion (0.2%) at 15 days interval. Soil application of phorate (2kg a.i./ha) at 2,6 and 10 weeks after transplanting; Grow resistant variety, Nasik red.

Onion maggot (Hylemia Sp.)

The maggot feeds upon onion plants of all ages from young seedlings to mature bulbs. They channel onion bulbs thus permitting bulb rotting organisms to enter. Where maggots are a problem, infected bulbs from storage-houses and packing sheds should be dumped to an isolated place and covered with at least 30 cm of soil in early spring before the adult emerges out. These burial places should then be dusted several times at 10 days intervals with 10 per cent parathion to kill the adults. if any, Perron *et al.* (1958) reported that the varieties of *Allium fistulosum* were more resistant than those of *A. cepa*.

Sandhu *et al.* (1975) reported that application of granular insecticides like phorate 1 kg, carbofuron 0.3 kg, carbaryl 0.5 kg, dimethoate 1 kg and Lindane 1 kg a.i./ha control the pest effectively when broadcast before light irrigation.

Several natural enemies in onion belong to *Orius* species : *O insidiosus, O. maxiedntex, Aelothrips collaris, Scymnus nibilus, Chrysopa* species and *C. orestes* were imported for *T. Tabaci* in India.

Diseases

Purple blotch (Alternaria sp.)

The disease is very serious and appears on the leaves, stems, bulbs and seed stalks as small, whitish, sunken lesions with purple center that rapidly enlarge. The leaves or seed stalks fall over gradually. The infection of bulbs occurs late in the season when the plants are nearly mature. For control, dry infected leaves may be clipped and sprayed with one kg of indofil M-45 or 1250 g of copper oxychloride in 500 litres of water/hectare using a sticker. Also spray trition (200 ml) or linseed oil (200 ml) in 200 litres of water per ha. This can be sprayed at 10 days interval for 3-4 times. Seed treatment with dithane M-45 (0.25%), dithane Z-78 (0.25%), difolatan (0.2%), thiram (0.2%) is recommended. Bulbs should be harvested carefully to avoid injury. The resistant varieties reported from other countries are Red Creole and Yellow Globes (Jones and Mann. 1963).

Bottom rot or basal rot (Fusarium oxysporum Schlecht)

This is a serious disease of bulb crop. Field symptoms usually do not appear until the soil has become warm. At first, there is a progressive yellowing and drying back from the tips of the leaves. Plants that are infected when young may continue to grow until harvest time. A semi-watery deceay affecting the fresh scales from the base and progresses upward and with early infection decay may be almost complete by harvest time. The fungus is able to cause the disease at termperatures of 15 to 30°C and above. As storage disease basal-rot is most active at or above room temperature. Use of healthy, disease-free planting material in disease-free plots is recommended. Sources of resistance have been indetified which can be grown (Sokhi *et al.*, 1974). Spaying with benlate (0.025%) and bulb treatment with potassium metabisulphate (100 pp), 2,4-D-100 ppm or acetic acid 100 ppm are the control measures.

Stemphylium blight (Stemphylium vesicarium)

This disease has become very serious like purple blotch and basal-rot. It causes burning tips of young seedlings and brown streaks on leaves and also develops brown streaks and blight on flowering stalks due to which seed crop fails even to the extent of 100 per cent. Dithane M-45 (0.25%) gives effective control (Anon., 1983). This disease occurs in eastern and nothern parts of the country.

Bacterial soft rot (Erwinia sp.)

This disease can cause considerable loss in storage. The rot usually begins at the neck of the bulb. Later the bulb lossens its firmness and it gives offensive smell through the neck when squeezed. It occurs most frequently in humid weather and specially in those varieties that have heavy green necks and do not cure well. Thorough and rapid drying at harvest time is essential and all the bulbs showing damage either mechanical or by insect pests may be discarded before storing.

Onion smut (Racists cepulae F.)

Smut appears as elongated dark slightly thickened area at the base of the seedling leaf as it eamerges from the soil. As new leaves are formed they become infected, swollen and bent downward. Raised black lesions appear near the base of the scales on plants starting to bulb. The fungus becomes inctive in soil temperature of 26.7°C and above. Most of the seedlings die within 3-5 days. The disease can be controlled by the application fo 0.5 litre of 40 per cent formaldehyde in 30 liters of water to be applied at the rate of 900 litres/ha (Jones and Mann, 1963). The disease may also be controlled by treating the seed with thiram (3 g/kg of seed). A *fistulosum* varieties are resistant to smut.

Downy mildew (Urocystis cepulae F.)

Smut appears as elongated dark slightly thickened areas at the base of the seedling leaf of seed stalk. Later the diseased portions become pale green. Yellow and finally the leaves collapse. After period of storage the infected bulbs become soft and shrivelled and the outer fleshy scale becomes partly or wholly amber in colour, wrinkled and watery. These bulbs must not be used for planting. However, the fungus in the bulbs can be destroyed by dry heating for four hours at 41°C (Yarwood, 1943). Regular spraying of difolatan (0.1%), cuman L (0.1%) or ridomil MZ (0.1%) at 10-12 days interval will control the diseases (Swarup, 2006). In the United States, the use fo sprays of dust has generally not given satisfactory control of onion mildew (Jones and Mann, 1963). In Holland the best results were obtained by spraying with 2 per cent zineb applied at the rate of 150 litres/ha with the addition of a commercial spreader (Doorn *et al.*, 1954).

Black mould (Aspergillus niger)

It is a serious disease during bulb storage. Rapid and thorough curing, good ventilation and temperature just above 0°C are important measures to control this disease.

Prominent bacterial diseases

Prominent becterial diseases in onion are seedling blight, soft rot and bacterial stalk rot of onion. Seedling blight is caused by *Pseudomonas siccata*. The infested leaves have streakes and dry from tip downwards.

Soft rot is caused by *Erwinia carotovora*. It affects the bulbs in the field as well as in storage. The disease enters the bulbs through wounds. Control include proper sanitation, curing of bulbs during storage and crop rotation with non-host crop.

Bacterial stalk rot

Caused by *pseudomonas gladiola* pv *allicola*. Symptoms are internal brown discolouration and dry rotting in the basal portion of the inflorescence stalk. Inflorescence dries up and falls down.

The disease infection can be reduced by soaking the bulbs in ceresan wet 2000 ppm. agrimycin-100 ppm and streptocyclin 200 ppm for 2 hrs.

Physiological Disorders

Tip burning

This is one of the most dangerous disorder of two bulb crops i.e. onion and garlic. The exact physiological stress of tip burning is still not known, because so far major research emphasis has been given on the disease like purple blotch and stemphyllium blight. However, it is believed that tip buring is the initial stage, which provide scope for the attack of thirps followed by occurrence of stemphyllium blight and purple botch. Therefore farmers should be make aware about this isorder and advised to examine attack of thrips and their control.

Bolting

Seedstalk initiation and development come under the general term 'bolting', which also includes premature bolting found in certain specific conditions. Onion is essentially a biennial crop, requiring one season for bulb development and another for seed production. This biennial nature is very specific and pronounced in case of long-day cultivars grown especially in temperate regions and also in rabi season in subtropical regions of North India. however the so called short day cultivars are grown during the summer to early winter in subtropical regions of India (Maharashtra) for both production and followed by a seed crop from winter to early summer, both the crops being completed within a year. This can be considered as an annual crop, even though the two specific seasons covering a year follow successively without bulb storage.

The growth of onion is charactrised by bulb development in the first year, followed by storage and seed production in the second year. In case of N-53 or Nasik Red, it appears that both these stage have been sacrified to produce the seeds, by growing kharif crop for bulbs and successively followed by rabi seed crop.

The initiation of floral formation is dependent mainly on lower temperature. In temperate countries, the onion are normally stored at a temperature of 0-2.2°C but it was found that bulbs kept at this temperature did not produce sufficient seed stalks compared to bulbs which were stored at 7.2-12.8°C. On the basis of these results it may be suggested that bulbs when stored at 0-4.5°C, should be lifted and put at a temperature of 10°C before planting for seed production. It is apparent that low temperature around 10-12°C is needed for initiation of floral primordia. Under north Indian conditions this temperature treatment is recieved by the bulbs, during their growing period in December- January, when the seedstalks are initiated. However, it appears that this temperature requiement varies amongst the cultivars, since cultivars like Bellary Red or N-53 and N-2-4-1 in Maharashtra produce seed stalks compartively at higher temperature.

It can be concluded that day length did not directly influence floral initiation and that under temperature, long days promoted the subsequent emergence of flowers and elongation of scapes. The multiplicity of effects of temperature at different seasons makes the lowering behaviour in onion variable rather unpredicatable in the field. High levels of N fertilizers reduced flowering in the field in over-wintered onion crops. Whereas low N nutrition during January and February, promoted subsequent flowering in plant, grown in culture solutions in greenhouse.

Inflorescence initiation is favoured by large set size or plant size. An onion plants should form at least 12-14 leaves before the appearance of flower but the evidence was not conclusive. However in case of premature bolting in the bulb crop in India. It is generally observed that some percentage of plants show seed stalks formation before the bulb development. The seed stalk development is the result of interaction of several factors, lie temperature, cultivar, age of seedlings etc. Hence it is necessary that the time of planting of seedlings for a particular cultivar in a specific location, should be worked out to avoid premature seed stalk development in the bulb crop. However, direct bolting without bulb formation can be utilized for resting and isolation of a line for mainataning male sterility in one year itself.

Splitting and Doubling of Bulbs

Under some adverse conditions and imbalanced manurial suchedule, splitting and doubling of bulbs occurred and affect bulb quality. Sometimes mechancial injury of plants, during hoeing and weeding initiate new growth and causes splitting and doubling of bulbs. According to choudhury (1967), water deficiency at the initial growth stage and irrigation after a long spell of drought may lead to splitting and doubling of bulbs. Higher doses of nitrogen are associated with the cause of splitting and doubling (Anon., 1984-85; Kale *et al.*, 1987; Pandey *et al.*, 1987-88). Pnadey *et al.* (1990) reported that in kharif, onion doubles increased with increase in the size of sets.

HARVESTING

The onion crops should be harvested depending upon the purposes for which it is planted. harvesting takes 45-90 days from field setting for green onion and 125-150 days (generally 4 months) for common bulb onions depending on the variety.

As the bulbs begin to mature, their tops fall over naturally and can be rolled off with a roller mounted onto a tractor. Discountinue irrigation and allow the soil to dry. Development of pigment and the characteristic pungency of the variety are also harvest indices of onion. Early harvest results in sprouting of the bulbs and late harvest gives rise to doubles or split bulbs. Harvest bulbs alongwith tops and wait till their neck begins to dry for 5-10 days after harvest. Clip both, the roots and tops with sharp shears leaving a 5cm neck before packing in bags for tranportation.

Bulbs for fresh market must be cured so that necks and outer scales dry up reducing decay. Onion may be cured almost entirely in the field during a dry weather in protected sheds, otherwise.

The green onions are harvested by pulling out be hand when there is little bulb formation and good, tender vegetative growth. The bulb crop is ready for harvest in three to five months after transplantig. When the bulb is ready for harvesting, about 70 per cent of the top shoots get dry and fall over even when part of the leaves are green. At this time the foliage should be trampled, i.e. 1-2 weeks before harvesting. At the time of harvesting the bulbs, it is necessary that the soil is loose which may be done by light irrigation some days earlier. The bulbs may be dug out by shovel or khurpi. For marketing, the tops are removed leaving bulbs only. If the season is mild the bulbs after harvesting are left in the field for curing which makes it firm and dry, whereas in hot weather the bulbs are removed to the shade for curing.

YIELD

Yield varied according to varieties and season. During rabi season yield range is 20-48 tonnes/ha which can be botained by adopting the modern agrotechinquies for raising of crops. A good variety under favourable conditions yields 250-400q/ha marketable bulbs. Yield of *Kharif* crop is comparatively low.

Due to poor storage facility farmers face problem for storage of produce harvested during late April-May and at that time there is glut in the market. Many times the prices crash down to the lowest, and therefore proper storage is essential. The highest price from onion is obtained usually after the rains. To solve the problems following measures are suggested :

(i) Stop irrigation at least 10 days before digging the crop. (ii) Keep the crop after harvest at a shady place for 5-6 days for proper curing of the bulbs. (iii) Only proper sized. well-cured, thin necked bulbs should be selected for storage. (iv) Use maleic hydrazide for checking the sprouting of the bulb during storage. Mix 750-1000 g maleic hydrazide in boiling water and later add it to 500 litre of water (1500-200 ppm solution) and any sticker also. Spraying this solution on the crop 15 days before harvest. The leaves of the crop should be not have dried up. Sprouting of the bulb is significantly reduced during storage by this treatment.

POST HARVEST HANDLING

It is estimated that 60-65% onion produced in India is consumed internally, 5% exported and 30-40% lost by post harvest damage.

Curing

Sprouting and rotting are common problems in storage since bulb contain high moisture. The bulb should be adequately cured for proper development of skin colour and to remove field heat before sotrage of bulb. It is done till the neck is tight and outer seales are dried. This will prevent infection of diseases and minimise shrinkage loss. Bulbs are cured either in field or in open shed or by artificial means before storage. During kharif season, bulbs are cured for 2-3 weeks along with top. In rabi, bulbs are cured in field for 3-5 days, tops are cut leaving 2.0-2.5 cm above bulb and again cured for 7-10 days to remove field heat.

The most common method of curing is by forced ventilation in the storage. Maximum safe temperature for onion curing at field has been reported to be 37-8°C for 3-5 days. Artificial curing of onion in crates are at 40°C for 16 hrs is formed to reduce rot losses in sotrage. Onions are considered cured when the neck is tight and the outer scales are dried. This condition is reached when onions have lost 3 to 5% of their weight. If not adequately cured, onions are likely to decay in storage. The most common form of decay is grey mould rot which occurs at the top of bulb. High temperature and high humidity during curing with good air circulation fovour development of desirable skin colour. Drying of superficial leaves of onion bulbs perfect them from microbial infection in storage and also gives skin colour to onion. It has been observed that if tops are cut before curing losses are more in handling than when tops are retained while curing bulbs with 2.0-2.5 cm tops when stored, losses are lower than with closed cut tops (Pandy, 1991).

Handling

Onion bulbs intended for storage must be as free as possible from cuts and bruises and so must be handled with extreme care. Onions should not be dropped on to a non-resilient surface from more than 1-2 m height which otherwise cause damage. When onions are dropped on to other onions, much less damage occur because of the cushioning effect between onions. Bruising in onion increased respiration rate, thus accelerating weight loss and may encourage infection by disease causing organisms. The depth to which onions can be stacked in bulk stored is reported to be between 2m and 3.5m. Excessive damage from pressure bruising will result at greater depths.

Sorting and grading

For avoiding loss due to rots, drying and sprouting and also to get a good price, the diseased, thick necked, damaged, bolted and doubles bulbs are removed from the lot. Medium size bulbs are better on storage point of view compared to small immature and bigger bulbs Sorted and graded bulbs have low storage losses than unsorted and ungraded (Pandey, 1991). Cured bulbs are graded based on size and depending on merket to which it is sent. Big size onion is in demand in New Delhi, medium size in Kolkata, Patna and small sized onion in North Eastern regions of the country.

Packing

Packing should be small for easy handling during transit and may vary according to merket demand. For cardboard boexs, it may be 15 kg and for jute bags it may be 40 kg. Thinly woven and low quality bags do not sustain in transport and rough handing 15 kg bamboo basket are also used for small onions in Karnataka and Tamil Nadu. For internal markets 40 kg packing is preferred and for export 8 to 25 kg packing sizes (nylon bags) are preferred depending upon the country. The size varies from 40-50 kg for consumer packs. Nylon netted and perforated plastic bags are also used. Packing should not be so tight as it causes injury to the bulbs by creating pressure among themselves or so loose resulting in deskinning of bulbs. Hooks should not be used to lift jute bags as it causes injury to bulbs.

Storage

The onion can be stored for several weeks under ordinary conditions. Under cold storage conditions it can be kept for about months at 0° to 1.7°C and 80 to 85 per cent relative humidity. At 60 per cent relative humidity it is possible to store even for 6 to 7 months. The bulbs are to be protected from direct sun light. Dampness and rain leackage are to be prevented completely. The essential of successful storage are thorough ventilation, uniform and comparatively low temperature and moderate dry atmosphere of 60-75 per cent R.H. Bulb harvested from kharif crop do not store well for long period.

The work at CFTRI, Mysore, revealed that if maleic hydrazide at 2500 ppm is sprayed 15 days before harvesting it checks the sprouting in the storage and the bulbs remain healthy for about 8 months. Bulbs from rabi season store better than those from Kharif season. Varieties Pusa Red, Arka Niketan, N-2-4-1 and Agrifound Light Red are good keepers and very popular among farmers. 75 kg of nitrogen per ha has increased the storage quality in Karnataka area (Nath *et al.*, 2008).

Various simple methods of storage of onions have been designed through ages and modified according to the local material available for their construction. Storage structures known as 'Chawl' in Maharashtra 'Kupe' or 'Tat' in Haryana, Delhi and U.P or tier system ventilated house know as 'Meda' in Gujarat. In Sudan, mud or straw cottages are used for storing onions. These have wooden stage inside which on the bulbs are piled up to about 1 m deep. They are constructed in such a way that they are ventilated by the prevailing wind passing through them. After 4 to 5 months storage by this method only 50-60% of the bulbs were marketable. In Korea, suquare straw cottages are built on upright poles driven into the ground. Nails are driven into the poles and bunches of onions, plaited together by their tops are suspended from these nails. Storage structure may vary from simple shelters, that protect from sun and rain to cold storage rooms that provide constant low temperature which prevent driage, sprouting and rottage.

In India onions are stored in bulk in special houses with thatched roof and sides of bomboo which are suitably well placed to allow good air circulation. In panipat district of Harayana and Jalalabad in U.P. the structures are made of bamboo/Sarkanda nets and roof is thatched one or made of sirki which is covered on top with Jute cloth. The size is 3.0×1.2×1.2 m and capacity is 40 quintals. The bottom net is fixed at about 15-20 cm height from ground level to have aeration. In Nasik and Pune area, the structures have side walls made of bamboo or wooden spaced at 1-2 cm apart. Roof is made of either thatch, asbestos sheet or tiles/tin sheet. There is however, no bottom ventilation but flooring is done with the help of soil, stone particles and sand and it is raised from ground level. Before loading onion, sticks are spread on the floor. The size of the structure varies from area to area. Normally width is kept as 1.5m and height is about 1.5m, the length may vary from 13.5-30 m and its capacity is in the range of 200-450 quintals.

In Bihar and Gujarat, the structures have 2-3 tiers and the floors are made of Bamboo pieces spaced at 1-2 cm apart. The dapth of loading is generally 30-60cm apart. Ventilation in the structure is provided by raised plateform and the windows on the side walls. Cold stores are rarely used for storage of onion in India.

Two tier onion storage sturcture

It is developed by NHRDF for storage of onion, its capcity in 40 quintals and the size of the stucture is 4.8 × 1.5 × 1.2 m. The side walls are made of bamboo splits with adequate support of wooden poles, the lower base is also made of bamboo splits with adequate structural support and its is raised from the ground level by about 75cm from bottom base. The second tier facilitates better aeration and therefore, reduces the losses. The total losses in this structures are about 36.6% over a period of four months.

Bottom ventilated onion storage godown

It is designed by NHRDF to hold 450 quitals of onion, it has a size of 15.8 ×4.8 × 2.4 m. Ventilation is porvided for free circulation from all sides including bottom ventilation. There is air gap of 60cm for this purpose below the floor. The Columns of the structures are made of birck masonry which supports the roof which is mangelore tiled on timber, trusses, purlins and rafters. There are two onion racks in each tructure of the size 15.×1.5m with a passage of 1.8m between them for loading/ unloading of onions. The onions racks are of wooden battons.

Refrigerated storage

It is often used for onions to be marketed late. Onions to be held in cold storage should be placed there immediately after curing. A temperature of 0.2°C will keep onions dormant and raesonably free from decay provided the onions are sound and well cured before storage. Sprout growth indicates high storage temperature, poorly cured bulb or immature bulbs. Root growth indicates high relative himidity. A comparatively low relative humidity (65 to 70%) is recommended for successful storage of onions. However, humidities as high as 85% and forced air circulation have given satisfactory results. Higher humidities expose onion to root growth and rot.

Bagged onion should be stacked to allow proper air circulation. Modern cold storage have forced ventilation system. Air is introduced through floor racks. Onion can be stored about 3m deep. However, soft onions at the bottom of the bin might be distorted in shape when stored in this manner. Onions should not be stored with other products which tend to absorb odours. When onions are removed from storage they are liable to sweet because of moisture condensation this may favour decay. Warming of onions gradually will avoid this difficulty (Maini and Chakrabarti, 2003).

Singh *et al.* (1982) studied the onion cultivars for storage qualities. Hisar-2, Arka Niketan, Arka Kalyan and Pusa Red cultivars were found to store 66.6, 65.2 and 65.1 per cent marketable bulbs respectively after 20 days of storage as compared to 48.6 per cent in VL-3 and 49.5 per cent in VL-1. Rotting was found maximum in Pusa White Round and sprouting was maximum in VL-3. Rampise *et al.* (2001) reviewed the factors affecting storage of onion. Out of total production of onion, 30-40 per cent produce, valued at about Rs. 600-700 crores, are lost every year due to desication, decay and sprouting during storage. The review suggests that to prolong the storage life of onion and to minimize losses owing to desication, rotting, sprouting and respiration, onion bulbs should be stored in single or two tier storage structures, constructed 30-40 cm above ground level having side walls of wooden splits having central and bottom ventilation facility. The optimum temerature of onion storage should be higher than 15°C along with 60-70 per cent relative humidity medium sized bulbs with 2-3 cm neck lenght are best for storage. The pre-harvest treatements of spraying MH-40, ethephone (0.2 to 0.3%) before 20-25 days of harvest and post harvest spray of 0.25 per cent mancozeb or 0.1 per cent carbandazim or benomyl will be helpful to reduce storage losses. Harvesting at proper maturity and sufficient curing in field and shed are useful to prolong the shelf life of onion bulbs.

Warade *et al.* (1995) reported that the performance of cultivars N-2-4-1 (49-57%) and AFLR (51.23%) was promising for storage for the period of 8 months during rainy season.

Maini *et al.* (1997) evaluated different structures used for storage of onions. During a storage period of 100 days, the physiological loss of weight was lowest in two tier wire system (26%) followed by single tier (38%) and conventional bamboo structure (47%). Maximum retention of quality was observed in two tier system. Sandhu *et al.* (2002) evaluated onion and garlic cultivars on ambient conditions. Punjab Naroya, PBR-1, ALR, Arka Niketan. PBR-2, PBR-4 and Punjab Red Round had good storage potential and had lost less than 20 per cent in weight after 150 days of storage. At room temperature they can be stored can be stored for several weeks but under cold storage they can be stored for 6-7 months. Sprouting at storage can be minimized by spraying maleic hydrazide@ 1500-2000 ppm.

Sagar and Maini (1997) reported the work on packaging and storage of onion powder. Onion powder prepared from 5 mm thick grated onion slices after drying in a cabinet dryer at 60°C for 10 hrs and grinding them in a laboratory powder mill and sieving with 30 mesh sieve, can be stored up to 6 months in 400 gauge bags at 7°C and four months at room temperature with better pungency. For storage of product havings 3.85 per cent moisture, the optimum RH was found to be 47.6 per cent and the critical and danger point was 9.14 per cent and 6.75 per cent moisture, respectively. The product was stable at low termperature as compared to high temperature in respect of colour and flavour.

Transportation

Transportation is the most important item for successful marketing. It plays an important role in marketing of onions both within the country and out side. Good transportation faicilities have made it possible to extend the area of production of Onion. For local markets onions are transported by head loads carts, tractor, tongas. For distant places, both railways and roadways are used. Care Should be taken while loading to avoid damage to the bulbs. Six to eight layers of bags are recommended. In railways wagons, while loading cushioning is provided at the bottom and side walls to avoid damage from heat. In fact, losses are more in wagons than in trucks but cost of transportation is high for Roadways. Rail transport takes about twice the time than road transport. Export consignments are generally shipped by ventilated ships as well as sailing vessels or mortor boats. Onions are also shipped in refrigerated vessels praticularly to for off countries. Sometimes, in absence of ventilations, cushioning or lining, damage to the bulbs is caused for 15-20 percent losses. And, therefore, care should be taken for all these aspects.

Marketing

Marketing of onions in the country is comparatively more organized than other vegatable crops. The main aim of the marketing is that the producers should relaize a suitable net return from his produce Timely and speedy delivery of produces with minimum damage and deterioration enroute at the lowest cost, are essential in marketing. Out of the total production in the country, it is estimated that the loss of about 30-40% occurs during various processes of handling after harvesting and in the course of maketing. About 50% of total supply is exported and remaining 60-65% is consumed in internally and used as seed material. There are number of agencies including producers villagers, merchants, whole sellers, commission agents, and co-peratives are involved in marketing. The co-operatives have been playing a significant role in the maketing of onions, particularly after NAFED'S (National Agricultural Co-operative Marketing Federation of India Limited) entry in this trade. In oder to provide onion at reasonable rates to consumers, NAFED had been purchasing onions on a large scale from the major onion markets in Maharashtra under the market intervention scheme and suppliying to the consumers at reasonable rates during off-season by maintaining buffer stocks. NAFED'S presence in the market had been maintaining price at reasonable level. Lasalgaon in Maharashtra is the biggest onion market in India. About 40% of the total production of onion of Nasik district arrives in this market. November to June is the peak period of onion in market.

Export

India occupies second position in onion production in the world and is traditional exporter of onion. The onion is an important foreign exchange earning crop in India. India is the third briggest exporter of onion, next to Netherlands and spain in world. Major export is to Gulf countires, Malaysia, Singapore, Sri Lanka and Bangladesh and East African countries. Export of onion is channelised through NAFED. Depending on preference of colour and size of bulbs, different varieties are exported. Middle East countries prefer light red to dark red bulbs. In Malaysia preference is for dark red bulbs. In America and Japan, demand is for yellowish or brown onion having mild pungency. Europe and Japan markets prefer large sized bulbs while in Singapore, demand is for small onions.

Onion accounts for major share (apporx. 75%) of vegetables exported from India. Export of 4.4 lakh tonnes of fresh onion worth of Rs. 332.43 crores during 2001-02 was increased to 8.33 lakh tonnes worth of Rs. 621.09 crores during 2004-05. Share of fresh onion export was 3.8% of the total export earning during 2004-05, from agriculture sector. Onion exports during 2006-07 touched 9.43 lakhs tonnes as against 7.7 lakh tonnes during 2005-06. Maharashtra accounts for 75 per cent of total onion exports. About 90 pecent of export from India is of big onion (4-6 cm diameter) and 10 per cent small onion (3-4 cm diameter) and multiplier onion. Large onion is exported to Malaysia, Gulf countries, Singapur, Sri Lanka, Bangladesh and small and multiplier onion to

Bangladesh, Singapore and Malayasia. Very small bulbs (2-3 cm diameter) with deep purple colour and high pungency are grown near Bangalore in Kolar district of Karnataka are exported to Malayasia and Hong Kong. Colour preference also varies with the country. Middle East countries demand light red to dark red, malaysia dark red,Sri Lanka dark red to light red and Bangladesh small onion of light red to dark red. European countries, Japan, and America prefer yellowish or brown onions having mild pungency.

In India yellow onions are not available for commercial cultivation and export. Only red onions are exported, restricting our export limited up to 20% of the world trade. Even among red onion group, varieties with superior quality and high bulb yield besides meeting specification of international trade are not available. There is a tremendous potential to increase the export by developing suitable varieties in yellow and red onion groups which cater the needs of the global market. Hence there is a definite need for developing yellow onion varieties and improving the existing red varieties to increase the export. Recently IIHR have developed the different varieties for export purpose. These are Arka Pitamber (Yellow), Arka Bindu and Arka Niketan (red), Bangalore rose (small onion) and Pusa madhavi from IARI, New Delhi.

Dehydration

To regulate the production and prices and also to earn more foreign exchange it is essential to develop dehydration indurstries in our country. Popularity of dehydrated onion is encreasing now-a-days. Advantages of dehydrated onions, are storage stability and ease of preparation. Commercial processing plants prescribes a shrinkage ratio of 7:1 to 17:1 with a moisture content of 4% in final product. Dehydrated onions are sold in many forms as slices, chopped, minced granulated and powdered. Bulbs for dehydration should be : (i) Devoid of any green patches so that it may not develop discolouration on drying. White onions are preferred to red or yellow onions. (ii) TSS should be 15-20 Brix in common onion and 25° in multiplier onion. (iii) Onion with small neck and root zone and those with tall globe shape are preferred than flat types to permit greater effeiciency in topping. (iv) Large bulbs are preferred due to economy in harvests. (v) Pungency should be high since dehydrated product is primarily used as flavouring agent. (vi) The bulb should have good stability with minimum shrinkage loss and rotting. Moisture content should be around 80%. South port white globe and white creosol are primarily used for dehydration in central California and central America. Varieties like Pusa white red, Pusa white flat, S-48 and N-257-1 and Udaipur-102 are suitable for dehydration purposes.

SEED PRODUCTION

Onion is a biennial for the purpose of seed production. All the varieties produce seed freely in the plains.

Seed-to-seed method

The seeds are sown in the nursery bed during during September to November in the plains for bulb production which get ready during summer. The selected bulbs for the seed purposes are usually planted in September-October in the plains which in turn produced seeds during summer next year. Thus seed-to-seed method take about 1½ year from the sowing of the seed from the new crop. In some regions like Bangalore with milder climate, the bulbs may be harvested from the monsoon crop (May to Septemeber) and may be planted immediaterly without storing, which in turn would produce the seed by next summer, i.e., within a year (Nath *et al.,* 2008).

Bulb-to-seed method

The selected bulbs are planted in well prepared field, the spacing of which would depend on the size of the bulb. Bulbs should be planted along one side of the ridge, 45cm wide giving a space of about 30 cm from plant to plant. In flat beds, the spacing of 45 cm x 30 cm may be given. Choudhury (1967) reported that Pusa Red spaced at 30 cm apart in the row produced more seeds than those plants which were spaced 45 cm apart both ways. Arakeri and Patil (1956) reported the highest seed yield from large-sized bulbs with a spacing of about 30 cm between the two plants in a row. Solomon and Patil (1959) also supported this view indicating that the bulbs weighing 81 to 100 g produced the maximum seed yield of 760 kg/ha. Mittal and Srivastava (1964) reported that the number of seed stalks influenced the seed yield and that the seed yield of Pusa Red was 50 per cent more than that of Early Grano. Sandhu and Korla (1976), Dadhlani and Bhagchandani (1978) and Singh *et al.* (1979) reported that plant and number of seed stalks per plant and also diameter of umbel by the latter increased the seed yield.

The after care and irrigation for seed crop are the same as those for onion raised for bulbs. Within three months the flowering stalks are produced and in another 6 weeks the seed ripens for harvest. Since the ripening of the seed is not uniform it is desirable to cut the heads before the seeds are shed. The heads are dreid in a well-ventilated shady place and then threshed. The seeds are then dried in the sun for a day or two before they are stored (Nath *et al.,* 2008).

Kulkarni (1963) recommended application of about 25kg of superphosphate at the time of planting of bulbs and about 45 kg of ammonium sulphate per hectare as top dressing after about 1½ months. A light earthing up should also be given after two months of planting to facilitate better growth of bulbs.

For nucleus and foundation seed production the bulb-to-seed method should be followed because it provides opportunity for selection and roguing. However, the seed-to-seed method may produce higher seed yield. Since it is a corss-pollinated corp, the two varieties should be isolated by about 1000 m in order to produce pure seeds (Anon., 1971a).

Vijay Kumar *et al.* (1991) reported the work of seed treatment and containers on the storability of onion seed. The results of the experiment showed that it would be beneficial to treat the Bellary onion seeds with captan 75 per cent WP at 2g/kg and packed at 8 per cent moisture content in 700 gauge polythene bag for maintaining viability.

On an average onion seed yield of 700-900 kg/ha can be obtained under good management practices. Seed yield of hybrids is very low (Singh, 2001).

Constraints in Onion Production

Production of onion limit due to following reasons :

i. There is need to develop high yielding varieties having improved quality of bulbs with longer storage life, suitable for different agro-ecological conditions.

ii. Non availability of suitable F_1 hybrids has been of the major constraints in improving the porductivity.

iii. Many of the diseases like purple blotch, basal rot, stemphyllium blight and storage rots, besides insect pests, like thirps take heavy toll of onion crop both in the field and store houses.

iv. Lack of suitable varieties for dehydration, adversely affecting the dehydration industry in a big way.

v. Post harvest life of onion depends on nutritional care i.e. more nitrogen decreases shelf life and more porash alongwith nitrogen increases the shelf life so there is need to apply balance dose of nutrition to the crop.

vi. Although India is exporting large quantities of onion, there exists a vast potential for increasing it by developing varieties suitable for export to the world market.

vii. Curing after harvesting of onion increase shelf life. It should be done in ambient condition.

viii. Onion follows paddy so, during November 15 farmers should sow seeds and it must be transplanted within December. It is the only vegetable in which seedlings of 6-8 weeks old are used.

ix. Too late sowing in nursery and hurridly planting in the field cause failure of the crop.

x. Non availability of varieties which are resistant to abiotic stresss especially moisture stress, high temperature, salinity and alkalinity.

Constraints in Onion Exports

i. Export industries preferred small and uniform size (1-2 cm diameter), less pungent as fresh onion. Bangalore rose and Arka bindu identified for export but production is very limited.

ii. Export as dehydrated form needed high T.S.S. (20%), where as Indian varieties have only 12% T.S.S.

iii. Alcohol insoluble solids more than 1% is desired if less than 1% quality will be poor. Indian varieties are inferior in this respect.

iv. The average per hectare yield in India is between 10-12 tonnes, so cost of production per unit is quite higher as compared to other countries. Indian onion thus gets out priced in international market.

v. The introduction of quota system in the export of onion has created uncertainity in the minds of importers because they are not sure if the Govt. of India would further allow export of onion after a particular quota is exhausted. As a result, other competing countries like Pakistan, Thailand, Phillipines, Holland have made significant in-roads in our traditional overseas markets.

vi. The export of onion from India have dwindled to some extent and our market have also shrunk in the recent past, so well thoughtout strategy needs to be worked out which not only ensures continuity in exports but also enables effecting supplied to importing countries at more competitive prices as compared to other competing countries.

vii. Post harvest losses are much higher as compared to other developed countries. This may be attributed to poor stransport system, lack of scientific storage, poor packing, poor handling etc. In order to minimise the post harvest losses, there is need to improve the transport system right from farm level to the ultimate destination.

viii. Present containerized-shipments has no proper system of ventilation, so many time losses are more. Modified containers with proper ventilation arrangement are thus needed.

ix. There are certain constraints in exports like non-availability of suitable railway wagons, no facility of storage at ports and non availability of vessels. Refrigerated/ventilated containers for shipment of yellow and white onions which are in demand in European, Japanese and others markets, are not available. It is, therefore, necessary to make available suitable insulated and ventilated railway wagons in sufficient numbers and also create facility of ventilated storage space at port.

x. It is also necessary to have refrigerated and ventilated containers so that yellow and white onions are produced and exported to European, Japanese and other markets which are in good demand. Further, with a view to remain in the market, it is necessary to arrange exclusive production of different types of onion meant for export every year (Singhal, 2003).

REFERENCES

Anonymous, (1971). Indian Minimum seed certification standards. Central Seed Committe, Ministry of Food and Agriculture, Community Development and Co-operation, New Delhi.

Anonymous, (1988). Vegetable Research Highlights. Project directorate of vegetable Research (ICAR) *Tech. Bulletin*, 2 : 18-22.

Anonymous, (1995). Manual for Agril Extension officers (1994-95 Rabi). Direcrorate of Agril. Food Production Orissa and OUAT, Bhubaneswar. pp. 70-71.

Anonymous, (2004). Report of the committee on agriculture - eighteen session : F.A.O., Rome, 9-10 February 2004.

Anonymous, (2005). Annual Report 2004-05, Indian Institute of Horticulture Research Bangalore, pp 80-81.

Anonynous, (1983). Annnual Report. Assciated Agricultural Development Foundation, New Delhi.

Arakeri, H.R. and Patil, S.S. (1956). Effects of bulb size, spacing and time of panting on the yield of onion seed. *Indian J. Agron*, 1 (2) : 75-80.

Aravadiya, L.K., Hiray, S.A. and Bhatt, R.I. (2002). Effect of planting dates, spacing and nitrogen levels on yield of onion bulbs (*Allium cepa* L.) cv N-53 under heavy rainfall zone of south Gujarat. *Intl. Conf. Veg.*, Bangalore, 11-14 November, 2002. *Abstr.* : pp. 186.

Arya, P.S. (2003). Vegetable breeding - production and seed production : Kalyani Publishers, New Delhi, 74.

Augusti, K.T. (1974). Effect of alloxan diabetes of allyl propyl disulphide obtained from onion. *Naturwissensechaften*, 61 : 172.

Augusti, K.T. (1976). Gas chromatographic analysis of onion principles and study of their hypoglycemic action. *Indian J. Exp. Biol.*, 14 :110.

Aykroyd, W.R. (1963). The nutritive value of Indian foods and the planning of satisfactory diets.. ICMR Special Report Series No. 42, New Delhi.

Brahmachari, H.D. and Augusti, K.T. (1962). Effects of orally effective hypoglycemic agents from plants on alloxan diabetes. *J. Pharm, Pharmacol*, 14 : 617.

Choudury, B. (1967). Vegetables. National Book Trust, India.

Currah, L. and Proctor, F.J. (1990). Onions in Tropical Regions. *National Resources Institute bulletine* No. 35, pp 323 NRI, UK.

Dadlani, N.K. and Bhagchandani, P.M. (1978). Variability and correlation studies in onion. *Indian J. Hort.*, 35 (2) : 118-22.

Dastane, H.G. and Joshi, R.S. (1964). Points to note about onions. *Indian Fmg.*, 14 (1) : 11-15.

Dhesi, N.S., Nandpuri, K.S. and Kumar, Jai Chand (1965). Onion cultivation in Punjab. *Agricultural Information Section*. Punjab Agric. University, Ludhiana.

Dixit, S.P. and K. Sharma Pritam (2002). Effect of phosphorus and lime application on the productivity uptake and build up of soil phosphorus taking onion (*Allium cepa*) as the crop. in mountain acidic lands of western Himalayas : *Indian Intl. Conf. Veg.*, Bangalore, 11-14 November, 2002. *Abstract* : pp 171.

Doorn, A.M., Van Koert. J.L. and Vander Vliet, M. (1954). Downy mildew in Onion and shallots. *Meded Dir. Tuinif.* 17 (16): 452-456. *Abstr. Rev. Appl. Mycol*, 34-277.

Hawthorn, L.H. and Pollared, L.H. (1954). Vegetable and Flower Seed Production. The Blakistan Co. INC, New York, pp. 135.

Hebbar, S.S. and Prabhakar, M. (2002). Studies on intergrated nutrient management and moisture conservation in rain fed onion. *Itnl. Conf. Veg., Bangalore*, 11-14 November, 2002. *Abstract* : pp. 219.

Jones, H.A. (1923). Pollination and self-fertility in the Onion. *Proc. Am. Soc. Hortic. Sci.*, 20: 191.

Jones, H.A. and Clarke, A. (1943). Inheritance of male sterility in the onion and the production of hybrids seeds, *Proc. Am. Soc. Hortic. Sci.*, 43: 189.

Jones, H.A. and Mann, L.K.A.E. (1963). Onions and their allies. World Crop Books. Interscience Publishers, Inc. New York.

Jones, H.A., Bailey, S.R. and Emsweller, S.L. (1934). Thrips resistance in the onion. *Hilgardia*, 8 : 215-232.

Katwale, T.R. and Saraf, R.K. (1991). Evaluation of onion (*Allium cepa* L.) varieties for rainy season in Satpura Zone of Madhya Pradesh. *Orissa J. Hort.*, 19(1&2): 6-10.

Kulkarni, C. (1963). Production of Onion seeds. *Indian Fmg.*, 10-11.

Kumar, H., Singh, J.V., Kumar Ajay and Singh, Mahek (1986). Studies on the effect of spacing on growth and yield of onion (*Allium cepa* L.) cv. Patna Red. *Indian J. Agric. Res.*, 32 (2) : 134-138.

Kumar, H., Singh, J.V., Kumar, Ajay and Singh, Mahak (1998). Studies on the effect of spacing on growth and yield of onion (*Allium cepa* L.) cv. Patna Red. *Indian J. Agric. Res.,* 32(2): 134-138.

Kumar, K. and Sahay, R.K. (1954). Effect of sulphur fertilization on the pungency of onion. *Curr. Sci.* 23 : 368-369.

Larenzon, M.C.A., Marfinho, M.R. and Freire, J.A.H. (1993). Effect of pollination by honey bees on onion hybrid seed production. *Pesquisa Agro Pecuaria Brasileira,* 28 (2) : 229-235.

Maini, S.B. and Chakrabarti, A.K. (2003). Post Harvest Management of Onion and Garlic, In : Approaches for sustainable development of Onion and Garlic. (Eds. Kirti Singh, K.E. Lawande, U.B. Pandey, Lallan Singh and S.R. Bhonde), Published by National Horticultural Research and Development Foundation (M.S.), India.

Maini, S.B., Sagar, V.R., Candon, S.S. and Kumar, Rajesh (1997). Evaluation of different structures for storage of onions. *Veg. Sci.,* 24 (1) : 73-74.

Manjunath, B.L. (1948). *Allium cepa* Linn. and *Allium sativum* Linn. in the wealth of India, Raw material Vol. 1, pp 56 (Manjunath B.L.) Ed., Council of Scientific and Industrial Res. Govt. India, Delhi.

Mc Collum, G.D. (1976) : Evolution of Crop Plants (Ed. Simmonds). Logman, London, New York. pp 186-190.

Mital, S.P. and Srivastava G. (1964). Seed yield in relation to bulb size and number of seed stalk in onion (*Allium cepa* L.) *Indian J. Hort.* 21 (3&6): 264-269.

Nadkarni, K.N. (1954). *Allium cepa* Linn and *Allium sativum* Linn. In : Indian Materia Medica, 3rd edn. (Part 1) (Puranik, M.V. and Bhatkal, G.R. Eds.) Popular Book Depot, Bombay, 63.

Narang, R.S. and Dastane, N.G. (1969). water use by *Allium cepa* L. (bulb onions) grown under different soil moisture regimes. *Indian J. Hort,* 26 (3 & 4) : 176-180).

Nath, Prem (1967). Major green leafy vegetables. Extn. Bull. 9 : Univ. Udaipur.

Nath, Prem; Srivastava, V.K., Dutta, O.P. and Swamy, K.R.M. (2008). Vegetable Crops – Improvement and Production. Dr. Prem Nath Agricultural Foundation, Bangalore, pp. 196-208.

Pandey R.C., Pandey, R.P., Dwivedi, S.K. and Khan, R.A. (1969). Manurial requirements of Onion, *Allium cepa* L. *Indian J. Agric. Sci.,* 39 (8) : 870-876.

Pandey, U.B. (1989). Onion (*Allium cepa* L.) *Indian Hort.,* 33-34 (4-1) : 58-62.

Pandey, U.B. (1991). Postharvest technology of onion. *Scientific Hort.* 2: 57-64.

Patel, J.A., Waghmari, D.R., and Patil, A.K. (1958) Age of seedlings in onion as a rabi crops. *Pune Agric. Collage Mag.,* 49 : 83-86.

Pathak, C.S. (1988). Studies on the inter-specific hybrids of *Allium cepa* x *allium fistulosum. Proc. Conf. Cytol. Genet.,* 1: 83-87.

Pathak, C.S., Aghora, T.S., Singh, D.P. and Deshpande, A.A. (1987). Exploitation of heterosis in onion (*Allium cepa* L.) by using indigenous source of male sterility. National symposium on heterosis: Exploitation Accomplishment and prospects. Marathwada Agric. Univ. Parbhani, (*Abstr.*) pp 53.

Pathak, C.S., Singh, D.P., Deshpande, A.A. (1987) A new type of cytoplasmic male sterility in onion (*Allium cepa* L.) First all India conference of cytology and genetics, Bangalore Univ., 1987. *Abstract*. pp 46.

Patil, A.K., Dod, V.N., Kale, P.B. and Kulwal, L.V. (1990). Performance of onion varieties with reference to seed production. *PKV. Res. J.*, 14 (2) : 123-125.

Patil, J.D., Rampise, S.A., Kakade, D.S. and Desai, U.T. (2002). Productivity and profitability of cabbage intercropped with leafy vegetables. *Intl. Conf. Veg.*, Bangalore, 11-14 November, 2002. Abstract, pp 197.

Perron, J.P., Jasmine, J.J. and Lafrance, J. (1958). Varietal resistance of seeded onion to the onion maggot, *Hylemya antequa* (Maci) (Deotera Anthomyiidae). *Candian Entomol.*, 90 : 653-656.

Purewal, S.S. (1957). Vegetable cultivation in North India. ICAR *Farm Bull* No. 26.

Purewal, S.S. and Dargan, K.S. (1962). Fertilizer and spacing experiment with onion crop. *India J. Agron.*, 7 (1) : 46-52.

Rampise, S.A., Birade, RM., Patil, B.T. and Sawant, S.V. (2001). Factors affecting the storage of onion : a review, *Orissa J. Hort.*, 29 (1) : 1-12.

Ray, P. and Gupta, H.N. (1980). 'Caraka Samhita' Scientific synopsis. Indian National Science Academy, New Delhi.

Rao, H. and Purewal, S.S. (1957). Onion and garlic cultivation in India. *Farm bulletin* No. 3, ICAR., New Delhi.

Rao, S.N., Rao, M.R. and Krishnamurti, G. (1967) Increasing yields and reducing storage losses of onion. *Indian Fmg.* 18 (4) : 14-16.

Roshania, B.D. and Agrawal, Y.K. (1981). N.P cholorphenyl substituted by droxamic acids and analytical reagents for the trace determination of vanadium (V). *Chem. Anal.* (Warsaw), 26: 191.

Sagar, V.R. and Maini, S.B. (1979). Studies on packaging and storage of onion powder. *Indian J. Hort.*, 54 (1) :L 75-79).

Saini, G.S. (1996). A Text Book of Vegetable Production. Aman Publishing House, Madhu Market Budhana Gate, meerut-250 002 (U.P.), India. 145.

Sandhu, G.S., Bhalla, J.S. and Singh, Balkaran (1995). Chemical control of Onion maggot with granular insecticides. *Haryana J. Hort. Sci.*, 4 (1&2) : 62-75.

Sandhu, J.S. and Korla, B.N. (1976). Interrelationship between economic characters and selection indices on onion seed crop. *Indian. J. Hort.* 33: 171-175.

Sankar, V., Aziz Qureshi A., Tripathi, P.C. and Lawande, K.E. (2002). Studies on Onion based cropping systems. *Intl. Conf. Veg. Bangalore,* 11-14 November, 2002. *Abstract*: pp 177.

Schweisguth, B. (1973). Study of a new type of male sterility in onion, *Allium cepa* L. *Ann. Amelior, Plant,* 23 : 221-253.

Selvaraj, K.V. (1993). Effect of irrigation and nitrogen fertilization on bulb yield of small onion. *Indian J. Hort.,* 50 (2) : 158-160.

Sharma, J.R. and Kumar, J.C. (1982-83). Growing Onion in Kharif season under North Indian conditions. Areview, *Punjab Veg. Grower,* 17-18 : 57-60.

Sharma, P.K., Yadav, G.L. and Kumar, Sudesh (2002). Effect of methods and dates of planting of onion sets on bulb yield of kharif onion. *Intl. Conf. Veg.,* Bangalore, 11-14 November, 2002. Abstract : pp 193.

Shrivastava, R.K., Verma, B.K., Dwivedi, S.K. and Dwivedi, Y.C. (1996). Response of onion (*Allium cepa* L.) vareities to plant spacing in relation to yield attributes and bulb productivity. *Veg. Sci.,* 23 (1) : 26-29.

Singh Samant, P.K., Singh, D.N and Sahu, S.K. (1982) Response of Onion (*Allium cepa* L.) to varying levels of N, P and K in Eastern Ghat High Land Zone of Orissa. *Orissa. J. Hort.* X. (1) : 33-37.

Singh, B., Babel, Y.S., Yadav, B.D. and Sharma D.R. (1995). Effect of size of onion sets and planting distance on yield and economies of kharif onion. *Indian J. Hort.,* 52(3) : 215-217.

Singh, Balraj (1996). Studies on weed control in Onion. *Veg. Sci.,* 33 (1) : 30-35.

Singh, D. (2001). Onion improvement in India. Kalyani Publishers, Ludhiana, 15-17.

Singh, D.N. (2000). Growth and yield of onion (*Allium cepa* L.) genotypes in winter season on the Entisol of Orissa. *Indian J. of Environment and ecoplanning.* 3 (3) : 677-700.

Singh, D.N. (2001). Path coefficient in onion (*Allium cepa* L). *Environment and Ecology.* 19 (14) : 980-982.

Singh, D.N. and Nath, Vishal (2011). Varieties and Hybrids of Vegetables. Satish Serial Publishing House, 403, Express Tower, Commercial Complex, Azadpur, Delhi-110033 (India). 86-93.

Singh, D.N., Nandi, A., Tripathy, P. and Sahu, A. (1995). Genetic variability and correlation in onion (*Allium cepa*). *Indian J. Agric Sci.,* 65 (11) : 793-6.

Singh, D.N., Tripathy, P. and Nandi, A. (1996). Periodical storage losses of Onion bulbs at ambient condition. *Environment and Ecology,* 14 (2) : 492-493.

Singh, D.P. and Joshi, M.C. (1975). Preliminary investigation on breeding onions for better keeping quality. *Haryana J. Hort. Sci.,* 4(3&4): 161-166.

Singh, D.P. and Joshi, M.C. (1978). Correlation studies of rotting and sprouting losses of onion with other traits during storage. *Veg. Sci.,* 5(1): 1-3.

Singh, D.P. and Singh, R.P. (1974). Studies on the effect of time of sowing and age of seedling on growth and yield of onion. *Indian J. Hort.,* 31 (1) : 69-73.

Singh, D.P., Padmaja Rao, S., Ramachandar, P.R. and Biswas, S.R. (1979). Inheritance, correlation and selection Index in onion. *Indian J. Agric. Sci.,* 49 (8) : 592-595.

Singh, D.P.; Swamy, K.R.M. and Joshi, M.C. (1976). A note on breeding onions suitable for dehydration. *Indian J. Hort.*, 33(3&4): 252-253.

Singh, Jitendra and Dhankar, B.S. (1989). Effect of nitrogen, potash and zinc on growth, yield and qualtity of onion. *Veg. Sci.*, 16(2) : 136-144.

Singh, Jitendra., Pandey, U.C. and Srivastava, V.K. (1992). Evaluation of onion cv. for storage quality, *Haryana J. Hort. Sci.*, 21 (1-2) : 83-85.

Singh, Kirti (1923). Current Status of Onion and Garlic Research in India and Future Prospects. In : Approaches for sustainable development of Onion and Garlic (Eds. Kirti Singh; K.E. Lawande, U.B. Pandey, Lallan Singh and S.R. Bhonde), Published by National Horticultural research and Development Foundation (M.S.), India.

Singh, M.P. and Singh, Ram (1969). Response of onion to different fertilizer application : *Indian J. Agric. Sci.*, 39 (1): 1026-1028.

Singh, N.P., Bhardwaj, A.K., Kumar A. and Singh, K.M. (2004). Modern Technology on Vegetable production. Interantional Book Distributing company, Lucknow.

Singh, R.P., Jain, N.K. and Poonia, B.L. (2001). Integrated nutrient management in rainy season Onion (*Allium cepa*) *Indian J. Agric. Sci.*, 71 (5) : 310-312.

Singh, S.P. (2001). Seed Production of Commercial Vegetables. Agrotech Publishing Academy, Udaipur-313 002. pp. 199-208.

Singh, V.B. and Singh, Kirti (1996). Spices. New Age International Publishers, New Delhi.

Singh, Vinay; Singh, J., Bisen, R.K. and Agrawal, H.P. (1997). A note on weed management in Onion. *Veg. Sci.*, 24 (2) : 157-158.

Singhal, S.C. (2003). Export of onion - Challenges and Strategies. In: Approaches for Sustainable Developement of Onion and Garlic, (Eds. Kirti Singh, K.E. Lawande, U.B. Pandey, Lallan Singh and S.R. Bhonde), Published by National Horticultural Research and Development Foundation (M.S.), India.

Sokhi, S.S., Sohi, H.S., Singh D.P. and Joshi, M.C. (1974). Sources of resistance to basal rot of onion caused by *Fusarium oxysporium*. *Indian J. Mycol. Plant Pathol.*, 4 (2) : 214-215.

Solomony, S. and Patil, J.A. (1959). Bulb size and seed yield in Onion. *Puna Agric. Coll. Mag.*, 50 : 30 33.

Swarup, Vishnu (2006). Vegetable Science and Technology in India. Kalyani Publishers, New Delhi, pp : 656.

Tackholm, V. and Drar, M. (1954). Flora of egypt. *Bull. Fac. Sci. Egypt. Univ.*, 3 : 1-644.

Thakral, K.K., Yadav, S.P.S., Khurana, S.C. and Nehra, B.K. (2002). Herbicidal control of weeds in onion nursery production. *Intl. conf. Veg.*, Bangalore, 11-14 November 2002, Abstract : pp 252-253.

Thamburaj, S. (1991). Biofertilizers for onion crop : Lesser the expense; healtheir the soil. *Indian Horticulature*, 36 (2) : 23.

Thompson, H.C. and Kelly, C.W. (1957). Vegetable crops. Mc graw Hill Book Co., Inc., USA.

Vavilov, N.I. (1951) The origin, variation, immunity and breeding of cultivated plants. *Chron. Bot.,* 13 : 1-6.

Vijay Kumar, A., Palanisamy, V., Jayraj, T. and Arumugam, R. (1991). Effect of seed treatments and container on the storability of onion seed. *South Indian Hort.,* 39 (5): 296-299.

Ways, S.B. (1967). Effect of N,P and K on yield and keeping quality of onion (*Allium cepa)* *Indian J. Agron.,* 12 : 349-382.

Yarwood, C.E. (1943). Onion downy mildew. *Hilgardia,* 14 : 595-161.

Yawalkar, K.S. (1969). Bulb vegetables, Vegetable crops of India. Agric. Hort. Pub. House, Nagpur.

Yawalkar, K.S., Kakata, P.N. and Srivastava, M.M.P. (1962). Commercial fertilizers in India. Agric. Hort Publishing House Nagapur.

❑❑❑

2. GARLIC

Botanical name	:	*Allium sativum* L.
Family	:	Alliaceae
Chromosome number	:	2n = 2x = 16

Garlic (*Allium sativum* L.), 2n = 2x = 16; also known as Lahsoon, Poodu, Lassan, Veluthull, Lasun, Rashun, Nohoyo, Vellulli, Lason, Shumandha, Bhutagna, Mahasuda; belongs to the family Alliaceae; is vegetatively propagated corp. Mostly local cloves are used on the basis of colour. The underground bulb consists of 10-50 segments called cloves and are enclosed in a thin sheath of white or pink colour (Singh and Nath, 2011).

Garlic is an important bulb crop and widely used as spice and a condiment. It is an important crop growing on commercial scale from which good foreign exchange is being earned by India. Garlic is the second most important among the cultivated *Alliums* after onion. It has long been recognized all over the world as a valuable spice for foods and a popular remedy for various ailments and physiological disorders.

Garlic oil is an item of commerce. Fresh garlic bulb and hole plant on steam distillation yield 0.1-0.2% and 0.05-09% respectively of essential oils. Up to 0.5% oil has been derived from the bulb. For recovery of the oil fresh garlic is the best. This essence consists primarily of diallyl, dimethyl and allyl methyl sulfide, disulfide and trisulphide and a few minior components including those containing an n-propyl group, all formed by thermal decomposition of allicin and its homologues (Singh and Pal, 2003).

ORIGIN

Garlic is well known, hardy, perennial plant of bulb crops. It is a native of Southern Europe. According to Choudury (1976), Central Asia is the prime origin and the mediterranean area is a secondary centre. Garlic has been cultivated for thousands of years. Central Asia is the primary origin and the Mediterranean area is a secondary. It was known in Egypt in pre-dynastic times before 3000 BC and also to ancient Greeks and Romans It has long been grown in India and china. Garlic was carried to western hemisphere by Spenish, Portuguese and French. It was used in England as early as frist half of 16th century (Shinde and Sontakke, 1986). It is said to be the native of Central Asia and Southern Europe, especially the Mediterranean region (Thompson and Kelly, 1975).

History

Garlic has long been known as cultivated plant. It is among the most ancient of cultivated vegetables giving pungency of the genus *Allium*. It was known in Egypt in predynastic times before 3000 B.C. and also to ancient Greeks and Romans. The anceint parent of the garlic now in cultivation was for a long time unknown. Regel (1875) followed by De candolle (1885) and considerably later Vavilov (1926), decribed the *Allium longicuspis* Rgl. subspecies as the ancient parent of garlic. It is found to be endemic to central Asia where it is distributed over the southern mountain slopes of Turkmenistan, northeast pamir etc. (Some authorities considered that *Allium longicuspis*) Regel, which is endemic to central Asia, is the wild ancestor and spread in ancient times to the mediterranean region. It has long been grown in India and China. Garlic was carried to the western hemisphere by the Spanish. Portuguese and French (Purseglove, 1975). Garlic was known to ancients and is said to have been disliked by Romans on account of strong odour and hence it was fed to their labourers and soldiers. It was used in England as early as the first half of the 16 the century.

The early domestication of garlic took a different turn from that of the largely seed propageted leek and onion; it became exclusively vegetatively propagated by cloves, and in those cultivars that still bolt, by infloresecence bulbils some modern cutivars may produce flowers mixed with the bulbils but the flowers never set seeds.

DISTRIBUTION

Garlic was known in Egypt in predynastic time, before 3000 B.C. and also to ancient greeks and Romans. It has long been grown in India. Besides India, garlic is grown in some other countries like China, the USA, (California, Texas, Loussiana), the Phillippines, Spain, Thailand, Korea Republic, Egypt, Italy, France, Morocco etc. China ranks first in garlic production followed by Turkey and India. The corp is grown throughout the plains of India and is consumed by most of the people. Madhya Pradesh is the largest state with an area of 32% and 28% of total production followed by Gujarat (18% in area and 17% in production). Orissa (14% in area and 11% in production) and Rajasthan (11% in area and 12% in production). M.P has 37.000 ha of land under garlic production with a production of 1.5 lakh tonnes, followed by Gujarat which has 21.000 ha area and 0.9 lakh tonnes production in 2001-2002 (Singh *et al.*, 2004).

BOTANY

It is a herbaceous annual for the bulb production and biennial for the seed production. Garlic is frost hardy bulbous perennial erect herb of 30-100 cm in height with narrow flat banes and bear small white flowers and bulbils. The scape of garlic is smooth, round and solid for its entire length unlike onion which is hallow. The edible underground stem is the composit-bulbs made up of numerous smaller bulbs known as cloves. If very small clove is planted or if growing condition is poor, a single small solid clove produced which is called round. Many colves of garlic do not produce flowerstalks. The inflorescence may be partially or not at all exerted, its bulbils forming a swelling some where within the false stems a few cm above the bulb (Purseglove, 1975).

There is a superficial adventitious root system. Bulbs composed of a disclike stem thin dry scales which are the bases of foliage leaves, and smaller bulbs or cloves formed from auxillary buds of the younger foliage leaves. The cloves are enclosed by the dry outer scales. Each cloves consists of a protective cylindrical sheath, a single thickened storage leaf sheath and a small central bud (Purseglove, 1975; Shinde *et al.,* 1993). Scape smooth, round solid coiled at first; spathe membranous, long-baked, splitting on one side, remaining attached to umbel. Small bulbils produced in inflorescence; flowers, variable in number and sometimes abscent, seldom open and may wither in bud. Flower on slender pedicels; perianth about 4 mm long, pinkish; anthers and style exerted. Seed seldom if ever produced (Purseglove, 1975).

Plant breeding

It is a cross pollinated crop usually propagated vegetatively by cloves. *Allium sativum* is a diploid with chromosome number 2n = 16. A good number of local strains are available in different regions of India. They vary in the number of cloves, which ranges from 16 to 50, and in the size of the bulbs. There is very little variation in shape. Most of the strains have white colour, some have reddish tinge also. Though its production and consumption are high, no special effort has been made to improve upon these strains. Garlic improvement is possible by clonal selection (Nath *et al.,* 2008).

Taxonomy

Garlic belongs to the family Alliaceae, genus *Allium* and spp. *sativum* L. The chromosome numbers is 2n = 2x = 16. Carl Linnaeus gave the name *Allium staivum* to garlic in 1754. At present two main classifications are accepted :

(I) *Allium sativum* L. spp. is divided in to three ssp.

(a) *A. sativum* var. *sativum* - non bolting

(b) *A. sativum* ophioscorodon - bolting, with snake like twisting flower stems.

(c) *A. sativum* var. scordoprasum-bolting with non-twisting-stems (includes rocambole and the other non-cultivable garlics).

(II) Kuznetsov (1948) presented a botanic and ecological classification of the species. He distinguished two ssp.

(a) *A. sativum* L; *A. sativum* var. *sagitatum* Kuz. (bolting garlic).

(b) *A. sativum* var. vulgarae Kuz. (non-bolting garlic). Bolting garlic is found in nearly all parts of the world. The sub-species can be divided into three ecotypes.

(1) *Asion ecotype* (multiflowered ecotype) : False stem (150 cm or more) inflorescence is large, number of sterile flowers 100 or more small aerial bulbils.

(2) *Caucasian ecotype* : Flower stem of 60-80 cm height, large inflorescence, small number of sterile flowers and aerial bulbils of about pea size. When such bulbils are planted in autumn, form bulbs (albeit smaller) during the next year.

(3) *Sea side ecotype* : Delayed formation of flower stem, which appears after the formation of three or four fertile leaves, at the base of which three circles of cloves are formed, causes prolonged vegetative period (12-15 days). Inflorescence consists 3-5 bulbils of diffrent size and shape.

Non-bolting subspecies

It includes great number of cultivars, distributed in all climatic zones. Non-formation of flower stem, causes only method of propagation by separation of the ground bulb into cloves. Kuznetsov subdivided the cultivars of the above-mentioned subsp into 3 ecophypes:

Continental ecotype

Longest vegetative period (30 days, more), false stem is thin, compact and about 40 cm high, the leaves are thin wide open, smooth and thicky waxed; the bulbs differ in size and numbers of sterile leaves (8-10), tightly arranged round the bulb. In many countries referred to as 'spring/summer' type and mostly grown without irrigaion.

South Russian ecotype

Mainly distributed in the steppe zones of Russia, Ukraine and some neighbouring countries. It has early maturity and good storability. The plants are x erophyte type, with thin, tall (up to 45 cm) false stems.

East Mediterranean ecotype

These cultivars grown along the northern shores of the mediterranean and other related areas. It has mesophyte structure influenced by moisture, usually grown in irrigated condition. False stems are thick, pulpy, with a softer consistancy. Cultivars vary with regard to the length of the false stem, which in itself is an important systematic feature (Batchvarov, 1966). The bulbs are large (upto 100g average weight) and consist of nemerous cloves (utpo 30) located more freely towards the bottom.

Garlic is a perennial sp., producing a much divided bulb, consisting of several cloves, covered by a thin white skin. It differs from *A. cepa* in that the leaf base does not store food, but matric as dry scales enclosing the cloves. The cloves are well developed axillary buds within the thin foliage leaves. Garlic is a sexually sterile diploid, unknown in the wild. Under domestication it has become exclusively propagated by vegetative methods, i.e. by cloves and in some cases inforescence bulbils (Jones, 1963).

COMPOSITION

Garlic contains a colourless, odourless, water-soluble amino acid (Allin) in uninjurd garlic. On the injury of the cell, enzyme alliinase changes it to Allicin. Allicin is the antibacterial substance of garlic and has typical odour of fresh garlic. It contains allyl radicle of disulphide. Garlic has been considered and a rich sources of carbohydrates, protein and potash, calcium, magnesium and phosphorus. Ascorbic content is very high in green garlic (Akroyd, 1963). It also contain fat and vitamin 'C' and sulphur. It has a higher nutritive value than other bulb crops (per 100 gm of edible protion).

Mositure	62.0g	Phosphorus	310 mg
Protein	6.3g	Iron	1.3 mg
Fat	0.1g	Vitamin 'A'	0 I.U
Minerals	1.0g	Thiamine	0.16 mg
Firbre	0.8g	Riboflavin	0.23 mg
Other carbohydrates	29.0g	Nicotinic Acid	0.4 mg
Calories	142	Vitamin 'C'	13.0 mg
Calcium	30 mg		

Uses

Garlic is consumed by most of the people in our country. It acts as flavouring and seasoning agent in various vegetable and meat preparations. Its preparation are administered as a cure against some long and lingering stomach diseases and sore eye and ear-ache. It has important medicinal values for digestive disorders. It is carminative and is a gastric stimulant and thus aids in digestion and absorption of food. It has insecticidal action and repellent property. A formulation of containing 1% garlic extract gave protection to persons agaisnt mosquitoes and black fly for about 8 hrs. Garlic has fungicidal action also. Use of garlic extract as seed treatment for control of foot rot of french bean. Garlic has some bactericidal properties. It has also been found that garlic constituents like allicin, allistains garlicin, dillyl disulphide, diallyl trisulphide and essential oils were active against certain becterial and other micro-organisms (Chouhan, 1972). The inhalation of garlic oil or garlic juice has generally been recommended by doctors in cases of pulmonary tuberculosis, rheumatism, sterility, impotency, cough and red eyes (Pruthi, 1979). A significant reduction in serum cholesterol level was observed by Mahanta *et al.* (1980), when 50g of raw garlic was fed daily to ten normal human volunteers for one month. Garlic oil and raw garlic reduced normal blood sugar significantly after 2 months of regular use. The blood sugar lowering effect of garlic was ascribed to allicin and related disulphide containing compounds. Deb-kiotanya *et al.* (1980) found garlic extracts possessing nematicidal and larvicidal properties, respectively.

VARIETIES AND HYBRIDS

Two types of varieties are available depending on their colour i.e., red and white. The white coloured varieties are more popular as they are less pungent. Red coloured varieties are mostly used for medicinal purposes. They are more pungent as compared to white varieties.

Amongst white varieties, some produce small-sized bulbs while some of them produce big-sized bulbs. The different varieties of these groups are as below :

1. Small sized varieties : T 56-4, Creole, Italian and Tabiti.
2. Big-sized varieties : Fawari, Rajalle Gaddi, Solan.

Some others are Madrasi, Jamnagar Local, Ladwa, Aru, Sweta etc.

Varieties developed by different Institutions/Organizations are given below :

MPKV, Rahuri : Godavari and Sweta

CCSHAU, Hissar : HG-1 and HG 6

IARI, New Delhi : Pusa Sel. 10

PAU, Ludhiana : LCC 1

VPKAS, Almora : ARU 52

NHRDF

Agrifound white (G-41), Yamuna Safed (G1), Yamuna Safed 2 (G 50), G 282 and Agrifound Parvati.

Acclimatise varieties

Early White or Medican varieties, Creole and Tatule and Italian.

Detail characteristics of varieties and hybrids described by Singh and Nath (2011) are given below :

Fawari

It has big size bulb with white colours and compact cloves. It is grown in Bellary region of Karnataka and high yielder with good keeping quality.

Rajalle Gaddi

It is grown in Karnataka and A.P. It gives high yield and has good keeping quality. It has medium size bulb with more pungency.

G-6

It has been found to be tolerant against soil salinity.

HG-7

It grows tall with high number of leaves/plant, best yielder with high number of cloves/bulb and weight of bulbs and cloves.

Godawari

It is good yielder but premature sprouting in the field is very high.

Madrasi

It is a desi/local variety and is very popular in Rajkot division of Gujarat and also a very high yielder.

Creole

It is earlier than Italian; has a flatten broader leaf a greener top, a shorter stem, fewer wrapper leaves and larger cloves. It is a heavier yielder, but its keeping quality is somewhat inferior.

Jahiti

Have the largest cloves and also are the darkest.

Agrifound white (G-41)

The bulbs are compact, silvery-white skinned with creamy flesh. Diameter of bulb is 3.5-4.5 cm and number of cloves/bulb is 20-25. The variety is susceptible to purple blotch and stemphylium blight which are common in northern India. TSS 41%, dry matter 43% and good storer. Average yield is 73 tonnes/ha.

Yamuna Safed (G-1)

The bulbs are compact, silvery white spinned with 25-30 sickle shaped cloves and creamy flesh. Diameter of bulb is 4.0-4.5 cm. It is tolerant to insect pests and diseases like thrips, purple blotch and stemphylium blight. Total sluble solids (TSS) 38%, dry matter 39.5%. The average yield is 15.0-17.5 tonnes/ha. It is recommended for cultivation all over India.

Yamuna Safed-2 (G-50)

The bulbs are compact, attractive with 35-40 cloves/bulb and white creamy flesh. Bulb diameter is 3.5-4.0 cm; TSS 38-40%; dry matter 40-41%. The average yield is 15-20 tonnes/ha. The variety is recommended for northern parts of India.

Yamuna Safed-3 (G-282)

The leaves are wider than other varieties. Bulbs are creamy-white and big size (5-6 cm diameter), having 15-16 cloves/bulbs. Total soluble solid 38-42%, dry-matter 39-43%, medium storer, yield 17.5-20.0 tonnes/ha; suitable for export purposes. It perform very well in northern and central India.

Agrifound Parvati (G-313)

A selection from the material collected from Hong-Kong market. It is a long day type and as such is suitable for cultivation in hill of northern states. Bulbs are of bigger size (5-6 cm diameter), creamy-white with Pinkish-tinge, 10-16 cloves/bulb, tolerant to common diseases. Average yield 17.5-22.5 tonnes/ha. It is medium storer and suitable for export.

Ooty 1

Bulbs bigger in size with dull white in colour, 20-25 cloves/bulb, yield 17 t/ha in 125-130 days. Resistant to thrips.

Pant Lohit

Moderately resistant to purple blotch. Cloves have pink tinge colour.

T 56-41

This variety bear medium size bulbs and is white in colour.

Jamnagar

It has bigger size white colour bulbs with compact cloves.

CLIMATIC INFLUENCE

Garlic is most hardy plant requiring cool and moist climate during growth and a relatively dry period during maturity of bulb. It prefers relatively moderate temperature in summer and winter. Bulbing takes place during longer days and a high temperatures, exposure to low termperature subsequent to bulb formation favours the process. Adequate vegetative growth promotes bulb formation. Bolting does not seems to be influenced by temperature and some cloves never produce flowers.

Mann (1952) found that long days and high termperatures encouraged bulb development in garlic. As soon as bulbing commences leaf initiation ceases. Becuase of this reason and high yield, garlic should be planted early to promote vegetative growth under short photoperiod and cool temperature. The yield potential depends on the amount of vegetative growth made before bulbing commences. The critical day length for bulbing is 12 hours. Exposure of dormant cloves or young plants around 20°C or lower depending upon the varieties for 1-2 months hastens subsequent bulbing. Late planting adversely affects vegetative growth and there by results in lower bulb yield.

Abdel (1973) reported that garlic bulbs planted on west side of North-South ridges where the temperature was lower, were larger than those on the east. Yield and survival were also higher at lower temperatures. Further, the survival rate bulb weight and yield of garlic planted on north side of West-East ridges was higher than that of plants grown on the South side or on the flat beds.

SOIL REQUIREMENT

Garlic can be grown on veriety of soils but thrives better on fertile, well drained loamy soils. According to Rao and Purewal (1957) and Joshi

(1961) garlic require medium black to well-drained loamy soils rich in humus with fairly good content of potash. The crop raised on sandy or loose soil does not keep for long and the bulbs too are lighter in weight. Heavy clay soils may result in mis-shapen bulbs and make harvesting difficult, bulbs are broken and bruished. It was found that pH range between 5-7 had little effects on growth and yield of garlic. Acidic soils are not good for clove development (Katyal 1985), and so they not keep well in storage.

When soil salinity exceeds 1.7ds/m, the mean yield starts declining by 1.68 per cent per unit increase in soil electrical conductivity and a 50 per cent reduction is occurred at 5.6-7-8 ds/m depending on cultivars (Mangal *et al.,* 1990).

SOWING / PLANTING

Medium land should be ploughed after the harvest of paddy for 4-5 times till a fine filth is obtained. The field is divided into small plots (strips) to facilitate irrigation and other cultural oprations.

The garlic is propagated by cloves which are carefully detached from the composite bulbs at the time of planting. It is planted from August in karnataka and Andhra Pradesh and November in the northern plains. In Nilgiris, it is planted in early May and October. On hills it is planted in March-April. The Planting can be done in follows ways:

Dibbling

Plot divided then cloves dibbled at given spacing and covered with loose soil tightly.

In furrow planting

Furrow are made then clove planted in furrow.

Broadcasting

Cloves are scattered evenly over a planked and levelled bed by hand. They are covered by harrowing and the field is divided into small plots for irrigations. In some localities the planting maybe done twice (May and October) if the climatic conditions are favourable. The cloves are dibbled in an upright position at a spacing of 15 cm from row to row and 7-8 cm from pant to plant. About 350 to 500 kg of cloves are required to plant one hectare of area. Hari Om and Srivastava (1977) observed that aerial bulbils can also be used as a planting material though maximum net return was obtained from cloves planted at closer spacing

(10 cm × 20 cm). Selvaraj *et al.* (1995) reported the effect of seasons and nutritional ingredients which indicated on the yield of garlic in Nilgirils. The experiment was carried out in two different seasons viz., April-July (main season) and October-January (autumn season). The results revealed that among two seasons, bulb yield from the main season's crop was much higher than the yield of autumn crop. This was due to the fact that the main crop season was shorter with uniformly moist weather throughout the crop season whereas, during autumn it was relatively very dry with more of evaporational loss. Mishra *et al.* (1997) studied the effect of different dates of planting on growth and yield of garlic under high altitude conditions of southern Orissa. The planting was done on 7 dates from 10 October to 11 December (at 10 days interval) at Regional Reserach Station, Semiliguda, Koraput for 3 years (Rabi 1991-92, 1992-93 and 1993-94). Planting of cloves between 21st October and 1st November restricted more number of cloves/bulb, higher bulb weight and higher total bulb yield. Delay in planting resulted in early harvest and low total bulb yield.

Nath *et al.* (2008) reported that planting of garlic in the main season i.e. April-June yield better than the early or late sowing. A planting distance of 15 cm X 10 cm is suitable for sowing garlic cloves. Storing bulb at 4°C for 1-2 months is helpful in breaking dormancy of cloves. Individual cloves are carefully separated from composite bulb without any injury. Bigger cloves of 8-10 mm diameter that too from outer side of bulb give higher yield.

NUTRITIONAL REQUIREMENT

Apply 40-50 tonnes of FYM/ha, at the time of field preparation and mix it thoroughly in the soil. Application of 120 : 60 : 120 kg N, P_2O_5, K_2O/ha give highest yield. ½ N, Full P_2O_5, ½ K_2O should given as basal dressing at the time planting. Remaining N should be top dressed in two split doses i.e. 3 weeks and 6 weeks after planting. Rest of the potash should be applied as 2nd top dressing along with 2nd split of nitrogen. At each application of fertilizers few quantity of farm yard manure should be mixed for proper utilization of fertilizers (Annon., 1995).

The recommended doses of garlic are FYM 50 tonnes. nitrogen 100 kg phosphorus 60 kg and potassium 60 kg/ha. The complete dose of FYM may be mixed up thoroughly in the soil at the time of field preparetion, whereas the complete dose of phosphorus and potassium and half of nitrogen may be applied as basal application before planting. Another half dose of nitrogen may be applied after one month of planting as a top dressing as reported by Nath *et al.* (2008).

The recommended doses of fertilizers and FYM in a few states of India, reported by Tandon (1991), are given below :

States	N	P_2O_5	K_2O	FYM
Andhra Pradesh	60	35	60	15
Arunachal Pradesh	125	60	100	20
Assam	50	60	100	25
Gujarat	100	50	50	10
Haryana	80	50	50	10
Himahcal Pradesh	75	40	40	10
Maharashtra	75-100	50	50	25
Punjab	125	60	0	50
Rajasthan	120	50	100	40
Uttar Pradesh	100	50	50	10

These doses may be raised and/or reduced according to climatic condition, soil type and cultivars grown (Singh, 2001).

Micronutrients are also effective in increasing the yield of garlic. Singh and Singh (1974) reported that growth of plant was inhibited and number of cloves reduced due to insufficient supply of boron. Application of boron at 0.50 to 1.00 pm was found optimum for high yield of garlic. Application of boron at 1.00 ppm was toxic. Choudhury (1960) mentioned that application of a mixture of 75 kg N,P,K in ratio of 18:2:3 plus 5.0 kg boron, zinc and molybdenum produced maximum amount of allyl propyl disulphide.

IRRIGATION

In early stage irrigate the crop at short interval and the frequency of irrigation incease with the increase in age of the crop and decrease with maturity. In general garlic need irrigation at an interval of 8 days during vegetative growth and 10-15 days during maturation. Moisture stress decrease yield.

In medium type of soils, during winter months, irrigation at 15 days interval is sufficient. In warm season it may be irrigated at 10 days interval which may eventually come to weekly intervals. Irrigation should be stopped at bulb maturity stage.

The first irrigation is given after planting that will provide proper sprouting. Irrigation has a beneficial effect on garlic yields and various yield components. Irrigation interval at 10-14 days causes the highest yield potential with the formation of heaviest (Singh, 1989 and Baten *et al.,* 1992) bulbs. The bulbs yield is expected to be higher at IW, CPI ratio of 1.2-1.4 (Panchal *et al.,* 1992) when crop approaches to maturity, interval of irrigation is increased and later on it stopped (Singh, 1989).

INTERCULTURE

Shallow cultivation is given between the rows to keep the soil around the plants weed free. Usually within 60 days two intercultural operations are reuired. Light, hoeing, weeding and earthing are rquired for the crop. Gernerally it is done after 20, 40 and 60 days after planting of the crop.

According to Joshi (1961), hoeing the crop just before the formation of bulbs (about two and a half months from sowing) loosens the soil and helps in setting of bigger compact bulbs. Heing in the later stage of bulb formation should be avoided to improve the quality of bulbs.

WEED CONTROL

Since garlic closely planted crop, manual weeding is tedious, expensive and often damages the plants. So the chemical weed control is economical and effective. Garlic can tolerate weeds only upto 3 weeks after seedlings or transplantation reported by Zimdahle (1980). Uncontrolled weeds can bring about yield reduction upto 60% as observed by sharma *et al.* (1980). Bhan (1972) and Bhan and Singh (1970) found promteryn 0.5-1.0 kg, nitrofen and BV 201 to be beneficial herbicides for garlic for controlling weeds and increasing crop yields over the weedy check. Sharma *et al.* (1983) got higher net return in garlic with a post-emergence spray of oxadiazon 1.0 kg/ha followed by a pre-plant application of fluchloralin 0.95 kg/ha.

Nandel *et al.* (2002) evaluated the effect of herbicides for weed control in galic. Herbicidal treatments applied at 1.0 kg a.i. ha^{-1} + HW (hand weeding) at 65 DAP (days after planting) gave the best results for maximum weed control (except fluchloralin). Highest garlic bulb yield (53.6q/ha) was recorded with pendimethalin applied @ 1.0 kg a.i. ha^{-1} + HW at 65 DAP. It was also revealed that nitrogen, phosphorus and potassium uptake by bulbs was directly correlated with yield and indirectly with weed population. Sandhu *et al.* (1997) studied the weed management in garlic under Punjab conditions. It was found that pendimethalin 30 EC @ 2.5 kg and 1.87 kg/ha and fluchlorlin 45 EC @ 2.5 kg and 1.87 kg/ha supplemented with one hand weeding having conducted after 105 days of sowing resulted in significant increase in bulb yield comparable to weed control. Linuron 50 EC @ 1.50 and 1.00 kg/ha tested for two years during 1995 and 1996 also gave comparable results. Tewari *et al.* (1998) studied the integrated weed management in garlic at Kanpur. Net monetary returns was maximized under 3 hand-weedings (Rs. 36,250/ha) followed by pendiumethalin @ 0.75 kg/ha + one hand weeding (Rs. 29,450/ha) and oxylfuorfen @ 0.07 kg/ha + 1 hand weeding (Rs. 26.694/ha). Pre-emergence application of oxylfurofen (0.15 kg/ha), pendimethalin (1.0 kg/ha) and alachlor (2.0 kg/ha) in combination with 1 normal weeding at 40 days stage were also reported to be the best choice for effective weed management in garlic grown in vertisols of southern Rajasthan (Porwal, 1995). In garlic. pre-emergence application of pendimethalin @ 1-1.5 kg a.i./ha, oxyflourfen (1.5-2.0 kg a.i/ha) and fluchloralin (1-1.5 kg a.i./ha) controls the weed population in the crop. These herbicidal sprays should be supplemented with two hand weedling at 30 days of sowing and another at 30 days after first weedings as reported by Nath *et al.* (2008).

Use of growth regulators

Sevlaraj *et al.* (1995) reported the effect of growth regulators on growth and yield of garlic. The results revealed that cycocel and maleic hydrazide significantly reduced sprouting and turberisation and increased the bulb yield. The highest yield of 18.0 t/ha was recorded with 1000 ppm of maleic hydrazide. Therefore growth regulators have a beneficial effect on the bulb yield. Application of GA3 @ 200 ppm or Ethrel 50 ppm at inflorescence stage enhanced bulblet formation and increased the number of cloves per bulb and delayed leaf formation in storage. Soaking of the cloves for 24 hours before planting in cycocel at 1000 ppm give maximum yield. Foliar spray of NAA 50 ppm, CCC 500 ppm at 60 and 90 days after planting gave maximum number of cloves and bulb yield. Spray of MH @ 300 ppm three weeks before harvest prolonged storage life of cloves and also reduced the weight loss and inhibit the sprouting in storage upto 300 days without any appreciable adverse effects on yield.

Crop Rotations

Garlic can be grown as companion crop with vegetables like beet root, tomato, lettuce and carrot. Growing of 2 more vegetable in the same field at the same times or season is called companions cropping. It is also known as double cropping. The different crop rotations can be adopted for more returns as suggested by Saini (1996).

1. Okra - Garlic - Okra (Summer)
2. Bitter gourd - Garlic- Cowpea
3. Cluster bean - Garlic - Cluster bean

4. Amaranthus - Garlic - Green Gram
5. Bottle gourd - Garlic - Maize
6. Sponge gourd - Garlic - Cowpea
7. Maize - Garlic - Okra
8. Cowpea - Garlic - Maize
9. Okra -Garlic - Cowpea
10. Black gram - Garlic - Green Gram
11. Green gram - Garlic - Green Gram
12. Maize - Garlic - Cluster bean

PLANT PROTECTION MEASURES

Insect Pests

Thrip (Thrips tabaci Lind.)

Nimphs and adults suck the sap from leaves and remain in the spaces of axils, leaf tips curled and plants dry up; bulb size is reduced. Spraying dimethoate 0.05% or acephate 0.05% to control the pest.

Aphids (Myzus persicae)

Though this insect is not so serious in garlic, and can easily be controlled by spraying of Malathion at 0.1 per cent (Lorini and Frizon, 1991 and Singh, 1989).

Mites (Rhizoglyphus echinopus)

Thses mites carry with the seed bulbs and prove very harmful for the growth and yield of garlic. These can be effectively controlled by the treatment of garlic bulbs with Hostathion at 0.05 per cent or Nemafos at 0.15 per cent for 10 to 15 minutes before planting or Dimethoate (0.03%) for *Petrobia latens* (Eastal *et al.*, 1992).

Cultivars like G-41, I.C. 49383 and G-1 are least infested with mite. *Aceria culipal* (Powar *et al.*, 1990). Nematodes and other pests are similar to onion.

Diseases

Most of the disesases (viz. Damping off, Bacterial soft rot, smut, smudge, pink root, Black mold, Aster yellow etc) affect both onion and garlic. The control measrues in onion are application to garlic also. Some other diseases alongwith their control measure have mentioned below :

Purple blotch (Alternaria porri (L) Gay

Small whitish water soaked areas on leaves and seed stalks, and soon they become brown in colour and later purplish in centre. Moist weather conditions are favourable. Severity of disease causes falling of leaves and seed stalks without forming seeds. The bulbs of the affected plants may rot later. It can be controlled by adopting long duration crop rotation. Make a provision of draining excess water from the field. Treat the seed with Thiram (2.5g/kg of seed) before sowing. Spray the crop with Zineb at 0.25 percent at 10 to 12 days interval (Quadri *et al.,* 1982) or Mancozeb at 0.25% + monocrotophos at 0.05% (Srivastava and Gupta, 1993) or Captafol (Upadhyay and Tripathi, 1995). Control thrips with regular spray of Malathion at 0.1-0.2% as infection becomes easy by *A. porri* through wounds made by thrips (Thind and Jhooty, 1982).

Stemphylium blight (Stemphylium vesicarium Wallroth)

This is an important disease of bulb crops particularly in northern and eastern parts of India, where it is responsible for yield loss of both in bulbs and seed crops. Infection appear on leaves in the form of small yellow to pale orange streaks in the middle of the leaves. The streaks develop into elongate spindle shaped ovate spots which later turn brown in colour and coalesce. Affected leaves dry prematurly. It causes poor development of bulbs and cloves and poor seed formation. It can be controlled by four sprays of Indofil M-45 at 0.25% (thiophanate methyl + mancozeb) at 10 days intervals 60 days after planting proved most effective (Lorini and Frizon, 1991).

Garlic variety 'HG-1' has the ability to tolerate purple blotch disease (*Alternaria porri*) and Stemphylium (Pandey and Singh, 1989). The last 2-3 weeks before harvesting of garlic crop, are the most critical period contributing the yield loss by purple blotch (Bisht *et al.,* 1991).

Pre-mature Sprouting

This is supposed to be the physiological/genetical disorder. In the field generally, before reaching to bulb maturity some cloves or almost all cloves of a garlic bulb sprout. Though, this melady is not permanent in occurrence, but the loss due to it may range from 0.5 to 50 per cent or so. Sprouting in white cultivars is seen more commonly than in pink or purple cultivars (Garcia, 1980). The sprouting may also be associated with excessive soil moisture and supply of nitrogen. Early planting in August-September initiate bulb formation during November and December, respectively. The low night temperatures during these months may held in breaking the dormancy and as a result there is heavy sprouting in August and September plantings in comparison to plantings done during October and November months (Lawande *et al.*, 1990). Premature sprouting of garlic cloves is less reported in sandy soils compared to heavy soils. This disorder seems to be governed by genetic constituents of the plant type, more injunct by environmental conditions those may be manipulated if not all, in such a way that could act against occurrence of clove sprouting. Presence of endogenous hormones(s) which becomes active under different environmental conditions need to be undestood in this regard.

Control

- Avoid excess soil moisture by withholding irrigations or increasing intervals of irrigations during rapid bulking and as soon as bulb formation is about to complete.
- Apply comparatively less quantity of nitrogen in the crop where this discorder becomes common and causes considerable loss.
- Plant the garlic crop at close distance (15× 7.5 cm) i.e. dense planting reduces the occurrence of sprouting (Garcia, 1980).
- Avoid raising garlic crop in heavy soils.
- Locate sprouting free seed material like purple colour type, if possible because sprouting is common in white type.

Storage Pests and Diseases of Garlic

It becomces essential to store garlic bulbs for home consumption and for planting in the next season in the field. For this purpose, garlic bulbs are stored. During storage certain pests and diseases cause harm to stored garlic bulbs.

Garlic cloves are eaten by the larvae of *Ephestia elutella* in storage which can be controlled by fumigation with posphine at 1-4 tablests per m^2 for 12 hours at 60 days interval (Bharadwaj and Thakur, 1974).

Macrophomina infection is seen in storage which is prevented by the treatment of garlic bulbs with 2 per cent boric acid just after harvest. Rhizopus rot is another disease in stored garlic bulbs; the bulbs which are stored for seed purpose i.e planting for next season crop; such bulbs/ cloves may be treated with mercuric chloride at 0.1% and with formalin at 0.3 per cent for consumption purpose. Bulbs affected with *Asperigilius niger* when planted in the field show poor performance. Treatment of bulbs with Dithane-M-45, Difolatan, Brassicol and Captan (0.1%) is very effective. Fusarium rot in the storage can be controlled effectively by fumigation with formalin (Rath and Mohanty, 1986).

Pre-Harvest Considerations

Garlic varieties differ considerably in their inherent duration of dormancy, as measured by the time to sprout either in storage or after planting in the field. However, very little information is available on this aspect in India. The stage of maturity of cloves at harvest, has significant influence on dormancy. Cloves lifted when the leaves are still green, require more time to sprout than mature cloves harvested when the foliage has dried. It appears that the more immature the cloves are, the deeper is the dormancy. It has also been reported that harvesting of garlic bulbs before the tops have fallen over indicating full maturity, results in poor storage, where as delayed harvesting results in splitting and sprouting of bulbs.

Sprouting is a major problems of garlic storage. Pre harvest application of growth regulator, maleic hydraizde (MH) can check sprouting. The chemical is to sprayed when the tops are green and before the neck fall. The recommended concentration is 1500 to 3000 ppm. It gives good effect if sprayed when the temperature is mild and during the cooler part of the day.

HARVESTING

The crop become ready for harvesting when the tops turn yellowish or brownish and show sign or drying up and bend over. The bulbs begin to mature in about 4-5 months.

Harvesting is done with the help of kudali. The bulbs are lifted and leaves are tied at the top. The top bulbs are cured for about 7-10 days or so under shade. A tractor drawn double-row harvester for garlic is useful for mechanical harvesting. The speed is approximately 0.22 hectare per hour and damage to the bulbs is 4.0 to 4.3 per cent (Gerasimchik *et al.*, 1989).

Like onion. the tops are removed and bulbs kept for storing under ordinary room temperature for months together. Sometimes green garlic is also harvested like green onion.

YIELD

The yield of bulbs is in betwen 100-200 q/ha. Recovery of clove in the bulb ranges from 86-96%.

In fact, the average productivity of garlic in India is very low 4.5 t/ ha. Whereas it is 18.8 t/ha in USA, 13 t/ha in Egypt and 9.5 t/ha in Republic of Korea (Singh, 2011).

POST HARVEST HANDLING

Curing

Curing of bulb is done for about a week in the field itself. Just after harvesting of bulbs they are left in the field for a couple of days in small groups of bunches. There is 2 stage and 3 stage dehydration process for garlic dehydration, which result in a saving of about 16 and 17 per cent of the total drying time respectively, compared with single stage drying at 60°C (Das and Bhatnagar, 1991). Artificial drying of garlic bulbs is quicker and does not adversely affect the bulbs if drying termperature does not exceed 42°C (Bharadwaj and Thakur, 1974).

Grading

All thick necked, split, injured, disease and insect infested bulbs are removed. Size grading is done manually or by a grader. Government of India has prescribed certain grades for different qualities of garlic for export which are given below Table 1 :

Table 1 :

Grade	Minimum diameter (mm)				General characteristics
	Nasik garlic	Madurai plain garlic	Madurai hill garlic	Jamnagar garlic	
Big	30	20	35	35	1. Bulbs shall be reasonably of uniform colour, shape and pungency characteristics of the variety type.
Medium	25	15	25	25	2. Bulbs shall be well cured and dried.
Small	10	10	10	12	3. Bulbs shall be compact with cloves well filled and fairly plump.
					4. Bulbs shall be free from damage caused by moulds, decay, split cloves, dirt, sunburn, dust, sprouting etc.

Note : Tolerance for size : For accidental error in sizing not more than 5 % by weight may be of the next lower grade.

Tolerance for requirement in respect of general characteristics : To allow for variation other than size, incidental error to proper grading and handling not more than 10% by weight of garlic in any container may be below the requirement specified under general characteristics but not more than a total of 1/5 th of this tolerance or 2% shall be allowed for garlic which is affected by decay.

Packing

In India, garlic is packed in open mesh jute bags of varying sizes normally between 40 kg and 60 kg. The garlic grading and packing rules provides for packing in 8 kg, 25 kg and 50 kg sizes. 18 kg and 25 kg packing is done in perforated 10 ply corrugated fibre board boxes for export. Nylon netted bags, when used for packing, have remeted in less storage loss because of good ventilation.

Storage

Well-cured garlic bulb can be kept for 1 to 1½ months in a ordinary well ventilated room. Sometimes garlic bulb with dried leaves are stored by hanging in well ventilated room. Under cold storage, condition it can be stored for many months at 0°C to 2.2°C temperature and 60% relative humidity. According to Pantastico (1975) garlic has an average freezing point of 25.4 and at 0°C with a relative humidity of 70-75% can be stored upto 28-36 weeks.

SEED PRODUCTION

Garlic

Garlic is entomophilous and some times self-pollination also occurs. The colour and the scent of the flower and also the nectar generally secreted by the septal glands of the ovary offer necessary attraction to the insects. Seed production is limited to an area of regions of low relative humidity, cool weather is rquired for its vegetative growth with moderately high temperature and dry atomosphere for the later part of the crop.

Time of planting

November, most of the variety transplanted to form seed stalk. Bulbs are form in the first year (leave sheath) and in the second year seeds are produced.

Method of seed prodcution

Bulb to seed method and seed to seed method. In the seed to seed method they are just like any common crop when the bulbs have matured and replanted for seed production.

It is normally propagated by cloves. The well-grown, uniform-sized composite bulbs of a particular variety are selected and the cloves are separated. Healthy and uniform cloves are selected and used for planting. The planting followed by the multiplication of bulbs is the same as that for the production of garlic bulbs for consumption. However, it has been recommended that the cloves may be sown at 7.5 to 12.5 cm in row which are spaced 40 cm apart.

Roguing

Rogue out off-type plants based on foliage characters and diseased plant. Prior to sowing, bulbs can also be sorted out which are distinct from the variety is to be grown. Under field condition, three roguing will solve the purpose.

Harvesting and curing of seed bulbs are similar as described in crop production chapter.

Seed yield

Generally, seed yield of garlic is 5 to 10 per cent less than the total yield because some oversiged and under sized bulbs, undesirable injured bulbs are removed. There is lot of variation in yield due to season of planting, cultivars, soil types, management practices etc. Normally 40 to 100 quintals per hectare bulbs are obtained with the recovery of clove in the bulbs 86 to 96 per cent (Shinde and Sontakke, 1988).

Storage

Storage of garlic is a serious problem both for growers and traders. The ultimate solution of this problem lies in the development of suitable storage structures, methods of storage, use of certain chemicals, selection of garlic cultivars having natural good keeping quality.

Properly cured garlic bulbs are stored. Common practice is to bunch 15 to 25 bulbs and hand in adequately ventilated shed or room. In this way, they can be stored for longer period of time. The storage life of garlic bulbs can further be increased in cold storage at 0.0°C to 1.6°C temperature and 65 to 75 per cent relative humidity. In cold storage bulbs without tops are kept in crates or in clean gunny bags of 40kg capacity. The sprouting during storage can be controlled up to great extent by spraying Maleic Hydrazide (MH) at 2500 ppm or so on foliage a fortnight before harvesting (Hari and Awasthi, 1978).

Twenty different garlic genotypes were studies for their strorability under ambient temperature, bulbs kept in basket, at Gujarat Agrcultural University, Junagadh during June to October 1990 (Kalola and Singh, 1994). The minimum physiological loss in weight, rotting, total loss in weight and maximum percentage of marketable bulbs were recorded in 'G-41' whereas these were highest in 'ARU-52' with lowest percentage of marketable bulbs (Table 2). Total soluble solids, dry matter and protein content of cloves varied significantly from genotype to genotype. The total soluble solids and dry matter content increased and protein content decreased at the termination of storage period (Table 3).

Garlic bulbs Cv. HG-1 when harvested at yellow leaves stage and stored in wooden boxes (60 x 30 x 45 cm size) with tops proved beneficial in extending shelf life because of more storability and acceptability due to lower quantitative losses. Keeping the bulbs in bundles proved to next best methods for storing garlic bulbs under room temperature for 5 months (Singh *et al.*, 1995). However, bulbs harveted at 100 per cent top fall and cured by the windrow method, stored for up to 5 and a half months in baksets under ambient conditions showed the total loss of 13.61 per cent which is very low and acceptable loss (Pandey *et al.*, 1996).

Dormancy of garlic cloves (cv. Caiyiapo) is broken and they start sprouting at 5°C, whereas there is no sprouting at 35°C. Ethylene production is high during the dormancy breaking stage, ABA increases during storage but the increase is least at 5°C. GA_3 and cytokinin levels are at peak during the early phase of dormancy. declines, then increases towards the end of dormancy (Liet *et al.*, 1996).

Table 2 : Percentage of various storage losses and marketable bulbs of different garlic genotypes after 150 days of stroage.

Genotypes	Physiological loss in weight (%)	Rotten bulbs (%)	Total loss in Weight (%)	Marketable bulbs (based on weight) %
IC-49682	18.07	29.13	37.50	62.54
IC-39373	19.84	19.84	34.46	65.54
Mota Ankadia	18.54	22.99	36.07	63.93
ARU-52	22.47	46.63	55.89	44.11
IC-35231	17.58	18.28	32.12	67.88
In. No-60	19.52	17.27	32.80	67.20
In. No.-15	17.74	22.30	33.50	66.50
GAUG-1	16.71	14.95	28.01	71.99
Lodhika	18.91	18.47	29.84	70.16
Thordi	18.28	22.72	33.14	66.86
GAUG-10	17.20	19.35	31.62	68.38
Vadla	18.32	32.27	39.80	60.20
G-41	14.98	11.12	20.78	79.22
Selection-2	16.13	30.59	37.99	62.01
IC-19084	19.77	20.49	33.05	66.95
Selection-10	18.33	19.59	33.06	66.94
Bawla	18.34	23.38	35.78	64.22
G-50	17.54	26.12	34.75	65.25
Pusa-10	18.19	22.17	33.60	66.40
G-1	18.19	12.97	25.67	74.33
CD at 5%	1.38	6.91	4.94	4.32

(Kalola and Singh, 1994)

Table 3 : Changes is qualitative characters of different gerlic genotypes before and after storage.

Genotyes	Total Soluble Solids (%)		Dry matter (%)		Protein (%)	
	Before storage	After storage	Before storage	After storage	Before storage	After storage
IC-49382	31.67	33.00	32.00	34.13	6.50	5.54
IC-49373	38.17	39.67	40.30	41.07	6.36	5.28
Mota Ankadia	34.50	35.83	36.16	37.24	6.33	5.22
ARU-52	27.83	28.83	29.10	32.33	7.18	4.99
IC-35231	33.67	34.33	36.03	37.36	6.24	5.28
In No.60	35.33	36.33	30.95	32.05	6.30	5.34
In.No.15	33.17	34.50	37.92	38.52	6.45	5.45
GAUG-1	34.00	36.33	36.06	37.73	6.27	5.45
Lodhika	38.67	40.00	41.05	42.44	6.42	5.48
Thordi	35.50	36.50	34.34	35.88	6.36	5.43
GAUG-10	34.83	35.83	33.23	34.85	6.33	5.40
Vadla	30.33	31.50	26.58	37.55	6.45	5.16
G-41	34.83	35.67	38.96	40.42	6.10	5.10
Selection-2	32.17	33.33	36.11	37.20	6.97	5.69
IC-19084	33.67	34.67	34.46	37.16	6.39	5.54
Selection-10	34.00	35.17	41.89	42.19	6.36	5.45
Bawla	35.00	36.00	30.84	31.59	6.68	5.43
G-50	33.33	34.33	37.19	38.34	6.13	5.40
Pusa-10	28.50	29.83	32.87	32.65	6.30	5.19
G-1	36.33	37.33	39.95	41.15	6.04	5.20
CD at 5%	0.99	1.05	0.76	1.90	0.17	0.29

(Kalola and Singh, 1994)

REFERENCES

Anonymous, (1995). Mannual for Agril. Extendion officers(1994-95 Rabi) Directorate of Agril. and Food Production. Orissa and OUAT, Bhubaneswar.

Aykroyd, W.R. (1963).The nutritive value of India Foods and the planning of satisfactory diets. ICMR Special Report Series No.42, New Delhi.

Batchvarov, S.I. (1966). Garlic, *Zemizdat*, Sofia.

Baten, M.A., Nahar, B.S., Sarkar, S. C. and Munsi, A.P. (1992). Effect of different intervals of irrigation on garlic growth and yield. *Pakistani J. Sci.*, 35 (7-8): 297-299.

Bhan, V.M. (1972). Weed control in vegetable crop. *Third International Symp. Subtrop. Hortic.*, Bangalore.

Bhan,V.M. and Singh, M. (1970). Weed control in field crop at Pantnagar, India. *Research Report, PANS,* 16 : 688-689.

Bhardwaj, A.K. and Thakur, R. C. (1974). *Ephestia elutella* (Walker) (phycitide): Lepidoptera, infesting, stored garlic (*Alliuu sativum* L.) *Curr. Sci.,* 43: 419-420.

Bisht, I.S., Agrawal, R.C. and Venketa shwaran, K. (1993). Yield loss in different genotypes of garlic due to purple blotch. *Indian Phytopathology,* 46 (1) : 89-91.

Candolle, De A., (1885). Origins of the cultivated plants, D. Appleton, New York.

Chaudhury, S.K. (1960). Allyl sulphide content of garlic. *Curr. Sci.,* 29: 351-352.

Chauhan, D.V.S. (1972). Vegetable production in India. Ram Prasad and Sons; Agra-3.

Choudhury, B. (1976). Vegetables. National Book trust, India, New Delhi.

Das, S.K. and Bhatnagar, S. (1991). Multistage dehydration for garlic. *Indian J. Agril. Engin.* 1, (1): 333-336.

Deb-Kiotanya, S., Ghos, M.R., Chaudhary, A.N. and Chaterjee, A. (1980). Extract of garlic as possible source of insecticides. *Indian J. Agric. Sci.,* 50: 507-509.

Eastal, P., Vinnela, E.,Jacas, J. and Macro, V. (1992). Evaluation of the efficiency of some pesticides. *Boletin de sanidad vegetable plagas,* 18 (2) : 331-335.

Garcia, A. (1980). Sprouting before harvest in garlic. *Communicado Techico, Empresa Brasileira de Resquisa Agropecuaria pelotas Res.,* 9:3.

Gerasimchik, V.G. and Kusaiko, V.S. (1989). A garlic harvester puller. *Khozyaistva,* 9 : 32-33.

Hari, Om and Srivastava, R.P. (1977). Influence of the planting material and spacing on growth and yield of garlic. *Indian J. Hort.,* 34 : 152-56.

Jones, M.A. and Mann, L.K. (1963). Onion and their Allies, Leonard Hill Books.

Joshi, S.N. (1961). Garlic is as good a crop as any. *Indian Farming,* 11(9): 33-34.

Kalola, R.H. and Singh, S.P. (1994). Storability of garlic (*Allium sativum* L.) genotypes under ambient temperature. *Indian J. Agric. Res.,* 28 (3): 197-200.

Katyal, S.L., (1985). Vegetable growing in India. Oxford and IBH pub. Co. New Delhi, pp 123.

Kuznetsov, A.V. (1948). Agrobiological studies of garlic, *phg, Krasnodar.*

Lawande, E.K. Pawar, R.D. and Kale, P.N. (1990). Preliminary studies on sprouting in garlic. *Haryana J. Hort. Sci.,* 19(3&4) :363-367.

Li, Xiaodong; Lu, Guoyi and Yu, Yan (1996). Effect of endogenous hormones on the dormancy of garlic (*Alllium sativum* L.) *Acta Hort.,* 23(2): 150-154.

Lorini, I. and Frizon, M.L. (1991). Evaluation of damage of thrips tabaci to garlic. *Anals de sociedade,* 20(2): 237-242.

Mahanta, R.K., Goswami, R.K., Kumar, D. and Goswami, P. (1980). Effect of *Allium sativum*, garlic on blood lipids, *Indian Med.* Gaz.; April, pp. 157.

Mangal, J.L., Singh, R.K. Yadav, A.C., Lal,S. and pandey , U.C. (1990). Evaluation of garlic cultivars for salinity tolerance *J. Hort. Sci.*, 65 : 657-58.

Mann, L.K. (1952). Anatomy of the garlic bulb and factors affecting bulb development. *Hilgardia*, 21:195-231.

Matrod, L., Faddoul, J., Elmeamar, A. and Alchaabi, S. (1991). The use of solar energy for controllling white rot disease of garlic. *FAO Pt. Prod. and Prote. Paper.* 109:118-128.

Mishra, H.N., Sahoo., S., Mohapatra, A.K. and Mohapatra, K.B. (1997). Effect of different dates of planting on growth and yield of garlic (*Allium sativum* L.) under high altitude conditions of southern Orissa. *The Orissa J. Hort.*, 25(1) : 53-54.

Nandel,T.R. , Shukla,Y.R. and Singh, Ravinder (2002). Evaluation and effect of herbicides for weed control in garlic (*Allium sativum* L.) *Int. Conf. Veg.*, Bangalore, 11-14 November, 2002. *Abstract*; pp 198.

Nath, Prem; Srivastava, V.K., Dutta, O.P. and Swamy, K.R.M. (2008). Vegetable Crops – Improvement and Production. PNASF, Bangalore. 209-212.

Om, H. and Awasthi, D.N. (1978). Effect of maleic hydrazide spray on the storage of garlic. *Progr. Hort.*, 9(4): 63-67.

Panchal, G.N., Modhwadia, M.M., Patel, J.C. and Sadaria, S.G. (1992). Response of garlic to irrigation, nitrogen and phosphorus. *Indian J. Agron*, 37(2): 397-98.

Pandey, U.B., Chauhan, K.P.S. and Singh, S.P. (1996). Effect of stage of harvesting and curing on storage quality of garlic. National Horticultural Research and Development Foundation *Newsletter*, 16(2): 1-3.

Pandey, V.C. and Singh, J. (1989). Performance of new varieties of garlic. *Hav. J. Ses.*, 19 (1): 69-71.

Pawar, D.B., Lawande, E.K. and Kale, P.N. (1990). Reaction of some garlic varieties to mite. *J. Maha. Agril. Univ.*, 15(5): 352-353.

Pruthi, J.S. (1979). Spices and Condiments. National Book Trust of India. New Delhi, pp. 125-132.

Purseglove, J.W. (1975). Tropical Crops : Monocotyledons. ELBS, Longman, London, pp. 52-56.

Qadri, S.M.H., Chougule, A.B. and Tripathi, S.P. (1987). Effect of different time of sowing on yield and quality of garlic. Agricultural Development Foundation, New Delhi, No. 17.

Rao, H. and Purewal S.S. (1957). Onion and garlic cultivation in India, *Farm Bulletin* No.3, ICAR, New Delhai, pp 18-22.

Rath, G.C. and Mohanty, G.N. (1986). Fusarium rot of stored garlic. *Indian phytopath;* 30: 614-615.

Regel, E. (1875). Alliorum adhuc cognitorum monographia, *Acta Hortic. Petrop.* 3: 266.

Saini, G.S. (1996). A Text Book of Vegetable Production. Aman Publishing House, Meerut (U.P.), India. pp. 145-147.

Sandhu, K.S., Singh, Daljit and Singh, Jaswinder (1997). Weed management in garlic (*Allium sativum* L.) under Punjab conditions. *Veg. Sci.*, 24(1): 7-9.

Selvaraj, N., Irulappan, I and Vedamuthu, P.G.B. (1995). Effect of growth regulators on growth and yield of garlic (*Allium sativum* L.) *South Indian Horticulture*, 43 (1&2) : 18-20.

Selvaraj, N., Irulappan, I.and Vedamuthu, Peter, G.B. (1994). Season and nutritional interaction on the yield of garlic (*Allium sativum* L.) *South Indian Horticulture*, 42(4): 251-258.

Selvaraj, N., Irulappn, I. and Vedamuthu, Peter, G.B.(1993). Effect of nitrogen, potassium and magnesium on foliage and bulb characters in garlic (*Allium sativum* L.) *South Indian Horticulture*, 41(6) : 350-354.

Sharma, A.B., Asrar, M.S. and Bhalla, B.L. (1980). Evaluation of effective, selective and economical herbicides for garlic (*Allium sativum* L.) In *Abstr. papers. Annu, Conf.* ISWS, Bhubaneshwar.

Sharma, A.B., Patel, R.K., and Tiwari, J.P. (1983). Chemical weed control in garlic. *Indian J. Weed Sci.*, 15: 17-22.

Shinde, N.N. and Sontakke, M.B. (1986). Bulb crops. In: vegetable crops (Ed. T.K. Bose *et al.*) .Naya Prokash, Calcutta.

Singh, Baldev (1982). Chemical weed control on garlic. M.Sc. thesis. Dept. of Vegetable Crops. HAU, Hisar, pp:70-72.

Singh, Baljeet and Pandey, U.C. (1989). Storage technique for onion and garlic. *Haryana farm*, 18:11.

Singh, D.N. and Nath, Vishal (2011). Varieties and Hybrids of Vegetables. Satish Serial Publishing House, Azadpur, Delhi (India). 78-80.

Singh, J., Vashist, D. and Khurana, S.C. (1995). Effect of harvesting stages and storage methods on storage life of garlic bulbs. *Scientific Hort.*, 4: 109-119.

Singh, Narendra and Netra, Pal (2003). Garlic In Vegetables, Tubercrops and Spices, (Eds. S. Thamburaj and Narendra Singh). ICAR, New Delhi.

Singh, N.P., Bhardwaj, A.K.,Kumar, A. and Singh K.M. (2004). Modern Technology on Vegetable Production. International Book Distributing company, Lucknow.

Singh, R.N. and Singh, J.R. (1974). Studies on the influence of boron nutrition on the growth characteristics of garlic (*Allium sativum* L.). *Indian J. Hort.*, 31: 255-258.

Singh, S.P. (1989). Production Technology of Vegetable Crops. Agricultural Research Communication centre, Kernal, Haryana (India), pp. 214-218.

Singh, S.P. (2001). Seed Production of Commercial Vegetables. Agrotech Publishing Academy, Udaipur. pp. 222-229.

Tandon, H.L.S. (1991). Fertilizer Recommendations for Horticultural Crops. A guidebook. Fertilizer Development and Consultation Organization, New Delhi.

Tewari, A.N., Rathi, K.S. and Singh, B. (1998). Integrated weed management in garlic (*Allium sativum*) *Indian J.Agric. Sci.*, 68 (5): 281-283.

Thompson, C.H. and Kelly, C.W. (1957). Vegetable Crops. Mc Graw Hill Book Co., Inc, USA.

Vavilov, N.I. (1926). Centres of the origin of cultivated plants, *Trudy Prikl. Bot, Genet. Sel.* 20: 1.

Zimdahl, L. (1980). Crop weed competition. A Review. Published in the United States by the International Plant Protection Centre, Oregon state University, Oregon, USA.

□□□

3. LEEK

Botanical Name	:	*Allium porrum* L. (Syn. *A. ampeloprasum* var. *porrum* L.)
Family	:	Allaceae
Chromosome number	:	2 n = 32

Leek (*Allium porrum* L.; syn. *A. ampeloprasum* var. porum L.): 2n = 32; belongs to the family Alliaceae is a non-bulb forming crop (Thomson and Kelly, 1957). It is cultigen derived from the wild, *A. ampeloprasum*. It is grown for its etiolated pseudostem formed by the leaf sheath. (Jones and Mann, 1963). The economic part is blanched stem and leaves which are highly nutritious and rich in medicinal properties, especially for heart ailments. It can be consumed as salad, in soup and in various cooked foods and can be quite relished for its onion flavour and nutritive value. Since. Indians are fond of green onions, therefore, leek thought to be good alternative leaving green onions to form bulbs.

The leek has greater cold resistance than the onion; bulbs are not formed. It does not have a rest period, and it is not influenced by photoperiod (Astley *et al.*, 1982). Plant is tall hardy biennial with white narrowly ovoid bulbs and broad leaves. Leek is highly cross-pollinated due to heteromorphism and male sterility. The production of inflorescences is induced by low temperatures. Growth is continuous and it can be harvested over a long period of time.

It is of ancient cultivation, and spread from its original home in the levant. It had spread to Europe in the middle ages and is now mainly a crop of northern Europe, where the largest production is in France. It is the national emblem of Wales.

ORIGIN

The leek is thought to be a native of Switzerland. Wild members of this family are found in the eastern Mediterranean region (Nonnecke, 1922). It was known to be the ancient Greeks and Romans (Thompson and Kelly, 1957). According to Harrison *et al.* (1969), leek is believed to be a native of eastern Mediterranean region. But Chauhan (1972) is of the opinion that leek is a native of Europe and West Asia. Reference to the leek can be found in Egyptian tombs, indicating that use has been made of this plant since early times. It has been known in northern Europe since the sixteenth century (Nonnecke, 1922).

DISTRIBUTION

Leek is grown in many countries like France, Britain, Northern Spain, Malaysia, Philippines, China. Belgium, Denmark and Holland, Sirya, Iraq,

Israel and many African countries (Harrison *et al.*, 1969). Recently it has been introduced to India. In India and Ceylon, it thrives well at higher altitudes but moist localities are adverse to its cultivation (Chauhan, 1972). It is not grown in India on a commercial scale but it is a favourite vegetable in the kitchen garden (Choudhury, 1967).

BOTANY

Leek is not a common bulb crop like onion and garlic. It is a non bulb forming, crop, grown for its blanched stem and leaves. Plant is tall hardy biennial with white, narrowly ovoid bulbs and broad leaves. It resembles the green onion but it is larger in size. The edible portion, which is cooked is the elongated base of the foliage leaves or false stem, which develop below ground in order to promote blanching. Lamina thin, keeled, folded lengthwise, 30-75 cm x 1.2-2.5 cm, broader than garlic; scape subtended by a single long pointed spathe; flowers numerous pinkish, with exserted anthers, (Purseglove, 1978). Leek is highly cross-pollinated due to heteromorphism and male sterility. Flowers are seldom produced in the tropics. The time of formation of the flower initially corresponds with the maxima in the curve showing percentage of carbohydrate and ash in the whole plant.

Taxonomy

Leek, *Allium ampeloprasum*, is a tetraploid with 2n = 32 (Mc Collum, 1976). According to Traub (1968), leek and garlic are put together in section *Allium*. Leek closely resembles wild *A. ampeloprasum*, differing mainly in a lesser tendency to form bulbs. Leek was known in Europe in the middle age and is still popular there, many cultivars having been selected for long, white, edible leaf bases and green tops, winter hardiness and resistance to bolting. Leek and Kurrat (*Allium kurrat*) are interfertile and it may be supposed that they would cross readily with wild *A. ampeloprasum* of the same (4x) ploidy level (Mc Collum, 1976).

COMPOSITION

Leek is rich in vitamin C, highly nutritious and having good amounts of minerals. An edible portions of 100 g leek contains (Aykroyd, 1941):

Moisture	78.9 g	Phosphorus	70 mg
Protein	1.8 g	Iron	2.3 mg
Fat	0.1 g	Vitamin A	30.IU.
Carbohydrate	5.0 g	Thiamin	0.23 mg
Minerals	0.7 g	Vitamin C	11.0 mg
Fibre	1.3 g	Calories	77

It is also rich in combined sulphur (0.66-0.072%).

Uses

The tender leek is eaten raw. The blanched, elongated stem and leaves are the parts used as a salad, for preparing soups and stems are used as vegetables, boiled or parboiled and fried in butter. (Harrison *et al.*, 1969). It is also used for flavouring in soup and stews (Choudhury, 1967). Leek has milder and more delicate flavour than onion, though of coarser texture. In medicine the bulbs are used as a renal stimulant (Shanmugavelu, 1989).

VARIETIES AND HYBRIDS

In leek many high yielding and promising varieties have been evolved (Robert Thompson, 1978). Borde Pearl' is one such high yielding line resistant to bolting. Bluevetia Early' is suitable for both summer and early autumn culture (Geriya, 1973). Copenhagen Market' and 'Elephant' are short-necked type and 'Gennivillies' and 'St. George' are long-necked types with resistant to frost.

Some other have been developed by pedigree methods are :

Miner

An autumn or early winter variety derived from crosses and family selection in the variety 'St George, has long stem and gives high yield.

Splendid

Cultivated throughout the season.

Renova

Derived by selection from the variety 'Regions' has darker leaves, longer stalk and less tendency to form bulb and is more uniform in size and shape.

Special 050

It is suitable for forcing.

American seed men varieties

London Flag, American Flag, Elephant or Monstrous Carentan, Giant Musselburgh. Lyon.

Belgium varieties

White Alberta, Blizzard and Carina are for winter. Acadia, Derrick, and Electra are suitable for autumn cultivation.

Canadan varieties

Longa, Odin, Kilima, Goliath, Siberia and Artico.

High yielding variety

Carina and Alberta.

Others

Regions Large Roven, Conqueror and Prize Taker.

Tropica seeds (Green Co.)

France, Large Long Summer.

Bejo Sheetal Seeds

Lincoln.

Nunhems Seeds

Carlton, Roxton, Parton.

Regins 930/57 (St. George x Special)

F_1 hybrid is good cropper especially in late autumn and early winter.

Detail characteristics of varieties and hybrids reported by Singh and Nath (2011) are given below:

London Flag

Tall with a thick stem, leaves broad, it is most generally cultivated.

Musselburg

It is very long and hardy; leaves broad and tall, stems long and thick.

Large Roven

Its leaves are very dark green, broad and thick. Stem rather short, but remarkably thick. It is said to grow as thick as man's arm. It is found to be the best kind of forcing as it acquires a sufficient thickness of stem sooner than any other varieties.

Lincoln

It matures in 80 days after transplanting. It is early Bulgarian giant type leek, which gives uniform long and thick stems. It is suitable for rabi season planting.

Carlton

This hybrid variety is very fast in growth, leaves are erect and green in colour. Long short length have excellent uniformity.

Roxton

Long short length with excellent uniformity. The growth of this hybrid is very fast with green to dark green irect leaf colour.

Parton

Medium growth of this hybrid produces medium shoft length of excellent uniformity. Leaves are erect and dark green in colour.

France

It shows medium maturity. The leaves are dark-blue green. Shanks are 18-20 cm long and 2.8 to 3.3 cm in diameter.

Large long summer

It is very early type and matures in four months. Foliage, medium green in colour. Plant long, medium sized shanks (20-25 cm long and 2.3 to 3.0 cm diameter), widely adapted and very fast growing.

Palam Paushtik

It has long, broad, flat garlic like bluish green leaves and long white pseudostem thickening cylindrically to attain 3-4 cm diameter at harvesting. It takes about 140-150 days to reach marketable stage. It has mild digestive flavour due to organosulphur compounds. It can be consumed like green onions after cooking with potatoes or in any other form of Indian dishes. Besides, it can be enjoyed raw as salad and in soup. It has healthful digester nutritive substances, especially protein, fibre, carbohydrates, vitamins and minerals. Its average yield is 30-35 tonnes/ha.

CLIMATIC INFLUENCE

Leek withstand heat and cold better than onion. Although leek is a cool season crop, temperature is more important than the length of day, in seed stalk development. However, like the onion, it is vernalized by cold temperatures, which cause undesirable premature bolting during the normal vegetative cycle. Leeks grown continuously at 15°C form flower stalks This is not likely to happen often in more temperate regions but might become a serious problem in the nineteen-to twenty-four-hour summer day length areas where leeks are grown-particularly in northern Europe.The tendency to bulb, an undesirable characteristic, is like wise temperature controlled, occurring between 15° and 18°C rather than at either 12°C or 21°C. Possibly due to these strict temperature parameters, the leek has not been promoted inNorth America as much as it might otherwise have been (Nonnecke, 1922). In India and Ceylone, it thrives well at higher altitudes and seeds produced at higher altitudes in the hills.

SOIL REQUIREMENT

Leek grows luxuriantly on medium soils which are rich in plant nutrient and organic matter. Loosening of the sandy clay loam subsoil which is helpful in penetration of roots up to 70 cm and thereby increase the yield. Sandy loam to loam soils are best suited for its cultivation. The most favourable range of soil pH for leek is 6.0-8.0 (Chauhan, 1972).

SOWING

Leek is grown either by direct seeding or transplanting the seedlings as in the case of onion except that greater spacing is maintained (Chaudhury, 1967). Sowing is done in August-October in raised nursery bed. About 5-7 kg seeds are sufficient to raise seedlings for one hectare land.The seedlings of 15 cm height are suitable for transplanting. Haworth (1960) recommended closest spacing 7.5 x 30 cm for obtaining the highest yields from large transplants although large transplants at the spacing 30 x 30 cm gave the highest mean weight per plant.

Germination and pre-treatment of seeds

Germination decrease with increasing in temperature. Pre-treatment at 3°C for 12-18 days of seed of Bulgarian and Elephant leek cultivars enabled germination to occur up to a temperature of 30°C. It is found that pre-chilled seeds in complete darkness at alternating temperature of 15°-28°C gave the highest percentage of germination.

PLANTING

Seedlings of height 10 cm to 15 cm and 3 to 4 weeks old are planted in well prepare field at spacing of 30 cm to 40 cm in row to row and 15 cm to 20 cm in plant to plant. Water is given immediate after planting of seedlings to establish quickly in the field. Sometimes transplanting is done in trenches about 30 cm deep and spaced 10 cm apart in rows 40 cm apart (Choudhury, 1967). By increasing plant density from 1,66,000 -2,5,000 plants/ha the yield increase. The best planting distance seems to be 40 cm x 10 cm (Singh and Singh, 2007).

NUTRITIONAL REQUIREMENT

Leek requires large amount of manures and fertilizers. It is more responsive to vigorous growth inducing N to encourage heavy foliage formation. Supplemental irrigation insures uninterrupted growth (Nonnecke, 1922). Wood and Robinson (1983) recorded that N in too low amounts causes old leaves to fade and shrivel, low P dwarfs the plant; low S leaves the foliage stiff and erect as well as chlorotic, low K causes dieback; low Mg typically produces yellowing in older leaves; and low Bo causes transverse cracks on the leaves. Choudhury (1967) recommended 100 kg N, 65 kg P_2O_5 and 130 kg K_2O per ha for obtaining 30 t of leek. Page and Williams (1977) reported optimum economic application of nitrogen for drilled leeks grown on the flat was 250 kg N per ha.

According to Laske (1962), nutrient absorption and growth are limited in the youthful stages. Both increased greatly at the commencement of the second half of the growing period. He recommended 86 kg N, 33 kg P_2O_5 and 100 kg K_2O besides 58 kg CaO for yielding 300 quintals of leeks per hectare.

Incorporate farm yard manure @ 20 t/ha at final land preparation. Apply nitrogen 80-100 kg, phosphorus 60-80 kg and potassium 80-100 kg for higher yield. Whole quantity of phosphorus and potassium along with half doses of nitrogen should be applied at the time of transplanting. Remaining half dose of nitrogen may be applied in two splits doses after 30 days and 45 days of transplanting for better crop yield (Singh and Singh, 2007). Use of heavy nitrogen give lower keeping quality product. High rates of potash appeared to be toughen the edible flower stalks, whereas high rates of nitrogen give more tender stalks (Shanmugavelu, 1989).

IRRIGATION

Leek requires large amount of water during its growth periods and very sensitive to moisture stress. Frequent light irrigations are required for growing of this crop. Without irrigation, fertilizer application is unprofitable to the crop. Irrigation has good effect in dry year on the crop growth and development. Geglareck (1978), reported that in a year of average rainfall supplementary irrigation at 160 mm result in 56.7% and 55.7% increase in total and commercial yields. In dry year, irrigation at 205 mm resulted in 147.8 and 148.2% increase in total and commercial yield respectively. It requires 375-400 mm of water for higher yield. Water requirement is low during the juvenile phase but subsequently increase. Sprinkler irrigation is beneficial to increase the yield of leek (Klatt, 1969). On a light clay soil, sprinkler irrigation applied at 65 per cent of field capacity increased the marketable yield by 10 to 30 per cent (Buczak *et al.*, 1976).

INTERCULTURE

Leek require to keep the ground cleans. Its root system is shallow, hence light hoeing and earthings are necessary for better growth of the plants. Generally 2-3 light hoeing and earthing is given to the crop. Earthing is done to blanch the stem of leek. If any flower stem appears, these should be nipped in the bud stage immediately (Shewell *et al.*, 1974).

Blanching

It is an essential operation in which leek plants are covered with soil from their ground level upwards to 15 cm to 20 cm in length and 2.5-5.0 cm in diameter nearer to the base of the leaves which blanch the covered portion and become yellowish white in colour due to lack of light thereby improving their quality and increasing acceptability and marketability. (Knott, 1975). Blanching can also be done by placing collars around stem up to the base of the leaves followed by drawing up soil. This will keep stems free of soil. Care should be taken to cover the plants with soil when these are young. It is quite possible to start blanching three weeks after planting and to continue by placing a little finely broken up soil around the stems once a fortnight. Blanching is also done by means of tubes of stiff brown paper also. The advantage of using brown paper is that it does prevent the soil getting in between the leaves (Shanmugavelu, 1989).

Crop-rotation

Frohlich (1962) stated that yields of three crops grown after leeks were higher than yields of the same crops grown after a lettuce or cauliflower crop. Buczak (1967) recommended the following vegetable crop rotation for maximum utilization of plant nutrients in organic and mineral manuring-celery, cauliflower, leek, cabbage, cucumber, tomato and onion. Pelletier (1975) found that four crop rotation-celeray, carrot, leek and celery are more profitable. The usual crop rotation for leek is cabbage and spinach (Kains *et al.*, 1976). It was also found that repeated treatments with linuron @ 0.75 kg/ ha caused no drop in yield in the above crop rotation.

WEED CONTROL

Since leek is a perennial crop, care should be taken to remove weeds from time to time from the field. Generally, three weed species dominating in the leek field. They are *Chenopodium album, Poa annua* and *Senecio vulgaris.* Singh and Singh (2007) suggested that weeds can be checked effectively by the use of following herbicides/weedicides, which details are given below:

a) Application of alachlor at 0.675 to 1.35 kg and propachlor at 1.80 to 3.60 kg/acre pre-emergence (both applied either as granules or spray) controlled a wide range of weeds.

b) Prynachlor at 3 kg per ha 12 days after planting control most of the weeds.

c) Gasagard 50 (prometryne 50%) at 2.5 to 3.0 kg per ha application as post emergence is effective to control weeds in planted leeks

d) Application of alicep at 50 g/acre or bladex and 10 plus tribunil at 20 g per acre give good weed control and only alicep enhance the yield also.

PLANT PROTECTION MEASURES

The important insect pests and diseases are leek moth, onion thrips, mites, white tip of leek, downy mildew, powdry mildew and purple blotch. Their details and control measures are given below:

Insect Pests

Leek moth (Acrolepiopsis assectella)

It is a serious pest of leek in Italy. The moth had 5 over lapping generation in a year. The third generation which occurred in July is most injurious. The application of insecticide like endosulfan @ 2 ml/litre of water control pest infection.

Onion thrips (Thrips tabacii) and mites (Rhizoglyphus echinopus)

These are most destructive insects of this crop. Application of malathion or nuvacron @ 0.03% has proven to be effective in controlling these pests.

White tip of leek

It is caused by the fungus *Phytophthora porri* Foister. It is a specific disease of leek. The tips of leaves become yellow, die and turn white for a distance up to 15 cm. Most affected leaves turn backward. Young severely infected plants remain stunted. Spraying with bordeaux mixture or copper oxychloride @ 2 g/litre of water is effective to control the disease.

Purple blotch

This disease is caused by by *Alternaria porri* and powdery mildew caused by *Leveillula taurica* can effectively be controlled by spraying dithane M-45 @ 2 g/litre of water.

HARVESTING

Leeks are long season crop and require about 120 days from seedling to harvest. Crop is harvested as tender stalks when stalks reaches the size of pencil thickness (2.0 to 2.5 cm). The plants are pulled by hand carefully then clean, trimmed properly and marketed in bunches. After bunching, the roots and stems are trimmed to the appropriate length required for shipping. The attractive green leaves with white base increase the marketability of this crop (Singh and Singh, 2007). After harvesting they should be kept under shade in cool condition.

YIELD

Yield depend upon several factors. However, an average yield of 200 q/ha can be obtained from a well-managed crop.

POST HARVEST HANDLING

Grading

After harvesting, cleaning and trimmings are done then plants are graded in 'super', 'fancy' and 'commercial' grades according to size of blanched stem and leaves. Then makes in bunches with different grades to facilitate marketing and to get high returns (Singh and Singh, 2007).

Packing

Singh and Singh (2007) suggested that bunches are packed in card board boxes with 5-6 kg capacity. Sometimes also pack in plastic crates and in paper packet in small quantities for proper marketing.

Storage

The green crop can be stored in good condition for 2-3 months at a temperature of 32°F (0°C) with a relative humidity of 85-90 per cent. Pre-drying the outer leaves before storage did not reduce moisture loss. Spraying of growth regulators one week before harvest for storage of leek is also beneficial in increasing shelf life (Herregods, 1971). Spraying of GA (100 ppm), benzimidazole (500 ppm) or benomyl (500 ppm) extended the shelf life of leek for 9-10 days.

SEED PRODUCTION

Leek does not set seed in the plains as the mature plants should be exposed to low temperature for a period before they form flowering stalks. The seed is produced at higher altitudes in the hills (Maity and Hazra, 1986). There are two methods for seed production (a) seed to seed method, (b) root to seed method.

Seed to seed method

Seeds are sown @ 2-3 kg per hectare 30 cm apart and 10 cm between plants. The young plants (stecklings) are lifted, trimmed and replanted. Two roguing stages are involved in this case one at proper establishment of plants and second at the flowering stage.

Root to seed method

The young plants are lifted after 10 weeks of sowing. They are graded and trimmed and those are of pencil thickness are replanted 10 cm deep, 8-10 cm apart, in rows 70 cm apart. Three stages of inspection and roguing are necessary. The first is when transplanting, second at establishment of transplanted steklings and third at the start of flowering in the second year.

The usual minimum isolation distance of 1000 m is required for growing its seed crop. Leek umbels are harvested by hand when approximately 5% of the capsules on individual heads have commenced shedding seeds. The mature umbles are individually cut from the mother plant with approximately 15 cm of seed stalk attached. The umbles are then spread on plastic sheets in open conditions. The harvested and dried umbels are ready for threshing as soon as seeds can be separated from the capsule by rubbing in the hand. The seed lot is dried down to 6 per cent moisture content and stored in vapour proof packets or containers.

REFERENCES

Astley, D., Innes, N.L. and Van der Meer, Q.P. (1982). Genetic resources of *Allium* species-global report. International Board for plant genetic Resources, Rome, 38.

Aykroyd, W.R. (1941). Health Bulletin, No.23. Published by Nutrition Research Lab. Koonoor.

Buczak, E.A., Hellwiig, R. and Motor, R. (1976). *Soils and Fertilizers*, 40(9): 5235.

Buczak, K. (1967). *Hort. Abstr.*, 38 (4): 7554.

Chauhan, D.V.S. (1972). Vegetable production in India. Ram Prasad and Sons.: Agra-3 (India), pp.240.

Choudhury, B. (1967). Vegetables. National Book Trust India, New Delhi.

Frohlich, H. (1962). *Hort. Abstr.*, 32 (3): 4981.

Geriya, R.F.(1973). *Plant Breed.Abstr.* 44(4): 2812.

Herregods, M. (1971). *Hort. Abstr.* 42(2): 3584.

Jones, H.A. and Mann, L.K. (1963) Onions and their allies. Leonard Hill (Books), London.

Kalloo, G. and Bergh, B.O. (1993). Genetic improvement of vegetables crops. Pergamon Press, New York.

Katyal, S.L. (1977). Vegetable growing in India. Oxford and IBH Publishing Co, New Delhi.

Klatt, F. (1969). *Hort Abstr.* 40(3): 6031.

Lorenz, O.A.and Maynard, D.N. (1980). Handbook for Vegetables (2nd ed: Knott), Wiley Interscience.

Maity, T.K. and Bazra, P.(1986). Leek. In Vegetable Crops in India, Ed. T.K. Bose and M.G. Som., Naya Prokash, Calcutta, 6, pp.602-605.

Mc Collum, G.D.(1976). Evolution of Crop Plants, Ed. N.W. Simmmonds, Longman, London and New York.

Nonnecke, I.B. Libner (1922).Vegetable Production. An AVI Book published by Van Nostr and Reinhold, Springer International Edition.

Page, E.R. and Williams, J.B. (1977). *Soil and Fertilizers*, 40(10): 5883.

Purseglove, J.W. (1978). Tropical crops: Monocotyledons. The English Language Book Society and Longman, London.

Robert Thompson (1978). *The gardener's Assistant*, pp.217-218.

Shanmugavelu, K.G.(1989). Production Technology of vegetable crops. Oxford and IBH Publishing Co. Pvt. Ltd, New Delhi.

Singh, D.N. and Nath, Vishal (2011). Varieties and Hybrids of Vegetables. Satish Serial Publishing House, Azadpur, Delhi (India), pp.82-84.

Singh, D.N. and Singh, S.P. (2007). Production Management of Underutilized Vegetables. Agrihortica Publications. Gandhidham, Junagadh (Gujarat) India, pp.171-179.

Thompson, H.C. and Kelly, W.C. (1957). Vegetable Crops . Tata Mc Graw Hill Publishing Company Ltd., New Delhi.

Traub, H.P. (1968). Plant Life, 24: 147-63.

Van der meer, Q.P. and Hanelt, P. (1990). Leek (*Allum ampeloprasum* var *porrum*) in Onions and allied crops, Vol.III. Biochemistry, food sciences and minor crops. Brewster, J.L. and Robinowitch,, H.D. eds, CRC Press, Boca Raton.

Wood, P. and Robinson, J.B.D. (1983). Diagnosis of mineral Disorders in Plants. Vol. 2 Vegetables, Her Majesty's. Stationery Office.

□□□

Chapter-8

Perennial Vegetables

ASPARAGUS

Botanical name	: *Asparagus officinalis* L.
Family	: Liliaceae
Chromosome number	: 2n = 20

Asparagus, a exotic hardy perennial and only popular vegetable of the family liliaceae which has more than 150 different species (Choudhury, 1967). It is known as Halgun, Halyun, Hillua and Shatawar and cultivated for its tender shoots, commonly known as spears, which has an agreeable flavour after it has been boiled. Asparagus is one of the most delicate, wholesome and appetizing products of the home garden (Thompson and Kelly, 1957). It gives economic yield after three years and continues to grow upto several years.

ORIGIN

Asparagus is originated in eastern Mediterranean lands and Asia minor. It is indigenous to temperate to Europe and Asia, where it has been in cultivation for over 2000 years. Thus, asparagus is a vegetable much esteemed by Europeans (Shanmugavelu, 1989). It has been introduced to many other parts of the world from there. Since the time of the ancient Greeks, who first gave the present name, "asparagus" has been used as vegetable.

DISTRIBUTION

Asparagus is widely distributed in America, Greece, France, Holland, Germany, Spain Hungry and England. The USA , Germany, Spain and France have large area under this crop. For Greek and Romans it is an important crop, specially for the medicinal properties. It is an important commercial crop in America (Kabir, 1993). It is newly emerged crop in India and its demand is gradually increasing. Recently grown in temperate places of the country to make it commercialized.

BOTANY

Asparagus is a perennial, dioecious herb with erect branched stem, growing 1 to 3 m tall. The roots are tuberous and the leaves narrow and needle like.The male flowers are yellowish green and conspicuous, while the female flowers (on separate plants) are less conspicuous. The female plants produced berries with one to three seeds per fruits. The seeds are large (3mm or less in diameter) rounded at the back but more or less flattened on one side, black in colour (Kabir, 1993).

Sawada *et al.* (1961) reported that longer the cutting season the greater are the number of spears and stalks produced. The number of crown buds and roots formed in the current year decreased with increase in the length of the cutting season. The fleshy roots reserved highest food in the plant where no spears removed.

Heigh (1960) reported that male plants are earlier, high yielding and live longer than female plants. An all male population may be produced by growing seeds from bisexual flowers on a male plant, crossing the male plants in the progeny with female plants to produce all male seeds.

COMPOSITION

Aspargus is a rich source of carbohydrates, proteins, calcium, iron, dietary fibre and ascorbic acid (Choudhury, 1967). White crystalline substance asparagine ($C_4H_8O_3N_2.H_2O$) is obtained from the juice of young shoots (Anonymous, 1943). The edible portion contains:

Moisture	93.0%	Ash	0.6%
Fat	0.2%	Total carbohydrates	3.9%
Calcium	0.21%	Sugar	1.3%
Fibre	0.7%	Iron	0.0012%
Starch	0.4%	Calories per 450 g	120
Protein	2.2%	Lignin	10.64%

Vitamin A 1400 I.U., B., 180 mgm, B_2 130 mgm, C 400 mgm, per 100 gms fresh leaves (Aykroyd, 1941).

Uses

The new succulent shoots which come every year constitute the asparagus of commerce and large quantities are canned (Chauhan, 1965). Asparagus is freshly cut and served as one of the most delicate items. It is used fresh as vegetable and also eaten green. The spears are treated as a delicacy in preparation of soup and other vegetable products. Large quantities of asparagus spears are canned and frozen. Asparagus is also preserved in cans or bottles, but the home grown products thus dealt with is far exceeded in quantity by the canned asparagus. It is considered to be a good diuretic and is used especially in cardiac dropsy and chronic gout (Anonymous, 1952).

VARIETIES AND HYBRIDS

The asparagus varieties are broadly divided into two groups : (a) Spears overcast with purple and (b) Spear tips almost free from purple overcast. The common varieties are: Perfection, Mary Washington, Martha-Washington, Colossal, Berr's Mammoth, Columbian Mammoth White, Donald's Elmira, Palmetta, Conovers Colossal, UC 72, Brock Imperial 84, Palmetto, Argentenil, Huchels Leistung Sausbese and Schewetzinger, Reading Giant, Gersey Queen,Jersey Giant, New Jersey, Improved Perfection, UC 66, UC 711, SL 831, Limbrass 22, Librass 126 and Lucullas 1813, Paradise Meisterschuss, Bergamo, Violet Dutch and White German are grown in different countries

Hybrids

HD-INDA, Gijnlim, Revolution, Lara and Mira, Faribo, Aneto and Desta.

Detail characteristics of varieties and hybrids described by Singh and Nath (2011) are given below:

Perfection

It is early, spears large, green, succulent, heavy cropper and uniform in maturity (Anonymous, 1998).

Conovers Colossal and Mammoth White

They produces light green or whitish spears. They are high yielder and commonly grown for marketing purpose.

Palmetts, Argentenil and Mary Washington

These are most important varieties and produces very high yield. The tips of spears before exposer to the light are purplish and the spears become dark green in sunlight (Krarup, 1989).

Violet Dutch

It is the oldest variety which may be regarded as the form from which all the modern types have been derived.

White German

It is selected in Germany and widely grown by Germans. It has rounded tip and good quality spears.

Palmetto

It has a little difference in appearance than other varieties and considerably resistant to rust and extensively grown throughout America.

Schewetzinger

It is a commercial variety and has a greater vigour when planted by clones.

Lara and Mira

These hybrids are early and high yielding evolved through breeding.

Aneto and Desta

These are inter-clone hybrid, high yielding and less heterogeneous. Evolved through *'in vitro'* tissue culture.

Faribo

It is a high yielder and produce good quality spears.

UC 711

It is a better hybrid than other and give maximum yield and high return.

Gijnlim

This asparagus hybrid is very early in production with medium thick, straight spears of excellent quality. A very good total yield by means of 100% male plants gives 20% higher production compared to female plants and gives the advantage of no production of seed berries and seedlings under normal conditions.

Paradise

It is a selection of Mary Washington developed in California are said to be superior to the standard strain in some regions.

Table 1 : Evaluation of Asparagus cultivars during 1993 and 1994.

Cultivar	Year	Spear yield / plant(g) (male plants)	Spear yield/plant (g) (female plants)	Average spear yield/plant	Number of spears/ plant	Spear weight (g)	Spear length (cm)	Spear diameter (cm)	Plant height (cm)
'Perfection'	1993	40355	280.45	342.00	14.2	24.09	23.56	1.82	123.33
	1994	416.39	297.18	356.78	16.5	26.27	23.30	2.20	131.52
'Brock Imperial 84'	1993	611.80	374.40	492.60	12.4	29.69	23.66	2.26	100.00
	1994	627.26	395.5	511.20	13.6	41.63	23.20	2.40	105.90
'Local'	1993	207.15	169.50	188.30	9.6	19.56	23.33	1.04	86.66
	1994	229.63	189.76	209.70	10.2	20.43	23.10	1.60	95.80
'Conovers Colossal'	1993	245.10	184.90	215.00	10.7	19.97	23.30	1.95	96.66
	1994	264.98	197.87	231.42	11.3	22.47	23.10	1.90	100.40
'UC 72'	1993	239.80	173.60	206.70	10.5	19.67	23.13	1.93	61.66
	1994	258.84	192.42	225.63	11.4	21.77	23.00	1.80	96.15

Jersey Queen

A clone cultivar selected from Mary Washington. It produce high quality spear and maximum yield (Ellison *et al.*, 1989).

MC Grady and Tilt (1990), in an evaluation trial, obtained highest number of 8-20 mm diameter spears in cv. Jersy Giant followed by UC 157.

Lucullus 1883 and Limbrass 126 are the best white type cultivars and limbrass 22 is the best green type cultivar grown in Poland (Knaflewski, 1988).

Performance of different varieties was evaluated by Sharma and Chaudhary (1996) and they found that Asparagus 'Brock Imperial 84' is highly suitable for commercial cultivation (Table 1).

CLIMATIC INFLUENCE

The temperate climate are highly suitable for asparagus cultivation (Sharma and Sharma, 1994). Asparagus can be grown successfully in temperate to sub tropical climates. It requires a cooler and humid climate for growth and yield for spear production. The upper portion dry up during winter and again sprout with the on-set of the spring season. The temperature should not be lower than 8°C for better germination of seeds and temperature of 15°-25°C require for good spear development (Sharma and Chaudhary, 1996).

A lower temperature is necessary for 60-90 days of the year when the plants are during dormant. It should be grown over 1000 m for economic yields (Rice *et al.*, 1990). Asparagus tolerates frost to some extent but continuous freezing temperatures, however, are harmful.

SOIL REQUIREMENT

Asparagus can be grown on nearly all kind of soils. Sandy, sandy loam and silt loam soils are used to a large extent in many regions. Deep, loose and well drained soils and well manured sandy loam soils are considered to be the best for cultivation of asparagus (Chauhan, 1965). For early crop sandy and sandy loam soils are advantageous. A pH range of 6.0-6.7 or neutral to slightly alkaline is optimum for higher yield. According to Sandhu (1963), this crop can grow well even on the soil which has a salt content too high for many other crops, but it prefers neutral soils.

SOWING

Asparagus is propagated through seeds or crown. About 2.5-4.0 kg of seeds are enough to raise seedlings (Young plants) or crowns (underground roots) for planting one hectare land. The sowing is done during March-May in the hills and July-November in plains.

Nursery Raising

The seeds should be sown in a well prepared raised nursery bed, seeds pre-soaked for 3-5 days, are sown in lines 15-20 cm apart at a depth of 1.5 to 2.0 cm. After germination, seedlings are thinned to a spacing of 5-7 cm between the plants within the row. The seeds sown during early spring take 3-4 weeks to germinate. Borthwick (1925) observed that seed soaked at 25° to 35°C germinated more quickly than that soaked at higher or lower temperatures. Soaking for 3-5 days at 30 to 35°C is recommended. After soaking, the seed should be spread out to dry and during the drying it should be stirred to hasten evaporation of water on the surface and to have it dry uniformly. Optimum temperature for seed germination is 20°C and germination is best at temperature of 25°-30°C (Kabir, 1993). Seedling emergence was more rapid from primed seeds in 0.6 MPa polyethylene glycol 8000 at 20°C than from water-soaked ones, but increased drying before planting led to slower emergence (Evans and Pill, 1989). In nursery beds seedlings are allowed to grow for about 7-12 months then transplanted in the main field.

Planting Distance

According to Choudhury (1967) the rows are spaced at about 1.5 to 2.0 metres and 45 to 60 cm distance is kept within rows. Kabir (1993) reported that ridges are formed about 75-90 cm apart and 50-60 cm between plants in row to get higher yield. Closer spacing results in a decrease in size of spears compared with wider spacing. Veeraragavathatham *et al.* (2001) suggested that row to row distance of 100-150 cm and plant to plant 45-60 cm is useful for planting of asparagus seedlings or crowns.

Selection of planting materials

Since asparagus is dioecious, has male and female flowers separately on different plants. Occurrence of perfect flower is rare. The male plants bear more spears than female plants. It is not possible to recognise the sex of the plants in the nursery as most of the plants do not flowers in a year. The plants should be selected according to their root growth. The crown having good growth and branching should be selected. Older crowns have more roots but a number of them may already be mutilated (Choudhury, 1967).

Land preparation

Preparation of soil thoroughly is essential since asparagus occupies the soil for many years. Adequate quantity of organic manure should be incorporated at the time of field preparation. Good drainage is essential as the rotting of crowns take place in wet soils. The main field is prepared by ploughing 2-3 times after incorporating 30-40 tonnes of farmyard manure per hectare. Ridges and furrows are formed at a given distance with a depth of 20-35 cm.

PLANTING

The seedlings are transplanted after 7-12 months or crown of one year old plants are divided to provide source of planting materials and are planted at a depth of 15 cm with same spacing as seedlings. After the furrows are made the crowns are dropped by hand and the furrows are filled with soil so as to cover the crowns to a height of 2-3 cm. The seeds can also be sown directly on the side of the ridges.

NUTRITIONAL REQUIREMENT

Since asparagus is a perennial crop, it is very necessary that the soil is made fertile and put in the best physical condition. It requires a good amount of humus. A yield of 17.5 quintals of asparagus removes 44 kg of nitrogen, 12.5 kg of phosphorus and 40 kg of potash (Choudhury, 1967). It is therefore, necessary to add about 50-60 kg of nitrogen, 25 kg of phosphorus and 50 kg of potassium every year. Application of 20 tonnes of farmyard manure, 60 kg N, 100 kg P and 100 Kg K/ha is recommended by Pandita and Bhan (1992). According to Mehwald (1988), application of 100-120 kg N, 50 kg P and 250-350 kg K per hectare resulted in a maximum yield of spears. Application of 20 tonnes of farm yard manure, 100 kg N, 100 kg P_2O_5 and 100 kg K_2O /ha gave better yield (Sharma and Chaudhary,1996). The farmyard manures is incorporated in the soil during land preparation. The fertilizers should be applied in two splits; first dose during early spring before the growth starts and the second when harvesting is completed. Application of common salt is often beneficial for asparagus cultivation.

IRRIGATION

Irrigation is a continuing operation to obtain high yield. The irrigation is given by means of flooding as per soil and climatic condition. The first irrigation is given at the time of planting and thereafter at 10-15 days intervals. The water is allowed to stand in furrows to facilitate percolation. Excess irrigation should be avoided and proper drainage provided. Irrigation is not common in humid region. Takatory *et al.* (1970) obtained highest yield in terms of weight and number of spears with the medium irrigation of 52.5 cm water annually. Under temperate conditions, the crowns became dormant during winter and formation of spears starts in the beginning of spring. In mild climatic conditions, the dormancy is not distinct.

INTERCULTURE

Asparagus is a perennial vegetable crop and hence requires different form of interculture. Hoeing should be done carefully in the crop to avoid any injury to the roots. It is desirable to disk and harrow thoroughly and apply fertilizers before growth starts. These operations should be done every year.

Inter cropping

When the rows are far apart, the inter space can be utilised for the first two year by growing some inter-crops in order to get some monetary benefit from the land every year, provided sufficient nutrients are available for both crops and a suitable crop is grown. Snap beans, early cabbage, lettuce or any other small growing crop may be planted between the rows of asparagus. Tall-growing or long season crop should not be grown with asparagus on account of shading and competition for moisture and nutrients. The cultivation required for good growth of asparagus will be sufficient for the companion crop and the return for the later should go a long way towards paying for the cost of growing both crops.

Blanching

Blanching is normally done to produce white asparagus. Mounding the soil to a height of 25-30 cm over the rows is practised to blanch the young spears, which are preferred for canning. The mound must be kept up as long as blanched spears are desired. After cutting season is over the mounds are levelled by means of a plough disk or other implements.

Mulching

It is found that mulching in asparagus bed is beneficial to control the weeds, to maintain the soil moisture, soil temperature for better growth and higher early yield with more total crop yield. (a) Coating of Bitumen mulch to one year old asparagus beds produce a crop twice large than non-mulch crop (Kempe, 1975). (b) Plastic covering to one year old crop also produce higher yield.(c) Mulching with polythene or PVC film raised soil temperature and asparagus yield by 25-30% more (Burn, 1969). (d) Use of a 0.05 mm thick transparent polythene film increased the soil temperature by 2°C to 3°C and give more early yield by 17% than non-mulch crop (Frenz and Munz, 1970).

WEED CONTROL

The weed control for asparagus is to be taken in timely and more attention should be given than other vegetable crops because its plants occupy the same area for many years. Weeds between the rows can be kept under control by shallow hoeing so that the young sprouts are not damaged. Use of herbicides are found effective to control weeds in asparagus crops, which details are given below :

(a) Pre-emergence application of metribuzin (1 kg/ha) in direct sown asparagus and early post emergence is recommended for control of broad-leaved weeds (Freeman and Maurer, 1989). (b) De Beni (1962) reported that application of monuron @ 3.0 kg per hectare before crop emergence gives best weed control in asparagus beds.(c) Monuron @ 0.8 kg and simazine @ 0.5 kg per hectare in 1000 litre of water are recommended for controlling annual weeds in seedbeds (Van Staalduire, 1962). (d) Spray of dalapon @ 8.5 kg per hectare immediately after harvest is very effective to control perennial grasses in production beds or sandy soil. (e) According to Ogg (1975) application of 2,4-D + dicamba at 1.1 + 0.6 kg per hectare, twice control Canada thistle (*Cirsium arvense*) and field bind weeds (*Convolvulus arvensis*).

PLANT PROTECTION MEASURES

The important insect pests and diseases are asparagus beetle, asparagus fly, garden centipede, asparagus rust, and fusarium wilt.

Insect Pests

Asparagus beetle

Asparagus is attacked by beetle (*Crioceris asparagi* L.) and the twelve-spotted asparagus beetle (*Crioeris duodecimpunctata* L.). Both the adults and larvae damage to the tender shoots, and to the plants later in the season. The larvae feeds on the tender portion of the tops, but the beetles gnaw the epidermis of the stem as well as the tender portions and weakening the roots. Later both the adults and larvae feed on the berries. Spraying with carbaryl 50 WP. (sevin) @ 3g per litre of water at least twice at 7 days interval control the beetles.

Asparagus Fly

It damage the tender portion of the plants and create problems for the crop. Spray with monocrotophos @ 2 ml per litre of water to control the fly. In case of severe infestation sprying should be repeated at 7 days interval.

Garden centipede (Scutigerella immaculata)

This pest makes large number of small , round holes in the spear below the surface of the ground. The fibres around the holes harden and make the spears unfit for canning. To control the pest flooding of water in scattered infested field is advisable. Soil application of furadon @ 15 kg per hectare is also effective to control the pest.

Diseases

Asparagus rust

This disease is caused by (*Puccinia asparagi*) and is the most serious disease of asparagus and appear on the plant as small reddish-yellow spots on the main stem and on the branches, enlarge into patches until the whole plant has a reddish brown orange colour, which becomes darker later in the season and reduces the yield by weakening the plants and often killing them. It can be controlled by dusting fine sulphur @ 24 kg per hectare even before infection take place. Spray benlate or bavistin @ 0.1 % or bordeaux mixture (5:5:50) at the early stage of the plant growth. The use of rust resistant cultivars has been suggested. Jersey Giant and UC 157 are slow rusting cultivars (Johnson and Lunden, 1992).

Fusarium wilt

This disease is caused by a fungus *Fusariium* sp. Wilt, affected spears show brown discolouration and become wilted and stunted. The fungus is soil borne and planting should be avoided in infested soil. The seed treatment and drenching of the plants with brassicol (0.2%) or bavistin (0.1%) help in controlling this disease. Grow tolerant varieties like Mary-Washington, UC 66 and UC 72.

HARVESTING

The spears (aerial shoots) arise from the buds on the rhizome six months after planting of seedlings or crowns. The spears are not harvested upto one year after planting the crown/seedlings, because of active growth of root system and rhizome during this period. The production of spears is at the expense of food stored in the roots during the previous year and as such sufficient time should be allowed for storage of food materials. The first cutting should be done two years after planting. From third year onwards spears are harvested regularly during spring-summer and again in autumn season. Normally spears should be harvested from March to July every year. They are cut by a special knife which cut 3 to 5 cm below the soil surface and at the length of 20-22 cm from the soil surface. Care should be taken while cutting so that the young adjacent sprouts are not damaged. Crooked and curved spears are rejected. Asparagus losses quality quickly after harvest, so it should be sent to market as soon as possible.

YIELD

Male plants give the higher total yield while female plants produces large individual spears. The plants remain productive up to 15-20 years. The spear yield increases gradually. The average yield per hectare every year is about 35-40 quintals (Rice *et al.*, 1990).

POST HARVEST HANDLING

Grading

Asparagus shoots are washed after harvesting then separated in to grades, based mainly on size. There are five different grades are used in International Markets. These are (a) very small less then 12.7 cm. (b) small 12.7-20.5cm, (c) medium 20.5-28.0 cm, (d) large 28.0-35.5 cm and (e) very large more than 35.5 cm. After being separated into the different grades shoots are made into bunches (Singh and Singh, 2007). Only one grade being put into a bunch and tied near each end after preparation of bunches. The butts are cut off evenly with a sharp knife and they packed immediately. The size of bunches ranges from 1-3 pounds depending on markets. Bunches of 2 ½ pounds are common in California.

Packing

Asparagus is sometimes packed in boxes or crates without bunching. Several types of packages are used for asparagus, the most popular one being a box having solid ends and made especially for this product. A pyramidal crate with two compartments are used in California. While packing in crates and during transport, the cut end is kept wet either by placing in a pan of water or a pad of wet moss. The bottoms of the crates usually are lined with paper on which is placed damp moss. The butts of the bunches are set on the moss and kept under good refrigeration. Asparagus can be shift long distances without serious loss in quality (singh and Singh, 2007).

Storage

The spears are cooled as soon as possible just after packed. During transit it is kept in pre cooled car at temperature of 45° F to 47° F. Refrigeration uses also for transportation. The spears can be stored at 0° to 2° C with 95% relative humidity for two to three weeks successfully in good condition. Spears stored in wet tissue paper may keep fresh and firm even after 13-16 days of storage (Poll, 1991). At room temperature, the spears become tougher if stored even for short time.

SEED PRODUCTION

Asparagus is a dioecious crop with male and female plants, cross-pollination is the rule and the cultivars should be grown in isolation to maintain the genetic purity. The ratio of male and female plant 1:4 in the field, need to maintain for seed production.

Isolation Distance

Isolation distance for the foundation seeds is1600m. and for certified seed 1000 m is maintained between two varieties for seed production.

Good crowns are selected and planted so that four female plants are grouped around one male to ensure good pollination.

Spacing

Spacing 120 cm from row to row and 50-60 cm from plant to plant is recommended for seed production. Ellison (1986), suggested the following important points for seed production of asparagus.

(i) Spears should not be harvested because it reduces stored reserves for maximum stalk growth and seed production. (ii) A large population of bee is necessary for good pollination. (iii) Asparagus beetles, should be controlled timely. (iv) Red ripe fruit should be harvested because seed from ripe berries are superior in germination to that from less mature bronzed coloured fruits.

Plantation of four to eight years old are chosen so they are said to yield the best seed. The branches of the fruiting plants are shortened, in order to promote early ripening and well nourished seeds. A seed measuring 3 mm in diameter and weight equivalent to 500 mg to 1 g is regarded as good. Germination is poor in seed harvested from late bush. Seeds from ripe berries are superior in germination to that from less mature bronze coloured fruit. Snope and Ellison (1963) reported that asparagus pollen withstand freeze-drying for up to 30 minutes without damage, and can be subsequently stored for up to six months at 1°C to 20°C in a nitrogen atmosphere with little loss of viability. Ripe berry should be collected for seeds extraction, its seeds give good germination. After harvest, seed should be separated from skin and pulp and dried in the sun. Seed production is better and faster under glass house than in open field.

REFERENCES

Anonymous (1943) .Food Values of Common Indian Food Stuffs, Published by Cal. Corporation.

Anonymous (1952). Wealth of India, Publ. C.S.I.R., New Delhi.

Anonymous (1998). Varieties and Hybrids. Division of Vegetable Crops, I.A.R.I., New Delhi.

Aykroyd, (1941). Health Bulletin, No.23, Published by Nutrition Research Lab, Koonoor.

Borthwick, H.A.(1925). Factores influencing the rate of germination of the seed of *Asparagus officinalis, Calif. Tech. Paper*, 18.

Burn, R. (1969). *Hort. Abstr.* XLI (1): 169

Chauhan, D. V. S.(1965). Vegetable Production in India, Ram Prasad and Sons : Agra-3

Choudhury, B .(1967). Vegetables. National Book Trust, India, New Delhi. (Edn, 7th reprint).

De Beni, P.V. (1962). *Hort. Abstr*, 32(4): 834.

Ellison, J.H. (1986) Asparagus Breeding. In: Breeding *Vegetable Crops* (Ed. M.J. Bassett)., The AVI Publishing Company, Inc. West Port, Connecticut..

Ellison, J.H., Kinelski, J.J. and Garrison, S.A. (1989). *Hort Science*, 24: 868.

Evans, T.A., and Pill, W.G. (1989). *J. Hort. Sci.*, 64: 275-82.

Freeman, J.A. and Maurer, A.R. (1989). *Canad, J. Pl. Sci.* 69: 265.

Frenz, F.W. and Munz, R. (1970) . *Hort. Abstr.*, 40(1): 151.

Heigh, J.C. (1960). *Hort. Abstr.*, 32(1): 116.

Johnson, D.A. and Lunden, J.D. (1992). *Plant Disease*, 76: 84-86.

Kabir, J. (1993). Asparagus . In Vegetable Crops in India. (Eds; T.K. Bose, M.G. Som and Kabir, J.) Naya Prokash Calcutta.

Kampe, W. (1975). *Hort. Abstr.*, 46(6): 500

Knaflewaski, M. (1988). *Gemuse*, 24: 238-239.

Krarup, A. (1989). *Asparagus Research News letter*, 6: 1-2.

Mc. Grady, J. and Tilt, P (1990). *Asparagus Research Newsletter*, 7: 12-13.

Mehwald, J. (1988) Gemuse, 24: 232-234.

Ogg, A.G., Jr. (1975). Weed Sci. 23(6): 458-461.

Pandita, P.N. and Bhan, M.K. (1992). *Indian Hort.* 36 (Oct.-Dec.): 23-26.

Poll, J.T.K. (1991). Asparagus Research News letter, 9: 6-9.

Rice, R.P., Rice, L.W. and Tindall, H.D. (1990). Fruits and Vegetable Production, Mac millan Publsihers Ltd., London and Basingstaoke, pp. 234-35.

Sandhu, A.S. (1963). Asparagus. *Intensive Agri.* Vol.1 No.X.

Shanmugavelu, K.G. (1989). Production technology of Vegetable Crops. Oxford and IBH Publishing Co. Pvt. Ltd., New Delhi.

Sharma, D.K. and Chaudhary, D.R. (1996).Cultivating Asparagus in Himachal Pradesh. *Indian Horticulture*, 41(2): 8-9

Sharma, D.K. and Sharma, P.P. (1994). Performance of asparagus cultivars in Himachal Pradesh. *Vegetable Science* 21(2): 118-121.

Singh, D.N. and Singh, S.P. (2007). Production Management of Underutilized Vegetables. Agrihortica Publications, "Shipra" I.V. Nagar Road, Girivihar Society, Gandhidham, Junagadh-362 001 (Gujarat), India.

Singh, D.N. and Nath, Vishal (2011). Varieties and Hybrids of Vegetables. Satish Serial Publishing House, Azadpur, Delhi (India).

Snope, A.J. and Ellison, J.H. (1963). *Hort. Abstr.* 34(1): 281.

Swada, E., Yakuwa, J. and Iwaki (1961). *Hort. Abstr.*, 32(2): 363.

Takatory, F.W., Cannell, G.W. and Asbell, C.W. (1970). *Calif. Agric.*, 24: 10-12.

Thompson, H.C. and Kelly, W.C. (1957). *Vegetable Crops* . Tata McGraw Hill Publishing Company Ltd., New Delhi.

Van Staalduir, D. (1962). *Hort. Abstr.* 32(2): 363.

□□□

About the Author

Dr. Dina Nath Singh is Professor of Horticulture, Orissa University of Agriculture and Technology, Bhubaneswar. He received B.Sc. (Ag.) degree from Udai Pratap College, Varanasi under Gorakhpur University, Gorakhpur (UP) in 1974 and M.Sc. (Ag.) Horticulture and Ph.D., Horticulture from Banaras Hindu University in 1976 and 1982, respectively. Dr. Singh has 31 years of Research experience in the field of Vegetable Science along with 13 years experience of teaching of Graduate, Post Graduate and Ph.D. students in the field of vegetable science. He has 80 research papers published in National and International Journals. Dr. Singh has guided several Post-Graduate and Ph.D. students in the field of vegetable science. He has authored two text books on vegetable science entitled "Production Management of Underutilized Vegetables" and "Varieties and Hybrids of Vegetables". He has developed a number of technologies for increasing production of vegetables in Orissa. Dr. Singh is a fellow of Society of Environmental Sciences. He is the life member of 4 scientific Societies. He has attended several workshops, seminars and symposia of national and international repute during his service careers related with Horticultural crops. Dr. Singh has also served as Associate Director of Research, RRTTS (WCTLZ) of Orissa; Professor of Vegetable Science; Professor of Horticulture and Officer-in-Charge of All India Coordinated Research Project on Medicinal & Aromatic Plants and Betelvine, OUAT, Bhubaneswar Centre; Associate Director of Research, RRTTS (CZ-1), Bhubaneswar and Joint Director (Distance Education), Directorate of Extension Education, OUAT, Bhubaneswar.

Prof. Vishal Nath, has completed his Master's degree from Narendra Dev University of Agricultural and Technology, Faizabad and Ph.D. from Kanpur. He worked as scientist (Horticulture) at Central Institute for Sub-tropical Horticulture, Lucknow and NRC for Arid Horticulture, Bikaner. He served as Senior Scientist at Central Horticultural Experiment Station (IIHR), Ranchi and joined as Professor (Horticulture) at Birsa Agricultural University, Ranchi. Presently, he is working as Principal Scientist & Head, Central Horticultural Experiment Station (IIHR), Bhubaneswar. Apart from his contribution to National Agricultural Research System (NARS), he has wider exposure to Hi-tech Horticulture being practiced in developed countries like Israel.

Prof. Vishal Nath has exclusively worked on fruits of major and minor importance being grown in arid and semi-arid regions and under waste land conditions. He has developed many technologies and varieties of different fruit crops suitable for resource poor areas. He has written several technical bulletins, extension folders and information brochures on these aspects. Prof. Vishal Nath has authored very important books on arid horticulture, Litchi and Fruit Production Technology for Plateau regions. Prof. Vishal Nath has more than 200 publications on different fruits which includes research papers, popular article, technical bulletins, research manuals and information bulletins in his credit.

About the Author